Mathematical Models in
Molecular and Cellular Biology

Mathematical models in molecular and cellular biology

EDITED BY
LEE A. SEGEL

Dean of the Faculty of Mathematical Sciences, The Weizmann Institute of Science, Rehovot, Israel

CAMBRIDGE UNIVERSITY PRESS

CAMBRIDGE

LONDON · NEW YORK · NEW ROCHELLE

MELBOURNE · SYDNEY

CAMBRIDGE UNIVERSITY PRESS
Cambridge, New York, Melbourne, Madrid, Cape Town,
Singapore, São Paulo, Delhi, Tokyo, Mexico City

Cambridge University Press
The Edinburgh Building, Cambridge CB2 8RU, UK

Published in the United States of America by Cambridge University Press, New York

www.cambridge.org
Information on this title: www.cambridge.org/9780521270540

© Cambridge University Press 1980

First published 1980
Re-issued 2011

A catalogue record for this publication is available from the British Library

ISBN 978-0-521-22925-8 Hardback
ISBN 978-0-521-27054-0 Paperback

Contents

Contents

List of authors

Bard, Jonathan L. B.

Ghozlan, Aline

Goldbeter, Albert

Hardt, Shoshana

Levitzki, Alexander

Liron, Nadav

Odell, Garrett M.

Parnas, Hanna

Perelson, Alan S.

Rapp, Paul E.

Western General Hospital, MRC Clinical and Population Cytogenetics Unit, Edinburgh, Scotland

Université de Paris XI, Institut d'Electronique Fondamentale, Orsay, France

Universite Libre de Bruxelles, Service de Chimie Physique II, 1050 Bruxelles, Belgium

The Weizmann Institute of Science, Department of Applied Mathematics, Rehovot, Israel

The Hebrew University of Jerusalem, Department of Biological Chemistry, Institute of Life Sciences, Jerusalem, Israel

The Weizmann Institute of Science, Department of Applied Mathematics, Rehovot, Israel

Rensselaer Polytechnic Institute, Department of Mathematical Sciences, Troy, NY 12181, USA

The Hebrew University of Jerusalem, Institute of Life Sciences, Jerusalem, Israel

*Brown University, Division of Biology and Medicine, Providence, RI 02912, USA
Now at:
University of California, Theoretical Division, Los Alamos Scientific Laboratory, Los Alamos, NM 87545, USA*

Medical College of Pennsylvania, Department of Physiology and Biochemistry, Philadelphia, PA 19129, USA

Rubinow, Sol I.   *Cornell University Graduate School of Medical Sciences, Biomathematics Division, New York, NY 10021, USA*

Segel, Lee A.   *The Weizmann Institute of Science, Department of Applied Mathematics, Rehovot, Israel*

Tolkovsky, Aviva M.   *The Hebrew University of Jerusalem, Department of Biological Chemistry, Jerusalem, Israel*

Wehrhahn, Christian   *Max-Planck-Institut für biologische Kybernetik, Tübingen, Germany*

Yagil, Gad   *The Weizmann Institute of Science, Department of Cell Biology, Rehovot, Israel*

Zeigler, Bernard P.   *The Weizmann Institute of Science, Department of Applied Mathematics, Rehovot, Israel*

Acknowledgments

The author of Chapter 3 takes pleasure in acknowledging the advice of Dr A. I. Mees. The presentation of linear control theory and catastrophe theory in Chapter 3 owes much to his lectures at the University of Cambridge, England. This author would also like to acknowledge a Research Fellowship from Gonville and Caius College, University of Cambridge.

The author of Section 4.1 is indebted to B. Hess, T. Erneux, L. A. Segel and D. Venieratos for stimulating discussions and for their collaboration in this work. He also wishes to thank Professors G. Nicolis and I. Prigogine for their continuous interest. The author is Charge de Recherches du Fonds National Belge de la Recherche Scientifique.

The author of Section 4.5 and Chapter 5 wishes to acknowledge that they are based on work performed under the auspices of the US Department of Energy. This work was also supported in part by BRSG grant S07 RR05664-11 awarded by the Biomedical Research Support Grant Program, Division of Research Resources, NIH.

The author of Chapter 7 would like to thank T. Poggio and W. Reichardt for advice and, together with K. Hausen, for criticism of the manuscript. He is also grateful to G. Palm for help with Section 7.6.

The authors of Appendix Sections A.1 and A.3 acknowledge the partial support of their contributions by the National Science Foundation.

The phase portraits that appear on the cover, in Section 6.7, and in Appendix A.3, were made using a Ramtek color graphics computer facility purchased by NSF grant MCS-77-27493. The curvature in these phase portraits is the result of a calibration fault in the computer graphics screen on which they were produced.

Particular thanks are due to Mrs Sandi Irvine for her painstaking and creative editorial work.

Introduction

What this book aims to convey

This book stems from a course on 'Mathematical models in biology' that was given in the spring of 1978 at the Weizmann Institute of Science, Rehovot, Israel. The course was sponsored in part by the Institute, and to a considerably larger extent by the European Molecular Biology Organization. The purpose of the course was to demonstrate to *experimental* biologists with a minimum of mathematical background the possible usefulness of mathematical models. It was hoped that participants would learn to 'read' various types of equations (to tell what assumptions they imply), to understand in broad outline a number of major theoretical concepts, to be aware of some of the difficulties connected with analytical and numerical solutions of mathematical problems, and thus to appreciate the significance of theoretical papers in their fields. It was also hoped that participants would acquire an increased awareness of when cooperative work with a theoretician might be useful to them, and a better ability to communicate in the course of such cooperation.

The primary aim of this book is identical to the aim of the course, but advantage has been taken of the precision of the written word to clarify and exemplify matters that could only be treated rather cursorily in the lectures. To enable a degree of comprehensiveness in coverage, topics treated are almost entirely limited to molecular and cellular biology, while the mathematics used is largely restricted to differential equations and related topics. The book can also serve applied mathematicians and other theorists as a source of mathematical applications, predominantly using differential equations, to central areas of modern biology.

Mathematical requisites

Those with a decent command of basic calculus should be able to understand a large portion of the theoretical material. Prospective readers can refer to the 'calculus refresher' in Section A.1 of the Appendix. If they are familiar with most of the major formulae, they are ready to begin with Chapter 1. Otherwise they are advised to use the Appendix or other more comprehensive sources to review the calculus. There are portions of the text that may be somewhat difficult even for individuals with a mathematical background that considerably surpasses the required minimum, but these portions can usually be skimmed with little resulting loss in general comprehension.

On the role of theory in biology

Implicit in the effort to include somewhat sophisticated mathematical ideas in this book is the conviction that there is a body of theoretical work, and the promise of more such work in the future, that bears directly on the professional concerns of experimental biologists. One basis for this conviction is the fact that explosive technological development is permitting the evermore comprehensive collection of quantified data, and the belief, based on the experience of the physical sciences, that organization and deep understanding of large bodies of such data require a theoretical framework. Moreover, present-day computer facilities greatly enhance our ability to study the large and complex systems that are characteristically encountered in the study of living organisms.

A major reason why mathematical modeling is required in biology is the fact that rather simple interactions can have consequences that are not predictable by intuition based on biological experience alone. Mathematics is a tool explicitly fashioned to aid intuition in drawing conclusions from a few assumptions. To cite one example from this book (Section 1.6), analysis of certain simple kinetic equations shows that various assumptions concerning the interaction of a few important regulatory chemicals result in quite different time courses of reaction. This permits the design of an experimental program to select a most likely mode of action from a number of possibilities. In contrast to this quantitative theory, the qualitative power of mathematics is illustrated in Section 4.1, for example, where certain kinds of cellular 'micro-differentiations' are identified with radical changes in the solutions of differential equations that can occur when parameters cross a threshold.

Biologists traditionally suspect that complex behavior results from what might be termed a concatenation of ratchets, each new twist of behavior resulting from some new molecular species. The action of complement provides an example of such a concatenation. But theoretical investigations have demonstrated time and again that complex organized behavior can also result from the interaction of a small number of factors. An example is the propagation of the action potential in nerves, shown by Hodgkin and Huxley to result from the fact that perturbations in ion concentrations bring about conductance changes that in turn influence ion concentrations.

It is noteworthy that there was mathematical input in both of the examples cited. Major mathematical calculations were required in Hodgkin and Huxley's demonstration that the system of differential equations that expressed their phenomenological findings indeed possessed a traveling wave solution with a shape and speed that conformed with observation (Section 6.7). Understanding of complement seems much more extensively based on pure experiment, but here too mathematical analysis has played a part in eliciting the mechanism (Section 5.5).

In addition to affording quantitative and qualitative analyses, mathematics plays another role in biology (and other sciences) by providing precisely defined concepts with which to view what otherwise might seem an almost unfathomable sea of experience. Autonomous oscillation, autocorrelation, balance equation, catastrophe, chaos, delay, diffusivity, feedback, flux density, optimization, power spectrum, stability, steady state, transfer function, traveling wave, white noise – these are a sample of the concepts that will be defined and discussed in a biological context.

Mathematical methodology

A certain lack of confidence in the ability of theoreticians to contribute to biology is summed up in the tale of a team of mathematicians who were given ten years by a group of biologists to contribute something to their subject. Summing up a decade's work by his group, the chief mathematician opened his lecture with the phrase, 'Consider a spherical elephant ... '. Biologists will not benefit much from theoretical work until they understand that this story is not really funny. Sometimes it is perfectly reasonable to regard an elephant as spherical, for the error is immaterial and a simple and insightful view of the situation is thereby obtained. (An example might be calculations that compare heat loss in elephants and mice.) Similarly it is 'wrong', but in certain circumstances it is wise, to regard tubes as infinitely long (Section 6.1) or to posit that certain adjustments occur infinitely fast (as in the quasi-steady state hypothesis of enzyme kinetics – see Section 1.1).

The elephant story illustrates the importance of judicious simplification, the key step in making a mathematical model of a biological situation. The next step is analysis of the equations or inequalities that comprise the model. As mathematics is far from a completed subject, the analysis may present problems whose solutions are currently infeasible. For such 'technical' reasons, further simplification of the problem may be required and/or a degree of scepticism may be advisable with respect to alleged approximate solutions. Nonetheless, skilled applied mathematicians usually come up with an adequate solution to the mathematical problem, so that the theoretician turns to the last step in the applied mathematical process, interpretation of the mathematical results in view of the original biological situation. The interpretation may lead to refinement of the original model, or to a substitution of a radically different model. Both cases require another iteration of the sequence – formulation, solution, interpretation.

Advances in computer science now make it possible for anyone who understands what a differential equation is to produce a numerical solution after a couple of hours' training. To illustrate this there follows essentially the complete computer program for the solution of a particular case of the Michaelis–Menten equation

$$\mathrm{d}s/\mathrm{d}t = 7s/(8+s), \qquad 0 < t \leqslant 10; \qquad s(0) = 15,$$

plus the tabulation and graphing of its solution:

```
F = 7 . * S / (8 . + S)
S = INTGRL (15 . , F)
TIMER FINTIM = 10 .
PRINT S
OUTPUT S
END
STOP
```

The program is written in the CSMP language. In the absence of specific instructions by the programmer, the computer itself selects a numerical integration method, suitably scaled axes for graphs, etc. Other such languages are available and even more powerful ones will doubtless be developed in the future. On the other hand, one must continually be aware of the fact that the possibility of programming errors plus subtleties of numerical analysis, as illustrated in Appendix Section A.5, imply that fully reliable computer solutions can only be obtained with the cooperation of experts. Nonetheless, modern biologists, and modern biological education, should take advantage of the fact that the capability to carry out exploratory calculations on relevant mathematical models can be quickly obtained by every student of science. (There were six hours of CSMP instruction and practice in the course on which this book is based, but this matter will not be pursued here. Interested readers can consult *A Guide to Using CSMP* by F. H. Speckhart and W. L. Green, Prentice-Hall, 1976.)

A good example of the use of CSMP can be found in Section 4.1. The first graphs like Figures 4.1.13 and 4.1.15 were produced by a CSMP program within hours of hitting upon the idea that relay and oscillation in slime mold might be alternative types of behavior for the same biochemical system. There followed considerable trial and error to determine the effect on the behavior of varying parameter values. A check on the computer results was obtained by verifying that numerically obtained points of transition between steady and oscillatory long-term behavior corresponded to stability–instability transitions according to linear stability analyses of the type discussed in Appendix Section A.3.

It is frequently the case that once a judicious combination of intuitively guided computer simulation and simple calculation reveals the presence of an interesting phenomenon, it becomes worthwhile to bring to bear heavier mathematical artillery. This is illustrated by calculations toward the end of Section A.3 that delineate (by previously unpublished calculations) the entire boundary of the oscillatory domain in an application of asymptotic analysis and the Hopf bifurcation theorem.

Topics considered in this book

Chapter 1 considers biochemical reaction theory, an area where mathematical analysis of the governing mass–action equations has played a central role from the beginning of the subject. An introduction to the classical theory is presented, together with more advanced material centered on the concept of cooperativity. The last sections of the chapter present examples of how the close coupling of theory and experiment permits deductions concerning the nature of various biochemical systems.

What does it mean to understand a large system, in addition to identifying its components? One path to understanding is simplification by combining similar entities. The issues involved in such 'lumping' are discussed in Chapter 2, with emphasis on biochemical systems. This chapter illustrates the type of work now being done by computer scientists with the aim of enabling the enormous capabilities of the computer to be of more service to subject matter specialists.

Control theory is the subject of Chapter 3. Apart from instruction in the standard Laplace transform method of analyzing linear systems, general qualitative principles are illustrated, such as the tendency of long chains of feedback or other types of delay to introduce oscillations into biochemical control systems.

Chapter 4 groups several kinetic studies, each with a different qualitative feature. The first section illustrates how sharp transitions in biochemical behavior can result from slow changes in the governing nonlinear differential equations. In the next section, saturation (another nonlinear effect) is suggested to have important consequences in synaptic release. A third section introduces the concept of optimization in the context of the use of storage materials. Next comes a 'textbook' instance where a combination of theory and experiment can discriminate between competing models, in this case inhibitory and stimulatory modes of liver regeneration. The final section shows how simple nonlinear kinetic laws can lead to results with every appearance of randomness.

With its numerous interacting cells and molecules, the immune system promises to be an area where mathematics can play a useful role in biology. Only recently, however, have enough basic facts been known to permit useful theoretical analyses. The current status of the theory is reviewed in Chapter 5.

Chapter 6 on partial differential equations in biology is unified by the general balance law, which states that the rate of change of any quantity in a small volume element in space is equal to the net creation rate of the quantity in the element plus the net rate at which the quantity flows into or out of the element. If 'creation' arises from the law of mass action and 'flow' comes from random motion, then the reaction–diffusion equation results. If the 'spatial' coordinate denotes a variable like age or maturity, then there result the equations that govern the proliferation of cell populations. A variety of biological phenomena is discussed in terms of these and similar equations. (Much of the chapter actually involves only ordinary differential equations, obtained for example, by considering solutions to the partial differential equations that are time independent or wave like.)

The material on fly visual behavior in Chapter 7 is in a sense outside the main stream of the book, for it is mainly concerned with phenomena on a supracellular level. On the other hand, this chapter brings out the important point that to attain the goal of understanding mechanisms on the cellular level, it may be wise to conduct rather extensive supracellular investigations first. Certainly the 'fly story' provides an excellent example of the use of mathematics in biology, since a central aim of the experimental program is the establishment of the fact that a certain stochastic differential equation underlies the dynamic behavior of the fly as it maneuvers in response to its visual environment. It is demonstrated that one cannot speak in profound terms about fly behavior without utilizing rather sophisticated mathematical concepts.

A mathematical appendix at the end of the book begins with a 'refresher' that is designed to make much of the book accessible to readers who have studied calculus for about a year but who have forgotten much of it. (The exponential function 'e' and the trigonometric functions sin and cos are defined as solutions to differential equations. The conventional approach is via continuously compounded interest and trigonometry, but these concepts have little relevance to the material of this book.)

There follows a brief treatment of topics such as determinants and complex numbers that are used in certain sections. Next there is a carefully crafted introduction to modern methods for determining the qualitative behavior of solutions to differential equations. This section is not easy reading, but it is accessible to the biologists who are the principle intended readers of this book; perseverance will generate comprehension of an elegant geometric way of thinking that is indispensable to the understanding of complex biological systems.

The remaining topics in the appendix are brief introductions to dimensional analysis and to the numerical methods used to transform differential equations into a form suitable for automatic computation.

A number of threads run through several chapters. The concepts of stability, control, and optimization, for example, are encountered several times. Dimensionless variables are repeatedly introduced. Several applications are made of the rule of thumb that in time T a tight cluster of particles will diffuse over a patch whose radius is of magnitude $\sqrt{(DT)}$, where D is the diffusivity. And, to give one more example, we see in several places how mathematical analysis can be used to extract information from bioassays: from assays in immunology (Section 5.5), from the capillary assay for cell motility (Section 6.1), and from observations of flow in affinity columns (Section 6.7).

Classroom use

Several different emphases are possible in using this book for a course in mathematical biology. Biologists with a minimum mathematical background will find a term's worth of material by concentrating on Chapters 1 and 4, some of the simpler material in Chapter 5, and the first three sections of Chapter 6. A presentation strongly focused on applications of mathematics in problems of molecular and cellular biology (leaving aside mathematical methods, more purely biochemical problems, and supracellular phenomena) can be obtained from Sections 1.6 and 1.7, Chapters 4 and 5, and highlights of Chapter 6.

Since this book is primarily intended for biologists, an effort has been made to confine detailed mathematical discussions to relatively elementary calculations and to stress concepts rather than technique when the mathematics becomes more advanced. Nonetheless, there is ample material for a course wherein students of applied mathematics can be introduced to much valuable material in the context of important applications. Courses of this nature can be planned with the aid of the list of Mathematical Applications that follows this introduction.

It will be seen that there is enough relatively elementary material and basic explanation to permit the teaching of considerable amounts of control theory, ordinary differential equations, and partial differential equations in the context of applications that are at once important and not difficult to understand. Probability theory, numerical analysis, computer simulation, stochastic differential equations, and even such currently fashionable topics as chaos and catastrophe theory are touched upon sufficiently to enrich a more conventional applied mathematics course with the introduction of topics on the frontiers of current practice.

Mathematical applications

Places in the book where various mathematical methods are applied are indicated by the number of chapter, section, equation or exercise (e.g. 3, 2.1, (4.2.3), Exercise 3.1.6). The list is not intended to be exhaustive, but to be helpful in gauging the general types of mathematics employed and in locating examples of particular techniques. Headings and subheadings are each arranged, very roughly, in order of increasing difficulty.

Algebra

Elementary: 1, 3.8, 4.2, 4.3, (6.3.7), (A.3.71–A.3.76)
Matrices: 3.8, (4.3.27)
Asymptotic methods: (A.3.80)

Calculus

Curvature and slope: 1.4, (4.2.7), 4.3, 5.3
Taylor series: (4.2.13), (6.1.6), (6.4.1)
Integration: (6.1.22)
Also see ordinary differential equations – separation of variables.

Ordinary differential equations

Separation of variables: (1.6.21), 1.7, (5.3.43), (6.3.11), (6.3.23), Exercises 1.1.7 and 6.6.1
Linear inhomogeneous: 1.7, Exercise 6.6.2
Quasi-steady state assumptions: 1.1, (4.1.6), (4.1.18), (5.3.57)
Laplace transforms: 3
Stability: (3.7.5), (4.1.9), Exercises 4.4.1 and 4.4.2, 6.4, (A.3.57)
By Routh–Hurwitz, Nyquist, and Gerschgorin theorems: 3
Global: Exercise 3.7.2, 3.8
Phase plane: 4.1, A.3
Singular perturbation method: 5.5, (6.3.25)
Stochastic: 7
Simulation: 2

Control theory: 3, 5, 4

Probability: 4.3, 5.2, 5.3, 5.5

Partial differential equations

Verification of a given solution: (6.1.15), (6.1.21)
Stability theory: 6.4, 6.5, 6.7
General balance equation: 6.1, 6.7
Diffusion equation: 6.1, 6.2
Other examples: 6
Generation expansion: (6.6.28)

Other topics

Least squares error: (1.2.35)
Delta function: Example 3.2.6, 5.5, (6.6.38)
Dimensionless variables: (3.8.2), (4.1.3), (6.3.25), (6.7.13), (6.7.42), (A.1.83)
Numerical analysis and simulation: 4.1, 4.2, 4.4, 6.4
Fourier series: (6.5.33)
Difference and delay equations: 3.4, 4.5
Chaos: 4.5
Catastrophe theory: 3.9

Conventions

Each chapter is divided into several sections (e.g. Section 2.3 is the third section of Chapter 2). Equations are numbered consecutively within each section. If the fourth numbered equation of Section 2.3 is referred to from another section, the notation '(2.3.4)' is employed, or (A.4.5) for the fifth numbered equation in Appendix 4. The designation 'Figure 3.2.1, 3.2.2, . . . ' is always employed for Figures.

In Chapters 2, 3, and 7, references are placed at the end of each chapter. Otherwise, references are placed at the end of each section.

Boldface is used when important words are defined. Important equations are indicated by a symbol on the left.

1 Biochemical reaction theory

1.1 FUNDAMENTAL CONCEPTS

Law of mass action

Consider a reaction in which a chemical A of concentration A combines reversibly with a chemical B of concentration B to yield **complex**, C, of concentration C. This reaction is symbolized by

$$A + B \underset{k_{-1}}{\overset{k_{+1}}{\rightleftharpoons}} C. \tag{1}$$

The **forward and backward rate constants** k_{+1} and k_{-1} are the proportionality factors in the **law of mass action** that is assumed to describe the process of the reaction. According to this law, the rate at which the species A reacts to form C is proportional to the mass of A, or equivalently, to the number of molecules of A available for reaction. In mathematical terms, the law takes the form of the following differential equations for the concentrations A, B, and C, at time t:

$$dA/dt = -k_{+1}AB + k_{-1}C, \qquad dB/dt = -k_{+1}AB + k_{-1}C, \tag{2a, b}$$

$$dC/dt = k_{+1}AB - k_{-1}C. \tag{2c}$$

In (2a), A is supposed to decrease at a rate jointly proportional to the concentrations of A and B. The idea behind this is again the law of mass action: doubling the concentrations of either A or B will double the rate of collision between these two molecules and hence will double the rate of 'successful' collisions that lead to the formation of C. Such an assumption is plausible as long as the concentrations are not too large. The break-up of an individual C molecule into its constituents is held to occur with a constant probability per unit time.

The phenomenological law of mass action can, in principle, be derived from statistical mechanics, or on a deeper level from quantum mechanics, but this law can be regarded as being well established because of experimental information on a wide variety of theories in the biological, chemical and physical sciences that assume it.

Enzyme–substrate complex system

Enzymes are large molecules that speed up the conversion of a chemical to an altered form. According to the theory of enzymatic reactions of Michaelis & Menten (1913), the enzyme accomplishes this in two steps. First the enzyme (concentration E) reacts reversibly with the chemical, called a **substrate** in this context, to form a **complex** (concentration C). Secondly, the complex breaks apart into an altered substrate or **product** and the original enzyme. This last reaction is often assumed to be irreversible, in which case one writes

$$E + S \underset{k_{-1}}{\overset{k_{+1}}{\rightleftharpoons}} C \overset{k_{+2}}{\longrightarrow} E + P.$$

The law of mass action for the concentrations $E(t)$, $S(t)$, $C(t)$, and $S(t)$ takes the form

$$dE/dt = -k_{+1}ES + k_{-1}C + k_{+2}C, \tag{3a}$$

$$dS/dt = -k_{+1}ES + k_{-1}C, \tag{3b}$$

$$dC/dt = k_{+1}ES - k_{-1}C - k_{+2}C, \tag{3c}$$

$$dP/dt = k_{+2}C. \tag{3d}$$

The above **system of differential equations** representing the enzymatic conversion of substrates to product was first put forward by Briggs & Haldane (1925). The equation must be supplemented by **initial conditions** that describe the system at some reference time. This time is conveniently designated $t = 0$. The standard initial conditions, which conform to the usual investigation of enzymatically controlled reactions, prescribe starting concentrations of enzyme and substrate, and assume that complex and product have had no opportunity to form:

$$E(0) = E_0, \qquad S(0) = S_0, \qquad C(0) = 0, \qquad P(0) = 0. \tag{4}$$

Addition of (3a) and (3c) yields

$$d(E + C)/dt = 0. \tag{5}$$

Consequently $E + C$ must be a constant, reflecting the fact that all enzyme molecules are either in their original form or bound in a complex at any time t. Using the initial conditions, the constant can be determined, so that we can write

$$E(t) + C(t) = E_0. \tag{6}$$

This equation may be used to eliminate E from (3a) and (3b), leaving two equations for the two unknown functions $E(t)$ and $S(t)$.

Pseudo-steady state: Michaelis–Menten equation

In laboratory experiments, it is typically the case that, at the start, many substrate molecules are present for each enzyme molecule. Under these circumstances one expects that after an initial short transient period there will be a balance between the formation of complex by the union of enzyme and substrate and the breaking apart of complex (either to enzyme and substrate, or to enzyme and product). Because there are so many substrate molecules, this balance will be achieved before there is perceptible transformation of substrate into product. One anticipates, therefore, that calculation of product formation can be carried out under the assumption that $dC/dt = 0$, or, from (3c),

$$k_{+1}ES = (k_{-1} + k_{+2})C. \tag{7}$$

This equation is said to result from a **quasi-** or **pseudo-steady state hypothesis**. If any quantity no longer changes with time it is said to be in a **steady state**. We add 'pseudo' or 'quasi' to the description of (7) as a steady state, since although C is fully adjusted to the instantaneous values of E and S, those values are changing slowly with time.

Upon substitution of (6) and (7) into (3b), we obtain the following equation for S:

$$dS/dt = k_{+1}k_{+2}E_0S/(k_{+1}S + k_{-1} + k_{+2}). \tag{8}$$

This equation can be solved, subject to the initial condition $S(0) = S_0$, to obtain the concentration S at any time t (Exercise 1). Biochemists are particularly interested in the velocity of reaction $V(t)$ defined variously as the rate of disappearance of substrate or the rate of appearance of product. In view of the steady state hypothesis, these definitions are equivalent; we have from (6), (7) and (3d) that

$$V(t) = dP/dt = k_{+2}C = |dS/dt|. \tag{9}$$

Biochemists determine this velocity at the beginning of the reaction. From (8) we can therefore write for this **initial velocity** $V_0 \equiv V(0)$,

◆◆ $$V_0 = \frac{VS_0}{K_m + S_0}, \tag{10a}$$

where

$$V = k_{+2}E_0, \qquad K_m = \frac{k_{-1} + k_{+2}}{k_{+1}}. \tag{10b, c}$$

Equation (10a) is called the **Michaelis–Menten** equation. Its graph starts from the origin, for the absence of substrate implies the absence of reaction, and approaches the asymptote $V_0 = V$ as S_0 becomes larger and larger (see Figure 1.1.1). Thus, when S_0 is large compared to K_m and $V_0 \approx V$, there is an abundance of substrate and the 'chemical factory' is working as fast as possible. In such cases the system is said to be **saturated**. Because the constant V is the maximum velocity that the reaction can attain, the term 'V-max' is used to describe it. (The term **Langmuir isotherm** is also associated with $(10a)$, which is said to have the form of a **rectangular hyperbola**.)

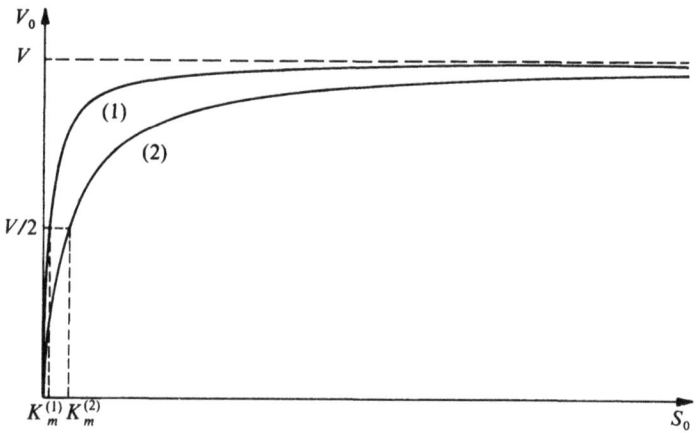

Figure 1.1.1. Graphs of the Michaelis–Menten equation $(10a)$ in two situations with the same maximum velocity V. The reaction represented by curve (1) is more specific than that of curve (2) because the Michaelis–Menten constant for it is smaller:

$$K_m^{(1)} < K_m^{(2)}.$$

The biochemical determination of the **Michaelis constant** K_m follows from the observation that when $S_0 = K_m$ then $V_0 = \frac{1}{2}V$. Thus K_m gives the concentration at which the reaction attains its half-maximal value. If this concentration is relatively low, then the reaction is said to be highly **specific**. A relatively low K_m means a relatively large k_{+1} and this in turn means that an enzyme–substrate collision is relatively likely to result in the formation of product, i.e. that the enzyme is specifically adapted to act on the particular substrate.

Biochemists frequently rearrange the Michaelis–Menten equation $(10a)$ into the **Lineweaver–Burk** or **double-reciprocal** form

$$\frac{1}{V_0} = \frac{1}{V} + \left(\frac{K_m}{V}\right)\frac{1}{S_0}. \tag{11}$$

The graph $1/V_0$ versus $1/S_0$ is thus a straight line, which simplifies the problem of fitting the theory to data. Then $1/V$ and $-1/K_m$ can be found at once as the intersection of this line with the vertical and horizontal axes, respectively (Figure 1.1.2).

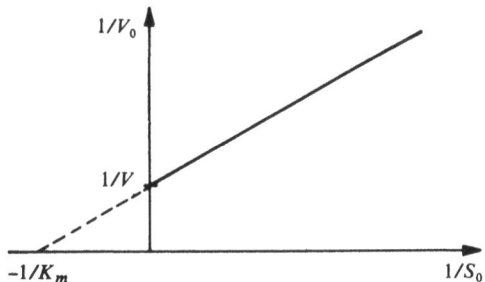

Figure 1.1.2. The Lineweaver–Burk plot, from which V and K_m can be readily determined. (The dashed part of the line corresponds to 'unphysical' negative substrate concentrations.)

Note from (10*b*) that V depends on the product of the initial enzyme concentration E_0 and the product formation rate constant k_{+2}. This reflects the fact that at high substrate concentrations the speed of reaction depends only on how many reaction units there are, and on how fast they can transform complex into product. Under such circumstances one says that the enzyme is the **rate-limiting chemical** and the complex–product conversion is the **rate-limiting step** in the conversion of substrate to product.

The back reaction for the conversion of complex to product can also be included in the theory. Further, Haldane (1930) has indicated that the reaction of product and substrate should be viewed symmetrically so that the complete set of reactions presumed to take place are represented as

$$S + E \underset{k_{-1}}{\overset{k_{+1}}{\rightleftarrows}} C \underset{k'_{-2}}{\overset{k'_{+2}}{\rightleftarrows}} C' \underset{k_{-2}}{\overset{k_{+2}}{\rightleftarrows}} P + E, \tag{12}$$

where C is an S–E complex, and C' is a P–E complex. The analysis of this reaction scheme does not alter the form of the Michaelis–Menten equation (10), although the meanings of V and K_m in terms of fundamental rate constants are more complicated than indicated by (10*b*) and (10*c*).

Our derivation of the steady state approximation has been informal. More careful mathematical treatment by means of singular perturbation theory is described elsewhere, for example in Lin & Segel (1973, Chapter 10) or in Rubinow (1975, Chapter 2). These treatments provide a careful analysis of the conditions under which the approximation is expected to be a good one and also give expressions for the concentration variations in the initial transient period. Such refinements may be deemed to add little to standard Michaelis–Menten theory, but the increased understanding gained has permitted extensions of the theory to such situations as enzyme–substrate–inhibitor systems (Rubinow & Lebowitz, 1970) and immunological systems where the characteristic time for reaching chemical equilibrium depends strongly on the back-reaction constant (Perelson & Segel, 1978).

Exercise

1   Solve (8) by the method of separation of variables, impose the initial condition, and obtain

$$S + \frac{k_{-1} + k_{+2}}{k_{+1}} \ln \frac{S}{S_0} = S_0 - k_{+2}E_0 t.$$

References

Briggs, G. E. & Haldane, J. B. S. (1925). A note on the kinetics of enzyme action. *Biochem. J.* **19**, 338–9.

Haldane, J. B. S. (1930). *Enzymes*, 2nd edn, London, Longmans, Green (reprinted by MIT Press, Cambridge, Mass. 1965), Chapter 5.

Lin, C. C. & Segel, L. A. (1973). *Mathematics Applied to Deterministic Problems in the Natural Sciences*, New York, Macmillan.

Michaelis, L. & Menten, M. L. (1913). Die kinetik der Invertinwirkung. *Biochem. Z.* **49**, 333–69.

Perelson, A. S. & Segel, L. A. (1978). On ligand–membrane binding reactions controlled by the ligand dissociation rate. *J. Math. Biol.* **6**, 75–85.

Rubinow, S. I. (1975). *Introduction to Mathematical Biology*, New York, Wiley.

Rubinow, S. I. & Lebowitz, J. L. (1970). Time-dependent Michaelis–Menten kinetics for an enzyme–substrate–inhibitor system. *J. Amer. Chem. Soc.* **92**, 3888–93.

1.2 EQUILIBRIUM BINDING OF MACROMOLECULES WITH LIGANDS

Theory of equilibrium dialysis

A basic and important method for studying the reaction of a protein P with a small molecule or ion C is **equilibrium dialysis**. In it, a known amount of the macromolecular protein is placed in solution inside a membrane bag which is suspended in a solution of the small molecule with which it is capable of reacting (the **ligand**). The membrane is permeable to the ligand, but impermeable to the macromolecule. Available membranes possessing such a permeability property require the molecular weight of the macromolecule to be greater than 10 000. The solution is allowed to stand for a sufficiently long time, of the order of one or two days, for the ligand to permeate the membrane and react with the protein so that equilibrium is established. At equilibrium, there exists both bound and unbound components of the ligand. Furthermore, the unbound concentrations on both sides of the membrane are equal to each other (if the ligand is an ion, electrical effects must be neutralized for the equality to hold true). By measuring the equilibrium concentrations outside the bag at the beginning and end of the equilibrium dialysis experiment, the bound ligand concentration is readily determined as the difference of those two quantities.

We present here the theory of this simple experimental procedure. We shall assume that the protein possesses n binding sites for the ligand, where n is an integer greater than unity in usual cases of interest. Let us denote by P_j the complex of a protein molecule with j ligand molecules attached, where $j = 0, 1, 2, \ldots, n$ (P_0 represents the bare protein). Then the reactions leading to equilibrium are as follows,

$$C + P_{j-1} \rightleftarrows P_j, \qquad j = 1, 2, \ldots, n. \tag{1}$$

Note that we have tacitly assumed that all complexes consisting of exactly j ligands attached are the same, regardless of the set of j attachment sites. We denote by italic lower case letters the concentrations of quantities represented by capital letters. Let us consider mathematically the first reaction above for $j = 1$. Employing the law of mass action, we express the time-dependent behavior of the concentration of one of these reactants, say p_0, as

$$dp_0/dt = -k_{+1} p_0 c + k_{-1} p_1. \tag{2}$$

At equilibrium, p_0, c, and p_1 are no longer time dependent and attain constant values which are interrelated, according to (2), by the relation

$$0 = -k_{+1} p_0 c + k_{-1} p_1. \tag{3}$$

It is customary to define the **association constant** K_a as

$$K_a \equiv k_{+1}/k_{-1}, \tag{4}$$

although for some purposes, it is found more convenient to utilize its inverse, the **dissociation constant** $K_d = 1/K_a = k_{-1}/k_{+1}$. Either one of these may be referred to as the **equilibrium constant**, although a certain ambiguity is thereby introduced by this usage if no further clarification is made. According to equations (3) and (4), at equilibrium

$$K_a = p_1/p_0 c. \tag{5}$$

To describe now the full set of reactions (1) at equilibrium, we introduce, for uniformity of notation, the n association constants K_j, $j = 1, 2, \ldots, n$, where

$$K_1 = \frac{p_1}{cp_0}, \quad K_2 = \frac{p_2}{cp_1}, \quad \ldots,$$

$$K_j = \frac{p_j}{cp_{j-1}}, \quad \ldots, \quad K_n = \frac{p_n}{cp_{n-1}}. \tag{6}$$

The quantities $p_0, p_1, \ldots, p_n$ are not usually experimentally determinable, but it is possible to find the average number of molecules of C associated with each macromolecule. This number is denoted by r and is defined as

$$r \equiv \frac{\text{total number of molecules of C combined with P}}{\text{total number of molecules of P}}. \tag{7}$$

We shall call r the **mean association function**. The numerator and denominator above are experimentally measurable quantities, as already indicated. Because there are j ligand molecules attached to each molecule P_j, r is expressible as

$$r = \frac{p_1 + 2p_2 + 3p_3 + \ldots np_n}{p_0 + p_1 + p_2 + \ldots + p_n}, \tag{8}$$

or, using (6),

$$\blacklozenge\blacklozenge \quad r = \frac{K_1 c + 2K_1 K_2 c^2 + 3K_1 K_2 K_3 c^3 + \ldots + nK_1 K_2 \ldots K_n c^n}{1 + K_1 c + K_1 K_2 c^2 + \ldots + K_1 K_2 \ldots K_n c^n}. \quad (9)$$

The above result is known as **Adair's equation** (Adair, 1925). A related quantity frequently utilized is the **saturation function** Y defined as the mean fraction of sites per protein molecule that are occupied, or

$$Y = r/n. \quad (10)$$

Identical independent sites

A significant simplification occurs when the binding sites in the protein are identical. Further, assume that binding at a given site is independent of the state of binding of all other sites. That is to say, let k_+ be the forward rate constant for attachment of a ligand molecule at a particular binding site, and let k_- be the associated backward rate constant. In terms of these rate constants, (3) assumes the form

$$0 = -nk_+ p_0 c + k_- p_1. \quad (11)$$

The factor n appears because there are n possible ways to form the state P_1 from the state P_0 (n available sites of attachment of the ligand molecule). Contrariwise, there is only one way for the ligand to be removed from the state P_1 to form the state P_0 (the ligand molecule is removed at its site of attachment). By a similar argument, we see that equilibrium between the states P_1 and P_2 is described as

$$0 = -(n-1)k_+ p_1 c + 2k_- p_2, \quad (12)$$

i.e. there are $n - 1$ empty sites available for ligand binding in the state P_1, and two ways to remove a ligand molecule from the state P_2. Hence, with the **intrinsic association constant** K defined as

$$K \equiv k_+/k_-, \quad (13)$$

we infer that $K_1 = nK$, $K_2 = (n-1)K/2$, and, in general (Bjerrum, 1941),

$$K_j = (n-j+1)K/j, \qquad j = 1, 2, \ldots, n. \quad (14)$$

Thus, the assumption that the binding sites of the protein are identical and independent is equivalent to the assertion that the intrinsic binding constant at one site is the same as at any other site, and moreover, is unaffected by the state of binding of the other sites. Using (14), Adair's equation assumes the particular simple Michaelis–Menten form (see, for example, the derivation in Rubinow, 1975, Section 2.4).

$$\blacklozenge\blacklozenge \quad r = \frac{nKc}{1 + Kc}. \quad (15)$$

A protein containing n binding sites which obey equation (15) is said to be **noncooperative** or to display **zero cooperativity**. By contrast, if the protein obeys (9) for $n > 1$ and the K_j are not related by (14), it is said to be **cooperative** or to display **cooperativity**.

Rather than repeat the cited derivation, we shall present here a simpler alternative derivation of (15) for the case of identical, independent binding sites. We focus on the individual binding sites and consider them as independent particles, as it were, because the fact that they are grouped in bundles of n on proteins has no bearing on their reaction properties. Let f be the equilibrium concentration of free binding sites, and b the equilibrium concentration of bound sites, which is the same as the equilibrium concentration of bound ligand. Then the intrinsic association constant of a binding site is

$$K = b/cf. \tag{16}$$

The total number of sites, whether free or bound, is expressed as

$$np = f + b. \tag{17}$$

According to the definition (7) of r, it is given as

$$r = b/p. \tag{18}$$

By dividing (17) by b, introducing (18), and eliminating f/b by means of (16), we see that (15) follows directly.

Hapten–antibody interactions and heterogeneity

The immunological system of higher organisms reacts to the stimulus of 'foreign' substances, called **antigens**, by producing certain large molecules, called **antibodies**, which combine with the antigens and thereby neutralize their harmful effect. The reaction of antibodies with antigens is chemical in nature. Antibodies are macromolecular proteins of blood serum of the order of 10^5 or 10^6 daltons in molecular weight, called **immunoglobulins**. Antigens normally have a molecular weight of 10^4 or larger. However, the formation of antibodies is a reaction to only a small part or chemical subunit of the antigen, called an **antigenic determinant**. Many low molecular weight chemical compounds have the ability to react with antibodies *in vitro*, and are called **haptens**, in the vocabulary of immunology. However, they lack the ability to elicit the production of antibody *in vivo*. Such an ability can be conferred on them by attaching them to protein carriers, thereby producing hapten–carrier complexes which are considerably larger than the hapten molecules.

Antibody molecules contain many binding sites for haptens. Moreover, it is well known that antibody is **heterogeneous** in its affinity for antigen. This means that, given an antigen containing a single antigenic determinant, or a hapten that is reacting with antibody, there will be a range of association constants characterizing the hapten–antibody complex. Here we shall concern ourselves with a particular type of heterogeneity, namely, that resulting from the macromolecule containing more than one set of binding sites for a given ligand. Such heterogeneity appears to be unique to the immune response, and is not seen for example, in the behavior of enzymes reacting with substrate molecules.

Let us assume, for definiteness, that the macromolecule contains two sets of binding sites, each set consisting of identical, independent binding sites. Let the first set consist of n_1 sites having an intrinsic association constant K_1, with the subscript 2 designating the second set of sites. In analogy with the derivation of equation (15) we can write that, at equilibrium,

$$b_1/p = n_1 K_1 c/(1 + K_1 c), \qquad b_2/p = n_2 K_2 c/(1 + K_2 c). \tag{19}$$

The ratio of the number of bound sites to the number of proteins, i.e. the mean association function is given by r,

$$r = (b_1 + b_2)/p. \tag{20}$$

Hence, from (19),

$$r = \frac{n_1 K_1 c}{1 + K_1 c} + \frac{n_2 K_2 c}{1 + K_2 c}. \tag{21}$$

More generally, when there are m subpopulations of identical, independent binding sites, it follows by the same argument (Tanford, 1961) that the average number of occupied sites per protein molecule is

$$r = \sum_{j=1}^{m} \frac{n_j K_j c}{1 + K_j c}. \tag{22}$$

The Scatchard plot

For the case of a macromolecule containing a single set of n identical, independent binding sites, we recognize that the functional relationship between r and c given by (15) is the same as that between the initial velocity v of the reaction between an enzyme and a substrate of concentration s, namely

$$v = Vs/(K_m + s), \tag{23}$$

where V and K_m are constants. The usual quantitative problem of characterizing such a reaction is to determine the constants V and K_m from the experimentally measured function $v(s)$. Similarly, we wish to determine n and K of (15), from the knowledge of $r(c)$. Analogous to the Lineweaver–Burk plot of enzyme kinetics (Lineweaver & Burk, 1934), we can plot $1/r$ as a function of $1/c$ (Klotz, 1946). Then if r satisfies (15), the relationship is linear, and the parameters n and K are readily found from the slope and intercept of the resulting straight line.

An alternative suggestion made by Scatchard (1949) is based on the rearrangement of (15) obtained by clearing the denominator and dividing by c so that (15) becomes

$$r/c = K(n - r). \tag{24}$$

Hence, a plot of the quantity r/c versus r is a straight line, as shown in Figure 1.2.1. Such a plot is called a **Scatchard plot**. For simplicity of

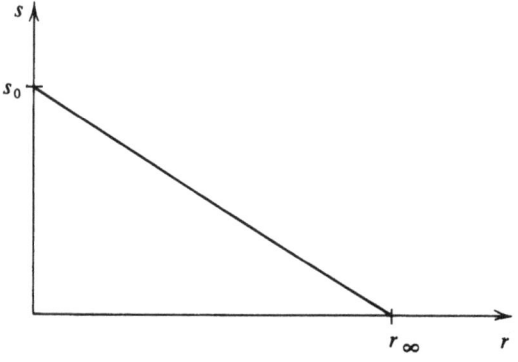

Figure 1.2.1. The Scatchard plot, based on (24) with $s_0 = nK$ and $r_\infty = n$.

notation, let us designate $s \equiv r/c$, a function of c, as the **Scatchard function**. We shall refer to the curve appearing in the Scatchard plot as the **Scatchard curve**. We note parenthetically that the relationship analogous to (24) for representing the initial velocity of an enzymatic reaction reads

$$v/s = (1/K_m)(V - v), \tag{25}$$

and the corresponding plot of v/s versus v, called an **Eadie plot** (Eadie, 1942), was originally suggested by Woolf (1932).

We ask, with respect to Figure 1.2.1, how is the straight line traversed as c increases from zero to infinity? We see from (15) that r increases monotonically as c increases, so that the straight line evolves from upper left to lower right in Figure 1.2.1. In fact, from (15),

$$r \to 0 \quad \text{as } c \to 0,$$
$$r \to n \quad \text{as } c \to \infty. \tag{26}$$

At the same time, we see from (15) after division by c that

$$s \to nK \quad \text{as } c \to 0,$$
$$s \to 0 \quad \text{as } c \to \infty. \tag{27}$$

By means of (26) and (27), the values of the intercepts $r_\infty = n$ and $s_0 = nk$ of the Scatchard curve shown in Figure 1.2.1 are deduced.

The fact that the entire variation of r with c is represented in the Scatchard plot by a curve of finite length implies that experiments conducted over a large range of variation of c receive equal representation in the curve, so to speak. This property, lacking in the curve $r(c)$ which tends to emphasize small values of c, or in a Lineweaver–Burk plot which tends to emphasize large values of c, is an attractive feature of the Scatchard plot that has helped enhance its popularity in recent times.

What is the form of the Scatchard curve when r satisfies (22)? The analog of the limits (27) and (26) are as follows:

$$s_0 \equiv \lim_{c \to 0} s = \sum_{i=1}^{m} n_i K_i, \qquad r_\infty \equiv \lim_{c \to \infty} r = \sum_{i=1}^{m} n_i. \tag{28}$$

However, the Scatchard curve is no longer a straight line. Figure 1.2.2 illustrates the Scatchard curve for the case $m = 2$, with representative values of the parameters for the binding of a ligand (1-iodonaphthalene-4-sulfonate) to IgM antibodies. We remark in passing that the Scatchard curve, when representing the functional form (22) with $n \geq 2$ always has the concave shape shown in the figure. By the phenomenological criterion of cooperativity (see Section 1.3), such curves indicate negative cooperative behavior. The solid circles in Figure 1.2.2 represent observations (Onoue, Grossberg, Yagi & Pressman, 1968) and point up a difficulty that arises in obtaining actual data, namely, that accurate experimental measurements are very difficult to carry out at either very small or very large values of the concentration c, so that the limiting values s_0 and r_∞ cannot readily be deduced. It is clear that for c sufficiently small, measurements will ultimately strain the limits of accuracy of the measuring apparatus. The same difficulty occurs at large values of c because the measurement of the bound ligand concentration and hence r requires determining the small difference between two large quantities, namely the concentration c at the beginning and end of the equilibrium dialysis experiment.

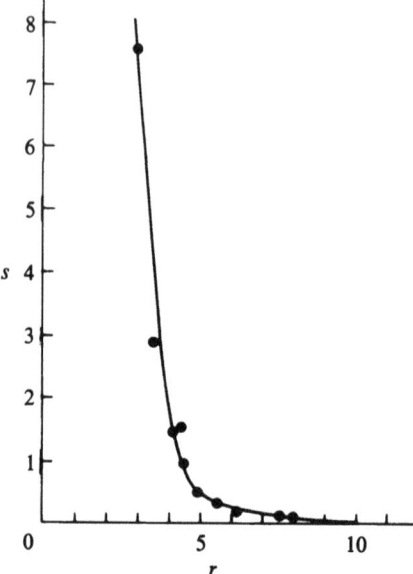

Figure 1.2.2. Scatchard plot representing the binding of a ligand (1-iodonaphthalene-4-sulfonate) by IgM antibodies. Solid circles are data points of Onoue *et al.* (1968). The curve is theoretical, and is based on (21) with parameter values given in Table 1.2.1 for $r_\infty = 10$. The ordinate unit is $(\mu\text{mole } l^{-1})^{-1}$. (From Rubinow, 1977.)

More generally, a mathematical problem of great practical importance is to determine how the parameters appearing in $r(c)$, namely K_i and n_i for $i = 1, 2, \ldots, m$ are to be inferred from observation. This problem arises also in conjunction with the Adair equation (9), and has attracted the attention of many investigators (for example, Pauling, Pressman & Grossberg, 1944; Karush & Sonenberg, 1949; Karush, 1956; Nisonoff & Pressman, 1958; Endrenyi, Chan & Wong 1971; Werblin & Siskind, 1972). Here we shall present a simple scheme for determining these parameters from the consideration of the Scatchard plot (Rubinow, 1977).

We restrict our attention to the case $m = 2$, so that r is assumed to be of the form shown in (21). Then r and s may be written as

$$r = \frac{a_1 c + a_2 c^2}{1 + b_1 c + b_2 c^2}, \qquad s = \frac{a_1 + a_2 c}{1 + b_1 c + b_2 c^2}, \qquad (29a, b)$$

where

$$a_1 = n_1 K_1 + n_2 K_2, \qquad a_2 = (n_1 + n_2) K_1 K_2,$$
$$b_1 = K_1 + K_2, \qquad b_2 = K_1 K_2. \qquad (30)$$

We clear the denominator of (29b) and use $r = sc$, so that it becomes

$$s + b_1 r + b_2 rc - a_1 - a_2 c = 0. \tag{31}$$

The unknown coefficients above can be determined from a set of L data points $(r_j, c_j), j = 1, 2, \ldots, L$, utilizing the method of least squares. In that case the algebraic equation for the unknown coefficients are linear, so the problem is reduced to 'linear regression'. A theoretical drawback of such a procedure is that the quantity a_2/b_2 which represents the total number of binding sites per molecule $r_\infty = n_1 + n_2$, will not in general be found to be an integer.

Alternatively, suppose for the moment that the quantity r_∞ is known from the Scatchard curve. Form the function

$$u \equiv (r_\infty - r)/s. \tag{32}$$

By explicit calculations, we find from (29) that

$$u = \frac{r_\infty + (r_\infty b_1 - a_1)c}{a_1 + a_2 c}. \tag{33}$$

Hence, u is precisely of the canonical form of r in (29a), with the advantage that the largest power of c appearing in either the numerator or denominator has been reduced to unity. This has the greatest significance in our example for which the largest power is 2, because we can consider u as a 'new' experimentally determined function r, and introduce a 'new Scatchard function' u/c. Thus, u and u/c are related as follows,

$$\frac{u}{c} = \frac{1}{a_1}\left[r_\infty\left(\frac{1}{c} + b_1\right) - a_1 - a_2 u\right], \tag{34}$$

which is a linear relationship similar to the one that motivated the introduction of the Scatchard plot. Again, linear regression can be utilized to determine the unknown coefficients in the above equation.

Inasmuch as r_∞ is known only approximately from the data, we suggest that an initial 'trial' integral value of r_∞ be chosen with which to define the function u. The procedure is repeated with neighboring integral values of r_∞, and the 'best' value of r_∞ is decided upon by the criterion that the error function E of the method of least squares is minimized. Thus, corresponding to (34), the error function E is defined as

$$E = \sum_{i=1}^{L} \left(\frac{u_i}{c_i}\frac{\alpha_1}{c_i} - \alpha_2 u_i - \beta_1\right)^2, \tag{35}$$

where the coefficients α_1, α_2 and β_1 are to be found by the minimization criterion. Then the a_i and b_i are found by comparison of (34) and (35) from the equations

$$a_1 = \frac{r_\infty}{\alpha_1}, \qquad a_2 = -\frac{r_\infty \alpha_2}{\alpha_1},$$

$$b_1 = \frac{1}{\alpha_1}(1 + \beta_1), \qquad b_2 = -\frac{\alpha_2}{\alpha_1}, \tag{36}$$

and the n_i and K_i in turn found from (30).

In this manner the data of Onoue *et al.* (1968) (see Figure 1.2.2) representing the binding of a ligand to IgM antibodies were considered, for the assumed values of r_∞ equal to 9, 10, and 11. Note that it would be very difficult to decide by extrapolation of the data points what the best value of r_∞ is, and it is useless to try to determine by extrapolation what s_0 is. The resulting parameter values of the minimization procedure are shown in Table 1.2.1. We note that the value $r_\infty = 10$ yields a local minimum of the error E and is therefore the value of choice. The solid line shown in Figure 1.2.2, based on (29) and the parameter values shown in the Table for $r_\infty = 10$, appears to give a satisfactory fit to the data points.

Table 1.2.1. *Representation of ligand binding to IgM antibodies (data of Onoue* et al., *1968)*

r_∞	n_1	n_2	K_1 (μmole l^{-1})$^{-1}$	K_2 (μmole l^{-1})$^{-1}$	E (μmole l^{-1})$^{-2}$
9	4.08	4.92	5.97	0.0293	0.310
10	4.07	5.93	6.24	0.0198	0.122
11	4.26	6.74	5.31	0.0131	0.219

Further support for the parameter choice $r_\infty = 10$ is derived from structural studies indicating that IgM is a pentameric molecule, and from the theoretical conception (Miller & Metzger, 1966) that IgM contains precisely 10 binding sites. It is likewise gratifying that the derived values of n_1 and n_2 are close to integers (see Table 1.2.1), as they should be on theoretical grounds. The fact that n_1 and n_2 are not both equal to 5, as might be expected, is of course unexplained by these considerations.

As a final cautionary word, it is worth remembering that the heterogeneous behavior of the macromolecular solution may be a consequence of the fact that it consists of a fraction γ of one homogeneous molecular species with n_1 binding sites, and a fraction $(1 - \gamma)$ of a second homogeneous species containing n_2 binding sites. Then r would still be represented by (21) with γn_1 replacing n_1 and $(1 - \gamma)n_2$ replacing n_2. In such a case the derived values of 'n_1' and 'n_2' would not be integers.

References

Adair, C. S. (1925). The hemoglobin system. VI. The oxygen dissociation curve of hemoglobin. *J. Biol. Chem.* **63**, 529–45.

Bjerrum, J. (1941). *Metal Ammine Formation in Aqueous Solution*, Copenhagen, P. Haase & Son.

Eadie, G. S. (1942). The inhibition of cholinesterase by physostigmine and prostigmine. *J. Biol. Chem.* **146**, 85–93.

Endrenyi, L., Chan, M.-S. & Wong, J. T.-F. (1971). Interpretation of non-hyperbolic behavior in enzymic systems. II. Quantitative characteristics of rate and binding functions. *Can. J. Biochem.* **49**, 581–98.

Karush, F. (1956). The interaction of purified antibody with optically isomeric haptens. *J. Amer. Chem. Soc.* **78**, 5519–26.

Karush, F. & Sonenberg, M. (1949). Interaction of homologous alkyl sulfates with bovine serum albumin. *J. Amer. Chem. Soc.* **71**, 1369–76.

Klotz, I. M. (1946). The application of the law of mass action to binding by proteins. Interactions with calcium. *Arch. Biochem.* **9**, 109–17.

Lineweaver, H. & Burk, D. (1934). The determination of enzyme dissociation constants. *J. Amer. Chem. Soc.* **56**, 658–66.

Miller, F. & Metzger, H. (1966). Characterization of a human macroglobulin. *J. Biol. Chem.* **241**, 1732–40.

Nisonoff, A. & Pressman, D. (1958). Heterogeneity and average combining constants of antibodies from individual rabbits. *J. Immun.* **80**, 417–28.

Onoue, K., Grossberg, A. L., Yagi, Y. & Pressman, D. (1968). Immunoglobulin M antibodies with ten combining sites. *Science* **162**, 574–6.

Pauling, L., Pressman, D. & Grossberg, A. L. (1944). The serological properties of simple substances. VII. A quantitative theory of the inhibition by haptens of the precipitation of heterogeneous antisera with antigens, and comparison with experimental results for polyhaptenic simple substances and for azoproteins. *J. Amer. Chem. Soc.* **66**, 784–92.

Rubinow, S. I. (1975). *Introduction to Mathematical Biology*, New York, Wiley. (1977). A suggested method for the resolution of Scatchard plots. *Immunochemistry* **14**, 573–6.

Scatchard, G. (1949). The attractions of proteins for small molecules and ions. *Ann. N.Y. Acad. Sci.* **51**, 658–66.

Tanford, C. (1961). *Physical Chemistry of Macromolecules*, New York, Wiley, p. 539.

Werblin, T. P. & Siskind, G. W. (1972). Distribution of antibody affinities: technique of measurement. *Immunochemistry* **9**, 987–1011.

Woolf, B. (1932). Quoted in Haldane, J. B. S. & Stern, K. G. *Allgemeine Chemie der Enzyme*, Dresden & Leipzig, Steinkopf Verlag, p. 119.

1.3 ALLOSTERIC AND INDUCED-FIT
THEORIES OF PROTEIN BINDING

Recall that ligand binding not obeying the hyperbolic law (1.2.15) is called cooperative. Many proteins display cooperative behavior, the most thoroughly studied example being hemoglobin, which possesses four binding sites for oxygen molecules. Moreover, the cooperative behavior of hemoglobin has a very important physiological significance in the transfer of oxygen from the blood of mammals to the tissue. By the same token, although the majority of enzymes obey Michaelis–Menten kinetics, a considerable number of them have been found to display cooperative behavior, and this behavior is likewise of great significance in the regulation and control of biosynthetic processes. Here we shall present the essential ideas underlying two important theoretical models of cooperative behavior of proteins: the **allosteric theory** of proteins put forward by Monod, Wyman & Changeux (1965), MWC theory for short; and the **induced-fit theory** of Koshland, Nemethy & Filmer (1966).

Allosteric theory

In MWC theory the protein is assumed to be an **oligomer**, that is to say, it is constructed from several identical subunits, or **protomers**, each of which contains one active site for binding with a ligand C. Furthermore, the subunits are assumed to be independent of each other, so that the intrinsic association constant of each and every binding site is the same. How then is cooperative behavior of the protein achieved? The answer lies in the assumption that each protomer can undergo a reversible **conformational change**, which alters its state from A to B, say. Further, the intrinsic affinity for the ligand at a binding site is different in each of the two states. Finally, it is assumed that the oligomer itself can exist in only two states, denoted by R and T, in which the protomers are either all in the A state or all in the B state.

Let $R_j(T_j)$ represent the protein state in which j ligand molecules are bound to the state R(T), in equilibrium. Then the possible reactions are expressed as

$$\left.\begin{array}{l} C + R_{j-1} \rightleftarrows R_j, \\ C + T_{j-1} \rightleftarrows T_j, \end{array}\right\} \quad j = 1, 2, 3, \ldots, n,$$

$$R_j \rightleftarrows T_j, \quad j = 0, 1, 2, \ldots, n. \tag{1}$$

The intrinsic dissociation constant in the states R and T are denoted by K_R and K_T, respectively. The dissociation constant representing the transition from R_0 to T_0 is denoted by L and called the **allosteric constant**, or more properly, the **isomerization constant**. The term 'allosteric' is meant to describe a reaction in which the binding of a ligand molecule to a protein at one site influences the binding of a second (identical or different) ligand, through the mediation of a conformation change in the protein. In practice, the term allosteric is used almost automatically to describe reactions exhibiting (positive or negative) cooperativity. Thus, as in the derivation of Adair's equation, which in fact was originally introduced to represent the binding of oxygen to hemoglobin, the equilibria of the reactions of (1) above are represented by (see (1.2.6) and (1.2.14))

$$L = \frac{t_0}{r_0}, \quad \frac{n-j+1}{j}\frac{1}{K_R} = \frac{r_j}{cr_{j-1}},$$

$$\frac{n-j+1}{j}\frac{1}{K_T} = \frac{t_j}{ct_{j-1}}, \quad j = 1, 2, 3, \ldots, n. \tag{2}$$

Here, lower-case italic letters again represent concentrations of quantities represented by associated capital letters.

It might be supposed that n additional parameters should be introduced above to represent the last set of reactions in (1) for j equal to 1 through n, but that is not the case, because of the **principle of detailed balancing** of chemical reactions in equilibrium (Onsager, 1931). According to this principle, the frequency of transitions from one molecular state to another is equal to that in the reverse direction, when the two states are in chemical equilibrium. This has the consequence that the mathematical expression of equilibrium between any two states which are connected to each other by more than one reaction pathway is independent of the pathway. Hence, for every cyclic set of reactions, there is a **loop relation** or equation of constraint on the equilibrium constants representing the elemental reaction steps making up the cycle or 'loop' of reactions. For example, for a set of three cyclic reactions in equilibrium, for which the equilibria among the concentrations of the reactants c_j, $j = 1, 2, 3$, are represented by the association constants K_j as $c_2 = K_1 c_1$, $c_3 = K_2 c_2$, $c_3 = K_3 c_1$, then the loop relation states that $K_1 K_2 = K_3$. In the case of the equilibrium reactions of (1), there are precisely n loops, hence n equations of constraint. Thus, it turns out that no error is made in the following derivation if the last set of reactions for j equal to 1 through n is simply ignored. (For further discussion, see Rubinow, 1975, Section 2.6.)

According to (1.2.8) and (1.2.10), the saturation function is expressed as

$$Y = \frac{\sum\limits_{j=1}^{n} j(r_j + t_j)}{n \sum\limits_{j=0}^{n} (r_j + t_j)}. \tag{3}$$

By substituting (2) into (3), it follows with the aid of some algebraic manipulation and the use of binomial coefficients (Exercise 1; or see Rubinow, 1975, Section 2.6) that

$$Y = \frac{Lcx(1+cx)^{n-1} + x(1+x)^{n-1}}{L(1+cx)^n + (1+x)^n}, \tag{4}$$

where

$$x = c/K_R \quad \text{and} \quad c = K_R/K_T. \tag{5}$$

This is the principal mathematical result of the MWC theory. Note that, if $L \to 0$, then (4) reduces to the noncooperative case equivalent to (1.2.15). We see that in addition to K_R, which only enters into the theory as a scale factor, these are two parameters, L and c. The function Y is able to represent positive cooperativity of the protein–ligand equilibrium binding behavior, but not negative cooperativity (see Section 1.4, below).

Induced-fit theory

Koshland *et al.* (1966) took issue with the assumption of MWC theory that the conformational change of state of the protein had to be an 'all-or-none' or 'concerted' change of state of the subunits. They proposed instead that the subunit changes its state from A to B only as a consequence of the binding of a ligand molecule (thus the description 'induced fit'). In other words, the ligand is only found bound to a subunit in the B state. Hence, proteins exist in hybrid conformational states composed of some subunits in A, and the remainder in B with attached ligand molecules.

These remarks are illustrated in Figure 1.3.1, for the special case in which the protein is a dimer, i.e. it consists of two subunits. Formally, we see that the induced-fit theory is the same as the Adair theory. What has been added is the concept that the equilibrium constants attain their

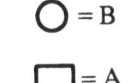

Figure 1.3.1. Allowed states of a dimer according to the induced-fit theory. Subunits in the A and B states are represented by squares and circles respectively. Binding of a substrate molecule to a subunit is represented by the letter S. It is assumed that such binding to a subunit in the state A causes it to be transformed into the state B.

values as a result of conformational change of the subunits. Consequently, the equilibrium constants depend on, and can be expressed in terms of, more fundamental equilibrium constants representing the three possible kinds of pairwise subunit interactions, namely, A–B, A–A, B–B. This dependence is a function of the number of each of the different kinds of subunit interaction pairs and therefore depends on the geometry of attachment of the subunits. For example, in the case of a tetramer (oligomer of four subunits) the induced-fit theory considers three different geometrical configurations, namely, 'linear', 'square', and 'tetrahedral' arrangements of the subunits. (The first attempt to relate the equilibrium constants of the Adair theory to the geometry of the protein was made by Pauling (1935) who assumed identical subunits, and therefore introduced only one pairwise interaction constant.)

Koshland *et al.* also considered a 'concerted' model, in which all the subunits can change their state reversibly from A to B in all-or-none fashion, but only the B state can bind ligand molecules. This can be recognized as the special case of the MWC model for which $c = 0$. The state of the protein which cannot bind ligand molecules is called a **dead-end complex**. They also considered the case of non-identical subunits. For example, it is known that the bare hemoglobin molecule consists of two types of subunits.

For purposes of comparison, the states of a dimer permitted in MWC theory are illustrated in Figure 1.3.2. Eigen (1967) has pointed out that both the MWC theory and the induced-fit theory can both be considered special cases of a more general theory that allows hybrid conformational states, as shown in Figure 1.3.3. It is assumed that in a transition from one state to another, either one subunit changes its state, or a ligand molecule becomes bound (or unbound), but not both. The MWC model (see Figure 1.3.2) permits only the left-most and right-most column of states. The induced-fit model permits only those states that lie along the diagonal line running from upper left to lower right.

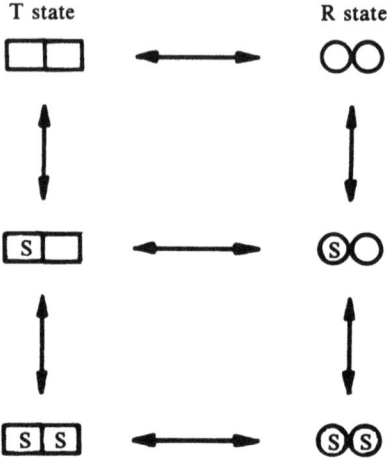

Figure 1.3.2. Allowed states of a dimer according to MWC theory. Only concerted changes of state of subunits are allowed, so that they are all in the state A (T state of the dimer), or in the state B (R state of the dimer).

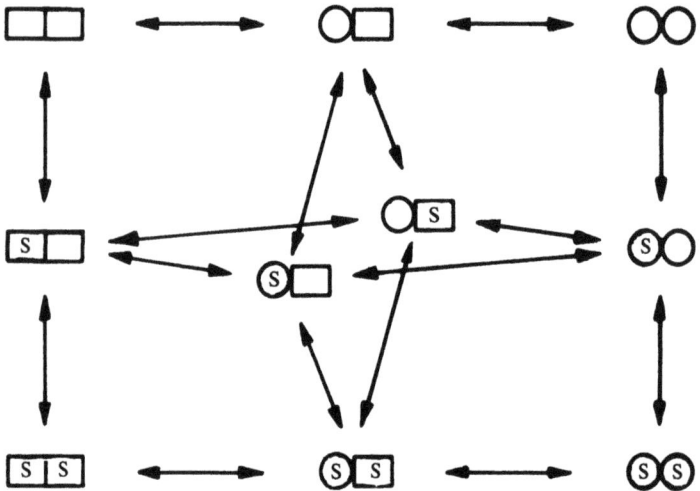

Figure 1.3.3. Representation of the states of a dimer in interaction with a ligand, according to the general formulation of the induced-fit model (Eigen, 1967). In each transition, either one subunit changes its state, or a ligand associates with (or dissociates from) a subunit, but not both.

Actually, as stated strictly, the model requires first, ligand binding, and, secondly, transition from state A to state B, so that the allowed states lie along a 'staircase' pathway just below the diagonal line. Moreover, application of the principle of detailed balancing to one of the cycles of Figure 1.3.3 is equivalent to removing one of the reversible transitions that form the cycle, because no additional information is supplied to the equilibrium conditions by its inclusion. One such transition can be removed for each loop that exists in a network of reactions in equilibrium. Hence, the set of equilibrium reactions represented by Figure 1.3.3 is completely equivalent to the subset of reactions shown in Figure 1.3.4. In

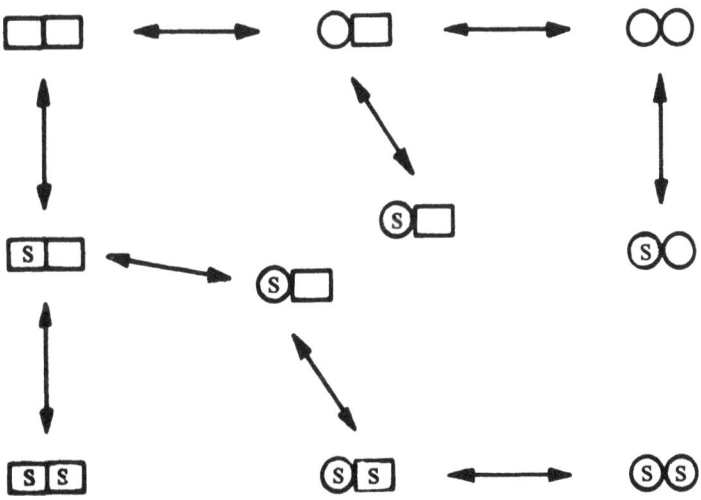

Figure 1.3.4. Subset of reactions equivalent to that of Figure 1.3.3 when the loop relations are exploited.

this form, the staircase pattern of reactions leading from upper left to lower right has been retained, in order to emphasize the formal similarity between induced-fit theory and its generalization. However, the equilibrium reaction system is also equivalent to the subset of reactions shown in Figure 1.3.5, chosen to emphasize the formal similarity of the general theory to MWC theory. From either Figure 1.3.4 or Figure 1.3.5, we infer that the general theory requires 9 equilibrium constants. As indicated by Eigen (1968), for the case of a tetramer the number of allowed states is 35. This number is even greater if the supposition is abandoned that the subunits in a given conformation are completely equivalent.

An example: aspartate transcarbamylase

Perhaps the most extensively investigated example of a protein that was discovered to exhibit cooperative behavior (Gerhart & Pardee, 1962) is aspartate transcarbamylase (ATCase), an enzyme that catalyzes the conversion of the substrates carbamyl phosphate and aspartate to the products carbamyl aspartate and phosphate. (For reviews of its

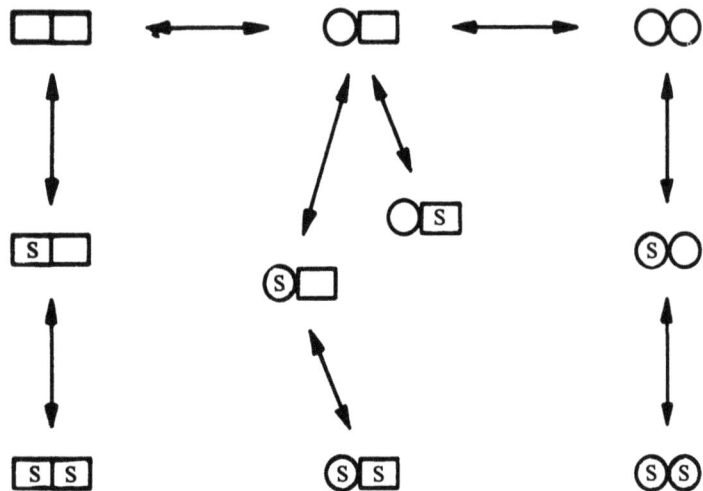

Figure 1.3.5. Subset of reactions equivalent to that of either Figure 1.3.3 or Figure 1.3.4.

properties, see Gerhart, 1970; Jacobson & Stark, 1973; Schachman, 1974.) This conversion is the first step in the biosynthesis of pyrimidines (cytosine, thymine, and uracil), some of the fundamental building blocks of the nucleic acids. ATCase is known to react also with cytosine triphosphate (CTP), one of the end products in the biosynthetic reactions initiated by ATCase. The binding of CTP to ATCase changes the catalytic behavior of the enzyme, making it more positively cooperative in its interaction with its substrates. This effect is nullified in the presence of adenosine triphosphate (ATP), presumably as a consequence of competition with CTP at its binding site. Moreover, ATCase can be made to dissociate into two kinds of subunits. One of them, called the **catalytic subunit**, possesses the catalytic function of ATCase, but does not react with CTP. The other one, called the **regulatory subunit**, reacts with CTP and does not possess any catalytic ability. Structural studies have shown that there are six catalytic sites (one site being able to react with a pair of molecules, carbamyl phosphate and aspartate) and six regulatory sites on each native ATCase molecule.

A theoretical model of ATCase has been recently introduced (Dembo & Rubinow, 1977) that is able to account for initial velocity studies, equilibrium binding studies of ATCase with substrate analogs that do not form products, and relaxation studies which yield quantitative information concerning changes of the enzyme molecule. No attempt is made to incorporate properties of the enzyme resulting from nucleotide (ATP, CTP) binding studies. The principal assumptions of the model are as follows.

(*a*) ATCase consists of three identical, noninteracting, cooperative dimers, or so-called 'allosteric units' (Markus, McClintock & Bussel, 1971).

(*b*) Ligand binding is highly ordered: carbamyl phosphate must bind before aspartate, and aspartate must vacate a binding site before carbamyl phosphate does.

(*c*) The enzyme molecule undergoes a slow concerted conformational change, such as postulated in MWC theory, but only after a carbamyl phosphate molecule is bound.

(*d*) The second active site is available for binding only after two conditions are fulfilled, namely, that the first site has been occupied by both a carbamyl phosphate molecule and an aspartate molecule, and that the slow conformational change of the enzyme molecule has taken place.

(*e*) When both sites are fully occupied, the dimer undergoes a fast conformational change, as postulated in induced-fit theory.

(*f*) The aspartate analog succinate can bind to the free dimeric subunit, and forms a dead-end complex unable to bind a single molecule of carbamyl phosphate.

Thus the enzyme ATCase, according to this model, exhibits some features of the induced-fit theory and some of the MWC theory. However, unlike the latter, the concerted conformational change occurs only after one of its ligands binds. Furthermore, even within a conformational state of the dimeric unit, cooperativity of ligand binding is permitted. Because all the complexity of conditions (*a*)–(*f*) above seem necessary to account for some of the ATCase data, it appears that MWC theory and induced-fit theory should be viewed as valuable simple constructs that serve to aid our understanding and thinking about the cooperative properties of proteins, even if it should turn out in the future that no proteins strictly obey these idealized models.

Exercise

1   Using (2), show that

(*a*) $\displaystyle\sum_{j=0}^{n} r_j = r_0 \sum_{j=0}^{n} b_j^n x^j = (1+x)^n$,

(*b*) $\displaystyle\sum_{j=1}^{n} jr_j = r_0 x \frac{\mathrm{d}}{\mathrm{d}x} \sum_{j=0}^{n} b_j^n x^j = r_0 x \frac{\mathrm{d}}{\mathrm{d}x}(1+x)^n$,

where

$$b_j^n = \frac{n!}{j!(n-j)!}$$

is the binomial coefficient, and $n! \equiv 1 \cdot 2 \cdot 3 \ldots (n-1) \cdot n$ is the factorial function. Here j and n are integers, with $j \leqslant n$. (By definition, $0! \equiv 1$.)

$$\left[\text{Hint: } (y+x)^n = \sum_{j=0}^{n} b_j^n y^{n-j} x^j.\right]$$

References

Dembo, M. & Rubinow, S. I. (1977). A kinetic model of cooperativity in aspartate transcarbamylase. *Biophys. J.* **18**, 245–67.

Eigen, M. (1967). Kinetics of reaction control and information transfer in enzymes and nucleic acids. In *Fast Reactions and Primary Process in Chemical Kinetics*, ed. S. Claesson, Nobel Symposium vol. 5, Stockholm, Interscience, Almqvist & Wiksell, pp. 333–69.

———— (1968). New looks and outlooks on physical enzymology. *Quart. Rev. Biophys.* **1**, 3–33.

Gerhart, J. C. (1970). A discussion of the regulatory properties of aspartate transcarbamylase from *Escherichia coli. Curr. Top. Cell Reg.* **2**, 275–325.

Gerhart, J. C. & Pardee, A. B. (1962). The enzymology of control by feedback inhibition. *J. Biol. Chem.* **237**, 891–6.

Jacobson, G. R. & Stark, G. R. (1973). Aspartate transcarbamylases. *Enzymes* **9**, 225–308.

Koshland, D. E. Jr, Nemethy, G. & Filmer, D. (1966). Comparison of experimental binding data and theoretical models in proteins containing subunits. *Biochemistry* **5**, 365–85.

Markus, G., McClintock, D. K. & Bussel, J. B. (1971). Conformational changes in aspartate transcarbamylase. *J. Biol. Chem.* **246**, 762–71.

Monod, J., Wyman, J. & Changeux, J. P. (1965). On the nature of allosteric transitions: a plausible model. *J. Mol. Biol.* **12**, 88–118.

Onsager, L. (1931). Reciprocal relations in irreversible processes. *Phys. Rev.* **37**, 405–26.

Pauling, L. (1935). The oxygen equilibrium of hemoglobin and its structural interpretation. *Proc. Nat. Acad. Sci., USA* **21**, 186–91.

Rubinow, S. I. (1975). *Introduction to Mathematical Biology*, New York, Wiley.

Schachman, H. K. (1974). Anatomy and physiology of a regulatory enzyme – aspartate transcarbamylase. *Harvey Lect.* **68**, 67–113.

1.4 POSITIVE AND NEGATIVE COOPERATIVITY

The importance of cooperativity has already been mentioned at the beginning of the previous section, and we shall see in later chapters several specific instances where the presence of cooperativity has important physiological consequences. Typically it is the shape of the saturation function that is of physiological importance. Strictly from the point of view of physiology, it makes little difference what molecular mechanism is responsible for the observed shape. But of course a deeper understanding of the physiological behavior requires knowledge of the underlying mechanisms.

In this section we study more closely the concept of cooperativity that has already been introduced and discuss some of the various molecular mechanisms responsible for its appearance. We adopt here the following precise molecular definition of cooperativity. We say that an oligomer in reaction equilibrium with a ligand exhibits 'positive cooperativity' when the fraction of bound sites at any given ligand concentration is larger than that expected for the case of identical, independent sites, and 'negative cooperativity' when the fraction of bound sites is smaller than expected. Moreover, we shall see that there is an unambiguous operational realization of this definition. That is to say, the definition leads to a unique identification of cooperativity.

We have already encountered some of the special graphs – Lineweaver–Burk plot, Scatchard plot – that are employed to make easier the process of estimating the parameters that govern biochemical reactions. We shall discuss in this section what can be deduced about the nature of cooperativity from the appearance of these plots as well as the Hill plot.

The identification of cooperativity in binding

Let us suppose we are studying the equilibrium binding of a ligand to an oligomer of unknown character. Let the saturation function Y be given as a function of the ligand concentration x by (1.2.9) and (1.2.10), i.e. $Y = Y(x)$. We shall call the curve $Y(x)$ the **standard curve** or **standard**

plot. As we have previously indicated, if the number of oligometric binding sites n is unity, or if the sites are identical and independent, then Y satisfies the equivalent of (1.2.15),

$$Y = Kx/(1+Kx), \tag{1}$$

where K is the intrinsic association constant at a binding site, defined by (1.2.13).

For our unknown oligomer, let us examine Y for very small values of x, and determine the quantity $\bar{K}$ defined as

$$\bar{K} = (dY/dx)_{x=0}. \tag{2}$$

If Y satisfied (1), then $\bar{K}$ would be equal to K. Hence, the meaning of $\bar{K}$ is that it represents the **average association constant** for the first ligand–oligomer binding process. Clearly, because $\bar{K}$ is determined from measurements of Y at very small concentrations, no higher order binding processes contribute to $\bar{K}$.

Using this quantity $\bar{K}$, we form the **reference saturation function** $\bar{Y}(x)$ defined as

$$\bar{Y} = \bar{K}x/(1+\bar{K}x). \tag{3}$$

In accordance with our previous discussion of cooperativity, we say that the binding represented by $Y(x)$ exhibits **positive cooperativity** if for all $x > 0$,

$$Y(x) > \bar{Y}(x). \tag{4}$$

Conversely, if for all $x > 0$,

$$Y(x) < \bar{Y}(x), \tag{5}$$

then the binding is said to exhibit **negative cooperativity**.

An additional definition will prove useful. We shall say that the binding curve is **sigmoidal** if the plot of Y starts near the origin with negative curvature (concave upward) and then makes a single change in curvature, so that the curve is concave downward for sufficiently large values of ligand concentration.

The reader should be warned that our definitions do not cover all interesting possibilities, e.g. the observed binding curve could start above the reference Michaelis–Menten curve and then fall below it, or the observed curve could exhibit several changes in curvature. These possibilities are explored in the following section. Also, our definitions of cooperativity are not uniformly agreed upon in the literature. The principal reason for this disagreement is that the definitions in the literature are usually made in terms of a particular plot, while we have proceeded here from a more fundamental molecular viewpoint.

Equilibrium ligand–dimer binding

We shall confine most of our detailed study of molecular mechanisms to the dimer, because it is the simplest example of an oligomer both to analyze and to understand. First we shall assume that the sites are initially identical with respect to binding of a single ligand molecule. Thus there are only three states of the dimer to consider – when no ligand is bound, when one ligand is bound, and when two ligands are bound. We shall denote these states by C_j, where j is the number of bound ligand molecules. The directed line segments of Figure 1.4.1 represent the

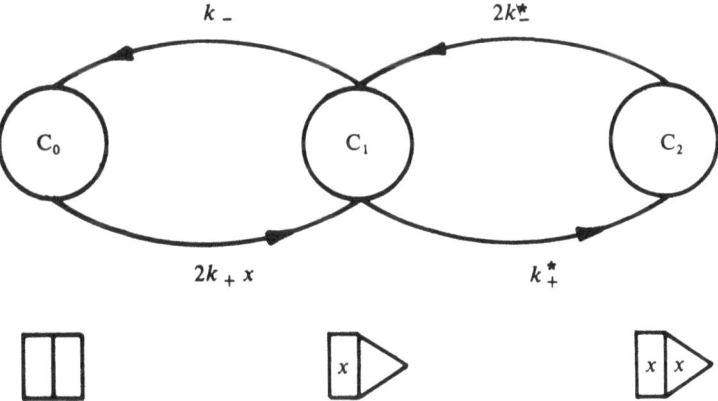

Figure 1.4.1. Graphical representation of transitions between the states of a dimer with initially identical sites. The three different states are also represented schematically, with the letter x indicating a bound ligand.

inter-state transitions, and the associated rate constants are indicated by their labels. Note in the figure that the overall rate constant for conversion from the state C_0 to the state C_1 has a factor 2, reflecting the fact that two sites are available for binding of a ligand molecule to the state C_0, and retaining for k_+ the meaning of an intrinsic forward rate constant at a single site. We denote by k_+^* and k_-^* the intrinsic forward and backward rate constants for the reaction between states C_1 and C_2; in general these constants are different from their unstarred counterparts. Comparison with the notation (1.2.4) used in deriving the Adair equation shows that

$$K_1 = 2K, \qquad K_2 = \tfrac{1}{2}K^*, \tag{6}$$

where $K = k_+/k_-$ and $K^* = k_+^*/k_-^*$ are the **intrinsic association constants** for binding at the first and second sites, respectively. With the present notation, the function $Y(x)$ is, from (1.2.9) and (1.2.10),

$$Y = (Kx + KK^*x^2)/(1 + 2Kx + KK^*x^2). \tag{7}$$

According to (2) and (3), the reference saturation function is

$$\bar{Y} = Kx/(1+Kx), \tag{8}$$

and the difference $Y - \bar{Y}$ is found, with the aid of a little algebra, to satisfy

$$Y - \bar{Y} = \frac{(K^* - K)Kx^2}{(1+Kx)(1+2Kx+KK^*x^2)}. \tag{9}$$

Hence, if $K^* > K$ then $Y > \bar{Y}$ and the binding is positively cooperative. Conversely, if $K^* < K$, $Y < \bar{Y}$, and the binding is negatively cooperative. In other words, binding is positively (negatively) cooperative when the intrinsic association constant for binding the second ligand molecule is greater (less) than that for the first ligand molecule. This result accords with the usual intuitive molecular meaning of positive and negative cooperativity in the literature (Cornish-Bowden & Koshland, 1975).

For our further investigations, the utilization of nondimensional variables and parameters facilitates and simplifies the mathematical investigation. Hence, we introduce the nondimensional quantities

$$\xi = Kx, \qquad \beta = K^*/K. \tag{10}$$

In terms of these, the nondimensional form of Y is

$$Y(\xi) = \frac{\xi(1+\beta\xi)}{1+2\xi+\beta\xi^2}, \tag{11}$$

while

$$\bar{Y}(\xi) = \xi/(1+\xi), \tag{12}$$

so that

$$Y - \bar{Y} = \frac{(\beta-1)\xi^2}{(1+\xi)(1+2\xi+\beta\xi^2)}. \tag{13}$$

Hence positive (negative) cooperativity requires $\beta > 1$ $(\beta < 1)$.

To investigate sigmoidality, we find by direct calculation (Exercise 1) that

$$\frac{dY}{d\xi} = \frac{1+2\beta\xi+\beta\xi^2}{(1+2\xi+\beta\xi^2)^2}, \tag{14}$$

and

$$\frac{d^2Y}{d\xi^2} = 2\frac{\beta-2-\beta\xi[3+3\beta\xi+\beta\xi^2]}{(1+2\xi+\beta\xi^2)^3}. \tag{15}$$

The quantity in the square brackets increases from three as ξ increases from zero. Thus the second derivative is always negative when $\beta < 2$, but the second derivative is positive in a region of ξ near the origin when $\beta > 2$. That is, Y is sigmoidal when $\beta > 2$. This result thus shows the inappropriateness of the assumption, sometimes made, that positive cooperativity is to be identified with sigmoidality, for Y exhibits positive cooperativity but is not sigmoidal when $1 < \beta < 2$.

Figure 1.4.2 shows some standard curves $Y(\xi)$ for various values of β. We see that sigmoidality is apparent only for rather strong positive cooperativity ($\beta \gg 1$), while negative cooperativity never shows qualitative characteristics that are markedly different from the noncooperative case.

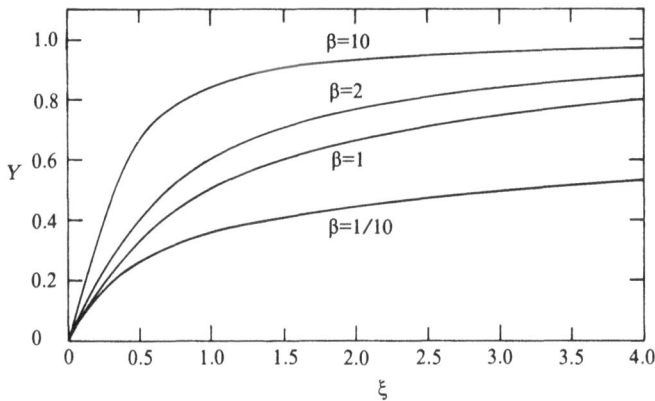

Figure 1.4.2. The standard plot of the nondimensional saturation function (11) for the dimer with initially identical sites.

Various special plots are used instead of the standard plot, in order to bring out different features of the kinetics. In Section 1.1 we mentioned that the Lineweaver–Burk plot of $1/Y$ versus $1/x$ gives a straight line for noncooperative binding. In the present case we see with the definition $\eta = 1/\xi$ that

$$\frac{1}{Y} = \frac{\beta + 2\eta + \eta^2}{\beta + \eta}. \tag{16}$$

Furthermore,

$$\frac{d(1/Y)}{d\eta} = \frac{\beta + 2\beta\eta + \eta^2}{(\beta + \eta)^2}, \tag{17}$$

and

$$\frac{d^2(1/Y)}{d\eta^2} = \frac{2\beta(\beta - 1)}{(\beta + \eta)^3}. \tag{18}$$

Hence, the slope of the Lineweaver–Burk curve is always positive and the curvature is positive (negative, zero) accordingly as β is greater than (less than, equal to) zero. Hence the curvature of the Lineweaver–Burk plot yields a direct indication of the cooperativity of the system. Figure 1.4.3 displays the Lineweaver–Burk curves corresponding to Figure 1.4.2.

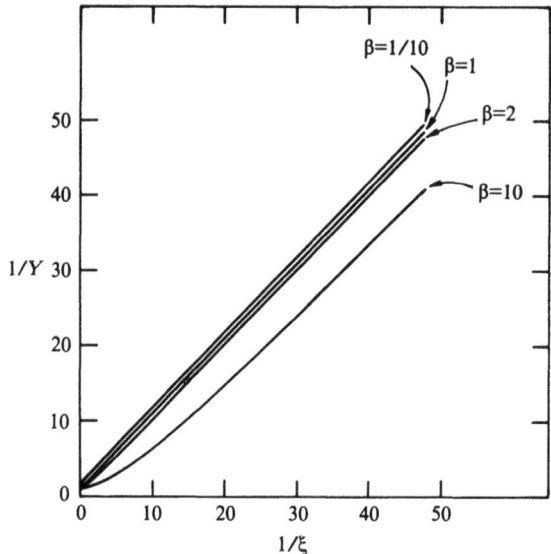

Figure 1.4.3. Lineweaver–Burk plots corresponding to the standard plots of Figure 1.4.2.

As pointed out in Section 1.2, in the Scatchard plot the function $S = Y/x$ is plotted as a function of Y, yielding a straight line for Michaelis–Menten kinetics. However, this linearity is lost when the system is cooperative. It may be readily calculated (Exercise 2) that, for $Y = Y(\xi)$ as given in (11), and $S = Y/\xi$,

$$\frac{\mathrm{d}S}{\mathrm{d}Y} = \frac{\mathrm{d}S}{\mathrm{d}\xi} \cdot \frac{\mathrm{d}\xi}{\mathrm{d}Y} = \frac{\beta - 1 - (1 + \beta\xi)^2}{1 + 2\beta\xi + \beta\xi^2}, \tag{19}$$

$$\frac{\mathrm{d}^2 S}{\mathrm{d}Y^2} = 2\beta(1 - \beta)\frac{(1 + 2\xi + \beta\xi^2)^3}{(1 + 2\beta\xi + \beta\xi^2)^3}. \tag{20}$$

Hence, the curvature of the Scatchard curve determines the cooperativity of the dimer, being always negative (positive, zero) accordingly as the cooperativity is positive (negative, zero). These observations are from Endrenyi, Chan & Wong (1971). Figure 1.4.4 shows the family of Scatchard curves corresponding to Figure 1.4.2.

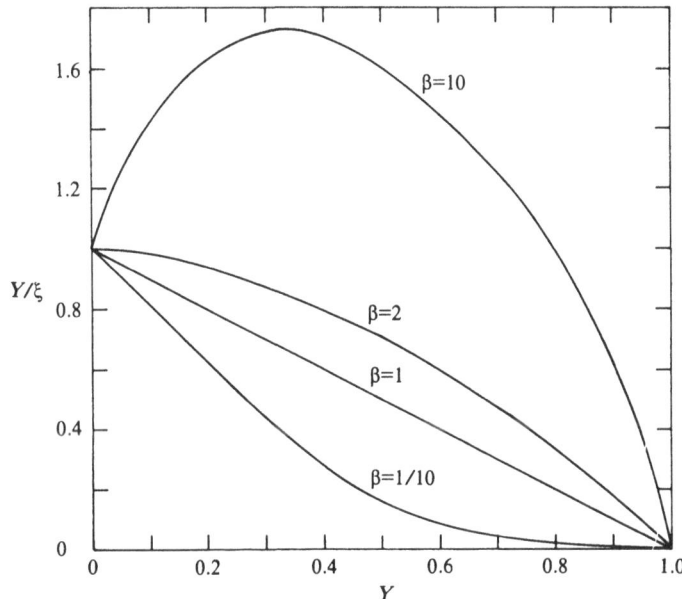

Figure 1.4.4. Scatchard plots corresponding to the standard plots of Figure 1.4.2.

The Hill plot (Hill, 1910) was originally introduced as a consequence of the following type of reasoning. Suppose that an oligomer of n subunits combines with substrate molecules so rapidly that all intermediate states between the bare and the fully occupied oligomer can be neglected. Then the reaction is symbolized as

$$P_0 + nx \underset{k_{n-}}{\overset{k_{n+}}{\rightleftharpoons}} P_n, \tag{21}$$

where P_n denotes the oligomer with n bound sites. The steady state and conservation equations for such a system are

$$-k_{n+}P_0x^n + k_{n-}P_n = 0, \qquad P_0 + P_n = \bar{P}, \tag{22}$$

where $\bar{P}$ is the total oligomer concentration. The saturation function in this case is

$$Y \equiv \frac{nP_n}{n\bar{P}} = \frac{K_n x^n}{1 + K_n x^n}, \tag{23}$$

where $K_n = k_{n+}/k_{n-}$. Since $Y/(1-Y) = K_n x^n$, we have

$$\ln[Y/(1-Y)] = \ln K_n + n \ln x, \tag{24}$$

and a straight line of slope n is obtained if the quantity in the left is plotted as a function of ln x. Such a plot is called the **Hill plot**. The quantity

$$n_H = \frac{d \ln [Y/(1-Y)]}{d(\ln x)} \tag{25}$$

is called the **Hill number**. A plot of the **Hill number** as a function of x is often routinely made, and the result used to estimate the number of binding sites on an oligomer in reaction equilibrium with a ligand. However, the above discussion shows that this interpretation is strictly valid only under very special circumstances.

In the case of a dimer obeying equation (11),

$$\ln \left(\frac{Y}{1-Y} \right) = \ln \xi + \ln \left(\frac{1+\beta\xi}{1+\xi} \right), \tag{26}$$

so that

$$n_H \equiv \frac{d}{d(\ln x)} \ln \left(\frac{Y(x)}{1-Y(x)} \right)$$

$$= \xi \frac{d}{d\xi} \ln \left(\frac{Y(\xi)}{1-Y(\xi)} \right) = 1 + \frac{(\beta-1)\xi}{(1+\xi)(1+\beta\xi)}. \tag{27}$$

Thus for the dimer, n_H is greater (less) than unity for $\xi > 0$ accordingly as β is greater (less) than unity, i.e. accordingly as cooperativity is positive (negative). In general, as this example shows, n_H will depend on concentration. Either the value of n_H at half saturation or an extreme value of n_H is frequently taken as an estimate of the number of sites involved in the reaction. In the present case (Exercise 3) analysis of (11) shows that $Y(\xi) = \frac{1}{2}$ when $\xi = \beta^{-\frac{1}{2}}$, and that Y has an extremum at this same value of ξ (a maximum if $\beta > 1$ and a minimum if $\beta < 1$). At $\xi = \beta^{-\frac{1}{2}}$, the half saturation value of n_H, which we denote as $\bar{n}_H$, is given as

$$\bar{n}_H = 1 + \frac{\beta^{\frac{1}{2}}-1}{\beta^{\frac{1}{2}}+1}. \tag{28}$$

Hence

$$\begin{aligned} \bar{n}_H < 1 \quad &\text{if } \beta^{\frac{1}{2}} < 1, \\ \bar{n}_H > 1 \quad &\text{if } \beta^{\frac{1}{2}} > 1, \end{aligned} \tag{29}$$

and, in fact, $\bar{n}_H \to 2$ from below if $\beta^{\frac{1}{2}} \to \infty$. The family of curves of Figure 1.4.2 are shown in a Hill plot in Figure 1.4.5.

We conclude that the Hill plot provides a sound basis for deciding on the nature of cooperativity for the dimer with identical sites. However, the maximum or half-saturation value of the Hill number has no meaning with respect to the true number of binding sites unless $\beta^{\frac{1}{2}}$ is large, when it approaches, but is less than, this number.

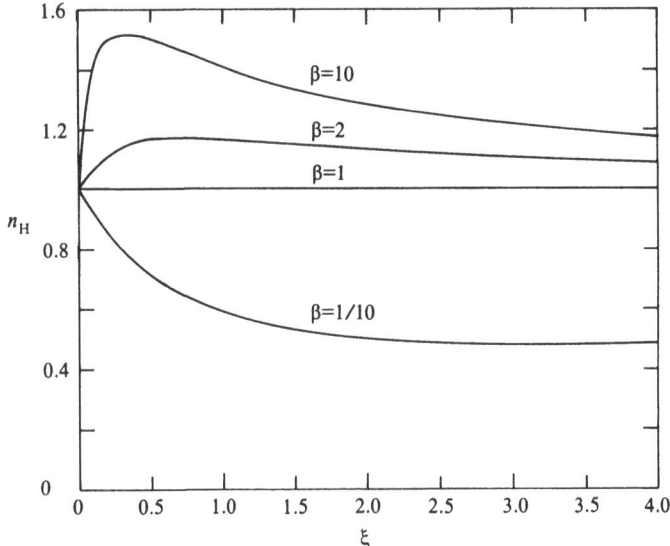

Figure 1.4.5. Hill plots corresponding to the standard plots of Figure 1.4.2.

Let us return to the question of when the Hill kinetic representation (23) is appropriate. As we have remarked, it is certainly permissible to construct a Hill plot irrespective of the validity of (23), which was used only for motivation. But our question is important in view of the fact that many authors use the Hill law as a simple representation of cooperative kinetics.

For the dimer, the Hill law will hold if the exact kinetic expression (7) can be approximated by

$$Y = KK^*x^2/(1 + KK^*x^2). \tag{30}$$

This approximation is *permitted* if $KK^*x^2 \gg Kx$, i.e. if $K^*x \gg 1$. But the approximation is only *useful* if the concentration for half-saturation is within the permitted region $x \gg (K^*)^{-1}$. Otherwise, although (30) closely approximates the true kinetic law, the law itself for the range of x under consideration is hardly distinguishable from the description of a saturated system wherein $Y \approx 1$. According to (30), half-saturation ($Y = \frac{1}{2}$) occurs when $KK^*x^2 = 1$, i.e. when $x = (KK^*)^{-\frac{1}{2}}$. This value of x will lie within the range of validity of (30) if $(KK^*)^{-\frac{1}{2}} \gg (K^*)^{-1}$, i.e. if

$$(K^*)^{\frac{1}{2}} \gg K^{\frac{1}{2}}. \tag{31}$$

Given our initial motivation, it is no surprise that a condition for the validity of the Hill approximation is that the association constant for the second binding be much larger than the corresponding constant for the first binding. (This condition also explains our observation that the half-saturation Hill number is close to the number of binding sites when $\beta \gg 1$.)

The approximation (30) will be inaccurate if x is smaller than, or comparable with, $1/K^*$. But for this range of x, the assumption $(K^*)^{\frac{1}{2}} \gg K^{\frac{1}{2}}$ guarantees that Kx and therefore Y will be very small, so that large percentage errors in Y are normally immaterial. For the dimer then, we have shown that the Hill law is a useful approximation to describe binding, for all ligand concentrations, provided that the condition (31) is satisfied.

Dimer binding: nonidentical sites

We now alter our assumption that the two sites of the dimer are initially identical with respect to binding of a single ligand molecule. We label the two different sites as 1 and 2, with associated rate constants $k_{\pm 1}$ and $k_{\pm 2}$, respectively. It follows that there are now four equilibrium states of the system, the bare dimer C_0, the state with site 1 bound C_1, the state with site 2 bound C_1', and the fully bound state C_2, as illustrated in Figure 1.4.6.

It turns out (Exercise 5(a)) that the saturation function is given by

$$Y = \tfrac{1}{2}(K_1 x + K_2 x + 2K_1 K_3 x^2)/(1 + K_1 x + K_2 x + K_1 K_3 x^2) \quad (32)$$

where

$$K_1 = k_{+1}/k_{-1}, \qquad K_2 = k_{+2}/k_{-2}, \qquad K_3 = k_{+3}/k_{-3},$$

and $k_{\pm 3}$ represents the rate constants associated with the transitions between the states C_1 and C_2. (When $K_3 = K_2$, (32) reduces to $\tfrac{1}{2}r$, where r is given by (1.2.21), as it should.) It might at first appear that the saturation function should depend on $K_4 = k_{+4}/k_{-4}$ as well (see Figure 1.4.6), but that is not the case because of the loop relation that follows from the thermodynamic principle of detailed balancing (see Section 1.3), which states that

$$K_1 K_3 = K_2 K_4. \quad (33)$$

Because K_1 and K_2 appear only in the combination $K_1 + K_2$, the expression (32) is only a two-parameter family, in fact of the same form as (7). This is readily seen by defining

$$J = \tfrac{1}{2}(K_1 + K_2) \quad \text{and} \quad J^* = 2K_1 K_3/(K_1 + K_2), \quad (34)$$

with which (32) becomes

$$Y = (Jx + JJ^* x^2)/(1 + 2Jx + JJ^* x^2). \quad (35)$$

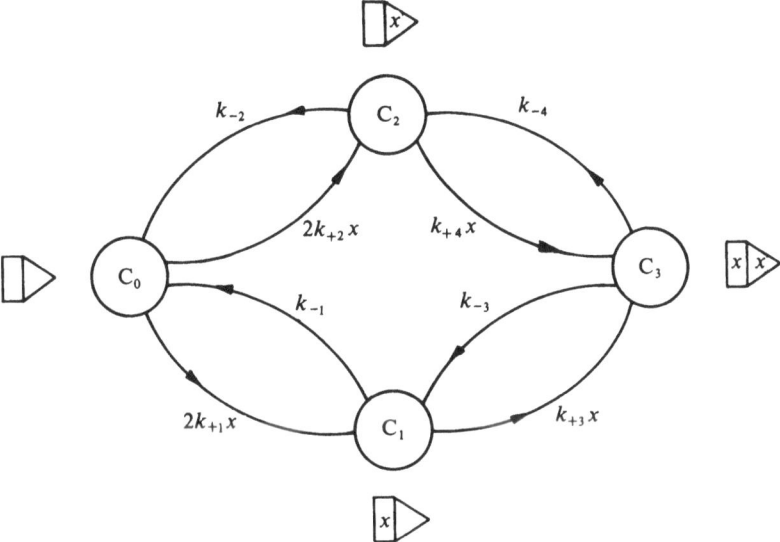

Figure 1.4.6. Graphical representation of transitions between the states of a dimer with two different sites. The four different states are also represented schematically, with the letter x indicating a bound ligand.

Hence, by introduction of $\xi = Jx$, Y is expressible in the form (11) with β replaced by β', where

$$\beta' = J^*/J = 4K_1K_3/(K_1+K_2)^2. \tag{36}$$

However β' now has a more general significance than previously ascribed to β. For example $\beta' = 1$ implies that the geometric mean of the equilibrium constants representing the transition from C_0 to C_2 equals the arithmetic mean of the equilibrium constants for the transition from C_0 to C_1 and C_0 to C_1', i.e. from (36),

$$\sqrt{(K_1K_3)} = \tfrac{1}{2}(K_1+K_2). \tag{37}$$

There are two important limiting cases of the system described in Figure 1.4.6. One occurs if $K_1 = K_2$. Then the states C_1 and C_1' are the same and the reaction scheme becomes identical with that of Figure 1.4.1, consisting of identical but dependent binding sites. The other important limiting case occurs if the binding sites are not identical so that $K_1 \neq K_2$, but the sites are independent, as occurs frequently for many immunoglobulins. Then $K_3 = K_2$, $K_4 = K_1$, and the saturation function is $\tfrac{1}{2}r$, where r is given by (1.2.21). More importantly, (36) can be rewritten as

$$\beta' = \frac{4K_1K_2}{4K_1K_2+(K_1-K_2)^2}, \tag{38}$$

from which we infer that it is always the case that $\beta' \leqslant 1$, with $\beta' = 1$ only when $K_1 = K_2$. From this there follows the well known observation that binding curves that exhibit negative cooperativity can arise either because the binding of one ligand molecule makes the second binding more difficult, or because the protein possesses two different independent binding sites for the ligand.

Cooperativity according to the MWC theory

Another mechanism for cooperativity is afforded by the allosteric theory of Monod *et al.* (see Section 1.3 above). The saturation function $Y(x)$ in this case is given by (1.3.4). The Michaelis constant for the first binding is obtained by examining (1.3.4) for small values of x. We obtain for $x \ll 1$, $Y(x) \approx (Lc + 1)x/(L+1)$, so that our reference saturation function is

$$\bar{Y} = \frac{(Lc + 1)x}{L(1 + cx) + (1 + x)}. \tag{39}$$

Some algebra shows that

$$Y(x) - \bar{Y}(x) = \frac{L(1 - c)x[(1 + x)^{n-1} - (1 + cx)^{n-1}]}{[L(1 + cx)^n + (1 + x)^n][L(1 + cx) + 1 + x]}. \tag{40}$$

It is now apparent that (except for an isolated noncooperative case when $c = 1$, i.e. when binding to both states is the same), it is always the case that $Y(x) - \bar{Y}(x) > 0$. This analysis justifies the frequently made remark in the literature that the allosteric model can only account for positive cooperativity.

Initial velocity studies

In reactions in which an enzyme catalyzes the formation of a product, there have been extensive experimental and theoretical studies of the cooperative behavior of the enzyme as revealed by initial velocity measurements. If the enzyme is an oligomer of n subunits, it can be shown that the initial velocity of the reaction, defined as the initial rate of product formation, is in general expressible as the ratio of two n-th order polynomials. We adopt here the same general procedure for defining cooperativity with respect to initial velocity studies as we did for binding studies.

Consider for reference an 'identical site' oligomer of n subunits. Let the sites be both identical and independent (so that the intrinsic rate constants both for binding and for product formation are the same at each site). Further, assume that the value of this intrinsic product formation rate constant is such that, at large ligand concentrations, the reference oligomer achieves the same maximum initial velocity as the unknown oligomer. Cooperativity of initial velocity behavior is defined now by comparison of the unknown oligomer with the above-defined identical site oligomer, each site of which represents in some sense the average behavior of the sites of the unknown oligomer; if the initial velocity of the unknown oligomer, measured as a fraction of the maximum velocity of the reaction, is greater (less) than that of the comparison identical site oligomer, the oligomer is said to exhibit positive (negative) cooperativity.

Although we have adopted here an unambiguous operational definition of cooperativity with respect to initial velocity studies, in analogy with the analysis of binding studies, we cannot claim that this definition has a clear meaning with respect to the individual enzyme molecules. The difficulties of interpretation will become apparent with our further discussion.

We shall assume that the initial velocity v approaches a finite limit V, as the substrate concentration becomes very large. We introduce the normalized initial velocity U, defined as the ratio of v to V:

$$U = v/V. \tag{41}$$

The function U has a formal resemblance to the saturation function Y. Thus, if all sites are identical and independent, then (as has been shown in (1.1.10)) the velocity obeys the Michaelis–Menten law

$$U = x/(K_m + x), \tag{42}$$

where K_m is the Michaelis constant.

In considering oligomeric enzymes of unknown character with normalized initial velocity functions $U(x)$, we define $\bar{K}_m$ to be the average Michaelis constant for a single ligand–oligomer reaction process, obtained from the behavior of $U(x)$ at small values of ligand concentration x:

$$\bar{K}_m = 1/[dU/dx]_{x=0}. \tag{43}$$

Note that (43) differs from the analogous equation (2) because for binding we employed the association constant while in velocity studies the Michaelis constant is primarily a measure of dissociation, although it involves the product formation process as well.

Using the $\bar{K}_m$ defined in (43) we construct the **reference initial velocity function** $\bar{U}$, based on the Michaelis–Menten law,

$$\bar{U} = x/(\bar{K}_m + x). \tag{44}$$

By analogy with our previous definition, we say that if the observed graph of U always lies above (below) the curve $\bar{U}$, then the initial velocity curve exhibits positive (negative) cooperativity.

We assume now that our enzyme is a dimer, with two identical sites, that catalyzes the conversion of the ligand into a product. All the transitions shown in Figure 1.4.1 are present. In addition, there is a transition from C_1 to C_0 with rate constant k_1 which transforms the ligand to a product molecule, and a transition from C_2 to C_1 with a rate constant designated by $2k_1^*$, also associated with the transformation of one of the bound ligand molecules to a product molecule. The resultant quasi-steady state initial velocity v of product formation is given as (see, for example, Rubinow, 1975, p. 67)

$$v = \frac{2e_0 x[k_1 k_m^* + k_1^* x]}{K_m K_m^* + 2K_m^* x + x^2}. \tag{45}$$

Here

$$K_m = (k_1 + k_-)/k_+, \qquad K_m^* = (k_1^* + k_-^*)/k_+^*, \tag{46}$$

and e_0 is the initial amount of enzyme present.

It is useful to introduce the nondimensional parameters and variables

$$\begin{aligned} \alpha &= k_1/k_1^*, \qquad \gamma = K_m/K_m^*, \qquad V = 2e_0 k_1^*, \\ \xi &= x/K_m, \qquad U = v/V. \end{aligned} \tag{47}$$

With these, (45) may be written as

$$U(\xi) = \xi(\alpha + \gamma\xi)/(1 + 2\xi + \gamma\xi^2). \tag{48}$$

Note that $U \to 1$ as $\xi \to \infty$, so that U is appropriately normalized. For small ξ, $U \approx \alpha\xi$. Consequently, our definition of cooperativity requires us to contrast (48) with the reference initial velocity function

$$\bar{U} = \alpha\xi/(1 + \alpha\xi). \tag{49}$$

We find easily that

$$U(\xi) - \bar{U}(\xi) = \frac{\xi^2(\alpha^2 - 2\alpha + \gamma)}{(1 + 2\xi + \gamma\xi^2)(1 + \alpha\xi)}. \tag{50}$$

Thus, cooperativity is identified as follows:

$$\begin{aligned} \gamma &> \alpha(2 - \alpha), \quad \text{positive cooperativity;} \\ \gamma &< \alpha(2 - \alpha), \quad \text{negative cooperativity.} \end{aligned} \tag{51}$$

Hence if $\alpha > 2$, positive cooperativity is assured, but if $\alpha < 2$, cooperativity depends on the relative values of γ and α as indicated above.

Because the criterion for the identification of cooperativity involves two parameters, α and γ, which in turn are intimately related to both the binding process (ligand association and dissociation) and the product formation process (product dissociation), it should perhaps not surprise us that it seems impossible to make a relatively simple interpretation of the molecular meaning of nonzero cooperativity as identified from initial velocity studies.

To see this better, let us introduce the binding association constants $K = k_+/k_-$, and $K^* = k_+^*/k_-^*$. On the basis of our molecular understanding of the binding process alone, we expect the enzyme to display positive cooperative behavior if $K^* > K$. With $\beta = K^*/K$ (as in (10)) we see from (46) and (47) that the parameter γ is expressed as

$$\gamma = \beta \frac{1 + (k_1^*/k_-)\alpha}{1 + (k_1^*/k_-^*)}. \tag{52}$$

Note that the expression multiplying β is greater (less) than unity if α is greater (less) than k_-/k_-^*. Thus either $\alpha > \delta$ and $\gamma > \beta$ or $\alpha < \delta$ and $\gamma < \beta$, where $\delta \equiv k_-/k_-^*$. The allowed parameter space in the α–γ plane is shown as the stippled region in Figure 1.4.7 with the criterion of cooperativity

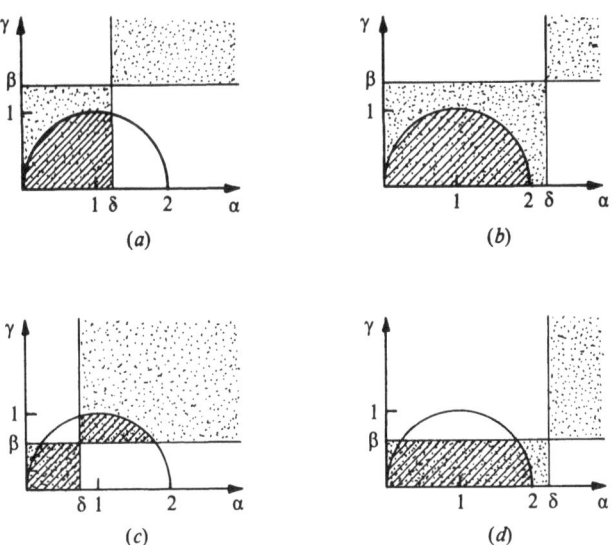

Figure 1.4.7. Stippled regions indicate permitted values of the parameters α and γ in (48) for the normalized initial velocity. The parts of the stippled regions that are hatched represent parameter domains corresponding to negative cooperativity. The remaining stippled regions correspond to positive cooperativity.
(a) $\beta > 1$, $\delta < 2$, (b) $\beta > 1$, $\delta > 2$, (c) $\beta < 1$, $\delta < 2$, (d) $\beta < 1$, $\delta > 2$.
Recall that $\beta = K^*/K = k_+^* k_-/k_+ k_-^*$, $\alpha = k_1/k_1^*$, $\gamma = K_m/K_m^* = [(k_1 + k_-)/k_+]/[(k_1^* + k_-^*)/k_+^*]$, $\delta = k_-/k_-^*$.

(51) superimposed. We see from these curves that four parameters must be considered in characterizing cooperativity. Further, we see that cooperative behavior defined exclusively in terms of the binding parameter β, i.e. β greater or less than unity, does not determine cooperativity in terms of initial velocity behavior, i.e. the criterion (51). We conclude that in spite of formal mathematical similarity between binding and velocity curves, the latter have a much more complicated molecular interpretation. The reason is that both association and dissociation processes interact in a complicated way to produce product. At the present time, no simple molecular interpretation of cooperativity in product formation has been given.

Exercises

1   Verify (14), (15), (17) and (18).
2   Verify (19), (20) and (23).
3   Find the maxima and minima of (11), for positive ξ. Also find the value of ξ at which $Y = \frac{1}{2}$.
4   Verify that (30) approximates (7) if $K^*K \gg 1$.
5   (a) Derive the saturation function (32) for the dimer with nonidentical sites.
       (b) Show that the function Y given in (32) can be written in the form (35), and hence in the form (11) with β as given in (36).
6   Verify (40) and (50).

References

Cornish-Bowden, A. & Koshland, D. E. (1975). Diagnostic uses of the Hill (Logit and Nernst) plots. *J. Mol. Biol.* **95**, 201–12.
Endrenyi, L., Chan, M.-S. & Wong, J. T.-F. (1971). Interpretation of non-hyperbolic behavior in enzyme systems. II. Quantitative characteristics of rate and binding functions. *Can. J. Biochem.* **49**, 581–98.
Hill, A. V. (1910). Possible effects of the aggregation of the molecules of haemoglobin on its dissociation curves. *J. Physiol.* **40**, iv–viii.
Rubinow, S. I. (1975). *Introduction to Mathematical Biology*, New York, Wiley.

1.5 GRAPHICAL REPRESENTATIONS FOR TETRAMER BINDING

The previous section provided a fairly detailed examination of the different types of cooperative behavior that are possible for a dimer. The basic mechanisms underlying the appearance of cooperative behavior are well revealed by a study of the dimer, but these mechanisms can exist in many combinations when oligomers have more than two sites. For dimers with identical sites, for example, binding of one ligand can make binding of the next easier (positive cooperativity), harder (negative cooperativity), or can have no effect (zero cooperativity). Already with trimers, the corresponding situation is much more complicated: binding of one ligand in principal can cause a conformational change that will make binding at both available sites easier, binding at one site easier and at the other harder, or binding at one site easier with the other site unaffected, etc.

In discussing n-mers, $n > 2$, an extended definition of cooperativity may be introduced with separate regions of positivity and negativity. Nonetheless the notions of cooperativity and sigmodality are not sufficient to describe with accuracy the variety of possible behaviors of occupancy and velocity curves. Moreover, graphs of binding and velocity data can exhibit a variety of shapes, and a given shape could be due to several different underlying molecular mechanisms.

In this section we shall illustrate in a study of tetramers some of the issues just mentioned. The reader is referred to the work of Kenett (1978) for a much more extensive theoretical treatment.

Scatchard plots for the tetramer

The basis of our discussion is the analog of (1.4.11) for a tetramer

$$Y = \frac{\xi + 3a\xi^2 + 3a^2b\xi^3 + a^3b^2c\xi^4}{1 + 4\xi + a\xi^2 + 4a^2b\xi^3 + a^3b^2c\xi^4}. \tag{1}$$

Equation (1) gives the saturation function Y as a function of the dimensionless ligand concentration ξ and dimensionless ratios of association constants a, b, and c, where

$$\xi = K_1 x \quad \text{and} \quad a = K_2/K_1, \quad b = K_3/K_2, \quad c = K_4/K_3. \tag{2}$$

In terms of the concentration of tetramer with i bound sites $E_i (i = 0, \ldots, 4)$, the intrinsic association constants K_i are given by

$$K_1 = \frac{E_1}{4E_0 x}, \quad K_2 = \frac{2}{3}\frac{E_2}{E_1 x}, \quad K_3 = \frac{3}{2}\frac{E_3}{E_2 x}, \quad K_4 = \frac{4E_4}{E_3 x}. \tag{3}$$

For the bulk of the discussions in this section, the parameters a, b, and c will be selected from the set of values 10, 1, and 0.1, giving 27 different possible combinations. These specific values represent a binding at the site in question that makes the next binding respectively easier, the same, or harder. Thus we shall make the association

$$10 \to +, \quad 1 \to 0, \quad 0.1 \to -.$$

We thereby describe individual binding events as displaying, relative to the previous binding event, a degree of positive, zero, or negative cooperativity. To give an example of the notation

$$[0 + -] \quad \text{means} \quad a = 1, \quad b = 10, \quad c = 0.1.$$

Let us now consider the 27 Scatchard plots of Y/ξ versus Y, corresponding to the 27 choices of a, b, and c (Figure 1.5.1, (a)–(d)). We shall analyze the results, stressing properties that allow one to distinguish at least to some extent the various sets of kinetic constants from qualitative observations of the nature of the curves.

Initial slope

The magnitude of the initial slope depends principally on the value of the parameter $a \equiv K_2/K_1$. In case $a = 1$, $b \equiv K_3/K_2$ becomes important. The situation is summarized in Table 1.5.1 (see p. 66). Note that if the initial slope is positive it is an increasing function of b, while initial negative slopes are independent of b. Initial slopes are uninfluenced by c.

Final slope

The value of the slope as $Y \to 1$, i.e. as $\xi \to \infty$, depends entirely on the product abc. Indeed a calculation shows that

$$\lim_{\substack{Y \to 1 \\ \text{(i.e. } \xi \to \infty)}} \frac{\mathrm{d}(Y/\xi)}{\mathrm{d}Y} = -abc. \tag{4}$$

This result can be verified from the graphs, for example the curves [+ 0 0] and [+ + −] in Figure 1.5.1*a* have the same final slope.

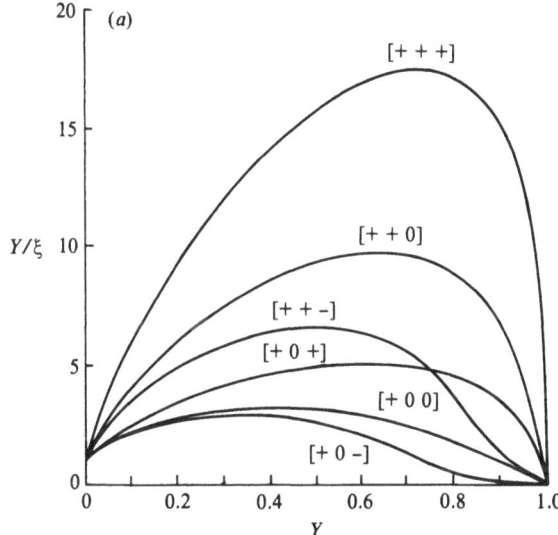

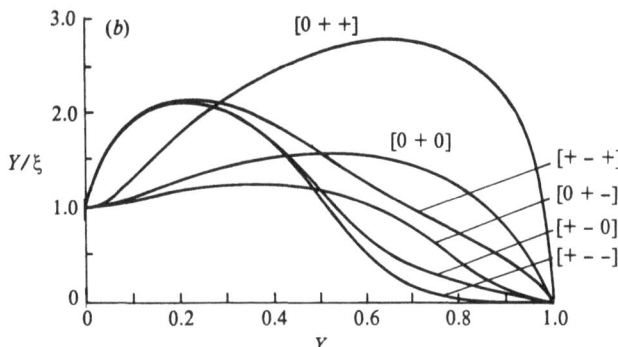

Figure 1.5.1. Scatchard plots for tetramers. Saturation function Y of (1) as a function of the dimensionless concentration $\xi \equiv K_1 x$. The curves are grouped into four families:
(*a*) [+ + ·], [+ 0 ·]; (*b*) [+ − ·], [0 + ·]; (*c*) [0 0 ·], [0 − ·]; (*d*) [− + ·], [− 0 ·], [− − ·].

There are 27 curves corresponding to all possible permutations of the binding coefficient ratios K_2/K_1, K_3/K_2, K_4/K_3 chosen from the values 10, 1, or 0.1. These values are symbolized by + (positive cooperativity), 0 (noncooperativity), − (negative cooperativity). The symbol · stands for all three alternatives.

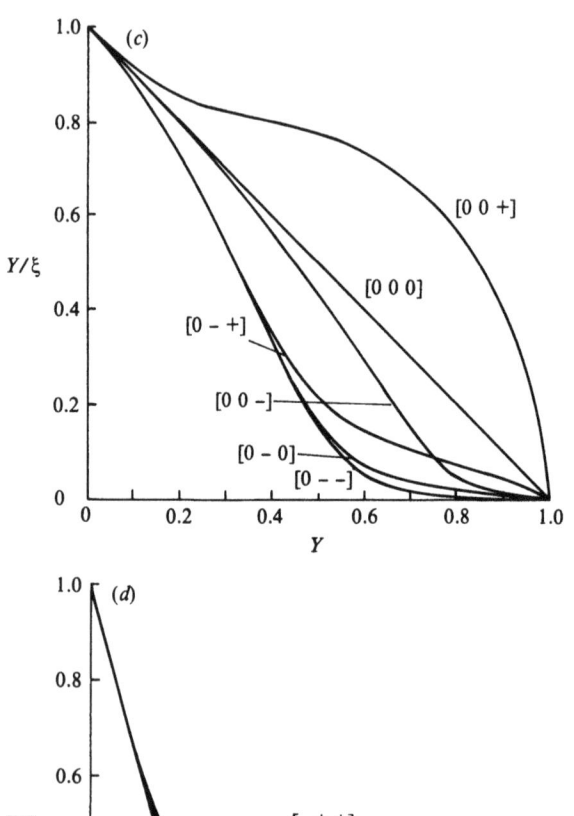

Figure 1.5.1. Continued.

Curvature

Most situations are summarized in Table 1.5.2 (p. 66) with a being the dominant influence on initial curvature and c on final curvature. The three curves $[- - \cdot]$ (where $\cdot$ denotes $+$, $-$, or 0) constitute an exceptional case in that the final curvature is essentially zero, whatever the value of c. The zero cooperativity case $[0\ 0\ 0]$ is a straight line devoid of curvature.

Position of maximum

Only curves of Figure 1.5.1a, b, that start with a non-negative initial slope possess a maximum. The location of the maximum is similar for curves with similar values of the product abc. The three curves $[+ - \cdot]$ form an exception for which the location of the maximum is independent of c.

Variation of ξ

We define ξ_L as the value of ξ necessary to reach $Y = 0.9$. Table 1.5.3 (p. 66) shows how this saturation concentration depends on a, b, and c. The product abc is the most important determining factor. When $c = 1$, ab is also important.

We emphasize that all the above discussion applies to a special subset of parameter values for which a, b, and c are selected from the restricted set 10, 1, 0.1. These values are in general representative, but there are some features of the curves that do depend rather sensitively on the values of a, b, and c. For example, for any curve whose initial slope is not large in magnitude, one would anticipate the possibility that a small change in coefficients could have a rather large effect. This is illustrated in the case $[0 + \cdot]$ of Figure 1.5.1b. If the value of b is changed from $b = 10$ to $b = 2$ then the Scatchard plot changes so that a minimum appears near the origin.

A paper by Gibson & Levin (1977) mentions a minimum like the one just described and presents data that support the existence of this minimum in the Scatchard plot for $[^3H]$acetylcholine. Gibson & Levin point out that such a minimum can be found according to the concerted model for allosteric enzymes, but not for the induced-fit model. Both these models are special cases of the Adair model (Edelstein, 1975) when the coefficients of (1) respectively have the forms

$$a = (1 + Lc^2)(1 + L)/(1 + Lc)^2,$$

$$b = (1 + Lc^3)(1 + Lc)/(1 + Lc^2)^2,$$

$$c = (1 + Lc^4)(1 + Lc^2)/(1 + Lc^3)^2,$$

and $a = b = c = K_{BB}/(K_{AB})^2$ where L, c, K_{BB}, and K_{AB} refer respectively to the constants involved in the allosteric and induced-fit models (see Section 1.3). Of course, one cannot rule out other possible special cases of Adair kinetics that could exhibit such a minimum.

The Hill plot and the Hill approximation for tetramers

We call attention to a valuable discussion of Hill plots, and discuss briefly the validity of the Hill approximation for tetramers. Cornish-Bowden & Koshland (1975) displayed graphs of saturation functions for essentially the same parameter sets used here, except that they did not use K_1 to nondimensionalize ligand concentration but rather chose K_1 to have values between $10^{-1.5}$ and $10^{1.5}$ in order to center the curves around the origin. These graphs are shown in Figure 1.5.2. Cornish-Bowden & Koshland's careful discussion of the curves includes such matters as the uses of the asymptotes in estimating K_1 and K_4 and the effects of random and systematic error. Previously we calculated the conditions required to assure that the Hill approximation will hold for the dimer. For the tetramer, the argument is analogous to that used before. Thus it is permitted to use the saturation function

$$Y = K_1 K_2 K_3 K_4 x^4 / (1 + K_1 K_2 K_3 K_4 x^4) \tag{5}$$

as an approximation to the correct function (the dimensional version of (1)) if

$$x \gg \max [K_4^{-1}, (K_3 K_4)^{-\frac{1}{2}}, (K_2 K_3 K_4)^{-\frac{1}{3}}].$$

This approximation is useful only if the half-saturation concentration $x = (K_1 K_2 K_3 K_4)^{\frac{1}{4}}$ lies within the permitted range of concentration. Hence the condition assuring that the Hill law may be used is

$$\max [K_4^{-1} (K_3 K_4)^{-\frac{1}{2}}, (K_2 K_3 K_4)^{-\frac{1}{3}}] \ll (K_1 K_2 K_3 K_4)^{-\frac{1}{4}} \tag{6}$$

i.e. if

$$\max [K_1^{\frac{1}{4}} K_2^{\frac{1}{4}} K_3^{\frac{1}{4}} K_4^{-\frac{3}{4}}, K_1^{\frac{1}{4}} K_2^{\frac{1}{4}} K_3^{-\frac{1}{4}} K_4^{-\frac{1}{4}}, K_1^{\frac{1}{4}} K_2^{-\frac{1}{12}} K_3^{-\frac{1}{12}} K_4^{-\frac{1}{12}}] \ll 1.$$

This condition requires that K_1 be sufficiently small compared to K_2, K_3, and K_4. As before, since errors at small concentrations are unimportant, it turns out that (5) may be used over the entire range of concentrations.

Standard and Lineweaver–Burk plots for tetramers

We close this section by presenting, for the same 27 cases as previously considered, tables summarizing the appearance of both standard plots (Figure 1.5.3) and Lineweaver–Burk plots (Figure 1.5.4). The philosophy of extracting information from such plots should be clear from the present discussion of Scatchard plots and from the discussion of Hill plots by Cornish-Bowden & Koshland (1975). It remains only to emphasize the remark that the different plots differently magnify some regions and compress others. If it is desired to extract maximum information from a given set of data, it is probably wise to make all four plots.

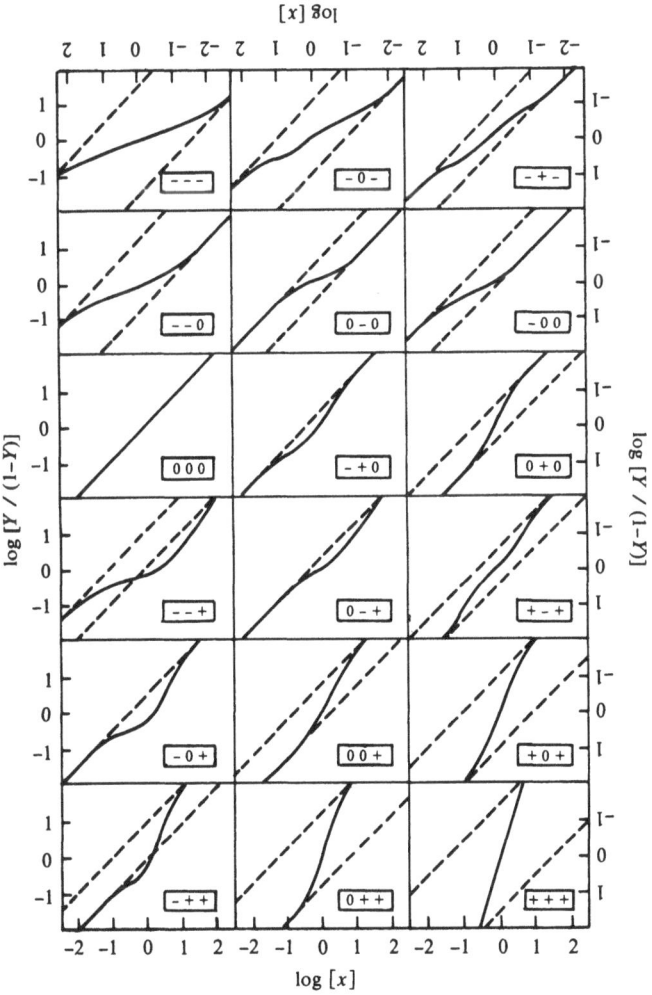

Figure 1.5.2. Hill plots (Figure 2 of Cornish-Bowden & Koshland, 1975). Dashed lines indicate asymptotes. Saturation function Y as a function of the dimensional concentration x. Nine new curves can be seen by viewing the figure upside down.

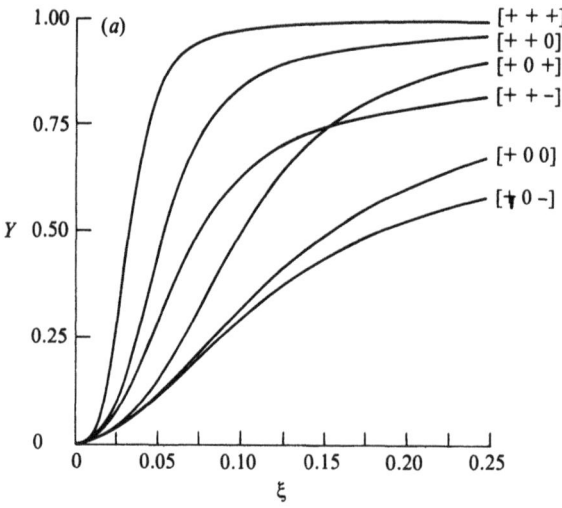

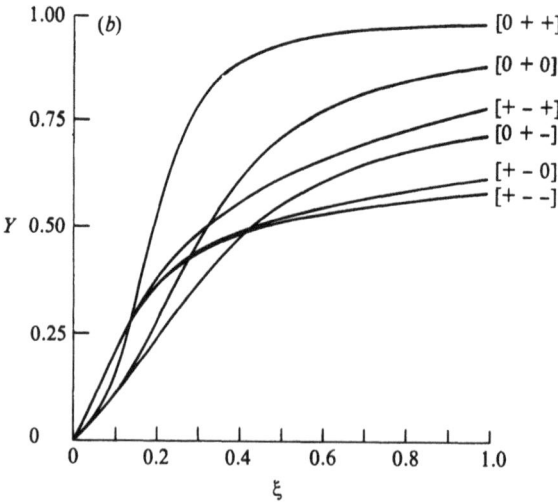

Figure 1.5.3. Standard plots. Saturation function Y as a function of the dimensionless concentration $\xi \equiv K_1 x$. The curves are here grouped into five families (a)–(e). The groups are the same as those in Figure 1.5.1 except that the family $[- - \cdot]$ appears in (e).

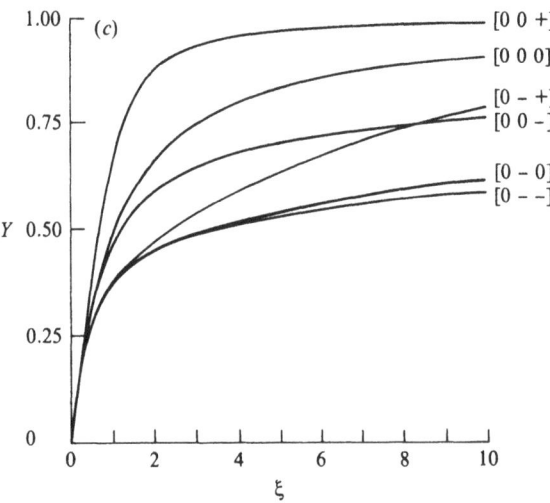

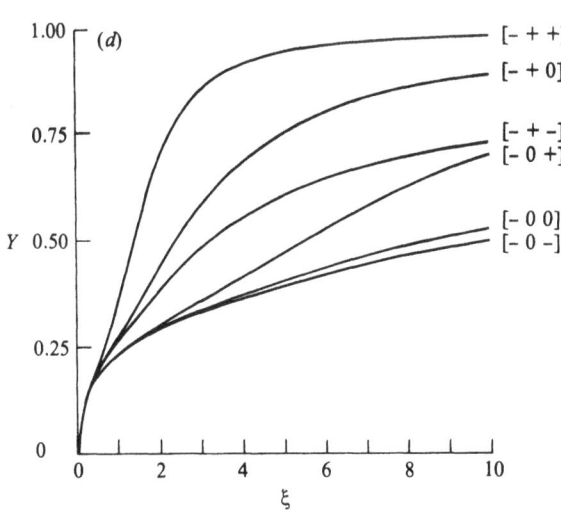

Figure 1.5.3. Continued.

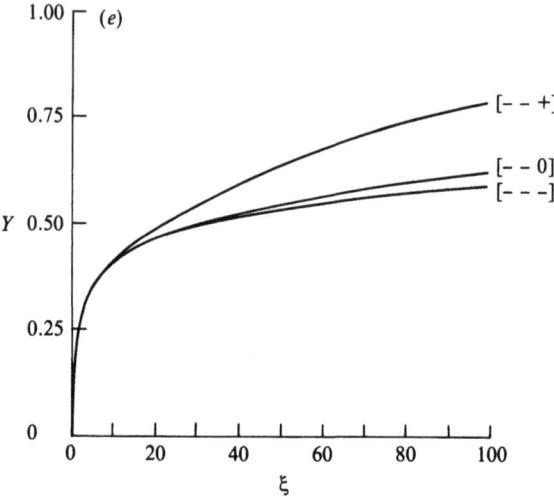

Figure 1.5.3. Continued.

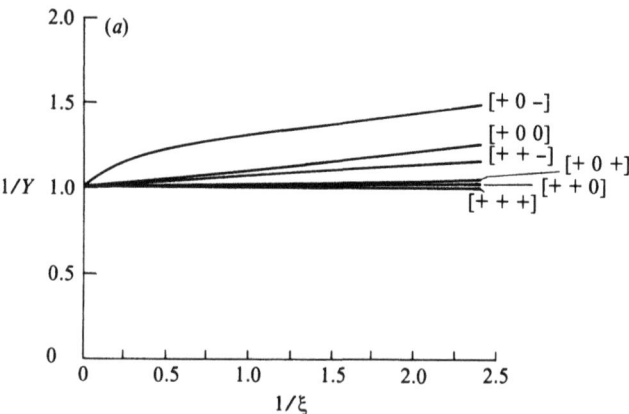

Figure 1.5.4. Lineweaver-Burk plots. Reciprocal of the saturation function Y as a function of the reciprocal of the dimensionless concentration ξ. The five groups (a)–(e) are as in Figure 1.5.3.

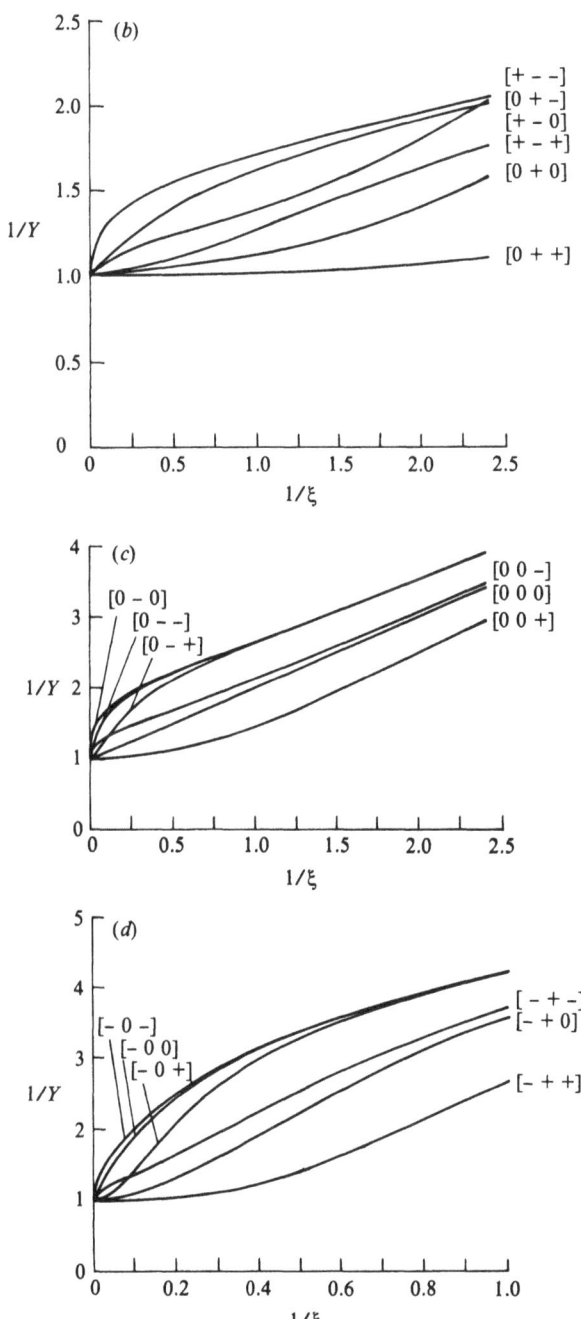

Figure 1.5.4. Continued.

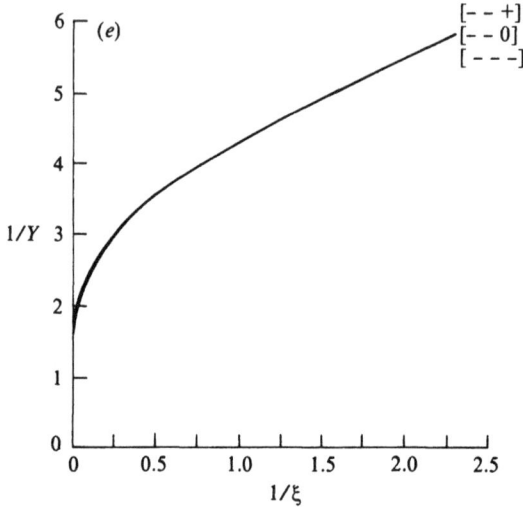

Figure 1.5.4. Continued.

Table 1.5.1. *Initial slopes*

Slope	Curves
Slightly positive	[0 + ·]
Definitely positive	[+ · ·]
Approximately −1	[0 0 ·], [0 − ·]
Less than −1	[− · ·]

Note: The symbol · is used whenever any one of the three symbols +, −, 0 is appropriate.

Table 1.5.2. *Curvatures at the boundaries of the Scatchard plots*

Boundary	Curvature	Curves
Left ($Y \to 0$)	Downwards	[+ · ·] [0 0 ·] [0 − ·]
	Upwards	[0 + ·] [− · ·]
Right ($Y \to 1$)	Downwards	[· · +]
	Upwards	[· · −]
	None	[− − ·]

Table 1.5.3. *Values of ξ_L where $Y(\xi_L) = 0.9$*

$abc = 10^3$ $abc = 10^2$	$\begin{cases} abc = 10 \\ ab \neq 10 \end{cases}$	$abc = 1$ $ab \neq 1$	$abc = 0.1$ $ab \neq 0.1$	
				$abc = 10^{-3}$
$\begin{cases} abc = 10 \\ ab = 10 \end{cases}$	$\begin{cases} abc = 1 \\ ab = 1 \end{cases}$	$abc = 0.1$ $ab = 0.1$	$abc = 10^{-2}$	
ξ_L: $0.06 \leqslant \xi_L \leqslant 1$	$1 \leqslant \xi_L \leqslant 10$	$10 \leqslant \xi_L \leqslant 10^2$	$10^2 \leqslant \xi_L \leqslant 10^3$	$\xi_L \approx 6 \cdot 10^3$

Exercises

1   Derive (4).
2   In the spirit of the discussion in this section of Scatchard plots, make a beginning on an analysis of the information contained in the standard plots; in the Lineweaver–Burk plots.

References

Cornish-Bowden, A. & Koshland, D. E. (1975). Diagnostic uses of the Hill (Logit and Nernst) plots. *J. Mol. Biol.* **95**, 201–12.
Edelstein, S. J. (1975). Cooperative interactions of hemoglobiñ. *Ann. Rev. Biochem.* **44**, 209–32.
Gibson, R. & Levin, S. (1977). Distinctions between the two-state and sequential models for cooperative ligand binding. *Proc. Nat. Acad. Sci., USA* **74**, 139–43.
Kenett, R. (1978). Studies in enzyme kinetics. Ph.D. Thesis, Faculty of Math. Sci., Rehovot, Israel, The Weizmann Institute of Science.

1.6 ENZYME INDUCTION

Introduction

We recognize today three principal mechanisms by which the flow of metabolites through the chemical and physical compartments of the cell is regulated.

(a) **Reversible association** of existing enzymes with effector molecules formed during metabolism. The effector may bind to the active site or to other sites, with or without allosteric change. This mechanism has already been discussed in this chapter.

(b) **Covalent modification** of existing enzyme molecules by specific phosphatases, kinases, adenylylating enzyme systems, etc. A quantitative analysis of this mode of regulation has recently been published by Stadtman & Chock (1977).

(c) **Induction–repression**, i.e. a change in amount of enzyme molecule participating in a particular process. This is achieved by the action of effector molecules on some site in the enzyme synthesis (or degradation) apparatus.

The genetic and molecular aspects of enzyme induction are well understood today, at least in selected prokaryotic systems (Lewin, 1974, 1975). Less attention has been paid to the quantitative aspects of the process, especially when compared with the considerable number of publications devoted to the other modes of regulation. An initial attempt to formulate the quantitative correlations involved is presented in a recent review (Yagil, 1975). The reader is referred to that review for a more complete treatment of the subject described in this section. Other treatment has been presented by Sanglier & Nicolis (1976), Savageau (1972), and Burns & Kacser (1977). Here we shall discuss a few elementary examples and treat in detail one situation not treated before, namely, a kinetic model for enzyme superinduction.

One important difference between the study of enzyme induction and that of the other regulating processes is that induction is mostly studied either in the intact cell or in multicomponent, often incompletely defined, 'cell-free' systems. This reflects the fact that quite a number of sub-processes are involved, each quite complex in itself. These include DNA transcription into messenger RNA (mRNA), mRNA processing, and mRNA translation, as well as several translocatory and degradative events. A quantitative treatment, if it is to be applicable to experimental results, requires a number of simplifying assumptions to be made in order to focus the treatment on those molecular events which are either rate limiting or subject to the regulatory interactions. We shall assume throughout that the rate of enzyme appearance is a direct measure of the amount of the principal controlled element (active operon or analogous component) active in enzyme synthesis. We shall also assume that the intracellular effector concentration is equal, or proportional, to its extracellular concentration. (This has been experimentally achieved only in the y , permease-less, *E. coli lac* systems or in cell-free systems.) When of importance, the intracellular effector concentrations can be experimentally determined.

It is of course possible, at least in principle, to construct quantitative models which include as many steps as are known to be present. This kind of approach is of considerable interest in the examination of models for complex biological processes, and has been carried out in relation to problems such as cell growth and multiplication (Goodwin, 1963; Heinmets, 1966), as well as cellular differentiation (Edelstein, 1972). Chapter 2 discusses some general principles that may be used in simplifying those necessarily very complex models.

In this discussion, the line taken is to set up a minimal number of equations required to describe the observed behavior of a system, because this seems the best way to relate to experiments. This should be borne in mind when consequences are drawn from a fit of experimental results to the prediction of a formula, or when the derived parameters are interpreted.

Effector concentration

The lactose (*lac*) operon of *E. coli*, which controls the formation of beta-galactosidase and two additional proteins, was the first genetic element whose regulatory circuits were established (Jacob & Monod, 1961), leading to many of our present concepts of regulation of enzyme synthesis (cf. Beckwith & Zipser, 1970). A scheme for the operation of the *lac* operon is shown in Figure 1.6.1. We shall examine the effect of the nonmetabolizable effector isopropyl-thiogalactoside (IPTG) as a basic example for evaluating the connection between effector concentration and the amount of enzyme formed in a system.

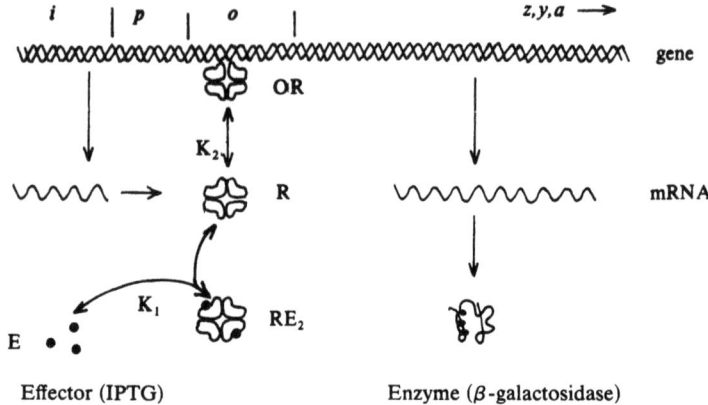

Effector (IPTG)   Enzyme (β-galactosidase)

Figure 1.6.1. Negative induction: effect of lactose analogs on the *lac* operon.

Two equilibrium relations govern the availability of the operator region for initiation by the transcribing machinery. The first equilibrium is between the protein termed the **repressor**, R, coded by the i gene, and the DNA of **free operators** O (at concentration O per unit culture volume), to form a complex OR (**blocked operator**):

$$O + R \rightleftarrows OR \qquad K_2 = O \cdot R/OR. \qquad (1)$$

The total number of operators O_t in a culture population is composed of blocked and free operators:

$$O_t = O + OR. \qquad (2)$$

In the second equilibrium, the repressor combines with n **effector** molecules E to form a complex RE_n with greatly reduced affinity to the operator DNA:

$$R + nE \rightleftarrows RE_n \qquad K_1 = R \cdot E^n/RE_n. \qquad (3)$$

In the simplest case, forms of the repressor bound to less than n molecules contribute little to the total amount of repressor R_t, or to the resulting effects, so that

$$R_t = R + RE_n. \qquad (4)$$

For a treatment where this is not the case, see Yagil & Yagil (1971). See also Section 1.4 for conditions under which the 'Hill approximation' used in (4) is valid.

In (4), we have not included OR because there are at most 4 operators per cell, i.e. $R_t \gg O_t$. Similarly, small effector molecules are, in *E. coli* at least, in large excess over the protein repressor, so that E represents practically total effector concentration ($E = E_t$). Combining (3) with (4), we find that

$$R = R_t \cdot K_1/(K_1 + E^n). \tag{5}$$

This can be inserted into (1) to give

$$O/OR = K_2(K_1 + E^n)/R_t K_1. \tag{6}$$

We shall further define

$$\beta = \frac{O/O_t}{1 - (O/O_t)} = \frac{O}{OR}. \tag{7}$$

β is the ratio of operons available for transcription to unavailable operons. β can be regarded as an experimental quantity in those cases where the fractional rate of enzyme formation is proportional to the fraction of available operons; this has been shown to be the case in most thoroughly studied bacterial operons. Introducing (7) into (6), we arrive at

$$\beta = (K_2/R_t) + (K_2 E^n/K_1 R_t). \tag{8}$$

When no effector is present ('basal conditions'), $E = 0$, so that

$$\beta_0 = K_2/R_t. \tag{9}$$

This leads to the final form of the **induction equation**

♦♦ $$\log[(\beta - \beta_0)/\beta_0] = n \log E - \log K_1. \tag{10}$$

This equation can be put to experimental test, because, as mentioned, β can be evaluated when we know the maximal rate of enzyme production either by saturating with effector or from constitutive strains. (For further elaboration, including the effect of cell growth, see Yagil & Yagil, 1971.) A straight line should result when $\log[(\beta - \beta_0)/\beta_0]$ is plotted versus $\log E$. If the model shown in Figure 1.6.1 is a sufficient description of the system, the slope of this line, n, gives the effective stoichiometry of repressor–effector molecule interaction. From the intercept, a value for K_1, the constant of effector affinity, to repressor, can be calculated. When only $\log(\beta - \beta_0)$ is plotted versus $\log E$, the intercept will yield an accurate measure of $\beta_0 = R_t/K_2$, i.e. for the ratio of total repressor concentration to its affinity to the operator region. The induction plot thus yields three parameters of the induction system.

An example is shown in Figure 1.6.2. Similar straight lines are obtained with many induction experiments in *E. coli*. In the *lac* system, quite a number of strains yield n values of $n = 2 \pm 0.2$ (Yagil & Yagil, 1971). This means that it is sufficient for two IPTG molecules to combine

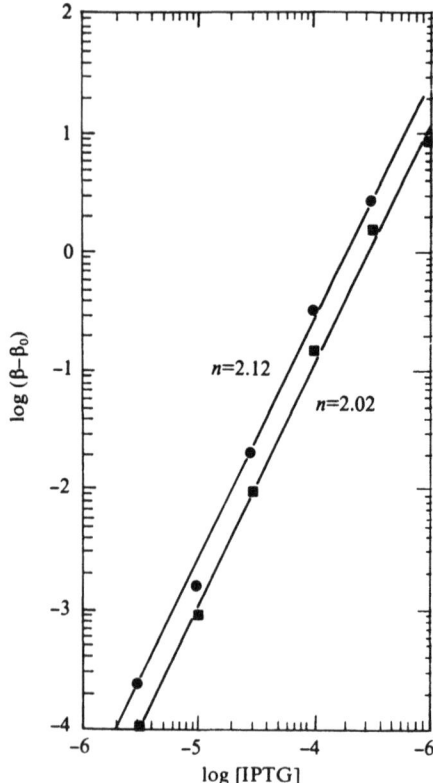

Figure 1.6.2. Induction of β-galactosidase in *E. coli* by IPTG. Data of Sadler & Novick (1972) are plotted according to (10). (●), Diploid strain W14D; (■), haploid strain W14. Values of $\log \beta_0 = -2.96$; 3.33 were subtracted from the ordinate; these differ by $\sim\log 2$ in accordance with the different gene dosage, and causes the two lines to be displaced by a similar factor.

with the tetrameric repressor to reduce drastically its affinity to the operator. This may well be a further example of half-site reactivity discussed by Levitzki & Koshland (1975). The values of K_1 obtained in the figures are 39 and 44 (μmole l^{-1})^2 respectively. The reader is encouraged to verify this from the data in the figure.

Positive induction

We have seen that the course of induction of the *lac* operon by IPTG and by other *lac* inducers conforms closely to the predictions derived from the simple Jacob–Monod regulatory circuit. The quantitative agreement can, however, be taken as a support of the model only if other circuits lead to different types of induction plots, and this, unfortunately, is not always the case. As an example, we may consider a case of positive induction, i.e. where the combination of repressor with effector converts it to a form which does associate with the operator to facilitate transcription.

In Figure 1.6.3, the action of cyclic AMP (cAMP) on the *lac* gene is depicted as an example of positive induction. The catabolite repression protein (CRP) has to be physically associated with the promoter region to

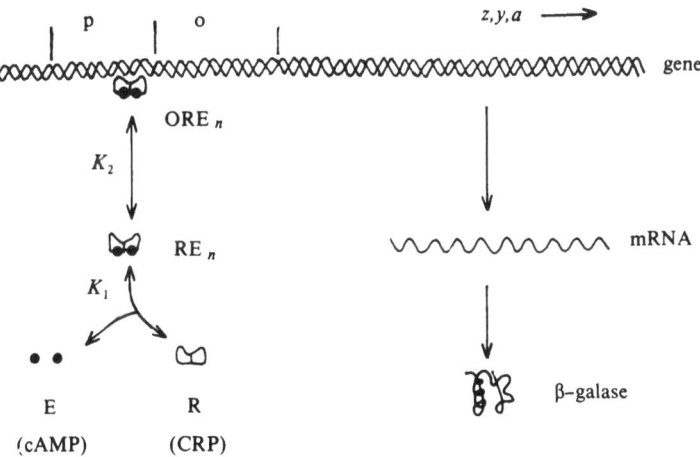

Figure 1.6.3. Positive induction: effect of cAMP and the catabolite repression protein (CAP) on the *lac* operon.

facilitate transcription and it does so preferentially when united with effector, cAMP. In other words, a ternary complex, to be termed ORE_n, has to be formed:

$$O + RE_n \overset{K'_2}{\rightleftharpoons} ORE_n \qquad K'_2 = O \cdot RE_n / ORE_n. \tag{11}$$

Here,

$$RE_n = R_t \cdot E^n / (K_1 + E^n). \tag{12}$$

There are two problems with positive induction not encountered in the negative case. The first is that maximal induction may not be experimentally achieved because, even with maximal amounts of effector, effector concentration may not be sufficient to cause association of all R with the operator. The maximal amount of active operative can be readily seen to be:

$$ORE_n^\infty = O_t \cdot R_t / (K'_2 + R_t). \tag{13}$$

Maximal rate of synthesis can be achieved only in cells in which $R_t \gg K'_2$.

We shall, therefore, in the simple case, express the ratio of transcribed to nontranscribed operon in the form of

$$\beta' = \frac{ORE_n/ORE_n^\infty}{1-(ORE_n/ORE_n^\infty)},\qquad(14)$$

and this leads to

◆◆ $\log \beta' = n \log E - \log (K_2' + R_t)/K_1 K_2,\qquad(15)$

which is the elementary induction equation for positive induction.

The other problem with positive induction is that no enzyme should be formed when no effector is present ($\beta = 0$ if $ORE_n = 0$). Nevertheless, in all enzyme systems examined, a basal rate of enzyme synthesis is encountered. One way to account for this is to assume that transcription of the operon can proceed to a limited extent, μ, without any repressor attached, or to an extent λ when associated with effector-free repressor. The fraction of operons available for transcription is then

$$\alpha' = (ORE_n + \lambda OR + \mu O)/(ORE_n^\infty + \mu O^\infty),\qquad(16)$$

and β' is now $\alpha/(1-\alpha')$. This can be developed, using (6) and (11)–(14), to give a final expression

$$\log \left(\frac{\beta' - \beta_0'}{1 + \beta_0'}\right) = n \log E - \log K_{app},\qquad(17)$$

with

$$\log K_{app} = K_1 \frac{K_2'(K_2 + R_t)}{(K_2' + R_t)K_2}.\qquad(18)$$

K_{app} is equal to K_1 only when both K_2 and K_2' are considerably larger than R_t. Equation (17) indicates that $\log (\beta' - \beta_0')$ should again be a linear function of the logarithm of effector concentration. This can be seen to be the case for cAMP induction of the *lac* gene in Figure 10 of Yagil (1975).[1]

A further example of induction plots is given in Figure 1.6.4. Here, the data of Whitlock and Gelboin (1974) on the induction of a mammalian liver enzyme, arylhydrocarbon hydroxylase, are plotted according to (10), as function of concentration of two of its inducers. It is seen that a straight line results. With one inducer, benzanthracene, the slope is quite close to unity, which means that a single benzanthracene molecule is

---

[1] Equations (15) and (18) are somewhat different from equations 88 and 91 in Yagil (1975), pp. 255–6 because of an unfortunate mistake in equation 81. This introduces no change, however, in the qualitative conclusion in that section, only the plot in Figure 10 and the constants derived are shifted by the factor $\beta_0'/(1 + \beta_0')$.

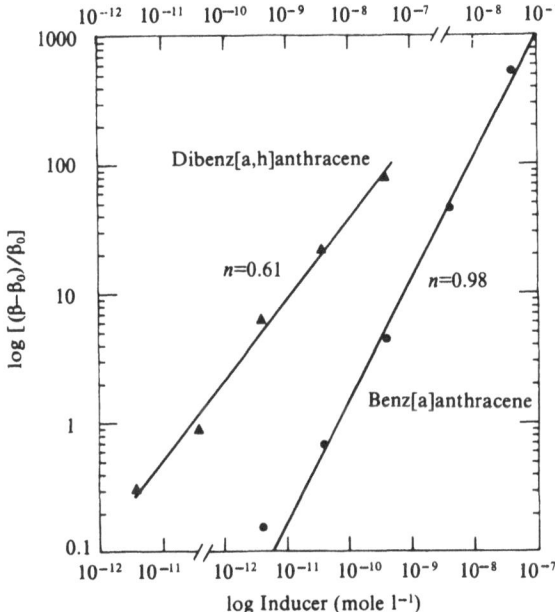

Figure 1.6.4. Induction of arylhydrocarbon(benzo[a]pyrene)hydroxylase by benz[a]anthracene (●) and dibenz[a,h]anthracene (▲) in cultured liver (BRL) cells. Data from Figure 3 of Whitlock & Gelboin (1974) are plotted according to (10), with $\beta_0 = 0.07$, 0.15, respectively. From the lines, values of $K_1 = 0.076$ nmol l^{-1} for benzanthracene and of 0.024 nmol l^{-1} for dibenzanthracene are calculated. To evaluate according to (17), these values have to be divided by $(1 + \beta_0)/\beta_0$, leading to $K_{app} = 1.17$, 0.16 nmol l^{-1}, respectively.

sufficient to convert the controlling element into the form which facilitates enzyme formation. The line with a slope of less than unity observed with dibenzanthracene might be due to the very low effector concentrations involved, in which case effector is not in excess of its receptor, as we have assumed.

We have no evidence yet, genetic or other, to establish whether a negative or positive type of control mechanism operates in the induction of any mammalian enzyme. Comparing (10) with (17) makes it apparent that they differ by the constant factor $\log [\beta_0'/(\beta_0' + 1)]$ (β' values were calculated using plateau enzyme values reached at high effector concentration; β' is, therefore, numerically equal to β, once positive induction is assumed). The linearity of the plots of Figure 1.6.4 thus do not permit us to determine whether induction is positive or negative. It is not even possible to conclude that the controlled element is indeed the structural gene. This element may also be a stable cytoplasmic messenger ('translation control'), as will be discussed in a following subsection.

The effect of time

Data from the arylhydrocarbon hydroxylase system can also be utilized to illustrate the time dependence of enzyme induction. In Figure 1.6.5 (left-hand panels), the experimental data describing the time course of the induction are shown. Arylhydrocarbon hydroxylase, like most mammalian proteins, undergoes continuous synthesis and degradation ('turnover'). The actually observed levels represent a steady state defined by the rate of formation, expressed in terms of translation frequency f_2

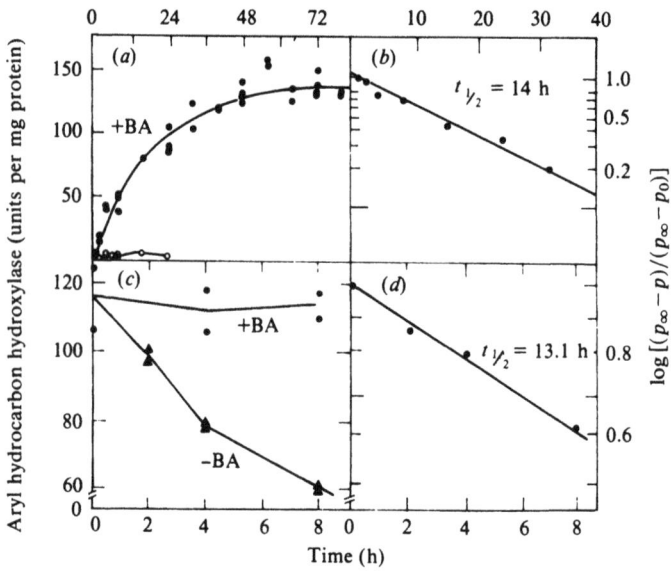

Figure 1.6.5. Time course of arylhydrocarbon hydroxylase induction by benz[a]anthracene (BA) in BRL cells. (*a*), (*b*), The accumulation phase. The experimental data of Whitlock & Gelboin (1973), Figure 1, are shown in panel (*a*) and replotted in panel (*b*) according to (21). (*c*), (*d*), The degradation phase. The data of Whitlock & Gelboin (1974), Figure 2, are shown in panel (*c*) and are replotted in panel (*d*) according to (2). Note the closeness of the values of $t_{\frac{1}{2}}$ obtained.

(Yagil, 1975) and by the rate of enzyme degradation (e.g. random, with time constant k_2). We shall discuss here the case when the message coding for the enzyme (protein, p) neither degrades nor accumulates, or when its formation is tightly coupled to the rate of protein synthesis. In both cases, the concentration of messenger, M, is constant:

$$\dot{p} = f_2 M - k_2 p, \tag{19}$$

and the steady state concentration of the induced protein (enzyme), p_∞, is then

$$p_\infty = f_2 M / k_2. \tag{20}$$

Integration of this equation leads (cf. Yagil, 1975, Equation 27) to

◆◆ $\quad (p_\infty - p)/(p_\infty - p_0) = \exp(-k_2 t). \tag{21}$

(p_0 is the starting concentration of protein.) In case of repression (i.e. a signal which either lowers f_2 or M, or increases k_2), p_0 will be larger than p_∞, as in panels c, d in Figure 1.6.5.

From the right-hand panels of Figure 1.6.5, it can be seen that a straight line results when the logarithm of the left-hand expression of (21) is plotted against the time. The half-times evaluated from the two plots, one representing induction following addition of inducer, the other degradation following withdrawal of inducer (or addition of an inhibitor which sets $f_2 = 0$), are indeed quite close. This demonstrates that the same mechanism operates in the rate-determining step of both the inductive and the repressive phases and brings forth an interesting property of (21), namely that the course of induction is determined by the *degradation* constant k_2 and, while the formation constant f_2 enters only implicitly, via p_∞. In other words, the kinetics of enzyme formation, like those of its disappearance, are determined solely by its degradation constant; the rate of synthesis is manifested in the level ultimately obtained. This was pointed out by Schimke (1969), as well as by Segal & Kim (1963), and serves as a routine basis for determination of enzyme 'half-lives' *in vivo* (Rechcigl, 1971; Goldberg & St John, 1976). Equation (21) has been found to be useful in the analysis of quite a number of inducible liver enzymes, particularly if allowance for an initial latent (lag) period is given. The next stage of complication comes when active messenger is not constant as in the case described, or when enzyme is not continuously turning over. The relevant equations can be set up with ease, although each experimental situation may have its own peculiarities. Several examples are discussed by Yagil (1975).

Superinduction

The term superinduction has been coined to describe situations where an inhibitor of transcription (or translation), instead of depressing increase in enzyme activity, causes this to increase above the levels attained without inhibitor. This effect was first discovered when actinomycin D was added at a late stage of infection of HeLa cells with pox virus, resulting in increased accumulation of the virus coded enzyme thymidine kinase (McAuslan, 1963). It has turned out that superinduction is quite a common phenomenon (Tomkins *et al.*, 1972; Steinberg, Levinson &

Tomkins, 1975), and it has recently been shown in one case (Kessler-Icekson & Yaffe, 1977) to be accompanied by an increase of mRNA translatable in 'cell-free' systems. Tomkins and coworkers (1969) proposed a model which attributes the superinduction of tyrosine aminotransferase in liver cells to the action of a cytoplasmic repressor capable of transforming active messenger into a form inactive in translation. The repressor and the messenger coding for it (if it is a protein) have to be much less stable than the messenger coding for the enzyme, so that when transcription inhibitor is added, this message and its products decay rapidly to permit the message for the enzyme to be converted into its active, enzyme-producing form.

A more complicated case of superinduction has been observed in the arylhydrocarbon hydroxylase system discussed above. Whitlock & Gelboin (1973, 1974) observed that when cycloheximide is added to the system and then removed 4 h later, arylhydrocarbon hydroxylase begins to appear even when no benzanthracene or analog is added. This phenomenon occurs even when actinomycin D is added upon removal of cycloheximide, which indicates that the messenger for the enzyme can survive the cycloheximide treatment. The amount of enzyme ultimately formed was found by Whitlock & Gelboin (1974) to depend on amount of cycloheximide given. When these data are plotted according to (10), the line shown in Figure 1.6.6 is obtained. This line may be interpreted as

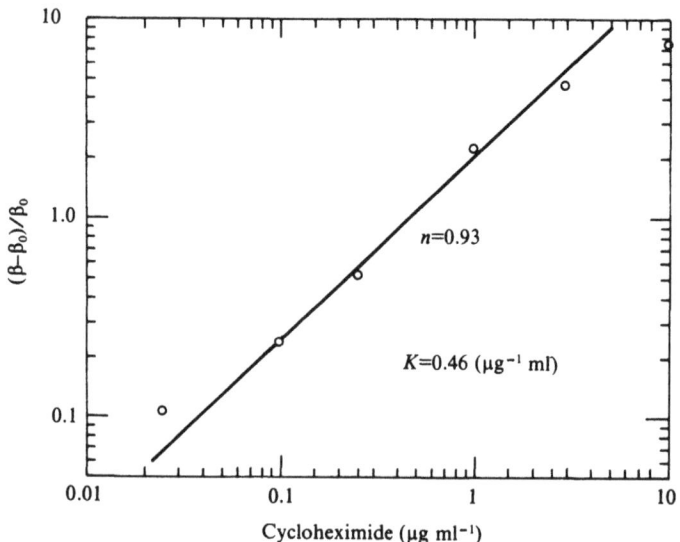

Figure 1.6.6. The induction of arylhydrocarbon hydroxylase by cycloheximide. The data of Whitlock & Gelboin (1974), Figure 6, are plotted according to (10).

meaning that cycloheximide is merely another effector of the system. However, a second translation inhibitor, puromycin, which is structurally very different from cycloheximide, has a similar superinducing effect. Further, the enzyme increase is observed quite a while after the drug is removed from the system.

We shall now show that the translation control model, depicted in Figure 1.6.7, which is in many features similar to the Tomkins *et al.* model (1969), can account for most features observed by Whitlock & Gelboin in this system. In this model, it is assumed that a factor α (possibly a

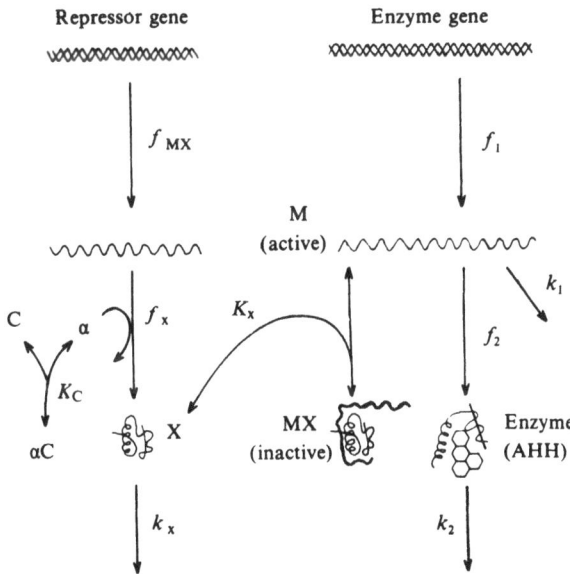

Figure 1.6.7. A model for superinduction by cycloheximide. Cycloheximide associates with a translation factor, α, to form an inactive complex αC. This shuts off the synthesis of cytoplasmic repressor X, and the remaining repressor decays exponentially with a rate constant k_X. The messenger is transformed from its inactive complex with X, MX, into its active form, M, (1.6.25), and begins to be translated into enzyme at a frequency f_2.

ribosomal protein) is required for translation to proceed. α may associate reversibly with cycloheximide (C) to form an αC complex inactive in translation:

$$\alpha + C \rightleftarrows \alpha C, \qquad K_C = \alpha \cdot C/\alpha C. \tag{22}$$

The frequency of formation of the cytoplasmic repressor, X, in this case a protein which rapidly reaches its steady state concentration (i.e. degrades rapidly with a constant k_X) is by (20)

$$X = f_X \cdot M_X/k_X. \tag{23}$$

f_X is the frequency of translation of M_X, the messenger for X. The relationship of the steady state concentration of X in presence of an inhibitor, to its concentration X_0 in absence of inhibitor will then be $[f_X = f_X^0 K_C/(K_C + C)]$

$$X = X_0 K_C/(K_C + C). \tag{24}$$

A second equilibrium exists between the active form M and the inactive form MX of the messenger coding for the enzyme:

$$M + X \rightleftarrows MX, \qquad K_X = M \cdot X/MX. \tag{25}$$

This leads to the fractional concentration of active message ($M_t = M + MX$):

$$M/M_t = K_X/(K_X + X). \tag{26}$$

Eventually, enzyme concentration too will reach a steady state (we know already from Figure 1.6.5 that its $k_2 = (-\ln 0.5)/14\ \text{h}^{-1}$):

$$E = \frac{f_2 M}{k_2} = \frac{f_2 M_t K_X}{k_2(K_X + X)} = \frac{f_2 M_t K_X (K_C + C)}{k_2(K_C(K_X + X_0) + K_X C)}, \tag{27}$$

(f_2 = frequency of translation of enzyme). This can be shown to lead to the by now familiar expression (with $n = 1$):

◆◆ $$\log\left[(\beta - \beta_0)/\beta_0\right] = \log C - \log K_C. \tag{28}$$

β is operationally taken as the ratio of enzyme activity at a certain effector concentration to the extent to which activity deviates from full induced activity: $\beta = (E - E_0)/(E_\infty - E)$, and $\beta_0 = E_0$.

The straight line seen in Figure 1.6.6 is, therefore, no less in accord with the superinduction model in Figure 1.6.7 than with the straight inducer role assigned to cycloheximide. We thus reach the conclusion that while quantitative models for superinduction based on elementary equilibrium and kinetic procedures can be set up, care has to be exercised in drawing conclusions from the observed quantitative correlation. The plot permits, however, the evaluation of affinity constants and thus produces a quantitative basis for comparing efficiency of inducers – carcinogens in the present case – as in Figure 1.6.4.

The quantitative treatment of the model presented permits the evaluation of one further parameter, namely the stability constant of the repressor protein X. This is done as follows. Before cycloheximide is added, a steady state level of the repressor, X_0, is present. This level is

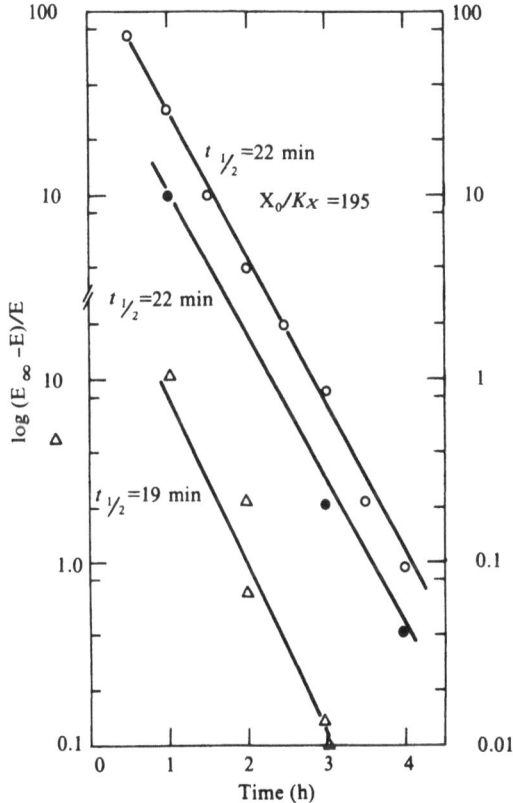

Figure 1.6.8. The rate of arylhydrocarbon reappearance after removal of cycloheximide form a culture of BRL cells. Data of Whitlock & Gelboin (1974) are plotted as function of the time the culture had been exposed to cycloheximide (i.e. the time the preexisting repressor X had been allowed to decay. (○), Data from *ibid*. Figure 6, p. 6118. Cycloheximide was followed by 4 h in actinomycin D; final enzyme levels $E_\infty = 155$ units per mg protein. (△), Data from Figure 10, p. 6120. Benzanthracene was present together with cycloheximide, followed by 4 h in actinomycin D: $E_\infty = 540$ units per mg protein. (●), Data from Figure 11, p. 6120; cycloheximide was followed by 4 h in actinomycin D and benzanthracene: $E_\infty = 300$ units per mg protein.

determined by the frequency of synthesis in absence of the cycloheximide, f_X^0 and rate of degradation, k_X, of X:

$$X_0 = f_X^0 M_X / k_X. \tag{29}$$

If the degradation of X is random and complete, then

$$X = X_0 \exp(-k_X t), \tag{30}$$

where t is the time the inhibitor is present. As before, we shall take the amount of enzyme formed (4 h out of inhibitor) to be

$$E = \frac{f_2 M}{k_2} = \frac{f_2 M_t K_X}{k_2(K_X + X)} = \frac{f_2 M_t K_X}{k(K_2 + X_0\, e^{-k_x t})}. \tag{31}$$

When t is sufficiently large, we have

$$E_\infty = f_2 M_t / k_2. \tag{32}$$

This will lead to

♦♦ $$\log\left[(E_\infty - E)/E\right] = \log\left(X_0/K_X\right) - k_X t. \tag{33}$$

A plot of the right hand quantity against time in cycloheximide should yield a straight line, the slope of which gives the decay constant of the cytoplasmic repressor k_X. The ratio of its steady state concentration to its affinity constant to enzyme messenger (or other critical component of the translation apparatus) can be calculated from the intercept (X_0/K_X).

The data of Whitlock & Gelboin (1974) on the effect of the time in cycloheximide *before* the enzyme is allowed to be induced in the presence or absence of actinomycin D, are plotted according to (33) in Figure 1.6.8. Straight lines result and lead to values for the half-life of repressor ranging between 19 and 22 min in the three experiments evaluated. A value for the ratio X_0/K_X can also be obtained.

We have thus demonstrated that in this particular case a quantitative model for a superinduction can be set up and correlated with experimental data in a way which leads to conclusions concerning the quantitative behavior of the system components. This should be of considerable help in the further experimental study of the molecular details of regulation mechanisms not only in classical induction systems but also in translation controlled systems.

References

Beckwith, J. R. & Zipser, D. (1970). *The Lactose Operon*, New York, Cold Spring Harbor Publ.

Burns, J. A. & Kacser, H. (1977). Allosteric repression, an analysis. *J. Theoret. Biol.* **68**, 199–213.

Edelstein, B. (1972). The dynamics of cellular differentiation and associated pattern formation. *J. Theoret. Biol.* **37**, 221–43.

Goldberg, A. L. & St John, A. C. (1976). Intracellular protein degradation in mammalian and bacterial cells. *Ann. Rev. Biochem.* **45**, 747–803.

Goodwin, B. C. (1963). *Temporal Organization in Cells*, New York, Academic Press.

Heinmets, F. (1966). *Analysis of Normal and Abnormal Cell Growth*, New York, Plenum Publ. Corp., Inc.

Jacob, F. & Monod, J. (1961). Genetic regulatory mechanisms in the synthesis of proteins. *J. Mol. Biol.* **3**, 318–56.

Kessler-Icekson, G. & Yaffe, D. (1977). Increased translatability in a cell-free system of RNA extracted from actinomycin D-treated cultures. *Biochem. Biophys. Res. Commun.* **75**, 62–8.

Levitzki, A. & Koshland, D. E. (1975). The role of negative cooperativity and half-of-the-sites reactivity in enzyme regulation. *Curr. Top. Cell. Reg.* **10**, 1–40.

Lewin, B. (1974). *Gene Expression*, vol. 1 *Bacterial genomes*, New York, Wiley.

—— (1975). *Gene Expression*, vol. 2 *Eukaryotic Chromosomes*, New York, Wiley.

McAuslan, B. R. (1963). The induction and repression of thymidine kinase in the pox-virus infected HeLa cell. *Virology* **21**, 383–9.

Rechcigl, M. (1971). *Enzyme Synthesis and Degradation in Mammalian Systems*, Basel, Karger.

Sanglier, M. & Nicolis, G. (1976). Sustained oscillations and threshold phenomena in an operon control circuit. *Biophys. Chem.* **4**, 113–21.

Savageau, M. A. (1972). The behavior of intact biochemical control systems. *Curr. Top. Cell. Reg.* **6**, 64–130.

Schimke, R. T. (1969). On the roles of synthesis and degradation in regulation of enzyme levels in mammalian tissues. *Curr. Top. Cell. Reg.* **1**, 77–124.

Segal, H. L. & Kim, Y. S. (1963). Glucocorticoid stimulation of the biosynthesis of glutamic alanine transaminase. *Proc. Nat. Acad. Sci., USA* **50**, 912–18.

Stadtman, E. R. & Chock, P. B. (1977). Superiority of interconnectable enzyme cascades in metabolic regulation. *Proc. Nat. Acad. Sci., USA* **74**, 2761–6.

Steinberg, R. A., Levinson, B. B. & Tomkins, G. M. (1975). 'Superinduction' of tyrosine aminotransferase by actinomycin D: a reevaluation. *Cell* **5**, 29–35.

Tomkins, G. M., Gelehrter, T. D., Granner, D., Martin, D., Samuels, H. H. & Thomson, E. B. (1969). Control of specific gene expression in higher organisms. *Science* **166**, 1474–80.

Tomkins, G. M., Levinson, B. B., Baxter, J. D. & Dethlefsen, L. (1972). Further evidence for post-transcriptional control of inducible tyrosine aminotransferase synthesis in cultured hepatoma cells. *Nature New Biol.* **239**, 9–14.

Whitlock, J. P. & Gelboin, H. V. (1973). Induction of aryl hydrocarbon (benzo[*a*]pyrene)-hydroxylase in liver cell culture by temporary inhibition of protein synthesis. *J. Biol. Chem.* **248**, 6114–21.

—— (1974). Aryl hydrocarbon (benzo[*a*]pyrene) hydroxylase induction in rat liver cells in culture. *J. Biol. Chem.* **249**, 2616–23.

Yagil, G. (1975). Quantitative aspects of protein induction. *Curr. Top. Cell. Reg.* **9**, 183–235.

Yagil, G. & Yagil, E. (1971). On the relation between effector concentration and the rate of induced enzyme synthesis. *Biophys. J.* **11**, 11–27.

1.7 MOLECULAR MODELS FOR RECEPTOR TO ADENYLATE CYCLASE COUPLING

Introduction

Numerous biochemical processes are triggered by the interaction of a specific membrane receptor with a specific ligand. In all of these cases, subsequent to the receptor-ligand binding step a post-receptor event is elicited. This post-receptor event may be either the activation of an enzyme such as adenylate cyclase or opening of a 'gate' to allow for specific ion fluxes. Little is known about the molecular events occurring subsequent to ligand binding to the receptor. Furthermore, it is not at all clear whether the receptor and the post-receptor entity responsible for generating the biochemical signal are pre-coupled to each other or are functionally uncoupled from each other and become coupled subsequent to ligand binding to the receptor. This question is the topic to which this study is devoted.

One of the means to reveal information about the state of coupling between the receptor and the enzyme is to perform a kinetic analysis of the rate of appearance of the active adenylate cyclase as a function of hormone concentration. Usually the accumulation of active cyclase is a very rapid process that is difficult to monitor. Fortunately, many adenylate cyclase systems can be activated to a permanently active state in the presence of hormone and Gpp(NH)p (guanylyl-imido-diphosphate), a non-hydrolysable analog of GTP (guanosine triphosphate) (Schramm & Rodbell, 1975; Pfeuffer & Helmreich, 1975; Sevilla *et al.*, 1976; Levitzki, Sevilla & Steer, 1976; Spiegel *et al.*, 1976). Although the highly active state is induced by hormone and therefore must be formed through interaction with the receptor, once formed this state is a permanent property of the enzyme and no longer involves receptor intervention.

We shall be concerned with the catecholamine-stimulated adenylate cyclase. Here propranolol, a specific β-adrenergic blocker of the catecholamine-stimulated adenylate cyclase system can no longer reduce in any way the activity of the highly active cyclase already formed but can block further activation of a partially activated enzyme (Sevilla *et al.*,

1976; Spiegel *et al.*, 1976). A detailed description of this phenomenon has already been presented (Sevilla *et al.*, 1976). Gpp(NH)p replaces the natural allosteric activator GTP. When GTP is present the active state of the enzyme is continuously reverted to its inactive state at a very rapid rate, concomitantly with the hydrolysis of GTP at the regulatory site (Cassel & Selinger, 1976). Thus under physiological conditions a steady state concentration of active enzyme is achieved (Levitzki, 1977; Sevilla & Levitzki, 1977). This steady state level is formed extremely fast and therefore the appearance of this state is impossible to monitor as a function of time. Only the steady state activity level can be measured. On the other hand, Gpp(NH)p activation is a rather slow process. Because of the irreversibility of the Gpp(NH)p activation process and its relative simplicity, it has become possible to probe into the state of enzyme–receptor coupling.

The approach used in this study was to analyze the kinetics of the process of enzyme activation induced by hormone and Gpp(NH)p in terms of different molecular models, allowing for different modes of enzyme–receptor coupling. These models also make specific predictions concerning the nature of the receptor itself, and the nature of hormone binding to the receptor. Therefore a detailed exploration of the different models of enzyme receptor coupling has been performed. Once this has been done, explicit experiments can be designed in order to test which of the different mechanisms of coupling is likely to be operating in the experimental system.

Indeed, such experiments have been carried out and are summarized at the end of this chapter. The more detailed experimental data are given elsewhere (Tolkovsky & Levitzki, 1978*a*, *b*, *c*). The theoretical analysis presented in this section is general and may be applied to receptor cyclase systems other than the *β*-adrenergic receptor dependent adenylate cyclase. In each of the models analyzed, expressions will be derived for (*a*) the binding of hormone, (*b*) the accumulation of the activated state of the enzyme in the presence of Gpp(NH)p, and (*c*) the steady state level of active enzyme in the presence of GTP. Also, the consequences are formulated of changing the receptor concentration or of the enzyme concentration, both on the binding of hormone and on the kinetics of enzyme activation.

Theory

The *β*-adrenergic-cyclase-activated system contains at least two basic components, the receptor and the adenylate cyclase. In order to elucidate the way by which the interaction of receptor and cyclase is dependent on hormone, a complete description of this system in terms of hormone–

receptor–cyclase interactions must be analyzed. Boeynams & Dumont (1975) have listed several possible models for receptor–enzyme interactions and their dependence on hormone concentration in general terms.

We shall now provide kinetic models of a system consisting of the three components hormone, receptor, and enzyme (cyclase) in kinetic terms, in order to describe the adenylate cyclase in turkey red blood cells. The analysis will be divided into five parts. A general derivation of each model in terms of active cyclase will be given along with the experimental predictions associated with each model.

In all models it will be assumed that the binding of hormone to the receptor is rapid and reversible. Such rapid equilibrium is well established for the β-adrenergic receptors of turkey erythrocytes analyzed in the present study (Levitzki, Steer & Atlas, 1974; Levitzki, Sevilla, Atlas & Steer, 1975). It is well documented that the interactions between the hormone and the β-adrenergic receptor are completely reversible since the introduction of a β-adrenergic blocker completely blocks the hormone activation in a competitive fashion (Levitzki *et al.*, 1974). It is also known that one can easily wash the hormone and lose the activating effect in the absence of Gpp(NH)p.

Another major assumption is that during the process of activation either the rate of formation of the hormone–receptor–enzyme complex (HRE) is fast compared to the rate of transition to the permanently active form or the transition step is fast compared to the rate of formation of HRE. This second assumption is justified on the basis of two sets of observations. (*a*) The accumulation of cyclic AMP (cAMP) in the adenylate cyclase assay, in the presence of GTP, is linear with time. (*b*) The appearance of the permanently active state of the enzyme in the presence of Gpp(NH)p is a purely exponential process without a lag period. (If the species HRE were to accumulate to a significant fraction of the total enzyme concentration at a rate comparable to the rate of transition to permanently active form, the appearance of the permanent state would occur nonexponentially with a lag time equal to the time it takes for HRE to reach a steady state concentration.)

The different models for receptor–enzyme interactions will now be listed. Because of our assumptions it can be presumed that either the concentration *HRE* is always small compared to total enzyme E_T because of the fast transition rate or that an equilibrium level of HRE is achieved rapidly compared to the step of enzyme activation. Swillens & Dumont (1976) suggested a few models where the binding of hormone to the receptor is not a rapid equilibrium process. This particular case is not treated here.

The precoupled receptor–cyclase model (I)

This model is described by the relationship

$$H + RE \underset{k_2}{\overset{k_1}{\rightleftharpoons}} HRE \overset{k_3}{\longrightarrow} HRE', \tag{1}$$

where H is hormone, RE the receptor–enzyme complex, k_1 and k_2 are the rate constants for hormone–receptor binding, and k_3 is the rate constant characterizing the conversion of the enzyme from the inactive state HRE to the stable active state HRE'.

In the absence of guanyl nucleotide, HRE accumulates and thus represents the bound complex of hormone to the receptor. In the presence of nucleotide, HRE' is also formed. If the nucleotide is non-hydrolysable, like Gpp(NH)p, HRE' accumulates until no more free receptor is available. In the presence of a hydrolysable nucleotide like GTP, HRE' is rapidly converted back to HRE.

Hormone binding

Using (1) one can obtain an expression for the dependence of *HRE* on *H*. (We use italic to denote concentration.) Note first that the conservation equation for RE is

$$RE_T = RE + HRE. \tag{2}$$

Since $H \gg RE_T$, $H \simeq H_T$,

$$HRE = \frac{k_1 H \cdot RE}{k_2}. \tag{3}$$

From (2) and (3) it follows that

$$RE_T = RE[1 + (k_1 H/k_2)]. \tag{4}$$

Consequently

$$HRE = \frac{k_1 H \cdot RE_T}{k_2[1 + (k_1 H/k_2)]} = \frac{H \cdot RE_T}{(k_2/k_1) + H}. \tag{5}$$

The ratio k_2/k_1 is in fact the hormone–receptor dissociation constant, which will be denoted by K_H:

$$K_H = k_2/k_1. \tag{6}$$

Equation (5) therefore takes the form

$$HRE = \frac{H \cdot RE_T}{K_H + H}, \tag{7}$$

which represents a noncooperative binding process (Figure 1.7.1).

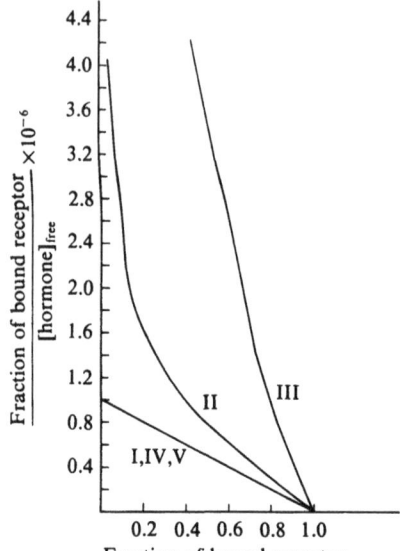

Figure 1.7.1. Simulation of hormone binding according to the different models of receptor to enzyme coupling. Simulation conditions presented in the figure are: $R_T = E_T = 10K_E$, $K_H = 1.0 \cdot 10^{-6}$ mole l^{-1}. Simulation was conducted with hormone concentrations from $1.0 \cdot 10^{-8}$ mole l^{-1} to $1.0 \cdot 10^{-5}$ mole l^{-1}. Changing the simulation conditions does not change the general characteristics of the curve. The Roman numerals signify the model according to which the simulation was conducted.

The formation of permanently active enzyme

Using (1) we now derive an expression for the rate of accumulation of the activated species of the enzyme HRE', in the presence of Gpp(NH)p. Since HRE' can now be formed, the conservation equation (2) is replaced by

$$RE_T = RE + HRE + HRE'. \tag{8}$$

Applying the quasi-steady state assumption (Section 1.1) for HRE we obtain

$$HRE = [k_1 H(RE_T - HRE - HRE')]/(k_2 + k_3). \tag{9}$$

Thus

$$HRE = \frac{H}{[(k_2 + k_3)/k_1] + H}(RE_T - HRE'). \tag{10}$$

The rate of accumulation of the active species HRE′ is given by

$$\frac{\mathrm{d}HRE'}{\mathrm{d}t} = k_3 HRE = \frac{k_3 H(RE_\mathrm{T} - HRE')}{[(k_2 + k_3)/k_1] + H}. \tag{11}$$

Upon integration, taking into account that $HRE' = 0$ at $t = 0$, one obtains

$$HRE' = RE_\mathrm{T}\left\{1 - \exp\left(-\frac{k_3 H}{[(k_2 + k_3)/k_1] + H}\right)t\right\}. \tag{12}$$

The steady state level of active enzyme in the presence of GTP

In the presence of GTP the scheme presented by (1) must be modified to

$$\mathrm{H + RE} \underset{k_2}{\overset{k_1}{\rightleftharpoons}} \mathrm{HRE} \underset{k_4}{\overset{k_3}{\rightleftharpoons}} \mathrm{HRE'}. \tag{13}$$

Here k_3 is the rate constant of enzyme activation and k_4 is the rate constant describing the conversion of the active enzyme to its inactive state, concomitantly with the hydrolysis of GTP.

Applying once again the steady state approximation for HRE we obtain

$$k_1 RE \cdot H + k_4 HRE' - (k_2 + k_3)HRE = 0. \tag{14}$$

Thus

$$HRE = [k_1 REH + k_4 HRE']/(k_2 + k_3). \tag{15}$$

Consequently

$$\frac{\mathrm{d}HRE'}{\mathrm{d}t} = \frac{k_3 k_1 RE_\mathrm{T}}{k_2 + k_3 + k_1 H} - HRE' \cdot \frac{k_1 H(k_3 + k_4) + k_2 k_4}{k_2 + k_3 + k_1 H}. \tag{16}$$

Integrating (16) and applying the initial condition $HRE = 0$ when $t = 0$, one obtains

$$HRE' = \frac{[k_3/(k_3 + k_4)]H \cdot RE_\mathrm{T}}{[k_2 k_4/k_1(k_3 + k_4)] + H}\left\{1 - \exp-\left(\frac{(k_2 k_4/k_1) + (k_3 + k_4)H}{[(k_2 + k_3)/k_1] + H}\right)t\right\}. \tag{17}$$

When steady state is achieved (t is large), (17) yields

$$HRE' = \frac{\{1/[1 + (k_4/k_3)]\}RE_\mathrm{T} \cdot H}{\{k_2/[k_1(1 + (k_3/k_4))]\} + H}. \tag{18}$$

If $k_4 \gg k_3$ (as in the case of turkey erythrocyte adenylate cyclase (Levitzki, 1977)), the affinity towards hormone measured in the presence of GTP is identical to that measured directly by binding, i.e. $k_2/k_1 = K_\mathrm{H}$. The maximal activity of the enzyme obtained in the presence of high hormone is given by

$$HRE'_{max} = \frac{RE_T}{1+(k_4/k_3)} \tag{19}$$

which indicates that $HRE'_{max} \ll RE_T$ when $k_4 \gg k_3$. This expression is similar to the one obtained by Levitzki (1977). A somewhat modified model which builds a steady state level of HRE' in the presence of GTP,

$$HR + E \underset{k_2}{\overset{k_1}{\rightleftharpoons}} HRE \overset{k_3}{\longrightarrow} HRE' \overset{k_4}{\longrightarrow} H + RE,$$

yields an equation for HRE' which is similar to (18). The expression for HRE'_{max} is identical for the two models. The rate governing the reversal process may be experimentally derived after achieving a steady state level of HRE' by the concomitant addition of the β-adrenergic antagonist PPL and labeled ATP. HRE' will decay rapidly to the ground state by a first order process and cAMP will accumulate exponentially (Tolkovsky & Levitzki, 1978b).

This model predicts that (a) there is no basal cyclase activity in the absence of hormone; (b) the rate at which the enzyme is activated depends on hormone concentration and is saturable according to a Michaelian function; (c) the affinity of the hormone to the receptor as measured kinetically should be identical or lower than that measured directly by binding experiments; (d) the decrease of receptor concentration or of enzyme concentration by irreversible inactivation should not affect the *rate* at which the activated species HRE' appears but should decrease the maximal level of HRE' attainable either in the presence of Gpp(NH)p or in the presence of GTP.

The dissociation receptor–cyclase model (II)

The model is described by the equation:

$$RE + H \underset{K_H}{\rightleftharpoons} HRE \underset{K_E}{\rightleftharpoons} HR + E; \qquad E \to E'. \tag{20}$$

The model postulates that E' is the active species of the cyclase. As in model I, the receptor is obligatorily coupled either to E or to R.

Hormone binding

Both R and RE bind hormone. In the absence of guanyl nucleotides, E' is not formed and one can write

$$HRE = \frac{REH}{K_H}, \tag{21}$$

and

$$HR = \frac{K_E HRE}{E} = \frac{K_E RE \cdot H}{K_H E}. \tag{22}$$

The conservation laws are

$$E_T = RE + HRE + E \tag{23}$$

and

$$R_T = RE + HRE + HR, \tag{24}$$

from which one obtains

$$E = E_T - RE - HRE = E_T - RE\left[1 + \frac{H}{K_H}\right]$$

$$= E_T - RE\left\{\frac{K_H + H}{K_H}\right\}, \tag{25}$$

$$B = HR + HRE, \tag{26}$$

where B is the concentration of the bound hormone. It follows from (22) and (23) that

$$B = \frac{RE \cdot H}{K_H}[1 + (K_E/E)]. \tag{27}$$

From (25), (26) and (27) one obtains

$$B = \frac{(R_T - B)H}{K_H}\left(1 + \frac{K_E K_H}{E_T K_H - R_T(K_H + H) + B(K_H + H)}\right), \tag{28}$$

which upon rearrangement yields

$$B^2 + \frac{B}{K_H + H}\left\{K_H\left(E_T + \frac{K_E H}{K_H + H}\right) - R_T(2H + K_H)\right\}$$

$$+ \frac{(R_T)^2 H(K_H + H) - R_T \cdot H(E_T + K_E)K_H}{(K_H + H)^2} = 0. \tag{29}$$

The dependence of B on H is not immediately apparent. However, simulation of (29) reveals that the binding is negatively cooperative (Figure 1.7.1). The extent of negative cooperativity becomes more pronounced as K_E is reduced relative to E_T. Reducing K_E results in the relative increase of HRE, the apparently low affinity form, relative to HR, the apparently high affinity species.

The rate of formation of the permanently active enzyme

The process of enzyme activation can be described by the scheme

$$RE + H \underset{k_2}{\overset{k_1}{\rightleftharpoons}} HRE \underset{k_4}{\overset{k_3}{\rightleftharpoons}} HR + E; \qquad E \overset{k_5}{\longrightarrow} E', \tag{30}$$

where $K_H = k_2/k_1$ and $K_E = k_3/k_4$. The accumulation of E' describes the activation of the enzyme to its permanently active state.
From the relationships

$$E_T = E + E' + RE + HRE, \tag{31}$$

$$R_T = RE + HRE + HRE, \tag{32}$$

$$HRE = \frac{E \cdot HR}{K_E} = \frac{H \cdot RE}{K_H} \quad \text{and} \quad RE = \frac{E \cdot HRK_H}{K_E H}, \tag{33}$$

one can write

$$E_T - E' = E + RE + HRE = E\left(1 + \frac{HRK_H}{K_E H} + \frac{HR}{K_E}\right). \tag{34}$$

From (34) it follows that:

$$HR = R_T - (E_T - E') + E. \tag{35}$$

Inserting (35) into (34) one obtains:

$$E^2 + E\left(\frac{K_E H}{K_H + H} + R_T - E_T + E'\right) - \frac{(E_T - E' K_E H)}{K_H + H} = 0. \tag{36}$$

When $k_5 > k_4$,

$$E \ll E' + RE + HRE, \tag{37}$$

and (36) simplifies to:

$$E = \frac{(E_T - E')(k_3/k_4)H}{(k_3/k_4)H + [R_T - (E_T - E')](K_H + H)}. \tag{38}$$

The rate of formation of E' is given by [see (30)]

$$\frac{dE'}{dt} = k_5 E. \tag{39}$$

Inserting (38) into (39) one obtains

$$\frac{dE'}{dt} = \frac{(k_5 k_3/k_4)(E_T - E')H}{(k_3/k_4)H + (R_T - (E_T - E')(K_H + H))}, \tag{40}$$

which can be rewritten as

$$\frac{dE'}{dt} = \frac{(E_T - E')}{k_5 + [k_4(K_H + H)/k_5 k_3 H][R_T - (E_T - E')]}, \tag{41}$$

or

$$\frac{dE'}{dt} = \frac{(E_T - E')}{k_5 + [k_4(K_H + H)/k_5 k_3 H]R_T - [k_4(K_H + H)/k_5 k_3 H](E_T - E')}. \tag{42}$$

Let us define

$$a = k_5 + \frac{k_4}{k_5 k_3} \cdot \frac{K_H + H}{H}(R_T - E_T), \tag{43}$$

$$b = \frac{k_4}{k_5 k_3} \cdot \frac{K_H + H}{H}. \tag{44}$$

From equations (42), (43) and (44) one obtains

$$\frac{dE'}{dt} = \frac{E_T - E'}{a + bE'}. \tag{45}$$

With these (42) takes the form

$$\frac{dE'}{(E_T/a) - (E'/a)} + \frac{E' \, dE'}{(E_T/b) - (E'/b)} = dt. \tag{46}$$

Upon integration one obtains the expression

$$\frac{\ln E_T}{E_T - E'} - \frac{bE'}{a + b^2 E_T} = \frac{t}{a + b^2 E_T}. \tag{47}$$

From (47) it is apparent that the kinetics of accumulation of E′ deviates from first-order kinetics. It can also be seen that reducing R_T or E_T by irreversible inactivation results in both the reduction of the maximal level of E′ attainable and a decrease in its rate of formation.

To summarize, the predictions of the dissociation model are as follows: (a) In the absence of hormone there might be basal activity; if not, all enzyme units are coupled to receptor units. (b) The binding of hormone will be negatively cooperative although one intrinsic binding constant governs the hormone binding. (This is because there are two hormone-bound species present in parallel, HR and HRE.) (c) The kinetics which govern the accumulation of E′ are not first order. (d) The effect of irreversibly reducing either R_T or E_T will slow down the rate of accumulation of E′ and decrease the maximal level of E′ attained.

The floating receptor–enzyme equilibrium model (III)

In this model it is assumed that equilibrium exists between the hormone-receptor complex and the enzyme. The scheme is:

$$H + R \underset{k_2}{\overset{k_1}{\rightleftharpoons}} HR; \qquad HR + E \underset{k_4}{\overset{k_3}{\rightleftharpoons}} HRE \overset{k_5}{\longrightarrow} HRE', \tag{48}$$

where $K_H = k_2/k_1$ and $K_E = k_4/k_3$. This type of model belongs to the class of models that allow the receptor to float in the membrane (Jacobs & Cuatrecasas, 1976).

Binding of hormone

In the absence of enzyme activation, the hormone binding species are R and RE. Thus

$$HRE = HR \cdot E/K_E = H \cdot R \cdot E/K_H K_E, \tag{49}$$

so that

$$HR = H \cdot R/K_H. \tag{50}$$

Consequently the total amount of bound hormone $B \equiv HR + HRE$ is given by

$$B = R[(H/K_H) + (E \cdot H/K_E K_H)]. \tag{51}$$

The conservation equations are

$$E_T = E + HRE, \tag{52}$$

$$R_T = R + HR + HRE. \tag{53}$$

From (53)

$$R = R_T - B. \tag{54}$$

Moreover (52) and (49) yield

$$E_T = E[1 + (HR/K_H K_E)], \tag{55}$$

that is

$$E = E_T K_H K_E/(K_H K_E + H \cdot R). \tag{56}$$

Inserting (51) into (53) and rearranging, one obtains

$$B^2 - B\left[\frac{K_E K_H H + 2H^2 R_T + K_H^2 K_E + (H \cdot R_T + H \cdot E_T K_H)}{H(K_H + H)}\right.$$

$$\left. + \frac{H \cdot R_T K_H K_E + H \cdot E_T \cdot R_T K_H + H^2 (R_T)^2}{H(K_H + H)}\right] = 0. \tag{57}$$

Simulation of (57) reveals negative cooperativity in binding (Figure 1.7.1). The tighter the interaction between HR and E, the further away are the two apparent binding constants derived for H from the thermodynamic dissociation constant.

The rate of HRE' formation

Using the scheme presented in (48) one can derive the rule of HRE' accumulation (in the presence of Gpp(NH)p) as a function of time. We see that

$$\mathrm{d}HRE'/\mathrm{d}t = k_5 HRE. \tag{58}$$

Applying the steady state conditions for HRE one obtains

$$\mathrm{d}HRE/\mathrm{d}t = 0 = k_3 HR \cdot E - (k_4 + k_5)HRE \tag{59}$$

or

$$HRE = k_3 HR \cdot E/(k_4 + k_5). \tag{60}$$

But since

$$E_T = E + HRE + HRE', \tag{61}$$

$$R_T = R + HR + HRE, \tag{62}$$

and since $HRE \ll E + HRE'$, (61) and (62) simplify to

$$E_T = E + HRE', \tag{63}$$

and

$$R_T = R + (H \cdot R/K_H) + HRE'. \tag{64}$$

Therefore

$$R = \frac{(R_T - HRE')K_H}{K_H + H}, \tag{65}$$

$$HR = \frac{H \cdot R}{K_H} = \left\{\frac{H(R_T - HRE')}{K_H(K_H + H)}\right\}K_H = \frac{H(R_T - HRE')}{H + K_H}, \tag{66}$$

and

$$HRE = \frac{k_3 H}{(k_4 + k_5)(K_H + H)}(R_T - HRE')(E_T - HRE'). \tag{67}$$

Since the rate of accumulation of HRE' is given by (48), one obtains

$$\frac{\mathrm{d}HRE'}{HRE'^2 + (R_T + E_T) \cdot HRE' - R_T \cdot E_T} = \frac{k_5 k_3 H}{(k_4 + k_5)(K_H + H)}\,\mathrm{d}t. \tag{68}$$

Integration of (68) can be carried out with the aid of any standard table of integrals. The result is, when $R_T < E_T$,

$$HRE' = \frac{E_T \cdot R_T\{1 - \exp\left(-[k_3 k_5 H(E_T - R_T)/(k_4 + k_5)(K_H + H)]t\right)\}}{E_T - R_T\{\exp\left(-[k_3 k_5 H(E_T - R_T)/(k_4 + k_5)(K_H + H)]t\right)\}}.$$

(69)

When $R_T > E_T$ one should interchange E_T and R_T in (69). From (69) it is apparent that the kinetics of HRE$'$ formation deviate strongly from first-order kinetics. When $k_5 < k_4$ HRE will rapidly accumulate *prior* to activation in an equilibrium fashion. We can therefore use scheme (20) to derive an expression for HRE:

$$HRE = \frac{H \cdot R_T(E_T - HRE' - HRE)}{K_E(K_H + H) + H(E_T - HRE' - HRE)}.$$

(70)

The full expression for *HRE* is then

$$HRE = \frac{E_T - HRE' + R_T}{2} + \frac{K_E(K_H + H)}{2H} - \sqrt{x},$$

(71)

where

$$x = \{[(E_T - HRE' + R_T)H + K_E(K_H + H)]^2/2H\} - (E_T - HRE')R_T.$$

We can now proceed to obtain an expression for HRE$'$ accumulation using (58).

$$dHRE'/dt = k_5 HRE.$$

(72)

Upon defining

$$C = [K_E(K_H + H) + (E_T + R_T)H]/2H,$$

(73)

we obtain

$$\frac{dHRE'}{dt} = k_5\left\{\left(\frac{-HRE'}{2}\right) + C - \sqrt{y}\right\},$$

(74)

where

$$y = [(HRE'^2/4) - HRE'(R_T + C) + C^2 - E_T \cdot R_T].$$

It can be seen that the rate law which governs the formation of the active species HRE$'$ deviates strongly from first-order kinetics. A solution identical to (69) is obtained when the activation follows the scheme

$$H + R \underset{k_2}{\overset{k_1}{\rightleftharpoons}} HR;$$

$$HR + E \underset{k_4}{\overset{k_3}{\rightleftharpoons}} HRE \overset{k_5}{\longrightarrow} HRE' \underset{k_7}{\overset{k_6}{\rightleftharpoons}} HR + E'.$$

(75)

Equation (69) (where $R_T \neq E_T$) describes a second-order process of accumulation of the activated species HRE'.

In most hormone adenylate cyclase systems $R_T = E_T$ (Levitzki *et al.*, 1975). When $R_T = E_T$ the formulation of the process is different but the nature of the process remains second order. Under these conditions (68) can be written in the form

$$\frac{dHRE'}{HRE'^2 + 2E_T \cdot HRE + (E_T)^2} = \frac{k_5 H}{K_E(K_H + H)} \, dt. \tag{76}$$

Integrating (76), we find that

$$\frac{1}{HRE'} = \frac{K_E(K_H + H)}{(E_T)^2 k_5 H t} + \frac{1}{E_T}$$

or

$$\frac{1}{HRE'} = \frac{K_E(K_H + H)}{(R_T)^2 + k_5 H t} + \frac{1}{R_T}. \tag{77}$$

Equation (77) reveals that the kinetics of HRE' accumulation are second order. In particular a plot of $1/HRE'$ versus $1/t$ should yield a straight line. It is therefore also clear that plotting the data according to first-order kinetics of activation, namely plotting $\ln[(HRE'_{max} - HRE'_t)/HRE'_{max}]$ versus t, will yield a nonlinear dependence. From (69) it is apparent that when $E_T \gg R_T$ the rate of appearance of HRE' approaches a first-order process. Similarly when $R_T \gg E_T$ the rate of accumulation of HRE' becomes first order.

In summary, the equilibrium floating model which assumes that the receptor–hormone complex and the enzyme are at equilibrium with the hormone–receptor–enzyme (HRE) species predicts that (*a*) the binding of hormone is negatively cooperative; (*b*) the overall rate process of appearance of the permanently active state of the enzyme is second order at every hormone concentration; (*c*) no explicit expression for the dependence of the apparent rate constant, characterizing the conversion of the inactive enzyme to its permanently active form can be formulated; (*d*) reduction in either R_T or E_T by chemical modification will reduce both the rate at which HRE' is produced and the maximal activation level attainable.

Another variation of the model presented in (75) is the following:

$$H + R \underset{K_H}{\rightleftharpoons} HR; \qquad HR + E \rightleftharpoons HRE \rightarrow HRE' \rightleftharpoons HR + E'$$

$$K_H \Big\updownarrow \tag{78}$$

$$H + RE'$$

This model is very similar to the one analyzed in detail and also predicts negative cooperativity in hormone binding and complex kinetics of accumulation of the active species of the enzyme (HRE', RE', and E'). In fact, the addition of any number of equilibrium steps subsequent to the irreversible step of activation does not fundamentally change the rate of activation. The receptor enzyme equilibrium model analyzed by Jacobs & Cuatrecasas (1976) is similar to the models discussed in this section, but less general. In their analysis, Jacobs & Cuatrecasas assumed that HRE is the active species of the enzyme, which is different from the more general case presented here. Also these authors included another simplifying assumption: that the hormone binds to the naked receptor and to the receptor–enzyme complex with identical affinities. The consequences of the latter assumption are (a) that the binding pattern of the hormone is noncooperative (derivation not shown) and (b) the kinetics of enzyme activation by hormone and Gpp(NH)p deviate from first-order kinetics (derivation not shown).

The collision coupling model (IV)

In this model, the mode of enzyme activation is described by the scheme

$$H + R \underset{K_H}{\rightleftharpoons} HR + E \underset{k_4}{\overset{k_3}{\rightleftharpoons}} X \overset{k_5}{\longrightarrow} HR + E' \tag{79}$$

where $K_H = k_2/k_1$ and X is the ternary encounter complex between hormone, receptor and enzyme. During the life time of this complex the enzyme is converted to its activated form E' such that $X = HRE \rightleftharpoons HRE'$.

The collision coupling model is in fact a special case of the floating equilibrium model, model III. In its formulation a few assumptions are made: (i) k_4 is very large, so that the affinity of HR for E is small, and thus k_3 becomes rate limiting; (ii) k_5 is also very rapid compared to k_3 and is of the same order of magnitude as k_4, $k_5 > k_4$; (iii) HRE never accumulates. In this case the irreversible step of activation (in the presence of Gpp(NH)p) is no longer between HRE and HRE', but describes the transition between HR and E since $k_5 \geqslant k_4$ and both are extremely fast. In analogy with the interaction of an enzyme with substrate, E is the substrate and R is the catalytic entity that is regenerated concomitantly with the formation of the 'product' E'.

Hormone binding

The binding pattern of hormone in this case is simple since the only bound hormone species that exists in significant amounts is HR. Therefore

$$HR = R_T H/(K_H + H), \tag{80}$$

and the binding of hormone is noncooperative (Figure 1.7.1).

The kinetics of appearance of permanently activated state

For the model presented in (79) one can write

$$E_T = E + X + E',\tag{81}$$

$$R_T = R + (HRE \rightleftarrows HRE') + HR.\tag{82}$$

Also

$$(HRE + HRE') \ll RE + E + E',\tag{83}$$

and

$$(HRE + HRE') \ll HR + RE + R.\tag{84}$$

One can write

$$dE'/dt = k_5 X,\tag{85}$$

so that

$$dX/dt = k_3 HR \cdot E - (k_4 + k_5)X.\tag{86}$$

At the steady state, which is rapidly established,

$$X = k_3 HR \cdot E/(k_4 + k_5).\tag{87}$$

Since

$$HR = R_T \cdot H/(K_H + H),\tag{88}$$

and

$$E = E_T - E',\tag{89}$$

one can rewrite (85) as

$$\frac{dE'}{dt} = \frac{k_5 k_3 R_T \cdot H \cdot E_T}{(k_4 + k_3)(K_H + H)} - \frac{k_5 k_3 R_T \cdot H \cdot E'}{(k_4 + k_3)(K_H + H)},\tag{90}$$

so that

$$\frac{dE'}{E_T - E'} = \frac{k_5 k_3 R_T \cdot H}{(k_4 + k_5)(K_H + H)}\,dt.\tag{91}$$

Integrating (91) one obtains:

$$E' = E_T\{1 - \exp\left(-[k_3 k_5 R_T \cdot H/(k_4 + k_5)(K_H + H)]t\right)\}.\tag{92}$$

In contrast to model III, the rate constant governing the appearance of the activated state depends solely on the total receptor concentration and not on both the total receptor and total enzyme concentrations. The maximal level of activated enzyme attainable is solely dependent on the total enzyme concentration and *independent* of the receptor concentration. The overall process of enzyme activation is first order and the apparent first-order rate constant is linearly dependent on total receptor concentration and increases as a function of hormone concentration in a noncooperative fashion.

Activation of the enzyme in the presence of GTP

In the presence of GTP the activated form of the enzyme is reverted to its inactive form concomitantly with the hydrolysis of GTP at the regulatory site (Cassel & Selinger, 1976; Levitzki, 1977; Sevilla & Levitzki, 1977). Under these conditions the collision coupling model is given by the scheme

$$H + R \underset{K_H}{\rightleftharpoons} HR + E \underset{k_4}{\overset{k_3}{\rightleftharpoons}} (HRE \rightleftharpoons HRE') \overset{k_5}{\longrightarrow} HR + E' \overset{k_6}{\longrightarrow} E. \tag{93}$$

The equation governing the rate of appearance of E' turns out to be

$$E' = \frac{E_T k_5 k_3 R_T \cdot H/(k_4 + k_5)(K_H + H)}{\{k_5 k_3 R_T \cdot H/(k_4 + k_5)(K_H + H)\} + k_6}$$

$$\times \left\{ 1 - \exp\left(-\left[\frac{k_3 k_5 R_T \cdot H}{(k_4 k_5)(K_H + H)} + k_6 \right] t \right) \right\}. \tag{94}$$

At high hormone concentration and if $k_5 \gg k_4$, (94) takes the form

$$E' = \frac{E_T}{1 + (k_6/k_3 R_T)} \{1 - \exp(-(k_3 R_T + k_6)t)\}. \tag{95}$$

After steady state conditions have been achieved, the concentration of enzyme in its active form is given by

$$E' = \frac{E_T}{1 + (k_6/k_3 R_T)}. \tag{96}$$

As in model I, it can be seen that only a fraction of the total enzyme is in its active form when the physiological regulator GTP is present. Also as in model I, when Gpp(NH)p is present as the allosteric activator, all the enzyme is converted to its active form after enough time has elapsed. This model predicts that (*a*) hormone binding is noncooperative; (*b*) the kinetics of activation of the enzyme in the presence of saturating Gpp(NH)p is first order; (*c*) the apparent first-order rate constant describing the process of activation is linearly dependent on total receptor concentration and is independent of enzyme concentration. Thus irreversible inactivation of the receptor will cause a proportional decrease in the rate of enzyme activation but not in the maximal activity attainable in the presence of Gpp(NH)p. Irreversible inactivation of the enzyme will cause a decrease in the maximal activity attainable but not affect the rate constant of activation in the presence of Gpp(NH)p. The model also predicts that (*d*) the apparent first-order rate constant of activation of the enzyme in the presence of Gpp(NH)p increases as a function of hormone concentration in a saturable fashion according to a noncooperative dependence curve.

Combination models (V)

One can combine the models discussed so far and construct combination models. One such model is presented in the following scheme:

$$
\begin{array}{ccc}
\mathrm{H+RE} \underset{K_{\mathrm{H}}}{\rightleftharpoons} \mathrm{X} \xrightarrow{k} \mathrm{HR+E'}, \\
K_{\mathrm{E}} \Updownarrow \qquad\qquad K_{\mathrm{H}} \Updownarrow \\
\mathrm{R+E} \qquad\qquad \mathrm{H+R}
\end{array}
\tag{97}
$$

where $\mathrm{X = HRE \rightleftharpoons HRE'}$. This model can be termed the precoupled-dissociation model.

Hormone binding

The conservation equations are

$$
R_{\mathrm{T}} = HRE + RE + HR, \tag{98}
$$

$$
E_{\mathrm{T}} = E + RE + HRE. \tag{99}
$$

The total amount of bound receptor is given by

$$
B \equiv HR + HRE = R_{\mathrm{T}} - RE - R, \tag{100}
$$

i.e.

$$
B = \frac{H \cdot R}{K_{\mathrm{H}}} + \frac{H \cdot RE}{K_{\mathrm{H}}} = \frac{H}{K_{\mathrm{H}}}(R_{\mathrm{T}} + RE). \tag{101}
$$

From (100) and (101) we obtain

$$
B = (H/K_{\mathrm{H}})(R_{\mathrm{T}} - B). \tag{102}
$$

The solution for B is

$$
B = R_{\mathrm{T}}H/(K_{\mathrm{H}} + H). \tag{103}
$$

Thus the binding of hormone under these conditions is noncooperative (Figure 1.7.1).

The kinetics of appearance of the activated enzyme

In the presence of Gpp(NH)p the scheme of enzyme activation takes the form

$$
\begin{array}{ccc}
\text{H+RE} \;\underset{k_2}{\overset{k_1}{\rightleftarrows}}\; \text{X} \;\xrightarrow{k_3}\; \text{HR+E}' \\[4pt]
K_E \Big\updownarrow \qquad\qquad K_H \Big\updownarrow \\[4pt]
\text{H+R+E} \qquad\qquad \text{H+R} \\[4pt]
K_H \Big\updownarrow \\[4pt]
\text{HR}
\end{array}
\qquad (104)
$$

where $K_H = k_2/k_1$. We have

$$E_T = E + RE + X + E', \qquad (105)$$

$$R_T = R + RE + X + HR, \qquad (106)$$

$$dE'/dt = k_3 X. \qquad (107)$$

When $k_3 \ll k_2$

$$X \ll E + RE + E', \qquad (108)$$

$$X \ll R + HR + RE, \qquad (109)$$

and one can write the steady state approximation

$$X = k_1 H \cdot RE/(k_2 + k_3). \qquad (110)$$

We wish to derive an expression for RE. To this end we use the relations

$$RE = R \cdot E/K_E, \qquad (111)$$

$$R = HRK_H/H, \qquad (112)$$

which together with (106) and (110) yield

$$R_T = RE + \frac{REK_E[(K_H + H)/K_H]}{(E_T - E' - RE)}. \qquad (113)$$

Upon rearrangement one obtains

$$RE^2 - RE\left(R_T + E_T - E' + \frac{K_E(K_H + H)}{K_H}\right) + R_T(E_T - E') = 0. \qquad (114)$$

An explicit analytical solution can be obtained for conditions where RE is relatively small, namely when

$$RE \ll E' + X + E \qquad (115)$$

at all times, so that the quadratic term can be neglected. This will be true when K_E is large relative to R and E and H is abundant. The approximate solution is

$$RE \approx \frac{R_T(E_T - E')}{R_T + E_T - E' + [K_E(K_H + H)/K_H]}, \tag{116}$$

since RE^2 is very small. Using (107) and (110) one obtains

$$\left\{ \frac{R_T + E_T + [K_E(K_H + H)/K_H] - E'}{E_T - E'} \right\} \frac{dE'}{dt} = \frac{k_3 k_1}{(k_2 + k_3)} H \cdot R_T. \tag{117}$$

Integrating (117) and setting $E' = 0$ when $t = 0$, we find that

$$\ln \left[\frac{(E_T - E')}{E_T} \right] = \frac{E'/R_T}{1 + [K_E(K_H + H)/R_T K_H]}$$

$$- \frac{k_3 k_1 H \cdot R_T t}{(k_2 + k_3)[R_T + (K_E/K_H)(K_H + H)]}. \tag{118}$$

The kinetics of appearance of E' deviate from a first-order process.

Since (118) was formulated for conditions where K_E is large relative to R_T and E_T we simulated the effect of the term

$$\frac{E'/R_T}{1 + [K_E(K_H + H)/R_T K_H]}, \tag{119}$$

which appears in (118), under conditions where $K_E = R_T$. At low hormone concentrations the contribution of the term (119) is very large at low E' and becomes progressively smaller as E' increases. At high hormone concentrations the contribution of (119) will be negligible at all E' levels. Thus deviations from first-order kinetics will be apparent at low hormone levels, while at saturating hormone levels first-order kinetics will be apparent. As in the collision model (model IV) the rate of E' appearance is governed by R_T and not by E_T, whereas the maximum level of activity attained is determined solely by E_T.

Experimental results

The turkey erythrocyte adenylate cyclase proved to be a very useful experimental system to investigate the possible modes of receptor to enzyme interactions (Tolkovsky & Levitzki, 1978a, b, c). The enzyme in this membrane system can be activated by either the β-adrenergic receptor upon binding of an *l*-catecholamine (Sutherland *et al.*, 1965), or by the adenosine receptor upon binding of adenosine (Tolkovsky & Levitzki, 1978c). In each of these cases we have followed (a) the binding of the hormone to the receptor; (b) the kinetics of adenylate cyclase activation by the hormone as a function of hormone concentration; (c) the change in the kinetic pattern of cyclase activation as a function of receptor and enzyme concentration.

These three types of experiments were shown above to be useful in deciphering the mode of coupling between the receptor and the enzyme.

Binding experiments

The binding of β-antagonists such as [^{3}H]propranolol and [^{125}I]-hydroxybenzylpindolol was shown to be noncooperative (Levitzki *et al.*, 1974; Atlas, Steer & Levitzki, 1974; Levitzki *et al.*, 1975; Tolkovsky & Levitzki, 1978*a*). Similarly the binding of agonists such as *l*-epinephrine, *l*-norepinephrine and *l*-isoproterenol was also found to be noncooperative. Thus, one can reject the dissociation model (model II, Tables 1.7.1 and 1.7.2) and the equilibrium floating model (model III, Tables 1.7.1 and 1.7.2) for receptor to cyclase coupling. Models II and III predict negatively cooperative binding (Figure 1.7.1).

Kinetics of adenylate cyclase activation

The kinetics of accumulation of the active species of adenylate cyclase was found to be pseudo first-order for both the *l*-catecholamine (Tolkovsky & Levitzki, 1978*a*, *b*) and the adenosine induced activation (Tolkovsky & Levitzki 1978*c*; Braun & Levitzki, 1979*a*, *b*). Furthermore, the pseudo first-order rate constant was found to depend on agonist concentration in a Michaelian fashion [see, for example, (12) and (92)]. These findings again exclude the dissociation model (model II, Tables 1.7.1 and 1.7.2) and the floating model (model III, Tables 1.7.1 and 1.7.2). On the basis of the finding that the kinetics of adenylate cyclase activation by either adenosine or *l*-catecholamines is first order even at very low hormone concentrations, the precoupled-dissociation model (model V, Tables 1.7.1 and 1.7.2) can also be rejected. On the same grounds the model proposed by Jacobs & Cuatrecasas (1976), which predicts non-first-order kinetics of activation, can also be rejected.

The kinetic pattern as a function of receptor concentration

Our findings so far can be accounted for by two diametrically opposed models: the precoupled model (model I, Tables 1.7.1 and 1.7.2) and the collision coupling model (model IV, Tables 1.7.1 and 1.7.2). Both models predict noncooperative binding of first-order kinetics of enzyme activation by the agonist. By further analysis of the two models, we have already established the different predictions of the two models as well as the

Table 1.7.1. *Mode of hormone binding as predicted by different models or receptor to enzyme coupling*

Model	Formulation	Mode of binding	Percent binding in each class of sites	Ratio of apparent hormone dissociation constant to K_H: K_{app}/K_H
I. Precoupled	$RE + H \underset{K_H}{\rightleftharpoons} HRE$	Noncooperative		1.0
II. Dissociation	$RE + H \underset{K_H}{\rightleftharpoons} HRE \underset{K_E}{\rightleftharpoons} RH + E$	Negatively cooperative. Cooperativity increases as K_E decreases relative to R_T or ET_T	1. 27% of one class and 73% of the second class when $K_E = E_T = R_T$ 2. 9% of one class and 81% of another class when $K_E = 0.1E_T = 0.1R_T$	0.017 0.30 0.013 0.64
III. Equilibrium floating	$H + R \underset{K_H}{\rightleftharpoons} HR + E \underset{K_E}{\rightleftharpoons} HRE$	Negatively cooperative	1. 70% of one class and 30% of another class when $K_E = E_T = R_T$ 2. 65% of one class and 35% of another class When $K_E = 0.1E_T = 0.1R_T$	0.33 0.31 0.062 0.25
IV. Collision coupling	$H + R \underset{K_H}{\rightleftharpoons} HR + E$	Noncooperative	100	1.0
V. Precoupled-dissociation	$H + RE \underset{K_H}{\rightleftharpoons} HRE$ $K_E \updownarrow$ $H + R + E \underset{K_H}{\rightleftharpoons} HR$	Noncooperative	100	1.0

Table 1.7.2. *Dependence of the rate of enzyme activation on hormone concentration as predicted by the different models of receptor to enzyme coupling*

Model	Overall kinetic features			Effect of reducing E_T	Effect of reducing R_T
	Rapid equilibrium prior to activation	Rapid activation compared to equilibrium	Type of dependence of apparent first-order rate constant on hormone concentration		
I. Precoupled	First order	First order	Noncooperative	1. Reduction of maximal activity attainable but no change in rate of activation	1. Reduction in maximal binding and in maximal activity attainable 2. No change in rate of activation
II. Dissociation	Deviates from first order	Deviates from first order	—*	1. Reduction of maximal activity attainable and in activation rate	1. Reduction in maximal binding and in maximal activity attainable
III. Equilibrium floating	Deviates from first order	Second order	—*	1. Reduction as in model II.	1. As in model II
IV. Collision coupling	Model excludes this condition	First order	Noncooperative	1. Reduction in maximal extent of enzyme activation with no change in rate constant for activation	1. Reduction in rate constant of enzyme activation with no change in maximal extent of enzyme activation
V. Precoupled-dissociation	Not examined	Deviates from first order at low hormone concentrations	—*	1. Reduction in extent of enzyme activation	1. As in model IV

* Due to deviations from first-order kinetics no dependence of the rate constant of enzyme activation on hormone concentration will be meaningful.

similarities between the two (see also Tables 1.7.1 and 1.7.2). We can summarize the predictions of the two models as follows:

The precoupled model (model I) predicts that the apparent rate constant, k_{obs}, is independent of the total receptor concentration, whereas the final concentration of attainable activated enzyme is proportional to the latter (12). The collision coupling model (model IV) in contrast, predicts that the apparent rate constant, k_{obs}, is linearly dependent upon the total receptor concentration, R_T, whereas the final concentration of attainable activated enzyme is independent of this quantity (92).

A diagnostic experiment was therefore designed in order to distinguish between the two possibilities.

The receptor concentration was progressively reduced by reacting the membrane preparation with an irreversible blocker which reacts covalently with the receptor. The progressive decrease in the concentration of the β-adrenergic receptor using a specific β-receptor directed affinity label resulted in a proportional decrease in the pseudo first-order rate constant whereas the maximal level of adenylate cyclase activation remains unaffected (Tolkovsky & Levitzki, 1978a, b). A summary of these experiments is shown in Figure 1.7.2. It is clear therefore that the mode of coupling of adenylate cyclase to the β-adrenergic receptor is of the 'collision coupling' type (model IV, Tables 1.7.1, 1.7.2).

In contrast, a similar experiment using an adenosine affinity label revealed that the adenosine receptor is tightly coupled to the adenylate cyclase (Braun & Levitzki, 1979a, b). The progressive destruction of the adenosine receptor by the adenosine affinity label results in a proportional decrease in the maximal level of enzyme activity attainable whereas the pseudo first-order rate constant of activation remains unchanged (Braun & Levitzki, 1979a, b). Other kinetic experiments corroborate the latter finding. For example, the pattern of adenylate cyclase activation by the combined ligands adenosine and epinephrine has been analyzed (Tolkovsky & Levitzki, 1978c). Knowing that the mode of coupling between the β-receptor and the cyclase is by the 'collision coupling' mechanism one can deduce the mode of coupling of the adenosine receptor to the enzyme from the kinetics of enzyme activation by the two ligands combined. This analysis (Tolkovsky & Levitzki, 1978c) revealed that the adenosine receptor is permanently coupled to the enzyme. In summary, the mode of coupling of adenosine to adenylate cyclase corresponds to the precoupled model.

Other experimental approaches

Since the activation of adenylate cyclase by the β-receptor is a bimolecular process, one expects that the kinetics of enzyme activation by the

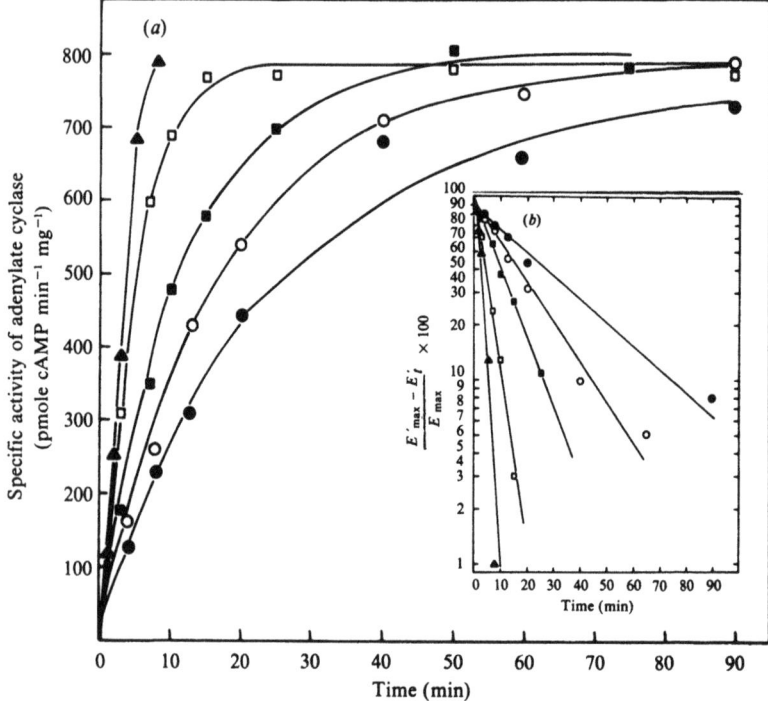

Figure 1.7.2. Time course of adenylate cyclase activation by saturating Gpp(NH)p and *l*-epinephrine subsequent to affinity labeling treatment. The membranes were treated with the affinity label as described elsewhere (Tolkovsky & Levitzki, 1978*b*). The rate of enzyme activation to its permanently active state was measured in the presence of $1.0 \cdot 10^{-4}$ mole l^{-1} *l*-epinephrine and $1.0 \cdot 10^{-4}$ mole l^{-1} Gpp(NH)p (saturating concentrations). (*a*) Progress curves for the accumulation of the active enzyme. E'_{max} is the maximal specific activity attained which was equal in this particular preparation to 790 pmoles cAMP mg^{-1} min^{-1}. (*b*) Data in (*a*) plotted on a semilogarithmic plot to demonstrate that the process of enzyme activation is first order. E'_{max} is the maximal specific activity attained, and E'_t the specific activity in each of the time points. $\blacktriangle$, untreated membranes; $\square$, $1.67 \cdot 10^{-5}$ mole l^{-1} affinity label; $\blacksquare$, $4.3 \cdot 10^{-5}$ mole l^{-1} affinity label; $\bigcirc$, $1.0 \cdot 10^{-4}$ mole l^{-1} affinity label; $\bullet$, $2.3 \cdot 10^{-4}$ mole l^{-1} affinity label.

hormone-bound receptor should depend strongly on the viscosity of the membrane matrix. Indeed, it has recently been shown (Hanski, Rimon & Levitzki, 1979) that the rate constant of enzyme activation by the hormone bound β-receptor depends linearly on the fluidity of the membrane. It appears that the bimolecular reaction between the receptor and the enzyme is diffusion controlled. As expected, on the basis of the experiments described in the previous section, the rate of adenylate cyclase activation by the adenosine receptor is independent of membrane fluidity. A summary of these results is given in Figure 1.7.3. In fact, one

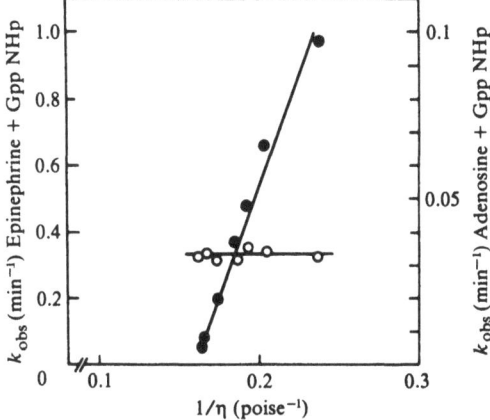

Figure 1.7.3. The dependence of k_{obs} on the fluidity of the membrane. Membrane fluidity was increased by the controlled insertion of the fluidizing agent *cis*-vaccenic acid. The fluidity of the membrane was determined by fluorescence polarization using 1,6-diphenyl-(1,3,5 all trans)-hextriene (DPH) as described elsewhere (Hanski *et al.*, 1978; Rimon *et al.*, 1978). In parallel, the kinetics of adenylate cyclase activation by adenosine and by epinephrine were determined in the presence of saturating concentration of Gpp(NH)p. k_{obs} was calculated from a computer fit of the data to (12) for adenosine and to (92) for *l*-epinephrine. ○, rate constant of cyclase activation by adenosine + Gpp(NH)p. ●, rate constant of cyclase activation by *l*-epinephrine + Gpp(NH)p.

can elucidate the mode of adenylate cyclase activation by hormone receptors by examining the kinetics of enzyme activation by the receptor as a function of membrane fluidity. The fluidity of the membrane can be increased by inserting fatty acids (Hanski *et al.*, 1979) and phospholipids (Hanski & Levitzki, 1978) or decreased by the insertion of cholesterol (Klein, More & Pastan, 1978).

Discussion

The mechanisms of receptor to adenylate cyclase coupling were explored in detail. Different modes of enzyme to receptor coupling within the membrane predict both different patterns of hormone binding and different mechanisms of activation of the receptor dependent enzyme. A summary of the important predictions of the different models is given in Tables 1.7.1 and 1.7.2 and in Figure 1.7.1. The possibility of generating conditions under which the process of hormone-dependent enzyme activation is essentially irreversible simplifies the mathematical analysis and enhances the possibility of designing experiments to probe the mode of receptor to enzyme coupling. Thus, performing binding experiments in conjunction with detailed kinetic analysis of enzyme activation may allow one to decide which of the possible modes of coupling operates *in vivo*. Since numerous hormone-dependent adenylate cyclases have been shown to depend also on GTP and to respond to GPP(NH)p, the analysis can be extended to other hormone-dependent adenylate cyclases. In some of the models explored, the response in the presence of GTP, instead of Gpp(NH)p, was also analyzed. Under such conditions not all the enzyme is converted to its activated form but a steady state concentration of active enzyme is achieved. This latter situation is the one encountered under physiological conditions. The predictions of the different models as to the behavior of the system under such conditions can also be tested against experimental findings.

The formulation of the coupling between enzyme and receptors within a membrane matrix is general and may apply to other receptor dependent responses where both response and ligand binding to the receptor may be studied simultaneously.

References

Atlas, D., Steer, M. L. & Levitzki, A. (1974). Stereospecific binding of propanolol and catecholamines to the β-adrenergic receptor. *Proc. Nat. Acad. Sci., USA* **71**, 4246–8.

Boeynams, J. M. & Dumont, J. E. (1975). Quantitative analysis of the binding of ligands to their receptors. *J. Cyclic Nucleotide Res.* **1**, 123–42.

Braun, S. & Levitzki, A. (1979a). The attenuation of epinepherine dependent adenylate cyclase by adenosine and the characteristics of the adenosine stimulatory and inhibitory sites. *Mol. Pharmacol.* **16**, 737–48.

——— (1979b). Adenosine receptor permanently coupled to turkey erythrocyte adenylate cyclase. *Biochemistry* **18**, 2134–8.

Cassel, D. & Selinger, Z. (1976). Catecholamine stimulated GTPase activity in turkey erythrocyte membranes. *Biochim. Biophys. Acta* **452**, 538–51.

Hanski, E. & Levitzki, A. (1978). The absence of desensitization in the beta adrenergic receptors of turkey reticulocytes and erythrocytes and its possible origin. *Life Sci.* **22**, 53–60.

Hanski, E., Rimon, G. & Levitzki, A. (1979). Adenylate cyclase activation by the β-adrenergic receptors as a diffusion-control process. *Biochemistry* **18**, 846–53.

Jacobs, S. & Cuatrecasas, P. (1976). The mobile receptor hypothesis and 'cooperativity' of hormone binding. Application to insulin. *Biochim. Biophys. Acta* **433**, 482–95.

Klein, I., More, L. & Pastan, I. (1978). Effect of liposomes containing cholesterol on adenylate cyclase activity of cultured mammalian fibroblasts. *Biochim. Biophys. Acta* **506**, 42–53.

Levitzki, A. (1977). The role of GTP in the activation of adenylate cyclase. *Biochem. Biophys. Res. Comm.* **74**, 1154–9.

Levitzki, A., Sevilla, N., Atlas, D. & Steer, M. L. (1975). Ligand specificity and characteristic of the β-adrenergic receptor in turkey erythrocyte plasma membranes. *J. Mol. Biol.* **97**, 35–53.

Levitzki, A., Sevilla, N. & Steer, M. L. (1976). The regulatory control of β-receptor dependent adenylate cyclase. *J. Supramol. Struct.* **4**, 405–18.

Levitzki, A., Steer, M. L. & Atlas, D. (1974). The binding characteristics and number of β-adrenergic receptors on the turkey erythrocyte. *Proc. Nat. Acad. Sci., USA* **71**, 2773–7.

Pfeuffer, T. & Helmreich, E. J. M. (1975). Activation of pigeon erythrocyte membrane adenylate cyclase by guanyl nucleotide analogues and separation of a nucleotide binding protein. *J. Biol. Chem.* **250**, 867–76.

Rimon, G., Hanski, E., Braun, S. & Levitzki, A. (1978). Mode of coupling between hormone receptor and adenylate cyclase elucidated by modulation of membrane fluidity. *Nature Lond.* **276**, 394–6.

Schramm, M. & Rodbell, M. (1975). A persistent active state of the adenylate cyclase system produced by the combined actions of isoproterenol and guanylyl imidophosphate in frog erythrocyte membranes. *J. Biol. Chem.* **250**, 2232–7.

Sevilla, N. & Levitzki, A. (1977). The activation of adenylate cyclase by *l*-epinephrine and guanylylimidophosphate and its reversal by *l*-epinephrine and GTP. *FEBS Lett.* **76**, 129–34.

Sevilla, N., Steer, M. L. & Levitzki, A. (1976). Synergistic activation of adenylate cyclase by guanylyl imidophosphate and epinephrine. *Biochemistry* **15**, 3493–9.

Spiegel, A. M., Brown, E. M., Fedak, S. A., Woodend, C. J. & Aurbach, G. D. (1976). Holocatalytic state of adenylate cyclase in turkey erythrocyte membranes: formation with guanylylimidophosphate plus isoproterenol without effect on affinity of β receptor. *J. Cyclic Nucleotide Res.* **2**, 47–56.

Sutherland, E. W., Øye, I. & Butcher, R. W. (1965). The action of epinephrine and the role of adenylate cyclase system in hormone action. *Rec. Prog. Horm. Res.* **21**, 623–46.

Swillens, S. & Dumont, J. E. (1976). The mobile receptor hypothesis in hormone action: A general model accounting for desensitization. *J. Cyclic Nucleotide Res.* **3**, 1–10.

Tolkovsky, A. M. & Levitzki, A. (1978*a*). Collision coupling of the β-adrenergic receptor with adenylate cyclase. In *Hormones and Cell Regulation*, vol. 2, ed. J. Dumont and J. Nunez, Amsterdam, Elsevier/North Holland, pp. 89–105.

—— (1978*b*). Mode of coupling between the β-adrenergic receptor and adenylate cyclase in turkey erythrocytes. *Biochemistry* **17**, 3795–810.

—— (1978*c*). Coupling of a single adenylate cyclase to two receptors: adenosine and catecholamine. *Biochemistry* **17**, 3811–17.

2 Simplification of biochemical reaction systems

2.1 INTRODUCTION

As more and more information accumulates about the components of biological systems it becomes increasingly important to fit the pieces together in a systematic way. Computer simulation models offer an attractive medium for such integration and it is tempting to conceive of progress in terms of an initial model which, by successive accretion of detail, exhibits more and more fidelity to its real system counterpart. Such a conception, however, is misguided in that as the model grows, it becomes more and more expensive in terms of space and time, and its program implementation is more and more difficult to verify, and to validate against the real system. Perhaps most crucially, it becomes less and less capable of providing insight into the key factors which govern the various behaviors which are of interest.

Integrated modeling (Zeigler, 1978, 1979) is a conception under development to deal with this problem. It advocates the development not of a single all-inclusive model but of a collection of models, each being a simplification oriented to answering a limited set of questions of interest. Methods of achieving valid simplification are thus of primary importance and have been studied in general and in various contexts (Zeigler, 1976).

In this chapter we study simplification of biochemical reaction systems. In particular, our motivation derives from attempts to develop manageable simulation models of the *Escherichia coli* cell (Weinberg & Zeigler, 1970; Goodman, 1972).

A glance at Figure 2.1.1 gives us an inkling of the problem. On the left are broad classes of biochemical constituents. As one proceeds to the right, these classes are refined into subclasses, and perhaps after a number of steps, finally into elementary biochemicals. Of course, the refinement might proceed even further into primitive chemical elements, atomic species, nuclear species, etc. Where should this refinement stop as far as its representation in a model is concerned? The answer to this depends on what has been called the **experimental frame** (Zeigler, 1976) – roughly, the set of questions toward which the model is addressed. In the case of Goodman (1972) and Weinberg & Zeigler (1970), the questions of

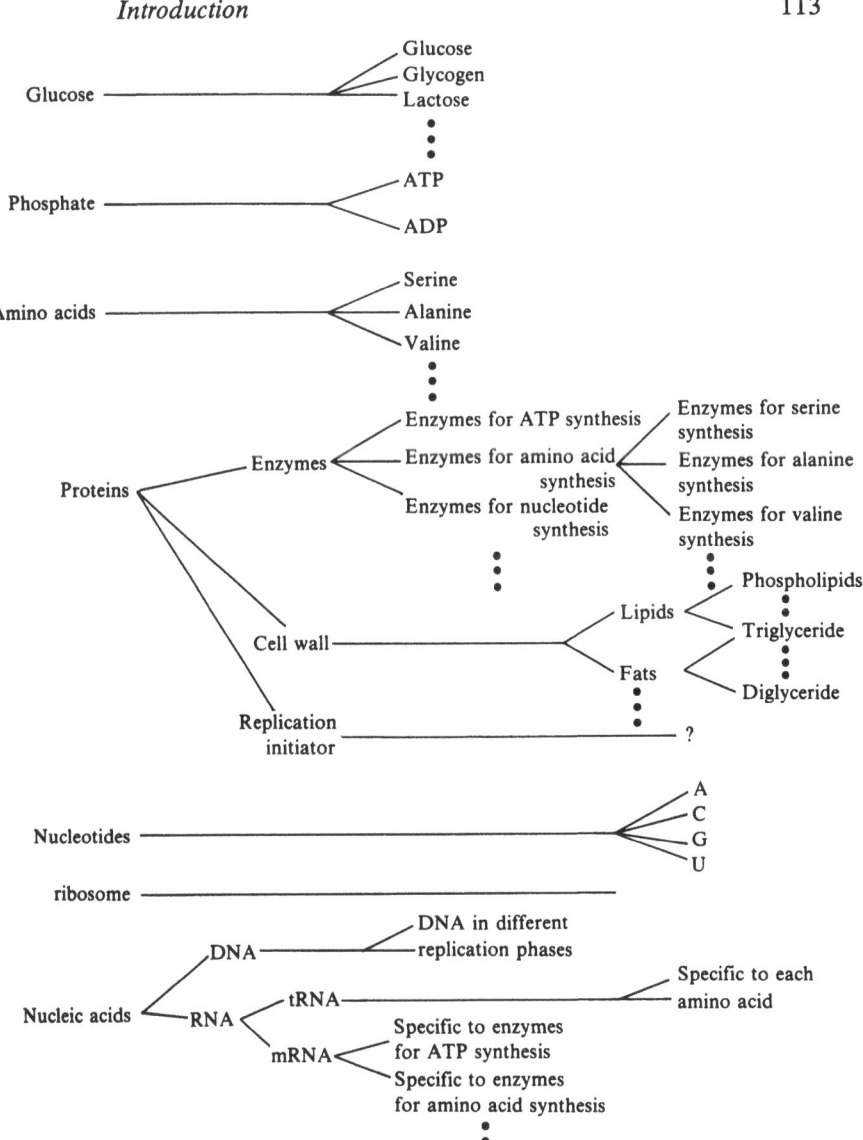

Figure 2.1.1. Refinement of biochemical classes.

interest related to the growth and replication response of the cell to shifts in its nutrient environment. In laboratory experiments, such things as number of cells in the colony, total mass, total DNA, total RNA etc. could be measured and their behavior in time recorded. It was such 'macroscopic' cell characteristics that were taken as model output variables to be compared with the experimental measurements. Since this experimental frame does not reflect any interest in the elementary biochemicals, it is reasonable to try for relatively simple models which employ only the coarser classes of Figure 2.1.1 as pools. Of course, the choice of pools is not entirely arbitrary. Certain choices may 'over simplify' in that they may not be capable of supporting valid models, i.e. models capable of reproducing the experimental data, even within the gross experimental frame of interest. For example, choice of the class 'proteins' for a pool would probably not work, since in order to account for metabolism one needs the enzyme fraction, to account for the cell size one needs the cell wall fraction, and to account for replication and cellular dimension one needs the initiator fraction. A model which does not distinguish at least these subclasses would have poor prospects for success.

On the other hand, such considerations may force ever finer partitioning of classes. Indeed, the partitioning of one class may force the partitioning of others, and these of others, and so on. For example, for each primary class (glucose, phosphate, etc.) that we distinguish, we may need to differentiate an enzyme to catalyze its synthesis. Splitting the primary class (e.g. amino acids into serine, alanine, etc.) requires further splitting of the enzymes for specificity to the new classes. For each enzyme pool we distinguish, we may need a messenger RNA (mRNA) to encode it. For each such mRNA, we may need to distinguish an enzyme to synthesize it, and so on.

Clearly this process must stop, either because it stops 'naturally', or because we overrun our complexity limits (mental or computational), or because we are willing to risk the consequences of 'prematurely' stopping.

Are there any criteria to help us make this decision?

There are, and their formulation and investigation constitutes a main concern of the theory of modeling and simulation (Zeigler, 1976). The relevant results take the form of theorems specifying the conditions under which a given simplification procedure will be valid. Necessarily, these conditions refer to a **base** model, whose structure must be known if these conditions are to be checked, and the procedure for reducing it to a **lumped** model are to be applied. If, as is usually the case, the base model is not known, these theorems take on a heuristic rather than an algorithmic cloak – they offer criteria whose plausibility may be assessible, if not fully ascertainable.

This chapter applies the theory of modeling and simulation to the simplification of chemical reaction systems.

Section 2.2 provides an exposition of the essential concepts of the theory of simplification. Section 2.3 develops a description of chemical networks which lends itself to simulation and simplification. Sections

2.4–2.7 develop the necessary and sufficient conditions for valid simplification, while Section 2.8 shows how these conditions may be employed in heuristic simplification procedures.

The reader interested only in this article's final product – the heuristics for simplification – is invited to skip directly to Section 2.8. The heuristics are intuitively reasonable and easily acceptable. The reader who wishes some preliminary justification for their necessity and sufficiency is invited to proceed to Section 2.2. The reader who wants a full understanding of the theory is invited to continue further into Sections 2.4–2.7.

The conditions developed in Sections 2.4–2.7 are quite stringent. Since they are shown to be necessary for justifying the simplifications under study (which are themselves reasonable), there is no getting away from them. The reader should thus want to know what the conditions are in order to understand when such simplification is indeed fully justified.

Although the conditions may not be fully satisfiable in most real systems of interest, it is reasonable to seek simplifications for which the conditions are approximately satisfied. In this case it becomes important to assess the effect of the error introduced. This issue is considered in Section 2.7.

2.2 ESSENTIALS OF LUMPING AND HOMOMORPHISM

Suppose we have a chemical system consisting of chemicals A_1, A_2, B_1 and B_2 and reactions

$$A_1 \xrightarrow{k_1} B_1,$$

$$A_2 \xrightarrow{k_2} B_2.$$

These reactions can be interpreted as determining the time course of the concentrations $A_1(t)$, $A_2(t)$, $B_1(t)$ and $B_2(t)$ (we shall use italic for the concentrations of chemicals) through the differential equations

$$dA_1/dt = -k_1 A_1, \qquad dB_1/dt = k_1 A_1,$$
$$dA_2/dt = -k_2 A_2, \qquad dB_2/dt = k_2 A_2. \tag{1}$$

Thus our base model consists of the chemicals A_1, A_2, B_1, B_2 and the equation set (1).

Suppose that we cannot distinguish between chemicals B_1 and B_2 and between A_1 and A_2 in a certain type of experiment. Let A and B refer to the pools as seen through such an experiment. In other words, a molecule will be counted as an A if it is either of type A_1 or of type A_2 and similarly for B. Thus by definition what we measure in the experiment is

$$A(t) = A_1(t) + A_2(t),$$
$$B(t) = B_1(t) + B_2(t). \tag{2}$$

Now let us ask whether it is possible to describe the time course of pools A and B by a simple model with just one reaction

$$A \xrightarrow{k} B.$$

In other words our lumped model consists of pools A and B whose time course is determined by the equation set

$$dA/dt = -kA,$$
$$dB/dt = kA. \tag{3}$$

We say that the lumped model is a valid simplification of the base model if the time courses of pools A and B predicted by each model are in complete agreement. In other words we require that $A(t)$ and $B(t)$ computed from (1) and (2) (the base model) agree with $A(t)$ and $B(t)$ computed from (3) (the lumped model).

Now consider the following diagram:

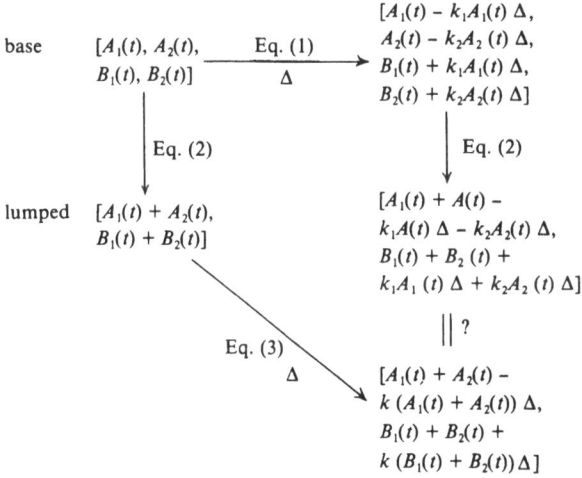

Starting in the upper left-hand corner and proceeding vertically down amounts to supposing that at time t both models give the same values to pools A and B. The question then is will they continue to produce values which agree at time $t + \Delta$, a short time later. The upper horizontal path represents what is predicted by the base model about the chemicals A_1, A_2, B_1 and B_2 using (1). Following this by the vertical path downwards yields the values of A and B predicted by the base model at time $t + \Delta$. Proceeding horizontally on the lower path yields the corresponding values predicted by the lumped model using (3). Thus the two models yield the same values of A and B at $t + \Delta$, if and only if

$$k_1 A_1(t) + k_2 A_2(t) = k(A_1(t) + A_2(t)). \qquad (4)$$

If it happens that $k_1 = k_2$, then (4) can be satisfied by choosing $k = k_1$. Thus a sufficient condition for lumped model validity is that the base model reactions have the same rate constants and that the lumped model rate constant is equal to either one of these. A brief restatement of this is that *you can lump two first-order reactions with the same rate constant into a single one of the same form which expresses the aggregate behavior of the substrates and products.*

As stated, the definition of validity implied that the two models produce trajectories which agree no matter what are the starting values for the individual concentrations. *Under these conditions, it can be seen that the requirement for equality of rate constants is also necessary.* To prove this, note that we can choose $A_1(t) = 0$, $A_2(t) \neq 0$ and then (4) reduces to $k_2 = k$ and similarly choosing $A_1(t) \neq 0$, $A_2(t) = 0$, we get $k_1 = k$. Hence $k_1 = k_2$ must hold if the proposed simplification is to be possible and there is to be no restriction on the starting values of the individual concentrations. In a moment we shall see that we may have to be content with validity for something less than all possible starting values.

When both paths in the above diagram lead to the same result it is said to **commute** and the aggregation relation expressed by $A = A_1 + A_2$, $B = B_1 + B_2$ is said to be a **homomorphism** (Zeigler, 1976). These concepts are fundamental in helping us to understand how and when simplification is possible and to design effective simplification procedures.

Now suppose that we have invented an instrument which will allow us to distinguish chemicals B_1 and B_2. We want to know if there is a lumped model which keeps track of B_1 and B_2 individually but still retains only the aggregate pool A. In other words the base model is the same as before (chemicals A_1, A_2, B_1 and B_2, reactions $A_1 \xrightarrow{k_1} B_1$ and $A_2 \xrightarrow{k_2} B_2$). But now the experimental measurements are A, B_1 and B_2 where

$$A(t) = A_1(t) + A_2(t). \tag{2'}$$

And the lumped model has pools A, B_1 and B_2 where we postulate the reactions

$$A \xrightarrow{K_1} B_1,$$

$$A \xrightarrow{K_2} B_2,$$

with equations

$$dA/dt = -K_1 A - K_2 A = -(K_1 + K_2)A,$$

$$dB_1/dt = K_1 A, \tag{3'}$$

$$dB_2/dt = K_2 A.$$

Let us note the problem of the lumped model here. It can at best keep track of the sum $A_1 + A_2$ and yet it needs the individual components A_1 and A_2 since the amount of A_1 determines the amount of B_1 produced (and similarly A_2 determines B_2) in the base model. We should expect then that either no lumped model can do this, or that, if such a model is possible, it will work only under rather special conditions. In fact we can guess (correctly) that the proportions of A_1 and A_2 in the pool A of the base model must remain fixed over time.

Using our commutative diagram for the relation (2') and the new lumped model (3') we find that we now have the following three equations to satisfy:

$$k_1 A_1(t) + k_2 A_2(t) = (K_1 + K_2)(A_1(t) + A_2(t)), \tag{5a}$$

$$k_1 A_1(t) = K_1(A_1(t) + A_2(t)), \tag{5b}$$

$$k_2 A_2(t) = K_2(A_1(t) + A_2(t)). \tag{5c}$$

While $(5a)$ can still be satisfied for all values of $A_1(t)$ and $A_2(t)$, this is not true for $(5b)$ and $(5c)$. In fact (assuming none of k_1, k_2, K_1, K_2 is zero), $(5b)$ requires that

$$A_1(t)/A_2(t) = K_1/(k_1 - k_1), \tag{6}$$

i.e. the ratio $A_1(t)/A_2(t)$ is constant in time. (Equation $(5c)$ also requires this by symmetry.) Thus we can only expect simplification under those conditions. But can they be realized in the base model? We check this as follows:

$$(A_1(t), A_2(t)) \xrightarrow[\Delta]{\text{Eq. (1)}} [A_1(t) - k_1 A_1(t)\Delta, A_2(t) - k_2 A_2(t)\Delta].$$

◆◆   If $\dfrac{A_1(t)}{A_2(t)} = c$ is $\dfrac{A_1(t) - k_1 A_1(t)\Delta}{A_2(t) - k_2 A_2(t)\Delta} = c$?

The answer is easily seen to be yes only if $k_1 = k_2$. Thus only a base model with equal rate constants will maintain fixed proportions of A_1 and A_2.

With $k_1 = k_2 = k$, the homomorphism conditions (5) now take the form

$$k[A_1(t) + A_2(t)] = (K_1 + K_2)[A_1(t) + A_2(t)], \tag{5'a}$$

$$kA_1(t) = K_1[A_1(t) + A_2(t)], \tag{5'b}$$

$$kA_2(t) = K_2[A_1(t) + A_2(t)]. \tag{5'c}$$

These relations can be satisfied only if K_1 and K_2 are such that $K_1 + K_2 = k$, $(5'a)$, and $A_1(t)$ and $A_2(t)$ such that $A_1(t)/A_2(t) = K_1/K_2$, $(5'c)$.

In other words if we choose rate constants K_1 and K_2 for the lumped model such that $K_1 + K_2 = k$, then we get a valid simplification of the base model, provided the latter starts off with chemicals A_1 and A_2 in the proportion K_1 to K_2 (it will maintain this proportion because of the equal component rate constants assumed).

The final base–lumped model pair can be displayed as

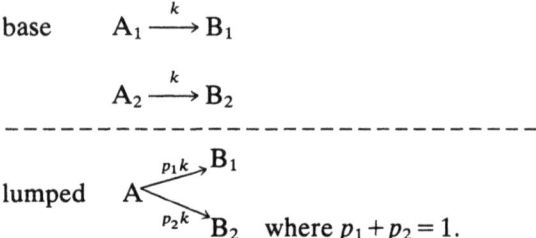

base $A_1 \xrightarrow{k} B_1$

$A_2 \xrightarrow{k} B_2$

----

lumped $A \overset{p_1 k}{\underset{p_2 k}{\lessgtr}} \begin{matrix} B_1 \\ B_2 \end{matrix}$ where $p_1 + p_2 = 1$.

(This is achieved by setting $p_1 = K_1/k$, $p_2 = K_2/k$.) It has the interpretation that *when we wish to distinguish products but lump reactants, we can only do so under the assumption that the base model reactants are maintained in a fixed proportion $p_1 : p_2$* and we employ the same p_1 and p_2 as probabilities of producing the products in the lumped model.

The next three sections show how the observations made here generalize to arbitrary chemical networks.

2.3 REPRESENTATION AND SIMULATION OF CHEMICAL NETWORKS

To understand more easily how chemical networks can be simplified, it will help first to have a clear idea of how such networks can be represented.

We shall find it convenient to specify a reaction system by means of a table listing the reactions and their characteristics. This table can be directly employed as the basis for a Monte Carlo simulation of the reaction network (Moebs, 1974). In this way the network is conceived as a stochastic model reflecting the chance collisions that bring about molecular reactions (Boudart, 1972). Assuming large numbers of molecules leads from the stochastic model to its deterministic counterpart (the form we shall be employing to discuss simplification) and finally to the standard differential equation presentation.

Consider for example the system

$$2A + E \underset{k_{-1}}{\overset{k_1}{\rightleftharpoons}} C \overset{k_2}{\longrightarrow} B + E$$

which might represent a conversion of substrate A to product B catalyzed by an enzyme E. In this reaction two molecules of A combine with an enzyme molecule to form a complex C which dissociates to form a molecule of B. This presentation is shorthand for the more complete and general form of presentation which says that we have a system consisting of chemicals A, E, C and B and reactions

$$\alpha : 2A + E \overset{k_1}{\longrightarrow} C,$$

$$\beta : C \overset{k_{-1}}{\longrightarrow} 2A + E,$$

$$\gamma : C \overset{k_2}{\longrightarrow} B + E.$$

Another still more complete means of representing these reactions is via the Table 2.3.1.

Table 2.3.1. *Example of reaction specification table*

Reaction	Reactants	Products	Rate	Effect of reaction on chemical			
				A	E	C	B
α	A, E	C	$k_1 AE$	-2	-1	1	0
β	C	A, E	$k_{-1} C$	2	1	-1	0
γ	C	B, E	$k_2 C$	0	1	-1	1

Thus for each reaction, we list its reactants, its products, the rate of the reaction, and the effect of the reaction. The latter is just the listing of stoichiometric coefficients of the reaction with a plus or minus sign according to whether the component is a product or reactant respectively (a zero signifies neither product nor reactant). The rate of reaction is assumed to be a function of the concentrations of its reactants in the form of a product of powers, where often but not always the powers are the associated stoichiometric coefficients (Boudart, 1972).

Such a table provides a complete specification of a chemical model being proposed. Its general appearance is illustrated in Table 2.3.2 which also displays the symbols to be used in denoting the various entities.

The specification table can be interpreted as a stochastic model which is simulated as follows:

Initialization

1. Initialize the concentrations of the pools to the desired starting values, $C_1, C_2, \ldots, C_n$, and the clock to initial time t.

State transition

2. Compute the rates of reactions $r_1, \ldots, r_m$ employing the current values of the concentrations.

$3(a)$. For each reaction α, $\alpha = 1, \ldots, m$, with probability $r_\alpha \cdot \Delta$ select this reaction to occur.

$3(b)$. For each reaction α, $\alpha = 1, \ldots, m$, if it has been selected in step 3, then carry out its effect on the pools: i.e. for each pool C_j, $j = 1, \ldots n$, set $C_j = C_j + n_{\alpha,j}$.

Time advance

4. Set $t = t + \Delta$.

Test for finish

5. If run is to continue, go to 2, else stop.

Notice that a reaction rate is interpreted as a probability per unit time that a reaction occurs (a collision of the reactants with energy exceeding the activation level (Boudart, 1972)). Also the effect of a reaction which occurs is to decrease the reactant concentrations and to increase the product concentrations by the stoichiometric coefficients. The algorithm above can actually be used for simulating chemical models but it may be improved by discrete event formulation.

By assuming that the concentrations are very large, the stochastic model can be represented by a deterministic model which well approximates the time course of the concentrations. The algorithm for simulating the deterministic model is the same as above, except that step 3 is replaced by the following:

3'. For each reaction, α, $\alpha = 1, \ldots, m$, carry out its expected effect on the pools: for each pool $j = 1, \ldots, n$, set $C_j = C_j + r_\alpha \cdot \Delta \cdot n_{\alpha,j}$.

From the above algorithm, it is clear that the net change in pool C_α from time t to $t + \Delta$ is

$$\Delta C_j = \sum_{\alpha=1}^{m} r_\alpha \cdot \Delta \cdot n_{\alpha,j}.$$

Letting $\Delta \to 0$, and writing $dC_j/dt = \Delta C_j/\Delta$ we obtain yet another model, this time a differential equation model:

$$\frac{dC_j}{dt} = \sum_{\alpha=1}^{m} r_\alpha n_{\alpha,j}, \quad j = 1, \ldots n.$$

This is the familiar formulation of chemical kinetics (Boudart, 1972). It says that the rate of change in pool C_j is obtained by multiplying the rate column by the appropriate effect column under pool C_j in Table 2.3.2 and summing down the column. Applied to Table 2.3.1 as an example, the procedure yields

$$dA/dt = -2k_1AE + 2k_{-1}C,$$

$$dE/dt = -k_1AE + k_{-1}C + k_2C,$$

$$dC/dt = k_1AE - k_{-1}C - k_2C,$$

$$dB/dt = k_2C.$$

While this is the standard formulation, it is not necessarily the most convenient or insightful. Indeed, we hope to have convinced the reader that the reaction rates and effects are primary and the expression of their combined effects through the above differential equations only secondary. We can deal with simplification much more conveniently by focusing on the reactions and their manifestation in the simulation algorithms than by looking at standard differential equations. We shall discuss simplification in terms of (a) lumping of reactions, (b) lumping of chemicals into pools, and then combine both procedures to form the most general simplification procedure.

Table 2.3.2. *Reaction specification table*

Reaction	Reactants	Products	Rate	Effect of reaction on chemical C_1,	C_2,	C_3,	$\ldots$,	C_n
α_1	$R_{\alpha 1}$	$P_{\alpha 1}$	$r_{\alpha 1}$	n_{11}	n_{12}	n_{13}	$\ldots$	n_{1n}
α_2	$R_{\alpha 2}$	$P_{\alpha 2}$	$r_{\alpha 2}$	n_{21}	n_{22}	n_{23}	$\ldots$	n_{2n}
$\vdots$	$\vdots$	$\vdots$	$\vdots$	$\vdots$	$\vdots$	$\vdots$	$\ldots$	$\vdots$
α_m	$R_{\alpha m}$	$P_{\alpha m}$	$r_{\alpha m}$	n_{m1}	n_{m2}	n_{m3}	$\ldots$	n_{mn}

Table 2.3.3. *Lumped model for base model of Table 2.3.1*

Reaction	Reactants	Products	Rate	Effect of reaction on pool A	E	C	B
α	A, E	C	$k_1 AE$	-2	-1	1	0
$\beta + \gamma$	C	A, E, B	$(k_{-1} + k_2)C$	$\dfrac{2k_{-1}}{k_{-1} + k_2}$	1	-1	$\dfrac{k_2}{k_1 + k_2}$

2.4 LUMPING OF REACTIONS

Consider lumping reactions β and γ together in the model of Table 2.3.1 to form the lumped model represented in Table 2.3.3. The lumped reaction, called $\beta + \gamma$, has as reactants (products) the union of the reactants (products) of β and of γ. Its rate is the sum of the individual rates. Finally its effect on a chemical is the sum of the effects of the individual reactions weighted according to their rates, e.g. the effect on A is

$$\frac{k_{-1} \cdot 2 + k_2 \cdot 0}{k_{-1} + k_2} = \frac{2k_{-1}}{k_{-1} + k_2}.$$

The general procedure is illustrated in Table 2.4.1. Suppose that we wish to lump a set of reactions which after reordering are listed in the first column of the table: $\alpha_1, \ldots, \alpha_k$. The resulting reaction is called α' (symbolically, it is the sum of $\alpha_1, \ldots, \alpha_k$), and its description is evident in the lumped model table.

To understand the justification for the procedure, we note that in our first simulation algorithm the probability of at least one reaction α_i, $i = 1, \ldots, k$ occurring in t to $t + \Delta$ is $\sum_{i=1}^{k} r_{\alpha i} \cdot \Delta$ (this neglects multiple reaction events, whose probability is of order Δ^2). Thus the lumped reaction is given the aggregated rate. The expected effect of reaction α' in the lumped model, according to our second simulation algorithm, is

$$r_{\alpha'} \cdot n_{\alpha',j} = r_{\alpha'} \cdot \sum_{i=1}^{k} n_{ij} \left(\frac{r_{\alpha i}}{r_{\alpha'}} \right) = \sum_{i=1}^{k} n_{ij} r_{\alpha i},$$

which is the expected effect of the disaggregated reactions in the base model. This is summarized in the commutative diagram

$$(\ldots, C_j(t), \ldots) \xrightarrow{\alpha_1, \ldots, \alpha_k} \left(\ldots, C_j(t) + \Delta \sum_{i=1}^{k} n_{ij} r_{\alpha i}, \ldots \right)$$

$$\updownarrow \qquad\qquad\qquad\qquad \updownarrow$$

$$(\ldots, C_j(t), \ldots) \xrightarrow{\alpha'} \left(\ldots, C_j(t) + \Delta \sum_{l=1}^{l} r_{\alpha'} \cdot n_{\alpha',j}, \ldots \right),$$

which the reader can verify.

Table 2.4.1. *Procedure for lumping of reactions*

Base

Reaction	Reactants	Products	Rate	Effect of reaction on chemical C_j　...
α_1	$R_{\alpha 1}$	$P_{\alpha 1}$	$r_{\alpha 1}$	n_{1j}　...
α_2	$R_{\alpha 2}$	$P_{\alpha 2}$	$r_{\alpha 2}$	n_{2j}　...
⋮	⋮	⋮	⋮	⋮　⋮
α_k	$R_{\alpha k}$	$P_{\alpha k}$	$r_{\alpha k}$	n_{kj}　...
α_{k+1}				...
⋮				
α_n				

Lumped

Reaction	Reactants	Products	Rate	Effect of reaction on pool C_j　...
$\alpha' = \sum\limits_{i=1}^{k} \alpha_i$	$R_{\alpha'} = \bigcup\limits_{i=1}^{k} R_{\alpha i}$	$P_{\alpha'} = \bigcup\limits_{i=1}^{k} P_{\alpha i}$	$r_{\alpha'} = \sum\limits_{i=1}^{k} r_{\alpha i}$	$n_{\alpha' j} = \sum\limits_{i=1}^{k} n_{ij}\left(\dfrac{r_{\alpha i}}{r_{\alpha'}}\right)$　...
α_{k+1}				
⋮				
α_n				

The double arrows indicate that, in this case, the homomorphism relation is in fact, a relation of identity. In such a case the base and lumped models are said to be isomorphic. To be more precise, the deterministic versions are isomorphic but the stochastic versions are not since the lumped model only reproduces the expected behavior of the base (this will be evident upon rereading the above justification).

Note that the lumped reaction effect $n_{\alpha' j} = \sum_{i=1}^{k} n_{ij}(r_{\alpha i}/r_{\alpha'})$ may be a function of the concentrations of its reactants $R_{\alpha'}$. A sufficient condition that $n_{\alpha j}$ is independent of these concentrations is that

$$\frac{r_{\alpha i}}{r_{\alpha'}} = \text{constant} = p_i, \qquad i = 1, \ldots, k,$$

i.e. the rate of reaction α_i relative to the block $\alpha_1, \ldots, \alpha_k$ is a constant over time, interpretable as p_i, the conditioned probability of α_i occurring given that one of $\alpha_1, \ldots, \alpha_k$ has occurred.

Also note that even in this case we can no longer enforce the constraint that the stoichiometric coefficients must be integers, which in any case is no longer demanded in modern formulations (Boudart, 1972).

A second sufficient condition for constancy of N_i is that

$$n_{1j} = n_{2j} = \ldots = n_{kj},$$

i.e. reactions $\alpha_1, \ldots, \alpha_k$ all have the same effect on concentration C_j. In this case,

$$n_{\alpha'j} = n_{1j}.$$

2.5 LUMPING CHEMICALS INTO POOLS

Table 2.5.1 exemplifies the procedure for lumping chemicals $C_1, \ldots, C_l$ into a pool C'. The effect of a reaction on the new pool is taken to be the sum of its effects on the pool constituents. That this is what is required is evident in the following commutative diagram:

$$(C_1, \ldots, C_l, \ldots) \overset{\alpha}{\to} (C_1 + \Delta r_\alpha \cdot n_{\alpha,1}, \ldots C_l + \Delta r_\alpha \cdot n_{\alpha,l}, \ldots)$$

$$\downarrow h \qquad\qquad\qquad\qquad \downarrow h$$

$$\left(\sum_{i=1}^{l} C_i, \ldots \right) \overset{\alpha}{\to} \left(\sum_{i=1}^{l} C_i + \Delta \cdot r_\alpha \cdot n'_{\alpha,1}, \ldots \right),$$

where the many to one correspondence h, taking the individual concentrations of pool constituents into their sum, is in this case a proper homomorphism.

There is something more than meets the eye here, since the reactants and products of α in the base model may have been transformed in the lumping process. More significantly, the rate r_α in the lumped model must now be expressed as a function of the concentrations of the lumped reactants.

Table 2.5.1. *Lumping of chemicals*

Base							
				Effect of reaction on chemical			
Reaction	Reactants	Products	Rate	$C_1,$	$\ldots,$	C_l	$\ldots$
$\vdots$	$\vdots$	$\vdots$	$\vdots$	$\vdots$	$\vdots$	$\vdots$	$\vdots$
α	R_α	P_α	r_α	$n_{\alpha 1}$	$\ldots$	$n_{\alpha,l}$	$\ldots$
$\vdots$	$\vdots$	$\vdots$	$\vdots$	$\vdots$	$\vdots$	$\vdots$	$\vdots$

Table 2.5.1. *Continued*

Lumped

Reaction	Reactants	Products	Rate	Effect of reaction on pool $C'_1 = \sum\limits_{i=1}^{l} C_i$	...
$\vdots$	$\vdots$	$\vdots$	$\vdots$	$\vdots$	$\vdots$
α	$R'_\alpha = R_\alpha / \pi$	$P'_\alpha = P_\alpha / \pi$	r_α	$n'_{\alpha,1} = \sum\limits_{j=1}^{l} n_{\alpha,j}$	...
$\vdots$	$\vdots$	$\vdots$	$\vdots$	$\vdots$	$\vdots$

Our example of Section 2.1, now represented in the following table, should make this clear.

Table 2.5.2

Base

				Effect of reaction on chemical			
Reaction	Reactants	Products	Rate	A_1	A_2	B_1	B_2
α_1	A_1	B_1	$k_1 A_1$	-1	0	1	0
α_2	A_2	B_2	$k_2 A_2$	0	-1	0	1

Lumped

				Effect of reaction on pool	
Reaction	Reactants	Products	Rate	A	B
α_1	A	B	$k_1 A$	-1	1
α_2	A	B	$k_2 B$	-1	1

Here we have lumped A_1 and A_2 to form pool A and B_1 and B_2 to form pool B. Note that to transform the base model table each chemical is replaced by the pool into which it is grouped. Thus the rate of α_1 in the lumped model must be expressed as $k_1 A$ rather than $k_1 A_1$. Since A and A_1 are not equal, r_α cannot be expressed as a function of the lumped concentrations in this example.

More generally, let π denote the partition of chemicals $C_1, \ldots, C_n$ into pools $C'_1, \ldots, C'_n$. That is π consists of a family of blocks (technically called equivalence classes); (C'_i) denotes the block of π which gets mapped into lumped pool C'_i. (Since π is a partition, the (C'_i) are pairwise disjoint, and their union is $C_1, \ldots, C_n$. In our example above, $(A) = \{A_1, A_2\}$ and $(B) = \{B_1, B_2\}$ are the blocks of π.

We shall use S/π to denote the set of pools into which a set S of chemicals is lumped by π. Thus in Table 2.5.1, the set of reactants of α, $R'_\alpha = R_\alpha/\pi$ and similarly for products.

The criterion for being able to express r_α in terms of the lumped concentrations is captured by the commutative diagram

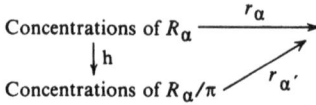

This says that there is a function r'_α which when applied to the pool concentrations as aggregated by h gives the same value as r_α does applied to the concentrations of R_α. Of course, in the aggregation by h, each pool receives the sum of the concentrations of the chemicals in R_α it represents. When this condition is satisfied we say that r_α is expressible in the lumped pools and we shall drop the prime in denoting the lumped model rate.

2.6 LUMPING OF REACTIONS AND OF CHEMICALS

It is not likely that rates are expressible in the lumped pools if we restrict ourselves just to lumping of chemicals, but prospects improve markedly if we lump both reactions and chemicals. Table 2.6.1 shows the general procedure which can be viewed as an application of the lumping of chemicals followed by the lumping of reactions. Table 2.6.2 illustrates the procedure applied to an example in which there are two types of substrates, A_1, A_2 and two types of enzymes, E_1, E_2, which combine to form four types of complexes, $C_1, \ldots, C_4$, and two types of products, B_1, B_2. The simplification lumps all substrates, enzymes, etc. (respectively) together.

The following commutative diagram ensures that the aggregation mapping, h (assigning to each pool the sum of the concentrations of its components), is a homomorphism:

$$
\begin{array}{ccc}
\overbrace{\ldots, C_{j_1}(t), \ldots, C_{j_m}(t), \ldots,}^{(j')} & \xrightarrow{(\alpha')} & \overbrace{\ldots, C_j(t) + \Delta \sum_{\alpha \in (\alpha')} n_{\alpha,j} r_\alpha}^{(j')} \\
\Big\downarrow h & & \Big\downarrow h \\
\ldots, \sum_{j \in (j')} C_j(t), \ldots, & \xrightarrow{\alpha'} & \ldots, \sum_{j \in (j')} C_j(t) + \Delta r_{\alpha'} n_{\alpha',j'}, \ldots,
\end{array}
$$

where recall that (j') is the set of chemicals to be lumped into pool j, and similarly (α') is the set of reactions to be lumped into reaction α'. The reader may check that the diagram commutes, noting the definition of $n_{\alpha'j}$ in Table 2.6.1.

Table 2.6.1

Base					Effect of reaction on chemical		
					$\overset{(j')}{\overbrace{\quad\quad}}$		
Reaction	Reactants	Products	Rate	$\ldots$	$\overset{\ulcorner}{C_{j1}},$	$\ldots$ $\overset{\urcorner}{C_{jm}}$	$\ldots$
$\vdots$	$\vdots$	$\vdots$	$\vdots$	$\vdots$	$\vdots$	$\vdots$	$\vdots$
$(\alpha')\begin{cases}\alpha_i \\ \vdots \\ \alpha_{i_n}\end{cases}$	$\begin{matrix}R_{\alpha i}\\ \vdots \\ R_{\alpha i}\end{matrix}$	$\begin{matrix}P_\alpha\\ \vdots \\ P_{\alpha i}\end{matrix}$	$\begin{matrix}r_\alpha\\ \vdots \\ r_\alpha\end{matrix}$	$\begin{matrix}\vdots\\ \\ \vdots\end{matrix}$	$\begin{matrix}n_{i1,j2},\\ \vdots \\ n_{in,j1},\end{matrix}$	$\begin{matrix}\ldots & n_{i1,jm}\\ \vdots \\ \ldots & n_{in,jm}\end{matrix}$	$\begin{matrix}\ldots\\ \vdots \\ \ldots\end{matrix}$
$\vdots$	$\vdots$	$\vdots$	$\vdots$	$\vdots$	$\vdots$	$\vdots$	$\vdots$

Lumped				
				Effect of reaction on pool
Reaction	Reactants	Products	Rate	$C_j,$ $\ldots$
$\vdots$	$\vdots$	$\vdots$	$\vdots$	$\vdots$ $\ldots$
α'	$R_{\alpha'} = \bigcup\limits_{\alpha\in(\alpha')} R_\alpha/\pi$	$P_{\alpha'} = \bigcup\limits_{\alpha\in(\alpha')} P_\alpha/\pi$	$r_{\alpha'} = \sum\limits_{\alpha\in(\alpha')} r_\alpha$	$n_{\alpha'j'} = \sum\limits_{\alpha\in(\alpha')}(r_\alpha/r_{\alpha'})$ $\times \sum\limits_{j\in(j')} n_{\alpha,j}$

We may summarize our previous results in a theorem.

Theorem 1

The aggregation mapping, h, is a homomorphism from the base model to the lumped model if
   (i) The reactions $r_{\alpha'}$ are expressible in the lumped concentrations and
   (ii) the coefficients $n_{\alpha',j'}$ are constants (independent of time).
Sufficient conditions for (ii) to hold are that
   (a) $\sum_{j\in(j')} n_{\alpha,j}$ is the same for each $\alpha \in (\alpha')$, or
   (b) $r_\alpha/r_{\alpha'}$ is independent of time, for all possible reactions and chemicals.
In words, (a) says that the compound effect on a pool is the same for each reaction in an equivalence class of reactions (reactions α in a block (α')). This is a natural criterion for lumping and is of the form characteristic of simplification conditions for other kinds of models (Zeigler, 1976). In this case $n_{\alpha',j'} = \sum_{j\in(j')} n_{\alpha,j}$, independent of $\alpha \in (\alpha')$. Condition (b) says that the relative rates of reactions in an equivalence class remain fixed in time. Although condition (b) may obtain, it puts much stronger constraints on the base model, as we shall see.

Let us note that in Table 2.6.2. condition (ii) is satisfied because of (a). But condition (i) is not satisfied, *unless* all $k_{ij} = \text{constant} = k$ for $i \in \{1, 2\}$ and all $l_i = \text{constant} = l$ for $i \in \{1, \ldots, 4\}$. In the latter case,

$$r_{\alpha'} = k \sum_{i=1}^{2} \sum_{j=1}^{2} A_i E_j = k(A_1 + A_2)(E_1 + E_2) = kAE$$

and

$$r_{\gamma'} = l \sum_{i=1}^{4} C_i = lC.$$

Thus the rates are indeed expressible in the lumped pools as required. Note that the resulting reactions α' and γ' are just the reactions α and γ of Table 2.3.1.

This example of Table 2.6.2. is an instance of a more general result. Suppose that we have a base model containing reactions

$$\alpha: n_1 A_{\alpha,1} + n_2 A_{\alpha,2} + \ldots n_m A_{\alpha,m} \to \mu_1 B_{\alpha,1} + \mu_2 B_{\alpha,2} + \ldots \mu_n B_{\alpha,n},$$

$$\alpha = 1, 2, \ldots, N_\alpha.$$

Suppose also that a partition π of chemicals is such that all the $A_{\alpha,i}$ are lumped to a pool A_i for $i = 1, \ldots, m$ and similarly all $B_{\alpha,j}$ form a pool B_j for $j = 1, \ldots, n$. Then all the reactions become identified with the reaction

$$\alpha': n_1 A_1 + n_2 A_2 + \ldots + n_m A_m \to \mu_1 B_1 + \mu_2 B_2 + \ldots + \mu_n B_n$$

under the partition π (substitute A_1 for $A_{\alpha,1}$, A_2 for $A_{\alpha,2}$, etc.).

We shall say that the reactions α all have the *same form* under π if the above holds.

Now notice that given the reaction α' and the partition π, we can generate a whole set of reactions of the same form. Symbolically this can be denoted

$$n_1 \begin{Bmatrix} A_{1,1} \\ A_{1,2} \\ \vdots \end{Bmatrix} + n_2 \begin{Bmatrix} A_{2,1} \\ A_{2,2} \\ \vdots \end{Bmatrix} + \ldots \to u_2 \begin{Bmatrix} B_{1,1} \\ B_{1,2} \\ \vdots \end{Bmatrix} + \ldots.$$

where every choice of the A's and B's from the bracketed possibilities generates a reaction of the same form under π.

We shall say that the base model contains a *complete set of reactions* if it contains reactions corresponding to all possible substitutions of the A's, i.e. the reactant classes. Note that we do not require that all combinations of the B's (products) appear. The base model in Table 2.6.2 contains complete sets of reactions of the form $2A + E \to C$ and $C \to B + E$ under the partition $\pi = \{\{A_1, A_2\}, \{B_1, B_2\}, \{C_1, C_2, C_3, C_4\}\}$.

Generalizing the result illustrated in Table 2.6.2, we obtain another theorem.

Table 2.6.2. *Lumping of chemicals and reactions*

Base

	Reaction	Reactants	Products	Rate	Effect of reaction on chemical									
					A		E		C				B	
					A_1	A_2	E_1	E_2	C_1	C_2	C_3	C_4	B_1	B_2
α'	α_1	A_1, E_1	C_1	$k_{12}A_1E_1$	−2		−1		1					
	α_2	A_1, E_2	C_2	$k_{12}A_1E_2$	−2			−1		1				
	α_3	A_2, E_1	C_3	$k_{21}A_2E_1$		−2	−1				1			
	α_4	A_2, E_2	C_4	$k_{22}A_2E_2$		−2		−1				1		
γ	γ_1	C_1	B_1, E_1	l_1C_1					−1				1	
	γ_2	C_2	B_1, E_2	l_2C_1						−1			1	
	γ_3	C_3	B_2, E_1	l_3C_2							−1			1
	γ_4	C_4	B_2, E_2	l_4C_4								−1		1

Lumped

Reaction	Reactants	Products	Rate	Effect of reaction on pool			
				A	E	C	B
α'	A, E	C	$r_{\alpha'} = \sum_{i=1}^{2} \sum_{j=1}^{2} k_{ij}A_iE_j$	−2	−1	1	0
γ'	C	BE	$r_{\alpha'} = \sum_{i=1}^{4} l_iC_i$	0	0	−1	1

Theorem 2

Suppose that a base model has a complete set of reactions of the same form under a partition of the chemicals into pools.

If all of the reactions α have rates of the form

$$r_\alpha(A_{\alpha,1}, A_{\alpha,2}, \ldots A_{\alpha,m}) = k_\alpha A_{\alpha,1} \cdot A_{\alpha,2} \ldots A_{\alpha,m},$$

with identical rate constant $k_\alpha = k$, then the reactions may be lumped into a single reaction of the common form α' (above) with rate

$$r_{\alpha'}(A_1, A_2, \ldots, A_m) = kA_1 \cdot A_2 \ldots A_m.$$

In other words, a lumped model constructed in this way would produce the same behavior of the pool concentrations over time as would the base model.

Proof. Both conditions of Theorem 1 are satisfied. (i) The total reaction rate is a sum of product terms. Since all combinations of variables appear with the same coefficient, this sum simplifies the product of the pool concentrations. (ii) All reactions have the same effect (condition (*a*) is satisfied) on the pools because they are of the same form under the partition. Thus the aggregation mapping, *h*, is a homomorphism from the base model to lumped model.

Now consider the lumpings of reactions in the second example of Section 2.1 and in Table 2.3.3. In these cases the lumped reactions do not have the same form under the relevant partition. What is still true, however, is that the reactant parts *do* have the same form but the product parts do not. This leads us to the following definition.

Reactions α and β have the **same reactant form** under π (a partition of the chemicals) if

$$\alpha: n_1 A_{\alpha,1} + n_2 A_{\alpha,2} + \ldots + n_m A_{\alpha,m} \to \text{product part of } \alpha$$

$$\beta: n_1 A_{\beta,1} + n_2 A_{\beta,2} + \ldots + n_m A_{\beta,m} \to \text{product part of } \beta$$

and $A_{\alpha,i} \underset{\pi}{\equiv} A_{\beta,i}$ (are in the same block of π) for each $i = 1, \ldots, m$.

Note that in this case reactions α and β need not have the same effect on the pool concentrations so Theorem 2 does not follow. However, the examples just mentioned lead us to look for conditions under which relative rates of reactions are constant in time. We should expect that such conditions will be more stringent on the base model and (as in the second example of Section 2.1) that the parameter of a valid lumped model will depend on the starting values of the base model. This is verified as follows.

Theorem 3

Suppose that a base model has a complete set of reactions having the same reactant form under a partition π of the chemicals into pools.

Suppose also that all the reactions α have rates of the form

$$r_\alpha (A_{\alpha,1}, \ldots, A_{\alpha,m}) = k_\alpha A_{\alpha,1} \cdot A_{\alpha,2} \ldots A_{\alpha,m}$$

with identical rate constants $k_\alpha = k$.

Finally, suppose that, for each block of π involved in the common reactant part, the following holds:

(*c*) Let the block of π be $\{C_1, C_2, \ldots, C_n\}$. Then there are

$$p_j \in [0, 1], \quad j = 1, \ldots, n, \quad \sum_{j=1}^{n} p_j = 1,$$

such that

$$C_j = p_j \sum_{i=1}^{n} C_i \Rightarrow \frac{\mathrm{d}C_j}{\mathrm{d}t} = p_j \sum_{i=1}^{n} \frac{\mathrm{d}C_i}{\mathrm{d}t} \qquad \text{for } j = 1, \ldots, n.$$

Then under these conditions, the base model is simplifiable to a valid lumped model.

Proof. As in Theorem 2, since the set of reactions is complete, the total rate $r_{\alpha'} = kA_1 A_2 \ldots A_m$ so that the relative rate $r_\alpha/r_{\alpha'} = (A_{\alpha,1}/A_1) \cdot (A_\alpha/A_2) \ldots (A_{\alpha,m}/A_m)$. Thus the relative rates are constant in time providing that all proportions $A_{\alpha,i}/A_i$ of reactant concentrations to pool concentrations remain constant. Condition (c) above, is equivalent to this requirement. Since $r_\alpha/r_{\alpha'}$ is independent of time, the rest follows as in Theorem 2.

We note that condition (c) can be checked directly (recalling that $\mathrm{d}C_j/\mathrm{d}t = \sum_\alpha r_\alpha n_{\alpha,j}$) and places $n - 1$ constraints on the base model parameters for each reactant pool of size n. Thus for a given base model, the number of constraints may well exceed the number of parameters with the implication that condition (c) cannot be satisfied for this model. Our second example of Section 2.1 and that of Table 2.3.3 show, however, that such constraints can be satisfied.

2.7 ASSESSMENT OF ERROR

Suppose that the conditions of Theorem 2 are all satisfied but that the rate constants k_α are not identical. We suspect that there will no longer be a lumped model valid for predicting the concentrations of the pools of the given partition. There remains the possibility that a finer partition might still satisfy the conditions of Theorem 2 and we shall return to this possibility soon. However, for now let us keep the partition fixed and try to relate the deviation between base and lumped model behavior resulting from the dispersion in the rate constants of the reactions being lumped together.

Examining the square diagram of Section 2.6 we find that it no longer commutes but that there is an error produced in each time step (Zeigler, 1976). It is not difficult to see that the error, e, as an absolute difference in the effect of reaction α' on pool j', is

$$e_{\alpha',j'} = \Delta r_{\alpha'} \left| 1 - \sum_{\alpha \in (\alpha')} \frac{r_\alpha}{r_{\alpha'}} \right| |n_{\alpha',j'}|.$$

Letting $k_{\alpha'}$ be the rate constant of α' in the lumped model and $r_{\alpha'}(A_1, \ldots, A_m) = k_{\alpha'} A_1 \cdot A_2 \ldots A_m$, we can see that the relative error for the effect of reaction α' on pool j' is

$$e_{\alpha',j'}^{\text{rel}} = \frac{e_{\alpha',j'}}{\Delta r_{\alpha'} |n_{\alpha',j'}|}$$

$$= \left| 1 - \sum_{\alpha \in (\alpha')} \frac{k_\alpha}{k_{\alpha'}} \frac{A_{\alpha,1}}{A_1} \frac{A_{\alpha,2}}{A_2} \cdots \frac{A_{\alpha,m}}{A_m} \right|. \tag{1}$$

Recalling that

$$A_i = \sum_{\alpha \in (\alpha')} A_{\alpha,i}, \qquad i = 1, \ldots m,$$

we see that the proportions $A_{\alpha,i}/A_i$ are formally identical to probabilities $p_{\alpha,i}$ and that the sum in (1) is formally identical to the expectation of $k_\alpha/k_{\alpha'}$ over the sample space (α') in which each has probability

$p_{\alpha,1} p_{\alpha,2} \cdots p_{\alpha,m}$ of occurring. It is well known that the maximum value of this expectation over all choices of probability assignments is $k_{max}/k_{\alpha'}$ where $k_{max} = \max \{k_\alpha | \alpha \in (\alpha')\}$ and the minimum value is $k_{min}/k_{\alpha'}$ where $k_{min} = \min \{k_\alpha | \alpha \in (\alpha')\}$. Thus

$$e^{rel}_{\alpha',j'} \leq \max \{|1 - (k_{min}/k_{\alpha'})|, |(k_{max}/k_{\alpha'}) - 1|\}.$$

Choosing $k_{\alpha'} = (k_{max} + k_{min})/2$ minimizes the upper bound and results in

$$e^{rel}_{\alpha',j'} \leq (k_{max} - k_{min})/(k_{max} + k_{min}).$$

Thus the greater the dispersion in rate constants, the greater will be the error introduced at each step in the lumped model prediction of base model behavior. Of course this error may also accumulate over time but whether it does so depends on the dynamics of the lumped model (Zeigler, 1976).

Note that the error bound applies even if the base model does not have a complete set of reactions. In this case we set $k_\alpha = 0$ for all missing reactions α, with the result

$$e^{rel}_{\alpha',j'} \leq (k_{max} - 0)/(k_{max} + 0) = 1.$$

Hence

$$e_{\alpha',j'} \leq \cdot \Delta r_{\alpha'} |n_{\alpha',j'}.$$

The maximum error is in fact achieved when $A_{\alpha,i} = A_i$ $(i = 1, \ldots, m)$ for a missing reaction α (all the reactants of α account for all the occupancy of their associated pools). In this case, the base model chemicals in (j') suffer no change while the lumped model pool j' is changed by an amount $\Delta r_{\alpha'} |n_{\alpha',j'}|$, which is indeed the maximum error. Returning to the essence of the simplification rationale – if all the k_α's of a complete set of reactions are equal, it does not matter to which reactants the base chemicals are concentrated; the effect of base and lumped reactions will be identical.

Let us note that demonstration that the error bound can be achieved also provides a proof of the *necessity* of the just mentioned conditions. Finally, we note that if, as in Theorem 3, the proportions $A_{\alpha,i}/A_i = p_{\alpha,i}$ are known, and can be assumed to be constant, the error is nullified (hence an exact homomorphism is obtained) with the choice $k_{\alpha'} = \sum_{\alpha \in (\alpha')} k_\alpha p_{\alpha,1} p_{\alpha,2} \cdots p_{\alpha,n}$.

2.8 THE SIMPLIFICATION PROCEDURE AND ITS APPLICATION TO THE *E. COLI* CELL

The essentials of the metabolic reaction system employed in the Weinberg–Goodman model of the *E. Coli* cell (Weinberg & Zeigler, 1970; Goodman, 1972) are shown in Figure 2.8.1. (The complete model contains, in addition, apparatus for DNA replication, cell wall synthesis,

1. External glycose + $enzyme_1$ $\longrightarrow$ glucose + $enzyme_1$
2. External lactose + $enzyme_2$ $\longrightarrow$ lactose + $enzyme_2$
3. External amino acids + $enzyme_3$ $\longrightarrow$ amino acids + $enzyme_3$
4. External nucleotides + $enzyme_4$ $\longrightarrow$ nucleotides + $enzyme_4$
- -
5. Lactose + $enzyme_5$ $\longrightarrow$ glucose + $enzyme_5$
- -
6. Glucose + ATP + $enzyme_6$ $\longrightarrow$ amino acids + ADP + $enzyme_6$
- -
7. Glucose + ATP + $enzyme_7$ $\longrightarrow$ nucleotides + ADP + $enzyme_7$
- -
8. Nucleotides + ATP + $enzyme_8$ $\longrightarrow$ ADP + $enzyme_8$
9. Glucose + ADP + $enzyme_9$ $\longrightarrow$ ATP + $enzyme_9$
- -
10-22. $mRNA_i$ + ATP + tRNA + ribosomes + amino acids + $enzyme_{10}$ $\longrightarrow$ $enzyme_i$ + ADP + $enzyme_{10}$ + $mRNA_i$

$$(i = 1,2,...,13)$$
- -
23-35. Nucleotides + DNA + amino acids + ATP + $enzyme_{11}$ + $enzyme_i$ $\longrightarrow$ $mRNA_i$ + DNA + ADP + $enzyme_{11}$

36-48. $mRNA_i$ $\longrightarrow$ nucleotides + amino acids

$$(i = 1,2,...,13)$$
- -
49. Nucleotides + amino acids + $enzyme_{12}$ + ATP $\longrightarrow$ ribosomes + $enzyme_{12}$
- -
50. Nucleotides + DNA + amino acids + $enzyme_{13}$ $\longrightarrow$ tRNA + DNA + $enzyme_{13}$
- -

Figure 2.8.1. Reactions in the *E. coli* cell model.

and division.) This reaction system, based on about 40 pools, is a simplification of a base model which might represent the finest level of constituents of a cell (about 3000 molecular species) illustrated in Figure 2.1.1. The ultimate test of such a lumped model is whether it is capable of reproducing the data observed for real cells within the experimental frame of interest (in this case, growth response to nutrient shifts). However, the theorems developed in this chapter help to test lumpings which may come to mind for *prior plausibility* and chances of success.

We shall state the information gained from the theorems in the form of a procedure. This procedure would be a true algorithm if we knew all the finest-level biochemicals in the cell, all the possible reactions involving these chemicals and their rate constants, assuming the rate expressions to be products of concentrations. Such knowledge would constitute a completely specified base model. In the absence of complete knowledge (the actual state of affairs) the following procedure becomes a set of heuristics whose truth may or may not be fully assessable.

Lumping procedure

1. Determine which are to be the input and output (to be compared with experimental results) chemicals or pools of the model. No lumping can be allowed to combine such pools further. For example, in the Weinberg–Goodman model, input pools are (external) glucose, lactose, amino acids and nucleotides; output pools are (internal) glucose, lactose, amino acids, nucleotides, proteins, ribosomes, DNA, transfer RNA and ATP.

2. Let the current set of pools consist of the chemical constituents of the base model considered as single element pools.

3. For the current set of pools consider a further lumping into a coarser set of pools (technically this is a partition of the current set) consistent with step 1 above.

4. Express the reactions based on the current set of pools in terms of the trial set by substituting for each pool in each reaction the pool into which it is being lumped. For example, if external glucose and external lactose are lumped into external GLUC; (internal) glucose and lactose are lumped into GLUC; and enzyme$_1$ lumped with enzyme$_2$ into enzyme$_{12}$, then reaction 7 of Figure 2.8.1 is rewritten:

$$\text{external GLUC} + \text{enzyme}_{12} \rightarrow \text{GLUC} + \text{enzyme}_{12}.$$

5. Is there a complete set of original reactions associated with each reactant form of the rewritten reactions? If not, try a new lumping (go to 3).

For example, the reactant form obtained in the example of step 4 above is: external GLUC + enzyme$_{12}$. Original reactions 1 and 2 of Figure 2.8.1 are associated with this reactant form since they both have this form after rewriting. But reactions 1 and 2 do not form a complete set which would also include the all possible reactant combinations. In this case, reactions

$$\text{external glucose} + \text{enzyme}_2 \rightarrow \text{glucose} + \text{enzyme}_2$$

$$\text{external lactose} + \text{enzyme}_1 \rightarrow \text{lactose} + \text{enzyme}_1$$

would complete the set.

6. For each of the complete sets of reactions discovered in step 5, is it the case that each of the reactions it contains has the same rate constant? If not, try a new lumping (go to 3).

7. For each of the rewritten reactions, does its reactant form appear in association with exactly one product form? If not, go to 8. If so, then to this rewritten reaction assign the common rate constant of the associated complete set of original reactions. If all reactions have been handled in this way, the resulting reaction system is a valid simplified version of the original one. If further lumping is desired, let the current set of pools be the ones just constructed, and go to 3.

For example, in Figure 2.8.1, 'nucleotides' is a reactant pool in reactions 8, 23–35, 49 and 50. One hypothesis justifying these reactions is that with each of these reactions is associated a set of four reactions obtained by substituting respectively A, C, G, U for nucleotides and each of the four reactions has the same rate constant.

8. For each reactant form which appears in association with more than one product form, can the proportions of the constituents of each of the pools be guaranteed to remain constant in time? If not, try a new lumping (go to 3). If so, the relative rates of the complete set of original reactions will also remain constant. These rates are used to weigh the stoichiometric coefficients of the original reactions' products to form the product coefficient of the lumped reaction. The rate constant of this reaction is the common one of the original set. If all reactions have been handled in steps 7 and 8, the resulting system is a valid simplified version of the original one, whose parameters may depend on the initial concentrations of the original.

For example, suppose we lumped the enzyme-specific {mRNA$_i$, $i = 1, \ldots, 13$} to form a single pool mRNA. The rewritten reactions 10–22 of Figure 2.8.1 would then all have the same reactant form but different product forms. Assuming that each of reactions 10–22 had the same rate constant and that the proportions of the various mRNA$_i$ remained fixed, we could justifiably replace reactions 10–22 by a single one in which each of the 13 enzymes appeared as a product with a coefficient proportional to that of its encoding mRNA type. Note that such fixed proportions would contradict the known capabilities of the cell to induce enzymes appropriate to its external nutrient source (modeled in reactions 23–35).

Allowance for error

Steps 5 and 6 can be replaced by 5' and 6': For each reactant form of the rewritten reactions, are the rate constants of the associated original reactions 'sufficiently close'? If so, proceed to step 7, with the recognition that an error proportional to the dispersion may be introduced at each step of the lumped model. If not, try a new lumping (go to 3). (In steps 7 and 8, the error may be minimized by choosing a lumped rate constant centered in the range of base rate constants.)

2.9 CONCLUSIONS

The conditions and procedures for simplification developed here are similar in nature to those developed in other domains such as neural networks (Zeigler, 1975, 1976) and queueing networks (Melamed, 1976). In particular, some known results in linear compartmental model simplification can be derived as a special case of the present ones (Zeigler, 1976).

While exact homomorphic simplification places strong constraints on the base model, there are some points to be noted which make the situation look more optimistic.

(*a*) *Simplifications may be quite robust with respect to deviations from the justifying conditions.* We have seen that the errors introduced at each step in the base–lumped relation are proportional to the dispersion in the rate constants of reactions being lumped. It is known that such errors accumulate or are damped out, according to whether mutually close trajectories tend to converge or diverge in the lumped model (Zeigler, 1976). Thus, properties of the lumped model (such as stability, memory span, etc. (Zeigler, 1976)) may act to reduce the sensitivity of the simplification to deviation from its necessary conditions.

(*b*) *The conditions justifying simplification require certain uniformities in structure that may often be expected on outside grounds.* Such grounds are information transmission limitation, spatial continuity, physio-chemical properties, balanced growth of cells, evolutionary design, hierarchical organization, etc.

(*c*) Most lumpings can be expected not to work (lead to valid models) because of the stringent conditions, *but those that do, then provide highly significant insight as to the most natural units of description of the system, and their coherent interaction* (Zeigler, 1971).

As a telling example, consider enzyme–substrate reactions. A lumping of all enzymes and substrates together (respectively) would probably not work because of the great dispersion in rates and affinities. But a lumping of enzymes and substrates according to specificity range might well work. (More precisely, the lumping would be such that most elements in enzyme class E_i would either react strongly with most elements in substrate class S_j, or not at all, for each pair E_i, S_j.)

We have earlier given the motivations for seeking simplification in the computer simulation context, and have now suggested a procedure for simplification via lumping of biochemical reaction systems. These results should now be placed within the more general context of model formulation and simplification. We note that lumping is but one simplification procedure commonly employed. An effective approach to complexity reduction should be cognizant of the variety of simplification procedures and the conditions for their applicability. The combined utilization of these procedures may yield valid simplifications far superior to those possible with any single technique. Particularly relevant to biochemical systems is the detection of pseudo-steady states, and the associated partitioning of aggregations according to the time scale of response (Park, 1974). Some other procedures are dropping of components, variables, and interactions (perturbation theory, see Chapter 1) coarsening of range sets, representation of multicomponent deterministic models by simpler stochastic models, and (conversely) representing stochastic models by deterministic models for their average behavior (Weinberg & Zeigler, 1970).

It is the task of the applied mathematician to develop these procedures and to characterize their domains of effective employment. The modeler of biological, or other, systems must necessarily employ such procedures, whether consciously or not, and whether guided by the existing theory or not. The task of the computer scientist is to develop software tools which more effectively aid in the modeling enterprise. Among the tools being developed are those intended to help the modeler state, and check, the validity of simplifications he is interested in (Hopfeld, 1978). While such technology is only now being developed, it may be expected rapidly to enter the mainstream of modeling practice.

References

Boudart, M. (1972). *Kinetics of Chemical Processes*, New Jersey, Prentice Hall.

Goodman, E. (1972). Adaptive behavior of simulated bacterial cells subjected to nutrional shifts, Doctoral dissertation, Ann Arbor University of Michigan.

Hopfeld, A. (1978). A software system for testing model simplification validity, Masters Thesis, Rehovot, Applied Mathematics Department, Weizmann Institute.

Melamed, B. (1976). Analysis and simplification of stochastic systems with application to Jackson queueing networks, Doctoral dissertation, Ann Arbor, University of Michigan.

Moebs, W. D. (1974). A Monte Carlo simulation of chemical reactions, *Math. Biosci.* **22**, 113–20.

Park, P. J. M. (1974). The hierarchical structure of metabolic networks and the construction of efficient metabolic simulators, *J. Theoret. Biol.* **46**, 31–74.

Weinberg, R. & Zeigler, B. P. (1970). Systems theoretic analysis of models: computer simulation of a living cell, *J. Theoret. Biol.* **29**, 35–56.

Zeigler, B. P. (1971). A note on system modelling aggregation and reductionism, *Int. J. Biomed. Computing* **2**, 277–80.

——— (1975). Statistical simplification of neural nets. *Int. J. Man-machine Studies* **7**, 371–93.

——— (1976). *Theory of Modelling and Simulation*, New York, Wiley.

——— (1978). Structuring the organization of partial models, *Int. J. Gen. Systems* **4**, 81–8.

——— (1979). Multi-level multi-formalism modelling, an ecosystem example, in *Theoretical Ecosystems*, ed. E. Halfon, New York, Academic Press, pp. 17–54.

3 Biological applications of control theory

3.1 METABOLIC REGULATION AS A CONTROL SYSTEM

This chapter introduces the concepts and basic techniques of control theory that have been useful in biological research. Being an introduction, it does not present a rigorous and detailed account of the theory. Particular attention has been paid to the definition of systems nomenclature, since, while most biologists are already accomplished at thinking about feedback control systems, they find the mathematician's language uncongenial. Mathematically sophisticated readers may be vexed by an explicit definition of, say, an input, but biological colleagues will be heartened to learn that it means exactly what they thought it did all along. Control theorists will also find these definitions unnecessarily restrictive and valid for only the simplest linear systems. This is necessary, however, since general definitions would assume knowledge of several topics, notably functions of complex variables, that a biological readership would find unfamiliar. In preparing this chapter, several references were found to be particularly valuable. Toates's (1975) *Control Theory in Biology and Experimental Psychology* and Milsum's (1966) *Biological Control System Analysis* are also directed to nonmathematicians and thus of interest to biologists. For the more determined novice, the Schaum Outline, *Feedback and Control Systems* (DiStefano, Stubberud & Williams, 1967) provides valuable drill work. Mathematical readers who require detailed information will find Hsu & Meyer (1968) and Atherton (1975) useful. For anyone considering active research in biochemical systems analysis, Savageau's (1976) book is an excellent introduction. A program that is similar in some respects to that developed by Savageau can be found in the article by Kacser & Burns (1973), who have investigated the control of multi-enzyme systems, and in that by Heinrich, Rapoport & Rapoport (1977), who have published an extended introduction to the mathematical modeling of metabolic regulation. Horrobin (1970) has written a useful nonmathematical introduction to biological control.

Twenty years ago a chapter on the applications of control theory to biology would begin with a brief essay on the importance of control theory techniques to biological research. This no longer seems necessary since these methods are now used in such a wide range of biological investigations. Control ideas can be examined by two approaches: simulation and analysis. Simulation techniques have been covered in Chapter 2. This paper will be concerned with analytical methods. While preparing and testing a computer model is usually the best first step in analyzing a complex physiological or biochemical process, a successful approach to any problem requires the application of both methods. The difficulties of analytical work has resulted in an overemphasis on computer simulations. The exclusive use of simulations as a means of studying biological systems is not fully satisfactory for several reasons. Any model of a biological process will contain several parameters such as reaction constants, transport rates and initial conditions. Often the numerical value of a parameter is uncertain. Since the behavior of a differential equation is sensitive to parameter variation it is necessary to run the program for several values of each parameter, otherwise important features of the system may be missed. A typical model may contain, say, 30 parameters. Suppose that 10 estimates of each were considered. This would result in 10^{30} separate simulations. This number could be compared with the estimated age of the universe (10^{17} s). Also the behavior of differential equations can be unstable with respect to variation in the form of the governing equations (for example if Michaelian enzyme equations are replaced by allosteric equations, the qualitative behavior of the solution can be very different). Ideally several different sets of governing equation should be tested. The preceding objections would be valid even if numerical methods were perfect. However, they are not and computer programs can yield very misleading results.

Before analytical methods can be applied, a complicated system must be reduced. The object is to produce a model that is simple enough to treat analytically but complicated enough to preserve essential biological behavior. This is the most taxing and important aspect of the biologist–mathematician interaction. The reduction process requires a sophisticated understanding of the biological problem and a good notion of available mathematical tools. No general procedure exists and usually each problem must be treated as a special case. A few general guidelines can be given. The most successful procedure is a reduction of the number of variables by separation of time scales. This is now discussed.

Decomposition by time scales

The recognition of different equilibrium times to effect a simplification of systems of differential equations is well known to those familiar with steady state enzyme kinetics, where two simplifying assumptions are made. First, it is assumed that total enzyme concentration is constant; this exploits the slow time scale on which enzyme concentrations actually do change. Secondly, it is assumed that the equilibrium between enzyme and substrate has been attained; this exploits the fast time scale for complex equilibration. It is supposed that during the course of reaction rate measurements dC/dt is very small (where C represents the concentration of enzyme substrate complex). The steady state condition is often expressed by the equation $dC/dt = 0$. Strictly speaking this is incorrect. Lin & Segel (1974) and Reich & Sel'kov (1974, Example 2.5) provide a careful derivation of the Michaelis–Menten equations from the point of view of singular perturbation theory. The separation of biochemical processes into different time regimes occurs naturally and is an important idea in the qualitative, as well as quantitative, understanding of metabolic regulation.

Following Waddington's (1957) format, Goodwin (1963) has classified biological systems according to their relaxation time. The **relaxation time** of a process can be loosely defined as the time required to return to equilibrium after a 'small' disturbance. The fast **metabolic time scale** consists of rapid events such as the enzymatic conversion of small molecules. The regulatory processes are enzyme activation and enzyme inhibition. Since a single enzyme molecule can convert 10 to 10 000 molecules of substrate per second the relaxation times will fall in the range of 0.1 s to 100 s. The second, slower time scale is the **epigenetic system**. Epigenetic events are defined as processes involving the synthesis and degradation of macromolecules and thus include regulation of enzyme concentration by enzyme induction and enzyme repression. The relaxation time would be on the order of minutes or hours. Goodwin's third time scale is the **genetic** level. The events include the evolutionary process involving the appearance and movement of novel genes in a population. This tertiary structure provides a general decomposition of biological processes. A more detailed breakdown is possible: electromechanical (neural firing and muscular contraction, milliseconds), metabolic (enzyme catalyzed reactions, seconds to minutes), epigenetic (short-term regulation of enzyme concentration, minutes to hours), developmental (control of gene activity during differentiation, hours to years), and evolutionary (movement of genetic material through a population, months or years).

A branch of control theory, called hierarchial systems theory is concerned with the structure and control of multi-level systems (Mesarovic', Macko & Takahara, 1970) but, because the relevant time scales frequently overlap, the decomposition of complex biological processes into smaller, manageable subunits is not always possible. A legitimate reduction can be effected only if the separation between time scales is sufficiently large. Often this is not the case. The ambiguity inherent in terms such as 'sufficiently large' almost precludes rigorous, general mathematical results. One of the few general theorems is Tikhonov's theorem, which specifies circumstances in which an n-dimensional system is effectively confined to a smaller dimensional subspace. A careful statement of the theorem in English has been given by Plant & Kim (1975). In practice each problem is examined separately. Reich & Sel'kov (1974, 1975) give several carefully presented biochemical examples of this process. The mathematical complexity of these problems unfortunately puts any further discussion of this topic out of the scope of this chapter.

Supposing that a complex system has been reduced to its essential subunits, the next step is to analyze each subunit. *Ab initio* the analysis of local control subunits would seem to present an unlimited number of special cases. However, examination of biological control systems shows that though the processes may be very different one finds that complicated systems are constructed by adding together subunits which are regulated by a comparatively small number of control circuit types. Therefore experience gained in analyzing representative members of each group can be applied to a wide range of problems. The next section gives examples of these archetypal control circuits.

Archetypal control circuits

A chemical system is **controlled** if one or more of the chemical components affects the rate of a reaction by some means other than just substrate availability. The simplest form of control is a single loop control system where only one chemical acts as a controller and only one reaction is regulated. The four possible types of single loop control (positive and negative feedback and feedforward control) are shown in Figure 3.1.1. Each type of archetypal control pattern has been found in biological systems.

Single loop negative feedback is the most commonly encountered form of biological regulation. As shown in Figure 3.1.1a, feedback metabolite x_n inhibits the x_0 to x_1 reaction; so an increase in x_n decreases the net rate of x_n production while a decrease in x_n deinhibits the first reaction thus, eventually, causing an increase in x_n synthesis. The first experimentally confirmed example of negative feedback control of a metabolic process was the inhibition of threonine deaminase by isoleucine reported by Umbarger (Figure 3.1.2; Umbarger, 1956). Several other examples were soon published. By 1963 enough cases had been established to make several general observations possible (Monod, Changeux & Jacob, 1963). Monod and his colleagues found that a general

(*a*)   Negative feedback control

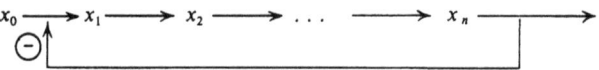

(*b*)   Positive feedback control

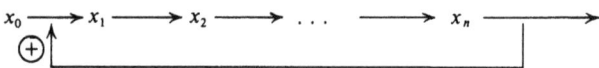

(*c*)   Negative feedforward control

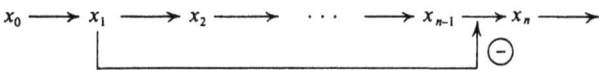

(*d*)   Positive feedforward control

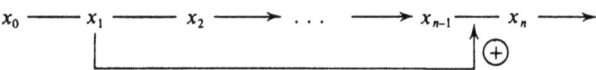

Figure 3.1.1. Four archetypal control loops. The symbol ⊖ indicates that the reaction is inhibited while ⊕ indicates that the reaction is accelerated.

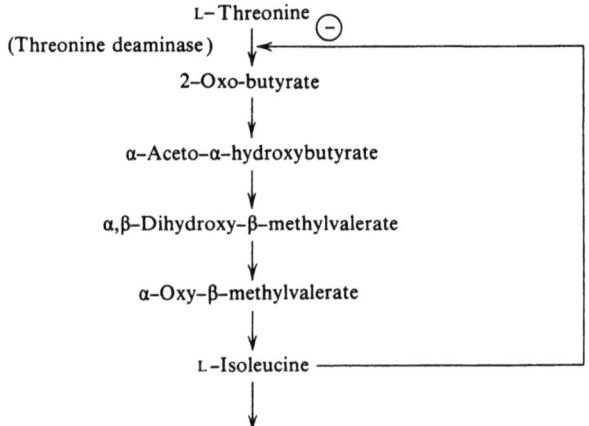

Figure 3.1.2. Feedback inhibition of threonine deaminase by L-isoleucine (Umbarger, 1956).

pattern was followed. Many experimentally investigated systems obeyed the following two rules:

(*a*) 'the regulatory enzymes (each of them acting immediately after a metabolic branching point) are all strongly and specifically inhibited by the terminal metabolite of the pathway in which each of them operates; intermediary metabolites in each pathway do not inhibit the regulatory enzyme;

(*b*) the enzymes which intervene after the regulatory one in each pathway are not significantly sensitive to inhibition by the terminal metabolite.' (Monod *et al.*, 1963).

Stated in the present nomenclature, all of the circuits considered by Monod are single loop negative feedback systems. The predominance of single loop systems is probably not accidental. Savageau (1976) has examined the optimal properties of several possible patterns of negative feedback control in unbranched pathways. He demonstrates that if all of the proposed networks have the same responsiveness to change in initial substrate, then the single loop alternative will be least sensitive to parameter variation. In considering input responsiveness, it is shown that a compromise between responsiveness and desensitization must be made, but, whatever balance is established, single loop end-product inhibition assures the maximum of both. These theoretical results suggest that single loop negative feedback would compete successfully in the natural selection process and accordingly the observed prevalence of this form of control is expected.

Negative feedback control is an intuitively satisfactory form of control since an increase (decrease) in output leads to a decrease (increase) in the rate of synthesis. A comparison with positive feedback (Figure 3.1.1*b*) indicates that positive feedback is unlikely to be useful since an increase in x_n leads to an increase in x_n synthesis. Indeed, Toates (1975, p. 118) suggests that positive feedback is only encountered in pathological conditions where normal homeostatic regulation has broken down. However, as shown in Section 3.8 below, positive feedback loops can be used to construct chemical on–off switches. The best-known examples of positive feedback are associated with processes where a rapid all-or-none switching response is required. For example the generation of a neuronal action potential is the result of a positive feedback loop (Horrobin, 1970; see also Section 7.7 below). A small initial membrane depolarization causes an increase in sodium permeability which permits inward sodium movement. This sodium in turn causes an increased depolarization. Eventually the fibre becomes positive and sodium influx stops (saturation of response). Active pumping moves sodium out of the cell. The saturation of response exhibited here is common in biological systems and explains why the response generated by a positive feedback loop is finite. Positive feedback in the control system regulating blood clotting provides a rapid response mechanism (Esnouf & MacFarlane, 1968; Davie & Kirby, 1973; Müller-Eberhard, 1975). A similar positive feedback switch generates the large sudden release of luteinizing hormone during ovulation (Figure 3.1.3; Horrobin, 1970).

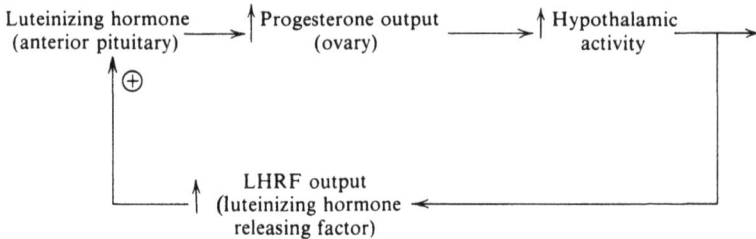

Figure 3.1.3. Positive feedback control of luteinizing hormone during the human female sex cycle (after Horrobin, 1970).

Besides forming chemical switches, positive feedback loops can be used to produce chemical oscillators. As shown in Section 4.1 below, positive feedback has been implicated in the cyclic AMP (cAMP) oscillation in *Dictyostelium* (Goldbeter & Segel, 1977) and in oscillations of the glycolytic pathway (Boiteux & Hess, 1974). It is interesting to speculate about the biological role of these oscillations. The cAMP oscillation in *Dictyostelium* has a well-understood purpose in controlling the aggregation process but the function, if any, of the glycolytic oscillation is unclear. Comparing the positive feedback loop in the glycolytic pathway with the preceding examples, it is possible to speculate that this control circuit primarily exists to fulfill the orthodox purpose of positive feedback, namely producing a chemical switch. The oscillations, which only exist for a comparatively narrow range of glucose entry rate, may be a switch defect. Switch 'chatter' in the form of oscillations is common in physical control systems.

Negative feedforward control is the forward control analog to positive feedback and few examples have been described in the biological literature. A common special case of negative feedforward control is the inhibition of an enzyme by high concentrations of its substrate. Examples include the inhibition of liver carboxylesterase by ethyl butyrate (Murray, 1930), the inhibition of xanthine oxidase by hypoxanthine (Dixon & Thurlow, 1924) and the inhibition of snake venom L-aminoacid oxidase by leucine (Dixon, Massey & Webb, 1964). The kinetic equations describing this inhibition have been published by Haldane (1930) and by

Dixon & Webb (1964, pp. 75–81). Inhibition of enzyme activity by high substrate concentrations prevents an unnecessary synthesis of intermediate compounds.

Positive feedforward control is the forward control analog of negative feedback. If a disturbance can be measured or predicted, this information can be passed by a forward control loop effecting necessary adjustments before a feedback controller would have time to operate. In metabolic systems, positive feedforward control adjusts the demand in the form of enzyme activity to meet the supply expressed as precursor concentration. A familiar class of examples is substrate activation of enzymes such as the activation of fumarate hydratase (Massey, 1953*a,b*) and the activation of phosphodiesterase by cAMP (Wang, Teo & Wang, 1972). Other examples include the control of cortisol secretion after hemorrhage (Blessner, 1969, pp. 528–31), cardiovascular and thermoregulatory response to exercise (Clynes & Milsum, 1970, p. 242) and in autonomic involuntary sensory motor systems (Talbot & Gessner, 1973, p. 222). The connection between prediction and feedforward control is especially apparent in man–machine coordination (Clynes & Milsum, 1970) and in the control of physiological processes by circadian rhythms. For example, petal movement in plants is necessarily slow. If the orientation to the sun did not begin until sunrise, several hours of sunlight would be lost. The circadian oscillator predicts dawn and a feedforward control loop begins the petal movement prior to sunrise. The generalization of this example is stated by Hoffman (1976): 'In general it is assumed that the adaptive significance of such endogenous rhythms lies in the fact that by proper phase adjustment, the organism is prepared to meet the challenge of cyclic changes in the environment; the organism can prepare in advance while a stimulus–response system might not allow sufficient time for the appropriate morphological, physiological and behavioral changes.'

If all biological control systems fell into one of these four classes of archetypal control system, the mathematical study of biological regulation would be far easier than it is. However, most biological control networks are multiple loop systems composed of archetypal subunits. Multiple loop systems are briefly considered in the next section.

Multiple loop biological control systems

In the context of metabolic regulation a **multiple loop system** is one in which several reactions are inhibited or activated and/or several metabolites are regulators. In Figure 3.1.4*a* a special case is shown in which several reactions are regulated by only one feedback metabolite and in Figure 3.1.4*b* the contrasting case is given where one reaction is regulated by several reactants. The usual biological case contains both forms of regulation (Figure 3.1.4c). Mathematical difficulties have limited theoretical investigations of multiple loop biochemical systems. A partial resolution for systems like Figure 3.1.4a has been published (Mees & Rapp, 1978) and work on the related system of Figure 3.1.4b is in progress. The details are beyond the scope of this chapter.

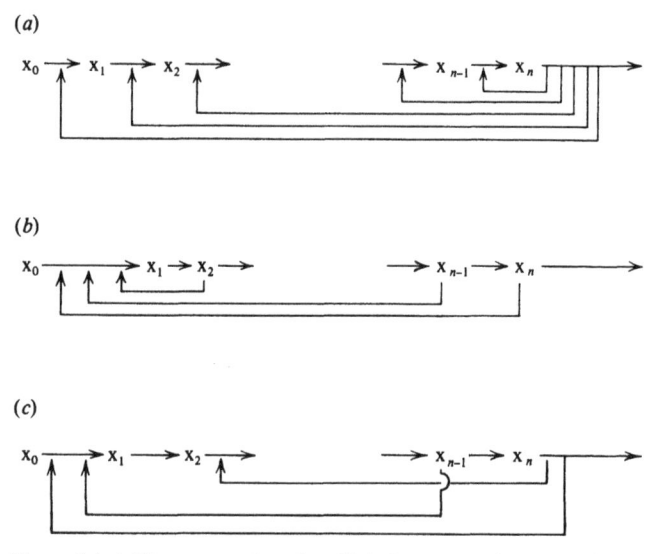

Figure 3.1.4. Three examples of multiple loop control systems: (a) The terminal metabolite either inhibits or activates several of the enzymes in the intermediate sequence. (b) Several different reactants affect the rate of the first reaction. (c) An example of a mixture of both of the previous forms of control (this is the most frequently encountered case).

Biological examples of multiple loop systems are commonplace. As a simple example, acetate in *Pseudomonas fluorescens* represses seven enzymes in the linear mandelate–acetate pathway (Mandelstam, 1968, p. 463). Examples of intermediate complexity include Toates's (1972) model of a division of the autonomic nervous system and the King-Smith & Morley model of granulocyte production (Figure 3.1.5, King-Smith & Morley, 1970).

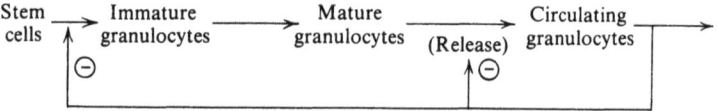

Figure 3.1.5. A specific example of a multiple loop control system is the King-Smith and Morley model of granulocyte differentiation and release (King-Smith & Morley, 1970; diagram redrawn from Tyson & Othmer, 1978).

Detailed models of biological processes inevitably involve highly interconnected multiple loop systems (e.g. the control of cortisol secretion: Gann, 1973, p. 224; Blessner, 1969, pp. 528–31). A particularly famous multiple loop system is Hill's hydrodynamic model of the nerve membrane which includes positive feedback and negative feedforward control (Talbot & Gessner, 1973, p. 82).

In practical terms additional loops complicate the analysis of control systems because their presence prevents the legitimate decomposition of a large system into smaller subunits. Parallel activation and inhibition between pathways (Figure 3.1.6) causes the same mathematical difficulties. Cross-pathway interactions coordinate the output of pathways that have a common source or converge to a common product. A biological example is provided by Stadtman's experimental analysis of aspartate metabolism in *Pseudomonas* (Stadtman, 1963).

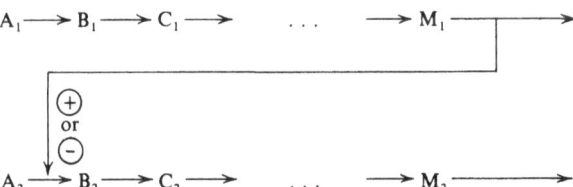

Figure 3.1.6. A hypothetical example of the cross coordination of two unbranched reaction sequences. A control system of this form would be expected if M_1 and M_2 ultimately formed a single product.

3.2 ANALYSIS OF A LINEAR, OPEN LOOP CONTROL SYSTEM

The methods of linear control theory are probably best presented by considering a specific example which is almost trivially simple but which will introduce techniques that generalize to more complicated problems. We consider Goodwin's model of the control of protein synthesis (Goodwin, 1965; Maynard Smith, 1968). Let $x_1(t)$ be the concentration of messenger RNA (mRNA), x_1, at time t. This messenger directs the synthesis of an enzyme, x_2, whose concentration is denoted by $x_2(t)$. This enzyme catalyzes a reaction producing a product, x_3, of concentration $x_3(t)$. The flow of information is mRNA to enzyme to enzyme product.

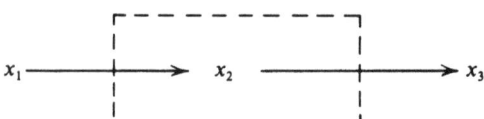

Figure 3.2.1. Diagrammatic representation of the conversion of compound x_1 to x_3 with a single intermediate x_2.

The **input** is the function of time $x_1(t)$. In general the input is a stimulus applied to a control system by an external agency. In this example we suppose that the concentration $x_1(t)$ will be controlled and that the concentration $x_3(t)$, the **output**, will be measured. It is common practice to represent control systems by block diagrams. A block diagram is a graphical representation of the relation between input, output and the components of a system. MacFarlane (1970; Section 3.3) gives the symbol conventions for more complicated systems. While the block

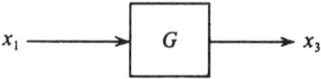

Figure 3.2.2. Block diagram representing the conversion of x_1 to x_3 as an open loop input–output relationship.

diagram is trivial for the example system, the technique is valuable for studying more complicated circuits where block diagrams can identify simpler equivalent circuits and facilitate the transformation of a control system to one that has already been analyzed. The block diagram has evolved into the **signal flow graph**. For a discussion of the well developed flow graph transformation techniques see DiStefano *et al.* (1967, Chapter 8).

The output x_3 depends on x_2, and hence on x_1. Neither x_2 or x_3 regulate their respective synthesis rates; there is no feedback. This is an **open-loop control system**. In the mathematical literature this is also referred to as components in a cascade. This can cause an unfortunate confusion with cascaded enzymatic mechanisms which can include feedback (Savageau, 1976, Chapter 13).

The x_1 to x_3 input–output relation is assumed to be **linear**. This means that the **law of superposition** (compare the discussion following (A.1.39)) is obeyed. Suppose input x_a gives output y_a and input x_b gives output y_b. If input $k_a x_a + k_b x_b$ gives output $k_a y_a + k_b y_b$ for all inputs x_a, x_b, and all constants k_a and k_b, then the system is linear (i.e. it obeys the law of superposition). In constructing a mathematical model of x_1 to x_3 conversion, it is supposed that the rate of enzyme synthesis is directly proportional to mRNA concentration and the synthesis of product x_3 is directly proportional to enzyme concentration. Also it is assumed that x_2 and x_3 leave the reaction system by transport or decomposition at a rate directly proportional to their concentrations. Translating these assumptions into differential equation form gives

$$dx_2(t)/dt = b_1 x_1(t) - b_2 x_2(t), \qquad dx_3(t)/dt = b_2 x_2(t) - b_3 x_3(t). \tag{1}$$

Here the initial conditions $x_2(0)$ and $x_3(0)$ are regarded as given, while b_1, b_2 and b_3 are *distinct positive* constants. Clearly this model is an enormous simplification of the truth, since translation actually involves thousands of separate nonlinear reaction steps and since the production of $x_3(t)$ should follow a nonlinear enzymatic rate law. The model should be regarded as a crude first approximation. (It would be possible to eliminate some of the parameters by transforming the problem to nondimensional variables; however, for this simple problem this process is not necessary.) This differential equation is simple enough to be solved explicitly by any of several techniques. The Laplace transform method is chosen because it will be useful in considering more complicated systems. Let $f(t)$ be a real function of time defined for $t > 0$; its Laplace transform, $L[f(t)]$, is defined as

$$\blacklozenge\blacklozenge \qquad L[f(t)] \equiv \hat{f}(s) \equiv \int_0^\infty \exp(-st)f(t)\,dt. \tag{2}$$

Properties of the Laplace transform

Example 3.2.1.   Find the Laplace transform for the step function, ramp input and the exponential.

Solution:
The step function is:

$$f_s(t) = \begin{cases} K & t \geqslant 0 \\ 0 & t < 0. \end{cases}$$

$$L[f_s(t)] = \hat{f}_s(s) = \int_0^\infty K \exp(-st) \, dt$$

$$= -\frac{K \exp(-st)}{s} \bigg|_0^\infty = K/s.$$

The ramp input is

$$f_r(t) = \begin{cases} Kt & t \geqslant 0 \\ 0 & t < 0. \end{cases}$$

Using the integration by parts formula (A.1.127), we find that

$$L[f_r(t)] = \hat{f}_r(s) = \int_0^\infty Kt \exp(-st) \, dt$$

$$= \frac{K[\exp(-st)](-st-1)}{s^2} \bigg|_0^\infty = \frac{K}{s^2}.$$

The exponential input is given by

$$f_e(t) = \begin{cases} \exp(-\alpha t) & t \geqslant 0 \\ 0 & t < 0. \end{cases}$$

$$L[f_e(t)] = \hat{f}_e(s) = \int_0^\infty \exp(-\alpha t) \exp(-st) \, dt$$

$$= \frac{-1}{s+\alpha} \exp[-(s+\alpha)]t \bigg|_0^\infty = \frac{1}{s+\alpha}.$$

In practice one can consult a table of Laplace transforms (Bateman, 1954) and thus avoid performing tiresome integrals. The connection between the Laplace transform and differential equations is established by the following three results (DiStefano *et al.*, 1967).

Example 3.2.2.   Show that the Laplace transform is a linear operator.

Solution:
We begin with the definition of a **linear operator** (compare (A.1.39)). An operator G acting on a class of objects $\{x_j\}$ is linear if it obeys the law of superposition, i.e.

$$G(k_a x_a + k_b x_b) = k_a G(x_a) + k_b G(x_b);$$

for all constants k_a and k_b and all objects x_a and x_b in the set $\{x_j\}$.

$$L[k_a f_a(t) + k_b f_b(t)]$$

$$= \int_0^\infty (k_a f_a(t) + k_b f_b(t)) \exp(-st) \, dt$$

$$= k_a \int_0^\infty f_a(t) \exp(-st) \, dt + k_b \int_0^\infty f_b(t) \exp(-st) \, dt$$

$$= k_a L[f_a(t)] + k_b L[f_b(t)],$$

so L is a linear operator.

Example 3.2.3.   Demonstrate that the Laplace transform of the derivative of a function is related to the Laplace transform of the function itself by

$$L[df/dt] = sL[f(t)] - f(0).$$

Solution:
This is most easily shown by using integration by parts

$$\int_0^\infty u \, dv = uv \Big|_0^\infty - \int_0^\infty v \, du.$$

$$L[df/dt] = \int_0^\infty \exp(-st) \frac{df}{dt} \, dt$$

$$= \exp(-st) f(t) \Big|_0^\infty + s \int_0^\infty f(t) \exp(-st) \, dt.$$

$$L[df/dt] = sL[f(t)] - f(0).$$

Example 3.2.4.   Show that the Laplace transform of the integral of a function is related to the Laplace transform of the function itself by

$$L\left[\int_0^t f(\tau) \, d\tau \right] = \frac{L[f(t)]}{s}.$$

Solution:
Again use integration by parts:

$$\int_0^\infty u\, dv = uv \Big|_0^\infty - \int_0^\infty v\, du.$$

$$u = \exp(-st), \qquad dv = f(t)\, dt, \qquad v = \int_0^t f(\tau)\, d\tau.$$

$$\int_0^\infty \exp(-st)f(t)\, dt = \left\{ \exp(-st) \int_0^t f(\tau)\, d\tau \right\} \Big|_0^\infty$$

$$- \int_0^\infty \int_0^t f(\tau)\, d\tau(-s) \exp(-st)\, dt.$$

$$L[f(t)] = s \int_0^\infty \left(\int_0^t f(\tau)\, d\tau \right) \exp(-st)\, dt$$

$$= sL\left[\int_0^t f(\tau)\, d\tau \right].$$

The Laplace transform will now be used to solve the differential equation (3.2.1). Taking the transform of both sides of each equation gives

$$s\hat{x}_2(s) - x_2(0) = b_1 \hat{x}_1(s) - b_2 \hat{x}_2(s),$$

$$s\hat{x}_3(s) - x_3(0) = b_2 \hat{x}_2(s) - b_3 \hat{x}_3(s).$$

Using the first equation it is possible to eliminate $\hat{x}_2$ and thus find $\hat{x}_3$ as a function of $\hat{x}_1$. *This demonstrates the principal advantage of the Laplace transform method: a differential equation problem has been reduced to an algebraic manipulation.*

$$\hat{x}_3(s) = \frac{b_1 b_2 \hat{x}_1(s)}{(s + b_2)(s + b_3)} + \frac{b_2 x_2(0)}{(s + b_2)(s + b_3)} + \frac{x_3(0)}{s + b_3}.$$

Input–output relations of this form are particularly common. Functions that can be expressed as the ratio of polynomials are called **rational functions**. If the degree of the denominator is greater than the degree of the numerator, they are termed **proper rational functions**. Having determined $\hat{x}_3(s)$, the next step is to find $x_3(t)$. This requires the inverse Laplace transform, L^{-1}.

$$L^{-1}[\hat{f}(s)] = f(t).$$

Like L, the inverse transform has an integral representation; however, its use requires an understanding of contour integration in the complex plane. The usual practice – using tables – will be followed here. L^{-1} is a linear operator, so the solution $x_3(t)$ naturally divides itself into two parts: one dependent on the input function x_1 and the other dependent on the initial conditions $x_2(0)$ and $x_3(0)$.

$$x_3(t) = L^{-1}[\hat{x}_3(s)]$$

$$= L^{-1}\left[\frac{b_1 b_2 \hat{x}_1(s)}{(s+b_2)(s+b_3)}\right] + L^{-1}\left[\frac{b_2 x_2(0)}{(s+b_2)(s+b_3)} + \frac{x_3(0)}{s+b_3}\right].$$

The contribution to the solution of a linear differential equation that depends on the input function is the **input response** (IR) or the **forced response**. The part that depends on the initial conditions is the **initial condition response** (ICR) or the **free response**.

The initial condition response is particularly simple and is examined first. Recalling that b_2 and b_3 are assumed to be unequal, the usual partial fraction expansion (see Appendix Section A.2) gives:

$$\frac{1}{(s+b_2)(s+b_3)} = \left(\frac{1}{b_3-b_2}\right)\left(\frac{1}{s+b_2}\right) + \left(\frac{1}{b_2-b_3}\right)\left(\frac{1}{s+b_3}\right).$$

Using the linearity of L^{-1} one finds the following restatement of the initial condition response:

$$\mathrm{ICR} = L^{-1}\left[\frac{b_2 x_2(0)}{(s+b_2)(s+b_3)} + \frac{x_3(0)}{s+b_3}\right]$$

$$= L^{-1}\left[\frac{b_2 x_2(0)}{(b_3-b_2)(s+b_2)} + \frac{b_2 x_2(0)}{(b_2-b_3)(s+b_3)} + \frac{x_3(0)}{s+b_3}\right]$$

$$= \frac{b_2 x_2(0)}{b_3-b_2} L^{-1}\left(\frac{1}{s+b_2}\right) + \frac{b_2 x_2(0)}{b_2-b_3} L^{-1}\left(\frac{1}{s+b_3}\right)$$

$$+ x_3(0) L^{-1}\left(\frac{1}{s+b_3}\right).$$

In Example 3.2.1 the Laplace transform of the exponential was derived.

$$L[\exp(-\alpha t)] = 1/(s+\alpha).$$

Applying the inverse Laplace transform to each side of the equation gives

$$\exp(-\alpha t) = L^{-1} L[\exp(-\alpha t)] = L^{-1}[1/(s+\alpha)].$$

The initial condition response can now be stated.

$$\mathrm{ICR} = \frac{b_2 x_2(0)}{b_3-b_2}\exp(-b_2 t) + \frac{b_2 x_2(0)}{b_2-b_3}$$

$$\times \exp(-b_3 t) + x_3(0)\exp(-b_3 t).$$

Note that the ICR tends exponentially to zero. For this reason the ICR is usually neglected by assuming zero initial values for all of the variables. We now state the final value theorem. Using this, it is possible to show that ICR → 0 without explicitly evaluating the inverse of the Laplace transform IĈR(s).

> *Example* 3.2.5.   Final value theorem.
> Let $f(t)$ have Laplace transform $\hat{f}(s)$. Show that if $\lim_{t \to \infty} f(t)$ exists, then
>
> $$\lim_{t \to \infty} f(t) = \lim_{s \to 0} s\hat{f}(s),$$
>
> Solution:
> Example 3.2.3 gives
>
> $$L[df/dt] = s\hat{f}(s) - f(0), \qquad \int_0^\infty \frac{df}{dt} \exp(-st) \, dt = s\hat{f}(s) - f(0).$$
>
> Take the limit $s \to 0$ on each side ($\exp(-st) \to 1$):
>
> $$\lim_{s \to 0} \int_0^\infty \frac{df}{dt} \exp(-st) \, dt = \int_0^\infty \frac{df}{dt} \, dt = \lim_{s \to 0} s\hat{f}(s) - f(0),$$
>
> $$f(t) \Big|_0^\infty = \lim_{t \to \infty} f(t) - f(0) = \lim_{s \to 0} s\hat{f}(s) - f(0),$$
>
> $$\lim_{t \to \infty} f(t) = \lim_{s \to 0} s\hat{f}(s).$$

The final value theorem can easily be applied to the initial condition response.

$$\lim_{t \to \infty} \text{ICR}(t) = \lim_{s \to 0} s\text{IĈR}(s)$$

$$= \lim_{s \to 0} \left[\frac{sb_2 x_2(0)}{(s+b_2)(s+b_3)} + \frac{sx_3(0)}{s+b_3} \right] = 0.$$

Because the initial condition response is short lived, attention is usually directed to the input response. As an example $x_3(t)$ is found for the case where $x_2(0) = x_3(0) = 0$ and $x_1(t)$ is a step function, i.e.

$$x_1(t) = \begin{cases} K & t \geq 0 \\ 0 & t < 0. \end{cases}$$

Since $x_2(0) = x_3(0) = 0$,

$$x_3(t) = L^{-1} \left[\frac{b_1 b_2 \hat{x}_1(s)}{(s+b_2)(s+b_3)} \right].$$

Using Example 3.2.1 we find that $\hat{x}_1(s) = K/s$ so

$$x_3(t) = L^{-1}\left[\frac{b_1 b_2 K}{s(s+b_2)(s+b_3)}\right].$$

Again a partial fraction expansion is used:

$$\frac{1}{s(s+b_2)(s+b_3)} = \frac{1}{b_2 b_3 s} + \frac{1}{b_2(b_2-b_3)(s+b_2)} + \frac{1}{b_3(b_3-b_2)(s+b_3)},$$

$$x_3(t) = \frac{Kb_1}{b_3}L^{-1}\left(\frac{1}{s}\right) + \frac{Kb_1}{b_2-b_3}L^{-1}\left(\frac{1}{s+b_2}\right) + \frac{Kb_1 b_2}{b_3(b_3-b_2)}L^{-1}\left(\frac{1}{s+b_3}\right),$$

$$x_3(t) = \frac{Kb_1}{b_3} + \frac{Kb_1}{b_2-b_3}\exp(-b_2 t) + \frac{Kb_1 b_2}{b_3(b_3-b_2)}\exp(-b_3 t).$$

The solution tends exponentially to a new final state Kb_1/b_3. Again use of the final value theorem would give the limiting value of $x_3(t)$ immediately:

$$\lim_{t\to\infty} x_3(t) = \lim_{s\to 0} s\hat{x}_3(s) = \lim_{s\to 0}\frac{sKb_1 b_2}{s(s+b_2)(s+b_3)} = \frac{Kb_1}{b_3}.$$

Transfer functions, the unit impulse response and stability of linear input–output systems

The definition of the transfer function is best motivated by recalling the simple example where

$$\hat{x}_3(s) = \frac{b_1 b_2 \hat{x}_1(s)}{(s+b_2)(s+b_3)} + \frac{b_2 x_2(0)}{(s+b_2)(s+b_3)} + \frac{x_3(0)}{s+b_3}.$$

The form of input–output relation obtained in the more general cases follows the same pattern

$$\hat{y}(s) = G(s)\hat{x}(s) + \text{terms due to initial conditions},$$

where $y(t)$ is the output and $x(t)$ is the input. For the special case where all initial values are zero

$$\hat{y}(s)/\hat{x}(s) = G(s).$$

Hence the definition: the **transfer function** of a linear system, $G(s)$, is the ratio of the Laplace transform of the output and the input for the special case where all initial conditions are zero. Thus for (1) the transfer function is:

$$G(s) = \frac{b_1 b_2}{(s+b_2)(s+b_3)}. \tag{3}$$

Work in systems analysis is often concerned with the experimental determination of transfer functions. Suppose there was an input $x(t)$ such that $\hat{x}(s) = 1$, then $\hat{y}(s) = G(s)$. The input function having this property is the **unit impulse** or **delta function**, $\delta(t)$, which is defined as the rectangular pulse function of height $1/\Delta t$ and width Δt taken in the limit $\Delta t \to 0$.

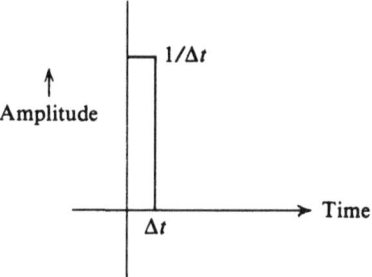

Figure 3.2.3. The rectangular pulse function of width Δt and height $1/\Delta t$. The unit impulse function is the mathematical idealization of a pulse input obtained in the limit $\Delta t \to 0$.

Example 3.2.6.   Show that $L[\delta(t)] = 1$.

Solution:
Let $u(t)$ be the unit step function. With this the definition of $\delta(t)$ becomes

$$\lim_{\Delta t \to 0} \left[\frac{u(t) - u(t - \Delta t)}{\Delta t} \right].$$

Thus

$$L[\delta(t)] = \int_0^\infty \lim_{\Delta t \to 0} \left[\frac{u(t) - u(t - \Delta t)}{\Delta t} \right] \exp(-st)\, dt$$

$$= \lim_{\Delta t \to 0} \left[\int_0^\infty \frac{u(t)}{\Delta t} \exp(-st)\, dt - \int_0^\infty \frac{u(t - \Delta t)}{\Delta t} \right] \exp(-st)\, dt.$$

The Laplace transform of $u(t)$ is $1/s$ (Example 3.2.1) so the first integral gives $1/(s\Delta t)$. A change in the variable of integration in the second integral gives

$$L[\delta(t)] = \lim_{\Delta t \to 0} \frac{1}{\Delta t} \left[\frac{1}{s} - \frac{\exp(-\Delta ts)}{s} \right].$$

The Taylor series has been defined in (A.1.30). Using this definition, we obtained the Taylor series (A.1.64) for the exponential function. In particular

$$\exp(-\Delta ts) = 1 - \Delta ts + \frac{(\Delta ts)^2}{2!} - \frac{(\Delta ts)^3}{3!},$$

so

$$L[\delta(t)] = \lim_{\Delta t \to 0} \frac{1}{\Delta t}\left[\Delta t - \frac{(\Delta t)^2 s}{2!} + \frac{(\Delta t)^3 s^2}{3!} - \ldots \right] = 1.$$

Thus the definition of the transfer function can be restated as the Laplace transform of the response to the input $\delta(t)$ for the case of zero initial conditions. $\delta(t)$ is a mathematical idealization but it can be approached experimentally. In actual practice there are more reliable techniques for measuring transfer functions. The response of the example system to an impulse response is determined in Example 3.2.7.

> *Example 3.2.7.* Find the response of the differential equation (3.2.1) to $\delta(t)$ for the case $x_2(0) = x_3(0) = 0$.

Solution:

$$\hat{x}_3(s) = \frac{b_1 b_2 \hat{x}_1(s)}{(s+b_2)(s+b_3)} = \frac{b_1 b_2 \hat{\delta}(s)}{(s+b_2)(s+b_3)} = \frac{b_1 b_2}{(s+b_2)(s+b_3)},$$

$$\hat{x}_3(s) = b_1 b_2\left[\left(\frac{1}{b_3-b_2}\right)\left(\frac{1}{s+b_2}\right) + \left(\frac{1}{b_2-b_3}\right)\left(\frac{1}{s+b_3}\right)\right],$$

$$x_3(t) = L^{-1}(\hat{x}_3(s)) = \frac{b_1 b_2}{b_3-b_2} \exp(-b_2 t) + \frac{b_1 b_2}{b_2-b_3} \exp(-b_3 t).$$

From Example 3.2.7. it is seen that the response to $\delta(t)$ decays. We shall refer to systems with this property as being stable, i.e. a linear system is **stable** if its response to an impulse input $\delta(t)$ approaches zero as time approaches infinity. (The reader is warned that there are several other definitions of stability in the literature; the present definition is essentially equivalent to definitions used elsewhere in this volume.)

A linear system has the Laplace transform representation

$$\hat{y}(s) = G(s)\hat{x}(s) + \hat{\mathrm{ICR}}(s).$$

Usually $\mathrm{ICR}(t) \to 0$ as $t \to \infty$, so the system is stable if

$$L^{-1}[G(s)\hat{\delta}(s)] = L^{-1}[G(s)] \to 0$$

as $t \to \infty$. For most simple systems $G(s)$ can be expressed in the form $G(s) = N(s)/D(s)$ where $D(s)$ is a polynomial. Suppose $D(s) = 0$ has roots $r_1 \ldots r_n$ (which may be complex, see Appendix Section A.2); then $D(s) = (s - r_1) \ldots (s - r_n)$. In Example 3.2.7, $D(s) = (s + b_2)(s + b_3)$: the roots $-b_2$ and $-b_3$ appear in the solution $x_3(t)$ in exponents $\exp(-b_2 t)$ and $\exp(-b_3 t)$. Similarly in the more general case roots $r_1 \ldots r_n$ appear in the solution in terms with $\exp(r_1 t) \ldots \exp(r_n t)$. The function $\exp(r_j t)$ decays with time only if $\mathrm{Re}(r_j) < 0$. This gives an important result, namely: *a necessary condition for stability is that the roots of* $D(s) = 0$ *have negative real parts.* The equation $D(s) = 0$ is usually called the **characteristic equation** and its roots $r_1 \ldots r_n$ are the **characteristic roots.** (Sometimes they are referred to as the **eigenvalues of the transfer function.**) The straightforward way of determining the stability of a linear system is to determine explicitly its characteristic roots. Unfortunately this is not usually possible; but there are several techniques for determining the stability properties of characteristic roots without solving the characteristic equation explicitly, e.g. the Routh–Hurwitz criterion and the continued fraction stability criterion. One of the most useful stability tests, the Nyquist criterion, will be presented in the discussion of closed loop feedback systems. Details of other tests can be found in any textbook on control theory.

3.3 FEEDBACK IN LINEAR PATHWAYS

The canonical negative feedback loop

Biologists have now established the broad outline of self-regulation in metabolic control systems, with the elucidation of some simple induction and repression mechanisms that regulate enzyme concentration. The allosteric regulation of enzyme activity has also been investigated. To use the analogy with mechanical control systems, we now understand something about the component parts of the control unit and something about their position in the circuit. The remaining problem – which is always much more difficult – is to understand the time-dependent response to a changing environment and to understand the way in which these 'local' control circuits are coordinated. By definition these are mathematical problems and unfortunately rather difficult ones. The subject will be introduced by considering some examples. Lengthy algebraic manipulations have been included with possibly excessive detail, and, while it could be argued that these details are not of interest to biologists, it is probably true that even a casual understanding of the methods discussed can be gained only by seeing a relatively complicated example worked through carefully.

In Section 3.2 we considered a process where an input in the form of mRNA produced a final product x_3. The rate of forward synthesis of x_3 was unaffected by its concentration (i.e. the process was not controlled). Clearly this is a potentially wasteful system, since, if x_3 was provided by some other source, say an external supply, the synthesis from mRNA and enzyme x_2 would still continue. We know in fact that biochemical processes are usually regulated and, in the case of negative control described in Section 3.1, an increase in x_3 concentration would decrease its net rate of synthesis. To provide a concrete example for mathematical examination, suppose x_3 exerts a negative feedback control at the translational level by inhibiting x_1. Previously the input to the synthesis process represented by transfer function $G(s)$ was $x_1(t)$; now suppose it is $x_1(t) - x_3(t)$. The control loop representation is given in Figure 3.3.1 with

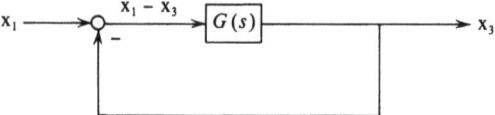

Figure 3.3.1. A specific example of a linear negative feedback system in which output x_3 exerts a negative control on its own synthesis.

$G(s)$ given in (3.2.3). $G(s)$ is referred to as the **direct transfer function** or the **forward transfer function**.

In order to define other concepts that frequently appear in the control literature, two complications will be added to this system: an explicit reference signal and a feedback controller. Often a reference signal y_{ref} equal to the desired output is available. This signal is subtracted from output before it is returned via the feedback loop. The output less the reference signal $(x_3 - y_{ref})$ is called the **error signal**. Frequently the error signal is modified before being subtracted from the input by a **feedback controller**. The feedback controller, which is also called the **output transducer**, has **feedback transfer function** $K(s)$. The resulting control system is shown in Figure 3.3.2. Circuits of this form are encountered so frequently in the control literature that they have been graced with the adjective canonical.

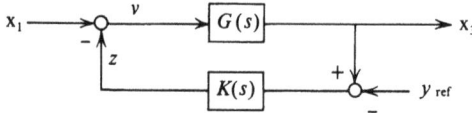

Figure 3.3.2. The canonical negative feedback loop.

Considering the complexity of biochemical control systems, the canonical feedback loop seems a simplistic caricature; however, a surprisingly large number of control systems can be transformed into this configuration. Also, since it is one of the simplest feedback systems it seems the best place to start. In analyzing a system of this kind three properties are examined. Accuracy: does the output follow the requirement specified by the reference signal? Stability: is the system reliable, or are there conditions which result in a radically unacceptable output? Speed: does the control system respond quickly enough to changes in input or reference signals? These questions will now be considered.

Accuracy of a simple linear negative feedback loop

The accuracy of the control loop in Figure 3.3.2 is considered for the following specific realization: the forward transfer function, $G(s)$, is given by (3.2.3). This transfer function described the forward synthesis process analyzed in Section 3.2. In this example we consider the simplest possible feedback controller which multiplies the feedback signal $(x_3 - y_{ref})$ by a *positive* constant; giving feedback transfer function $K(s) = k$. The constant k could of course be $k = 1$ in which case the feedback signal is unmodified. Systems where $K(s) = k$ are called **proportional feedback systems**, and k is a gain constant. The **gain** of a control component can be most simply defined as the transmission function of the component when it is a simple multiplier. A more general definition would be the ratio of a change in output to a change in input, i.e. if a change of z units in the input results in a change Kz in the output, then the gain of the process is K. In this more general sense gain can be a very complicated function depending on the form of the input:

$$\text{Gain} = \frac{\Delta \text{ output}}{\Delta \text{ input}}.$$

As in Section 3.2 the analysis is conducted in Laplace transform space. Setting initial conditions to zero gives

$$\hat{x}_3(s) = G(s)\hat{v}(s),$$

$$\hat{v}(s) = \hat{x}_1(s) - \hat{z}(s) = \hat{x}_1(s) - k[\hat{x}_3(s) - \hat{y}_{ref}(s)],$$

$$\hat{x}_3(s) = G(s)\hat{x}_1(s) - kG(s)\hat{x}_3(s) + kG(s)\hat{y}_{ref}(s).$$

This equation is solved, determining the output transform $\hat{x}_3(s)$:

$$\hat{x}_3(s) = \frac{G(s)}{1 + kG(s)}[\hat{x}_1(s) + k\hat{y}_{ref}(s)]. \tag{1}$$

In the case of the open loop system, the process was described by the transfer function $G(s)$; here it is described by the **closed loop transfer function** $G(s)/[1 + kG(s)]$.

A control system is accurate if the output closely tracks the reference setting. It is useful to define a function of time called the **offset** as the difference output − reference signal, which for this example would be $[x_3(t) - y_{ref}(t)]$. The Laplace transform of the offset is

$$\hat{x}_3(s) - \hat{y}_{ref}(s) = \frac{1}{k}\left[\frac{G(s)\hat{x}_1(s) - \hat{y}_{ref}(s)}{G(s) + (1/k)}\right].$$

The aim is to have the smallest possible offset. Since superficial examination of this equation suggests that $|\hat{x}_3(s) - \hat{y}_{ref}(s)|$ appears to approach zero as $k \to \infty$, it would seem that the offset could be decreased by increasing the gain of the feedback loop. Indeed, this is frequently the case and provides a simple and successful method for improving control accuracy. However, if increasing k causes $G(s)+(1/k)$ to approach zero, the offset can rapidly increase; the accuracy of control would dramatically decrease with an increase in gain. This is a specific example of a general phenomenon termed the control–stability conflict (or tightness of control–stability trade-off). We can now see that the problem of accuracy of control is intimately connected with the problem of stability.

Stability of a linear negative feedback loop

The closed loop transfer function (3.3.1) has denominator $1+kG(s)$. If this function has a zero, $s=r_j$ with $\mathrm{Re}\,(r_j)>0$, then $x_3(t)$ contains a growing exponential, because partial fraction expansions of $G\hat{x}_1/(1+kG)$ and $kG\hat{y}_{\mathrm{ref}}/(1+kG)$ give terms $\rho_j/(s-r_j)$ (ρ_j being the expansion coefficient) and hence the inverse transform $\rho_j \exp{(r_jt)}$. So a necessary condition for the stability of the control system of Figure 3.3.2 is that all of the zeros of $1+k\hat{G}(s)=0$ have negative real parts. This equation is the characteristic equation of the closed feedback loop and the roots are the characteristic roots.

For the example case, the characteristic equation is simple enough to be solved explicitly.

$$1+kG(s)=1+\frac{kb_1b_2}{(s+b_2)(s+b_3)}=0,$$

$$(s+b_2)(s+b_3)+kb_1b_2=0,$$

$$s^2+(b_2+b_3)s+b_2b_3+b_1b_2k=0, \tag{2}$$

$$2r_{1,2}=-(b_2+b_3)\pm[(b_2+b_3)^2-4(b_2b_3+b_1b_2k)]^{\frac{1}{2}}.$$

The problem is to determine the stability properties of the system not for a single value of feedback gain k but for the entire range of positive values. The plot of the roots r_1 and r_2 in the complex plane as k increases from zero is the **root locus diagram**. For stability, the roots have to be on the left side of the complex plane (i.e. they must have negative real parts). A summary of the basic properties of complex numbers is given in Appendix A.2.

Example 3.3.1.   Draw the root locus diagram as a function of k for the system of Figure 3.3.2 for the case $b_1=0.5$, $b_2=1$, $b_3=2$.

Solution:
For these parameter values the two roots are

$$r_1=-\tfrac{3}{2}+[\tfrac{1}{4}-(k/2)]^{\frac{1}{2}},\qquad r_2=-\tfrac{3}{2}-[\tfrac{1}{4}-(k/2)]^{\frac{1}{2}}.$$

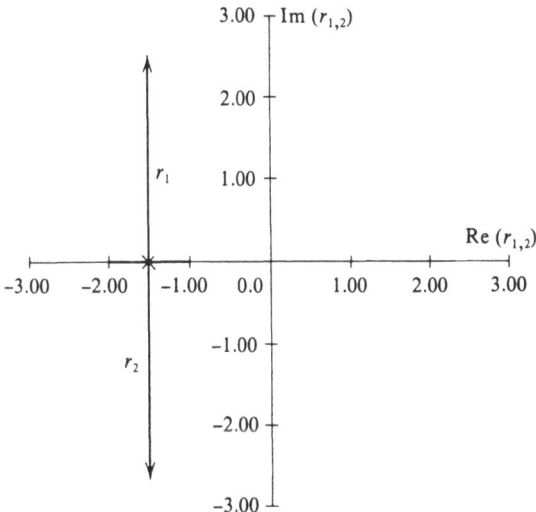

Figure 3.3.3. Root locus diagram for characteristic equation (2) when $b_1 = 0.5$, $b_2 = 1$ and $b_3 = 2$. Arrows indicate the direction of increasing feedback gain, k.

The root locus diagram is simple enough to draw by hand, but following usual practice a simple computer program was written to produce the diagram.

For all positive values of k the roots have negative real parts, so the system appears to be stable. For more complicated systems the root locus method is not completely satisfactory. Usually the characteristic equation cannot be solved explicitly. Also the example produced the root locus diagram for a specific set of parameters b_1, b_2 and b_3; however, this does not generally ensure that it is impossible to find a set of positive b's that give instability. Let us next consider a more powerful technique, the Routh–Hurwitz criterion, which shows that the roots have negative real parts for any set of positive b's and positive k.

If stability is the only concern it is only necessary to show that the characteristic roots all have negative real parts. Thus a test like the Routh–Hurwitz criterion that shows this without actually finding values of r_j's is usually acceptable. In its general case the criterion considers polynomial equations of order n,

$$a_n s^n + a_{n-1} s^{n-1} + \ldots + a_1 s + a_0 = 0,$$

where it is assumed that $a_n > 0$. (If this is not the case, divide the equation by -1. Usually one divides through by the coefficient of s^n so that a_n can be taken equal to unity.) The theorem states that all roots have negative real parts if and only if $\Delta_1 > 0$, $\Delta_2 > 0$, ..., $\Delta_n > 0$, where the Δ's have a rather complicated form. (A summary of the definitions and elementary properties of determinants is given in Appendix A.2.)

$$\Delta_1 = a_{n-1},$$

$$\Delta_2 = \begin{vmatrix} a_{n-1} & a_{n-3} \\ a_n & a_{n-2} \end{vmatrix},$$

$$\Delta_3 = \begin{vmatrix} a_{n-1} & a_{n-3} & a_{n-5} \\ a_n & a_{n-2} & a_{n-4} \\ 0 & a_{n-1} & a_{n-3} \end{vmatrix},$$

$$\Delta_4 = \begin{vmatrix} a_{n-1} & a_{n-3} & a_{n-5} & a_{n-7} \\ a_n & a_{n-2} & a_{n-4} & a_{n-6} \\ 0 & a_{n-1} & a_{n-3} & a_{n-5} \\ 0 & a_n & a_{n-2} & a_{n-4} \end{vmatrix},$$

and so on. If a_{n-r} does not appear in the polynomial equation (or $n - r < 0$) then a_{n-r} is set equal to zero. There are several equivalent ways of defining Δ's; this convention has been followed by DiStefano *et al.* (1967).

> *Example 3.3.2.* Use the Routh–Hurwitz criterion to show that the roots of the characteristic equation
>
> $$(s + b_2)(s + b_3) + b_1 b_2 k = 0$$
>
> have negative real parts if k and the b's are positive.

Solution:
In the previous example the root locus was drawn for the special case $b_1 = \frac{1}{2}$, $b_2 = 1$ and $b_3 = 2$. Now the analysis is performed for arbitrary positive k and positive b's (indicating the value of analytical methods).

$$a_n = a_2 = 1, \qquad a_{n-1} = a_1 = b_2 + b_3,$$

$$a_{n-2} = a_0 = b_2 b_3 + b_1 b_2 k.$$

Expressions for Δ_1 and Δ_2 follow from the previous definitions.

$$\Delta_1 = a_{n-1} = a_1 = b_2 + b_3.$$

Since b_2 and b_3 are both positive it follows that Δ_1 is positive.

$$\Delta_2 = \begin{vmatrix} a_{n-1} & a_{n-3} \\ a_n & a_{n-2} \end{vmatrix} = \begin{vmatrix} a_1 & 0 \\ a_2 & a_0 \end{vmatrix},$$

$$\Delta_2 = \begin{vmatrix} b_2 + b_3 & 0 \\ 1 & b_2 b_3 + b_1 b_2 k \end{vmatrix} = (b_2 + b_3)(b_2 b_3 + b_1 b_2 k).$$

Since each factor in Δ_2 is positive, $\Delta_2 > 0$. The roots of the closed loop characteristic equation have negative real parts even if the feedback gain is arbitrarily large.

While the Routh–Hurwitz criterion satisfactorily disposed of the problem in Example 3.3.2, some experimentation with the formula for Δ indicates that the method becomes very tedious as the dimension n increases. This is a major drawback for biological work where the dimension is frequently very large. Fortunately the Routh–Hurwitz criterion is complemented by the Nyquist criterion which is often easier to apply to large-dimension systems. An exact statement of the Nyquist theorem requires an understanding of elementary complex analysis. We propose to make a substantial compromise with rigor and attempt to give a purely operational presentation of the methods used by considering again Example 3.3.2. The feedback system in Figure 3.3.2 has a closed loop transfer function

$$\hat{x}_3(s) = \frac{G(s)}{1 + kG(s)}(\hat{x}_1(s) + k\hat{y}_{\text{ref}}(s)), \qquad G(s) = \frac{b_1 b_2}{(s + b_2)(s + b_3)}. \tag{3}$$

In Figure 3.3.4 the complex function $G(i\omega)$, $i = \sqrt{(-1)}$, is plotted for $\omega = 0$ to $\omega = \infty$, for the parameters $b_1 = 0.5$, $b_2 = 1$ and $b_3 = 2$. This curve is the **Nyquist locus**.

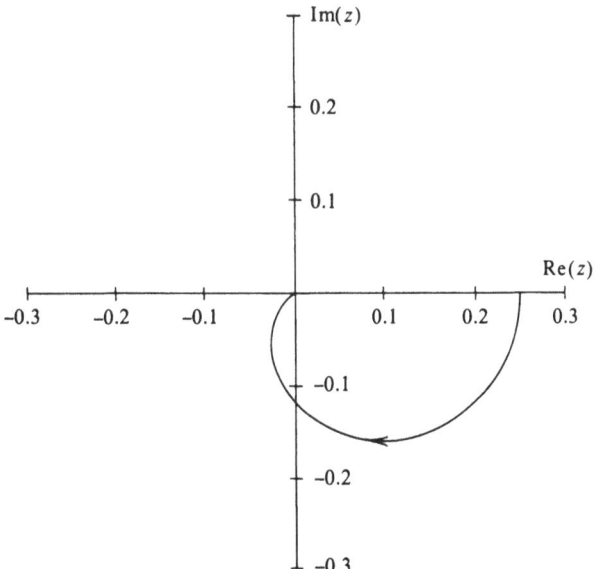

Figure 3.3.4. Nyquist locus, $G(i\omega)$, for the transfer function of equation (3) for $b_1 = 0.5$, $b_2 = 1$ and $b_3 = 2$. The arrow indicates direction of increasing ω.

The Nyquist theorem shows that the characteristic equation has a root with positive real part if and only if the point $-1/k$ is encircled by the Nyquist locus. The difficulty in understanding the criterion centers on the ambiguity of the word 'encircled'. Consider the Nyquist locus and its reflection through the real axis (Figure 3.3.5). The resulting figure is a closed curve with a direction defined by the direction of increasing ω. The interior of the curve is shaded. (If you walk around the boundary of a directed closed curve, your right shoulder is on the inside.) A point is encircled by the Nyquist locus if it is in the shaded region.

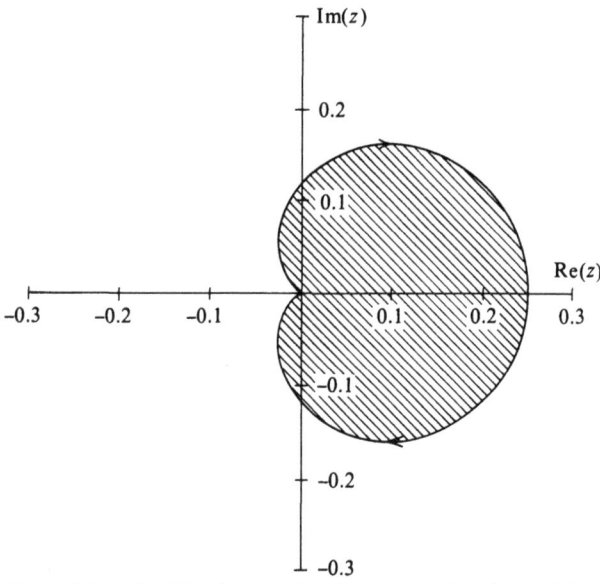

Figure 3.3.5. The Nyquist locus for the transfer function of (3) and its reflection through the real axis with the interior shaded. The arrows indicate direction of increasing ω.

Using the Nyquist locus in Figure 3.3.5 and this definition of encirclement, the stability properties of the loop can be determined.

Example 3.3.3.   Using the Nyquist criterion, show that the roots of the characteristic equation of the closed loop transfer function of (3.3.1) have negative real parts when $b_1 = 0.5$, $b_2 = 1$ and $b_3 = 2$ and when k is an arbitrary positive gain constant.

Solution:
Figures 3.3.4 and 3.3.5 show that no point on the negative real axis is encircled by the Nyquist locus. Gain k is a positive real number so $-1/k$ is a negative real number and hence never encircled. The Nyquist criterion demonstrates that the roots of the characteristic equation always have negative real parts.

In this example the Nyquist criterion demonstrated stability for $b_1 = 0.5$, $b_2 = 1$, $b_3 = 2$ and for feedback gain k, while the Routh–Hurwitz criterion showed stability for a completely general set of positive b's. In fact, it is fairly easy to generalize the Nyquist criterion result. This is done in Example 3.3.4.

> *Example* 3.3.4. Using the Nyquist criterion, show that the roots of the characteristic equation (3.3.2) have negative real parts for all positive b_1, b_2, b_3 and positive k.

Solution:
The argument has six steps. (Listing these steps may make the solution seem more complicated than it really is. The reader can convince himself or herself of the simplicity of the method by sketching a few Nyquist loci by hand.)
1. Instability is indicated only if the point $-1/k$ is encircled by the Nyquist locus.
2. k is a positive real number, so $-1/k$ is a negative real number.
3. From steps (1) and (2) instability can occur only if the Nyquist locus encircles some interval of the negative real axis.
4. The Nyquist locus can encircle an interval of the negative real axis only if it intersects the negative real axis.
5. An intersection of $G(i\omega)$ and the real axis takes place for ω such that Im $G(i\omega) = 0$ (i.e. the imaginary part of $G(i\omega)$ is zero).
6. If the only solution of Im $G(i\omega) = 0$ is $\omega = 0$, then it follows from steps (4) and (5) that the negative real axis is never intersected and thus the system is stable.

$$G(i\omega) = \frac{b_1 b_2}{(b_2 + i\omega)(b_3 + i\omega)} \left[\frac{(b_2 - i\omega)(b_3 - i\omega)}{(b_2 - i\omega)(b_3 - i\omega)} \right],$$

$$G(i\omega) = \frac{b_1 b_2 (b_2 - i\omega)(b_3 - i\omega)}{(b_2^2 + \omega^2)(b_3^2 + \omega^2)}$$

$$= \frac{b_1 b_2 [-\omega^2 + b_2 b_3 - i\omega(b_2 + b_3)]}{(b_2^2 + \omega^2)(b_3^2 + \omega^2)},$$

$$\text{Im } G(i\omega) = \frac{-b_1 b_2 (b_2 + b_3)\omega}{(b_2^2 + \omega^2)(b_3^2 + \omega^2)}.$$

Using the expression for Im $G(i\omega)$ it is seen that the equation Im $G(i\omega) = 0$ has only the solution $\omega = 0$. The system is stable.

The tightness of control–stability trade-off has been mentioned and it was claimed that increasing feedback gain can cause a loss of stability. The claim is not confirmed by the above example, since the system is stable for arbitrarily large feedback gain, k. The next example shows that this is atypical.

Instability due to high feedback gain

In this control circuit the forward operator describes a three-step chemical reaction.

$$dx_2(t)/dt = b_1 x_1(t) - b_2 x_2(t),$$

$$dx_3(t)/dt = b_2 x_2(t) - b_3 x_3(t),$$

$$dx_4(t)/dt = b_3 x_3(t) - b_4 x_4(t),$$

where b_1, b_2, b_3 and b_4 are positive reaction constants. Assuming initial values are zero, the transfer function of the process is found by taking the Laplace transform of each differential equation.

$$s\hat{x}_2(s) = b_1 \hat{x}_1(s) - b_2 \hat{x}_2(s),$$

$$s\hat{x}_3(s) = b_2 \hat{x}_2(s) - b_3 \hat{x}_3(s),$$

$$s\hat{x}_4(s) = b_3 \hat{x}_3(s) - b_4 \hat{x}_4(s).$$

The second and third equations are used to eliminate $\hat{x}_2(s)$ and $\hat{x}_3(s)$, giving $\hat{x}_4(s)$ as a function of $\hat{x}_1(s)$:

$$\hat{x}_4(s) = \frac{b_1 b_2 b_3}{(s+b_2)(s+b_3)(s+b_4)}\hat{x}_1(s) = G(s)\hat{x}_1(s). \tag{4}$$

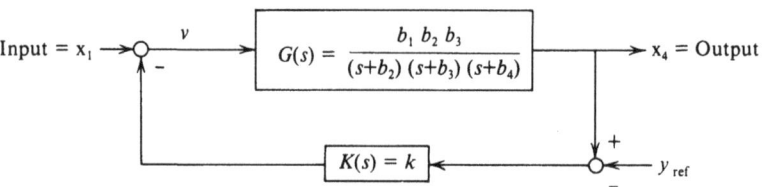

Figure 3.3.6. The canonical control loop in which the forward operator models a three-step chemical reaction and the feedback controller is a gain constant.

As before let us consider proportional feedback where the feedback transfer function is again constant, $K(s) = k$. The same form of closed loop transfer function is obtained.

$$\hat{x}_4(s) = \frac{G(s)}{1+kG(s)}[\hat{x}_1(s) + k\hat{y}_{ref}(s)].$$

Stability is determined by the roots of the characteristic equation.

$$1 + kG(s) = 1 + \frac{kb_1b_2b_3}{(s+b_2)(s+b_3)(s+b_4)} = 0. \tag{5}$$

Example 3.3.5. Using the Routh–Hurwitz criterion, examine the stability properties of characteristic equation (5) for positive b's and positive feedback gain k.

Solution:
The characteristic equation is

$$s^3 + s^2(b_2+b_3+b_4) + s(b_2b_3+b_2b_4+b_3b_4)$$
$$+ b_2b_3b_4 + kb_1b_2b_3 = 0.$$

$a_n = a_3 = 1,$

$a_{n-1} = a_2 = b_2 + b_3 + b_4,$

$a_{n-2} = a_1 = b_2b_3 + b_2b_4 + b_3b_4,$

$a_{n-3} = a_0 = b_2b_3b_4 + kb_1b_2b_3.$

Expressions for Δ_1, Δ_2 and Δ_3 must be derived. There is a root with positive real part if (and only if) there is a negative Δ.

$\Delta_1 = a_{n-1} = a_2 = b_2 + b_3 + b_4.$

Since all b's are positive, Δ_1 is positive.

$$\Delta_2 = \begin{vmatrix} a_{n-1} & a_{n-3} \\ a_n & a_{n-2} \end{vmatrix} = \begin{vmatrix} a_2 & a_0 \\ a_3 & a_1 \end{vmatrix} = a_1a_2 - a_0a_3,$$

$\Delta_2 = (b_2+b_3+b_4)(b_2b_3+b_2b_4+b_3b_4) - b_2b_3b_4 - kb_1b_2b_3,$

$\Delta_2 = b_2(b_2b_3+b_2b_4) + (b_3+b_4)(b_2b_3+b_2b_4+b_3b_4) - kb_1b_2b_3.$

The first two terms are positive; the third is negative. The magnitude of the third term increases with k, so if k is big enough Δ_2 is negative and the system is unstable (tight control (increasing k) causes instability (Δ_2 negative)). The inequality determining the minimum k for *instability* is

$0 > \Delta_2 = b_2(b_2b_3+b_2b_4) + (b_3+b_4)(b_2b_3+b_2b_4+b_3b_4) - kb_1b_2b_3,$

$kb_1b_2b_3 > b_2(b_2b_3+b_2b_4) + (b_3+b_4)(b_2b_3+b_2b_4+b_3b_4),$

$$k > \frac{b_2(b_2b_3+b_2b_4) + (b_3+b_4)(b_2b_3+b_2b_4+b_3b_4)}{b_1b_2b_3}.$$

The system is unstable if k is greater than the expression on the right, which is defined as k_c (c = critical). Analysis of Δ_3 gives the same result.

$$\Delta_3 = \begin{vmatrix} a_{n-1} & a_{n-3} & a_{n-5} \\ a_n & a_{n-2} & a_{n-4} \\ 0 & a_{n-1} & a_{n-3} \end{vmatrix} = \begin{vmatrix} a_2 & a_0 & 0 \\ a_3 & a_1 & 0 \\ 0 & a_2 & a_0 \end{vmatrix},$$

$$\Delta_3 = a_2 \begin{vmatrix} a_1 & 0 \\ a_2 & a_0 \end{vmatrix} - a_0 \begin{vmatrix} a_3 & 0 \\ 0 & a_0 \end{vmatrix},$$

$$\Delta_3 = a_0(a_1 a_2 - a_0 a_3).$$

From the characteristic equation, $a_0 = b_2 b_3 b_4 + k b_1 b_2 b_3$. Since all of these constants are positive, it follows that a_0 is positive. This means that the sign of Δ_3 is the same as the sign of its second factor:

$$\text{sgn } \Delta_3 = \text{sgn } (a_1 a_2 - a_0 a_3).$$

But $a_1 a_2 - a_0 a_3 = \Delta_2$, so

$$\text{sgn } \Delta_2 = \text{sgn } \Delta_3.$$

Thus Δ_3 gives the same stability condition as Δ_2, namely:

$k < k_c$ roots stable (negative real parts);

$k > k_c$ roots unstable (positive real parts).

The special case $k = k_c$ gives pure imaginary roots. This would result in periodic solutions of the differential equation exactly as in the case of simple harmonic motion. However, it must be stressed that linear systems cannot generally successfully model chemical oscillations. The reason can be easily seen using the example. If $k = k_c$ *exactly*, then oscillations result, but there is always noise in any real system so k will move above or below k_c. If k is even an ε above k_c (ε being an *arbitrarily small* positive number), the system is unstable. If k is ε below k_c, the oscillations damp out. The inevitable variation from k_c will destroy periodic motion.

Using the Routh–Hurwitz criterion, it has been shown that increasing feedback gain can cause instability. In Example 3.3.6 this result is reproduced via the Nyquist criterion.

> *Example* 3.3.6.   Determine the stability properties of the system of Figure 3.3.6 by the Nyquist criterion for parameters $b_1 = \frac{1}{2}$, $b_2 = 1$, $b_3 = 1.05$ and $b_4 = 1.1$.

Solution:
The Nyquist locus spirals in a counterclockwise direction from $G(0)$.

$$G(0) = b_1b_2b_3/b_2b_3b_4 = b_1/b_4 = 0.4545\ldots.$$

There is an intersection with the negative real axis at $\omega \cong 1.82$. Instability is indicated if $-1/k$ is encircled by the Nyquist locus. The points on the negative real axis between the origin and the intersection of $G(i\omega)$ and the negative real axis correspond to feedback gain that can produce instability. (This is the region marked by the heavy line in Figure 3.3.7.) For $1/k$ small enough

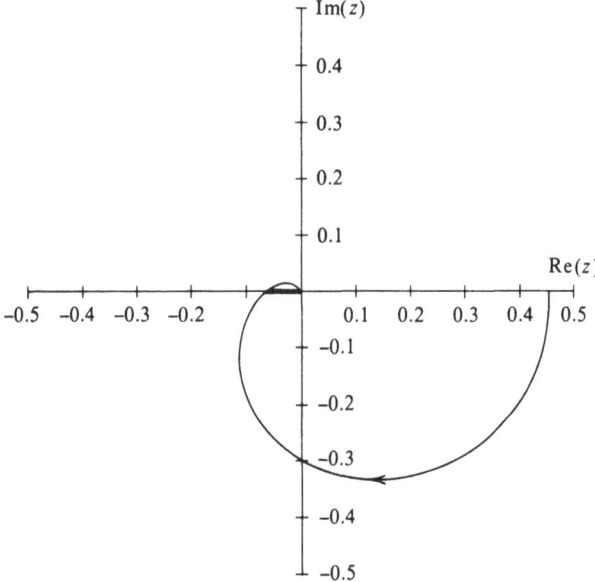

Figure 3.3.7. The Nyquist locus, $G(i\omega)$, for the transfer function of (4) for $b_1 = 0.5$, $b_2 = 1$, $b_3 = 1.05$ and $b_4 = 1.1$. The arrow indicates the direction of increasing ω. The overlined segment of the negative real axis is encircled by the Nyquist locus.

(i.e. k big enough), $-1/k$ is encircled. The point of intersection gives the minimum gain for instability. The intersection occurs at

$$-1/k_c = -0.0568, \qquad k_c = 17.6.$$

This is the same k_c found by the formula of Example 3.3.5.

The three step reaction system gives a useful example of a general control system phenomenon; as the number of separate steps in a process increases, instability becomes more likely. In a five-step reaction scheme, for example, k_c would be smaller and so on. A detailed analysis of this effect is given in Rapp (1976). Similarly a time delay between reaction steps can destabilize a control loop. The destabilizing effect of time delays is briefly considered in Section 3.4.

3.4 TIME DELAY AND CONTROL LOOP STABILITY

The example considered is a generalization of the two-step reaction scheme. x_1 is transformed to x_2. After a time delay τ, x_2 is transformed to y_2 where it is assumed that for time less than τ, $y_2 = 0$ (i.e. y_2 is initially zero and thus stays at zero until after a time delay τ). Mathematically this relationship is represented by $y_2(t) = x_2(t - \tau)$. Biologically, this time delay could be the result of transport processes from one cellular compartment to another. The differential equations that describe this process are analogous to those previously encountered.

$$dx_2(t)/dt = b_1 x_1(t) - b_2 x_2(t),$$

$$y_2(t) = x_2(t - \tau),$$

$$dx_3(t)/dt = b_2 y_2(t) - b_3 x_3(t).$$

In the previous examples the solution procedure was to take Laplace transforms of each equation and solve algebraically for $\hat{x}_3(s)$ as a function of $\hat{x}_1(s)$. For this example, $L[x_2(t - \tau)]$ must be determined. Fortunately a time-delayed function has a particularly simple Laplace transform.

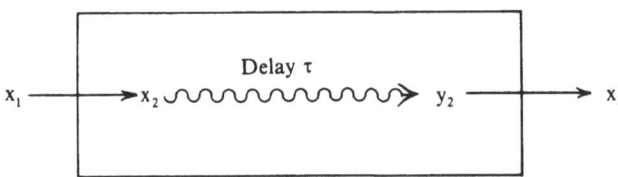

Figure 3.4.1. Diagrammatic representation of time delayed two-step reaction.

Example 3.4.1. Let $f(t)$ be a Laplace transformable function of time. Establish the relation between $L[f(t)]$ and $L[f(t - \tau)]$ where $\tau > 0$ and $f(t - \tau) = 0$ for $t < \tau$.

Solution:
Start with the definition of the Laplace transform.

$$L[f(t-\tau)] = \int_0^\infty f(t-\tau)\exp(-st)\,dt,$$

$$L[f(t-\tau)] = \int_0^\tau f(t-\tau)\exp(-st)\,dt + \int_\tau^\infty f(t-\tau)\exp(-st)\,dt.$$

Given the assumption $f(t-\tau) = 0$ for $t < \tau$,

$$\int_0^\tau f(t-\tau)\exp(-st)\,dt = 0,$$

$$L[f(t-\tau)] = \int_\tau^\infty f(t-\tau)\exp(-st)\,dt.$$

The desired result is obtained by transforming the variable of integration. Let $\theta = t - \tau$. Since τ is a constant $d\theta = dt$, when $t = \tau$, $\theta = 0$ and when $t = \infty$, $\theta = \infty$. Thus

$$L[f(t-\tau)] = \int_0^\infty f(\theta)\exp[-s(\theta+\tau)]\,d\theta$$

$$= \exp(-s\tau)\int_0^\infty f(\theta)\exp(-s\theta)\,d\theta.$$

But the symbol used for the variable of integration is immaterial, so that it can be replaced by t:

$$\int_0^\infty f(\theta)\exp(-s\theta)\,d\theta = \int_0^\infty f(t)\exp(-st)\,dt = L[f(t)]$$

thus

$$L[f(t-\tau)] = \exp(-s\tau)L[f(t)] = \exp(-s\tau)\hat{f}(s).$$

Using this result the transfer function of the time-delayed process can be established. Assuming zero initial conditions,

$$\frac{dx_2(t)}{dt} = b_1 x_1(t) - b_2 x_2(t), \qquad s\hat{x}_2(s) = b_1 \hat{x}_1(s) - b_2 \hat{x}_2(s),$$

$$y_2(t) = x_2(t-\tau), \qquad \hat{y}_2(s) = \exp(-s\tau)\hat{x}_2(s),$$

$$\frac{dx_3(t)}{dt} = b_2 y_2(t) - b_3 x_3(t), \qquad s\hat{x}_3(s) = b_2 \hat{y}_2(s) - b_3 \hat{x}_3(s).$$

Using these three equations one can obtain the transfer function

$$\hat{x}_2(s) = \frac{b_1 \hat{x}_1(s)}{s + b_2},$$

$$\hat{x}_3(s) = \frac{b_2 \hat{y}_2(s)}{s + b_3} = \frac{b_2 \exp(-s\tau)\hat{x}_2(s)}{s + b_3} = \frac{b_1 b_2 \exp(-s\tau)\hat{x}_1(s)}{(s + b_2)(s + b_3)}.$$

Thus

$$G(s) = \frac{b_1 b_2 \exp(-s\tau)}{(s + b_2)(s + b_3)}. \tag{1}$$

We now consider the proportional feedback loop in Figure 3.3.2 where the forward transfer function is the time-delayed process of (3.4.1). In Example 3.4.2, the effect of different values of τ on loop stability is determined by the Nyquist criterion. (This is an example of a problem that can be solved by the Nyquist criterion but not by the Routh–Hurwitz criterion; the latter cannot treat a characteristic equation containing an exponential such as $(s + b_2)(s + b_3) + kb_1 b_2 \exp(-s\tau) = 0$.)

> *Example* 3.4.2. For the transfer function in (3.4.1) and the parameters $b_1 = 0.5$, $b_2 = 1$, and $b_3 = 2$, show (*a*) that even though the system describes a two step reaction scheme, it is possible to destabilize the loop with sufficiently high feedback gain if $\tau > 0$; and show (*b*) that as τ increases, the minimum value of gain causing instability, k_c, decreases.

Solution:
For $\tau = 0$, the transfer function becomes

$$G(s) = b_1 b_2 / (s + b_2)(s + b_3).$$

This is the same as the transfer function of Example 3.3.3. No points on the negative real axis are encircled, so the system is stable for arbitrary feedback gain. The Nyquist loci for $\tau = 0, 1$, and 3 are given in Figure 3.4.2. For $\tau = 1$ and $\tau = 3$ (and indeed any positive τ) points on the negative real axis are encircled, i.e. for feedback gain k greater than some lower bound, k_c, the system is unstable. The first intersection of $G(i\omega)$ for $\tau = 3$ is to the left of that for $\tau = 1$. Further numerical testing or an analytic demonstration can show this to be true for any positive τ. This means that k_c for $\tau = 3$ is less than k_c for $\tau = 1$, since

$$k_c = -1/(\text{value of } x\text{-intersection}),$$

for example

$$\tau = 0 \qquad k_c = \infty,$$
$$\tau = 1 \qquad k_c = 9.1,$$
$$\tau = 2 \qquad k_c = 6.2,$$
$$\tau = 3 \qquad k_c = 5.3,$$
$$\tau = 8 \qquad k_c = 4.4.$$

It can be shown that, as $\tau \to \infty$, $k_c \to 4$.

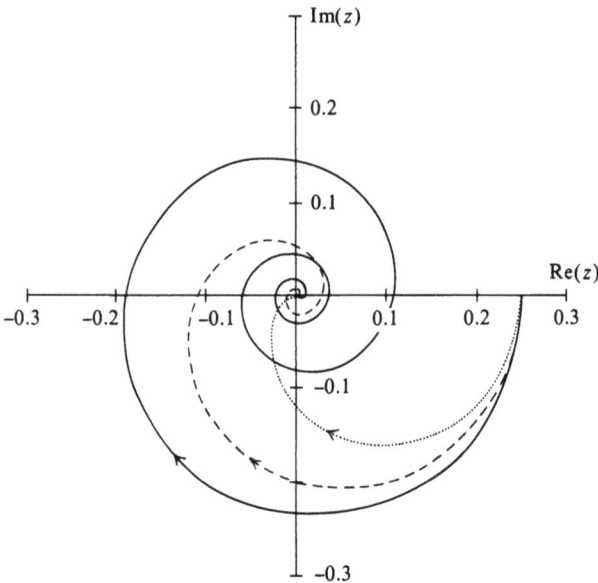

Figure 3.4.2. Nyquist loci for the transfer function of (1). $\omega > 0$ for $\tau = 0$ ($\cdots$), $\tau = 1$, ($----$) and $\tau = 3$ (———). Arrows indicate the direction of increasing ω.

It should be noted that the discrete delay considered here, where the value of a variable depends on the value of another variable *at a specific time*, is a mathematical artifice. Real systems contain distributed lags in which the value of a variable depends on the previous history of another variable. However, an investigation of discrete delays is simple and often successfully approximates the actual system. The possible importance of distributed and delayed lags on the stability of biological regulatory systems has only recently been fully appreciated. For a detailed mathematical treatment see MacDonald (1978).

3.5 THE EFFECT OF DIFFERENT CONTROLLERS

In all of the previous examples a proportional feedback loop was considered where the controller was a positive gain constant, $K(s) = k$. Two other forms of controller are common enough to deserve a brief mention. The first is an integrating controller, which is also called a phase lag controller. Here

$$K(s) = 1 + 1/sT_L$$

where T_L is some positive constant. The second is a differentiating phase advance controller for which

$$K(s) = 1 + sT_A,$$

where T_A is a positive constant. Both of these components are abstractions since it is impossible to produce a pure integrator or a pure differentiator. It is common practice to put these components in a cascade so that $K(s)$ becomes

$$K(s) = k(1 + 1/sT_L)(1 + sT_A).$$

The constants k, T_L, and T_A can be adjusted to give desired performance. Integration (phase lag) gives good low frequency control (long-term performance) while proportional control (no lag or advance) gives good medium frequency control. Differentiation (phase advance) gives good high frequency control (transient, fast responses). The selection of controllers (either by a design engineer or by 'evolution') is determined by the conditions that the system will encounter.

3.6 NONLINEAR MODELS

Experience suggests that most biological systems are inherently nonlinear and that linear analysis is at best only an approximation which is valid for a possibly restricted range near dynamic equilibrium. There are two classes of behavior of particular biological importance unique to nonlinear systems: switching and oscillations.

Nonlinear chemical systems can have more than one steady state. The selection of a particular steady state will depend on the parameter values and on the past history of the system. As shown in Section 3.8, it is possible to cause a chemical system to move abruptly from one state to another in response to a small change in a parameter, i.e. the system responds as a switch. Several authors have suggested that switching behavior could provide a basis for cell differentiation or for the transformation from normal to neoplastic growth (Simon, 1965; Grigorov, Polyakova & Chernavskii, 1967; Babloyantz & Nicolis, 1972; van Cauter & Dumont, 1978).

As indicated in Section 3.3, the linear oscillator cannot successfully model biological oscillations which are known to be insensitive to fairly large variations in parameters and initial conditions. A chemical system can oscillate only if it is nonlinear, thermodynamically open and far from equilibrium (Prigogine, 1967, 1969). Biochemical oscillations in positive feedback loops are exemplified by the glycolytic oscillator (Section 4.1). Oscillations in positive feedback loops are intuitively plausible since one would expect that positive feedback could destabilize a steady state. However, since the response of biological systems saturates, the output should ultimately be bounded. A stable oscillation would be the expected 'compromise' in bounded systems with an unstable steady state (for an example see Goldbeter & Segel, 1977). Unforced oscillations (i.e. oscillations that are produced by the system itself and are not due to external periodic forcing) in negative feedback loops seem less likely because the action of the control is to direct the output to the reference signal. Indeed at one time it was commonly supposed that oscillations in negative feedback *biochemical* systems were impossible. However, experience in engineering control systems has indicated that negative feedback systems are quite likely to have stable oscillations unless an effort is made to prevent them. Indeed in Example 3.6.1 it is shown that even very simple negative feedback loops can oscillate.

> *Example* 3.6.1.   The Chancellor of the Exchequer Oscillator (so named because the control element always does the right thing but too late and in an extreme way). Consider the control loop in Figure 3.6.1. The object of the control is to hold the value of x to zero. If x becomes negative, then after a time delay, τ, the value of $\dot{x}$ is $+a > 0$, i.e. x will increase and return to zero. If x becomes positive then, after the delay, its derivative is $-b < 0$, so x decreases returning to zero. Thus the negative sign on the return loop appears because the output of the nonlinear control element is $+a$ if $x(t-\tau) < 0$ and $-b$ is $x(t-\tau) > 0$. The component $1/s$ appears because the output of the nonlinearity is equal to $\dot{x}$ and not x. The input is the constant zero, so if oscillations appear it is not as a result of periodic forcing. Suppose that initially $x = 0$ and the nonlinear controller is set at $-b$. The existence of oscillations will be demonstrated by explicitly following the subsequent movement of x and $\dot{x}$ and showing that the system returns to this initial state, i.e. a repeating cycle is established.

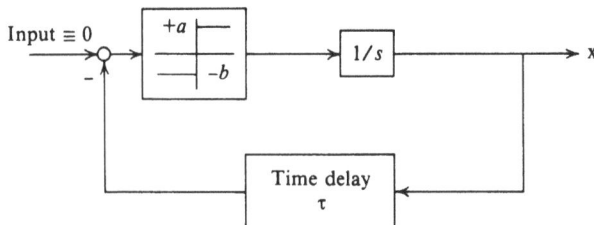

Figure 3.6.1. The Chancellor of the Exchequer Oscillator: the derivative of x is either $+a$ or $-b$. The instability results from the time delay τ. Details are in the text.

Solution:

Initially $x = 0$ and $\dot{x} = -b$; because of the time delay element, x will decrease for a time interval τ. It will then increase and eventually become positive, but again $\dot{x}$ will not change sign until after a delay interval τ; x then decreases to zero reproducing the initial conditions $x = 0$ and $\dot{x} = -b$. Figure 3.6.2 shows the variation of x and its derivative. The argument is made clearer by explicitly calculating the period and amplitude.

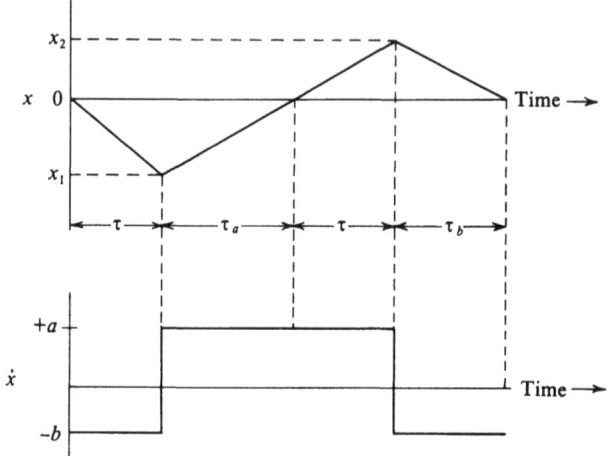

Figure 3.6.2. Transitions in x and its derivative in the control loop of Figure 3.6.1. Initially $x = 0$ and $\dot{x} = -b$. The transition to $\dot{x} = +a$ follows after a time delay τ.

(*a*) Calculation of x_1.
Start at $x = 0$ and decrease at rate $-b$ for time τ. Thus $x_1 = -\tau b$.

(*b*) Calculation of τ_a.
Start at $x = -\tau b$ and increase at rate $+a$.

distance = (rate) · (time),   i.e. $\tau b = (a)(\tau_a)$, $\tau_a = \tau b/a$.

(*c*) Calculation of x_2.
Start at $x = -\tau b$ and increase at rate $+a$.

$x_2 = a\tau$.

(*d*) Calculation of τ_b.
Start at $x_2 = a\tau$ and decrease at rate $-b$.

$\tau_b = a\tau/b$.

Thus

period $= \tau + \tau_a + \tau + \tau_b = 2\tau + \tau[(b/a) + (a/b)]$,

amplitude $= x_2 - x_1 = \tau(a + b)$.

The system in Example 3.6.1 is clearly an extreme case in having a mechanical plus–minus nonlinearity and in containing an idealized pure delay. However, it does show that oscillations can occur in a negative feedback system that appears always to decrease displacements from a constant reference setting. A model of a biochemical negative feedback system is discussed in Section 3.7.

3.7 A NEGATIVE FEEDBACK BIOCHEMICAL SYSTEM: THE GOODWIN EQUATIONS FOR $N = 3$

Derivation of the equations and translation to nondimensional form

In Section 3.2 we introduced a three-variable model of a biochemical process where x_1 is an mRNA species, x_2 is the enzyme coded by the mRNA and x_3 is the product of the enzyme catalyzed reaction. In the previous analysis we treated x_1 as an input variable. Now we consider the case originally considered by Goodwin (1965) where the input consists of a constant nucleotide pool necessary for mRNA synthesis and where x_3 directly represses the synthesis of x_1. In this simple model it is assumed that ρ molecules of x_3 combine with the gene coding for the mRNA to form a nonfunctional operator–inhibitor complex. The rate of x_1 synthesis is assumed to be directly proportional to the fraction of time that the operator is not repressed. To calculate this fraction as a function of x_3, suppose that there is a large population of identical gene copies. Let G_T be the total number of genes in the population, G_A, the number of active genes and G_R the number of repressed genes. The fraction of time that a single gene is unrepressed is equal to the fraction of unrepressed genes in a large population, G_A/G_T. This ratio is found by using an argument analogous to those of classical enzyme kinetics. It is supposed that the repressor–operator reaction can be approximated as the direct combination of ρ molecules of repressor. This is clearly not the case chemically since the possibility of higher-order molecular collisions is vanishingly small. Strictly speaking one should consider the sequential addition of repressor molecules. However, the reaction of (1) below is taken as a first approximation. The approximation is surprisingly successful for a wide range of concentrations under certain circumstances; in particular, when binding the first repressor is the rarest step.

$$G_A + \rho x_3 \underset{h_2}{\overset{h_1}{\rightleftharpoons}} G_R;$$

$$dG_A/dt = -h_1 G_A x_3^\rho + h_2 G_R.$$

(1)

This reaction is assumed to be close to equilibrium so

$$G_T = G_A + G_R = G_A(1 + \alpha x_3^\rho), \quad \text{where } \alpha \equiv h_1/h_2.$$

$$G_A/G_T = 1/(1 + \alpha x_3^\rho).$$

The rate of x_1 synthesis is taken as being directly proportional to this function. Let k_0 be the constant of proportionality. It is also assumed that the rate of mRNA destruction is directly proportional to its concentration. As in Section 3.2 the synthesis and destruction of x_2 and x_3 are assumed to be first order. (Note that (2) is slightly more general than (3.2.2) since it admits the possibility $b_2 \neq g_2$.)

$$dx_1/dt = [k_0/(1 + \alpha x_3^\rho)] - b_1 x_1,$$

$$dx_2/dt = g_1 x_1 - b_2 x_2, \tag{2}$$

$$dx_3/dt = g_2 x_2 - b_3 x_3.$$

The approximation associated with assuming a $(\rho + 1)$-order reaction has already been discussed, but there is another approximation implicit in modeling gene control processes by continuous ordinary differential equations, namely that the law of mass action (large numbers) applies. However, in typical systems there may be very few copies of repressor molecules (Gilbert & Müller-Hill, 1966; Lin & Riggs, 1975). Berg & Blomberg (1977) have compared the binding laws derived for a system containing a discrete number of molecules with the classical mass action equation and found that the difference between them increases as the number of molecules in the system decreases. However, it is fair to conclude that the order of error introduced by this assumption is probably small compared to the others involved in the model, notably reducing mRNA synthesis and enzyme synthesis to a single equation. It is possible to produce (2) by an argument involving feedback inhibition in a metabolic system where the problem of limited numbers of molecules is less severe. In any case the initial purpose of the present exercise is to learn something in general about the behavior of continuous negative feedback systems.

Equation (2) contains seven parameters (k_0, α, b_1, b_2, b_3, g_1 and g_2). Substantial simplification is achieved by transforming the problem to non-dimensional coordinates. This is done in Example 3.7.1.

Example 3.7.1. Find scaling factors for concentrations and time to simplify (2) as much as possible; namely find scaling factors for t, x_1, x_2 and x_3 and hence new variables τ, z_1, z_2 and z_3 and new parameters k_1, k_2 and k_3 so that (2) becomes

$$\frac{dz_1}{d\tau} = \frac{1}{1 + z_3^\rho} - k_1 z_1, \tag{3}$$

$dz_2/d\tau = z_1 - k_2 z_2,$

$dz_3/d\tau = z_2 - k_3 z_3.$

Solution:
Clearly the first step is to eliminate α from the denominator of the nonlinearity. Define intermediate variables $y_i = \alpha^{1/\rho} x_i$. With these,

$$\frac{dy_1}{dt} = \frac{\alpha^{1/\rho} k_0}{1 + y_3^\rho} - b_1 y_1,$$

$$\frac{dy_2}{dt} = g_1 y_1 - b_2 y_2,$$

$$\frac{dy_3}{dt} = g_2 y_2 - b_3 y_3.$$

Define scale factors w and h_j such that $\tau = wt$ and $z_j = h_j y_j$. The differential equation becomes

$$\frac{dz_1}{d\tau} = \left(\frac{\alpha^{1/\rho} k_0 h_1}{w}\right)\left(\frac{1}{1 + z_3^\rho/h_3^\rho}\right) - \left(\frac{b_1}{w}\right) z_1,$$

$$\frac{dz_2}{d\tau} = \left(\frac{g_1 h_2}{w h_1}\right) z_1 - \left(\frac{b_2}{w}\right) z_2,$$

$$\frac{dz_3}{d\tau} = \left(\frac{g_2 h_3}{h_2 w}\right) z_2 - \left(\frac{b_3}{w}\right) z_3.$$

The desired simplification results if

$$\frac{\alpha^{1/\rho} k_0 h_1}{w} = 1, \qquad \frac{g_1 h_2}{w h_1} = 1, \qquad \frac{g_2 h_3}{h_2 w} = 1, \qquad h_3 = 1.$$

Taking the product of the first three conditions fixes w:

$$w = (g_1 g_2 \alpha^{1/\rho} k_0)^{\frac{1}{3}}.$$

Solving the first three conditions sequentially we find

$$h_1 = \frac{(g_1 g_2 \alpha^{1/\rho} k_0)^{\frac{1}{3}}}{\alpha^{1/\rho} k_0}, \qquad h_2 = \frac{(g_1 g_2 \alpha^{1/\rho} k_0)^{\frac{2}{3}}}{g_1 \alpha^{1/\rho} k_0}, \qquad h_3 = 1.$$

So, as is necessary, the fourth condition is automatically satisfied. The transformation equations can be generalized to arbitrary dimension (Rapp, 1976).

It is important to consider now the significance of the parameter ρ in (3). As ρ increases the synthesis versus inhibitor function becomes steeper (Figure 3.7.1). In the terminology of Section 3.3, control becomes

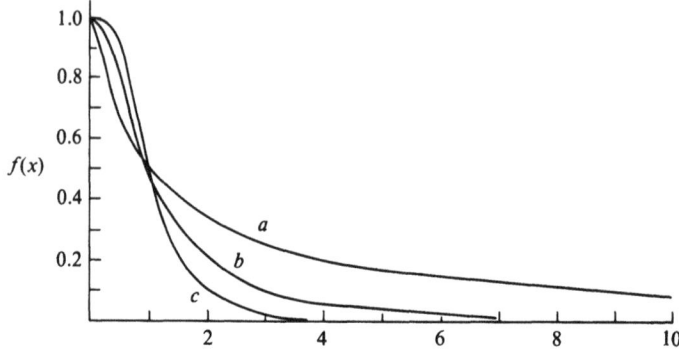

Figure 3.7.1. The effect of increasing ρ on the nonlinearity $f(x) = 1/(1 + x^\rho)$. In Curve a, $\rho = 1$; in Curve b, $\rho = 2$ and in Curve c, $\rho = 3$. As ρ increases the transition from high synthesis rate (large f) to low synthesis rate (small f) becomes sharper.

tighter as ρ increases. Accordingly we would expect that increasing ρ might destabilize the system. We begin the analysis of (3) by establishing two simple global properties of the equation.

Global properties

We now show that all biologically significant behavior of (3) is contained in a finite region of positive concentration space. In addition it is shown that this region contains a single steady state.

Example 3.7.2 (Griffith, 1968). Suppose that initially z_1, z_2 and z_3 are positive or zero. Since these variables are concentration functions this must be the case. Show that after a sufficiently large time the following inequalities are satisfied.

$$0 \le z_1(\tau) \le \max\left(z_1(0), (1/k_1)\right),$$

$$0 \le z_2(\tau) \le \max\left(z_2(0), (1/k_1 k_2)\right),$$

$$0 \le z_3(\tau) \le \max\left(z_3(0), (1/k_1 k_2 k_3)\right).$$

Solution:
For $j = 1$, 2 and 3, dz_j/dt is positive when $z_j = 0$ so if $z_j(0)$ is non-negative then $z_j(t)$ must be positive for all time $\tau > 0$. This establishes the lower bound. Consider the z_1 governing equation

$$\frac{dz_1}{d\tau} = \frac{1}{1 + z_3^\rho} - k_1 z_1.$$

Since $z_3 \geq 0$, we know that $1 + z_3^\rho \geq 1$, so that

$$\frac{\mathrm{d}z_1}{\mathrm{d}\tau} = \frac{1}{1+z_3^\rho} - k_1 z_1 \leq 1 - k_1 z_1.$$

Thus $\dot{z}_1$ is negative for all $z_1 > (1/k_1)$. Accordingly if $z_1(0)$ is less than $1/k_1$ we know that it will never exceed $1/k_1$. If $z_1(0)$ is greater than $1/k_1$, the derivative will be negative and so z_1 will decrease until after some finite time it is less than $1/k_1$. Hence

$$0 \leq z_1(\tau) \leq \max[z_1(0), (1/k_1)].$$

The bound on z_2 is established by a similar argument. After a sufficient time $z_1 < 1/k_1$. Consequently

$$\mathrm{d}z_2/\mathrm{d}\tau = z_1 - k_2 z_2 \leq (1/k_1) - k_2 z_2.$$

For all sufficiently large time and for all $z_2 > 1/k_1 k_2$, we now know that $\dot{z}_2$ is negative, i.e.

$$0 \leq z_2(\tau) \leq \max[z_2(0), (1/k_1 k_2)].$$

The bound on z_3 is established by an analogous argument. Let B denote the set in z_1–z_2–z_3 space formed by the rectangular box with vertices $(0, 0, 0)$ and $(1/k_1, 1/k_1 k_2, 1/k_1 k_2 k_3)$ with sides parallel and perpendicular to the coordinate axes. If an initial point is in set B, the solution of the differential equation remains in B. Sets with this property are called **invariant**.

In Example 3.7.3 the next property of the system is established: it has only one physical steady state (where here we use the word 'physical' to mean non-negative, as is required of concentration functions).

Example 3.7.3. Show that there is only one steady state in positive concentration space and that it is in set B.

Solution:
Let the superscript '0' denote steady state values. They are defined by the steady state equation $\mathrm{d}z_j/\mathrm{d}t = 0$.

$$0 = \frac{1}{1+(z_3^0)^\rho} - k_1 z_1^0, \qquad 0 = z_1^0 - k_2 z_2^0, \qquad 0 = z_2^0 - k_3 z_3^0.$$

Using the second and third equation to eliminate z_1^0 and z_2^0 and the definition $c_0 = k_1 k_2 k_3$ we find that

$$c_0 z_3^0 = 1/[1+(z_3^0)^\rho] \equiv f(z_3^0). \tag{4}$$

As shown in Figure 3.7.2 the left-hand side of this equation increases from zero to infinity while the right-hand side

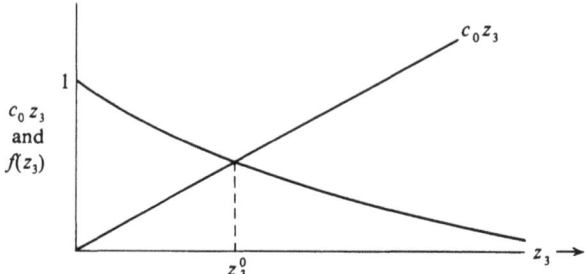

Figure 3.7.2. Solution of the steady equation $f(z_3) = c_0 z_3$ for z_3^0. The uniqueness of the solution for positive z_3 follows from the monotonicity of the functions (they are increasing or decreasing for all positive z_3).

decreases from one to zero so there must be an intersection (and hence a solution to the steady state equations). Because both $c_0 z_3^0$ and $f(z_3^0)$ are strictly monotone (i.e. always increasing or always decreasing) there is only one intersection. Thus there is only one positive z_3^0 and hence only one positive z_1^0 and z_2^0 satisfying the steady state equation.

To show that this steady state is in invariant set B, we make the following manipulations of the steady state equation.

$$c_0 z_3^0 = 1/[1 + (z_3^0)^\rho],$$

$$c_0 z_3^0 + c_0 (z_3^0)^{\rho+1} = 1,$$

$$f_a(z_3^0) \equiv c_0(z_3^0)^{\rho+1} = 1 - c_0 z_3^0 \equiv f_b(z_3^0).$$

The functions f_a and f_b are plotted in Figure 3.7.3. Their unique intersection occurs at the positive solution of the steady state

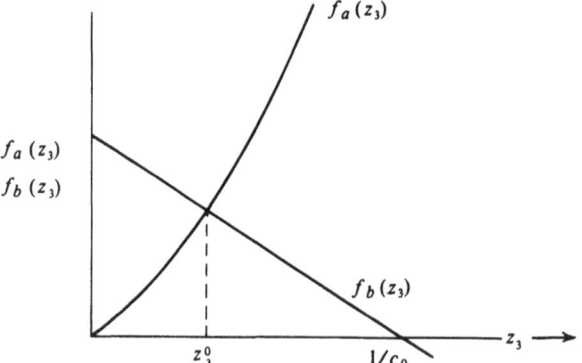

Figure 3.7.3. Proof that $0 < z_3^0 < (1/c_0)$. In this diagram $f_a(z_3) = c_0(z_3)^{\rho+1}$ and $f_b(z_3) = 1 - c_0 z_3$.

equation. Since f_b is zero at $1/c_0$ we have

$$z_3^0 < 1/c_0 = 1/k_1 k_2 k_3.$$

Similarly

$$z_2^0 = k_3 z_3^0 < 1/k_1 k_2, \qquad z_1^0 = k_2 z_2^0 < 1/k_1,$$

so (z_1^0, z_2^0, z_3^0) is in set B.

Both of these results are fairly easy to prove, but are more valuable than they first appear since they can be used in conjunction with some recent mathematical results to identify conditions which ensure the existence of oscillations. Examples 3.7.2 and 3.7.3 have been generalized to a nonlinear n-dimensional multiple loop system that contains (3) as a special case (Mees & Rapp, 1978).

Local stability analysis

The analysis of the stability properties of steady states is the essential first step in the analysis of a nonlinear system. A steady state is said to be locally stable if all sufficiently close initial points possess solutions that tend to the steady state as $t \to \infty$. The local stability analysis begins by translating the original nonlinear differential equation (here (3)) to an approximate linear differential equation that describes the behavior of small variations around the steady state. Let z_j' be the variation from z_j^0, i.e. $z_j = z_j^0 + z_j'$. The equation for z_1' is found by substituting these new variables into the equation for z_1:

$$dz_1/d\tau = f(z_3) - k_1 z_1,$$

$$\frac{d}{d\tau}(z_1^0 + z_1') = \frac{d}{d\tau} z_1' = f(z_3^0 + z_3') - k_1(z_1^0 + z_1').$$

Here we have used the fact that z_1^0 is a constant and hence has a zero derivative. Using Taylor's theorem (A.1.30) we have the following expansion of $f(z_3)$:

$$f(z_3) = f(z_3^0 + z_3')$$

$$= f(z_3^0) + f'(z_3^0)z_3' + \frac{f''(z_3^0)(z_3')^2}{2!} + \dots.$$

Recall that we are considering only small displacements from the steady state so z_3' is small and thus

$$z_3' \gg (z_3')^2 \gg (z_3')^3,$$

so we have a linear approximation

$$f(z_3^0 + z_3') = f(z_3^0) + f'(z_3^0)z_3'.$$

Substituting this into the differential equation gives

$$dz_1'/d\tau = f(z_3^0) + f'(z_3^0)z_3' - k_1(z_1^0 + z_1').$$

The steady state condition is

$$f(z_3^0) - k_1 z_1^0 = 0,$$

so the differential equation becomes

$$dz_1'/d\tau = f'(z_3^0)z_3' - k_1 z_1'.$$

The translation process for z_2 and z_3 is simpler since these equations are already linear:

$$dz_2/d\tau = z_1 - k_2 z_2,$$

$$\frac{d}{d\tau}(z_2^0 + z_2') = z_1^0 + z_1' - k_2(z_2^0 + z_2').$$

Since $z_1^0 - k_2 z_2^0 = 0$ this becomes

$$dz_2'/d\tau = z_1' - k_2 z_2'.$$

Similarly for z_3 we find

$$dz_3'/d\tau = z_2' - k_3 z_3'.$$

The set of equations collected in (5) below is referred to as the linear variational equations taken about the steady state:

$$\begin{aligned}
dz_1'/d\tau &= f'(z_3^0)z_3' - k_1 z_1', \\
dz_2'/d\tau &= z_1' - k_2 z_2', \\
dz_3'/d\tau &= z_2' - k_3 z_3'.
\end{aligned} \tag{5}$$

If this system of differential equations is stable, i.e. $z_j'(\tau) \to 0$ as $\tau \to \infty$ for $j = 1$, 2 and 3, then the steady state is stable. The stability of the linear variational equations will be determined by the Nyquist criterion in a manner identical to the analyses in Sections 3.3 and 3.4. Equation (5) could be investigated directly, without transforms, but the transform method can be used for more general situations such as those with time delays. It is assumed that initially the system is at the steady state, so $z_j'(0) = 0$. Taking Laplace transforms we find

$$\begin{aligned}
s\hat{z}_1' &= f'(z_3^0)\hat{z}_3' - k_1\hat{z}_1', \\
s\hat{z}_2' &= \hat{z}_1' - k_2\hat{z}_2', \\
s\hat{z}_3' &= \hat{z}_2' - k_3\hat{z}_3'.
\end{aligned}$$

The second and third equation can be used to eliminate $\hat{z}'_1$ and $\hat{z}'_2$:

$$(s + k_1)\hat{z}'_1 = f'(z_3^0)\hat{z}'_3,$$

$$(s + k_2)\hat{z}'_2 = \hat{z}'_1,$$

$$(s + k_3)\hat{z}'_3 = \hat{z}'_2,$$

$$(s + k_1)(s + k_2)(s + k_3)\hat{z}'_3 = f'(z_3^0)\hat{z}'_3,$$

$$\hat{z}'_3 = f'(z_3^0)[1/(s + k_1)(s + k_2)(s + k_3)]\hat{z}'_3.$$

We define $G(s)$ as

$$G(s) = 1/(s + k_1)(s + k_2)(s + k_3), \tag{6}$$

so

$$\hat{z}'_3 = f'(z_3^0)G(s)\hat{z}'_3.$$

The control loop corresponding to this differential equation is shown in Figure 3.7.4.

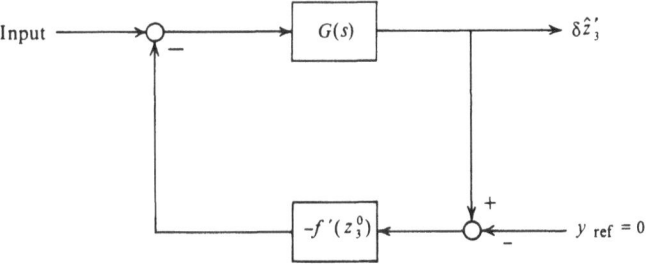

Figure 3.7.4. Local stability analysis by application of the Nyquist criterion follows from a restatement of the stability equations as a feedback system.

The stability of the variational equations, and hence of the steady state, is now found by application of the Nyquist criterion.

Example 3.7.4.   Let $G(s)$ be given by (6). The Nyquist locus has the same form as that in Figure 3.3.7; $G(0)$ is a positive real number and the locus spirals into the origin in a clockwise direction intersecting the negative real axis once. Let ω_1 be the value of ω at this intersection. Show that the control system of Figure 3.7.4 is unstable if and only if

$$W_1(k_1, k_2, k_3) = \rho(1 - c_0 z_3^0)|G(i\omega_1)/G(0)| > 1,$$

where $c_0 = k_1 k_2 k_3$.

Solution:
The analysis follows the program of Example 3.3.6. Here the gain constant analogous to k is $-f'(z_3^0)$ (recall that since f is

monotone decreasing; i.e. decreasing for all positive z_3, $f'(z_3^0)$ is negative and thus $-f'(z_3^0)$ is positive). The control loop is unstable if the point $1/f'(z_3^0)$ (analogous to $-1/k$) is encircled by $G(i\omega)$. Since $1/f'(z_3^0)$ is a real number we conclude that this can happen only if

$$G(i\omega_1) < 1/f'(z_3^0) < G(0).$$

Since $1/f'(z_3^0)$ is negative and $G(0)$ is positive, the inequality on the right is satisfied automatically so the instability condition is

$$G(i\omega_1) < 1/f'(z_3^0).$$

Again using the known sign properties of these numbers this condition can be restated:

$$-|G(i\omega_1)| < -1/|f'(z_3^0)|,$$

$$|G(i\omega_1)| > 1/|f'(z_3^0)|,$$

$$|f'(z_3^0)||G(i\omega_1)| > 1.$$

The function $|f'(z_3^0)||G(i\omega_1)|$ is a function of the parameters k_1, k_2 and k_3. Defining W_1 as shown, we can conclude that $W_1 > 1$ is a necessary and sufficient condition for instability:

$$W_1(k_1, k_2, k_3) \equiv |f'(z_3^0)||G(i\omega_1)| > 1.$$

It now only remains to transform W_1 into the desired final form. By definition

$$G(0) = 1/k_1 k_2 k_3 = 1/c_0,$$

so

$$W_1(k_1, k_2, k_3) = \frac{|f'(z_3^0)|}{c_0}|G(i\omega_1)/G(0)|.$$

It is necessary to show that $|f'(z_3^0)|/c_0 = \rho(1 - c_0 z_3^0)$. But

$$f(z_3) = \frac{1}{(1 + z_3^\rho)}, \qquad f'(z_3) = \frac{-\rho z_3^{\rho-1}}{(1 + z_3^\rho)^2}.$$

At the steady state

$$c_0 z_3^0 = 1/(1 + (z_3^0)^\rho),$$

so

$$f'(z_3^0) = -\rho c_0^2 (z_3^0)^{\rho+1}.$$

Again using the steady state equation we find

$$c_0 z_3^0 = 1/(1 + (z_3^0)^\rho),$$

$$c_0 z_3^0 (1 + (z_3^0)^\rho) = c_0 z_3^0 + c_0 (z_3^0)^{\rho+1} = 1,$$

$$c_0 (z_3^0)^{\rho+1} = 1 - c_0 z_3^0,$$

$$f'(z_3^0) = -\rho c_0 (1 - c_0 z_3^0),$$

and thus

$$\frac{|f'(z_3^0)|}{c_0} = \rho |1 - c_0 z_3^0|.$$

In Example 3.7.3 it was shown that $z_3^0 < 1/c_0$ so

$$|1 - c_0 z_3^0| = 1 - c_0 z_3^0,$$

and the desired result is obtained:

$$W_1(k_1, k_2, k_3) = \rho(1 - c_0 z_3^0)|G(i\omega_1)/G(0)|.$$

It should be noted that this result is immediately valid for a system of arbitrary dimension n where

$$G(s) = \prod_{j=1}^{n} \frac{1}{s + k_j}.$$

The behavior of $W_1(k_1, k_2, k_3)$ should now be examined. Even for $n = 3$ this is a comparatively complicated function. However, most of the more important information can be obtained by considering the special case of its maximum value. We begin by evaluating the maximum of $|G(i\omega_1)/G(0)|$. It should be realized that because $c_0 = k_1 k_2 k_3$, the factors $\rho(1 - c_0 z_3^0)$ and $|G(i\omega_1)/G(0)|$ are coupled so that *ab initio* one should not determine the maximum of the two factors separately. However, it can be shown that the maximum value of $|G(i\omega_1)/G(0)|$ is obtained when $k_1 = k_2 = k_3$ (Rapp, 1975). The common value of the k's (and hence of c_0) is immaterial so the two factors can be maximized separately.

$$\max |G(i\omega_1)/G(0)| = [\cos(\pi/3)]^3,$$

so

$$\max W_1(k_1, k_2, k_3) = \max [\rho(1 - c_0 z_3^0)] \cos(\pi/3)^3.$$

It can be shown (Rapp, 1975) that

$$\max \rho(1 - c_0 z_3^0) = \lim_{c_0 \to 0} \rho(1 - c_0 z_3^0) = \rho.$$

The maximum of W_1 has been found

$$\max W_1(k_1, k_2, k_3) = \rho \cos(\pi/3)^3 = \rho/8.$$

For $\rho = 1, \ldots 7$, max $W_1 < 1$; so for any positive set of k's the steady state is locally stable. For $\rho = 8$, max $W_1 = 1$, exactly, and since the maximum is only reached in the limit $c_0 \rightarrow 0$, we conclude that also for $\rho = 8$ the steady state is stable for any set of k's. However, for $\rho > 9$ it is possible to find finite positive k's that produce a system with $W_1 > 1$, i.e. that produce a system that has an unstable steady state. It is interesting to note that both W_1 and max W_1 are directly proportional to ρ. Thus as predicted, increasing ρ tends to destabilize the steady state. As ρ increases the synthesis versus inhibitor function becomes steeper and there is a more sharply defined on–off control (i.e. tightness of control is increased approaching the on–off controller in the Chancellor of the Exchequer oscillator). The implications of local instability for global behavior are considered next.

Oscillations: biological examples of tightness of control and local instability

It can be demonstrated that if the steady state of (3) is unstable (i.e. $W_1 > 1$) then the system possesses a periodic solution (Hastings, Tyson & Webster, 1975). The proof given by Hastings, Tyson & Webster proves this proposition for a much more general n-dimensional system. A simpler proof of the three-dimensional case of (3.7.3) has been published separately (Tyson 1975). However, even the proof of the simple case is beyond the scope of this book. An intuitive justification of the result can be argued along the lines previously presented for the case of oscillations in positive feedback loops. As shown in Example 3.7.2 the solution will always remain in a finite region. This is true even if the steady state is unstable. So again an oscillation results as a compromise between a bounded response and an unstable steady state.

However, granting that the mathematical result is intuitively acceptable, it does not necessarily follow that it is obvious that negative feedback systems can form the basis of biological oscillations. Indeed since local instability can result only if $\rho \geqslant 9$, systems analogous to (3) might seem unlikely candidates as biochemical oscillators. However, several considerations suggest that this is not necessarily the case.

First, it should be pointed out that the number of intermediate reactions in a biochemical system would often be larger than the $n = 3$ case of (3). A more satisfactory model would be one with a larger number of reactants:

$$\frac{dz_1}{d\tau} = \frac{1}{1 + z_n^\rho} - k_1 z_1,$$

$$dz_j/d\tau = z_{j-1} - k_j z_j; \qquad j = 2, \ldots n.$$

The local instability condition is again

$$W_1(k_1, \ldots k_n) \equiv \rho(1 - c_0 z_n^0)|G(i\omega_1)/G(0)| > 1,$$

where

$$G(s) = \prod_{j=1}^{n} 1/(s + k_j).$$

The maximum of W_1 has been established for this more general system (Rapp, 1975):

$$\max W_1 = \rho \cos(\pi/n)^n.$$

As n increases, $\cos(\pi/n)^n$ increases, and the minimum value of ρ admitting the possibility of local instability decreases rapidly (Table 3.7.1).

Table 3.7.1. *Instability and tightness of control*

For a specified value of n, $\cos(\pi/n)^n$ is given. The minimum integer value of ρ such that $\max W_1 > 1$ is found from the equation $\max W_1 = \rho \cos(\pi/n)^n$.

n	$\cos(\pi/n)^n$	$\min \rho$
3	0.125	9
4	0.250	5
5	0.347	3
6	0.422	3
7	0.482	3
8	0.531	2
9	0.571	2
10	0.605	2

Time delays between reaction steps have the same destabilizing effect as increasing the number of chemical intermediates. As the sum of all separate delays approaches infinity $|G(i\omega_1)/G(0)| \to 1$ so an equation with arbitrary dimension n, and any value of ρ except $\rho = 1$ can be unstable if the time delays are sufficiently large. (In fact it is possible for these systems to oscillate even if $n = 1$ and $\rho = 2$ (Rapp, 1974). This example is a mathematical curiosity, and does not correspond to any chemical analog but the example does make the point particularly vividly.)

Thus it is seen that the requirement for a large value of ρ is an artifact resulting from considering the system for the special case of $n = 3$ with no time delays. However, it should be pointed out that ρ can be fairly large. The DNA of bacteriophage λ has two operators recognized by the same repressor. Each operator binds six repressor molecules sequentially

(Maniatis & Ptashne, 1973). Here $\rho = 6$ but because the binding is sequential a more accurate nonlinearity would be the function

$$(z_n^6 + \alpha_5 z_n^5 + \ldots \alpha_1 z_n + \alpha_0)^{-1}.$$

Also the possibility of oscillations in negative feedback metabolic systems becomes more plausible when it is noted that the measured activity versus inhibitor functions of allosteric enzymes can be very steep. Specific examples are found in the enzymes that regulate cyclic AMP (cAMP) concentration. The inhibition of adenylate cyclase (the enzyme that synthesizes cAMP from ATP) by calcium (Marcus & Aurbach, 1971) and EGTA (Bradham, Holt & Sims, 1970; Johnson & Sutherland, 1973) and the activation of this enzyme by magnesium (Drummond & Duncan, 1970; Hepp, Edel & Wieland, 1970) and by epinephrine (Birnbaumer & Rodbell, 1969) all follow functions that are so steep that they approach the form of a mechanical on–off switch. This is also true of the activation of the cAMP catabolic enzyme phosphodiesterase by calcium (Kakiuchi, Yamazaki, Teshima & Uenishi, 1973; Teo & Wang, 1973). This suggests that if these components were incorporated into negative feedback loops oscillations could result. It has been argued that this is indeed the case (Rapp & Berridge, 1977) and that oscillations in calcium–cAMP control loops form the basis of several high frequency biological rhythms including potential oscillations in pancreatic β-cells, smooth muscle and the cardiac pacemaker. The feedback loop proposed for the cardiac pacemaker is shown in Figure 3.7.5. It is proposed that, as in the case of brain tissue, calcium, in conjunction with a calcium-dependent receptor protein, activates adenylate cyclase (Brostrom, Huang, Breckenridge & Wolff, 1975) and produces an increase in cAMP levels. In cardiac muscle cAMP stimulates the sequestration of calcium in the sarcoplasmic reticulum (Tada, Kirchberger & Katz, 1975). The existence of an oscillation in intracellular calcium is well established and calcium is the contraction trigger in all types of muscle (Morad & Goldman, 1973). An oscillation in cAMP concentration has been measured directly (Brooker, 1973, 1975; Wollenberger *et al.*, 1973). Because cAMP and calcium directly affect the membrane, oscillations in internal calcium and cAMP would produce

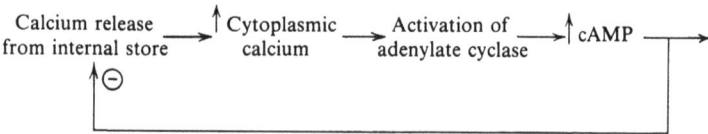

Figure 3.7.5. The interrelation of calcium and cAMP in cardiac muscle. Calcium increases the concentration of cAMP which in turn stimulates the sequestration of cytosol calcium.

observed oscillations in membrane potential. An alternative and more generally accepted model of the basis of pacemaker activity is based on the nonlinear properties of the ion-conducting channels of the cardiac sarcolemma (Jack, Noble & Tsien, 1975).

It has been suggested that other biological rhythms may be the result of oscillations in negative feedback loops. These include oscillations in cyclic enzyme synthesis (Masters & Donachie, 1966; Knorre, 1968, 1973; Donachie & Masters, 1969) and circadian rhythms (Benson & Jacklet, 1977).

3.8 POSITIVE FEEDBACK IN THREE DIMENSIONS

Derivation of the governing equations and transformation to dimensionless coordinates

In this section we consider the three-dimensional system with positive feedback analogous to the negative feedback system considered in Section 3.7. In this system, x_3 acts as its own inducer by combining with a repressor R:

$$R + \rho x_3 \underset{h_2}{\overset{h_1}{\rightleftharpoons}} (Rx_3^\rho).$$

In its active form, repressor R combines with the gene that directs the synthesis of mRNA x_1:

$$G_A + R \underset{h_2'}{\overset{h_1'}{\rightleftharpoons}} G_I.$$

G_A represents an active gene and G_I an inhibited gene. It is assumed that the total repressor concentration R_T is fixed. As in Section 3.7, the rate of x_1 synthesis is taken as being proportional to the fraction of the time that gene G_A is unrepressed. It is supposed that the inducer–repressor reaction and the repressor–gene reaction are near equilibrium, so that

$$dR/dt = -h_1(R)(x_3^\rho) + h_2(Rx_3^\rho),$$

$$h_1/h_2 \equiv \alpha_1 = (Rx_3^\rho)/(R)(x_3^\rho),$$

and similarly

$$\alpha_2 \equiv h_1'/h_2' = G_I/(G_A)(R).$$

The repressor conservation relation is

$$R_T = R + (Rx_3^\rho) + G_I.$$

Because of the comparatively low concentration of repressor-specific sites on the gene, it is reasonable to assume that

$$R_T \cong R + (Rx_3^\rho) + R = \alpha_1(R)(x_3^\rho),$$

$$R = R_T/(1 + \alpha_1 x_3^\rho).$$

The conservation relation for the gene is

$$G_T = G_A + G_I = G_A(1 + \alpha_2 R).$$

The rate of x_1 synthesis is taken as being directly proportional to G_A/G_T:

$$\text{rate} \sim \frac{G_A}{G_T} = \frac{1}{1 + \alpha_2 R} = \frac{1}{1 + \alpha_2\left(\dfrac{R_T}{1 + \alpha_1 x_3^\rho}\right)} = \frac{1 + \alpha_1 x_3^\rho}{K + \alpha_1 x_3^\rho},$$

where

$$K = 1 + \alpha_2 R_T > 1.$$

The equations for x_2 and x_3 are the same as those in Section 3.7.

$$\frac{dx_1}{dt} = \frac{b_0(1 + \alpha_1 x_3^\rho)}{K + \alpha_1 x_3^\rho} - b_1 x_1,$$

$$\frac{dx_2}{dt} = g_1 x_1 - b_2 x_2, \tag{1}$$

$$\frac{dx_3}{dt} = g_2 x_2 - b_3 x_3.$$

A transformation process introduces the following new variables and constants:

$$(b_0 w) = (g_1 g_2 \alpha_1^{1/\rho} b_0)^{\frac{1}{3}}, \qquad \tau = (b_0 w)t, \qquad z_1 = \left[\frac{(b_0 w)}{b_0}\right] x_1,$$

$$z_2 = \left[\frac{(b_0 w)^2}{g_1 b_0}\right] x_2, \qquad z_3 = \left[\frac{(b_0 w)^3}{g_1 g_2 b_0}\right] x_3, \qquad k_j = \left(\frac{b_j}{b_0 w}\right).$$

In terms of the new variables the differential equation becomes

$$\frac{dz_1}{d\tau} = \frac{1 + z_3^\rho}{K + z_3^\rho} - k_1 z_1,$$

$$dz_2/d\tau = z_1 - k_2 z_2, \tag{2}$$

$$dz_3/d\tau = z_2 - k_3 z_3.$$

Throughout this section $f(z)$ refers to the nonlinearity

$$f(z) = \frac{1 + z^\rho}{K + z^\rho},$$

where it is to be understood that K is positive and greater than 1. The function is increasing for all positive z. This property is established in Example 3.8.1.

> *Example 3.8.1.* Show that $f(z)$ is increasing for all positive z and that it tends to a maximum value of 1.

Solution:
This is shown by differentiating $f(z)$ and demonstrating that the derivative is always positive. We have

$$f(z) = \frac{1 + z^\rho}{K + z^\rho},$$

$$f'(z) = \frac{\rho z^{\rho-1}}{K + z^\rho} - \frac{(1 + z^\rho)(\rho z^{\rho-1})}{(K + z^\rho)^2} = \frac{\rho z^{\rho-1}(K - 1)}{(K + z^\rho)^2}.$$

Since $K > 1$, $K - 1 > 0$. The derivative is always positive and thus the function is always increasing. The maximum of the function will be reached in the limit $z \to \infty$

$$\lim_{z \to \infty} f(z) = \lim_{z \to \infty} \frac{1 + z^\rho}{K + z^\rho} = \lim_{z \to \infty} \left[\frac{(1/z^\rho) + 1}{(K/z^\rho) + 1} \right] = 1.$$

The effect of increasing ρ is shown in Figure 3.8.1. The effect is analogous to that seen in Section 3.7. As ρ increases the function becomes steeper.

A global stability result like that of Example 3.7.2 can be easily established. Since $f(z)$ is less than one for positive z, we have

$$\frac{dz_1}{d\tau} = f(z_3) - k_1 z_1 \leq 1 - k_1 z_1,$$

so

$$0 \leq z_1(t) \leq \max{(z_1(0), 1/k_1)}.$$

Bounds on z_2 and z_3 are then established and the same invariant set B is constructed; namely, the box with vertices $(0, 0, 0)$ and $(1/k_1, 1/k_1 k_2, 1/k_1 k_2 k_3)$. At the steady state

$$z_1^0 = f(z_3^0)/k_1 \leq (1/k_1),$$

so z_1^0 and hence z_2^0 and z_3^0 are such that (z_1^0, z_2^0, z_3^0) is in B. However, establishing the number of positive steady states is more difficult for the system with positive feedback. First, it is shown in Example 3.8.2 that there is always at least one steady state point in B.

> *Example 3.8.2.* Show that (2) has at least one steady state point in invariant set B.

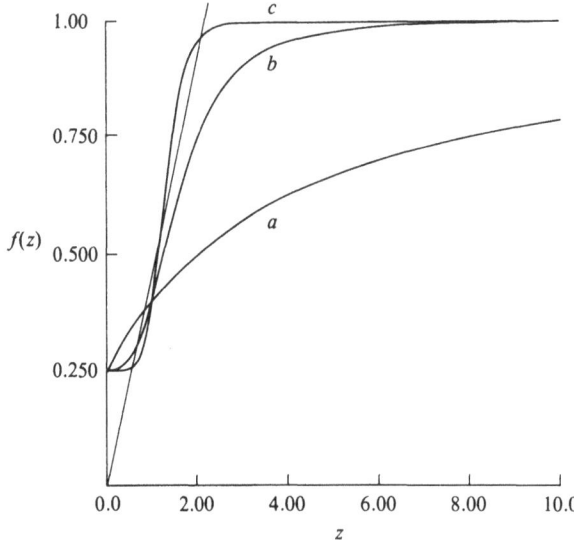

Figure 3.8.1. The effect of increasing ρ on the nonlinearity $f(z) = (1 + z^\rho)/(K + z^\rho)$. In Curve a, $\rho = 1$; Curve b, $\rho = 2$; and in Curve c, $\rho = 6$. As in the previous section, increasing ρ increases the rate of the transition from high synthesis to low synthesis.

Solution:
The steady state equation is

$$k_1 z_1^0 = k_1 k_2 z_2^0 = k_1 k_2 k_3 z_3^0 = c_0 z_3^0 = f(z_3^0).$$

Thus a steady state exists if the lines $c_0 z_3$ and $f(z_3)$ intersect at a positive value of z_3. Clearly $c_0 z_3$ starts at zero and approaches infinity; $f(z_3)$ starts at $1/K$ and, as shown in Example 3.8.1, tends to a finite limit so the $c_0 z_3$ line must 'overtake' $f(z_3)$; i.e. the two lines must intersect.

Variations in the number of steady states

As shown in Figure 3.8.2, it is sometimes possible for there to be three roots. Assuming that the diagram is accurate, there can be one or three roots for $\rho = 6$ depending on the value of c_0 (and for the very exceptional case where $c_0 z_3$ and $f(z_3)$ are tangent there can be two solutions). The analysis of the number of solutions to the steady state equation begins with an application of Descartes' Rule.

Descartes' Rule: (DiStefano et al., 1967). Consider the polynomial equation $P(x) = 0$ where $a_n > 0$ and all a_i are real:

$$P(x) = a_n x^n + a_{n-1} x^{n-1} + \ldots + a_1 x + a_0 = 0.$$

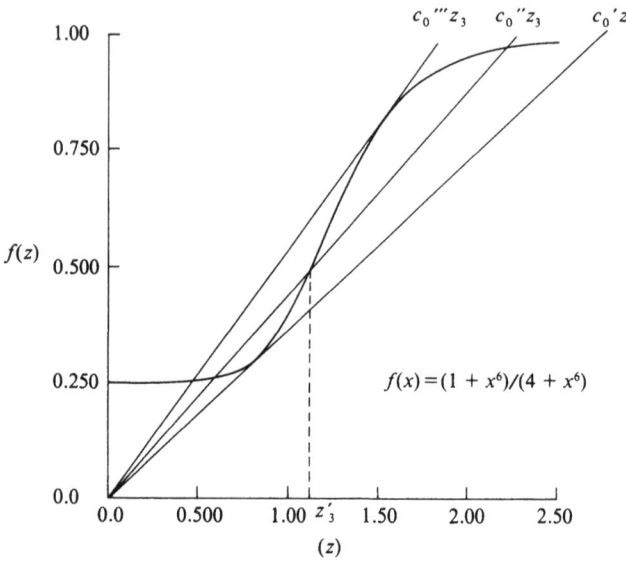

Figure 3.8.2. Multiple solutions of the steady state equation can appear for intermediate values of c_0. For large values, c_0''', and small values, c_0', only one solution is possible. In this diagram $f(z)$ is given by $(1 + z^6)/(K + z^6)$.

The number of positive real roots of $P(x) = 0$ *cannot exceed* the number of variations of sign in the coefficients in the polynomial $P(x)$, where the nonappearance of a given coefficient does not constitute a sign change (i.e. if $a_j = 0$ there is no sign change between a_{j+1} and a_j) and where the number of a root is counted according to its multiplicity (for example, in the equation $(x - 1)(x - 2)^3(x - 7)^2 = 0$, the root 1 has multiplicity 1, $x = 2$ has multiplicity 3 and $x = 7$ has multiplicity 2). As an example of the theorem consider

$$x^7 + 5x^3 - 2x^2 + 3 = 0.$$

There are two changes, one in going from the x^3 to the x^2 term and one in going from the x^2 to the x^0 term. So the maximum number of positive real roots is two. Similarly the number of negative real roots cannot exceed the number of changes in sign of the coefficients of $P(-x)$.

> *Example 3.8.3.* Show that there are at most three positive real roots of the steady state equation and that for the particular case $\rho = 1$ there is at most one.
>
> Solution:
> Consider the $\rho = 1$ case first:
>
> $$c_0 z_3^0 = \frac{1 + z_3^0}{K + z_3^0},$$

$(z_3^0)^2 + [K - (1/c_0)]z_3^0 - (1/c_0) = 0.$

There are three cases depending on whether $(K - 1/c_0)$ is positive, negative or zero, but in each case there is only one sign change in the coefficients of the equation so there is at most one positive real root. However, according to Example 3.8.2 there is at least one positive real root. Hence there is exactly one root.

For $\rho \geq 2$ the steady state equation becomes:

$(z_3^0)^{\rho+1} - (1/c_0)(z_3^0)^\rho + Kz_3^0 - (1/c_0) = 0.$

Recalling that both c_0 and K are positive, we conclude there are three sign changes and thus at most three positive roots.

The application of Descartes' Rule specifies an upper bound on the number of positive roots but it does not establish conditions that ensure the existence of more than one root. This condition is determined in Example 3.8.4 (Othmer, 1976).

Example 3.8.4. Consider the steady state equation

$c_0 z_3^0 = f(z_3^0) = \dfrac{1 + (z_3^0)^\rho}{K + (z_3^0)^\rho},$

where $c_0 > 0$ and $K > 1$. Show that if

$$\rho > \frac{K + 1 + 2K^{\frac{1}{2}}}{K - 1}, \tag{3}$$

then there exist positive real numbers c_a and c_b $(c_a < c_b)$ such that if

$c_0 < c_a$ there is one positive real root,
$c_0 = c_a$ these are two positive real roots,
$c_a < c_0 < c_b$ there are three positive real roots,
$c_0 = c_b$ there are two positive real roots,
$c_b < c_0$ there is one positive real root.

Solution:
From Figure 3.8.2 it is clear that for very large and very small c_0 there is only one positive root. If there is to be more than one solution, there must be an intersection of $f(z_3)$ and $c_0 z_3$ like that at z_3'. At that point, the slope of f, namely $f'(z_3')$, is greater than the slope of the $c_0 z_3$ line which is $\Delta y/\Delta x = f(z_3')/z_3'$. Therefore we conclude that if there is a positive z_3 such that

$$f'(z_3) > [f(z_3)/z_3], \tag{4}$$

then there can be multiple roots. Thus the problem is to find a condition on ρ that ensures satisfying condition (4). The possibility of more than three positive roots is precluded by Descartes'

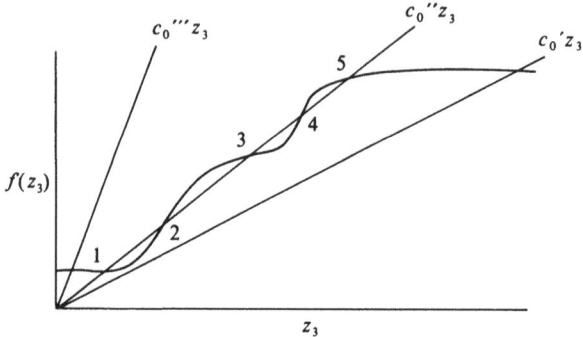

Figure 3.8.3. A hypothetical example of a bounded, monotone increasing, nonlinearity having more than three solutions. The fourth and fifth solutions are possible because there are two distinct intervals of z_3 where $f'(z_3) > [f(z_3)/z_3]$.

Rule. However, in the process of establishing (3) we will reproduce this result by another argument. Given the requirement that f is increasing, more than three roots could result only from a nonlinearity like that of Figure 3.8.3. So, excepting the peculiar cases where $c_0 z_3$ and $f(z)$ are tangent, the number of roots changes by two. Note from Figure 3.8.3 that for five roots to exist there must be two separate regions where (4) is satisfied. More generally, if there are j distinct regions where this condition is satisfied then the *maximum* number of roots is $2j + 1$. The explicit qualification specified by the word maximum should be stressed since there could be fewer roots as with c_0' and c_0''' in Figure 3.8.3. We will show the following:

1. If (3) is satisfied, there is a region of positive z_3 such that $f'(z_3) > f(z_3)/z_3$ and hence there is an interval of c_0 giving three roots.

2. There is at most only one region of z_3 such that (4) is satisfied, so there are at most three roots (thus confirming the application of Descartes' Rule) and there is only one interval of c_0 giving three roots, i.e. there are c_a and c_b with the properties specified in the statement of the problem.

We start with:

$$f'(z_3) > [f(z_3)/z_3].$$

This becomes

$$\frac{\rho z_3^{\rho-1}}{K+z_3^\rho} - \frac{(1+z_3^\rho)\rho z_3^{\rho-1}}{(K+z_3^\rho)^2} > \frac{1}{z_3} \frac{1+z_3^\rho}{K+z_3^\rho}. \tag{5}$$

Defining a new variable $y = z_3^\rho$ and the function $F(y)$ as shown, condition (3.8.5) becomes

$$F(y) = y^2 + y(1 + K - \rho K + \rho) + K < 0.$$

$F(0) = K$, which is positive, and the limit of $F(y)$ for large y is positive. Also $F(y)$ is quadratic so there are *at most* two positive values of y giving $F(y) = 0$. Thus if $F(y)$ is ever zero it must look something like the function in Figure 3.8.4.

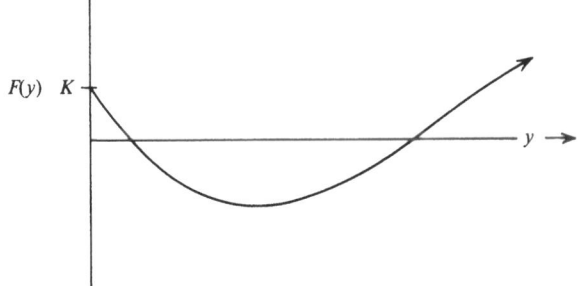

Figure 3.8.4. Since $F(y)$ is a quadratic function, it has at most two intersections with the positive real axis. $F(0)$ and $\lim_{y\to\infty} F(y)$ are positive, so if $F(y)$ is ever negative it is negative only for a single interval of y values.

There is only one region of y such that $F(y) < 0$ which means that there is at most only one region of z_3 such that (4) is satisfied; and so there is only one interval of z_3 such that multiple roots are possible, giving a maximum of three roots. By the preceding construction, multiple roots can exist only if the minimum of $F(y)$ is negative. A necessary, but not sufficient, condition for this to happen is for $F'(0)$ to be negative. Since

$$F'(0) = 1 + K - \rho K + \rho < 0,$$

this condition is

$$\rho > (K + 1)/(K - 1). \tag{6}$$

Condition (6) ensures the existence of a minimum to $F(y)$ at $y > 0$. The value of y giving the minimum of $F(y)$, y_m, satisfies $F'(y) = 0$. Thus

$$y_m = -\tfrac{1}{2}(1 + K - \rho K + \rho).$$

We require $F(y_m) < 0$, i.e.

$$F(y_m) = -\tfrac{1}{4}(1 + K - \rho K + \rho)^2 + K < 0.$$

Upon collecting terms and dividing through by $(K-1)^2 > 0$, this becomes

$$\rho^2 - \frac{2\rho(K+1)}{K-1} + 1 > 0.$$

Thus ρ must be larger than the positive root of this quadratic, which gives

$$\rho > (K + 1 + 2K^{\frac{1}{2}})/(K-1),$$

a condition that includes (6).

The proposition that there is only one interval of c_0 giving multiple roots is demonstrated by showing that there are at most two values of c_0 such that $f(z_3)$ and $c_0 z_3$ are tangent. This is left as an algebraic exercise.

Example 3.8.4 indicates how many positive steady states will be found for a given value of ρ and set of constants k_1, k_2, k_3 and K. A local stability analysis of these steady states is given below.

Local stability properties of steady states in the positive feedback system

The linear variational equations derived in Section 3.7 are unchanged. The only difference is that $f'(z_3^0)$ has changed form. In particular it should be remembered that since $f(z)$ is increasing (positive feedback), $f'(z_3^0)$ is positive. Let z_i' be a small displacement from a steady state. Then

$$dz_1'/d\tau = -k_1 z_1'(\tau) + f'(z_3^0) z_3'(\tau),$$

$$dz_2'/d\tau = z_1'(\tau) - k_2 z_2'(\tau), \tag{7}$$

$$dz_3'/d\tau = z_2'(\tau) - k_3 z_3'(\tau).$$

As before it is assumed that the system is initially at the steady state $z_i'(0) = 0$. Taking Laplace transforms we obtain the same result as before

$$\hat{z}_3'(s) = f'(z_3^0) \left(\frac{1}{(s+k_1)(s+k_2)(s+k_3)} \right) \hat{z}_3'(s)$$

$$= f'(z_3^0) G(s) \hat{z}_3'(s).$$

The control loop corresponding to this differential equation is shown in Figure 3.7.4. The closed loop transfer function is

$$\frac{G(s)}{1 - f'(z_3^0)G(s)}.$$

The stability of the control loop, and hence of the steady state point, is determined by the roots of the characteristic equation (eigenvalue equation)

$$1 - f'(z_3^0)G(s) = 0. \tag{8}$$

The steady state is stable if all numbers satisfying this equation have negative real part. The stability of (7) can also be investigated by first supposing exponential solutions $z'_1 = \alpha_1 \exp(\lambda\tau)$, $z'_2 = \alpha_2 \exp(\lambda\tau)$ and $z'_3 = \alpha_3 \exp(\lambda\tau)$. Substituting these into (7) we find

$$\lambda\alpha_1 \exp(\lambda\tau) = -k_1\alpha_1 \exp(\lambda\tau) + f'(z_3^0)\alpha_3 \exp(\lambda\tau),$$

$$\lambda\alpha_2 \exp(\lambda\tau) = \alpha_1 \exp(\lambda\tau) - k_2\alpha_2 \exp(\lambda\tau),$$

$$\lambda\alpha_3 \exp(\lambda\tau) = \alpha_2 \exp(\lambda\tau) - k_3\alpha_3 \exp(\lambda\tau).$$

Dividing by $\exp(\lambda\tau)$ and rearranging, we find that

$$(\lambda + k_1)\alpha_1 = f'(z_3^0)\alpha_3, \quad (\lambda + k_2)\alpha_2 = \alpha_1, \quad (\lambda + k_3)\alpha_3 = \alpha_2.$$

Combining these equations we conclude that λ must satisfy

$$(\lambda + k_1)(\lambda + k_2)(\lambda + k_3) - f'(z_3^0) = 0.$$

Differential equation (7) is stable only if all λ satisfying this equation have negative real parts. It is possible to generalize this result using matrices. The matrix format is introduced here because some stability tests (notably Gerschgorin's theorem) can be more simply expressed in matrix terms.

$$\frac{\mathrm{d}}{\mathrm{d}\tau}\begin{pmatrix} z'_1 \\ z'_2 \\ z'_3 \end{pmatrix} = \begin{pmatrix} -k_1 & 0 & f'(z_3^0) \\ 1 & -k_2 & 0 \\ 0 & 1 & -k_3 \end{pmatrix}\begin{pmatrix} z'_1 \\ z'_2 \\ z'_3 \end{pmatrix} \equiv A\begin{pmatrix} z'_1 \\ z'_2 \\ z'_3 \end{pmatrix},$$

where A is the 3×3 matrix. The stability of the differential equation can be specified in terms of the eigenvalues of matrix A; λ is an eigenvalue of A if it satisfies the equation $|A - \lambda I| = 0$ where the vertical bars denote the determinant and I is the 3×3 identity matrix. The connection between stability expressed in terms of the closed loop transfer function and in terms of matrix A is made explicit in Example 3.8.5.

> *Example* 3.8.5. Show that the characteristic equation of the closed loop transfer function given by (8) is the same as the eigenvalue equation of matrix A.
>
> Solution:
> The characteristic equation of the feedback loop is
>
> $$1 - f'(z_3^0)G(s) = 0, \quad \text{or}$$
>
> $$(s + k_1)(s + k_2)(s + k_3) - f'(z_3^0) = 0.$$

The eigenvalue equation of matrix A is

$$|A - \lambda I| = \begin{vmatrix} -(k_1 + \lambda) & 0 & f'(z_3^0) \\ 1 & -(k_2 + \lambda) & 0 \\ 0 & 1 & -(k_3 + \lambda) \end{vmatrix}$$

$$= -(k_1 + \lambda) \begin{vmatrix} -(k_2 + \lambda) & 0 \\ 1 & -(k_3 + \lambda) \end{vmatrix}$$

$$+ f'(z_3^0) \begin{vmatrix} 1 & -(k_2 + \lambda) \\ 0 & 1 \end{vmatrix}.$$

Multiplying through by -1 we find that

$$(k_1 + \lambda)(k_2 + \lambda)(k_3 + \lambda) - f'(z_3^0) = 0.$$

Thus the two equations are the same.

We begin the local stability analysis by stating the final result.

1. If three positive steady states exist, the intermediate is unstable and the upper and lower steady states are locally stable.

2. If there is only one positive steady state, it is locally stable.

We neglect consideration of steady states established by the tangential intersection of $f(z_3)$ and $c_0 z_3$, since they exist for only very exceptional sets of parameter values. In order to derive the maximum educational benefit from this exercise, we will try to produce the stability from four different techniques:

1. Direct examination of the eigenvalue equation.
2. Gerschgorin's theorem.
3. The Routh–Hurwitz criterion.
4. The Nyquist criterion.

Previous experience suggests that direct examination of the eigenvalue equation usually fails to provide much useful information in return for a substantial amount of algebraic effort. However, as shown in Example 3.8.6, the present problem is an exception to this rule.

> *Example* 3.8.6.   Establish the stability properties of the steady states by direct examination of the eigenvalue equation.

Solution:

We begin with the recapitulation of an essential piece of information extracted from the analysis of Example 3.8.4. If the steady state in question is the intermediate one in a system containing three steady states, $f'(z_3^0)$ is greater than the slope of the $c_0 z_3$ line, i.e.

$$f'(z_3^0) > c_0.$$

If the steady state z_3^0 is unique, or is the upper or lower steady states of a set of three, the opposite is true:

$$f'(z_3^0) < c_0.$$

We can now show that an intermediate steady state is unstable. Let $P(\lambda)$ denote the eigenvalue equation.

$$P(\lambda) = (k_1 + \lambda)(k_2 + \lambda)(k_3 + \lambda) - f'(z_3^0) = 0.$$

By the above, $P(0)$ is negative for intermediate steady state

$$P(0) = c_0 - f'(z_3^0) < 0.$$

Consider the derivative of $P(\lambda)$ with respect to λ.

$$dP/d\lambda = 3\lambda^2 + 2(k_1 + k_2 + k_3)\lambda + (k_1 k_2 + k_1 k_3 + k_2 k_3).$$

Since the k's are positive $dP/d\lambda$ is positive for $\lambda > 0$. Thus $P(\lambda)$ starts at a negative value and increases monotonically to infinity. Therefore it has a unique real positive root. There is an eigenvalue with a positive real part, so the steady state is unstable.

We now consider the case of steady states where $c_0 > f'(z_3^0)$. Suppose that λ' has positive real parts; then since $k_j > 0$

$$\text{Re } (\lambda' + k_j) > k_j,$$

$$\text{Re } (\lambda' + k_1)(\lambda' + k_2)(\lambda' + k_3) > c_0,$$

$$\text{Re } [(\lambda' + k_1)(\lambda' + k_2)(\lambda' + k_3) - f'(z_3^0)] > c_0 - f'(z_3^0),$$

$$\text{Re } P(\lambda') > c_0 - f'(z_3^0) > 0.$$

For every λ' such that Re $\lambda' > 0$, we have shown that the real part of $P(\lambda')$ is strictly positive; thus it cannot be a root of $P(\lambda) = 0$. It should be noted that this analysis immediately generalizes to a system containing an arbitrary number of intermediate reactions.

The other three methods used to examine the local stability of the steady states are often more effective than direct examination of the eigenvalue equation. The analysis continues with a statement of Gerschgorin's theorem.

Gerschgorin's theorem (Minc & Marcus, 1964)

1. Every eigenvalue of matrix A lies on the complex plane in at least one of the circular discs with centers a_{ii} and radii r_i equal to the sum of the magnitudes of the off-diagonal elements of the i-th row, i.e.

$$r_i = \sum_{j \neq i} |a_{ij}|.$$

2. If m of the circular discs of (1) form a connected region which is isolated from the other discs, then there are precisely m eigenvalues of A in this connected region.

3. Alternatively, in calculating radii one could sum off-diagonal elements in columns, obtaining

$$r_i = \sum_{j \neq i} |a_{ji}|.$$

For a 3×3 matrix there would be three discs. A hypothetical example of three possible cases is shown in Figure 3.8.5. In case (a), all of the discs are completely in the left-half plane so all eigenvalues are in the left-half plane. They have negative real part and the system is stable. In the third case one of the discs is contained entirely in the right-half plane so the system is unstable. In case (b), one of the discs intersects the imaginary axis. The stability of the system is undetermined.

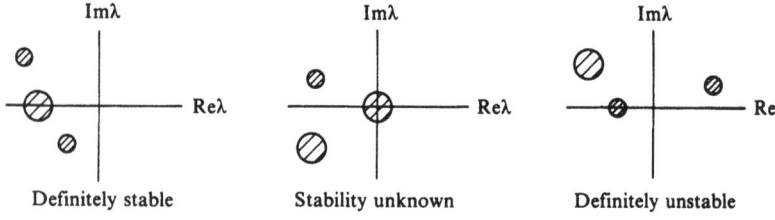

Definitely stable   Stability unknown   Definitely unstable

Figure 3.8.5. Three examples of the application of Gerschgorin's theorem for a three dimensional system. In the first case all of the discs are in the left-half plane, so the system is known to be stable. In the third example one disc is in the right-half plane so there is one eigenvalue with a positive real part and the system is unstable. In the intermediate case the application of Gerschgorin's theorem is indeterminant since one disc is partly in the left-half plane and partly in the right-half plane.

Example 3.8.7.   Apply Gerschgorin's theorem to matrix A.

Solution:

$$A = \begin{pmatrix} -k_1 & 0 & f'(z_3^0) \\ 1 & -k_2 & 0 \\ 0 & 1 & -k_3 \end{pmatrix}.$$

Discs have centers at $-k_1$, $-k_2$ and $-k_3$. These are negative so if the radii r_j are less than k_j, the disc is contained in the left-half plane. If radii are produced by summing along *rows* we find that the following conditions ensure stability:

$$k_1 > |f'(z_3^0)| = f'(z_3^0), \qquad k_2 > 1, \qquad k_3 > 1.$$

In general we do not know if the conditions are met so the test is indeterminate. If radii are produced by summing down *columns* we produce the following stability conditions:

$$k_1 > 1, \qquad k_2 > 1, \qquad k_3 > f'(z_3^0).$$

Again the test is indeterminate. (It should be understood that these are sufficient conditions and not necessary. For example one could easily have $k_2 < 1$ and still have a stable steady state.)

Unfortunately Gerschgorin's theorem is too conservative to resolve the stability question for this system. For other problems it can be more successful; an example has been given by Othmer (1976) who considers the special case of (1) where $g_j = b_j$.

Besides giving a sufficiency condition for stability, the application of Gerschgorin's theorem provides a useful – albeit expected – piece of qualitative information. The conditions $k_1 > f'(z_3^0)$ and $k_3 > f'(z_3^0)$ suggest that increasing f' could destabilize the steady state. This is exactly analogous to the results of Section 3.7 for negative feedback. A complete analysis of local stability for $n = 3$ can be produced by application of the Routh–Hurwitz criterion.

Example 3.8.8. Apply the Routh–Hurwitz criterion to the characteristic equation $P(\lambda)$.

Solution:

$$P(\lambda) = \lambda^3 + (k_1 + k_2 + k_3)\lambda^2$$
$$+ (k_1 k_2 + k_1 k_3 + k_2 k_3)\lambda + k_1 k_2 k_3 - f'(z_3^0) = 0.$$

The characteristic equation is stable if and only if $\Delta_j > 0$ for $j = 1, 2$ and 3.

$$\Delta_1 = a_{n-1} = k_1 + k_2 + k_3 > 0,$$

$$\Delta_2 = \begin{vmatrix} a_{n-1} & a_{n-3} \\ a_n & a_{n-2} \end{vmatrix} = \begin{vmatrix} a_2 & a_0 \\ a_3 & a_1 \end{vmatrix} = a_1 a_2 - a_0 a_3$$

$$= (k_1 + k_2 + k_3)(k_1 k_2 + k_1 k_3 + k_2 k_3) - [c_0 - f'(z_3^0)]$$

$$= k_1(k_1 k_2 + k_1 k_3) + (k_2 + k_3)(k_1 k_2 + k_1 k_3 + k_2 k_3) + f'(z_3^0).$$

Δ_2 is clearly positive since the k's are positive and $f'(z_3^0)$ is positive for the case of positive feedback.

$$\Delta_3 = \begin{vmatrix} a_{n-1} & a_{n-3} & a_{n-5} \\ a_n & a_{n-2} & a_{n-4} \\ 0 & a_{n-1} & a_{n-3} \end{vmatrix} = \begin{vmatrix} a_2 & a_0 & 0 \\ a_3 & a_1 & 0 \\ 0 & a_2 & a_0 \end{vmatrix} = a_0(a_1 a_2 - a_0 a_3)$$

$$= a_0 \Delta_2.$$

Δ_2 has been shown to be positive so sgn (Δ_3) = sgn (a_0). Thus, a steady state is stable if a_0 is positive and unstable if a_0 is negative, where

$$a_0 = c_0 - f'(z_3^0).$$

Recall that $f'(z_3^0) > c_0$ for intermediate steady states in systems with three steady states. So for these steady states, $\Delta_3 < 0$ and the point is unstable. However, for unique steady states and the upper and lower steady states of a triplet, the opposite is true. This gives the desired result.

The fourth investigation of local stability uses the Nyquist criterion.

Example 3.8.9. Apply the Nyquist criterion to the stability analysis of a system with closed loop transfer function $G(s)/[1 - f'(z_3^0)G(s)]$ where

$$G(s) = 1/(s + k_1)(s + k_2)(s + k_3).$$

Solution:
Fortunately the Nyquist locus $G(i\omega)$ is unchanged from Figure 3.3.7. The steady state is unstable if the point $1/f'(z_3^0)$ is encircled by $G(i\omega)$. Since $f'(z_3^0)$ is positive, $1/f'(z_3^0)$ is encircled only if

$$0 < [1/f'(z_3^0)] < G(0).$$

Recall that $G(0) = 1/c_0$. Thus instability occurs only if

$$0 < \frac{1}{f'(z_3^0)} < \frac{1}{c_0},$$

i.e. only if

$$c_0 < f'(z_3^0).$$

So, as before, the intermediate steady state is unstable and all others are locally stable.

The implications of these local results on global behavior are now considered.

Global behavior: hysteresis

Present knowledge of the global behavior of (2) is incomplete. There is no theorem for positive feedback systems analogous to the Hastings–Tyson–Webster theorem for negative feedback systems. Local stability analysis, computer simulation and the results of analytic methods of nonlinear control theory (Othmer, 1976) suggest (but do not rigorously demonstrate) the following summary of global behavior.

1. The system never oscillates.
2. If the steady state is unique all solutions converge on the steady state.
3. If there are three solutions, the upper solution $(z_3)_u$ and the lower solution $(z_3)_l$ are regional attractors. Solutions will approach one of these two solutions depending on the initial conditions. A solution will stay in the region of its attracting steady state unless:

    (a) The system is perturbed over the boundary that divides the attracting regions of the two steady states, or

    (b) the number of steady states changes because of a change in parameter c_0. This is related to switching behavior.

It should be noted that in describing the behavior of solutions attracted by a steady state point we have used the phrase 'in the region of its attracting steady state'. This reflects an important mathematical point. Though the solution may come very close to the point, it will never actually reach it in a finite time. To do so would violate the property of uniqueness of solutions. However, in practical terms one generally expects that the solution will be so close to the steady state after its approach from the initial point that the difference is negligible.

Figure 3.8.6 shows the number of steady states as a function of c_0. Assuming the hypothesized global behavior to be correct, consider the effect of varying c_0. Suppose initially c_0 is less than c_a. The only steady state is $(z_3)_u$ and the solution should be near this point. Now suppose that c_0 is increased so that $c_a < c_0 < c_b$. The point $(z_3)_u$ still exists and is stable so the solution remains near this point. If the parameter c_0 is increased still further, so that c_0 is greater than c_b, $(z_3)_u$ no longer exists; the solution makes a transition to the neighborhood of $(z_3)_l$ and remains near this point for all $c_0 > c_b$. Next we consider what happens if c_0 is decreased. The

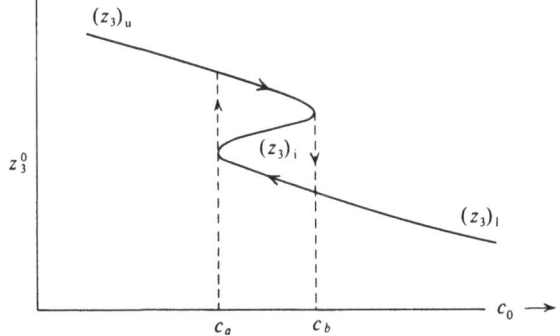

Figure 3.8.6. The number of steady states as a function of c_0, where $(z_3)_u$ denotes the upper equilibrium, $(z_3)_l$ the lower and $(z_3)_i$ the unstable intermediate point.

solution remains near $(z_3)_l$ until $c_0 = c_a$ when a transition is made to $(z_3)_u$. Switching that follows this pattern is said to display **hysteresis**. For a value of c_0 in the range c_a, c_b, the steady state value of z_3 depends on the past history of the system. It shows a primitive form of memory. (Indeed a nonlinear switch with a hysteresis characteristic is referred to in the control literature as a **nonlinearity with memory**. Conversely, a single-valued nonlinearity is frequently called a **memoryless non-linearity**.) Besides providing a chemical mechanism for constructing a trivial memory unit, hysteresis also has the useful property of reducing switch chatter. This can be understood by comparing the response of the switch shown in Figure 3.8.6 and in Figure 3.8.7 to noise. If in the step transition of Figure 3.8.7, c_0 increases through c_a, a step down in the steady state is observed. If c_0 were then to decrease, even slightly, the reverse transition would occur. It is easy to see that in a noisy system with $c_0 = c_a$, frequent on–off step transitions would result. However, consider the hysteresis transition from $(z_3)_u$ to $(z_3)_l$ effected by increasing c_0 through $c_0 = c_b$. If c_0 decreases slightly after the transition, a return to $(z_3)_u$ does not occur and will not occur unless c_0 decreases all the way to c_a. Reverse transitions occur only in response to finite and not infinitesimal changes in c_0.

Examples of hysteresis have been found experimentally. Naparstek, Romette, Kernevez & Thomas (1974) found a pattern in the number of steady states similar to that in Figure 3.8.7 in the hydrolysis of benzoyl ethyl ester by papain and in the oxidation of uric acid by uricase. In the first case, nonlinear behavior is the result of an autocatalytic effect of the product while in the second it is the result of inhibition by an excess of substrate. Hysteresis has also been observed in the reversible denaturation of collagen, lysozyme and ribonuclease. Kozak & Benham (1974) consider cases where two parameters, salt concentration and tempera-

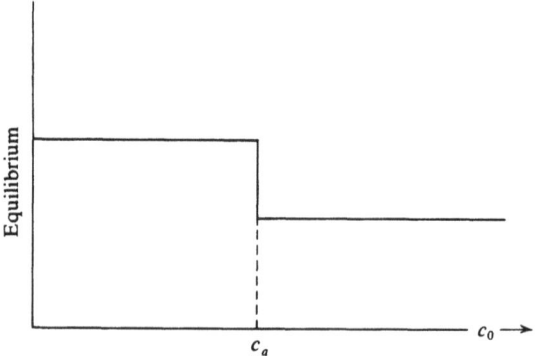

Figure 3.8.7. Example of a transition at c_a without hysteresis. Reverse transitions can occur as the result of possibly infinitesimal changes in c_0.

ture, are varied, and thus deal with a parameter plane and not a parameter axis, as in Figure 3.8.6.

The best-known biological examples of hysteresis are found in the mechanical properties of biological materials such as the pressure–volume curves in the lung (Johnson, 1966), the heart (Rothe, 1966) and veins (Caro, Pedley, Schroter & Seed, 1978) and in the length–tension relation of muscle (Rothe, 1966). Several more complicated models of chemical systems with multiple steady states have been published (Bunow & Colton, 1975; this reference includes a useful bibliography). Pavlidis (1973) has produced a model of circadian rhythms which shows hysteresis not in transitions of steady states but in the change of oscillator frequency that results from a continuous change in light intensity. In Cowan's (1974) model of large-scale neural networks, multiple hysteresis transitions are seen in the variation of activity as a function of external stimulation.

The study of transitions in complex nonlinear systems still requires a great deal of research. However, for the case of a special class of systems, advances have recently been made. These results are outlined in the next section.

3.9 CATASTROPHE THEORY

It is certainly unusual and possibly unprecedented for a piece of pure mathematics to cause the interest in the general press that catastrophe theory has generated. At the center of the controversy is a theorem that even the most outspoken anti-catastrophists describe as 'beautiful' (Zahler & Sussmann, 1977). The purpose of this section is to provide a qualitative description of it and to try to suggest why it has been the cause of such an emotional response.

Conjugacy, catastrophes and catastrophe sets

One of the problems considered in the preceding sections was an investigation of how the qualitative behavior of the solutions of a differential equation can change as a parameter is varied. For example, a small change in a parameter may cause the appearance of periodic solutions (Section 3.7) or it may cause a change in the number of equilibrium points (Section 3.8). A catastrophe is said to occur if there is a major change in the form of the solution. To make this definition more precise we must first define **conjugacy**. Suppose there is a family of differential equations defined over a set of possible parameter values. For example in differential equations modeling metabolic systems the parameters would include reaction constants, temperature and the concentrations of enzyme, if these are constants. Let $x(a, t)$ be the solution corresponding to parameter value a and similarly let $x(b, t)$ be the solution corresponding to parameter value b. The solutions are said to be **conjugate** if there is a continuous function that superimposes one solution on the other and has a continuous inverse that reverses the process. If the solutions are thought of as curves on a graph this means that they can be stretched and twisted until they superimpose, but the rules of this superpositioning game require that the graphs cannot be folded or broken. If the solutions $x(a, t)$ and $x(b, t)$ are conjugate then the transition from parameter value a to parameter value b can be made smoothly without any discontinuous change in solution behavior. If changing the parameter value from a

through another value, say r, causes some abrupt and discontinuous change in the shape of the solutions a **catastrophe** is said to have occurred and parameter value r is an element of the **catastrophe set**. A more formal statement of these definitions would be the following: consider a family of differential equations defined on a parameter space P. Let K denote the catastrophe set. It is a subset of P: $K \subset P$. A parameter value $r \in P$ is an element of the catastrophe set K if it is impossible to find a range of values around r in which all of the corresponding solutions to the differential equation are conjugate. Note that this range of values need only be infinitesimally small. A **catastrophe** is the variation of a parameter through a value in the catastrophe set, K. The solutions of the differential equation before and after the catastrophe are not conjugate, i.e. there has been a nontrivial change in the topology of the solutions. (Technical point: it should be noted that the word catastrophe is also used in another related sense when it is used to refer to a particular type of catastrophe set, e.g. the cusp catastrophe or swallowtail catastrophe (Zeeman, 1977).) The catastrophe set of the Goodwin system with negative feedback (Section 3.7) would be the parameter values corresponding to the appearance of periodic solutions. These are the values of k where

$$W_1(k_1, k_2, k_3) \equiv |f'(z_3^0)||G(i\omega_1)|.$$

Similarly the catastrophe set for the Goodwin system with positive feedback (Section 3.8) consists of the values of k's (and hence of c_0) where the number of solutions to the steady state equation changes, i.e. at the values c_a and c_b of Example 3.8.4. An even simpler example that is important to the development of catastrophe theory is the differential equation

$$\dot{x} = 3x^2 - r. \tag{1}$$

If r is negative there are no real solutions to the equation $\dot{x} = 0$; there are no steady states. If r is equal to zero, there is a single steady state at $x = 0$. If $r = 0$ and the initial value of x is negative, $x(t)$ approaches zero as time approaches infinity. If the initial value of x is positive the solution becomes unbounded as time approaches infinity. If r is positive there are two steady states $x = +\sqrt{(r/3)}$ and $x = -\sqrt{(r/3)}$.

The behavior of the differential equation is established by finding the sign of the derivative $\dot{x}$ on the three intervals of the real axis:

$x < -\sqrt{(r/3)}, \qquad \dot{x} > 0;$

$x = -\sqrt{(r/3)}, \qquad \dot{x} = 0;$

$-\sqrt{(r/3)} < x < +\sqrt{(r/3)}, \qquad \dot{x} < 0;$

$x = +\sqrt{(r/3)}, \qquad \dot{x} = 0;$

$x > +\sqrt{(r/3)}, \qquad \dot{x} > 0.$

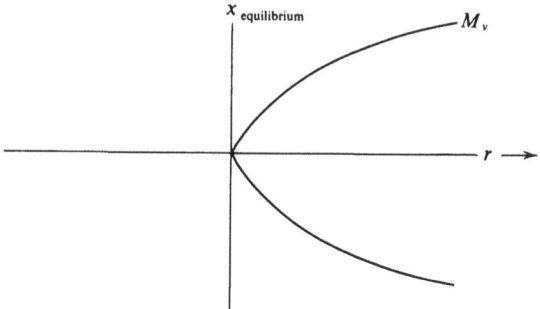

Figure 3.9.1. The steady states of differential equation (1) as a function of parameter r. The curve formed by the equilibrium points is called the equilibrium manifold, M_v.

If the initial value of x, x_0, is equal to one of the equilibrium values then $x(t)$ remains at that point for all time. If x_0 is less than $+\sqrt{(r/3)}$ the solution tends to $-\sqrt{(r/3)}$, and if x_0 is greater than $+\sqrt{(r/3)}$ the solution is unbounded as time approaches infinity. Clearly a catastrophe occurs when the number of steady states changes at $r = 0$. The catastrophe set is $K = \{0\}$. This is called the **fold catastrophe**.

The steady states are plotted as a function of r in Figure 3.9.1. The steady state values form a curve which is labelled M_v. M_v will be referred to as the **equilibrium manifold** (Technical note: in the differential equation literature, the word equilibrium is used interchangeably with steady state and, accordingly, we use the conventional nomenclature in referring to the equilibrium manifold. It should be understood that the word equilibrium is not used here in a thermodynamic sense.) For systems with a single parameter, M_v is a curve that can be drawn on a graph. If there were two parameters, for example a and b, then depicting the equilibrium manifold would require a three-dimensional picture. M_v would be a surface defined above, or possibly below, the (a, b)-plane.

Looking at Figure 3.9.1 we note that the value of r in the catastrophe set is the value at which M_v is vertical. Said another way the catastrophe is the singularity of the projection of M_v onto the r-axis (where here the word **singularity** refers to a discontinuous change since the projection of M_v on to the r-axis begins at $r = 0$). This description of the catastrophe set seems pretentious and unnecessarily complicated for the simple system under consideration, but it is precisely this formulation of the catastrophe set that generalizes to higher dimensional systems.

A more complicated equilibrium manifold is depicted in Figure 3.9.2. The corresponding differential equation is

$$\dot{x} = x^3 + ax + b. \tag{2}$$

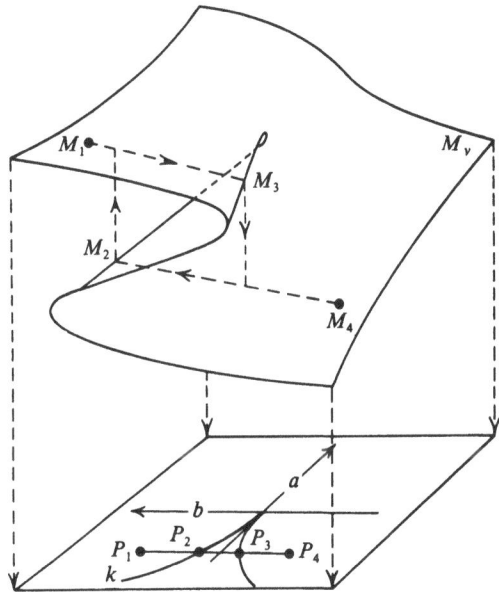

Figure 3.9.2. The equilibrium manifold for differential equation (2) and its projection on the (a, b) plane. Points P_1, P_2, P_3 and P_4 are the projections of M_1, M_2, M_3 and M_4. This is the cusp catastrophe.

M_v is the surface of points defined by $\dot{x} = 0$. For most values of a and b this equation has a unique solution; however, in a manner reminiscent of the positive feedback loop of Section 3.8, some (a, b) pairs admit three real solutions. The equilibrium manifold is shown in Figure 3.9.2.

The catastrophe set K is shown in the projection of M_v on to the (a, b)-plane. It consists of points where the number of solutions changes from one to three. As seen in Figure 3.9.3 this curve has two branches:

$$b = -(-a/3)^{\frac{3}{2}} - a(-a/3)^{\frac{1}{2}},$$
$$b = +(-a/3)^{\frac{3}{2}} + a(-a/3)^{\frac{1}{2}}.$$

A derivation of these equations is given in the next section.

The parameter a is called the **splitting factor**, since it divides the parameter space into two regions: one in which catastrophes are impossible ($a > 0$) and one in which catastrophes are possible ($a < 0$). The parameter b is referred to as the **normal factor**. Catastrophes of this form are referred to as **cusp catastrophes**.

The qualitative dynamics of the cusp catastrophe are analogous to the hysteresis transitions involving a single parameter seen in the case of the positive feedback loop (Section 3.8). If there is only one steady state it is locally stable and it attracts nearby trajectories. If there are three, the

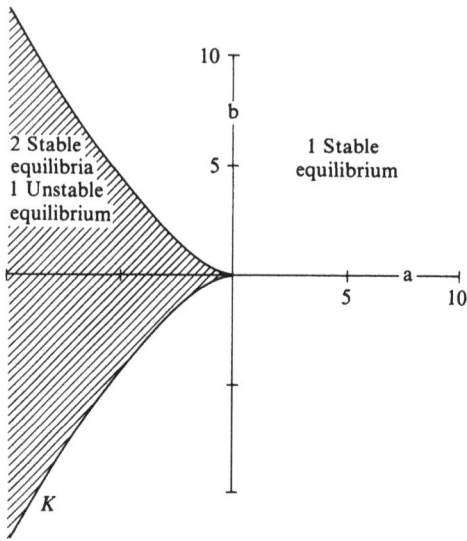

Figure 3.9.3. The catastrophe set for differential equation (3.9.2).
Differential equations corresponding to parameter values (a, b) inside the
shaded region have three steady states. Otherwise the differential equation
has one steady state.

steady states on the upper and lower sheet are locally stable and the
intermediate steady state on the folded sheet of M_v is unstable. Consider
the points M_1, M_2, M_3 and M_4 on the equilibrium manifold and their
projections P_1, P_2, P_3 and P_4 on the (a, b)-plane. Suppose the system is
originally in the vicinity of point M_1 and parameter b increases. The
solution remains on the upper sheet until the point M_3 is reached (when
the corresponding point on the (a, b)-plane is P_3, an element of the
catastrophe set). After passing through M_3 a rapid transition is made to
the lower sheet. As b continues to increase the solution approaches point
M_4. If b is decreased from point P_4 the transition occurs at point P_2. A
hysteresis characteristic like that of Section 3.8 is seen.

We have considered the form of the catastrophe set for two particularly simple cases: the fold catastrophe and the cusp catastrophe.
Catastrophe theory rests on a classification theorem that says that catastrophe sets are made of shapes that are equivalent to a small number of
prototype catastrophes. However, it should be stressed that this
classification is valid only for the very special class of differential equations called *potential systems*. These systems are defined and described in
the next subsection.

Potential systems and the analytic location of catastrophe sets

Consider a general form of differential equation

$$\frac{d\mathbf{x}}{dt} = \mathbf{f}(\mathbf{x}, \mathbf{r}). \tag{3}$$

The independent variable is time and the dependent variable, $\mathbf{x}(t)$, is a vector function of n real components:

$$\mathbf{x}(t) = \begin{pmatrix} x_1(t) \\ x_2(t) \\ \vdots \\ x_n(t) \end{pmatrix}.$$

For example the x's could be chemical concentration functions in a system of n chemical components. We summarize this with $\mathbf{x} \in \mathbb{R}^n$ where $\mathbb{R}^n$ denotes the real variables of n distinct components. In practice $\mathbf{x}(t)$ does not vary throughout all of $\mathbb{R}^n$, it is usually confined to a subset $S \subseteq \mathbb{R}^n$. For example if x's are concentrations they can never be negative and S is the orthant of where all components of the vector are positive, i.e. $S = \mathbb{R}^n_+$.

S is called the **state space**, the **behavior space** or the **phase space**. The vector $\mathbf{r} = (r_1, r_2 \ldots r_m)$ in the differential equation is the parameter vector. In a chemical system it would consist of reaction constants. Vector $\mathbf{r}$ is an element of set P, a subset of $\mathbb{R}^m$; P is called the **control space** or the **parameter space**. The function $\mathbf{f}$ is a function of $\mathbf{x}$ and $\mathbf{r}$ and has a value that has the same dimension as $\mathbf{x}$. Writing out the components of (3) we have

$$\frac{dx_1(t)}{dt} = f_1(x_1, \ldots x_n, r_1, \ldots r_m),$$

$$\frac{dx_2(t)}{dt} = f_2(x_1, \ldots x_n, r_1, \ldots r_m),$$

$$\vdots$$

$$\frac{dx_n(t)}{dt} = f_n(x_1, \ldots x_n, r_1, \ldots r_m).$$

Potential systems are a special case of (3), which is a **potential system** if there is a function $V(x_1 \ldots x_n, r_1 \ldots r_m)$ so that

$$\frac{dx_1(t)}{dt} = \frac{\partial}{\partial x_1} V(\mathbf{x}, \mathbf{r}),$$

$$\frac{dx_2(t)}{dt} = \frac{\partial}{\partial x_2} V(\mathbf{x}, \mathbf{r}),$$

$$\vdots$$

$$\frac{dx_n(t)}{dt} = \frac{\partial}{\partial x_n} V(\mathbf{x}, \mathbf{r}).$$

More concisely, $\dot{x} = \mathbf{f}(\mathbf{x}, \mathbf{r})$ is a potential system if there is a V such that $\mathbf{f} = \mathrm{grad}_x V$. $V(\mathbf{x}, \mathbf{r})$ is called the **potential function** of the differential equation. A potential function for (1) is

$$V(x, r) = x^3 - rx,$$

and a potential function for (2) is

$$V(x, a, b) = (x^4/4) + (ax^2/2) + bx.$$

Two points should be stressed. First, the potential function of a differential equation is not unique since they could always vary by an added constant. For example $V(x, r) = x^3 - rx + 25$ is also a potential function of (1). Secondly, not all differential equations have potential functions. Indeed most do not. (Though a differential equation of a single variable always has a potential function; namely the integral.) Since, strictly speaking, catastrophe theory is only valid for potential systems, the range of applicability of the theory is severely limited. (Technical note: the theorem is valid for gradient systems, a generalization of potential systems. However, for every gradient system there is a conjugate potential system so little is gained by this added complication.)

Potential systems are a particularly simple class of differential equation; for example, they can never oscillate. Also, the only catastrophes that can occur in a potential system are those associated with a change in the number of steady states. If an explicit statement of V is available then finding the catastrophe set is a purely algebraic exercise (though possibly difficult). For systems with only one dependent variable (i.e. $S \subseteq \mathbb{R}^1$), the catastrophe set consists of those parameter values that satisfy

$$dV/dx = 0, \qquad d^2V/dx^2 = 0.$$

Since $\dot{x} = dV/dx$, the first equation is the steady state equation. It defines the equilibrium manifold. The second equation is satisfied when the number of solutions to the equilibrium equation changes.

Example 3.9.1.   Find the catastrophe set for the fold and the cusp catastrophe.

Solution:
For the fold catastrophe $V(x, r) = x^3 - rx$, so the catastrophe set

consists of those values of r satisfying

$$dV/dx = 3x^2 - r, \qquad d^2V/dx^2 = 6x.$$

These equations have the unique solution $r = 0$ found before. For the cusp catastrophe the potential function is

$$V(x, a, b) = \tfrac{1}{4}x^4 + \tfrac{1}{2}ax^2 + bx.$$

The catastrophe set satisfies

$$dV/dx = x^3 + ax + b = 0, \qquad d^2V/dx^2 = 3x^2 + a = 0.$$

The second equation has the solution $x = \pm(-a/3)^{\frac{1}{2}}$ so the catastrophe set has two branches depending on the sign taken. Substituting these values of x into the first equation and solving for b gives

$$b = +\left(\frac{-a}{3}\right)^{\frac{3}{2}} + a\left(\frac{-a}{3}\right)^{\frac{1}{2}}, \qquad b = -\left(\frac{-a}{3}\right)^{\frac{3}{2}} - a\left(\frac{-a}{3}\right)^{\frac{1}{2}},$$

as claimed in the previous section.

For cases with more than one dependent variable, the catastrophe set condition is more complicated. For the general case $\mathbf{x} \in \mathbb{R}^n$, $\mathbf{r} \in \mathbb{R}^m$ the catastrophe set is determined by the simultaneous equations

$$\mathrm{grad}_x\, V(\mathbf{x}, \mathbf{r}) = 0, \qquad \mathrm{Hess}_x\, V(\mathbf{x}, \mathbf{r}) = 0,$$

where $\mathrm{Hess}_x\, V(\mathbf{x}, \mathbf{r})$ denotes the Hessian of the potential function

$$\mathrm{Hess}_x\, V(\mathbf{x}, \mathbf{r}) \equiv \begin{vmatrix} \dfrac{\partial^2 V}{\partial x_1^2} & \dfrac{\partial^2 V}{\partial x_1\, \partial x_2} & \cdots & \dfrac{\partial^2 V}{\partial x_1\, \partial x_n} \\[2mm] \dfrac{\partial^2 V}{\partial x_2\, \partial x_1} & \dfrac{\partial^2 V}{\partial x_2^2} & \cdots & \dfrac{\partial^2 V}{\partial x_2\, \partial x_n} \\[2mm] \vdots & \vdots & & \vdots \\[2mm] \dfrac{\partial^2 V}{\partial x_n\, \partial x_1} & \dfrac{\partial^2 V}{\partial x_n\, \partial x_2} & \cdots & \dfrac{\partial^2 V}{\partial x_n^2} \end{vmatrix}.$$

The proof is given in Zeeman (1976). A potential system with three variables is given in Example 3.9.2.

> *Example* 3.9.2. (A. I. Mees, unpublished). Find the catastrophe set for the system of differential equations corresponding to the potential function
>
> $$V(x, y, z, a, b, c) = \tfrac{1}{3}x^3 + ax^2 + y^2 + z^2 + bx + cy,$$
>
> where x, y and z are dependent variables and a, b and c are constants.

Solution:

The problem will first be solved by direct examination of the steady state corresponding to the appropriate potential system

$$dx/dt = \partial V/\partial x = x^2 + 2ax + b,$$

$$dy/dt = \partial V/\partial y = 2y + c,$$

$$dz/dt = \partial V/\partial z = 2z.$$

Since this is a potential system, the only catastrophes are those resulting from a change in the number of steady states. At a steady state x, y and z are

$$x_0 = -a \pm (a^2 - b)^{\frac{1}{2}}, \qquad y_0 = -c/2, \qquad z_0 = 0.$$

The number of roots changes when $(a^2 - b)$ changes sign, i.e. when $b = a^2$. The catastrophe set can also be found by solving the simultaneous equations

$$\mathrm{grad}_x\, V = 0, \qquad \mathrm{Hess}_x\, V = 0.$$

We find that

$$
\mathrm{Hess}_x\, V =
\begin{vmatrix}
\dfrac{\partial^2 V}{\partial x^2} & \dfrac{\partial^2 V}{\partial x\,\partial y} & \dfrac{\partial^2 V}{\partial x\,\partial z} \\[2ex]
\dfrac{\partial^2 V}{\partial x\,\partial y} & \dfrac{\partial^2 V}{\partial y^2} & \dfrac{\partial^2 V}{\partial y\,\partial z} \\[2ex]
\dfrac{\partial^2 V}{\partial z\,\partial x} & \dfrac{\partial^2 V}{\partial z\,\partial y} & \dfrac{\partial^2 V}{\partial z^2}
\end{vmatrix}
=
\begin{vmatrix}
2x + 2a & 0 & 0 \\
0 & 2 & 0 \\
0 & 0 & 2
\end{vmatrix}.
$$

$\mathrm{Hess}_x\, V = 4(2x + 2a) = 0, \quad if\; x = -a.$

Substituting $x = -a$ into the first steady state equation $\partial V/\partial x = 0$ we find

$$a^2 - 2a^2 + b = 0, \qquad b = a^2,$$

as before.

Thom's theorem and the classification of catastrophes

We precede a statement of the theorem with a recapitulation of the definition of the differential equation. Consider the equation

$$\dot{x} = \mathrm{grad}_x\, V(\mathbf{x}, \mathbf{r}),$$

where $\mathbf{x} \in S \subseteq \mathbb{R}^n$ is a state space of dimension n and where $\mathbf{r} \in C \subseteq \mathbb{R}^m$ is a control space of dimension m. M_v, the equilibrium manifold, is the set of steady states as a function of parameter values. It is defined by the simultaneous equations $\dot{x} = 0$.

K is the catastrophe set, containing the parameter points that correspond to a change in the number of steady states. It consists of the singularities of the projection of M_v onto the control space and is defined by the simultaneous equations $\dot{x} = \text{grad}_x \, V = 0$ and $\text{Hess}_x \, V = 0$. Over most of its content, K is 'smooth'; it is composed of unbroken curves or surfaces. At some points K loses 'smoothness'; for example, at the single point of the fold catastrophe or at the point where the two branches of K meet in the cusp catastrophe. Regions where K loses smoothness are called 'singularities' of K. A singularity is said to be equivalent to another if there is an infinitely differentiable function that can map one singularity to another (and back again by its inverse) continuously. Thom's theorem catalogs the possible singularities of K.

Theorem (Thom, 1972; Trotman & Zeeman, 1976). If $m \leqslant 5$ (where m is the dimension of the control space, the number of distinct parameters), any singularity of catastrophe set K is equivalent to one of a finite number of types of singularities called **elementary catastrophes**. If $m > 5$ the classification becomes infinite.

The number of catastrophe types varies with m. A list of the names of the catastrophes according to the dimension of the control space is given in Table 3.9.1. For example if the potential function contains three parameters then singularities of the catastrophe set will be equivalent to one of five different elementary catastrophes (fold, cusp, swallowtail, hyperbolic umbilic and elliptic umbilic).

Table 3.9.1. *Identification of elementary catastrophes*

$m = \dim C$†	1	2	3	4	5
Number elementary catastrophes	1	2	5	7	11
Names of the catastrophes	Fold	Fold Cusp	Fold Cusp Swallowtail Hyperbolic umbilic Elliptic umbilic	Fold Cusp Swallowtail Butterfly Hyperbolic umbilic Elliptic umbilic Parabolic umbilic	

† dim C denotes the dimension of the control space C.
* 7 previous catastrophes plus 4 others.

There are model potential functions that give catastrophe sets for each of the elementary catastrophes. These are listed in Table 3.9.2. It should be recalled that the potential functions are not unique. Different potential functions have equivalent catastrophe sets. Two of these potential functions have already been seen. The fold catastrophe set consists of a single point and can appear in a system with a one dimensional parameter space. The cusp catastrophe can only appear if there are at least two parameters.

(Technical note: Table 3.9.2 indicates that the model potential functions need only have one canonical variable x or two variables x and y. This may seem surprising. To state the case very imprecisely, it can be shown that all of the elementary catastrophes have manifolds M_v of dimension k that can be embedded in $\mathbb{R}^{k+1}$ or $\mathbb{R}^{k+2}$ and thus at most only two canonical variables are required.)

Table 3.9.2. *Model potential functions for elementary catastrophes*

		dim S	dim C	Model potential function V
Cuspoids	Fold	1	1	$x^3 - rx$
	Cusp	1	2	$\frac{1}{4}x^4 + \frac{1}{2}ax^2 + bx$
	Swallowtail	1	3	$\frac{1}{5}x^5 + \frac{1}{3}ax^3 + \frac{1}{2}bx^2 + cx$
	Butterfly	1	4	$\frac{1}{6}x^6 + \frac{1}{4}ax^4 + \frac{1}{3}bx^3 + \frac{1}{2}cx^2 + dx$
Umbilics	Hyperbolic	2	3	$x^3 + y^3 + ax + by + cxy$
	Elliptic	2	3	$x^3 - 3xy^2 + ax + by + c(x^2 + y^2)$
	Parabolic	2	4	$x^2y + y^4 + ax + by + cx^2 + dy^2$

dim S is the dimension of the state space. dim C is the dimension of the control space.

Visualization of the higher catastrophes is extremely difficult. The swallowtail catastrophe set is a surface in the three dimensional parameter space. It is shown in Figure 3.9.4. The interested reader is referred to Woodcock & Poston's (1974) computer graphics study of the geometry of more complicated catastrophes.

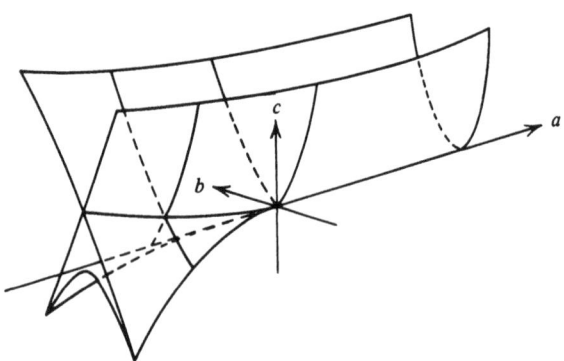

Figure 3.9.4. The butterfly catastrophe (from Thom, 1972).

The theorem says that the geometry of the catastrophe set depends on the dimension of the control space and is independent of the dimension of the state space. Thus a very complicated system can have very simple dynamic behavior. This astonishing result has motivated several controversial applications of catastrophe theory. We now consider some virtues and defects of these models.

The applications of catastrophe theory

The publicly displayed emotional response to catastrophe theory is probably unprecedented in mathematics history. Certainly it is unusual for topology to arouse such strong feelings. Goodwin (1973) reviewing Thom's (1972) book wrote: 'Despite certain shortcomings, the book gave me a sense of liberation and enlightenment akin to what I imagine Ptolemaic astronomers may have felt when offered Copernican heliocentric geometry as an alternative to the endless epicyclic calculations required by the geocentric picture.' Similarly Stewart (1977) says: 'Properly understood and exploited this ever expanding web of concepts promises mankind a unique weapon against ignorance and a profound insight into the universe.' However, contrary views have been expressed, notably by Zahler & Sussman (1977):

'Our conclusion is that the claims made for the theory are greatly exaggerated and that its accomplishments at least in the biological and social sciences are insignificant. This is because catastrophe theorists have misused the basic mathematics in ways that lead to incorrect reasoning; they have offered models which are based on unreasonable assumptions and which lead to erroneous conclusions; and they have made predictions which are either vacuous, tautologous, vague or impossible to test experimentally.'

It would be unimaginative to suggest that the truth lies between these extreme views, but nevertheless we wish to suggest that this is indeed the case. The excitement is the result of Thom's demonstration that the singularities of the catastrophe sets of potential systems with a sufficiently small number of parameters are locally equivalent to a small number of model types, and that this number is independent of the dimension of the behavior space. This unexpected result suggested to some that even very complicated systems, for example a developing organism, could be described by a small number of geometrically expressed models.

This point of view fails for several reasons. (1). The theorem is strictly valid for potential systems and there is no reason *a priori* to suppose that complicated biological and social systems should have potential functions. In fact this seems unlikely to be the case. (2). The classification becomes infinite if there are more than five parameters varying simultaneously. Complex systems would have many parameters and it is usually very difficult to construct an experimental environment which allows only two or three to vary. (3). The local nature of the theorem is frequently unappreciated. The equivalence holds only in neighborhoods close to the singularities of the catastrophe set. The theorem does not give any information about the global properties of the catastrophe set or about the global dynamical behavior of the differential equation. Certainly there is no reason to conclude that the global structure of the catastrophe set can be established from a possibly very limited set of data points.

With these observations in mind we would suggest that many of the biological applications of catastrophe theory are fallacious. However, this need not always be the case. The extreme Zahler & Sussman (1977) view that 'the many researchers now being attracted to catastrophe theory stand to gain nothing but disappointment and wasted time,' is probably unduly pessimistic and almost certainly premature. It would be quite unprecedented if a theorem of the importance of Thom's did not eventually lend itself to important applications. Certainly the interest in catastrophe theory has already had the valuable effect of redressing the overemphasis on nongeometric forms of analysis and reintroducing mathematicians, pure and applied, to the importance of geometry. One of Zahler & Sussman's criticisms of Zeeman's biological papers is that the predictions are independent of catastrophe theory. However, if catastrophe theory encourages people to think in terms of simple geometrical analogs which, independently of Thom's theorem itself, lead to useful insights, then in our view the 'theory', if not its 'mathematics', has been a success.

3.10 CONCLUSIONS

Metabolic pathways are self-regulating. For this reason ideas of control theory can provide biologists with a conceptual structure that leads to valuable insights in attempts to understand both the normal functions, and abnormal behavior, of metabolic systems. At the beginning of the discussion of control theory in this chapter, we examined the distinction between numerical simulation and mathematical analysis. In order to apply these techniques, it is necessary to reduce large-scale processes to more tractable subsystems. The decomposition into archetypal control systems and the separation of time scales provides part of the solution.

The presentation of mathematical control theory began with an introduction to linear systems theory. As well as being an essential prerequisite to an understanding of nonlinear methods, linear methods are surprisingly useful in their own right. Laplace transform methods have the virtue of reducing a calculus problem to an algebraic manipulation and the Nyquist criterion used in stability tests of idealized linear models is also useful in the examination of more realistic nonlinear systems. Several of the ideas introduced by linear systems have important generalizations. For example the canonical control system has several features found in biochemical examples. Also the definition of stability and the relation between tightness of control and stability are of general importance.

Nonlinear systems display two forms of behavior that are important in biological contexts: switch-like behavior between steady states and self-sustained oscillations. Switching, in some cases with hysteresis, has long been observed in biochemical and biological systems and is important in understanding development and certain pathologies of metabolic regulation. Oscillations can appear in both negative and positive feedback loops. Biochemical oscillators have taken central roles in models of the control of metabolism (Goodwin, 1963, 1965), the control of development (Turing, 1952; Goodwin & Cohen, 1969; Cooke & Goodwin, 1972), the self-organization of prebiotic systems (Eigen, 1971;

Prigogine & Nicolis, 1971) and as the basis of biological rhythms (Pavlidis & Kauzmann, 1969; Vanden Driessche, 1973).

The Goodwin system has been an important example in the development of mathematical models of negative feedback biochemical oscillators. It appears to be complicated enough to mimic important biological behavior but simple enough to be analyzed. The analysis in Section 3.7 demonstrated several important techniques: (*a*) the transformation to nondimensional coordinates, (*b*) the construction of a bounded invariant set, (*c*) the local stability analysis of the equilibrium, and (*d*) an application of a general theorem showing that local instability can lead to oscillations. Two important factors contribute to the destabilization of the equilibrium, the tightness of control (as expressed by ρ the stoichiometric coefficient of the feedback metabolite) and the length of the control loop (which depends on the number of intermediate reactions and on the magnitude of intervening time delays). This is an important and general result which is useful in understanding oscillatory behavior in a much larger class of control system.

Analysis of the positive feedback loop of Section 3.8 follows the same pattern as the investigation of the negative feedback system. The major departure here is the possibility of multiple equilibria and the possibility of transitions between these steady states. The question of changes in the number of equilibria is the central concern of catastrophe theory. This theory provides an elegant classification of the structure of transitions for potential systems with no more than five independent parameters. However, the usefulness of the theory in applied problems is still very much a matter of dispute.

The mathematical study of metabolic regulation began in the 1950s using the concepts of control theory and simulation by computers. Research of this period had the useful result of providing biochemists with a set of paradigms that were helpful in thinking about regulation. The second 'generation' of mathematical biology combines analysis and experimental work with computer simulations. This work is now in hand and is of help in understanding what might be called the 'local circuits' of metabolism. Third 'generation' developments await the discovery of a usable theory of hierarchical control systems.

Exercises

1   A circuit takes input $x(t)$ and gives output

$$Y(t) = (dx/dt) + 5x(t) + \int_0^T x(t)\,dt.$$

Show that it is linear. A second circuit gives output $y(t) =$

$x^2(t)$. Show that it is nonlinear (hint: use the law of superposition).

2 Find the Laplace transform of

$$f(t) = \begin{cases} 5t + 6 \exp(-3t) & t \geq 0 \\ 0 & t < 0, \end{cases}$$

and of

$$f(t) = \begin{cases} 1 + 4 \exp(-st) & t \geq 0 \\ 0 & t < 0. \end{cases}$$

3 Let $\hat{y}(s)$ be the Laplace transform of $y(t)$. Show that

$$L(d^2y/dt^2) = s^2\hat{y}(s) - sy(0) - (dy/dt)|_0,$$

$$L(d^3y/dt^3) = s^3\hat{y}(s) - s^2y(0) - s\,dy/dt|_0 - (d^2y/dt^2)|_0.$$

4 Let output $y(t)$ be related to input $x(t)$ by the differential equation

$$(d^2y/dt^2) + 4(dy/dt) + 3y(t) = [dx(t)/dt] + 2x(t).$$

Find the Laplace transform of the output as a function of the Laplace transform of the input and the initial data.

5 Find the partial fraction expansions of the following:

(a) $\dfrac{s^2+2}{s^3 - 9s^2 + 26s - 24}$

(b) $\dfrac{s+3}{s^3 - s}.$

(c) $\dfrac{1}{s^3 - 2s^2 + s}.$

6 The input $x(t)$ and the output $y(t)$ of a linear system are related by the differential equation

$$(d^2y/dt^2) + 5(dy/dt) + 6y(t) = (dx(t)/dt) - 3x(t).$$

(a) Find the Laplace transform of the output as a function of the transform of the input.

(b) Suppose $x(t) = \begin{cases} 2t - 3 & t \geq 0 \\ 0 & t < 0. \end{cases}$

Find $\hat{x}(s)$, the Laplace transform.

(c) Suppose that y and its derivatives are zero at $t = 0$, and the input (b) is applied. Find the Laplace transform of the output.

(d)   Taking inverse Laplace transforms find $y(t)$ (hint: determine the partial fraction expansion of the answer to (c)).

7   The output $y(t)$ is related to the input $x(t)$ by the differential equation

$$(d^2y/dt^2) + 5(dy/dt) + 4y = x(t).$$

$y(0)$ and $y'(0)$ are assumed to be nonzero. Let $x(t)$ be

$$x(t) = \begin{cases} 3 & t \geq 0 \\ 0 & t < 0. \end{cases}$$

Find $\hat{y}(s)$ and $y(t)$ and thus show that the contribution to $y(t)$ containing initial value terms decays exponentially. (This is a specific example of a general result for stable linear systems that often justifies neglecting initial value terms.)

8   Using the final value theorem find $\lim_{t \to \infty} z(t)$ when

(a) $\hat{z}(s) = \dfrac{s^2}{7s^4 + 3s + 2}$,

(b) $\hat{z}(s) = \dfrac{s^2}{6s^2 + 9}$.

9   Output $y(t)$ is related to input $x(t)$ by the differential equation

$$\frac{d^3y}{dt^3} - 3\frac{d^2y}{dt^2} + 3\frac{dy}{dt} - y(t) = x(t).$$

Find $\lim_{t \to \infty} y(t)$ when

$$x(t) = \begin{cases} -\exp(-4t) + \exp(-2t) & t \geq 0 \\ 0 & t < 0. \end{cases}$$

10   Find the transfer functions for systems where input $x(t)$ is related to output $y(t)$ by the following differential equations:
(a) $(dy/dt) + y(t) = x(t)$,
(b) $(d^3y/dt^3) + 7(dy/dt) = d^2x/dt^2$,
(c) $(d^4y/dt^4) + 6(d^3y/dt^3) + y(t) = 9(d^2x/dt^2)$.

11   Let $\hat{y}(s) = G(s)\hat{x}(s)$. Find the unit impulse response for the following transfer functions:

$$G(s) = (s^2 + 2)/(s^3 + 9s^2 + 26s - 24),$$

$$G(s) = (s + 3)/(s^3 - s).$$

12 Let z_1 and z_2 be the complex numbers $z_1 = 1 - i$, $z_2 = 3 + i$.
Find $z_1 + z_2$, $z_1 z_2$, $|z_1|$, $|z_2|$, $|z_1 z_2|$.

13 Let z_1 and z_2 be arbitrary complex numbers. Show that
$|z_1||z_2| = |z_1 z_2|$.

14 Consider a simplified canonical feedback loop of the form in
Figure 3.3.2 where $k(s) = 1$ and $y_{ref} = 0$. Let

$$G(s) = \frac{N(s)}{D(s)} = \frac{a_m s^m + \ldots a_1 s + a_0}{b_n s^n + b_{n-1} s^{n-1} + \ldots + b_1 s + b_0}.$$

Find the linear differential equation corresponding to this
system.

15 Find a canonical feedback loop corresponding to the
following differential equations:

(a) $(dy/dt) + y = x(t)$,
(b) $(d^2 y/dt^2) + (4 \, dy/dt) + 3y(t) = (dx/dt) + 2x$,
(c) $(d^4 y/dt^4) + (6d^3 y/dt^3) + y = 9(d^2 x/dt^2)$.

16 For the canonical feedback system of Figure 3.3.2 find the
roots of the closed loop characteristic equation when $k(s) = 1$
and

(a) $G(s) = 1/s$,
(b) $G(s) = 3/(s + 4)$,
(c) $G(s) = 5/(s + 2)(s + 3)$,
(d) $G(s) = 1/s(s^3 - s^2 + s - 1)$.

17 Evaluate the following determinants

(a) $\begin{vmatrix} 2 & 3 \\ 1 & -4 \end{vmatrix}$,

(b) $\begin{vmatrix} -1 & 0 & 3 \\ 2 & 3 & 1 \\ 6 & 0 & 2 \end{vmatrix}$,

(c) $\begin{vmatrix} 2 & 3 & 0 & -1 \\ 4 & 2 & 1 & 1 \\ 5 & 3 & 2 & 1 \\ 0 & 1 & -2 & -3 \end{vmatrix}$.

18 Establish the stability properties of the control loops of
problem 16 by application of the Routh–Hurwitz criterion.

19 Draw the root locus diagram for the system of Figure 3.3.2,
when $K(S) = k$, $G(s) = 5/(s + 3)$, for positive feedback gain k.

20   Draw the Nyquist locus for

    (a) $G(s) = 1/(s+1)$,

    (b) $G(s) = 1/(s+1)(s+2)$,

    (c) $G(s) = 1/(s+1)(s+2)(s+3)$.

21   Use the Nyquist criterion to show that the control loop of problem 19 is stable for any positive k.

22   Using the Nyquist criterion find the maximum gain $K(s) = k$ that gives a stable system when

$$G(s) = 1/(1+s)^3.$$

23   Consider a time delayed negative feedback system $G(s) = \exp(-s\tau)$, $K(s) = k$. What is the maximum k that gives a stable system?

24   Consider the four dimensional Goodwin system with negative feedback

$$dz_1/dt = [1/(1+z_n^\rho)] - k_1 z_1,$$

$$dz_j/dt = z_{j-1} - k_j z_j; \qquad j = 2, 3, 4.$$

Suppose initially

$$z_1(0) < 1/k_1 \equiv B_1,$$

$$z_2(0) < 1/k_1 k_2 \equiv B_2,$$

$$z_3(0) < 1/k_1 k_2 k_3 \equiv B_3,$$

$$z_4(0) < 1/k_1 k_2 k_3 \equiv B_4.$$

Show that $z_j(t) < B_j$ ($j = 1, 2, 3, 4$) for all later time t.

25   Let $(z_1^0, z_2^0, z_3^0, z_4^0)$ denote the positive steady state of Problem 24. Show that this point is unique and that the set consisting of only this point is invariant.

26   Consider the one-dimensional analog of the Goodwin system

$$dz_1/dt = [1/(1+z_1^\rho)] - k_1 z_1.$$

Show that all solutions with non-negative initial values approach the equilibrium state.

27   Consider the two-dimensional analog of the Goodwin equations

$$dz_1/dt = [1/(1+(z_2)^\rho)] - k_1 z_1,$$

$$dz_2/dt = z_1 - k_2 z_2.$$

    (a)   Show there is a unique positive steady state.

    (b)   Derive the linear variational equations about that point.

(c) Express these equations in terms of a closed loop negative feedback system.

(d) Use the Nyquist criterion to show that the steady state is locally stable for all positive constants k_1, k_2 and any positive integer ρ.

(e) Derive the characteristic equation of the closed loop feedback system and use the Routh–Hurwitz criterion to show that the steady state is always locally stable.

28   Consider a two-dimensional system with positive feedback.

$$\frac{dx_1}{dt} = \frac{b_0(1 + \alpha_1 x_2^\rho)}{K + \alpha_1 x_2^\rho} - b_1 x_1,$$

$$dx_2/dt = g_1 x_1 - b_2 x_2.$$

Make the following definitions:

$$(b_0 w) = (g_1 \alpha_1^{1/\rho} b_0)^{\frac{1}{2}}, \qquad \tau = (b_0 w) t, \qquad z_1 = \frac{(b_0 w)}{b_0} x_1,$$

$$z_2 = \frac{(b_0 w)^2}{g_1 b_0} \cdot x_2, \qquad k_1 = \left(\frac{b_1}{b_0 w}\right), \qquad k_2 = \left(\frac{b_2}{b_0 w}\right).$$

Show that

$$\frac{dz_1}{d\tau} = \frac{1 + z_2^\rho}{K + z_2^\rho} - k_1 z_1, \qquad \frac{dz_2}{d\tau} = z_1 - k_2 z_2.$$

29   Use Descartes' Rule to establish the maximum number of positive and negative real roots of the following polynomial equations:

(a) $x^2 + 5x = 0$,

(b) $x^4 - 7x^3 + 7x^2 + 5 = 0$,

(c) $x^6 - 3x^5 - 4x^4 + 2x^2 - x + 15 = 0$.

30   What results does application of Gerschgorin's theorem yield for the following matrices?

(a) $\begin{pmatrix} -3 & 1 \\ \frac{1}{2} & -1 \end{pmatrix}$,   (b) $\begin{pmatrix} -3 & 4 \\ \frac{1}{2} & -1 \end{pmatrix}$,   (c) $\begin{pmatrix} 3 & -1 \\ \frac{1}{2} & -1 \end{pmatrix}$.

31   Find the system of differential equations corresponding to the following potentials:

(a) $V(x_1, x_2, x_3, x_4) = x_1^2 + 3x_2 + 4x_3^3 - \sin x_4$,

(b) $V(x_1, x_2) = \exp(-x_1) + x_2^{\frac{1}{2}}$.

32   Find the catastrophe set of the differential equation defined by the following potentials:

(a) $V(x) = x^3 - ax + b$,

(b) $V(w, x, y, z) = \frac{1}{4}x^4 + \frac{1}{2}ax^2 + bx + \frac{1}{4}y^4 + \frac{1}{2}cy^2 + dy + z^3 + w^4$.

33   Derive the family of equations that defines the catastrophe set for the hyperbolic, elliptic and parabolic umbilics.

References

Atherton, D. P. (1975). *Nonlinear Control Engineering: Describing Function Analysis and Design*, London, Van Nostrand Reinhold Company.

Babloyantz, A. & Nicolis, G. (1972). Chemical instabilities and multiple steady state transitions in Monod–Jacob type models. *J. Theoret. Biol.* **34**, 185–92.

Bateman, H. (1954). *Tables of Integral Transforms*, vol. 1, compiled by the staff of the Bateman Manuscript Project, New York, McGraw-Hill.

Benson, J. A. & Jacklet, J. W. (1977). Circadian rhythm of output from neurones in the eye of *Aplysia*. IV. A model of the clock: differential sensitivity to light and low temperature pulses. *J. Exp. Biol.* **70**, 195–211.

Berg, D. G. & Blomberg, C. (1977). Mass action relations in vivo with application to the *lac* operon. *J. Theoret. Biol.* **67**, 523–33.

Birnbaumer, L. & Rodbell, M. (1969). Adenyl cyclase in fat cells. II. Hormone receptors. *J. Biol. Chem.* **244**, 3477–82.

Blessner, W. B. (1969). *A Systems Approach to Biomedicine*, New York, McGraw-Hill.

Boiteux, A. & Hess, B. (1974). Oscillations in glycolysis. Cellular respiration and communication. In *Faraday Symposium, No. 9 Physical Chemistry of Oscillatory Phenomena*, pp. 202–14.

Bradham, L. S., Holt, D. A. & Sims, M. (1970). The effect of Ca on the adenyl cyclase of calf brain. *Biochim. Biophys. Acta* **201**, 250–60.

Brooker, G. (1973). Change in myocardial adenosine 3′,5′-cyclic monophosphate during the contraction cycle. In *Myocardial Metabolism*, ed. N. S. Dhalla, Baltimore, University Park Press, pp. 207–11.

────── (1975). Implications of cyclic nucleotide oscillations during the myocardial contraction cycle. *Adv. Cyclic Nucleotide Res.* **5**, 435–52.

Brostrom, C. D., Huang, Y.-C., Breckenridge, B. Mc. L. & Wolff, D. J. (1975). Identification of a calcium-binding protein as a calcium-dependent regulator of brain adenylate cyclase. *Proc. Nat. Acad. Sci., USA* **72** 64–8.

Bunow, B. & Colton, C. K. (1975). Substrate inhibition kinetics in assemblages of cells. *Biosystems.* **7**, 160–71.

Caro, C. G., Pedley, T. J., Schroter, R. C. & Seed, W. A. (1978). *The Mechanics of the Circulation*, Oxford, Oxford University Press.

Clynes, M. & Milsum, J. H. (1970). *Biomedical Engineering Systems*, New York, McGraw-Hill.

Cooke, J. & Goodwin, B. C. (1972). Periodic wave propagation and pattern formation: application to problems in development. In *Some Mathematical Questions in Biology*, vol. 2, ed. J. D. Cowan, Providence, R.I., Amer. Math. Soc., pp. 33–60.

Cowan, J. D. (1974). Mathematical models of large-scale nervous activity. In *Some Mathematical Questions in Biology*, vol. 5, ed. J. D. Cowan, Providence, R.I., Amer. Math. Soc., pp. 99–133.

Davie, E. W. & Kirby, E. P. (1973). Molecular mechanisms in blood coagulation. *Curr. Top. Cell. Reg.* **7**, 51–86.

DiStefano, J. J., Stubberud, A. R. & Williams, I. J. (1967). *Theory and Problems of Feedback and Control Systems*, Schaum's Outline Series, New York, McGraw-Hill.

Dixon, M., Massey, V. & Webb, E. C. (1964). Unpublished observations cited in Dixon, M. & Webb, E. C., *Enzymes* (1964), London, Longman, p. 80.

Dixon, M. & Thurlow, W. (1924). Studies on xanthine oxidase. II. The dynamics of the oxidase system. *Biochem. J.* **18**, 976–88.

Dixon, M. & Webb, E. C. (1964). *Enzymes*, London, Longman.

Donachie, W. D. & Masters, M. (1969). Temporal control of gene expression in bacteria. In *The Cell Cycle, Gene–Enzyme Interactions*, ed. G. M. Padilla, G. L. Whitson and I. Cameron, New York, Academic Press, pp. 37–76.

Drummond, G. & Duncan, L. (1970). Adenyl cyclase in cardiac tissue. *J. Biol. Chem.* **245**, 976–83.

Eigen, M. (1971). Molecular self-organization and the early stages of evolution. *Quart. Rev. Biophys.* **4**, 141–212.

Esnouf, M. P. & MacFarlane, R. G. (1968). Enzymology and the blood clotting mechanism. *Adv. Enzymol.* **30**, 255–315.

Gann, D. S. (1973). The control of adrenal secretion of cortisol. In *Engineering Principles in Physiology*, vol. 1, ed. J. H. U. Brown and D. S. Gann, New York, Academic Press, pp. 213–26.

Gilbert, W. & Müller-Hill, B. (1966). Isolation of lac repressor. *Proc. Nat. Acad. Sci., USA* **56**, 1891–8.

Goldbeter, A. & Segel, L. A. (1977). Unified mechanism for relay and oscillation of cyclic AMP in *Dictyostelium discoideum. Proc. Nat. Acad. Sci., USA* **74**, 1543–7. Also see Section 4.1.

Goodwin, B. C. (1963). *Temporal Organization in Cells: A Dynamic Theory of Cellular Control Processes*, New York, Academic Press.

——— (1965). Oscillatory behavior in enzymatic control processes. In *Advances in Enzyme Regulation*, vol. 3, ed. G. Weber, pp. 425–38.

——— (1973). 'Mathematical metaphor in biology: A review of structural stability and morphogenesis' by R. Thom. *Nature, Lond.* **242**, 207–8.

Goodwin, B. C. & Cohen, M. H. (1969). A phase shift model for the spatial and temporal organization of developing systems. *J. Theoret. Biol.* **24**, 49–107.

Griffith, J. (1968). Mathematics of cellular control processes. I. negative feedback to one gene. *J. Theoret. Biol.* **20**, 202–8.

Grigorov, L. N., Polyakova, M. S. & Chernavskii, D. S. (1967). Model investigation of trigger schemes and the differentiation process. *Molekulyarnaya Biologiya* **1**, 410–18.

Haldane, J. B. S. (1930). *Enzymes*, London, Longman.

Hastings, S., Tyson, J. J. & Webster, D. (1975). Existence of periodic solutions for negative feedback cellular control systems. *J. Diff. Eqn* **25**, 39–64.

Heinrich, R., Rapoport, S. M. & Rapoport, T. (1977). Metabolic regulation and mathematical models. *Prog. Biophys. Mol. Biol.* **32**, 1–82.

Hepp, K. D., Edel, R. & Wieland, D. (1970). Hormone action on liver adenyl cyclase activity: The effects of glucagon and fluoride on a particulate preparation from rat and mouse liver. *Eur. J. Biochem.* **17**, 171–7.

Hoffman, K. (1976). The adaptive significance of biological rhythms corresponding to geophysical cycles. In *The Molecular Basis of Circadian Rhythms*, ed. J. W. Hastings and H.-G. Schweiger, Berlin, Dahlem Konferenzen, pp. 63–76.

Horrobin, D. F. (1970). *Principles of Biological Control*, Aylesbury, Medical and Technical Publishing.

Hsu, J. C. & Meyer, A. U. (1968). *Modern Control Principles and Applications*, New York, McGraw-Hill.

Jack, J. J. B., Noble, D. & Tsien, R. W. (1975). *Electric Current Flow in Excitable Cells*, Oxford, Clarendon Press.

Johnson, P. C. (1966). The dynamics of respiratory structures. In *Physiology*, 2nd edn, ed. E. E. Selkurt, Boston, Little, Brown and Co., pp. 427–43.

Johnson, R. A. & Sutherland, E. W. (1973). Detergent dispersed adenylate cyclase from rat brain. *J. Biol. Chem.* **248**, 5114–21.

Kacser, H. & Burns, J. A. (1973). The control of flux. In *Symposia of the Society for Experimental Biology, No. 27 Rate Control of Biological Processes*, London, Cambridge University Press, pp. 65–104.

Kakiuchi, S., Yamazaki, R., Teshima, Y. & Uenishi, K. (1973). Regulation of nucleoside cyclic $3',5'$-monophosphate phosphodiesterase activity from rat brain by a modulator and Ca. *Proc. Nat. Acad. Sci., USA* **70**, 3526–30.

King-Smith, E. A. & Morley, A. (1970). Computer simulation of granulopoiesis: normal and impaired granulopoiesis. *Blood* **36**, 254–62.

Knorre, W. A. (1968). Oscillations of the rate of synthesis of beta-galactosidase in *Eschericha coli* ML 30 and ML 308. *Biochem. Biophys. Res. Comm.* **31**, 812–17.

—— (1973). Oscillations in the epigenetic system: biophysical model of the beta-galactosidase control system. In *Biological and Biochemical Oscillations*, ed. B. Chance, E. K. Pye, A. K. Ghosh and B. Hess, New York, Academic Press, pp. 449–55.

Kozak, J. J. & Benham, C. J. (1974). Denaturation: an example of a catastrophe. *Proc. Nat. Acad. Sci., USA* **71**, 1977–81.

Lin, C. C. & Segel, L. A. (1974). *Mathematics Applied to Deterministic Problems in the Natural Sciences*, New York, Macmillan.

Lin, S.-Y. & Riggs, A. D. (1975). The general affinity of *lac* repressor for E. coli DNA: implications for gene regulation in procaryotes and eucaryotes. *Cell* **4**, 107–11.

MacDonald, N. (1978). *Time Lags in Biological Models*, Lecture Notes in Biomathematics, vol. 27, Berlin, Springer-Verlag.

MacFarlane, A. G. J. (1970). *Dynamical Systems Models*, London, Harrap.

Mandelstam, J. (1968). Coordination: induction, repression and feedback inhibition. In *Biochemistry of Bacterial Growth*, 1st edn, ed. J. Mandelstam and K. McQuillen, Oxford, Blackwell Scientific Publications.

Maniatis, I. & Ptashne, M. (1973). Structure of the lambda operators. *Nature, Lond.* **246**, 133–6.

Marcus, R. & Aurbach, G. D. (1971). Adenyl cyclase from renal cortex. *Biochim. Biophys. Acta* **242**, 410–21.

Massey, V. (1953*a*). Studies on fumarase. 2. The effects of inorganic anions on fumarase activity. *Biochem. J.* **53**, 67–71.

—— (1953*b*). Studies on fumarase. 4. The effects of inhibitors on fumarase activity. *Biochem. J.* **55**, 172–7.

Masters, M. & Donachie, W. D. (1966). Repression and the control of cyclic enzyme synthesis in *Bacillus subtilis. Nature, Lond.* **209**, 476–9.

Maynard Smith, J. (1968). *Mathematical Ideas in Biology*, London, Cambridge University Press.

Mees, A. I. & Rapp, P. E. (1978). Periodic metabolic systems. Oscillations in multiple-loop negative feedback biochemical control networks. *J. Math. Biol.* **5**, 99–114.

Mesarovic', M. D., Macko, D. & Takahara, Y. (1970). *Theory of Hierarchical Multilevel Systems*, New York, Academic Press.

Minc, H. & Marcus, M. (1964). *A Survey of Matrix Theory and Matrix Inequalities*, Boston, Prindle, Weber and Schmidt.

Milsum, J. H. (1966). *Biological Control System Analysis*. New York, McGraw-Hill.

Monod, J., Changeux, J.-P. & Jacob, F. (1963). Allosteric proteins and cellular control systems. *J. Mol. Biol.* **6**, 306–29.

Morad, M. & Goldman, Y. (1973). Excitation-contraction coupling in heart muscle: membrane control of development of tension. *Prog. Biophys. Mol. Biol.* **27**, 257–313.

Müller-Eberhard, H. J. (1975). Complement. *Ann. Rev. Biochem.* **44**, 697–724.

Murray, D. R. P. (1930). Inhibition of esterases by excess substrate. *Biochem. J.* **24**, 1890–6.

Naparstek, A., Romette, J. L., Kernevez, J. P. & Thomas, D. (1974). Memory in enzyme membranes. *Nature, Lond.* **249**, 490–1.

Othmer, H. G. (1976). The qualitative dynamics of a class of biochemical control circuits. *J. Math. Biol.* **3**, 53–78.

Pavlidis, T. (1973). *Biological Oscillators: Their Mathematical Analysis*, New York, Academic Press.

Pavlidis, T. & Kauzmann, W. (1969). Toward a quantitative biochemical model for circadian oscillators. *Arch. Biochem. Biophys.* **132**, 338–48.

Plant, R. E. & Kim, M. (1975). On the mechanism underlying bursting in the Aplysta abdominal ganglion R15 cell. *Math. Biosci.* **26**, 357–75.

Prigogine, I. (1967). *Introduction to the Thermodynamics of Irreversible Processes*, 3rd edn, New York, Wiley Interscience.

—— (1969). Structure, dissipation and life. In *Theoretical Physics and Biology*, ed. M. Marois, Amsterdam, North Holland, p. 23.

Prigogine, I. & Nicolis, G. (1971). Biological order, structure and instabilities. *Quart. Rev. Biophys.* **4**, 107–48.

Rapp, P. E. (1974). Discussion note on delays in biochemical control loops. *Faraday Symposium of the Chemical Society, No. 9 The Physical Chemistry of Oscillatory Phenomena*, 215–17.

—— (1975). A theoretical investigation of a large class of biochemical oscillators. *Math. Biosci.* **25**, 165–88.

—— (1976). Analysis of biochemical phase shift oscillators by a harmonic balancing technique. *J. Math. Biol.* **3**, 203–24.

Rapp, P. E. & Berridge, M. J. (1977). Oscillations in calcium–cyclic AMP control loops form the basis of pacemaker activity and other high frequency biological rhythms. *J. Theoret. Biol.* **66**, 497–525.

Reich, J. G. & Sel'kov, E. E. (1974). Mathematical analysis of metabolic networks. *FEBS Lett.* **40**, *Suppl.* S119–S127.

——— (1975). Time hierarchy, equilibrium and nonequilibrium in metabolic systems. *Biosystems* **7**, 39–50.

Rothe, C. F. (1966). Cardiodynamics. In *Physiology*, 2nd edn, ed. E. E. Selkurt, Boston, Little Brown and Co., pp. 311–31.

Savageau, M. A. (1976). *Biochemical System Analysis: A Study of Function and Design in Molecular Biology*, Reading, Mass., Addison Wesley Advanced Book Program.

Simon, Z. (1965). Multi-steady-state model for cell differentiation. *J. Theoret. Biol.* **8**, 258–63.

Stadtman, E. R. (1963). Symposium on multiple forms of enzymes and control mechanisms. II. Enzyme multiplicity and function in the regulation of divergent metabolic pathways. *Bacteriol. Rev.* **27**, 170–81.

Stewart, I. N. (1977). *Encyclopedia Britannica Book of the Year*, New York, Encyclopedia Britannica.

Tada, M., Kirchberger, M. A. & Katz, A. M. (1975). Phosphorylation of a 22,000 dalton component of the cardiac sarcoplasmic reticulum by adenosine $3',5'$-monophosphate dependent protein kinase. *J. Biol. Chem.* **250**, 2640–7.

Talbot, S. A. & Gessner, U. (1973). *Systems Physiology*, New York, Wiley Interscience.

Teo, T. S. & Wang, J. H. (1973). Mechanism of activation of a cyclic adenosine $3',5'$-monophosphate phosphodiesterase from bovine heart by calcium ions. *J. Biol. Chem.* **248**, 5950–5.

Thom, R. (1972). *Stabilité structurelle et morphogénése essai d'uni théorie générale de modéles*, ed. A. S. Wightman, New York, W. A. Benjamin, preface by C. H. Waddington, pp. 363. English Translation (1975) by D. H. Fowler, Reading, Addison-Wesley Advanced Book Program.

Toates, F. M. (1972). A model of an autonomic effector control loop. *Measurement and Control* **5**, 354–7.

——— (1975). *Control Theory in Biology and Experimental Psychology*, London, Hutchinson Education.

Trotman, D. J. A. & Zeeman, E. C. (1976). The classification of elementary catastrophes of codimension less than or equal to five. In *Structural Stability, The Theory of Catastrophes and Applications in the Sciences*, Lecture Notes in Mathematics, vol. 525, Berlin, Springer-Verlag, pp. 263–327.

Turing, A. M. (1952). The chemical basis of morphogenesis. *Phil. Trans. Roy. Soc.* **237**, 37–72.

Tyson, J. J. (1975). On the existence of oscillatory solutions in negative feedback cellular control processes. *J. Math. Biol.* **1**, 311–15.

Tyson, J. J. & Othmer, H. G. (1978). The dynamics of feedback control circuits in biochemical pathways. *Prog. Theoret. Biol.* **5**, 1–62.

Umbarger, J. E. (1956). Evidence for a negative-feedback mechanism in the biosynthesis of isoleucine. *Science* **123**, 848.

van Cauter, E. & Dumont, J. E. (1978). Cross inhibition models for the transmission of hormone signals. *J. Theoret. Biol.* **73**, 657–77.

Vanden Driessche, T. (1973). A population of oscillators: a working hypothesis and its compatibility with the experimental evidence. *Int. J. Chronobiol.* **1**, 253–9.

Waddington, C. H. (1957). *The Strategy of the Gene*, London, Allen and Unwin.

Wang, J. H.-C., Teo, T. S. & Wang, T. H. (1972). Hysteretic substrate activation of bovine c-AMP phosphodiesterase. *Biochem. Biophys. Res. Commun.* **46**, 1306–11.

Wollenberger, A., Babskii, E., Krause, E., Genz, S., Blohm, D. & Bogdanova, E. (1973). Cyclic changes in levels of cyclic AMP and cyclic GMP in frog myocardium during the cardiac cycle. *Biochem. Biophys. Res. Commun.* **55**, 446–52.

Woodcock, A. E. R. & Poston, J. (1974). *A Geometrical Study of the Elementary Catastrophes*, Lecture Notes in Mathematics, vol. 373, Berlin, Springer-Verlag.

Zahler, R. S. & Sussman, H. J. (1977). Claims and accomplishments of applied catastrophe theory. *Nature, Lond.* **269**, 759–63.

Zeeman, E. C. (1976). The umbilic bracelet and the double-cusp catastrophe. In *Structural Stability, The Theory of Catastrophes and Applications in the Sciences*, Lecture Notes in Mathematics, vol. 525, Berlin, Springer-Verlag, pp. 328–66.

——— (1977). *Catastrophe Theory. Selected Papers, 1972–1977*, Reading, Mass., Addison-Wesley Advanced Book Program.

4 Case studies in kinetics

4.1 MODELS FOR OSCILLATIONS AND EXCITABILITY IN BIOCHEMICAL SYSTEMS

Introduction

In response to a change in external conditions, such as substrate input or effector levels, open biochemical systems commonly reach a stationary state. Monotone evolution to a stable time-independent regime is, however, by no means the rule. Even in spatially homogeneous conditions, such diverse phenomena as oscillations, excitability, or transitions between multiple stationary states are possible in metabolic pathways. Space-dependent structures such as chemical waves can be obtained when diffusion is coupled to chemical reaction. Nonequilibrium thermodynamics shows that these phenomena constitute various modes of self-organization in systems close to or beyond instability of a stationary state (Nicolis & Prigogine, 1977). The conditions for the occurrence of such *dissipative structures* (Prigogine, 1969) are (i) a critical distance from equilibrium, (ii) a flow of matter or energy through the system, and (iii) appropriate nonlinear kinetic laws. The three conditions for self-organization in nonequilibrium systems are often satisfied in biology. Oscillatory processes indeed occur at all levels where regulation is exerted. This ranges from periodicities in predator–prey populations (May, 1972) to oscillations in metabolic pathways and in the nervous system (Hess & Boiteux, 1971; Chance, Pye, Ghosh & Hess, 1973; Goldbeter & Caplan, 1976).

Excitability is the capability of a system, initially at a stable steady state, to amplify suprathreshold perturbations in a pulsatory manner. Experiments show that excitable chemical systems can give rise, in slightly different conditions, to sustained autonomous oscillations (Winfree, 1972; De Kepper, 1976). The link between the two phenomena has been studied theoretically in models for the nerve membrane (Fitzhugh, 1961), for an enzyme with autocatalytic pH-dependent kinetics (Hahn, Nitzan, Ortoleva & Ross, 1974), and for the Belousov–Zhabotinsky reaction which is the best-known example of a chemical clock (Tyson, 1977). The purpose of this section is to analyze the generation of excitable and oscillatory behavior in biochemical systems. A great deal of theoretical work has been devoted to oscillatory enzyme reactions (see the recent reviews by Goldbeter & Caplan, 1976, and by Tyson & Othmer, 1979). In the following, we shall restrict ourselves to the analysis of two biochemical systems for which experimental evidence exists for excitable and/or oscillatory behavior. This will allow comparison of the predictions of theoretical models with the available experimental data.

We shall consider in turn a soluble and a membrane-bound enzymatic system. The first is that of glycolytic oscillations. These periodicities, observed in intact yeast cells as well as in yeast and muscle extracts, are the prototype of metabolic oscillations (Hess & Boiteux, 1971; Goldbeter & Caplan, 1976). They have been studied extensively and their mechanism is relatively simple: they result from the positive regulation exerted on phosphofructokinase by a product of the enzyme, ADP, or by AMP which is linked to ADP through the adenylate kinase reaction. The model studied for the phenomenon is that of an allosteric enzyme regulated by positive feedback. The analysis of this model accounts for many experiments on glycolytic oscillations and predicts conditions in which the pulsatory amplification of suprathreshold ADP pulses should be observed in yeast extracts.

In the second part of this section, we shall analyze the cyclic AMP (cAMP) signaling system which controls periodic aggregation in the cellular slime mould *Dictyostelium discoideum*. As demonstrated in cell suspension experiments, this membrane-bound signaling system is capable of amplifying extracellular cAMP pulses and of generating autonomously periodic cAMP signals. Here also the mechanism appears to be based on the positive control exerted on adenylate cyclase by the product of the enzyme, cAMP. The study of the phosphofructokinase and adenylate cyclase models throws light on the closely related conditions in which both excitable and oscillatory behavior arise in regulated bio-chemical systems.

Excitability and oscillations in glycolysis

Experimental observations

Glycolytic oscillations were first discovered in 1964 and have since been extensively reviewed (Pye, 1969; Hess & Boiteux, 1971; Chance *et al.*, 1973; Goldbeter & Caplan, 1976). Here we shall only recall the salient features of these periodicities. The most convenient system in which to study the oscillations is that of cell extracts of yeast or muscle. In yeast extracts, which are most commonly utilized, the period of the phenomenon is of the order of five minutes (Pye, 1969; Hess & Boiteux,

1971); in muscle extracts, the period approaches twenty minutes (Frenkel, 1968; Tornheim & Lowenstein, 1974). When yeast extracts are subjected to a constant input of glycolytic substrate, oscillations are observed for substrate injection rates lying between two critical values (Hess *et al.*, 1969). Similar experiments have been performed in suspensions of intact yeast (Von Klitzing & Betz, 1970). All glycolytic intermediates oscillate with a unique frequency but with various phases; the periodicities can be conveniently observed by recording the fluorescence of NADH.

That the oscillations are produced by phosphofructokinase (PFK) has been demonstrated in numerous ways. An obvious proof is the suppression of periodicity when the PFK step is bypassed in glycolyzing yeast extracts by the injection of fructose-1,6-biphosphate (FDP) which is the product of PFK (Hess & Boiteux, 1968). Moreover, positive as well as negative effectors of PFK are potent inhibitors of glycolytic oscillations (Frenkel, 1968; Hess, Boiteux & Krüger, 1969). The reconstitution *in vitro* of an oscillating system comprising the complete sequence of glycolytic enzymes from hexokinase to alcohol dehydrogenase confirms that the periodicities originate from this metabolic pathway (Hess & Boiteux, 1968). An oscillating system consisting of a shorter sequence of reactions centered around the PFK step has recently been described (Hofmann, 1978).

As shown by phase-shift experiments, it is the couple ATP/ADP which controls the periodic change in PFK activity (Hess & Boiteux, 1968; Pye, 1969). The second substrate, fructose-6-phosphate, does not play any significant regulatory role; FDP also activates the enzyme in muscle extracts (Tornheim & Lowenstein, 1974). Models based on the activation of PFK by ADP or FDP have been proposed by Higgins (1964) and Sel'kov (1968); these models account qualitatively for several features of oscillating glycolysis. A model for the oscillatory PFK reaction explicitly taking into account the allosteric properties of the enzyme was subsequently developed (Goldbeter & Lefever, 1972; Goldbeter & Nicolis, 1976). This model yields insight into the molecular mechanism of glycolytic oscillations and allows prediction of the conditions in which excitable behavior should be observed in glycolyzing yeast or muscle extracts.

Allosteric model for the oscillatory phosphofructokinase reaction

The allosteric model for PFK is based on the concerted transition mechanism proposed by Monod, Wyman & Changeux (1965) for multisubunit enzymes (see Section 1.3). The results obtained in a concerted model could also be obtained with the sequential mechanism proposed by

Koshland, Némethy & Filmer (1966). Besides positive feedback, the existence of cooperative interactions between the enzyme subunits is necessary for a nonequilibrium instability in the PFK reaction. This cooperativity may arise from either a concerted or a sequential transition of the enzyme protomers between multiple conformational states upon binding of the substrate or of the positive effector.

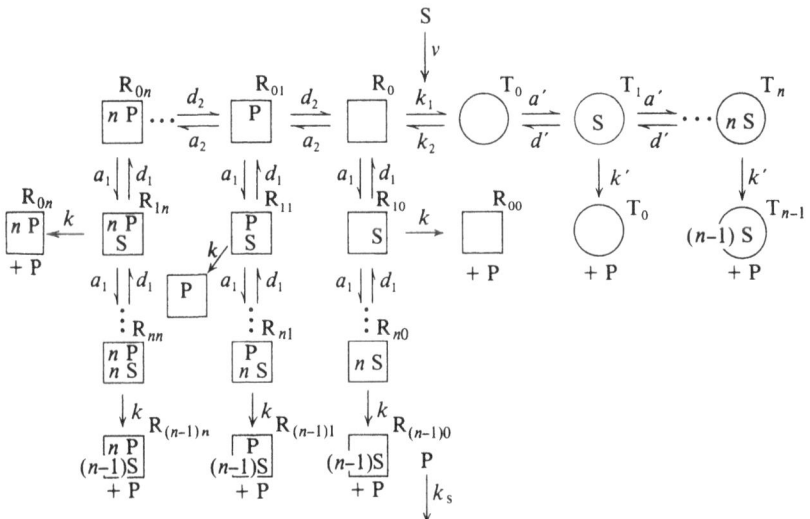

Figure 4.1.1. Concerted allosteric model for the oscillatory phosphofructokinase (PFK) reaction. The enzyme contains n protomers, each of which possesses a catalytic site for the substrate S and a regulatory site for the product P which is a positive effector (see text for details).

The model is represented in Figure 4.1.1 in the general case of an enzyme containing n protomers. Basic assumptions are as follows.

(i) The substrate is injected into the system at a constant rate v.

(ii) Each protomer can exist in two states, R and T, which differ in their affinity for the substrate (K effect) and/or in their catalytic activity (V effect). The protomers undergo a concerted transition between the forms R_0 and T_0 which are free of ligands.

(iii) The product of the reaction binds exclusively to the R state, which has the largest affinity for the substrate and/or the largest catalytic activity.

(iv) The reaction product leaves the system at a rate proportional to its concentration (this amounts to the assumption of a nonsaturated Michaelian sink).

In the following, we shall treat the situation of a system that is homogeneous in space; this situation holds in the experiments carried out in continuously stirred extracts of yeast or muscle. A review of the theoretical results on the occurrence of space-dependent structures in the PFK allosteric model has been given by Goldbeter & Nicolis (1976). In homogeneous conditions, the time evolution of the metabolite and enzyme concentrations is governed by the following differential equations (see also Dalziel, 1968):

$$\mathrm{d}R_0/\mathrm{d}t = -k_1 R_0 + k_2 T_0 - na_2 PR_0 + d_2 R_{01} - na_1 SR_0 + (d_1 + k)R_{10},$$

$$\vdots$$

$$\mathrm{d}R_{0n}/\mathrm{d}t = a_2 PR_{0(n-1)} - n\,\mathrm{d}_2 R_{0n} - na_1 SR_{0n} + (d_1 + k)R_{1n},$$

$$\vdots$$

$$\mathrm{d}R_{nn}/\mathrm{d}t = a_1 SR_{(n-1)n} - n(d_1 + k)R_{nn},$$

$$\mathrm{d}T_0/\mathrm{d}t = k_1 R_0 - k_2 T_0 - na'ST_0 + (d' + k')T_1,$$

$$\vdots$$

$$\mathrm{d}T_n/\mathrm{d}t = a'ST_{n-1} - n(d' + k')T_n,$$

$$\mathrm{d}S/\mathrm{d}t = v - na_1 S\Sigma_0 - (n-1)a_1 S\Sigma_1 - \ldots$$

$$\qquad - a_1 S\Sigma_{n-1} + d_1\Sigma_1 + 2d_1\Sigma_2 + \ldots$$

$$\qquad + nd_1\Sigma_n - na'ST_0 - (n-1)a'ST_1 - \ldots$$

$$\qquad - a'ST_{n-1} + d'T_1 + 2d'T_2 + \ldots$$

$$\qquad + nd'T_n,$$

$$\mathrm{d}P/\mathrm{d}t = -na_2 PR_0 - (n-1)a_2 PR_{01} - \ldots$$

$$\qquad - a_2 PR_{0(n-1)} + d_2 R_{01} + 2d_2 R_{02} + \ldots$$

$$\qquad + nd_2 R_{0n} + k\Sigma_1 + 2k\Sigma_2 + \ldots$$

$$\qquad + nk\Sigma_n + k'T_1 + 2k'T_2 + \ldots + nk'T_n$$

$$\qquad - k_s P, \tag{1}$$

with the conservation relation

$$R_0 + R_{ij} + T_0 + T_i = D_0 \qquad (i, j = 1, \ldots n). \tag{2}$$

We have denoted by S and P the concentrations of substrate and product, and by R_{ij} the concentration of the enzymatic form in the R state carrying i molecules of S and j molecules of P; T_i refers to the form in the T state carrying i molecules of S. Moreover,

$$\Sigma_i = \sum_{j=0}^{n} R_{ij} \qquad (i = 0, \ldots n).$$

The various rate constants are defined in Figure 4.1.1.

When the concentration of the enzyme is smaller than that of the metabolites, the former varies on a faster time scale. A quasi-steady state hypothesis (Section 1.1) can then be made for the enzyme (Reich & Sel'kov, 1974). Let us define the dimensionless concentrations (see Appendix Section A.4)

$$\alpha = S/K_R, \qquad \gamma = P/K_P, \tag{3}$$

with

$$K_R = d_1/a_1, \qquad K_P = d_2/a_2. \tag{4}$$

The algebraic equations $\dot{R}_0 = 0$, $\dot{R}_{ij} = 0$, $\dot{T}_i = 0$ yield the following expressions relating the enzymatic forms to the dimensionless concentrations of substrate (α) and product (γ):

$$\Sigma_0 = R_0(1+\gamma)^n, \quad \Sigma_1 = n\alpha e \Sigma_0, \quad \ldots, \quad \Sigma_n = (\alpha e)^n \Sigma_0.$$

$$T_0 = LR_0, \quad \ldots, \quad T_n = (\alpha ce')^n T_0,$$

with

$$R_0 = D_0/[L(1+\alpha ce')^n + (1+\alpha e)^n(1+\gamma)^n]. \tag{5}$$

Insertion of these relations into the kinetic equations for the metabolites yields the final equations:

$$d\alpha/dt = \sigma_1 - \sigma_M \Phi,$$

$$d\gamma/dt = k_s(\lambda \Phi - \gamma), \tag{6}$$

with $\lambda = (q\sigma_M/k_s)$ and

$$\phi = [\alpha e(1+\alpha e)^{n-1}(1+\gamma)^n + L\theta\alpha ce'(1+\alpha ce')^{n-1}]/[L(1+\alpha ce')^n$$

$$+ (1+\alpha e)^n(1+\gamma)^n]. \tag{7}$$

The parameters σ_1 and σ_M appearing in (6) are the normalized input of substrate and maximum enzyme reaction rate:

$$\sigma_1 = v/K_R, \qquad \sigma_M = nkD_0/K_R = V_M/K_R. \tag{8}$$

Furthermore, $q = K_S/K_P$; $e = (1+\varepsilon)^{-1}$ and $e' = (1+\varepsilon')^{-1}$ where $\varepsilon = k/d$ and $\varepsilon' = k'/d'$;

$$\theta = k'/k; \qquad c = K_R/K_T; \qquad L = k_1/k_2.$$

Parameters c and L are, respectively, the nonexclusive binding coefficient of the substrate and the allosteric constant of the enzyme (Monod *et al.*, 1965). The latter is a *perfect K system* when $\theta = 1$, i.e. when the T and R states have the same catalytic activity and differ only by their affinity for the substrate ($c < 1$). When $c = 1$ and $\theta < 1$, the enzyme represents a *perfect V system*. Both parameters L and c play a prominent role in determining the degree of cooperativity of the enzyme response to changes in metabolite concentrations.

Sustained oscillations and excitability in the PFK reaction

To analyze sustained oscillations and excitability, it is particularly convenient to examine the dynamics of the system governed by (6) in the phase plane (α, γ) determined by the substrate and product concentrations. (Phase plane methods are discussed in Appendix Section A.3.) Similar phase-plane analyses have been conducted in previous studies of excitable systems (Fitzhugh, 1961; Hahn *et al.*, 1974; Tyson, 1977). In the phase plane, two curves are of primary importance: these are the *nullclines* $\sigma_1 = \sigma_M \Phi$ and $\gamma = \lambda \Phi$ corresponding to $(d\alpha/dt) = 0$ and $(d\gamma/dt) = 0$, respectively (see Fig. 4.1.2).

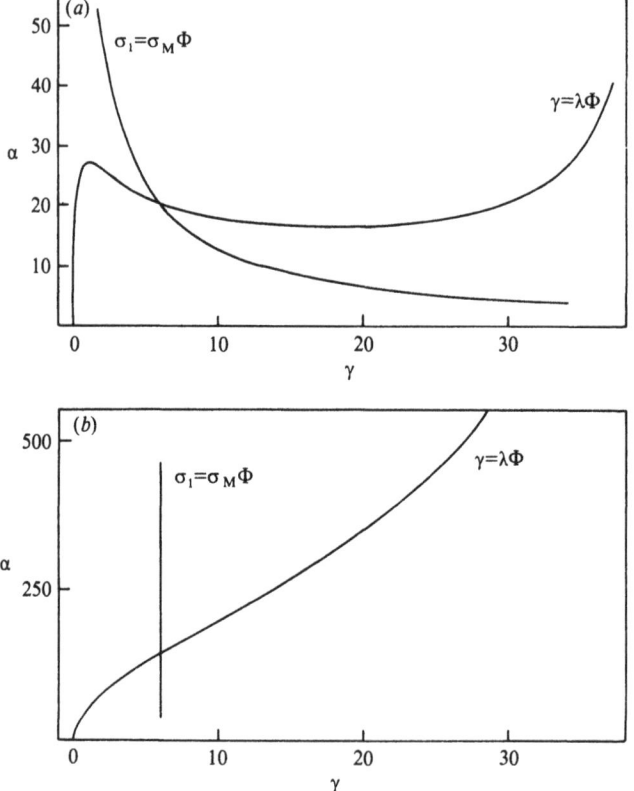

Figure 4.1.2. The two nullclines of system (6) (*a*) with and (*b*) without positive feedback. Parameter values are $n = 2$, $q = \theta = 1$, $k_s = 0.1\,\mathrm{s}^{-1}$, $\sigma_1 = 0.6\,\mathrm{s}^{-1}$, $\sigma_M = 4\,\mathrm{s}^{-1}$, $L = 10^5$, $c = 10^{-5}$, $\varepsilon = \varepsilon' = 0.1$.

A necessary condition for excitable and oscillatory behavior is that the nullcline $\gamma = \lambda \Phi$ should be an S-shaped sigmoid (see below). This happens in a well-defined parameter range, i.e. for large values of L, and above a critical value of λ (T. Erneux & A. Goldbeter, unpublished). Then the curve $\gamma = \lambda \Phi$ passes successively through a maximum and a minimum in α as γ increases (Fig. 4.1.2a). In the absence of regulatory feedback, the nullcline $\gamma = \lambda \Phi$ can never be S-shaped (Fig. 4.1.2b).

The intersection of the two nullclines defines the steady state (α_0, γ_0); for the present model, this steady state is always unique. The stability properties of the steady state depend on its location on the sigmoid nullcline. Stability is always ensured when the steady state lies to the left of the maximum or to the right of the minimum on the sigmoid (Goldbeter & Erneux, 1978). This can be easily seen by means of an analysis in which the time evolution of infinitesimal perturbations around the steady state is determined. Our analysis will be a special case of that discussed in section A.3.

Let us denote the perturbations in α and γ by x and y, respectively, so that $\alpha = \alpha_0 + x$, $\beta = \beta_0 + y$. Substitution of these equations into (6) and linearization of (6) yields

$$dx/dt = -\sigma_M (\partial \Phi / \partial \alpha)_0 x - \sigma_M (\partial \Phi / \partial \gamma)_0 y,$$

$$dy/dt = k_s \lambda (\partial \Phi / \partial \alpha)_0 x + k_s [\lambda (\partial \Phi / \partial \gamma)_0 - 1] y, \qquad (9)$$

where the subscript zero refers to the steady state. The system (9) being linear admits solutions of the form $x = a \exp (\omega t)$ and $y = b \exp (\omega t)$. Substituting back into (9) yields a homogeneous algebraic system of first degree for a and b. For this system to admit nontrivial solutions requires that its determinant be zero. This condition yields the characteristic equation

$$\omega^2 + \omega [\sigma_M (\partial \Phi / \partial \alpha)_0 + k_s - k_s \lambda (\partial \Phi / \partial \gamma)_0] + \sigma_M (\partial \Phi / \partial \alpha)_0 = 0. \quad (10)$$

Determining the stability properties of the steady state then reduces to analyzing the real part of the solutions of (10).

In the absence of substrate inhibition, i.e. when $c \leqslant 1$, the independent term of the characteristic equation is always positive. Then the condition for ω to have a positive real part is

$$\sigma_M (\partial \Phi / \partial \alpha)_0 + k_s [1 - \lambda (\partial \Phi / \partial \gamma)_0] < 0, \qquad (11)$$

or

$$k_s > \sigma_M (\partial \Phi / \partial \alpha)_0 [\lambda (\partial \Phi / \partial \gamma)_0 - 1]^{-1}. \qquad (12)$$

To relate the condition of instability (12) to the position of the steady state on the sigmoid nullcline, we note that derivation with respect to γ of the relation $\gamma = \lambda \Phi$ yields

$$1 = \lambda[(\partial\Phi/\partial\alpha)(d\alpha/d\gamma) + (\partial\Phi/\partial\gamma)]. \tag{13}$$

Hence,

$$d\alpha/d\gamma = -[\lambda(\partial\Phi/\partial\gamma) - 1]/\lambda(\partial\Phi/\partial\alpha). \tag{14}$$

The relation (14) holds on $\gamma = \lambda \Phi$, and in particular on the intersection point of the two nullclines. Again using the subscript zero to denote this point, we insert (14) into (11) to obtain, as a necessary and sufficient condition for instability,

$$(d\alpha/d\gamma)_0 < -(1/q). \tag{15}$$

It is now clear that the steady state is stable on the ascending branches of the S-shaped curve, where $d\alpha/d\gamma$ is positive, and that unstable steady states, if any, will lie on sufficiently steeply sloping portions of the descending branch. The precise domain of instability can be determined by obtaining the explicit form of the partial derivatives $(\partial\Phi/\partial\alpha)_0$ and $(\partial\Phi/\partial\gamma)_0$ and by the subsequent evaluation of condition (12) on a digital computer (Goldbeter & Lefever, 1972; Goldbeter & Nicolis, 1976). As indicated by (15), the boundaries of the unstable domain tend to coincide with the extrema on the sigmoid as the value of parameter q increases.

Sustained oscillations.   When condition (12) is satisfied, the system governed by (6) undergoes sustained oscillations around the unstable stationary state. In the phase plane, these oscillations correspond to a unique closed trajectory, the limit cycle, which is reached regardless of initial conditions (Figure 4.1.3). The existence of a limit cycle can be proved mathematically by the construction in the phase plane of a closed curve enclosing the unstable steady state across the boundary of which solutions always head inward (Erle, Mayer & Plesser, 1979).

Experimental data for the parameters of the PFK reaction are available for the enzyme from *Escherichia coli* (Blangy, Buc & Monod, 1968), and are currently being obtained for yeast. Inserting the values obtained for the bacterial enzyme in (6), one obtains oscillations whose period is of the order of several minutes. This range agrees with that observed for the oscillations in yeast or muscle. The large values of the allosteric constant L (of the order of 10^5–10^6) used in the simulations are taken from the *E. coli* data; for yeast, these values reflect the existence of regulatory sites for the inhibition of PFK by the substrate ATP (Laurent *et al.*, 1978).

One of the main results obtained in the experiments with yeast extracts has been the demonstration that glycolytic oscillations occur in a finite domain of substrate injection rates comprised between 20 and

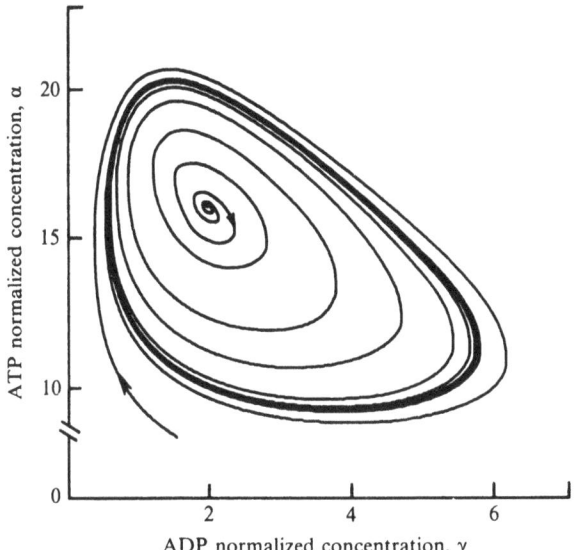

Figure 4.1.3. Limit cycle in the allosteric model for the PFK reaction. The curve is obtained by integration of (6) on an analog computer for $n = 2$, $\sigma_1 = 0.2 \, \text{s}^{-1}$, $\sigma_M = 10^3 \, \text{s}^{-1}$, $k_s = 0.1 \, \text{s}^{-1}$, $q = 1$, $L = 7.5 \cdot 10^6$, $c = 10^{-2}$, $\varepsilon = 0.1$, $\varepsilon' = \theta = 0$ (the enzyme represents a K-V system); α and γ are the concentrations of ATP and ADP divided by the dissociation constant $K_R = 5 \cdot 10^{-2} \, \text{mmole} \, \text{l}^{-1}$. The limit cycle can be reached from the outside, or from the unstable steady state ($\alpha = 16$, $\gamma = 2$). The period of sustained oscillations is 145 s (from Goldbeter & Caplan, 1976).

160 mmole $\text{l}^{-1} \, \text{h}^{-1}$ (Hess *et al.*, 1969). This result is easily explained by discussion of the phase portrait in Figure 4.1.2*a*. As shown above, the domain of sustained oscillations around an unstable steady state lies on the region of negative slope on the sigmoid nullcline $\gamma = \lambda \Phi$. Moreover, the location of the steady state on this curve as a function of the substrate injection rate σ_1 is known, since at the steady state the equations of (6) yield the value $\gamma_0 = q\sigma_1/k_s$ for the product concentration. For low values of σ_1, the steady state lies on the left branch of the sigmoid and is stable. An increase in the substrate injection rate brings the steady state into the region of negative slope; above a critical value of σ_1 the steady state then becomes unstable and limit cycle oscillations develop. Further increase in σ_1 brings the steady state across the instability domain into the right limb of the sigmoid. There thus exists a second, larger critical value of the substrate injection rate above which sustained oscillations disappear and the system evolves toward a stable stationary state. In the model, the oscillatory range of substrate injection rates ranges, typically, from 19 to 246 mmole $\text{l}^{-1} \, \text{h}^{-1}$ (Boiteux, Goldbeter & Hess, 1975).

The theoretical characteristics of the oscillations as a function of the substrate input compare with experimental observations both qualitatively and quantitatively for the variation in period and amplitude (Boiteux *et al.*, 1975; Goldbeter & Nicolis, 1976). The model does not account, however, for the double periodicity which is sometimes observed at low substrate injection rates in cellular extracts (Hess *et al.*, 1969). The activity of PFK in the middle of the oscillatory domain determined in the model undergoes a periodic on–off variation between 1 and 75% of the maximum reaction rate, with a mean activity over a period of close to 17% V_M (Figure 4.1.4); these data match those obtained in yeast extracts (Hess *et al.*, 1969; Boiteux *et al.*, 1975).

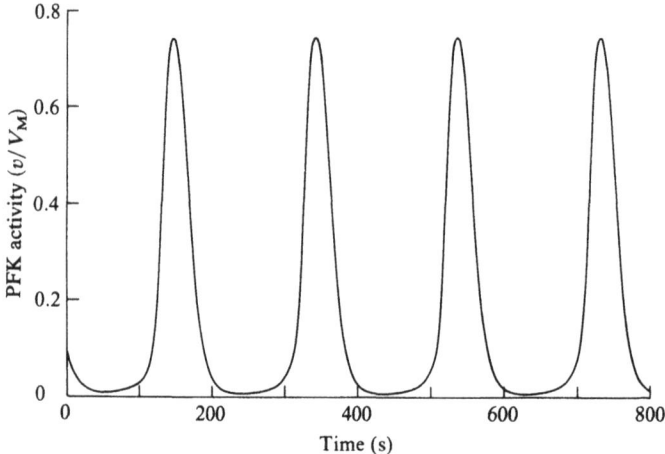

Figure 4.1.4. Periodic variation of PFK activity. The curve is obtained by integration of (6) on a digital computer for $\sigma_1 = 0.7 \text{ s}^{-1}$, $L = 10^6$ (other parameter values as in Figure 4.1.2). The ratio v/V_M is equal to function Φ given by (7) (from Goldbeter & Caplan, 1976).

Phase shift experiments have shown that the addition of adenine nucleotides at certain times over the period causes a phase advance or a phase delay of glycolytic oscillations (Pye, 1969). Theoretical experiments carried out on a digital computer show, in agreement with experimental observations, that the addition of 0.7 mmole 1^{-1} ADP at the minimum of ADP oscillations induces a delay of 1 to 2 min (Goldbeter & Nicolis, 1976). The fact that the addition of ADP has a much smaller influence at other times over the period is easily explicable: the positive effector can appreciably shift the equilibrium between the two enzyme conformations from the T to the R state only when it is added at a minimum of ADP, i.e. when the enzyme is predominantly in the T state and is thus most sensitive to activation by the product.

Entrainment of glycolytic oscillations by a periodic source of substrate has been demonstrated both in experiments with yeast extracts and in the model (Boiteux *et al.*, 1975; Goldbeter & Nicolis, 1976). Two kinds of entrainment are possible: the enzyme can be locked either to the fundamental frequency of the external input or to a subharmonic frequency. The latter type of entrainment is of special interest as it demonstrates the nonlinear nature of the glycolytic oscillator. It is also reminiscent of the frequency demultiplication that takes place in circadian rhythms which can be altered from a 25 h to a 24 h period by an oscillatory input of 6 h period. Finally, a double periodicity occurs both in the model and the experiments when the period of the external input is much larger than that of the oscillatory enzyme.

Most preceding data have been obtained for a dimeric allosteric enzyme. Phosphofructokinase from many sources is known to be a tetramer (Mansour, 1972). In yeast, recent studies show that the enzyme contains two types of subunits, probably catalytic and regulatory (Laurent *et al.*, 1978); the number of subunits of each type is either three (Laurent *et al.*, 1978) or four (Tamaki & Hess, 1975; Hofmann, 1978). The yeast enzyme would thus appear as a trimer or a tetramer with dimer subunits. In view of this observation, it is interesting to know the effect of an increasing number of enzyme subunits on metabolic oscillations.

The number of protomers constituting the oscillatory enzyme either has or does not have significant effect on the periodicity, depending on whether or not there exists two distinct time scales in the system of (6). Two time scales will exist for large values of k_s at constant λ, i.e. for large values of the parameter q that reflects the differential affinity of the reaction product and of the substrate for the enzyme. When $q \gg 1$ the amplitude of the oscillations in the product concentration remains practically unchanged as the number of protomers increases from 2 to 8. This is illustrated in Figures 4.1.5 and 4.1.6 where the limit cycles obtained for $n = 2$ and $n = 6$ are shown, with $q = 100$. In these conditions, the oscillations are of 'relaxation' type, because the transit times from one branch of the sigmoid to the other (the horizontal parts on the limit cycle) are short, of the order of several seconds, whereas the other portions of the limit cycle are covered in minutes. Such relaxation oscillations give rise to periodic pulses in the product concentration, and have been observed both in a reconstituted glycolytic system (Hess & Boiteux, 1968) and in intact yeast cells (Von Klitzing & Betz, 1970).

A totally different picture obtains when the two equations of (6) have similar time scales. Then the amplitude of the oscillations in the product concentration markedly decreases as the number of protomers passes from 2 to 8. The question arises as to whether this phenomenon is the result of a reduction in either the amplitude of the limit cycle or the width

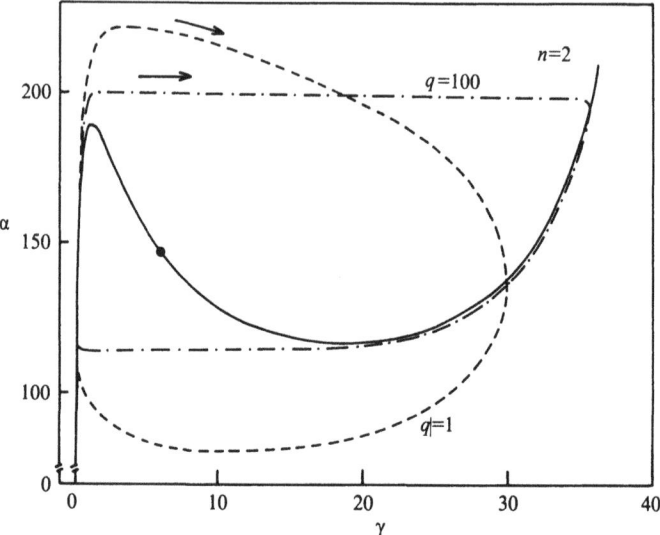

Figure 4.1.5. Limit cycles for a dimeric enzyme. The closed trajectories
($---$, $-\cdot-$) are obtained at constant λ for $q = 1$, $k_s = 0.1 \ s^{-1}$ (period = 366 s)
and $q = 100$, $k_s = 10 \ s^{-1}$ (period = 186 s); $\sigma_1 = 0.6 \ s^{-1}$, $L = 5 \cdot 10^6$; other
parameters are as in Figure 4.1.2. The solid line represents the sigmoid
nullcline $\gamma = \lambda \Phi$ on which the dot denotes the unstable steady state. The
value taken for σ_1 yields the largest amplitude of oscillation for $n = 2$.

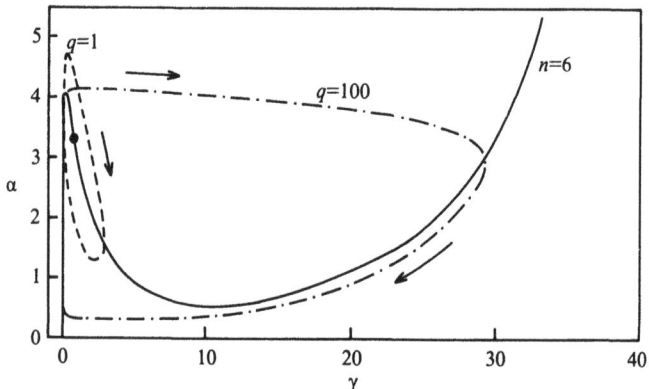

Figure 4.1.6. Limit cycles for a hexamer. The closed trajectories ($---$, $-\cdot-$)
are obtained at constant λ for $q = 1$, $k_s = 0.1 \ s^{-1}$ (period = 78 s) and $q = 100$,
$k_s = 10 \ s^{-1}$ (period = 58 s); $\sigma_1 = 0.07 \ s^{-1}$; other parameters as in Figure 4.1.5.
The solid line represents the sigmoid nullcline on which the dot indicates the
unstable steady state. The value taken for σ_1 yields the largest amplitude of
oscillation for $n = 6$.

of the sigmoid nullcline as n increases. A phase plane analysis of the effect of n shows that the former explanation holds, as indicated in Figures 4.1.5 and 4.1.6 by the curves drawn for $q = 1$ at constant λ. A further effect of the protomer number concerns the substrate level and the period, which both diminish as n increases. The dependence of glycolytic periodicities on the number of enzyme subunits and on the time scale structure of system (6) has been analyzed in more detail in recent publications (Venieratos & Goldbeter, 1979; Goldbeter & Venieratos, 1980).

Thinking of the possible physiological significance of metabolic oscillations, one sees that the number of protomers constituting an oscillatory enzyme can have a decisive influence when metabolic variables in the oscillatory reaction evolve on similar time scales. Only a low number of protomers ($n = 2$ to 4) then allows the large amplitude changes in the product concentration that could initiate a specific metabolic or cellular response. The above results on the role of n are obtained in systems regulated by positive feedback, such as the PFK reaction. Different results are obtained in the analysis of oscillatory enzyme reactions controlled by end-product inhibition (Walter, 1970).

Excitability. To examine the phenomenon of excitability in the PFK reaction, it is convenient to resort to a phase plane analysis of the model governed by (6). As shown in the previous section, the steady state of this system is stable when located on the left or right branches of the sigmoid nullcline. Any oscillation can only be damped, but the system becomes excitable as it can amplify perturbations whose amplitude exceeds a threshold (Goldbeter & Erneux, 1978). In the situation described in Figure 4.1.7, the steady state is located just to the left of the maximum on the sigmoid.

Let us consider the response of the system, initially at the stable steady state, to a small instantaneous increase in ADP corresponding to the addition of a pulse of this metabolite. When the amplitude of the pulse is sufficiently low, the system returns immediately to the stable steady state following a small-amplitude trajectory in the (α, γ)-plane (Figure 4.1.7). When the ADP pulse exceeds a threshold, the system follows a large-amplitude trajectory in the phase plane before returning to the steady state (Figure 4.1.7). This far-ranging trajectory corresponds to the synthesis of a pulse of ADP followed by a monotone evolution toward the steady state level (Figure 4.1.8). The above results demonstrate the capability of the enzyme to function as an amplifier of suprathreshold signals, for parameter values close to those for which it generates autonomous oscillations. When the parameter values yield a period of several minutes in the oscillatory domain, the time for reaching the

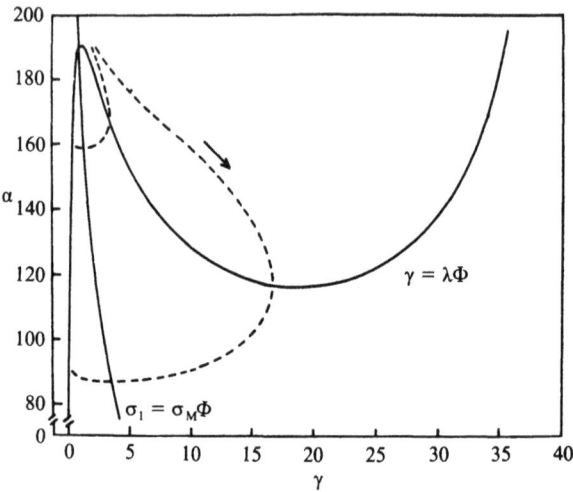

Figure 4.1.7. Excitability in the phase plane of the PFK model. The stable steady state lies at the intersection of the $\gamma = \lambda \Phi$ and $\sigma_1 = \sigma_M \Phi$ nullclines. Dashed lines represent two trajectories in response to an ADP pulse. The curves are obtained by integration of (6) for $\sigma_1 = 0.1 \, s^{-1}$, $L = 5 \cdot 10^6$ (other parameters are as in Figure 4.1.2). Initial conditions are $\gamma = 2$ for the small-amplitude trajectory and $\gamma = 2.4$ for the large-amplitude curve; the initial substrate concentration is equal to the steady-state value $\alpha = 189.82$. The excitation threshold is $\gamma = 2$ (see Figure 4.1.9) (from Goldbeter & Erneux, 1978).

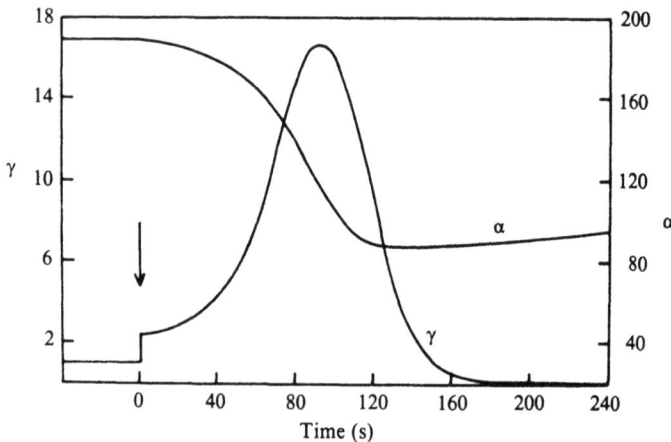

Figure 4.1.8. Synthesis of an ADP pulse in the PFK reaction. The time evolution of the substrate (α) and product (γ) concentrations is shown in response to a suprathreshold addition of ADP at time zero (arrow). The curves correspond to the large-amplitude trajectory in Figure 4.1.7 (from Goldbeter & Erneux, 1978).

maximum response to a signal just above threshold in the near excitable domain is of the order of 100 s (Figure 4.1.8).

The magnitude of the amplification depends on the parameters of the model and, more specifically, on the value of the allosteric constant. This is well indicated in the dose–response curve in Figure 4.1.9. Both the amplification and the discontinuous nature of the curve increase with the value of the allosteric constant. Large values of L thus favor excitable behavior just as they favor the onset of sustained oscillations in the PFK reaction (Goldbeter & Lefever, 1972; Goldbeter & Nicolis, 1976). The results of Figure 4.1.9 demonstrate the possibility of a seven-fold amplification of ADP signals above a threshold. These results are obtained in the absence of any time scale difference between the two equations of (6). As shown in the section devoted to excitability in the adenylate cyclase reaction in *Dictyostelium discoideum*, larger amplification factors and sharper thresholds are obtained when the metabolic variables evolve on different time scales.

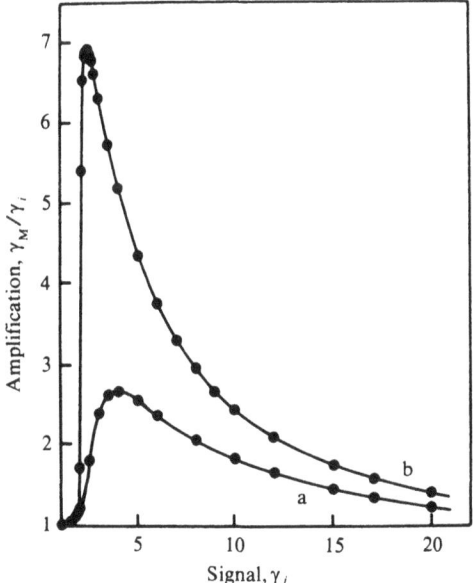

Figure 4.1.9. Amplification of the ADP signal as a function of stimulation in the PFK reaction. The curves are obtained by successive integrations of (6) for two values of the allosteric constant, $L = 10^6$ (a) and $5 \cdot 10^6$ (b). The amplification is defined as the ratio of the maximum of the synthesized ADP peak, γ_M, divided by the initial ADP concentration, γ_i ($\gamma = 1$ at the steady state). Parameter values are those of Figure 4.1.7 (from Goldbeter & Erneux, 1978).

The predictions on excitability in the PFK reaction could be checked experimentally in glycolyzing yeast or muscle extracts. As already mentioned, in yeast the technique of a constant rate of substrate injection has shown that sustained glycolytic oscillations occur when the input of fructose or glucose ranges from 20 to 160 mmol $l^{-1} h^{-1}$ (Hess *et al.*, 1969). The model suggests that excitation by ADP pulses should occur for steady states located to the left of the oscillatory domain in the immediate vicinity of the maximum on the sigmoid nullcline (Fig. 4.1.7). Experimentally, this corresponds to substrate injection rates slightly below 20 mmole $l^{-1} h^{-1}$. Using such input rates, the amplification of ADP pulses should be observed above a threshold; the phenomenon should correspond to a burst in NADH fluorescence which is the most suitable marker of oscillating glycolysis. The variations in NADH reflect an interplay between the lower and upper parts of the glycolytic pathway (Hess *et al.*, 1969), and it is thus difficult to extract from them quantitative information on a transient activation of PFK. The determination of CO_2 production, which is the end-process of glycolysis, or the assay of metabolites such as fructose diphosphate (FDP) or the adenine nucleotides should bring more clear-cut evidence for excitability in the PFK reaction. It would be useful to interpret variations in the adenine nucleotide pool in the light of a three-enzyme model for glycolytic oscillations proposed by Plesser (1977) who explicitly took into account the activation of PFK by AMP and the formation of the latter metabolite via the adenylate kinase reaction.

When the dissociation constant of ADP for PFK is taken as 5.10^{-5} mole l^{-1} (Blangy *et al.*, 1968), the value predicted in Figure 4.1.9 for the threshold of stimulation by ADP is 0.1 mmole l^{-1}. As the amplification of the signal becomes larger when the value of L increases, addition of inhibitors of PFK, such as citrate in muscle, should enhance the excitable response, since negative effectors augment the apparent value of the allosteric constant. Besides ADP, other positive effectors of PFK such as FDP in muscle or AMP could be used to demonstrate excitability of the enzyme.

Until now, excitability has been discussed only as a response to the addition of ADP pulses. When the value of the substrate injection rate is such that the steady state lies to the right of the oscillatory domain on the sigmoid nullcline, excitability can occur as a response to a transient decrease in ADP concentration (see Figures 4.1.10 and 4.1.11). Such a 'negative' signal is followed first by a further decrease in ADP and, subsequently, by the synthesis of a pulse of this metabolite before the system returns to steady state; a pulse of substrate is also produced in response to the signal. Experimentally, this could be observed in yeast extracts subjected to a substrate injection rate larger than

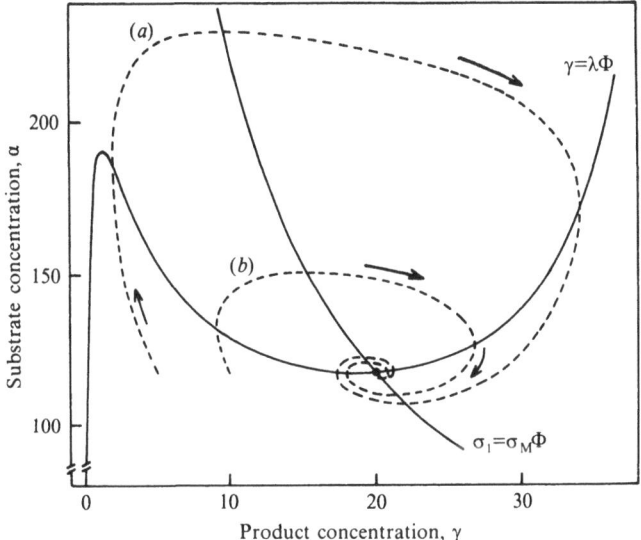

Figure 4.1.10. Excitation by 'negative' ADP signals in the PFK reaction. The stable steady state ($\alpha = 117.13$, $\gamma = 20$) lies at the intersection of the $\sigma_1 = \sigma_M \Phi$ and $\gamma = \lambda \Phi$ nullclines. The trajectories (dashed lines) follow a 50% (*a*) and a 75% (*b*) reduction in ADP (γ); they are obtained by integration of (6) for $\sigma_1 = 2 \, \text{s}^{-1}$. Other parameter values are as in Figure 4.1.7.

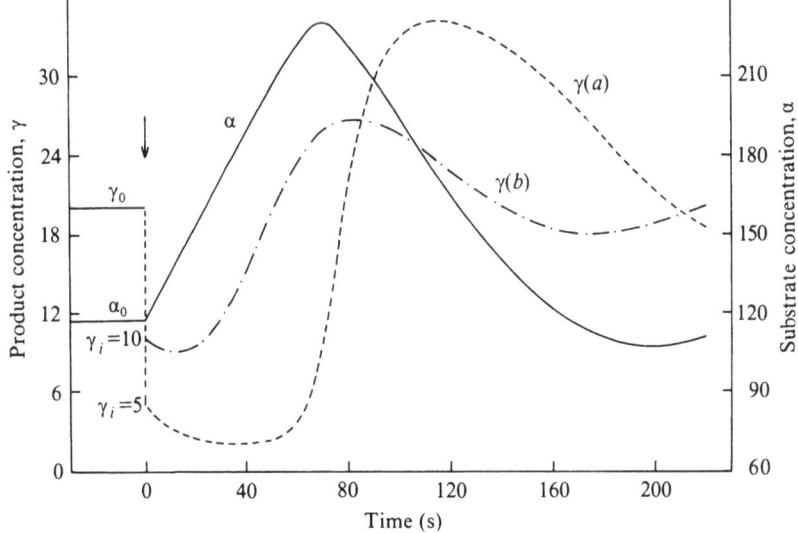

Figure 4.1.11. Pulsatory response to 'negative' ADP signals in the PFK reaction. The time evolution of the ADP concentration corresponding to trajectories (*a*) and (*b*) of Figure 4.1.10 is shown, with the time evolution of the substrate concentration for (*a*). α_0 and γ_0 are the steady state levels.

$160 \, \text{mmole} \, l^{-1} \, h^{-1}$, upon a sudden decrease in the activator (ADP or AMP) concentration.

Phosphofructokinase shares the other properties of excitable systems, such as the existence of a refractory period and the increase of the threshold of excitation as the steady state moves further away from the oscillatory domain (Fitzhugh, 1961). These two properties are discussed in greater detail in the following section devoted to the generation of cAMP pulses in *D. discoideum*. Before turning to this system, the question can be asked as to whether all-or-none transitions between multiple steady states are possible in the model for the PFK reaction. Figure 4.1.2*a* shows that such a situation can indeed occur in an oscillatory system which loses one degree of freedom: if the substrate level is maintained constant, there exist substrate levels for which the horizontal nullcline $\dot{\alpha} = 0$ will have three intersections with the $\gamma = \lambda \Phi$ sigmoid. Two stable steady states will then be separated by an unstable state. All-or-none transitions will occur from one stable state to the other. These transitions differ from excitable trajectories, since in the latter case the system returns to the original steady state after the pulsatory amplification of the initial perturbation.

The cAMP signaling system in *Dictyostelium discoideum*

In contrast with glycolysis, where the physiological significance of the oscillations remains unclear, the cellular slime mould *D. discoideum* provides an example of an integration of excitable and oscillatory behavior in the life cycle of an organism. *D. discoideum* cells grow as solitary amoebae until they exhaust their food supply. Then, about eight hours after the beginning of starvation, they collect around centers by a chemotactic response to cAMP signals (Konijn, van de Meene, Bonner & Barkley, 1967). As many as 10^5 amoebae may aggregate around a center; the multicellular body thus formed migrates and finally transforms into a fruiting body consisting of only two types of differentiated cells, those that belong to the stalk and those that become spores. This simple pattern of differentiation in an eucaryotic organism, as well as the existence of a mechanism of intercellular communication during the aggregation phase, explains why *D. discoideum* is a major model in developmental biology (Bonner, 1967; Loomis, 1975).

The process of aggregation in *D. discoideum* has a periodicity of several minutes: waves of inward amoeboid movement appear to propagate outward from the center to the periphery of the aggregation field. To account for this observation, Shaffer (1962) postulated the existence of a dual mechanism for the periodic release of chemotactic attractant by the centers and for the relay of the chemotactic signals by aggregating cells. The chemotactic factor was later identified as cAMP and, more recently, direct evidence for both the autonomous oscillations of cAMP and the relay of cAMP pulses was obtained in suspensions of *D. discoideum* cells (see below).

The existence of a mechanism for relay and oscillation of the chemotactic signal in *D. discoideum* allows the formation of large aggregation territories (Gerisch, 1968; Cohen & Robertson, 1971; Alcantara & Monk, 1974). Species such as *D. minutum* in which the chemotactic factor is propagated by simple diffusion indeed form much smaller aggregation territories (Gerisch, 1968). In addition to their role as chemotactic signals, periodic cAMP pulses promote and synchronize cell differentiation in *D. discoideum* during the interphase which separates starvation from aggregation (Darmon, Brachet & Pereira da Silva, 1975; Gerisch & Malchow, 1976). This second physiological role consists of inducing the synthesis of specific proteins, most of which belong to the cAMP signal machinery (Klein & Darmon, 1977).

Experimental observations in cell suspensions

The studies of *D. discoideum* suspensions throw light on the cAMP signaling mechanism. Several hours after starvation, the cells exhibit spontaneous oscillations with a period of about seven minutes. These oscillations, revealed by periodic changes in light scattering (Gerisch & Hess, 1974), correspond to sustained oscillations in the concentration of intra- and extracellular cAMP (Gerisch & Wick, 1975). The oscillations can be phase shifted upon addition of cAMP pulses.

The phenomenon of relay is demonstrated in suspensions of cells that have not yet begun – or have ceased – to oscillate. The addition of a pulse of extracellular cAMP (10^{-8} to 10^{-6} mmole 1^{-1}) elicits the synthesis of a pulse of intracellular cAMP of much larger magnitude (Roos, Nanjundiah, Malchow & Gerisch, 1975; Shaffer, 1975). The intracellular pulse occurs generally after 1–2 min and is followed by a peak in extracellular cAMP.

Do the phenomena of relay and oscillation originate from a common mechanism or are they caused by two distinct signaling processes? The model analyzed below suggests that the relay of cAMP signals reflects the excitability of adenylate cyclase in *D. discoideum*. It is proposed that the adenylate cyclase reaction in this species of slime mold belongs to the class of excitable and oscillatory chemical systems (Goldbeter, Erneux & Segel, 1978).

Model for the cAMP signaling system

The signaling system consists of a cell surface cAMP receptor and of a functionally coupled adenylate cyclase (Gerisch & Malchow, 1976;

Klein, Brachet & Darmon, 1977). The latter enzyme transforms ATP into cAMP. Taking this into account, Goldbeter & Segel (1977) proposed a model based on the observation (Roos & Gerisch, 1976) that binding of extracellular cAMP to the receptor activates adenylate cyclase which faces the inner side of the plasma membrane (Farnham, 1975). This model is a modification and an extension of that proposed by Goldbeter (1975) for cAMP oscillations on the basis of the results of Rossomando & Sussman (1973) for the intracellular regulation of adenylate cyclase. Here, the control of the enzyme by extracellular cAMP is analyzed explicitly.

How the activation of adenylate cyclase proceeds is still unclear. The mechanism might involve allosteric regulation either directly by extracellular cAMP or by an intracellular intermediate whose concentration rapidly changes upon reception of the signal. The enzyme could also be regulated by covalent modification (Greengard, 1978). The observation that a pulse of intracellular cyclic GMP (cGMP) precedes the synthesis of a cAMP pulse (Mato *et al.*, 1977) suggests a possible control of adenylate cyclase by a cAMP-dependent protein kinase (Gerisch *et al.*, 1977*a*). Experimental evidence for such a process is, however, still lacking. Moreover, the cGMP response could be linked to the chemotactic process only, since it is observed in amoeboid species that lack the mechanism for relay and oscillation (Mato & Konijn, 1977). Calcium could be the intracellular effector (Rapp & Berridge, 1977), since a rapid calcium influx follows cAMP binding to the cell surface receptor (Wick, Malchow & Gerisch, 1978). Calcium ions seem, however, to affect the basal activity of adenylate cyclase rather than the oscillations (Roos, Scheidegger & Gerisch, 1977).

To represent the regulation by cAMP in the model (Figure 4.1.12), the simplest assumption is made that the cAMP receptor behaves as a regulatory subunit of adenylate cyclase. Furthermore, the catalytic and regulatory parts of the enzyme are treated as dimers, and the resulting complex is considered as obeying the concerted allosteric model (Monod *et al.*, 1965) with exclusive binding of both substrate and positive effector to the R state. As in the modeling of oscillating glycolysis in yeast extracts, we treat here the case of a spatially homogeneous system. Most experiments on relay and oscillation are indeed performed in continuously stirred suspensions of *D. discoideum* cells. Given the above assumptions, the behavior of the cAMP signaling system is governed by three ordinary differential equations (Goldbeter & Segel, 1977; Goldbeter *et al.*, 1978):

$$d\alpha/dt = v - \sigma\Phi,$$

$$(1/q)\, d\beta/dt = \sigma\Phi - (k_t\beta/q),$$

$$d\gamma/dt = (k_t\beta/h) - k\gamma, \tag{16}$$

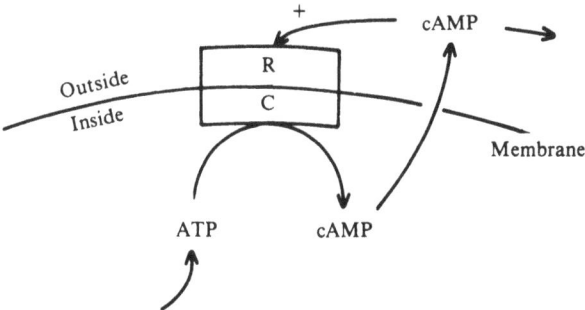

Figure 4.1.12. Model for the cAMP signaling system in *D. discoideum*. R and C denote the cAMP receptor and the functionally coupled adenylate cyclase, respectively. Cyclic AMP (cAMP) is synthesized intracellularly from ATP and transported into the extracellular medium where it is hydrolysed by phosphodiesterase. The + sign refers to the activation of adenylate cyclase that follows cAMP binding to R (from Goldbeter & Segel, 1977).

where

$$\Phi = \alpha(1+\alpha)(1+\gamma)^2/[L+(1+\alpha)^2(1+\gamma)^2]. \tag{17}$$

The three metabolic variables α, β and γ denote the concentrations of intracellular ATP, intracellular cAMP and extracellular cAMP divided by K_S, K_P and K_P, respectively, where K_S and K_P are the Michaelis constant of adenylate cyclase for ATP and the dissociation constant of the cAMP receptor. Parameters v and σ relate to the constant ATP input and to the maximum cyclase activity, divided by K_S. (An ATP input is needed to maintain the system far from equilibrium, since the latter condition is the thermodynamic prerequisite for sustained oscillatory behavior.) Also, $q = K_S/K_P$; k_t and k are apparent first order rate constants for the cAMP transport across the cell membrane and for the phosphodiesterase reaction, which are both considered as linear processes; L is the allosteric constant of adenylate cyclase; the dilution factor h is the ratio of extracellular fluid volume to cell volume in the experiments with cell suspensions. Parameter k accounts for both the membrane-bound and the extracellular forms of phosphodiesterase which hydrolyse extracellular cAMP (Farnham, 1975). The equations of (16) have been obtained on the assumption of a quasi-steady state for the enzymatic forms; they are derived in the way outlined for the PFK reaction in the section on glycolytic oscillations.

Sustained oscillations and excitability in the adenylate cyclase reaction

The system (16) admits a single steady state solution whose stability properties can be determined by a linearized stability analysis. To discuss excitable and oscillatory behavior, it is, however, convenient to resort to a phase plane analysis of (16). This is facilitated by the observation that $q \gg 1$. Indeed, $q = K_S/K_P$, where K_S is of the order of 10^{-4} mol l^{-1} (Klein, 1976; Gerisch & Malchow, 1976) whereas K_P values range from 10^{-9} to 10^{-7} mol l^{-1} (Henderson, 1975; Gerisch & Malchow, 1976; Mullens & Newell, 1978). The system (16) can thus be approximated by the reduced system (Goldbeter *et al.*, 1978)

$$\beta = (q\sigma/k_t)\Phi, \qquad d\alpha/dt = v - \sigma\Phi, \qquad d\gamma/dt = k(\lambda\Phi - \gamma), \quad (18)$$

by means of a quasi-steady state assumption for β, in the limit $q \to \infty$ with $\lambda = (q\sigma/hk)$ and (k_t/q) remaining finite. The dynamics of the cAMP signaling system can now be studied in the (α, γ) phase plane. (Also see the concluding portion of Appendix Section A.3 where oscillations of the full system (16) is studied as an example of Hopf bifurcation.)

Sustained oscillations of cAMP. The results obtained in the phase plane analysis of the PFK model hold for the adenylate cyclase system, since the equations of (6) are formally identical to the reduced system (18). As in the case of the PFK reaction, the existence of sustained oscillations is linked to the form of the nullclines $v = \sigma\Phi$ and $\gamma = \lambda\Phi$ and the position of their point of intersection which defines the steady state. A necessary condition for oscillation and excitability is that the nullcline $\gamma = \lambda\Phi$ should be an S-shaped sigmoid.

As in the PFK system, therefore, sustained oscillations around an unstable steady state occur when the intersection of the two nullclines lies in the region of negative slope on the sigmoid $\gamma = \lambda\Phi$ (the instability condition is given by (12) or (15)). Limit cycle oscillations similar to those of Figures 4.1.3, 4.1.5 and 4.1.6 develop, that correspond to the autonomous synthesis of periodic pulses of intra- and extracellular cAMP (Figure 4.1.13). These oscillations are accompanied by a periodic variation in ATP, the amplitude of which can be restricted to a 10% variation around the mean ATP level by appropriate parameter choices. Such a reduced amplitude can be related to the observation that ATP remains practically constant in the course of cAMP oscillations (Roos *et al.*, 1977; Geller & Brenner, 1978a). In the model, the oscillations are generally accompanied by variations of larger amplitude in the substrate concentration. A possible compartmentation of the membrane-bound adenylate cyclase reaction *in vivo* could also explain the lack of observable ATP oscillations.

The behavior of the model qualitatively changes when the ATP level is held constant. Oscillations are no longer possible, but the system governed by (18) becomes able to switch between two different stable steady states for a given set of parameter values. Such a situation can be visualized in the (α, γ) phase plane where the nullcline $\dot{\alpha} = 0$ becomes a

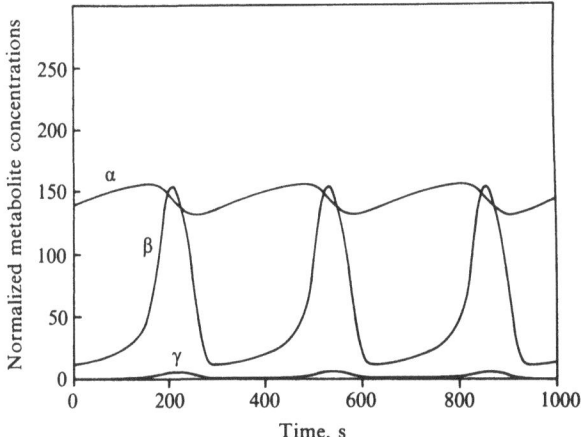

Figure 4.1.13. Sustained oscillations of intracellular ATP (α), intracellular cAMP (β), and extracellular cAMP (γ). The curves are obtained by integration of the equations of (16) for $v = 0.2\,\text{s}^{-1}$, $k = 1\,\text{s}^{-1}$, $\sigma = 1.2\,\text{s}^{-1}$, $k_t = 0.4\,\text{s}^{-1}$, $q = 100$, $L = 10^6$, $h = 10$.

horizontal line that has one or three intersections with the sigmoid nullcline $\gamma = \lambda\Phi$, depending on the ATP level. The steady state is still unstable when located in the region of negative slope on the sigmoid, but oscillations do not develop as the system evolves toward either one of the two stable steady states. Some variation in ATP is thus needed for oscillatory behavior in the system described by (16) or (18).

Relay of cAMP signals. The phase plane analysis of the cAMP signaling system is particularly helpful for the comprehension of the relay mechanism, as the response to a pulse of extracellular cAMP can be directly visualized in the (α, γ) plane. When the intersection of the two nullclines is located on the left of maximum A or to the right of minimum D on the sigmoid (see Figure 4.1.14), the steady state is stable and excitability is mathematically possible. Since relay consists of the amplification of a pulse of extracellular cAMP, the physiological conditions for excitability correspond to a steady state located on the left branch of the sigmoid, near maximum A (Figure 4.1.14).

In the phase plane, the effect of adding a pulse of extracellular cAMP is simulated by an instantaneous increase in γ, i.e. by a horizontal displacement to the right of the steady state. The terminal point of this displacement defines the initial condition (α_i, γ_i) from which the behavior of the system is determined by integration of (18). The trajectory marked (a) in Figure 4.1.14 shows the response to a suprathreshold stimulation by extracellular cAMP. The system undergoes a large excursion in the

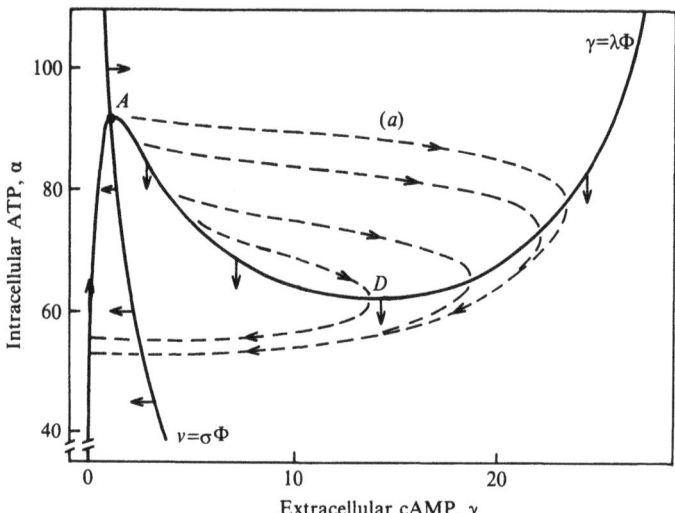

Figure 4.1.14. Excitability in the adenylate cyclase reaction. In the (α, γ) phase plane of the reduced system (18), the steady state (dot) is located at the intersection of the nullclines $v = \sigma \Phi$ and $\gamma = \lambda \Phi$ on which arrows indicate the local direction of the solution trajectory. Several trajectories (dashed lines) are shown, corresponding to different suprathreshold initial conditions. Trajectory (a) shows the response of the system, initially at steady state, to a pulse of extracellular cAMP. Parameter values are $v = 0.04 \text{ s}^{-1}$, $k = 0.4 \text{ s}^{-1}$; other parameters are as in Figure 4.1.13. A and D denote a maximum and a minimum, respectively, on the sigmoid (from Goldbeter *et al.*, 1978).

phase plane across the right branch of the sigmoid before returning to the stable steady state. This wide-ranging trajectory in the (α, γ)-plane corresponds to the synthesis of a substantial pulse of intracellular cAMP. As shown by Figure 4.1.15(a), the relay response obtained by integration of the equations of (16) yields good agreement with the phase plane analysis of the reduced system (18).

The other trajectories in Figure 4.1.14 indicate that the signaling system can be reexcited on its way back to the steady state. The threshold for these signals decreases and the amplitude of the relay response increases as the system approaches the steady state. The above properties demonstrate the existence of a relative refractory period for relay. During the synthesis of the pulse itself, the refractoriness is absolute and no second signal can be generated on top of the other upon further stimulation. Both types of refractory period have been described by Fitzhugh (1961) in his theoretical study of excitation in nerve membrane.

When the steady state lies in the immediate vicinity of maximum A, the dose–response curve that links the amplitude of the relay response to the magnitude of the external signal exhibits a sharp threshold followed

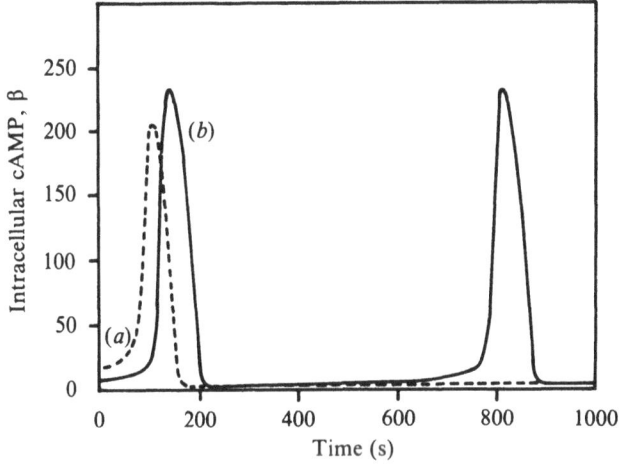

Figure 4.1.15. Relay of a cAMP signal (*a*) and autonomous oscillations of cAMP (*b*). The curves are obtained by integration of the equations of (16) for $v = 0.04\,\text{s}^{-1}$ (*a*) and $v = 0.1\,\text{s}^{-1}$ (*b*); other parameter values are those of Figure 4.1.14 (redrawn from Goldbeter & Segel, 1977).

by a plateau (Figure 4.1.16*a*). Dose–response curves become less discontinuous when the steady state lies further away from maximum *A*. Once the extracellular cAMP signal becomes so large that it brings the system initially across the right limb of the sigmoid nullcline, no amplification of the signal occurs since the system returns directly to the steady state (see Figure 4.1.14; this remark also holds for the PFK system, as shown in Figure 4.1.9). No pulse of extracellular cAMP is thus produced above a certain level of stimulation, as observed experimentally by Grutsch & Robertson (1978). Simulations of system (16) indicate that a rapid pulse of intracellular cAMP can nevertheless be obtained in these conditions.

The theoretical characteristics of relay compare semi-quantitatively with the experimental observations in *D. discoideum* suspensions (Roos *et al.*, 1975; Gerisch *et al.*, 1977*b*), viz. the half-width of the response (of the order of 1 min), amplification factor (of the order of 10–20), and time for maximal response (about 100 s) (Goldbeter & Segel, 1977). The delay of 30–40 s between the peaks in extra- and intracellular cAMP (Roos *et al.*, 1975) can be matched in the model provided phosphodiesterase is treated as a Michaelian enzyme that does not function only in the linear regime (A. Goldbeter, unpublished). It should be noted that this delay could be the result of storage of intracellular cAMP into vesicles prior to transport across the plasma membrane (Maeda & Gerisch, 1977).

The parameter values taken for the above simulations are in part arbitrary. This is the case for L, v and k_t which are difficult to evaluate.

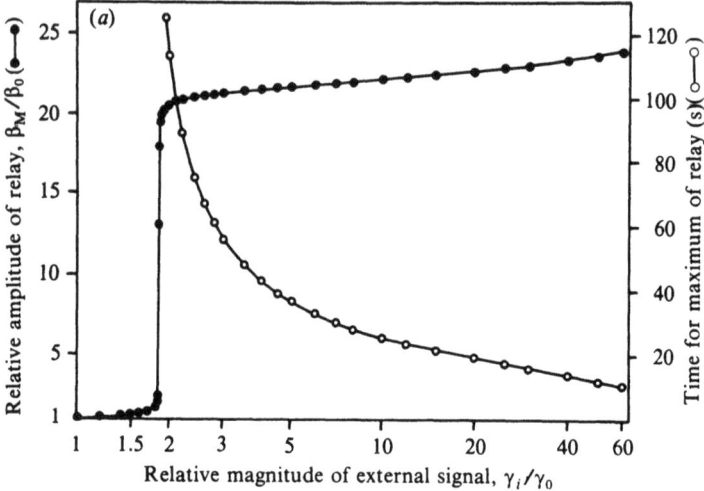

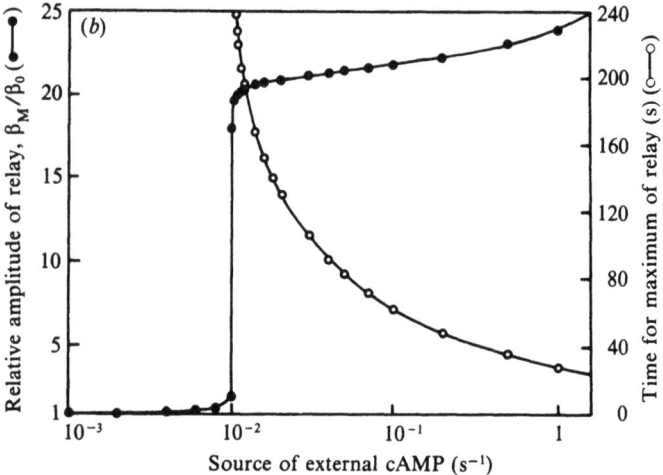

Figure 4.1.16. Excitation of the cAMP signaling system by a pulse (a) and by a constant source (b) of extracellular cAMP. The amplification factor (●) is given as the maximum of the intracellular cAMP peak, β_M, divided by the steady state level β_0. The second curve in each graph shows the time at which β_M is reached after beginning of stimulation (○). The curves are obtained by integration of the equations of (16) for the parameter values of Figure 4.1.14. In (b) a constant source term divided by K_P has been inserted in the evolution equation for γ. In (a), the amplitude of the external cAMP signal is given as the initial extracellular cAMP concentration γ_i divided by the steady state level γ_0 which is equal to unity (from Goldbeter & Segel, 1977, and Goldbeter *et al.*, 1978).

The value of these parameters has been adjusted so as to give the experimentally observed steady state levels of ATP and cAMP. Experimental data for the adenylate cyclase and phosphodiesterase reactions in *D. discoideum* (Gerisch & Malchow, 1976; Klein, 1976) suggest that parameters k and σ are of the order of 10^{-2} s^{-1}, whereas q is in the range 10^3–10^5. Excitability and oscillations still occur for such parameter values, with metabolite levels in the physiological range, but the time scale of both phenomena becomes too slow. Larger values of the adenylate cyclase activity, of the order of those used in our simulations, have recently been reported (Brenner, 1978). The isolation and subsequent study of the signaling system in *D. discoideum* ghosts or extracts should facilitate its quantitative analysis.

Until now, excitability has been considered in response to cAMP pulses only. A continuous signal of extracellular cAMP can also elicit the synthesis of a pulse of intracellular cAMP. Dose–response curves obtained in the model for such a constant signal resemble those obtained for the response to cAMP pulses (Figure 4.1.16b). They exhibit both a threshold and a plateau, and the time for maximal relay decreases as the amplitude of the cAMP source increases. What happens if a constant suprathreshold signal is maintained once the relay response is completed? Depending on the stability properties of the new steady state admitted by (16) in which a constant source term is inserted in the evolution equation for γ, the system will evolve to this steady state or further intracellular cAMP pulses will be synthesized, in which case the modified system (16) admits a limit cycle solution.

The theoretical predictions that excitability can occur in the cAMP signaling system as a response to both constant or pulsatory signals above a threshold agree with the observations of Robertson & Drage (1975) on relay in *D. discoideum* fields on agar. The value of the threshold predicted for a pulsatory signal (Figure 4.1.16a) is of the order of K_P, i.e. in the range 10^{-9}–10^{-7} mole l^{-1} found experimentally for the dissociation constant of the cAMP receptor (Henderson, 1975; Gerisch & Malchow, 1976; Mullens & Newell, 1978). Such a range for the threshold agrees with that found by Robertson & Drage (1975) on agar (see also Grutsch & Robertson, 1978) as well as with the amplitude of cAMP pulses used in the experiments on relay in cell suspensions (Roos *et al.*, 1975; Shaffer, 1975; Gerisch *et al.*, 1977b). For a constant signal (Fig. 4.1.16b), the value of the threshold is of the order of $10^{-2}K_P$ s^{-1}, i.e. in the range 10^{-11} to 10^{-9} mole l^{-1} s^{-1} depending on the value taken for K_P. This prediction could be tested experimentally in cell suspension experiments.

Excitability can also occur in the model as a response to a 'negative' cAMP signal when the steady state lies on the right branch of the sigmoid nullcline, near minimum D (see Figure 4.1.14). Then a transient decrease in extracellular cAMP could result in the synthesis of a peak of intracellular cAMP in the way predicted in Figures 4.1.10 and 4.1.11 for the PFK reaction. It is difficult to test this prediction experimentally in the slime mold, since the cAMP signaling system has not yet been isolated in cell ghosts or extracts; until then, it will be difficult to control the location of the steady state on the sigmoid nullcline, in contrast with the situation that prevails for glycolysis in yeast where the oscillatory and excitable domains can be controlled by the technique of constant substrate injection.

The phase plane analysis of the cAMP signaling system yields a qualitative explanation for several observations made on relay in *D. discoideum*, e.g. the existence of a threshold for excitation by extracellular cAMP (Cohen & Robertson, 1971) and the existence of a refractory period (Shaffer, 1962; Gerisch, 1968; Robertson & Drage, 1975). The analysis substantiates the description of the aggregation fields as excitable media (Durston, 1973). According to the model, relay and oscillation of cAMP are two phenomena that are necessarily linked and occur in closely related conditions. The fact that the time for maximal relay in the model varies from 100 to 10 s depending on the magnitude of the constant or pulsatory signal (Figure 4.1.16) could explain why cells in suspensions relay after 100 s (Roos *et al.*, 1975; Shaffer, 1975) whereas those on agar respond after some 15 s (Cohen & Robertson, 1971; Alcantara & Monk, 1974); the latter might receive a larger signal locally, because of the absence of stirring. The model also explains why the half-width and the waveform of the cAMP peaks during relay and oscillations are similar (see Figure 4.1.15), as observed experimentally (Gerisch *et al.*, 1977*b*): the phase plane trajectories in both conditions follow proximate paths dictated by the form of the sigmoid nullcline.

In support of a common mechanism is the observation that relay and oscillations in *D. discoideum* have a similar temperature dependence (Gross, Peacey & Trevan, 1976). The observation that 2,4-dinitrophenol can suppress the oscillations by rendering the cells excitable (Geller & Brenner, 1978*b*) does not necessarily imply that the two phenomena are caused by two different mechanisms. 2,4-dinitrophenol affects the ATP level and is likely to modify parameter v in the model. If initially v is such that the nullcline $v = \sigma\Phi$ intersects the sigmoid nullcline in the region of negative slope, i.e. in the oscillatory domain, a decrease in v can shift the steady state to the left of maximum A (see Figure 4.1.14), thus suppressing the oscillations and bringing the system into the excitability domain.

Developmental control of the signaling system

During the hours that follow starvation, *D. discoideum* amoebae are first able to respond chemotactically to cAMP signals; then they become capable of relaying these signals, before being able to generate autonomously periodic cAMP pulses (Robertson, Drage & Cohen, 1972). Goldbeter & Segel (1977) suggested that such a sequence of developmental transitions can be explained in the model for the signaling system by supposing that some parameters are drifting during interphase. If the system is initially in a state where neither relay nor oscillations occur, such a move in the parameter space can bring the cells first into a region where the system is excitable and later into the domain of autonomous oscillations.

To illustrate in a simple way such a sequence of developmental transitions, let us consider the effect of a variation in the substrate injection rate v on the behavior of the system in the phase plane (α, γ). The steady state value of γ is (qv/hk). Starting from a low value of v that corresponds to a steady state located on the left branch of the sigmoid nullcline (see Figure 4.1.14), an increase in v brings the steady state near maximum A, i.e. in the domain where excitation occurs in response to a suprathreshold pulse of extracellular cAMP. A further increase in v then brings the system into the region of limit cycle oscillations; if continued, the increase in input rate brings the system into a stable steady state which is no longer excitable.

Experimentally, no evidence exists for a variation in the rate of ATP input to adenylate cyclase during interphase. The transitions discussed as a function of v can also take place in the model as a result of a variation in σ or k, which relate to the activities of adenylate cyclase and phosphodiesterase. Experimental evidence indeed exists for a developmental control of these two enzymes during the hours that follow starvation (Klein, 1976; Gerisch *et al.*, 1977*a*; Klein & Darmon, 1977). As these changes in enzyme activity take place over a time scale which is much larger than the period of cAMP oscillations, the dependence of the behavior of system (16) on σ and k can be determined assuming that these parameters are constants.

A theoretical path in the adenylate cyclase–phosphodiesterase space can be found that would account for the observed sequence no-relay–relay–oscillations (Goldbeter & Segel, 1980). Qualitatively, the path would traverse four regions in this parameter space, in the sequence *ABCD* (see Figure 4.1.17). Region *B* is the relay domain which is always located to the right of the oscillatory domain *C*. Cells following this path in the course of development would start, at the onset of starvation, from a state *A* with low phosphodiesterase and adenylate cyclase activity. This initial state would be devoid of signaling properties and would correspond to a small constant leakage of cAMP into the extracellular medium. Cells entering *B* would become able to relay. Those cells that are the first to enter *C* would become aggregation centers, being able to generate periodic pulses of cAMP. Finally, the cells that leave the

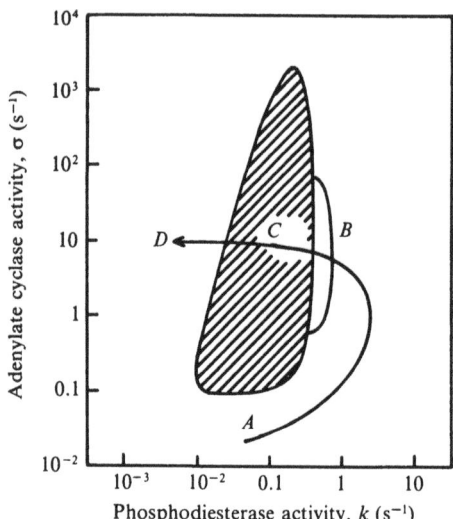

Figure 4.1.17. Developmental path for the cAMP signaling system in *D. discoideum* as a function of adenylate cyclase and phosphodiesterase activities (see text for details). The oscillatory domain (*C*) and the relay domain (*B*) were determined by linear stability analysis and computer simulation; see text for *A* and *D*. For relay, simulations were performed with the initial condition $\gamma_i = 5$. Parameter values are as in Figures 4.1.13 and 4.1.14.

oscillatory domain to enter the stable steady state region *D* would secrete cAMP at a high constant rate. This transition, first discussed by Cohen (1977), is thought to occur in the tips of late aggregates. In agreement with experimental observations, the path of Figure 4.1.17 suggests that cells in such a state *D* are likely to be the cells that previously were aggregation centers in *C*. Other cells could leave the oscillatory domain by returning to *B*, in which case they would again become excitable. Such a process has been observed experimentally in cell suspensions (Gerisch *et al.*, 1977*a*). Let us note here that the behavior of cells in suspensions can differ from the behavior of cells on agar, because the pattern of enzyme synthesis – including that of phosphodiesterase – differs in the two situations (Town & Gross, 1978).

The developmental path suggested in Figure 4.1.17 fails, however, on two accounts when compared with experimental data. First, the decrease in phosphodiesterase activity required for entering the relay domain prior to the oscillatory regime does not correlate with the time course of this enzyme during aggregation. A decrease in phosphodiesterase activity is indeed observed, but it follows rather than precedes the appearance of autonomous centers. Second, state *D* in Figure 4.1.17 does not correspond to a constant level of extracellular cAMP larger than in state A, for a given value of k. This is because, in the model described by (16), the steady state concentration of extracellular cAMP is simply $\gamma_0 = qv/hk$. Thus, contrary to expectations, γ_0 does not depend on the adenylate cyclase activity σ. The reason for this paradoxical result lies in the form of the kinetic equation for α, which at steady state yields the relation $v = \sigma\Phi$. The dependence on σ thereby vanishes at steady state in the equations for β and γ.

In order to remedy this unwanted simplification, one has to modify the evolution equation for ATP. The new system of equations to be considered is

$$d\alpha/dt = v - \sigma\Phi - k'\alpha,$$

$$(1/q)\, d\beta/dt = \sigma\Phi - (k_t\beta/q),$$

$$d\gamma/dt = (k_t\beta/h) - k\gamma, \tag{19}$$

where Φ is given by (17). Comparing with system (16), the only difference is that a term $(-k'\alpha)$ has been added in the kinetic equation for α. As shown below, this single modification overrides the difficulties encountered with the developmental path in Figure 4.1.17. The motivations for the alteration are twofold. In system (16), the only ATP-consuming process is the reaction catalyzed by adenylate cyclase. This assumption certainly represents an oversimplification since ATP is utilized in other metabolic processes as well. The term $-k'\alpha$ can thus be related to the disappearance of ATP in reactions other than that catalyzed by adenylate cyclase. Alternatively (or simultaneously), this term can be combined with the input rate v and be rewritten as $k'(\alpha_0 - \alpha)$, with $v = k'\alpha_0$. Such a description would correspond to the exchange of ATP between two compartments, one being a constant intracellular pool (α_0), and the other being the layer near the membrane where ATP (α) is immediately available to adenylate cyclase for the synthesis of cAMP. A similar alteration of the kinetic equation for α has been suggested independently by E. L. Coe (personal communication) on the grounds of a comparison between the theoretical and experimentally observed time evolution of ATP.

The dynamic behavior of the model described by (19) in the σ–k plane is shown in Figure 4.1.18. When k' is less than $10^{-4}\ \mathrm{s}^{-1}$, the effect of the term $-k'\alpha$ is negligible and the diagram obtained is practically identical to that of Figure 4.1.17. The diagram of Figure 4.1.18 has been established for the intermediary value $k' = 10^{-3}\ \mathrm{s}^{-1}$. It shows that the relay domain B has been shifted and lies below the oscillatory domain C. Hence the observed sequence no relay–relay–oscillations can now be

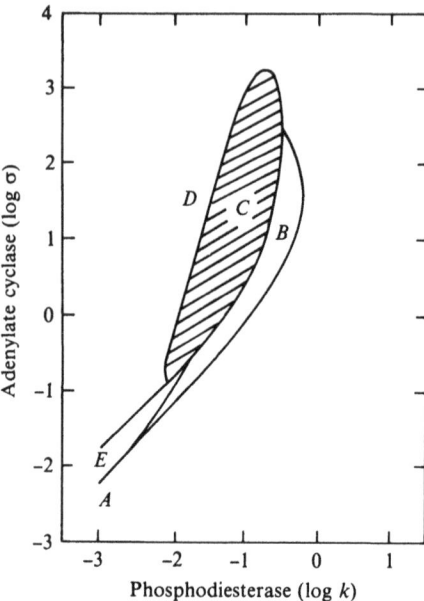

Figure 4.1.18. Stability diagram in the σ–k plane, for the system governed by eqs (19). Region C denotes the domain in which autonomous oscillations occur around an unstable steady state. In region B, the system is capable of relaying a cAMP signal. This region was determined by computer simulations, taking α and β at steady state and setting the initial concentration of extracellular cAMP as $\gamma = 10$; relay was obtained when the pulse produced in response to the signal exceeded the initial value of γ. Regions A and D refer to stable steady states corresponding, respectively, to low and large levels of extracellular cAMP. In region E, multiple steady states occur (see Figure 4.1.19). The diagram is established for the parameter values of Figure 4.1.17, with $k' = 10^{-3}$ s^{-1} (see Goldbeter & Segel, 1980).

obtained in the model without a prior decrease in phosphodiesterase activity, along the path *ABC*. In addition, the system may leave the oscillatory regime by reentering the relay domain or by going into region *D*. In the latter case, the steady state level of extracellular cAMP will be larger than at the initial point *A* since γ_0 now increases with σ as illustrated by Figure 4.1.19 below.

A further difference with respect to the graph of Figure 4.1.17 concerns the existence of a domain of multiple stationary states (region *E* in Figure 4.1.18). Two different situations are encountered in this domain, as shown in Figure 4.1.19. At low values of *k*, two of the three steady states obtained as a function of σ are stable. For larger values of *k*, the upper branch of steady states becomes unstable. Let us consider the latter situation, illustrated by the curve for $k = 0.008$ s^{-1} in Figure 4.1.19.

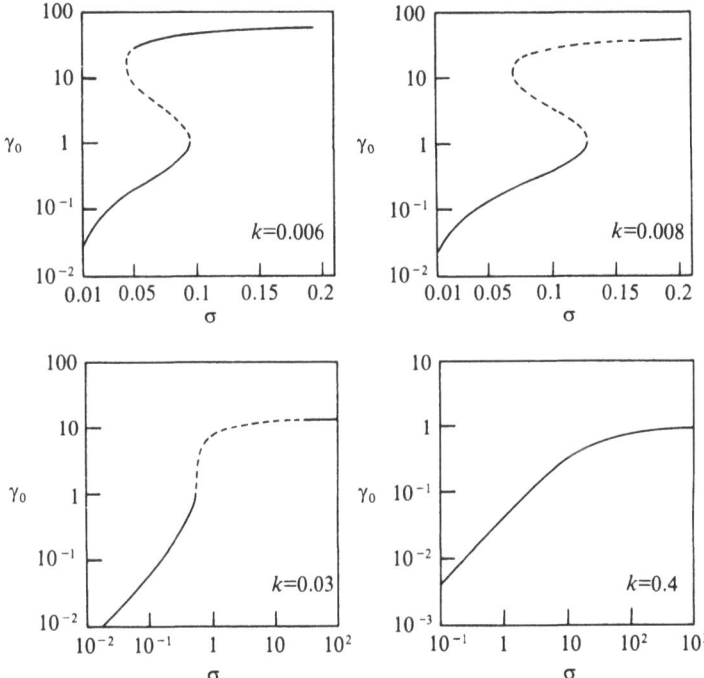

Figure 4.1.19. Patterns for the dependence of the steady state concentration of extracellular cAMP (γ_0) as a function of the maximum adenylate cyclase activity (σ) obtained for different values of the phosphodiesterase rate constant k in the system governed by equations (19). Multiple steady states obtain for low values of k. A dashed line denotes unstable steady states. The graphs correspond to four different cuts through the diagram of Figure 4.1.18.

When the value of σ is so large (e.g. $\sigma = 0.15$ s^{-1}) that only the upper branch of steady states exists, a limit cycle encloses the steady state when it is unstable. For lower values of σ (e.g. $\sigma = 0.1$ s^{-1}) corresponding to three stationary states, the question arises as to whether limit cycle oscillations take place on the upper branch of unstable steady states. Such oscillations have not been found in numerical integrations performed for several initial conditions. If the oscillations exist, starting from another portion of the three-dimensional phase space, a phenomenon of hard excitation could occur on the lower branch of stable steady states. Pulses of extracellular cAMP (γ) could then bring the system from such a steady state to an oscillatory regime around another, higher cAMP level. Such a process could account for the acceleration of development observed upon addition of periodic cAMP pulses (Darmon *et al.*, 1975; Gerisch & Malchow, 1976). Evidence for hard excitation has, however, not been found in the simulations of system (19).

In the situation of one stable and two unstable steady states, numerical integration always showed the system evolving toward the stable steady state. This state is excitable. When perturbed above a threshold value of γ, the system undergoes a large excursion in the phase space, above the upper unstable steady state before returning to the original stable state. Such a behavior gives rise to the synthesis of a cAMP pulse (curve *b* in Figure 4.1.20). This type of excitable response differs from that encountered in the presence of a single stationary state (Figure 4.1.14).

When two out of three steady states are stable in region *E*, all-or-none transitions between them can be observed (curve *c* in Figure 4.1.20). For a given value of the maximum adenylate cyclase activity σ, the system can then function with two different effective rates for the enzyme reaction. These two rates correspond to two cAMP steady state levels, one low and the other a hundredfold larger. Upon application of cAMP pulses, the

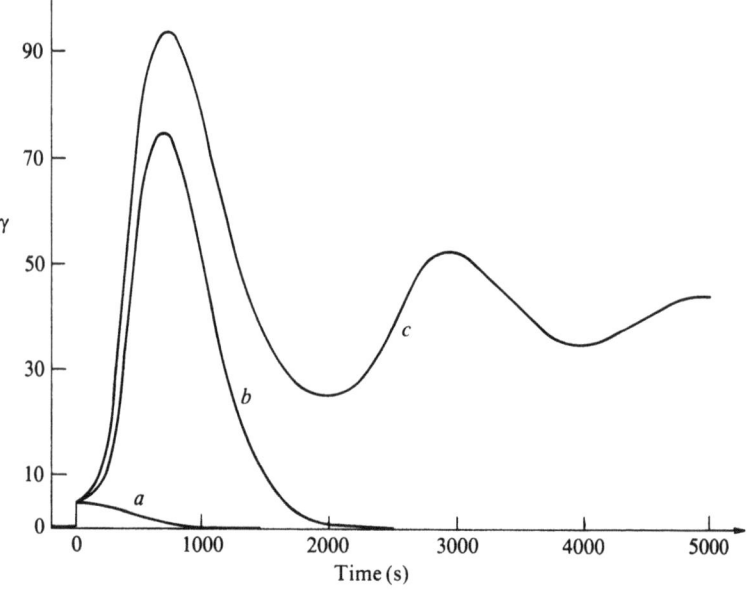

Figure 4.1.20. Time evolution of the system governed by equations (19) illustrating a transition between two stable steady states (curve *c*, $k = 0.006\,\text{s}^{-1}$, $\sigma = 0.08\,\text{s}^{-1}$), a response to a subthreshold perturbation (curve *a*, $k = 0.006\,\text{s}^{-1}$, $\sigma = 0.05\,\text{s}^{-1}$), and excitation in response to a suprathreshold perturbation (curve *b*, $k = 0.008\,\text{s}^{-1}$, $\sigma = 0.09\,\text{s}^{-1}$). Initial conditions for α and β correspond to steady states located on the lower branch of the curves of multiple steady states in Figure 19. Perturbation of the steady state is realized by increasing at time zero the concentration of extracellular cAMP to the value $\gamma = 5$.

system can leave the lower state and evolve toward the upper branch (Figure 4.1.20c). Such a behavior would correspond to a transition from an 'inactive' to an 'active' form of adenylate cyclase. Evidence for such a transition in *D. discoideum* amoebae at the beginning of interphase in response to cAMP pulses has been obtained by Juliani & Klein (1978).

The domain of multiple stationary states is small with respect to the oscillatory and relay domains in Figure 4.1.18. In the course of development, the signaling system may or may not cross the region of multiple steady states, depending on the initial values of parameters k and σ. If it does, the aforementioned phenomena such as transitions between two stable steady states, excitability, or transitions from a stable steady state to an oscillatory regime around a higher mean level of cAMP might occur.

The changes in enzyme activity on the path ABC in Figure 4.1.18 compare qualitatively with the changes occurring during interphase in the activity of phosphodiesterase and adenylate cyclase (Goldbeter & Segel, 1980). Although the hypothesized path in a parameter space formed by two essential enzyme activities already yields a qualitative explanation for the development of the signaling system, it should be stressed that the real path takes place in a multi-dimensional parameter space. Around fifty genes are essential for the completion of aggregation in *D. discoideum* (Coukell, 1975; Williams & Newell, 1976). The fraction of these genes that are required for the operation of the cAMP signaling system itself is, however, not known.

More generally, any change in any parameter that enters into λ, which is defined as $(q\sigma/hk)$, is likely to alter signaling properties. The latter are indeed linked to the S-shaped sigmoidicity of the nullcline $\gamma = \lambda \Phi$ which obtains only above a critical value of λ. Particularly amenable to experimental manipulation is the dilution parameter h; dilution of the cell suspension can result in the suppression of oscillation and excitability by decreasing λ through an increase in h.

An analogous explanation for the development of the cAMP signaling system has been proposed in a model based on a different regulation of adenylate cyclase (Cohen, 1977). In this, the sequence of developmental changes does not result from a variation in enzyme activities but from a variation, resulting from the decrease of reserve material after starvation, in the level of a catabolite that controls cAMP metabolism. The phase plane dynamics of the cAMP signaling system in that model and in the present one well illustrates how continuous changes in one or two parameters can cause a series of discontinuous developmental transitions.

Conclusions

Sustained oscillations and excitability are closely associated in chemical systems. The conditions for their occurrence in biochemistry have been studied here in models for two well-known oscillatory systems; glycolysis in yeast or muscle and the cAMP signaling system in *Dictyostelium discoideum*. In both models, the dynamic behavior is governed by a set of nonlinear differential equations which describe the control of an enzyme – phosphofructokinase or adenylate cyclase – by positive feedback. Much information on the dynamics of these reactions can be obtained even when the number of metabolic variables that control the enzymatic system is reduced to two. The analysis of the models then shows how excitability and sustained oscillations occur in contiguous domains of parameter values. A necessary condition for both phenomena is that one of the two nullclines of the system should be an S-shaped sigmoid in the phase plane determined by the metabolite concentrations.

In the two models, the enzyme is treated as an allosteric protein which obeys a concerted transition mechanism (Monod *et al.*, 1965). Experimental evidence indicates that this assumption holds for phosphofructokinase. In all sources investigated so far – except *D. discoideum* – PFK is an oligomeric protein that exhibits cooperative allosteric interactions (Mansour, 1972). In many instances, the data fit a concerted *K* or *K–V* system, as in *E. coli* (Blangy *et al.*, 1968), muscle (Goldhammer & Hammes, 1978), or yeast (Tamaki & Hess, 1975; Hofmann, 1978; Laurent *et al.*, 1978). The allosteric model for the oscillatory PFK reaction could be extended to take explicitly into account the existence of two substrate–product couples as well as the existence of regulatory sites for the inhibition of the enzyme by ATP. It appears, nevertheless, that the simple monosubstrate allosteric model based on the product activation of PFK already yields agreement with a large number of experiments on glycolytic oscillations in yeast and muscle. In addition, the phase plane analysis of this model shows the possibility of a pulsatory amplification of ADP pulses beyond a threshold. The conditions for observing this phenomenon of excitability in yeast extracts have been determined: the theory predicts that it should occur for substrate injection rates just below those that produce glycolytic oscillations.

In *D. discoideum*, the assumption that adenylate cyclase is an allosteric enzyme still lacks experimental support. The enzyme could in fact be regulated in a different way, e.g. by covalent modification. Currently under investigation (J. L. Martiel & A. Goldbeter, unpublished) is the question as to whether excitability and oscillations are possible in enzyme cascades (Stadtman & Chock, 1978) controlled by positive feedback. Such a situation would apply if adenylate cyclase was regulated by a protein kinase itself under control by cAMP or cGMP (In *D. discoideum* the latter metabolite is synthesized in response to an extracellular cAMP signal prior to adenylate cyclase activation). Preliminary results suggest that nonequilibrium instabilities could occur in these systems as they occur in reactions controlled by allosteric regulation. The main reason is the nonlinear nature of the control that cyclic nucleotides exert on protein kinases (Ogez & Segel, 1976). Although the model for adenylate cyclase suggests that regulation of this enzyme is the primary cause of periodicity and excitation, a role for regulation at the cAMP receptor level (King & Frazier, 1977) cannot be excluded in these two processes.

Recent experiments by Devreotes & Steck (1979) show that amoebae in the course of relay adapt to a given cAMP stimulus. Whereas the signal remains constant, the response is still pulsatory and corresponds to a transient activation of adenylate cyclase. Devreotes & Steck suggested that this result can be explained in terms of the response regulator model proposed by Koshland (1977) for the adaptation to chemical gradients observed in bacterial chemotaxis. This model postulates that the signal controls the rates of two antagonistic processes whose interplay produces a transient response. In contrast, in the afore-analyzed model for the cAMP signaling system it is the depletion of the substrate ATP which causes termination of the response to a pulsatory or constant cAMP signal. Experimental evidence indicates that ATP depletion is not, however, the main factor involved in the decrease in adenylate cyclase activity (Devreotes & Steck, 1979; M. Brenner, personal communication). As an alternative, a recent analysis (J. L. Martiel & A. Goldbeter, unpublished results) shows that the ATP concentration can be replaced as a variable by the cAMP receptor concentration, provided the receptor can undergo a reversible transition between an active and a desensitized state. Termination of the response to a constant cAMP signal is then caused by receptor desensitization. The comparison of such a mechanism with a response regulator model is currently under investigation.

Whatever the precise mechanism of adenylate cyclase activation, the fact that in *D. discoideum* cAMP itself acts as an extracellular hormone which elicits the synthesis of cAMP suggests that this positive feedback is ultimately responsible, as in many other chemical systems (Nicolis & Prigogine, 1977), for the occurrence of excitable and oscillatory behavior. It should be noted that transient responses to hormonal or chemotactic stimuli, which reflect the existence of an adaptation process, do not generally represent manifestations of excitable behavior. As in *Dictyostelium*, the latter phenomenon is associated with the capability for self-sustained oscillations and implies the existence of a sharp threshold for excitation.

It is the conjunction of cooperativity and feedback regulation that leads to the formation of dissipative structures in biochemical systems (Goldbeter & Nicolis, 1976). The examples treated here are based on positive feedback. Most models of metabolic oscillations are based on the more frequent control by end-product inhibition (see Walter, 1970; Tyson & Othmer, 1978; and Chapter 3 in this volume). Sel'kov (1972) has discussed a variety of regulations that give rise to the phase plane characteristics needed for oscillatory and excitable behavior. Among the oscillatory enzyme reactions presently known, it appears that positive feedback is the mechanism most commonly responsible for the onset of metabolic periodicity (Goldbeter & Caplan, 1976). Besides the regulation by allosteric effectors or by covalent modification, the bell-shaped dependence of enzyme reaction rates upon pH can also lead to instability as a result of autocatalysis in reactions that produce an acid or a base (Hahn *et al.*, 1974). Such is the mechanism of the papain membrane oscillator (Caplan, Naparstek & Zabusky, 1973).

The sources for nonlinearity in biochemistry are manifold. It is probable, therefore, that regulated biochemical systems frequently possess the capacity for nonequilibrium self-organization of which excitable and oscillatory behavior are two of the most common modes.

References

Alcantara, F. & Monk, M. (1974). Signal propagation during aggregation in the slime mould *Dictyostelium discoideum. J. Gen. Microbiol.* **85**, 321–34.

Blangy, D., Buc, H. & Monod, J. (1968). Kinetics of the allosteric interactions of phosphofructokinase from *Escherichia coli. J. Mol. Biol.* **31**, 13–35.

Boiteux, A., Goldbeter, A. & Hess, B. (1975). Control of oscillating glycolysis of yeast by stochastic, periodic, and steady source of substrate: A model and experimental study. *Proc. Nat. Acad. Sci. USA* **72**, 3829–33.

Bonner, J. T. (1967). *The Cellular Slime Molds.* Princeton, NJ, Princeton University Press.

Brenner, M. (1978). Cyclic AMP levels and turnover during development of the cellular slime mold *Dictyostelium discoideum. Develop. Biol.* **64**, 210–23.

Caplan, S. R., Naparstek, A. & Zabusky, N. J. (1973). Chemical oscillations in a membrane. *Nature, Lond.* **245**, 364–6.

Chance, B., Pye, E. K., Ghosh, A. K. & Hess, B., eds. (1973). *Biological and Biochemical Oscillators.* New York, Academic Press.

Cohen, M. H. & Robertson, A. (1971). Wave propagation in the early stages of aggregation of cellular slime molds. *J. Theoret. Biol.* **31**, 101–18.

Cohen, M. S. (1977). The cyclic AMP control system in the development of *Dictyostelium discoideum. J. Theoret. Biol.* **69**, 57–85.

Coukell, M. B. (1975). Parasexual genetic analysis of aggregation-deficient mutants of *Dictyostelium discoideum. Mol. Gen. Genet.* **142**, 119–35.

Dalziel, K. (1968). A kinetic interpretation of the allosteric model of Monod, Wyman and Changeux. *FEBS Lett.* **1**, 346–8.

Darmon, M., Brachet, P. & Pereira da Silva, L. H. (1975). Chemotactic signals induce cell differentiation in *Dictyostelium discoideum. Proc. Nat. Acad. Sci., USA* **72**, 3163–6.

De Kepper, P. (1976). Etude d'une réaction chimique périodique. Transitions et excitabilité. *C. R. Heb. Acad. Sci., Paris, Série C* **283**, 25–8.

Devreotes, P. N. & Steck, T. L. (1979). Cyclic 3',5'-AMP relay in *Dictyostelium discoideum.* Requirements for the initiation and termination of the response. *J. Cell Biol.* **80**, 300–9.

Durston, A. J. (1973). *Dictyostelium discoideum* aggregation fields as excitable media. *J. Theoret. Biol.* **42**, 483–504.

Erle, D., Mayer, K. H. & Plesser, T. (1979). The existence of stable limit cycles for enzyme catalyzed reactions with positive feedback. *Math. Biosci.* **44**, 191–208.

Farnham, C. J. M. (1975). Cytochemical localization of adenylate cyclase and 3',5'-nucleotide phosphodiesterase in *Dictyostelium. Exp. Cell Res.* **91**, 36–46.

Fitzhugh, R. (1961). Impulses and physiological states in theoretical models of nerve membrane. *Biophys. J.* **1**, 445–66.

Frenkel, R. (1968). Control of reduced diphosphopyridine nucleotide oscillations in beef heart extracts. 1. Effects of modifiers of phosphofructokinase activity. *Arch Biochem. Biophys.* **125**, 151–6.

Geller, J. & Brenner, M. (1978a). Measurements of metabolites during cAMP oscillations of *Dictyostelium discoideum. J. Cell. Physiol.* **97**, 413–20.

—— (1978b). The effect of 2,4-dinitrophenol on *Dictyostelium discoideum* oscillations. *Biochem. Biophys. Res. Commun.* **81**, 814–21.

Gerisch, G. (1968). Cell aggregation and differentiation in *Dictyostelium.* In *Current Topics in Developmental Biology*, ed. A. A. Moscona and A. Monroy, vol. 3, New York, Academic Press, pp. 157–97.

Gerisch, G. & Hess, B. (1974). Cyclic-AMP controlled oscillations in suspended *Dictyostelium* cells: their relation to morphogenetic cell interactions. *Proc. Nat. Acad. Sci., USA* **71**, 2118–22.

Gerisch, G., Maeda, Y., Malchow, D., Roos, W., Wick, U. & Wurster, B. (1977a). Cyclic AMP signals and the control of cell aggregation in *Dictyostelium discoideum.* In *Developments and Differentiation in the Cellular Slime Moulds*, ed. P. Cappuccinelli and J. M. Ashworth, Amsterdam, Elsevier/North-Holland Biomedical Press, pp. 105–24.

Gerisch, G., Malchow, D., Roos, W., Wick, U. & Wurster, B. (1977b). Periodic cyclic-AMP signals and membrane differentiation in *Dictyostelium.* In *Cell Interactions in Differentiation*, ed. M. Karkinen-Jääskeläinen, L. Saxén and L. Weiss, New York, Academic Press, pp. 377–88.

Gerisch, G. & Malchow, D. (1976). Cyclic AMP receptors and the control of cell aggregation in *Dictyostelium.* In *Advances in Cyclic Nucleotide Research*, ed. P. Greengard and G. A. Robison, vol. 7, New York, Raven Press, pp. 49–68.

Gerisch, G. & Wick, U. (1975). Intracellular oscillations and release of cyclic AMP from *Dictyostelium* cells. *Biochem. Biophys. Res. Commun.* **65**, 364–70.

Goldbeter, A. (1975). Mechanism for oscillatory synthesis of cyclic AMP in *Dictyostelium discoideum. Nature, Lond.* **253**, 540–2.

Goldbeter, A. & Caplan, S. R. (1976). Oscillatory enzymes. *Ann. Rev. Biophys. Bioengin.* **5**, 449–76.

Goldbeter, A. & Erneux, T. (1978). Oscillations entretenues et excitabilité dans la réaction de la phosphofructokinase. *C. R. Heb. Acad. Sci., Paris, Série C* **286**, 63–6.

Goldbeter, A., Erneux, T. & Segel, L. A. (1978). Excitability in the adenylate cyclase reaction in *Dictyostelium discoideum*. *FEBS Lett.* **89**, 237–41.

Goldbeter, A. & Lefever, R. (1972). Dissipative structures for an allosteric model. Application to glycolytic oscillations. *Biophys. J.* **12**, 1302–15.

Goldbeter, A. & Nicolis, G. (1976). An allosteric enzyme model with positive feedback applied to glycolytic oscillations. In *Progress in Theoretical Biology*, ed. F. Snell and R. Rosen, vol. 4, New York, Academic Press, pp. 65–160.

Goldbeter, A. & Segel, L. A. (1977). Unified mechanism for relay and oscillation of cyclic AMP in *Dictyostelium discoideum*. *Proc. Nat. Acad. Sci., USA* **74**, 1543–7.

—— (1980). Control of developmental transitions in cyclic AMP signalling system of *Dictyostelium discoideum*. *Differentiation*, in press.

Goldbeter, A. & Venieratos, D. (1980). Analysis of the role of enzyme cooperativity in metabolic oscillations. *J. Mol. Biol.* **138**, 137–44.

Goldhammer, A. R. & Hammes, G. G. (1978). Steady-state kinetic study of rabbit muscle phosphofructokinase. *Biochemistry* **17**, 1818–22.

Greengard, P. (1978). Phosphorylated proteins as physiological effectors. *Science* **199**, 146–52.

Gross, J. D., Peacey, M. J. & Trevan, D. J. (1976). Signal emission and signal propagation during early aggregation in *Dictyostelium discoideum*. *J. Cell Sci.* **22**, 645–56.

Grutsch, J. F. & Robertson, A. (1978). The cAMP signal from *Dictyostelium discoideum* amoebae. *Develop. Biol.* **66**, 285–93.

Hahn, H. S., Nitzan, A., Ortoleva, P. & Ross, J. (1974). Threshold excitations, relaxation oscillations, and effect of noise in an enzyme reaction. *Proc. Nat. Acad. Sci. USA* **71**, 4067–71.

Henderson, E. J. (1975). The cyclic adenosine 3′:5′-monophosphate receptor of *Dictyostelium discoideum*. *J. Biol. Chem.* **250**, 4730–6.

Hess, B. & Boiteux, A. (1971). Oscillatory phenomena in biochemistry. *A. Rev. Biological Membranes*, ed. J. Järnefelt, Amsterdam–London–New York, Elsevier Publishing Company, pp. 148–62.

Hess, B. & Boiteux, A. (1971). Oscillatory phenomenal in biochemistry. *A. Rev. Biochem.* **40**, 237–58.

Hess, B., Boiteux, A. & Krüger, J. (1969). Cooperation of glycolytic enzymes. In *Advances in Enzyme Regulation*, vol. 7, Oxford and New York, Pergamon Press, pp. 149–67.

Higgins, J. (1964). A chemical mechanism for oscillation of glycolytic intermediates in yeast cells. *Proc. Nat. Acad. Sci. USA* **51**, 989–94.

Hofmann, E. (1978). Phosphofructokinase – a favourite of enzymologists and of students of metabolic regulation. *Trends in Biochem. Sci.* **3**, 145–7.

Juliani, M. H. & Klein, C. (1978). A biochemical study of the effect of cAMP pulses on aggregateless mutants of *Dictyostelium discoideum*. *Develop. Biol.* **62**, 162–72.

King, A. C. & Frazier, W. A. (1977). Reciprocal periodicity in cyclic AMP binding and phosphorylation of differentiating *Dictyostelium discoideum* cells. *Biochem. Biophys. Res. Commun.* **78**, 1093–9.

Klein, C. (1976). Adenylate cyclase activity in *Dictyostelium discoideum* amoebae and its changes during differentiation. *FEBS Lett.* **68**, 125–8.

Klein, C., Brachet, P. & Darmon, M. (1977). Periodic changes in adenylate cyclase and cAMP receptors in *Dictyostelium discoideum*. *FEBS Lett.* **76**, 145–7.

Klein, C. & Darmon, M. (1977). Effects of cyclic AMP pulses on adenylate cyclase and the phosphodiesterase inhibitor of *D. discoideum*. *Nature, Lond.* **268**, 76–8.

Konijn, T. M., van de Meene, J. G. C., Bonner, J. T. & Barkley, D. S. (1967). The acrasin activity of adenosine-3′,5′-cyclic phosphate. *Proc. Nat. Acad. Sci., USA* **58**, 1152–4.

Koshland, D. E., Jr (1977). A response regulator model in a simple sensory system. *Science*, **196**, 1055–63.

Koshland, D. E., Némethy, G. & Filmer, D. (1966). Comparison of experimental binding data and theoretical models in proteins containing subunits. *Biochemistry* **5**, 365–85.

Laurent, M., Chaffotte, A. F., Tenu, J. P., Roucous, C. & Seydoux, F. J. (1978). Binding of nucleotides AMP and ATP to yeast phosphofructokinase: Evidence for distinct catalytic and regulatory subunits. *Biochem. Biophys. Res. Commun.* **80**, 646–52.

Loomis, W. F. (1975). *Dictyostelium discoideum: A Developmental System*, New York, Academic Press.

Maeda, Y. & Gerisch, G. (1977). Vesicle formation in *Dictyostelium discoideum* cells during oscillations of cAMP synthesis and release. *Exp. Cell Res.* **110**, 119–26.

Mansour, T. E. (1972). Phosphofructokinase. In *Current Topics in Cellular Regulation*, ed. B. L. Horecker & E. R. Stadtman, vol. 5, New York, and London, Academic Press, pp. 1–46.

Mato, J. M. & Konijn, T. M. (1977). Chemotactic signal and cyclic GMP accumulation in *Dictyostelium*. In *Developments and Differentiation in the Cellular Slime Moulds*, ed. P. Cappuccinelli and J. M. Ashworth, Amsterdam, Elsevier/North-Holland Biomedical Press, pp. 93–103.

Mato, J. M., Van Haastert, P. J. M., Krens, F. A., Rhijnsburger, E. H., Dobbe, F. C. P. M. & Konijn, T. M. (1977). Cyclic AMP and folic acid mediated cyclic GMP accumulation in *Dictyostelium discoideum*. *FEBS Lett.* **79**, 331–6.

May, R. M. (1972). Limit cycles in predator–prey communities. *Science* **177**, 900–2.

Monod, J., Wyman, J. & Changeux, J. P. (1965). On the nature of allosteric transitions: a plausible model. *J. Mol. Biol.* **12**, 88–118.

Mullens, I. A. & Newell, P. C. (1978). cAMP binding to cell surface receptors of *Dictyostelium*. *Differentiation* **10**, 171–6.

Nicolis, G. & Prigogine, I. (1977). *Self-Organization in Nonequilibrium Systems*, New York, Wiley-Interscience.

Ogez, J. R. & Segel, I. H. (1976). Interaction of cyclic adenosine 3′ : 5′-monophosphate with protein kinase. Equilibrium binding models. *J. Biol. Chem.* **251**, 4551–6.

Plesser, T. (1977). Dynamic states of allosteric enzymes. In *Proceedings of the VII Internationale Konferenz über Nichtlineare Schwingungen, Abhandlungen der Akademie der Wissenschaften der DDR N6*, vol. 2, Berlin, Akademie-Verlag, pp. 273–80.

Prigogine, I. (1969). Structure, dissipation and life. In *Theoretical Physics and Biology*, ed. M. Marois, Amsterdam, North-Holland Publishing Company, pp. 23–52.

Pye, E. K. (1969). Biochemical mechanisms underlying the metabolic oscillations in yeast. *Can. J. Bot.* **47**, 271–85.

Rapp, P. E. & Berridge, M. J. (1977). Oscillations in calcium–cyclic AMP control loops form the basis of pacemaker activity and other high frequency biological rhythms. *J. Theoret. Biol.* **66**, 497–525.

Reich, J. G. & Sel'kov, E. E. (1974). Mathematical analysis of metabolic networks. *FEBS Lett.* **40**, S119–27.

Robertson, A. & Drage, D. J. (1975). Stimulation of late interphase *Dictyostelium discoideum* amoebae with an external cyclic AMP signal. *Biophys. J.* **15**, 765–75.

Robertson, A., Drage, D. J. & Cohen, M. H. (1972). Control of aggregation in *Dictyostelium discoideum* by an external periodic pulse of cyclic adenosine monophosphate. *Science* **175**, 333–5.

Roos, W. & Gerisch, G. (1976). Receptor-mediated adenylate cyclase activation in *Dictyostelium discoideum. FEBS Lett.* **68**, 170–2.

Roos, W., Nanjundiah, V., Malchow, D. & Gerisch, G. (1975). Amplification of cyclic-AMP signals in aggregating cells of *Dictyostelium discoideum. FEBS Lett.* **53**, 139–42.

Roos, W., Scheidegger, C. & Gerisch, G. (1977). Adenylate cyclase activity oscillations as signals for cell aggregation in *Dictyostelium discoideum. Nature, Lond.* **266**, 259–61.

Rossomando, E. F. & Sussman, M. (1973). A 5′-adenosine monophosphate-dependent adenylate cyclase and an adenosine 3′:5′-cyclic monophosphate-dependent adenosine triphosphate pyrophosphohydrolase in *Dictyostelium discoideum. Proc. Nat. Acad. Sci. USA* **70**, 1254–7.

Sel'kov, E. E. (1968). Self-oscillations in glycolysis. 1. A simple kinetic model. *Eur. J. Biochem.* **4**, 79–86.

Sel'kov, E. E. (1972). Nonlinearity of multienzyme systems. In *Analysis and Simulation of Biochemical Systems*, ed. H. C. Hemker and B. Hess, Amsterdam, North-Holland Publishing Company, pp. 145–61.

Shaffer, B. (1962). The acrasina. In *Advances in Morphogenesis*, vol. 2, New York, Academic Press, pp. 109–82.

Shaffer, B. M. (1975). Secretion of cyclic AMP induced by cyclic AMP in the cellular slime mould *Dictyostelium discoideum. Nature, Lond.* **255**, 549–52.

Stadtman, E. R. & Chock, P. B. (1978). Superiority of interconvertible enzyme cascades in metabolic regulation: analysis of monocyclic systems. *Proc. Nat. Acad. Sci. USA* **74**, 2761–5.

Tamaki, N. & Hess, B. (1975). Purification and properties of phosphofructokinase (EC 2.7.1.11) of *Saccharomyces carlsbergensis. Hoppe-Seyler's Zeitschrift für Physiologische Chemie*, **356**, 399–415.

Tornheim K. & Lowenstein, J. M. (1974). The purine nucleotide cycle. IV. Interactions with oscillations of the glycolytic pathway in muscle extracts. *J. Biol. Chem.* **249**, 3241–7.

Town, C. & Gross, J. (1978). The role of cyclic nucleotides and cell agglomeration in postaggregative enzyme synthesis in *Dictyostelium discoideum. Develop. Biol.* **63**, 412–20.

Tyson, J. J. (1977). Analytic representation of oscillations, excitability, and traveling waves in a realistic model of the Belousov–Zhabotinskii reaction. *J. Chem. Phys.* **66**, 905–15.

Tyson, J. J. & Othmer, H. G. (1978). The dynamics of feedback control circuits in biochemical pathways. In *Progress in Theoretical Biology*, ed. F. Snell and R. Rosen, vol. 5, New York, Academic Press, pp. 2–62.

Venieratos, D. & Goldbeter, A. (1979). Allosteric oscillatory enzymes: influence of the number of protomers on metabolic periodicities. *Biochimie* **61**, 1247–56.

Von Klitzing, L. & Betz, A. (1970). Metabolic control in flow systems. 1. Sustained glycolytic oscillations in yeast suspensions under continual substrate infusion. *Archiv. Mikrobiol.* **71**, 220–5.

Walter, C. (1970). The occurrence and the significance of limit cycle behavior in controlled biochemical systems. *J. Theoret. Biol.* **27**, 259–72.

Wick, U., Malchow, D. & Gerisch, G. (1978). Cyclic-AMP stimulated calcium influx into aggregating cells of *Dictyostelium discoideum. Cell Biol. Int. Rep.* **2**, 71–9.

Williams, K. L. & Newell, P. C. (1976). A genetic study of aggregation in the cellular slime mould *Dictyostelium discoideum* using complementation analysis. *Genetics* **82**, 287–307.

Winfree, A. T. (1972). Spiral waves of chemical activity. *Science* **175**, 634–6.

4.2 LINEAR VERSUS SATURATED RATES IN SYNAPTIC RELEASE

Introduction

The subject of this section is a quantitative analysis of the synaptic release of transmitter. We shall begin with the essential biological background. There are basically two ways by which signals are transferred from a nerve to another nerve or to a muscle: one electrical and the other chemical. We will concentrate on the latter. Chemical signals are transferred through the **synapse**, the region that connects the nerve to the target cell. The organization of a typical synapse is shown in Figure 4.2.1. The main components of a synapse are (*a*) the **nerve terminal** whose membrane is the **presynaptic membrane** and in which the **synaptic vesicles** are seen; (*b*) the **postsynaptic membrane**, which is the membrane of the target cells; and (*c*) a **gap** between the membranes whose thickness is typically somewhat greater than 50 nm. In Figure 4.2.2, one can see the steps involved in signal transmission as described by Katz (1965). (*a*) The nerve impulse reaches the nerve terminals. (*b*) The impulse causes release of transmitter from the terminals into the gap. (*c*) The transmitter diffuses through the gap to the postsynaptic membrane. (*d*) At the postsynaptic membrane the transmitter binds to receptors. This causes permeability changes in the postsynaptic membrane and thus brings about an **end-plate potential** (e.p.p.).

A typical e.p.p. is seen in Figure 4.2.3(*b*). If this potential reaches a threshold level, it evokes a muscle (nerve) impulse that propagates along the muscle (nerve). The size of the e.p.p., if it is not too high, is the experimental indication of how much transmitter has been released. The most common neurotransmitter presently known is acetylcholine (Ach).

Transmitter is released from the nerve terminal very rapidly, once the impulse reaches it. This 'synchronized release' has been shown to *require* external calcium (Del Castillo & Katz, 1954*a*) and to be *antagonized* by external magnesium (Katz & Miledi, 1967*a*, *b*). Even if the nerve is not stimulated, end-plate potentials are still seen across the post-synaptic membrane, but they are small and appear spontaneously and randomly.

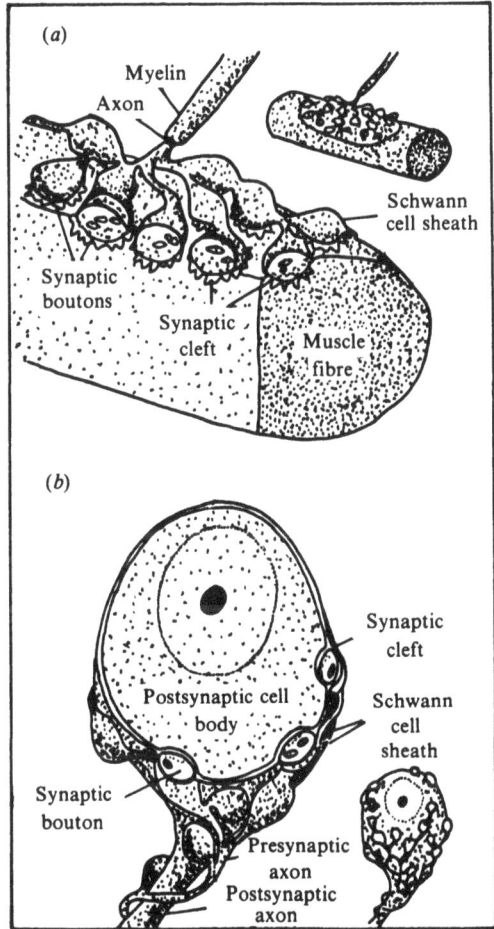

Figure 4.2.1. Organization of a typical synapse. (*a*) Nerve–muscle synapse; (*b*) nerve–nerve synapse (from Kuffler & Nicholls, 1976).

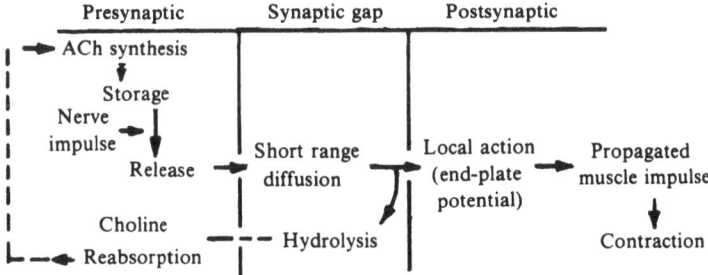

Figure 4.2.2. Chemical signal transmission (from Katz, 1965).

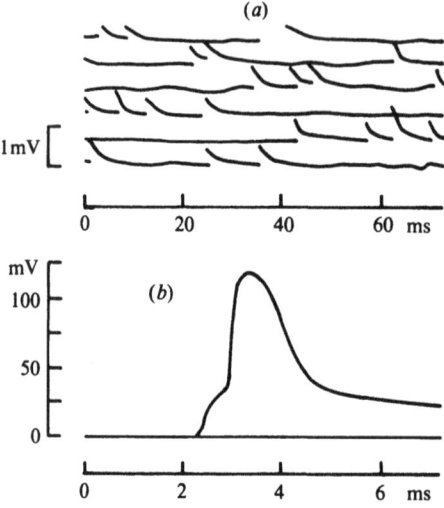

Figure 4.2.3. (*a*) A typical miniature end-plate potential (m.e.p.p.); (*b*) a typical end-plate potential (e.p.p.) (from Kuffler & Nicholls, 1976).

These potentials are called **miniature end-plate potentials** (m.e.p.p.); a typical picture of them is seen in Figure 4.2.3*a*. The m.e.p.p. do not require external calcium, but their frequency increases greatly after an e.p.p. (Del Castillo & Engback, 1954; Miledi & Thies, 1967). This increased frequency is antagonized by external magnesium, in a way similar to the antagonism by Mg of **evoked release** (the release that follows nerve impulse) (Silinsky, Mellow & Phillips, 1977).

The m.e.p.p. represents quantal release, probably the content of one vesicle, and the evoked release is built of many such units. The average number of units that is released after an impulse is several hundred, during about 1 ms. Since in frog the mean frequency of m.e.p.p. is $1\,s^{-1}$, (Del Castillo & Katz, 1954*b*), there is an increase of 10^5 in the frequency of transmitter release after an impulse.

Evidence has accumulated that both evoked and spontaneous release might share a common mechanism. Because of differences in the experimental methods by which the two types of release are measured and because the relation between them is still not absolutely clear, we will concentrate here on evoked release.

It was found many years ago that, if an impulse is given presynaptic-ally, a certain number of quanta are released. If 5, 20 or even 50 ms later a second impulse is given, the amount that is released after the second impulse is normally higher than after the first (Eccles, Katz & Kuffler, 1941). The release continues to increase with successive shocks until it

Figure 4.2.4. Facilitation during and following 5 impulses (from Mallart & Martin, 1967).

eventually saturates. This increase in the amount of transmitter released with successive impulses is called **facilitation** if one or few impulses are given, and **potentiation** if a train of impulses is given. Typical examples of facilitation are seen in Figure 4.2.4.

The basic theory of facilitation was put forward by Katz & Miledi (1967*b*) who suggested that when an impulse is given, Ca becomes attached to active sites on the membrane and that the number of quanta that are released reflects the number of active sites to which Ca is bound. Ca is continuously removed from those sites. When the second impulse arrives, if there are still some sites occupied by Ca then the new number of occupied sites is higher than after the first impulse and more transmitter will be released. This is the 'residual theory'.

The residual theory was further developed by Dodge & Rahamimoff (1967). They showed that the amount of transmitter being released follows a sigmoid curve as a function of external Ca concentration, the slope of which is approximately 4 when Ca is small. They suggested therefore that only sites which are occupied by four molecules of Ca are effective for release. When the second impulse arrives, if there are sites which have two or three Ca molecules, it will be easier to reach the required 'four-state' and thus more transmitter will be released.

The 'residual theory' that describes accumulation of a Ca complex can easily be extended to describe accumulation of Ca in the nerve terminal if a connection between release and internal Ca can be shown. Relevant evidence was obtained when Miledi (1973) injected Ca into nerve terminals of a squid and found elevated transmitter release without nerve stimulations and with no external Ca. There are important differences between normal evoked release and the elevated release that followed injection. If one consults Figures 4.2.5 and 4.2.3, one can see that normal e.p.p. is both higher and much shorter than the release as a result of injection of Ca. Still Miledi's experiments support the notion that release reflects the level of free Ca in the nerve terminals and that facilitation is the result of accumulation of Ca in the terminals. Further support comes from the work of Alnaes & Rahamimoff (1975) which shows that inhibition of the uptake of Ca into mitochondria increases release.

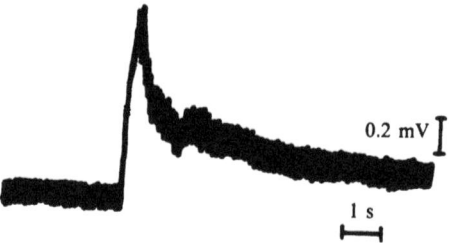

Figure 4.2.5. Transmitter release following injection of Ca into squid nerve terminals (from Miledi, 1973).

One can thus visualize facilitation as arising from three processes: **entry**, **removal**, and **release**.

(*a*) During nerve stimulation, conductance changes occur in the pre-synaptic membrane and Ca enters the nerve terminal. Thus the level of free Ca is temporarily raised from a resting level to a new higher level.

(*b*) Internal Ca is continuously removed into internal Ca stores and the level of free Ca decreases back toward the resting level.

(*c*) Transmitter is rapidly released following the nerve impulse in an amount that reflects the concentration of internal free Ca.

When two pulses are given, let L_1 and L_2 denote the amounts of transmitter released after the first and second pulses. The facilitation (F) is quantitatively defined by

$$F = L_2/L_1. \tag{1}$$

Experimentally, the decay of facilitation is measured by giving the second (test) pulse after increased durations of time. In this context F is regarded as a function of the time t between the two pulses.

Previous models to describe F

Facilitation has been treated quantitatively by several authors. Some of the equations as well as the assumptions that led to them will now be briefly described.

$$\text{(i)} \quad F(t) = F_0 \exp(-\alpha t), \quad \text{Mallart \& Martin (1967).}$$

Mallart and Martin characterized facilitation as being composed of two components, early and late, according to the time course of their decay. These authors proposed that the magnitude and time course of both components of facilitation are the same for every shock in a short train of repetitive stimulations, and that the individual facilitatory effects sum linearly. They also assumed that both components of facilitation decay exponentially. Equation (i) describes the decay of the early component.

(ii) $F = [1 + \exp(-\alpha t)(1 - A)]^4$,   Rahamimoff (1968).

Rahamimoff defined A as the fraction of sites occupied by Ca and thus obtained the results $L_1 = KA^4$ and $L_2 = KA^4(1 - A \exp(-\alpha t) + \exp(-\alpha t))^4$. He assumed cooperativity in release and also assumed that the removal of Ca from the active sites follows an exponential decay curve.

(iii) $F(t) = F_0 \exp(-t/\tau)$,   Balnave & Gage (1974).

Similar to (i).

(iv) $f(t) = 0.8 \exp(-t/50) + 0.12 \exp(t/300)$

$\qquad + 0.025 \exp(-t/3000)$,   Magleby (1973).

Here $f(t)$ is the facilitation contributed by each impulse. According to Magleby, this facilitation has three components, each of which decays exponentially with a different time course.

(v) $A \underset{k_{-1}}{\overset{k_1}{\longleftrightarrow}} B \underset{k_{-2}}{\overset{k_2}{\longleftrightarrow}} C$,   Balnave & Gage (1977).

Here the first reaction represents influx of Ca and the second conversion of Ca from B to an activated form C. The decay of C is responsible for the decay of facilitation and this is given by $C = C_0 \exp(-k_2 t)$, again an exponential decay.

Without separately discussing each of the above models, and others like them that have not been mentioned, one can see that some of the models in the literature do not suggest any physical mechanism for the release and facilitation process, others do suggest a mechanism, but all of them postulate that facilitation or its cause *decays exponentially*. Recall now the picture that was described earlier for release, which assumed that release reflects the internal free Ca concentrations. According to this picture an exponential decay of facilitation or its cause means that the internal Ca concentration declines exponentially after being raised temporarily after the pulse.

The purpose of this section is to look critically at the assumption of exponential decay, using the framework mentioned above in which facilitation arises from the interplay of entry, removal, and release.

Before turning to the details of our theoretical analysis, we shall make some general remarks about notation. As has already been stated, our models will be built of three elements – *e*ntry, re*m*oval, and re*l*ease. The italicized letters provide mnemonics for the notation that we shall employ. The letters 'E' and 'ε' will be associated with *e*ntry, 'μ' with re*m*oval, and 'L' and 'λ' with re*l*ease. In particular, when functions of saturation type are assumed, the 'Michaelis constants' that give the concentrations for half-saturation will be denoted by K_ε, K_μ, and K_λ

respectively. In connection with these saturation functions, we point out that it is the *general* saturation phenomenon that will prove of central importance; the true saturation functions could well be different from the Miahaelis–Menten functions used in our theory, but this would not change the main lines of our argument.

We shall employ the notation $C = C(t)$ to denote the internal Ca concentration at time t, and C_e to denote the external Ca concentration (assumed constant).

Entry and release equations

Dodge & Rahamimoff (1967) found, as was mentioned earlier, that a sigmoid curve relates the amount of transmitter released after an impulse to the external Ca concentration. In particular, this curve is observed to saturate as C_e becomes large. Moreover, Rahamimoff (1968) showed that facilitation was a decreasing function of C_e in experiments where the time interval between the two pulses was very short. We shall now demonstrate that these experimental facts mandate the assumption that release saturates at high internal Ca concentrations.

The observed saturation in release could occur if entry were a saturable function but release itself was linear. Let us work out what facilitation would be in this situation. To do this we first note that entry is expected to be a function of the difference between the external Ca concentration C_e and the internal Ca concentration C (at the time of the impulse). Initially C has a resting value $C_r = 1 \ \mu\text{mole l}^{-1}$, a value that is small compared even to the lowest values of C_e normally used $(0.2 \ \text{mmole l}^{-1})$. After the first entry, C is still small compared to C_e, so that the second entry will be virtually the same as the first. If entry causes internal Ca concentration to increase by an amount E, then after the first entry $C = C_r + E$. The time between pulses is so short that removal is negligible. After the second entry, then, $C = C_r + 2E$. If release is proportional to C, then

$$F = (C_r + 2E)/(C_r + E) \approx 2. \tag{2}$$

There is no dependence on C_e. Thus the experimental results cannot be explained with the assumption that only entry is a saturable function.

Given the results of Dodge & Rahamimoff, it has been shown that we have no alternative to the assumption that release is a saturable function of external Ca. We will now work out what facilitation would be in this situation. We note that release (and entry too) are processes that occur over a period of time. To treat these processes with some degree of precision, one should describe the flow of chemical by an equation wherein the rate at some time t depends on conditions at that time. In fact, both processes are so fast that they can be regarded as taking place instantaneously. This means that the amount of chemical that is released, or that enters the cell, is completely determined by the state of the cell at the time of triggering.

We shall assume that a release that is triggered by an impulse when the internal concentration is C is given by

$$L(C) = \lambda C^{n_\lambda} / [(K_\lambda + C)^{n_\lambda}]. \tag{3}$$

Note that when $C \ll K_\lambda$, $L \approx \lambda C^{n_\lambda}$. For small concentrations, then, release appears to require 'collision' of n molecules. Thus if data-fitting requires $n_\lambda > 1$, one expects that cooperativity will be found in the molecular mechanism for release. (Section 1.4 contains a somewhat detailed discussion on the relation between molecular cooperativity and a phenomenological relationship such as (3). Note that (3) – which was employed by Dodge & Rahamimoff (1967), agrees with the Hill assumption (1.4.23) when $C \ll K_\lambda$.)

Let us assume for the present that entry has a linear functional dependence on C_e only,

$$E = \varepsilon (C_e - C) \approx \varepsilon C_e. \tag{4}$$

(This simple entry assumption will have to be modified later when repetitive stimulation is considered.) Remembering that removal is negligible in the experiments under consideration at the moment, we see that the release in the first two pulses is given by

$$L_1 = \lambda \left(\frac{\varepsilon C_e}{K_\lambda + \varepsilon C_e} \right)^{n_\lambda} \quad \text{and} \quad L_2 = \lambda \left(\frac{2\varepsilon C_e}{K_\lambda + 2\varepsilon C_e} \right)^{n_\lambda} \tag{5a, b}$$

so that

$$F = \frac{L_2}{L_1} = \left[\frac{2(K_\lambda + \varepsilon C_e)}{K_\lambda + 2\varepsilon C_e} \right]^{n_\lambda}. \tag{6}$$

Note that $F \to 1$ as $C_e \to \infty$. Moreover it is easy to show that F is a decreasing function of C_e, for

$$\frac{\partial F}{\partial C_e} = n_\lambda \left[\frac{2(K_\lambda + \varepsilon C_e)}{K_\lambda + 2\varepsilon C_e} \right]^{n_\lambda - 1} \frac{\partial}{\partial C_e} \left[\frac{2(K_\lambda + \varepsilon C_e)}{K_\lambda + 2\varepsilon C_e} \right]$$

$$= \frac{-2\varepsilon K_\lambda n_\lambda [2(K_\lambda + \varepsilon C_e)]^{n_\lambda - 1}}{[K_\lambda + 2\varepsilon C_e]^{n_\lambda + 1}} < 0. \tag{7}$$

A summary of the foregoing considerations and the corresponding experimental results are shown in Figure 4.2.6. This figure demonstrates that the assumption of saturation in release at higher internal Ca concentrations fits the experimental finding of decline in F as C_e increases, when the time interval between pulses is short.

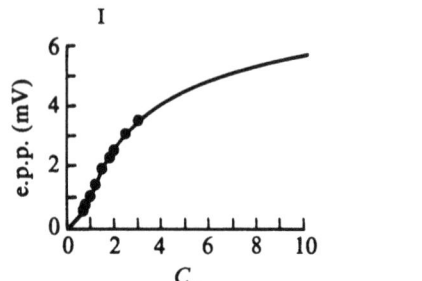

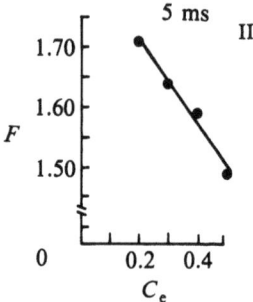

Theoretical analysis

Assume   Predict

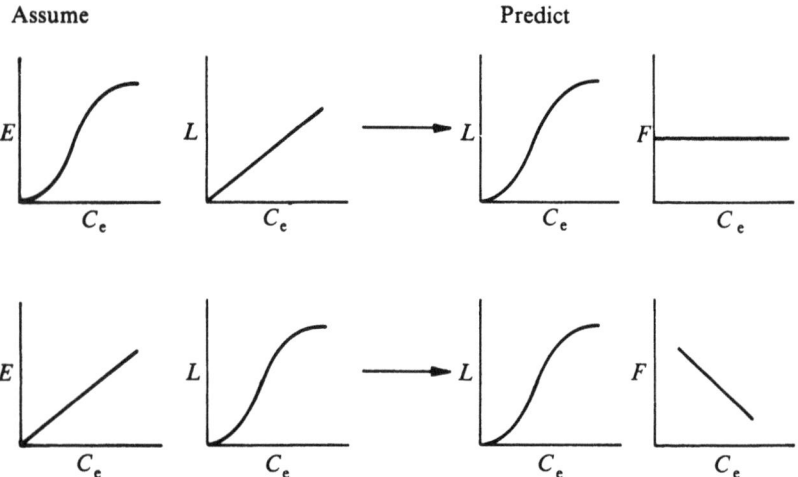

Figure 4.2.6. Experimental and theoretical considerations in choosing the release equation. *Experimental results*: I. The amount of transmitter released following an impulse relates by a sigmoid curve to the external concentration (from Dodge & Rahamimoff, 1967). II. Facilitation declines as external Ca concentration (C_e) is raised, when the time interval between impulses is 5 ms (from Rahamimoff, 1968). For symbols see text.

Theoretical considerations Saturable entry but linear release predicts only I, while linear entry, but saturable release predicts both I and II.

The general entry equation will not be fully discussed here, first because there is not yet a clear way to decide upon it, and secondly because, as will be shown later, its exact nature is not important to the discussion of the main points to be treated here. Because Ca is deemed to enter through channels and because of the competition that occurs between Ca and Mg, it is reasonable to assume the following saturable form for the entry function E:

$$E(C) = \varepsilon(C_e - C)/[K_\varepsilon + (C_e - C)]. \tag{8}$$

No cooperativity is assumed in entry.

Finally, for the moment we shall make the simplest reasonable assumption concerning removal, namely that the removal rate is a linear function of concentration.

$$dC/dt = -\mu C, \quad \text{i.e.} \quad C = C_0 \exp(-\mu t). \tag{9a, b}$$

The crucial experiments

The main subject of this investigation is whether an exponential removal function can account for certain experimental results that concern the relation between facilitation and the external Ca concentration, C_e, and, if not, what the removal function should be. The relevant experimental results are summarized in Figure 4.2.7a. One can see that when there are short time intervals between pulses then F declines if C_e is raised, as expected from the saturation of the release function L at higher C. On the other hand, when the time interval between pulses is increased to 50 ms, then F *increases* as C_e is raised. This result cannot be predicted on the basis of saturation in release. One can also see in Figure 4.2.7 that the time interval over which the test pulses still show facilitation (the duration of facilitation) increases as C_e is raised, both in two-pulse experiments (Rahamimoff, 1968) and after a train of repetitive stimulation (Rosenthal, 1969). Finally, if the decay of facilitation with time is investigated at two different values of C_e, the curves have an intersection point. This result follows from previous observations – at larger values of C_e, facilitation is lower when the test pulse is given shortly after the control pulse, but duration is longer.

Let us consider those experiments for which the interval between pulses is sufficiently long to make it necessary to take removal into account. Using the exponential removal law (9b), we see that (5b) and (6) must be modified to

$$L_2 = \lambda \left[\frac{\varepsilon C_e(1 + \exp(-\mu t))}{K_\lambda + \varepsilon C_e(1 + \exp(-\mu t))} \right]^{n_\lambda},$$

$$F = \left[\frac{(1 + \exp(-\mu t))(K_\lambda + \varepsilon C_e)}{K_\lambda + \varepsilon C_e(1 + \exp(-\mu t))} \right]^{n_\lambda}. \tag{10a, b}$$

The quantitative conclusion of (7) remains true, however, for once again $\partial F/\partial C_e < 0$ for all t. That is, in contradiction to experiment, the present theory predicts that F will *always* decrease as C_e is raised, with any time interval between pulses. For time intervals long compared to μ^{-1}, this decrease will not be detectable. Graphs of F as a function of C_e at various time intervals between pulses are presented in Figure 4.2.7.

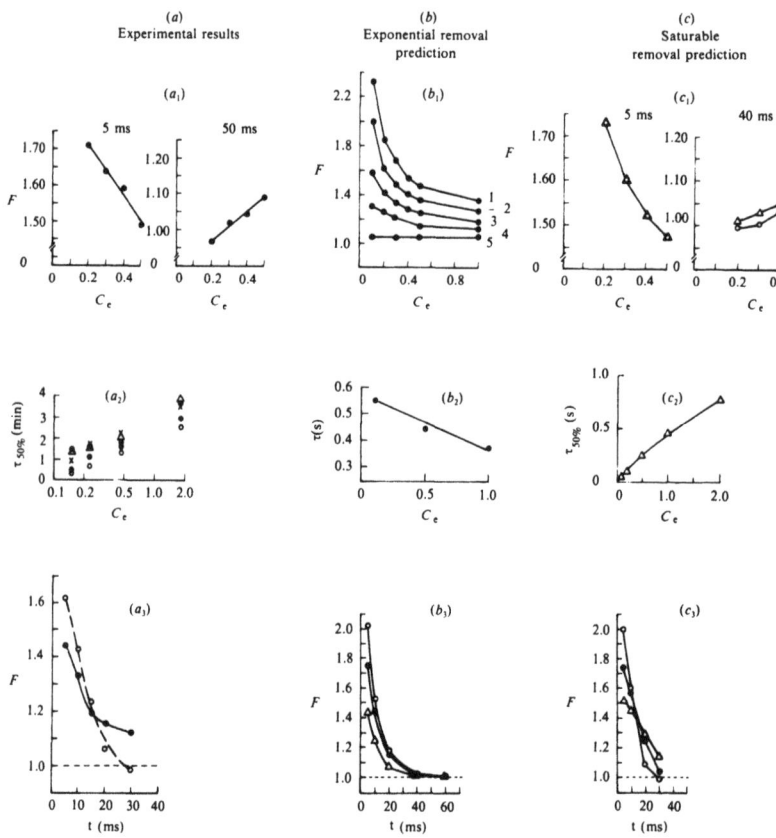

Figure 4.2.7. Experimental and theoretical considerations in choosing the removal equation. Experimental results (a): (a_1), facilitation declines as C_e is raised when the time interval between impulses is 5 ms, but increases, as C_e is raised, when the time interval is 50 ms (from Rahamimoff, 1968); (a_2), the duration of facilitation increases as C_e is raised following a train of repetitive stimulations (from Rosenthal, 1969); (a_3), decay curves of facilitation at $\bigcirc$, $C_e = 0.2$ mmole l^{-1} and $\bullet$, $C_e = 0.4$ mmole l^{-1} (Rahamimoff, 1968).

Theoretical considerations. Given saturable functions for entry and release, and exponential removal (b) fails to predict either (a_1), (a_2), or (a_3). See (b_1), (b_2) and (b_3). In (b_1) 1–5 are 5, 20, 40, 60 and 80 ms, respectively. In (b_3) $C_e = \bigcirc$, 0.1 mmole l^{-1}; $\bullet$, 0.2 mmole l^{-1}; $\triangle$, 0.4 mmole l^{-1}. Saturable removal (c), however, predicts correctly (a_1), (a_2) and (a_3). See (c_1), (c_2) and (c_3). In (c_1), $\triangle$ and $\bigcirc$ are $n_\mu = 1$ and 2. In (c_3), $\bigcirc$, $\bullet$ and $\triangle$ are $C_e = 0.1, 0.2$ and 0.4 mmole l^{-1}, respectively.

Duration of facilitation

We provide a quantitative definition for the **duration of facilitation** τ_q as the time that it takes for F to revert to the value q, where q will normally be a few percent above unity. (The letter τ itself will be used in discussions where the precise value of q is immaterial.) From (10b), τ_q can be found as the solution of the equation

$$q^{1/n_\lambda} = \frac{[1 + \exp{(-\mu\tau_q)}][K_\lambda + \varepsilon C_e]}{K_\lambda + \varepsilon C_e[1 + \exp{(-\mu\tau_q)}]}. \tag{11}$$

A little manipulation shows that one can write

$$1 + \exp{(-\mu\tau_q)} = q^{1/n_\lambda}/[1 - \varepsilon C_e K_\lambda^{-1}(q^{1/n_\lambda} - 1)]. \tag{12}$$

It is already clear that an increase in C_e will decrease the denominator on the right side, and therefore will decrease τ_q. Let us exploit the fact that $q \approx 1$, by employing the approximations $(1 - x)^{-1} \approx 1 + x$ and $\ln(1 - x) \approx x$, where x is any quantity satisfying $|x| \ll 1$. (These approximations are just the first terms of the appropriate Taylor series.) By this means, if $q \approx 1$ we can write

$$\tau \approx \mu^{-1}\ln{[1/(q^{1/n_\lambda} - 1)]} - (\varepsilon C_e/\mu K_\lambda)q^{1/n_\lambda}. \tag{13}$$

This formula explicitly shows that duration decreases as C_e increases, in contrast to the experimental findings depicted in Figure 4.2.7a.

We have demonstrated that the exponential removal function fails to predict correctly *any* of the experimental results which deal with the effect of C_e on facilitation. If one examines the discrepancies between the exponential removal predictions and the experimental results, one can see that the removal process should 'compensate' for the decrease in the derivative of the release function as a function of C_e. Let us demonstrate this assertion in connection with the observation that F is an increasing function of C_e when the time interval between pulses is approximately 50 ms, but is a decreasing function when this interval is considerably less than 50 ms. If 50 ms are enough to remove *all* the Ca that entered at low C_e, but are not enough to remove all the Ca that entered at higher C_e, then $F \approx 1$ at lower C_e and $F > 1$ at higher C_e. Consequently the rate of removal must saturate at higher Ca concentrations, instead of increasing linearly as assumed by the hypothesis of exponential removal. Indeed we will see that this simple assumption of saturation in the rate of removal leads to correct predictions for all the above-mentioned experiments.

It is important to mention that saturation in removal is not only a mathematical requirement which emerged from our theoretical study, but it is also a logical assumption considering the biological facts known about removal. Since it has been pointed out by Alnaes & Rahamimoff (1975) that mitochondria are involved in the removal of internal free Ca, that Ca is extruded through the presynaptic membrane by an active process (Blaustein & Hodgkin, 1969), and that the kinetics follow a Michaelis–Menten curve (Dipolo, 1973), the assumption of saturation seems eminently reasonable.

Final basic model

In our final basic model, the *entry* and *release* processes are governed as before by (8) and (3), respectively. *Removal* will be described by a simple form of saturation with cooperativity. (It will be shown later that cooperativity in removal is not essential to predict correctly the experimental results.) The following equations thus constitute our final basic model.

$$\text{Entry:} \quad E(C) = \frac{\varepsilon(C_e - C)}{K_\varepsilon + C_e - C}. \tag{14a}$$

$$\text{Release:} \quad L(C) = \lambda\left(\frac{C}{K_\lambda + C}\right)^{n_\lambda}. \tag{14b}$$

$$\text{Removal:} \quad \frac{\mathrm{d}C}{\mathrm{d}t} = -\mu\left(\frac{C}{K_\mu + C}\right)^{n_\mu}. \tag{14c}$$

A numerical solution of (14a, b, c) with the parameters listed in Table 4.2.1 is shown in Figure 4.2.7(c). One can see that all the qualitative experimental results are obtained with the saturable rate of removal. By comparing the results in Figure 4.2.7b and 4.2.7c one can also conclude that the dependence of τ on C_e and hence the intersection point at $F > 1$

Table 4.2.1. *List of standard numerical values for the parameters*

ε	0.1 mmole l^{-1}	μ	0.0016 mmole l^{-1} ms^{-1}
K_ε	0.5 mmole l^{-1}	K_μ	0.004 mmole l^{-1} when $n_\mu = 2$,
λ	1 mmole l^{-1}		or 0.008 mmole l^{-1} when $n_\mu = 1$
K_λ	0.015 mmole l^{-1}	n_μ	1 or 2
n_λ	4 or 1	C_r	0.001 mmole l^{-1}

when decay of F is measured at different C_e, as well as the increase in F as C_e increases at longer time intervals – all these results stem from the saturation in removal and not from saturation in any other process. The reason is that the entry equation (14a) and the release equation (14b) are the same in Figures 4.2.7b and 4.2.7c, only the removal equation differs.

It can be concluded that the three equations of (14) provide a correct qualitative description of the phenomenon of facilitation and its dependence on C_e. Therefore the saturability in removal should be accepted and its consequences should be kept in mind by those who attempt to explain other aspects of facilitation. The main consequence of saturability in removal is that the time course of facilitation depends on C_e, or to be more precise, on the internal Ca concentrations at the end of stimulation.

Before we move to a study of facilitation seen after a train of repetitive stimulations (PTP), two points must be made. First, as can be seen in Figure 4.2.9, cooperativity in removal is not essential to explain the major

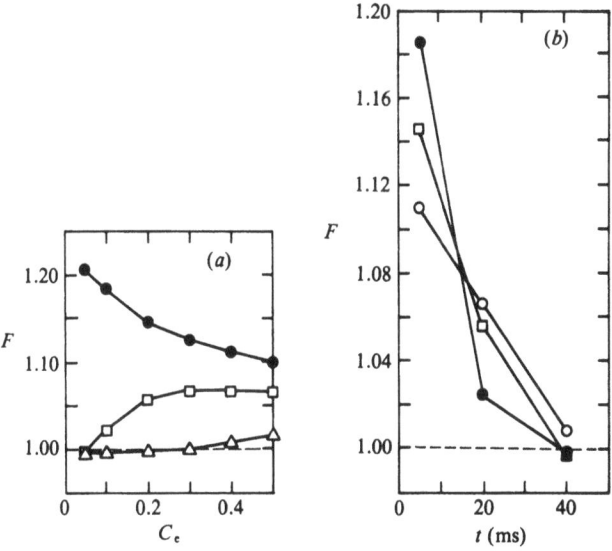

Figure 4.2.8. Various aspects of facilitation when no cooperativity in release is assumed. (*a*) Facilitation is a function of C_e, at time intervals of ●, 5, □, 20 and △, 40 ms. (*b*) Decay of facilitation at C_e = ●, 0.1; □, 0.2 and ○, 0.4 mmole l^{-1}.

effects of saturability in removal. In this figure, F as a function of C_e is seen at 5 and 40 ms intervals between pulses. The results are virtually the same for $n_\mu = 1$ and $n_\mu = 2$; the only thing that differs is the numerical value that some of the parameters must be assigned. Another point is that cooperativity in *release* is also not important and the same basic results are obtained when $n_\lambda = 1$ as long as the removal process is a saturable one. This result can be seen in Figure 4.2.8 in which $n_\lambda = 1$ and we still see the intersection point and the decrease in F as C_e is raised at short time intervals, and the increase in F at longer time intervals. The exact value of

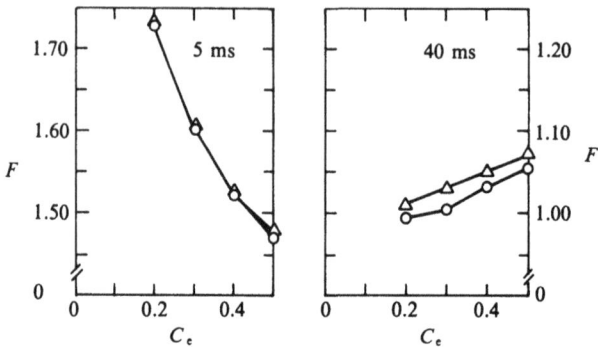

Figure 4.2.9. Aspects of facilitation with and without cooperativity in removal. $\triangle$, $n_\mu = 1$; $\bigcirc$, $n_\mu = 2$.

n_λ determines the magnitude of facilitation but not its behavior. See Parnas & Segel (1980) for a further study of the basic facilitation phenomena.

Are early and late facilitation, augmentation and post-tetanic-potential different processes?

One of the most important differences between the results of exponential and saturable removal is that while the first assumption predicts independence of the time course of facilitated release on the initial level of internal Ca (immediately following the stimulation), the assumption of saturability in removal predicts that the time duration increases as the initial level of internal Ca increases. Because of this difference, advocates of the exponential removal approach must explain the different time courses observed in different experimental conditions as caused by different processes; while advocates of saturable removal will look for different initial concentrations of internal Ca caused by the different experimental conditions.

The exponential removal approach is the only one that has been used until now. Thus it is not surprising that the phenomenon of increased transmitter release following a test pulse has been given at least four names, each corresponding to a different process which is believed to take place. Figure 4.2.10 summarizes the situation.

One can see that the durations of facilitated release differ by orders of magnitude when the experimental conditions differ. Thus when one pulse is given before the test pulse, C_e is 0.2 or 0.5 mmole l^{-1}, and when the external Mg level, M_e, is 1 mmole l^{-1}, then the duration of facilitation is 30–50 ms. If a *few* preliminary pulses are given, if C_e is 1.8 mmole l^{-1} and if M_e is greatly increased to 12–17 mmole l^{-1}, the duration is already

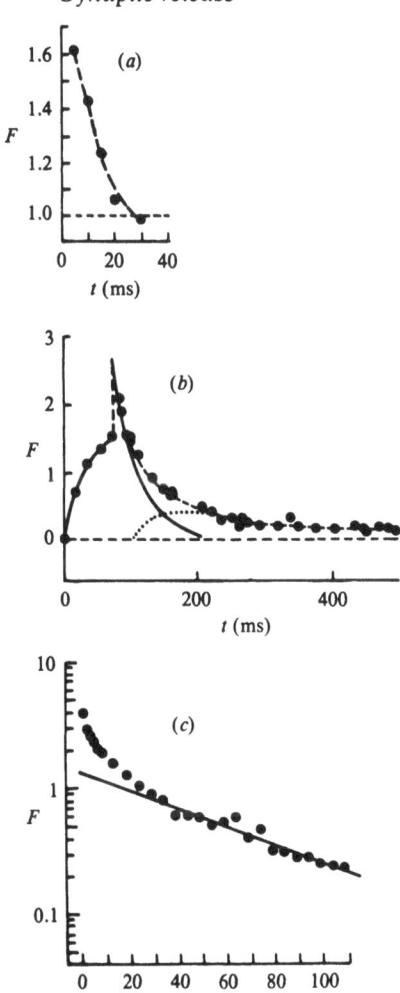

Figure 4.2.10. Typical (exponential) early and late facilitation (as well as augmentation and potentiation. (*a*) Facilitation with a time course of 30–50 ms (from Rahamimoff, 1968). (*b*) Early and late facilitation (from Mallart & Martin, 1967). (*c*) Augmentation and potentiation (Magleby & Zengel, 1976).

several hundreds of ms. Because these observed phenomena could not be described by a single exponent, Mallart & Martin (1967) proposed that **early** and **late facilitation** were two processes occurring at different times, with total facilitation being the sum of the two individual facilitations.

When a long train of repetitive stimulation is given, the duration is seconds, or even minutes in other experiments, and here too the whole period of decay cannot be fitted by one exponent. As a result, two more processes have been suggested. **Augmentation** follows the repetitive stimulation and has a time course of several seconds, while **PTP** happens later with a time course of tens of seconds or even minutes.

It should be emphasized that the 'different processes' postulated by the various authors do not necessarily require that a factor different from internal Ca is involved, but that at different times different processes regulate the internal Ca concentration (Erulkar & Rahamimoff, 1978) or that more than one pool of Ca is involved (Magleby, 1973). According to the common view, then, we have at least four processes that contribute at different times to facilitated release, and the decay of facilitation represents the sum of those processes.

Is saturation in removal enough to unify facilitation and PTP?

If we are to adhere to the saturation model we must try to explain the long time course of PTP with the equations of (14). Moreover it is seen experimentally (Figure 4.2.10) that both the decay of facilitation after one pulse and the decay of PTP sometimes has two time constants. Could saturation in removal account for this? We shall now examine this question.

According to the rules so far, PTP and facilitation both depend on the same pool of free internal Ca and the decay of both depends on the initial level of internal Ca and the rate of its removal. In the case of repetitive stimulation, the initial internal Ca concentration is normally higher than after one pulse, therefore a longer time course for PTP is expected. However, if the initial level of internal Ca is set equal to the external concentration (Ringer: $C_e = 1.8$ mmole l^{-1}) one can calculate the longest possible duration τ that can be expected with the given parameters, on the basis of saturation in removal only. The maximum τ is found to be around 1 s, which is much too short to account for PTP.

Another shortcoming of the present theory is revealed by simulating a train of five pulses with the equations of (14). Figure 4.2.11a, shows the experimentally observed increase in facilitation during a train of five pulses and its decay after stimulation (Mallart & Martin, 1967). There are three basic differences.

(a) The computed increase in F is too small. This discrepancy is even larger than it appears since Mallart & Martin (1967) define F as $(L_2/L_1) - 1$.

(b) The initial decay rate is too slow in the computed results.

(c) The duration of facilitation is too short.

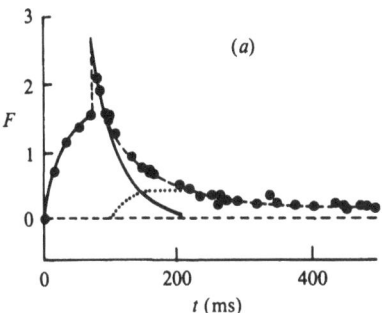

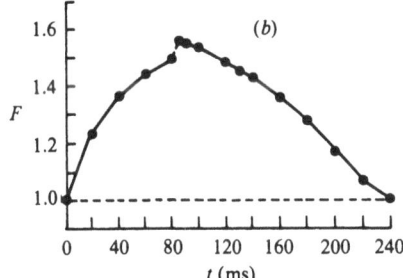

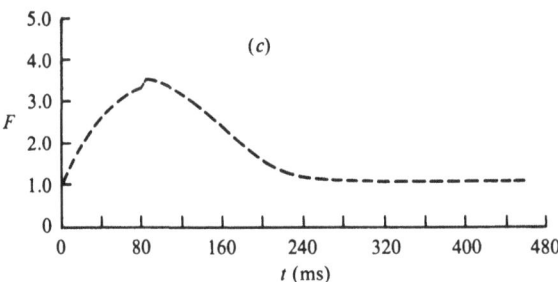

Figure 4.2.11. Experimental and calculated build up and decay of facilitation. (a) Experimental build up and decay of facilitation (Mallart & Martin, 1967). (b) Build up and decay of facilitation calculated with the equations of (14) and the set of parameters listed in Table 4.2.1. (c) Build up and decay of facilitation calculated with (15)–(17), parameters: $\bar{K}_\varepsilon = 2$ mmole l^{-1}, $C_e = 1.8$ mmole l^{-1}.

It is apparent that saturation in removal, even though it does explain correctly all types of dependence of facilitation on Ca concentration, is not enough to account for some of the results seen after a train of stimulations. Examination of the three differences between the computed and experimental results shown in Figure 4.2.11 reveals that both the low

F which is attained during stimulation and its lower decay immediately following stimulation might result from the same source, namely higher $C(0)$ in the computations than in the experiment. One can see from Figure 4.2.7 that when $C_e = 0.2$ mmole l^{-1}, both F and its initial decay rate were higher than when $C_e = 0.4$ mmole l^{-1}. This means that in the calculations, too much Ca was allowed to enter.

Why, then, is τ too short? It may be that the calculated rate of removal is too fast, but what can account for this? If one compares the experimental conditions of Rahamimoff (1968), when one pulse is given, to those of Mallart & Martin (1967), one can see that there is a major difference. While in Rahamimoff's experiments the external concentration of Mg was 1 mmole l^{-1}, it was 12–17 mmole l^{-1} in the five pulse experiments of Mallart & Martin (1967).

As was mentioned earlier, the role of Mg is to block transmitter release. When a train of repetitive stimulation is given, transmitter may be released so frequently and in such large amounts that the internal pools of transmitter become exhausted. In such an event, the measured facilitation would be an artifact of the depletion of internal stores of transmitter. In addition to this, if a nerve–muscle preparation is used then high release can cause a strong contraction of the muscle and thus a jump of the microelectrode used to measure the e.p.p. High external concentrations of Mg are therefore employed in many of the experiments in which repetitive stimulation is given. (It should be mentioned that Mg is not the only means used to overcome the two problems just referred to, but it is a very common one.)

Although the ability of Mg to reduce the amount of transmitter released has often been exploited, its mode of action has not been investigated until recently. In order to inhibit release, if release indeed reflects the internal Ca concentration, a substance must either inhibit the entry of Ca into the nerve terminal during stimulation, or inhibit release, or both. If Mg does inhibit release, it should itself enter the nerve terminal. If Mg competes with Ca on entry, this will explain why the calculated concentration of Ca that entered was too high (in the 5-pulse experiments) and thus F was too low and its initial decay too slow. The shorter calculated duration cannot be explained by that competition.

In recent years it has been found that Mg indeed inhibits the entry of Ca (Baker, 1975). Furthermore Rojas & Taylor (1975) showed that Mg does enter the nerve terminals together with Ca. According to their calculations, more Mg enters than Ca. Miledi (1973) injected Mg together with Ca into squid axon. He compared release with injected Ca alone to that with Ca and Mg together, and found that release was slightly inhibited in the presence of Mg. The main inhibition of release caused by Mg seems thus to be the result of competition on entry of Ca. Carafoli &

Crompton (1975) showed that Mg competes with Ca on uptake into heart mitochondria. On the basis of these facts, it is suggested here that Mg competes with Ca in each of the three processes: entry, release and removal.

The competition of Ca and Mg on removal is an extension of the saturation in removal rate. We suggest that this competition results in the long durations of facilitation that occur after a train of stimulation, during which the internal concentration of both Ca and Mg are raised. It is not the intention here to say that only Mg is able to cause long durations of facilitation. It is suggested that any causes of saturation in removal together with a decrease in the rate of removal can result in prolonging the duration τ by orders of magnitude.

There are other ways to reduce the rate of removal such as lowering the temperature, or by competition between Ca and other ions that are known to enter during nerve stimulation. The result will be the same – prolongation of τ. Mg is chosen for further theoretical investigation since it is such a common component of the experiments in which a long train of stimulations is given. We shall use $M(t)$ to denote the internal concentration of Mg, and M_e to denote the external concentration. An overbar will be employed to distinguish the various constants and functions pertaining to Mg from the corresponding constants and functions for Ca.

Competition between Ca and Mg on entry, release, and removal

We shall take into account the competition between Ca and Mg by generalizing the equations of (14) as follows:

$$\text{Entry of Ca:} \quad E(C, M) = \cfrac{\varepsilon}{\cfrac{K_\varepsilon}{C_e - C}\left(1 + \cfrac{M_e - M}{\bar{K}_\varepsilon}\right) + 1}. \tag{15a}$$

$$\text{Entry of Mg:} \quad \bar{E}(C, M) = \cfrac{\varepsilon}{\cfrac{\bar{K}_\varepsilon}{M_e - M}\left(1 + \cfrac{C_e - C}{K_\varepsilon}\right) + 1}. \tag{15b}$$

Note that if $C_e - C \gg K_\varepsilon$ and $M_e - M \gg \bar{K}_e$ then $E + \bar{E} \approx \varepsilon$, so that ε can be regarded as the maximum capacity of the common Ca–Mg channels.

$$\text{Release:} \quad L(C, M) = \cfrac{\lambda}{\left[\cfrac{K_\lambda}{C}\left(1 + \cfrac{M}{\bar{K}_\lambda}\right) + 1\right]^{n_\lambda}}. \tag{16}$$

Mg is not effective in release (Miledi, 1973), but it can bind to the release sites with a dissociation constant $\bar{K}_\lambda$.

Removal of internal Ca:
$$\frac{dC}{dt} = \frac{-\mu}{\left[\dfrac{K_\mu}{C}\left(1+\dfrac{M}{\bar{K}_\mu}\right)+1\right]^{n_\mu}}. \tag{17a}$$

Removal of internal Mg:
$$\frac{dM}{dt} = \frac{-\bar{\mu}}{\left[\dfrac{\bar{K}_\mu}{M}\left(1+\dfrac{C}{K_\mu}\right)+1\right]^{n_\mu}}. \tag{17b}$$

For simplicity, the degree of cooperativity has been assumed to be the same for Ca and Mg. Such an assumption seems justified in view of our past experience that the details of the equations have little effect.

If we reexamine the previously found defects of the model, we see that they are abolished by introducing competition in removal in addition to saturation. The duration τ increases by orders of magnitude as both $C(0)$ and $M(0)$ increase, as can be seen in Table 4.2.2. The effect of saturation

Table 4.2.2. *Durations of facilitated release* (τ) *under various conditions**

C_e mmole l^{-1}	M_e mmole l^{-1}	No. of impulses	$C(0)$ mmole l^{-1}	$M(0)$ mmole l^{-1}	τ ms
0.2	1	1	0.028	1	116
0.5	1	1	0.049	1	137
0.2	1	50	0.177	1	237
0.2	20	50	0.15	19.9	5600
1.8	20	10	0.148	4.9	724

* See text for discussion of parameters

is seen if one compares results when $C_e = 0.2$ mmole l^{-1} for 1 and 50 pulses; τ increased from 116 to 237 ms. A more striking change is caused by the competition in removal. If one compares $C_e = 0.2$ mmole l^{-1} in the presence of 1 and 20 mmole l^{-1} of external Mg, when 50 pulses are given τ changes from 237 to 5600 ms. These results are preliminary. The numerical values of the parameters have not yet been studied so that it is not possible at present to give a fully meaningful comparison between these and the previously described results (Ca alone).

At the moment we are content to show that τ can change by orders of magnitude if the experimental conditions are such that both Ca and its competitor on removal can accumulate internally. Such conditions exist in repetitive stimulation. In Figure 4.2.12, a decay of facilitated release is

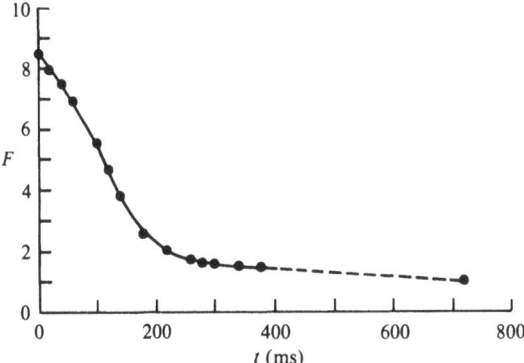

Figure 4.2.12. Decay of facilitated release following a train of 10 impulses with high M_e. Calculated with (15)–(17) and parameters as in Figure 4.2.11c, except that $M_e = 20$ mmole l^{-1}.

seen which corresponds to $C_e = 1.8$ mmole l^{-1} and $M_e = 20$ mmole l^{-1} when 10 pulses are given. Two distinct decay rates, i.e. two 'time constants' are clearly seen, with the first much larger than the second. Because of the high value of M_e and the relative small number of pulses, $C(0)$ is relatively low while $M(0)$ is relatively high, which results in a decay curve with two time constants. This result shows that *more than one time constant in a decay curve does not necessarily mean more than one process. With appropriate parameter choices, competition on decay can result in the same shaped curve.*

Further support to the idea that the shape of the decay curve reflects $C(0)$ and the initial rate of removal can be obtained from the experimental results of Rosenthal (1969). In Figure 4.2.13, we present decay curves of PTP following tetani (pulse trains) of different frequencies. One can

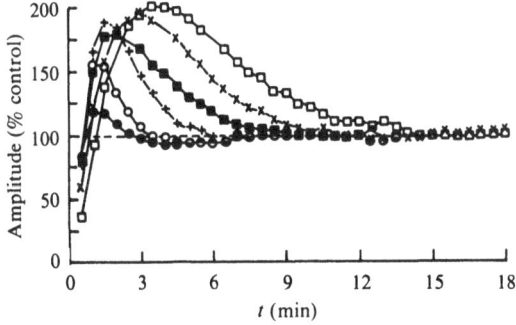

Figure 4.2.13. Decay curves of PTP following tetani of different frequencies (Rosenthal, 1969). ●, $40\,s^{-1}$; ○, $50\,s^{-1}$; +, $62.5\,s^{-1}$; ■, $77\,s^{-1}$; ×, $100\,s^{-1}$; □, $125\,s^{-1}$.

see that the higher the frequency, which means higher initial internal Ca concentration $C(0)$, the slower is the initial decay and the less distinct is the separation into two time constants. At the very high frequency of $125\ \mathrm{s}^{-1}$, one can even observe a decay curve which is slower in the beginning and increases later on when C is reduced. Finally, if competition between Ca and Mg is used in the three processes and the simulation of Mallart & Martin's (1967) experiment is repeated, one can see in Figure 4.2.11 that the results of the computation fit very nicely with the experimental findings in all three aspects: size of F, its initial decay rate, and the duration of facilitation.

The following additional experimental evidence supports the suggestion made here that all results derive from a single set of processes.

(i) As was shown in Figure 4.2.7, both F and PTP show an increase in τ as C_e is raised. If PTP results even in part from a process that does not connect internal to external Ca, such similar dependence would be difficult to explain.

(ii) The duration of both F and PTP have the same temperature dependence. Their Q_{10}'s have the same value, namely around $Q_{10} \approx 4$ (Eccles *et al.*, 1941; Magleby & Zengel, 1976).

Preliminary simulations of (15)–(17) indicate that a set of parameters that show good agreement with two-pulse experiments will enable a τ of seconds following a train of repetitive stimulation, but not minutes. This difficulty is expected since the model presented here is incomplete. The postulated removal process is a lumping of several processes. The reduction in the rate of removal that is suggested here is only because of an increase in the K_m of uptake into the mitochondria, caused by accumulation of internal Mg. It is known, however, that external Mg slows the efflux of internal Ca through the presynaptic membrane. This reduction is because of a decrease in the V-max of the removal, which obviously brings about a longer τ than is calculated here. However, this prolongation is independent of the number of pulses, while the reduction caused by internal accumulation depends on this number. It is also known that Na enters during stimulation and it has been suggested that internal Na slows down the uptake of internal Ca into the mitochondria. Under normal conditions, external Na is present and accumulates in the nerve terminals during repetitive stimulation. Its internal accumulation causes reduction in the rate of removal of internal Ca. Therefore the rate of removal of internal Ca, following tetanic stimulation is even lower than expected on the basis of inhibition by internal Mg only. There is thus no intention to say that only internal Mg causes PTP, and that (15)–(17) fully describe facilitation and PTP. But the theory does unify all the suggested processes involved in facilitated release into one common process.

The conclusions are as follows:

(a) In many cases in which high external Mg was present and repetitive stimulation was given, the accumulation of internal Mg during the stimulation and its competition on removal are the causes for the very long durations of facilitation.

(b) The initial concentration of internal Ca together with its *rate of removal* explains the whole period of facilitated release.

(c) Early and late facilitation as well as augmentation and PTP do not necessarily differ from each other. They may just represent different experimental conditions, and, because of this, different initial concentrations and rates of removal. Both higher external concentrations (of Ca and Mg) and (more importantly) repetitive stimulation increase the initial concentrations of internal Ca and Mg, and thus prolong the duration of F and also change its decay shape.

References

Alnaes, E. & Rahamimoff, R. (1975). On the role of mitochondria in transmitter release from motor nerve terminals. *J. Physiol.* **248**, 285–306.

Baker, P. F. (1975). The regulation of intracellular calcium. *Soc. Exp. Biol. Sym.* **30**, 67–88.

Balnave, R. J. & Gage, P. W. (1974). On facilitation of transmitter release at the toad neuromuscular junction. *J. Physiol.* **239**, 657–75.

—— (1977). Facilitation of transmitter secretion from toad motor nerve terminals during brief trains of action potentials. *J. Physiol.* **266**, 435–51.

Blaustein, M. P. & Hodgkin, A. L. (1969). The effect of cyanide on the efflux of calcium from squid axons. *J. Physiol.* **200**, 497–527.

Carafoli, E. & Crompton, M. (1975). Calcium ions and mitochondria. *Soc. Exp. Biol. Sym.* **30**, 89–114.

Del Castillo, J. & Engbaek, L. (1954). The nature of the neuromuscular block produced by magnesium. *J. Physiol.* **124**, 370–84.

Del Castillo, J. & Katz, B. (1954a). Quantal components of the end-plate potential. *J. Physiol.* **124**, 560–73.

—— (1954b). Statistical factors involved in neuromuscular facilitation and depression. *J. Physiol.* **124**, 574–85.

Dipolo, R. (1973). Calcium efflux from internally dialyzed squid giant axons. *J. Gen. Physiol.* **62**, 575–89.

Dodge, F. A., Jr & Rahamimoff, R. (1967). Cooperative action of calcium ions in transmitter release at the neuromuscular junction. *J. Physiol.* **193**, 419–32.

Eccles, J. C., Katz, B. & Kuffler, S. W. (1941), Nature of the 'end plate potential' in curarized muscle. *J. Neurophysiol.* **4**, 362–87.

Erulkar, S. D. & Rahamimoff, R. (1978). The role of calcium ions in tetanic and post-tetanic increase of miniature end-plate potential frequency. *J. Physiol.* **278**, 501–11.

Katz, B. (1965). *Nerve, Muscle and Synapse*, New York, McGraw-Hill Publishing Company.

Katz, B. & Miledi, R. (1967a). A study of synaptic transmission in the absence of nerve impulses. *J. Physiol.* **192**, 407–36.

——— (1967b). The timing of calcium action during neuromuscular transmission. *J. Physiol.* **189**, 535–44.

Kuffler, S. W. & Nicholls, J. G. (1976). *From Neuron to Brain*, Sunderland, Mass., Sinauer Assoc., Inc.

Magleby, K. L. (1973). The effect of tetanic and post-tetanic potentiation on facilitation of transmitter release at the frog neuromuscular junction. *J. Physiol.* **234**, 353–71.

Magleby, K. L. & Zengel, J. E. (1976). Augmentation: a process that acts to increase transmitter release at the frog neuromuscular junction. *J. Physiol.* **257**, 449–70.

Mallart, A. & Martin, A. R. (1967). An analysis of facilitation of transmitter release at the neuromuscular junction of the frog. *J. Physiol.* **193**, 679–94.

Miledi, R. (1973). Transmitter release induced by injection of calcium ions into nerve terminals. *Proc. R. Soc. Lond.* B **183**, 421–85.

Miledi, R. & Thies, R. E. (1967). Post-tetanic increase in frequency of miniature end-plate potentials in calcium-free solutions. *J. Physiol.* **192**, 54–55P.

Parnas, H. & Segel, L. A. (1980). A theoretical explanation for some effects of calcium on the facilitation of neurotransmitter release. *J. Theoret. Biol.*, **84**, 3–29.

Rahamimoff, R. (1968). A dual effect of calcium ions on neuromuscular facilitation. *J. Physiol.* **195**, 471–80.

Rojas, E. & Taylor, R. E. (1975). Simultaneous measurements of magnesium, calcium and sodium influxes in perfused squid giant axons under membrane potential control. *J. Physiol.* **252**, 1–27.

Rosenthal, J. (1969). Post-tetanic potentiation at the neuromuscular junction of the frog. *J. Physiol.* **203**, 121–33.

Silinsky, E. M., Mellow, A. M. & Phillips, T. E. (1977). Conventional calcium channel mediates asynchronous acetylcholine release by motor nerve impulses. *Nature, Lond.* **270**, 528–30.

4.3 OPTIMAL STRATEGIES FOR THE METABOLISM OF STORAGE MATERIALS IN UNICELLULAR AND MULTICELLULAR ORGANISMS

Introduction

It is widely accepted that the function of energy-storage materials in the form of carbohydrates and lipids is to provide energy when there is a shortage in its normal source. Deficiency in energy can result from various factors. A few examples follow, in which the cause of the deficiency and the role of storage materials in overcoming this deficiency are specified.

(a) There may be a shortage in an external source of energy, as in a period of starvation. Cook (1966) showed that algae which grow in natural light–dark cycles store energy in the form of the starch **paramylum** during the day and use it during the night when no external source of energy is available. Microorganisms are known to break down their glycogen, or PHB (poly-β-hydroxy butyrate) during starvation periods.

(b) Higher than normal demand for energy might take place during migration periods. It has been shown that birds and whales break down their reserve materials during migration (Odum, Marshall & Marples, 1965; Brodie, 1975).

(c) Lower capability of using the external source may occur in relatively poor conditions, usually winter. Schindler, Clark & Gray (1971) showed that storage materials are used by some animals during the winter.

Given that the role of storage materials is to enable the organism to survive during periods in which there is a temporary shortage in energy, the unavoidable conclusion is that these materials must be synthesized prior to the time of demand for them. This conclusion is almost self-evident when the following conditions are fulfilled: (i) the appearance of a period of shortage can be predicted; (ii) the exact time of its appearance is certain; and (iii) the same organism both prepares and uses the storage materials according to these conditions.

An organism that faces unpredictable starvation periods does not fulfill these conditions. Nor do microorganisms, which normally have a short life cycle. Indeed the common view is that microorganisms synthesize storage materials *if and only if* there is a surplus of external energy over that required for present growth (Wilkinson, 1959; Palmstierna, 1956; Holme, 1957; Cook, 1963.) This 'excess theory' leads to situations in which there is no connection between the preparation of the storage materials and their use. Since a high external level of energy is not normally followed by immediate starvation, a bacterium with life cycle of 20 min, say, will prepare glycogen when the external carbon concentration is high, and since it is high this specific bacterium will not use the glycogen. Daughter cells facing a decreasing external carbon concentration will not prepare glycogen and will not be ready for the starvation that is likely to come after a gradual decrease in the external carbon source.

Even in metazoa the ideas are not very clear. Slobodkin (1962) asserted that a high energy content per unit weight is favored under circumstances such as preparation for fasting or stress, but he also said that natural selection normally favors production of the maximum number of progeny rather than high energy content. This idea means that if the demand for energy and the time of its appearance are certain (conditions (i) and (ii)) then the organism should prepare itself. However, it does not draw any correlation between the variable natural ecological conditions of the organism and its policy in respect to metabolism of storage materials. Calow & Jennings (1974) demonstrated that there is such a correlation.

The purpose of this section is to describe a quantitative way to determine the policy as well as the level of storage materials in an organism given its natural ecological conditions. The model is developed for microorganisms, but some of its conclusions can be extrapolated to metazoa. Much of the material, with extensive quotations, is taken from Cohen & Parnas (1976) and Parnas & Cohen (1976).

Analogy between storage materials in living systems and inventory in firms

There is a striking resemblance between storage materials in a living organism and inventory in firms. Both are stored for future use, and in the future both can be used for production instead of external supply if there is a shortage. Moreover, in both cases storage in the present time is likely to be at the expense of current production when external supply is limiting.

In economic systems inventory control is determined by an **optimal policy**. The policy is to maximize the **profit**, which is the difference between the extra income and the extra expenses caused by storage.

There is an analogy between economic and biological systems that follows from the belief that natural selection as well as the economic system favors maximization of 'profit'. Consequently, it is suggested here that we employ the same mathematical approach which is used to solve inventory control in order to determine the policy of synthesis and the level of storage materials in a living organism. We propose that storage materials are synthesized at the present time in an amount which results from optimization of the 'profit', taking into consideration both the loss and the gain arising from storing reserve materials. This optimal amount thus answers the demand in the future for storage materials in the natural environment of the organism.

The profit function in biological systems

We define the profit function in microorganisms as the **long term growth rate** in their natural environment. This growth rate is the long-term average of growth in periods when growth is possible and of zero growth (survival) during starvation periods. The profit function in multicellular organisms in a similar way is the long-term number of progeny. Therefore the loss at the present time because of storage, the gain in the future due to the stored materials, and the profit which is the difference between the gain and the loss, should be expressed in units of growth rate in the case of microorganisms and in the number of progeny in multicellular organisms. For a population of microorganisms we denote the loss function as L, the gain function as G and the profit as π.

The natural environment of microorganisms

There are many factors that determine the natural environment and hence the long term growth rate of a population of microorganisms. However, we only consider here the level (or rate) of supply of energy to the cells, since this is the environmental factor which determines the level of storage materials. In this respect one can distinguish between two extreme environments: deterministic and random. Deterministic environments will be characterized by an external energy (carbon) source concentration c_1 that lasts for t_1 units of time, followed by concentration c_2 for time t_2. Various cases can be considered.

(*a*) Concentration c repeatedly alternates between c_1 and c_2, $c_1 > c_2 > 0$. c_2 is lower than c_1 but is larger than zero so that there is no starvation period. This environment is rather an artificial one but can elucidate many of the general factors that determine the optimal policy. We will not discuss this environment here. A full treatment of such a pattern was carried out by Parnas & Cohen (1976).

(*b*) Same as (*a*) except that $c_1 > c_2 = 0$. This environment constitutes a regular alternation of concentration c_1 for a time t_1, followed by starvation for a period t_2. Such a pattern cannot be studied as a special case of pattern (*a*), since the metabolic behavior during starvation is completely different from that during growth, even when growth is very slow. A general study of such an environment is provided by Parnas & Cohen (1976).

(*c*) A deterministic environment in which the carbon source appears in regular pulses but the concentration of the substrate gradually falls. This fall may depend on population growth, or may occur independently. In either case we assume that a starvation period of a fixed duration t_2 always occurs only after the low concentration at the end of the pulse.

A natural situation that corresponds to pattern (*c*) is a photoperiodic environment for photosynthetic organisms. A detailed solution for the policy and level of storage materials in photosynthetic algae (as in Cohen & Parnas, 1976) is given below. The theoretical predictions are examined and confirmed experimentally. (See also Parnas & Cohen, 1976.)

It is important to construct the model for a natural deterministic environment which can be easily repeated experimentally, and only then to extend the model to a random environment. Therefore we will show the approach and the solution for photosynthetic algae and then briefly describe a random environment.

A deterministic environment in which $c_1 > c_2 = 0$

A photoperiodic environment is best described by pattern (*c*) wherein dark (starvation) follows a decline in light intensity. It might very well be that the decline in light intensity is the signal for the starvation. However, in our treatment we assume that the light intensity during the day is constant and that a day of a given length is followed by a fixed period of night. We also do not base our calculations on the possibility that the decline in light intensity is a signal for starvation. Therefore the photoperiodic environment falls into category (*b*).

The temporal distribution of the increase in weight and of cell division

We begin with some definitions. First we assume that the number of new cells produced, N, is proportional to the increase in weight of the cell materials excluding the storage materials,

$$N = (W_{PT} - W_{P0})/W_M \equiv \Delta W_P / W_M, \tag{1}$$

where W_M is the average weight of a daughter cell, W_{PT} is the weight of the cell materials W_P at the end of the day, and W_{P0} is its weight at the beginning of the day. The creation of a new cell requires a quantity of energy E_d in carbohydrate equivalents. The total energy needed for division, expressed as a fraction of the total weight, is called E.

$$E = E_d N / W_{PT}. \tag{2}$$

If the cells grow and divide at the same time, we may assume that photosynthesis is depressed during the time needed for the production of the necessary energy (Spectorov, Slobodskaya & Nichiporovich, 1963). We therefore see that when division takes place during the day, less time will be left for the production of cell components and the number of daughter cells will therefore be reduced.

Conclusion 1
In order to maximize the number of daughter cells per day, the cell should increase its weight, and store the necessary energy during the day, and divide during the night.

At very high light intensities and long days, the cells may reach their upper limit of size during the day, and will then divide. Since the specific rate of photosynthesis falls when the cells grow very large, the cells should divide every night, provided that their weight has reached twice the minimal weight.

The optimal timing of synthesis of storage materials during the day

We have just seen that during the day the cells have to store the energy for division. Energy can be derived from specific storage materials, such as starch, or from the breakdown of the cellular proteins which make the synthetic machinery of the cells.

Each cell starts the day with quantity W_{P0} of cell materials, i.e. materials capable of synthesis, and a quantity of W_{S0} of storage materials. At the end of a day of length T these become W_{PT} and W_{ST}. S is the ratio between them, i.e.

$$S = W_{ST} / W_{PT}. \tag{3}$$

(W_{S0} may not be zero, but we shall neglect the small contribution stemming from this source.) The ratio of energy required during the night (for division and respiration) to the synthetic machinery is called S_{min}.

$$S_{min} = [E_M(24 - T) W_{PT} + E_d N] / W_{PT}, \tag{4}$$

where E_M is the maintenance energy per hour per unit weight of cell material. Substituting N from (1) gives

$$S_{min} = E_M(24 - T) + (E_d/W_M)[(W_{PT} - W_{P0}/W_{PT})].\qquad(5)$$

At the beginning of the night it is possible to have

$$S = S_{min}, \qquad S < S_{min}, \qquad S > S_{min}.\qquad(6)$$

If $S = S_{min}$, S is completely used up during the night. If $S < S_{min}$ part of W_P must be converted to energy, equal to

$$W_{PT} \cdot \frac{1}{\alpha'}(S_{min} - S)\qquad(7)$$

where α' is the coefficient for the conversion of W_P to energy; or more accurately, the ratio between the conversion coefficients for W_P and W_S. If $S > S_{min}$, the excess of W_S will be converted to W_P with an efficiency β. The increase in W_P will be

$$W_{PT} \cdot \beta(S - S_{min}).\qquad(8)$$

Note that $\alpha' < \beta < 1$.

The optimal timing of the synthesis of W_P and W_S during the day

Since N is proportional to ΔW_P, by (1) the optimal timing should minimize the loss in W_P, while reaching the required level of W_S. At limiting light intensities, the production of W_S during any time interval during the day will decrease the final level of W_P. Since the rate of synthesis of W_S is proportional to W_P we obtain

$$\Delta W_S = W_{Pt}R_P\Delta t,\qquad(9)$$

where Δt is the time required for the synthesis of an increment ΔW_S at the specific photosynthesis rate R_P, and W_{Pt} is the level of W_P at the time t of synthesis of W_S. It is clear that Δt for a given W_S is smallest when W_{Pt} is largest. Since W_{Pt} always increases with t, Δt will be minimal when the synthesis of ΔW_S takes place as late as possible. As this is true for any increment ΔW_S and Δt, it is also true for any sum of such increments. Thus conclusion 2 follows.

> *Conclusion 2.* For any given W_S, W_{PT} is maximized if only
> W_P is synthesized at the beginning, and only W_S is
> synthesized at the end of the day. Since this is true for any
> W_S, it is also true for the optimal W_S. The time it takes to
> synthesize W_S is called τ. It is given by

$$\tau = S/R_P.\qquad(10)$$

Calculation of the profit function

The gain function G

G is independent of the timing of synthesis of W_S during the day, i.e. G is the saving in the breakdown of W_P to supply S_{min}. When $S > S_{min}$, G is the additional gain by the conversion of the excess S to W_P. Thus, when $S < S_{min}$

$$G = (1/\alpha')SW_{PT}, \tag{11}$$

and when $S > S_{min}$

$$G = (S_{min}W_{PT}/\alpha') + \beta W_{PT}(S - S_{min}). \tag{12}$$

The loss function L

By Conclusion 2, W_S is synthesized at the end of the day and τ denotes the amount of time devoted to this synthesis of storage material. L will be the loss in W_{PT} caused by the synthesis of W_S. We assume exponential growth at rate R_P, so that

$$L = W_{P0}\{\exp(R_P T) - \exp[R_P(T - \tau)]\}. \tag{13}$$

Substituting τ from (10) and rearranging we obtain

$$L = W_{P0}\exp(R_P T) - W_{P0}\exp(R_P T - S). \tag{14}$$

We see that L always increases with S.

The profit function π

This is given by $G - L$. When $S > S_{min}$,

$$\pi = W_{PT}[(S_{min}/\alpha') + \beta(S - S_{min})]$$
$$- [W_{P0}\exp(R_P T) - W_{P0}\exp(R_P T - S)]. \tag{15}$$

Substituting $W_{PT} = W_{P0}\exp(R_P T - S)$, we find that

$$\frac{\partial \pi}{\partial S} = W_{P0}\exp(R_P T - S)\left[\beta - \frac{S_{min}}{\alpha'} - (S - S_{min}) - 1\right], \tag{16}$$

which is always negative.

We see that the optimal S, S_{op}, must either be equal to or less than S_{min}. When $S < S_{min}$

$$\pi = (W_{PT}S/\alpha') - W_{P0}\exp R_P T + W_{P0}\exp(R_P T - S), \tag{17}$$

and

$$\frac{\partial \pi}{\partial S} = W_{P0} \exp (R_P T - S)[(1 - S - \alpha')/\alpha'].$$ (18)

If $S_{min} < (1 - \alpha')$ the derivative is a positive decreasing function of S, so that S_{op} must be at the boundary of S_{min}. If $S_{min} > (1 - \alpha')$, the derivative becomes zero at an intermediate value of $S_{op} < S_{min}$. In this case,

$$(1 - S - \alpha')/\alpha' = 0;$$ (19)

i.e. $S_{op} = 1 - \alpha'$.

> *Conclusion 3.* S_{op} is equal to S_{min} when $S_{min} < 1 - \alpha'$, and equal to $1 - \alpha'$ when $S_{min} > 1 - \alpha'$.

S_{op} as a function of the rate of photosynthesis R_P

The contribution of a model is tested not only by its ability to predict and explain experimental results, but also by its ability to distinguish between various theoretical possibilities, in this particular case between the 'excess' and the 'optimal policy' theories. According to the 'excess' theory if the light intensity is limiting in such a way that the cell cannot even double its dry weight during the day, then the cell will synthesize proteins only and will not synthesize any storage materials (Cook, 1966). The predicted policy according to the 'optimal policy' theory regarding synthesis of storage materials at very low light intensities can be studied in two ways. The first way requires examination of the derivative of S_{min} (which was found to be equal to S_{op} under most conditions) with respect to the rate of photosynthesis (R_P). The other way is to examine the number of hours during the day that cells will synthesize W_S at various light intensities, i.e. to study $\partial S_{min}/\partial R_P$ and $\partial \tau/\partial R_P$.

To study S_{op}, we substitute $S_{op} = S_{min}$, and $W_{PT} = W_{P0} \exp (R_P T - S)$, and differentiate S_{min}, (5), with respect to R_P at constant T. This gives

$$\frac{\partial S_{min}}{\partial R_P} = \frac{T(E_d/W_M)}{\exp (R_P T - S_{min}) + (E_d/W_M)}.$$ (20)

The derivative is positive and decreasing for all R_P. When R_P is very high, $W_{PT} \gg W_{P0}$. Thus, from (5), as $R_P \to R_{max}$

$$\text{maximal } S_{min} = E_M(24 - T) + (E_d/W_M).$$ (21)

However, when R_P is very low, and W_{PT} does not differ very much from W_{P0} one can see from (5) that as $R_P \to 0$

$$\text{minimal } S_{min} = E_M(24 - T).$$ (22)

The effect of R_P on τ

The derivative of τ, (10), assuming $S = S_{op} = S_{min}$, gives

$$\frac{\partial \tau}{\partial R_P} = \frac{R_P(\partial S_{min}/\partial R_P) - S_{min}}{R_P^2}. \tag{23}$$

Substituting $\partial S_{min}/\partial R_P$ (20), and S_{min}, (5), in the numerator of (23) gives

$$\frac{\partial \tau}{\partial R_P} = \frac{E_d}{R_P^2 W_M}\left[\frac{TR_P}{\exp(R_P T - S_{min}) + (E_d/W_M)} + \frac{1}{\exp(R_P T - S_{min})} - 1\right]. \tag{24}$$

To see whether this is positive or negative we have employed numerical values of the parameters taken from *Euglena*. The value of S_{min} reaches a limiting value of $E_M(24 - T)(E_d/W_M)$ at saturating light intensities. Under these conditions paramylum makes up 20% of the dry weight of the cells (Cook, 1963), which is equivalent to $S_{min} = 0.25$ according to our definition. E_M is about $0.005\ h^{-1}$ (Cook, 1966), so that E_d/W_M is about 0.2 for a day length of 14 h.

At zero growth, $R_P T = S_{min}$. In *Euglena* one such combination is of $R_P \sim 0.02\ h^{-1}$, $T \sim 5$, so that S_{min} is then ~ 0.1. Since $\partial S_{min}/\partial R_P$ decreases exponentially with R_P, it is sufficient to prove that

$$(\partial S_{min}/\partial R_P)R_P < S_{min} \tag{25}$$

at this low R_P. Substituting the numerical values in

$$\frac{\partial \tau}{\partial R_P}(R_P T - S_{min})$$

shows that this indeed is the case.

> *Conclusion 4.* The number of hours during which storage materials are synthesized, τ, *decreases* when the rate of photosynthesis *increases*.

Conclusion 4 also holds when $S_{op} = 1 - \alpha'$, since then

$$\partial \tau/\partial R_P = (\alpha' - 1)/R_P^2, \tag{26}$$

which is always negative. The numerical dependence of τ on R_P has been obtained by iteratively solving (5) for S_{min}, and then setting $\tau = S_{min}/R_P$. The results are given in Figure 4.3.1, in which we see S_{min} increasing and τ decreasing as a function of R_P, as predicted by the theoretical analysis.

The effect of the length of the day (T) on S_{min} and τ is not given here since both S_{min} and τ depend very little on T under natural conditions. A full mathematical description of $\partial S_{min}/\partial T$ and $\partial \tau/\partial T$ is provided by Cohen & Parnas (1976).

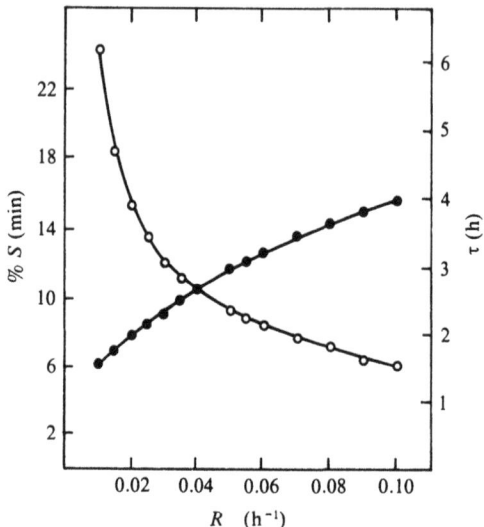

Figure 4.3.1. A numerical solution of the model showing S_{min} and τ as a function of R_P. Length of day, T, = 14 h. Note the two different scales: $\bigcirc$, τ; $\bullet$, S_{min}. From Cohen & Parnas (1976).

Experimental validation of the model

The main predictions of the model are as follows:
1. Cells grow during the day and divide at night.
2. They first synthesize proteins and then the reserve material paramylum.
3. The duration of paramylum synthesis increases as the light intensity decreases.

These predictions were tested by growing a culture of *Chlamydomonas reinhardii* at different conditions of light intensities, and light–dark cycles.

The changes in starch, protein, and cell number in a culture growing in 14 h light followed by 10 h dark cycle at $10.23 \cdot 10^2$ lx are shown in Figure 4.3.2. The pattern is typical for all synchronized cultures. We see that cell division takes place only during the dark period. This is in complete agreement with the first prediction of the model. This pattern of cell division has also been found in other algae (Cook, 1961; Hase, Morimura & Tamiya, 1957; Leedale, 1959; Tamiya *et al.*, 1953).

We also see that protein is synthesized throughout the day, although at a somewhat slower rate toward the end of the light period, while no starch is synthesized for the first 8.5 h of the light period and is then rapidly synthesized for the remaining 5.5 h of the light period. The starch level falls to its initial value during the night. The beginning of starch synthesis

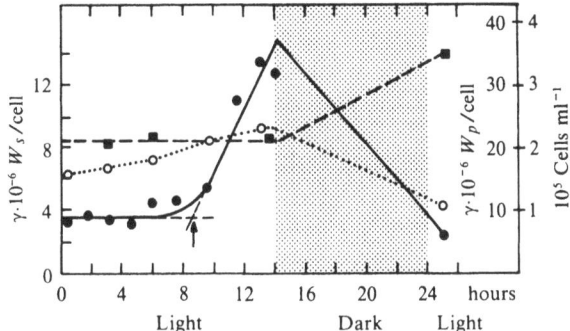

Figure 4.3.2. Cell division and protein and starch synthesis (γ) in a 14 h light to 10 h dark cycle, at $10.23 \cdot 10^2$ lx. γ is amount of ●, starch per cell; ○, protein per cell, ■, cells per ml. Notice the three different scales. Arrow indicates the beginning of starch synthesis. From Cohen & Parnas (1976).

was always estimated by extrapolation of the slope of the starch level back to its initial basal level. This finding agrees with the second prediction of the model. A similar pattern has been found in *Chlorella* (Hase *et al.*, 1957) and in *Euglena* (Cook, 1966). In *Euglena* in $14.21 \cdot 10^2$ lx, protein synthesis completely stopped after 10 h of light. Since protein is ~55% of the dry weight in *Chlamydomonas* (Mayer, 1963), the observed ratio of $W_{ST}/W_{PT} = 0.46$, means that starch makes 32% of the total dry weight. There is a good correspondence between the relative increase in protein per cell during the light period, 1.4, and the relative increase in cell number during the dark period, 1.5. This validates the assumption that N is proportional to the relative increase in cell proteins.

Examination of the dependence of the duration of starch synthesis on light intensity is a strong test for the model since, according to the hypothesis of Cook (1966), increasing the light intensity would be expected to make the synthesis of starch start *earlier*, as all the other energy requirements of the cell would be satisfied earlier, and allow the excess supply to go into storage. The prediction of our model is the opposite, as shown above both analytically and numerically (Figure 4.3.1), i.e. that starch synthesis should start earlier when the light intensity is *decreased*.

The results of the experiments are shown in Figure 4.3.3, in which it can be seen that, in both light–dark cycles, increasing the light intensity decreased the duration of starch synthesis.

Our model predicts that the duration of starch synthesis will not be altered much by changes in the duration of the light period T. Comparing the values for $10.23 \cdot 10^2$ lx in $T = 14$ (Figure 4.3.3a) and $T = 10$ (Figure 4.3.3b), it is clear that the values of τ are essentially the same for the two day lengths. As a result of this, the beginning of starch synthesis is delayed

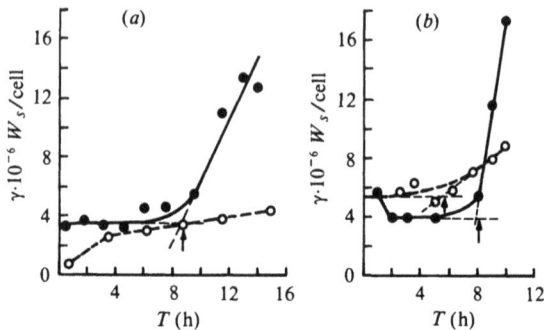

Figure 4.3.3. Starch levels, (γ) in cells grown at different light–dark cycles and at different light intensities. (a) 14 h light, 10 h dark. ●, light intensity, $10.23 \cdot 10^2$ lx, $\tau = 5.5$ h. ○, light intensity, 7×10^2 lx, $\tau = 13.5$ h. (b) 10 h light, 14 h dark. ●, light intensity, $13.46 \cdot 10^2$ lx. $\tau = 2$ h; ○, light intensity, $10.23 \cdot 10^2$ lx, $\tau = 5$ h. Arrows indicate the beginning of starch synthesis, γ is starch per cell. From Cohen & Parnas (1976).

in the longer day length by the difference between the longer T and the shorter T.

A randomly varying environment

Although normal cycles of light and darkness give a periodic deterministic environment, there is always the possibility that the cells will find themselves in the dark for long and unpredictable periods. A random component is characteristic of most natural environments. Randomness is very pronounced for example in the environments of soil and water bacteria. The intestinal environment of *E. coli* is also variable, although to a lesser extent. It has a large component of regular periodic changes.

A full treatment of the dependence of S_{op} on the length of the starvation period, as well as on other important factors is given by Parnas & Cohen (1976) for nonphotosynthetic microorganisms. One of the principal conclusions is that S_{op} *increases as the duration of starvation increases* until the marginal loss resulting from synthesis of storage materials is greater than the marginal gain resulting from their synthesis.

The effect of the probability of having starvation following any level of substrate concentration can be demonstrated by examining a randomly varying environment in which substrate levels change in discrete pulses that appear randomly according to a particular distribution. Both the heights and the frequencies of the pulses may change.

Let us suppose that there are n concentrations C_i, and that any one concentration may appear after any other concentration with probability P_i. Let us also consider the concentrations in an ascending order, such that $C_n > C_{n-1} > \ldots, C_1 \geqslant 0$.

The profit function on transition between any of the concentrations appears in matrix $\mathbf{A}$, where

$$
\mathbf{A} = \begin{pmatrix}
0 & 0 & 0 & 0 & \ldots & 0 \\
\pi(C_1 \to C_0) & 0 & 0 & 0 & \ldots & 0 \\
\pi(C_2 \to C_0) & \pi(C_2 \to C_1) & 0 & 0 & \ldots & 0 \\
\pi(C_3 \to C_0) & \pi(C_3 \to C_1) & \pi(C_3 \to C_2) & 0 & \ldots & 0 \\
\vdots & & & & & \\
\pi(C_n \to C_0) & \pi(C_n \to C_1) & \pi(C_n \to C_2) & \pi(C_n \to C_3) & \ldots & 0
\end{pmatrix}. \tag{27}
$$

The zeros in the matrix are for the transitions from any concentration to a concentration higher or equal to it, for which the profit function is zero (as proved in Section (A) (i), Parnas & Cohen, 1976). Each term in matrix $\mathbf{A}$ is equivalent to a profit function in a certain environment, with its associated gain and loss functions.

The probabilities for the occurrence of the various concentrations are given by a vector $\mathbf{P} = [P_0, P_1, P_2, \ldots, P_n]$. The expected profit functions are defined as a vector $\mathbf{B} = \mathbf{A} : \mathbf{P}$

$$
\mathbf{B} = \mathbf{A} : \mathbf{P} = \begin{pmatrix}
0 \\
\pi(C_1 \to C_i) \cdot P_i \\
\sum_{i=0}^{i=1} \pi(C_2 \to C_i) \cdot P_i \\
\sum_{i=0}^{i=2} \pi(C_3 \to C_i) \cdot P_i \\
\vdots
\end{pmatrix}. \tag{28}
$$

The optimal storage function will be a vector $\mathbf{S}_{lop}$ which maximizes all the terms in $\mathbf{B}$.

It is clear that the optimal value of any S_i depends only on the S values of lower concentrations. It is therefore possible to solve $\mathbf{S}_{lop}$ by an iterative procedure, starting from the lowest concentration, and going in steps to next highest concentration.

First define $\qquad\qquad\qquad\qquad\qquad\qquad S_0 = 0$

$\qquad\qquad\qquad\qquad\qquad\qquad\qquad\qquad\qquad\qquad\qquad \downarrow$

Then solve for S_{1op} when $S_0 = 0$ $\qquad\qquad S_{1op} = \max \mathbf{B}_1$

$\qquad\qquad\qquad\qquad\qquad\qquad\qquad\qquad\qquad\qquad\qquad \downarrow$

Then solve for S_{2op} when $S_0 = 0$ and $S_1 = S_{1op}$ $\qquad S_{2op} = \max \mathbf{B}_2$

$\qquad\qquad\qquad\qquad\qquad\qquad\qquad\qquad\qquad\qquad\qquad \downarrow$

In the end solve for S_{nop} when $\qquad\qquad\qquad\qquad \downarrow$

$\qquad\qquad\qquad\qquad\qquad\qquad\qquad\qquad\qquad\qquad\qquad \downarrow$

$(S_0 = 0;\ S_1 = S_{1op},\ S_{n-1} = S_{n-1op})\qquad\qquad \downarrow$

$\qquad\qquad\qquad\qquad\qquad\qquad\qquad\qquad\qquad\qquad S_{nop} = \max \mathbf{B}_n$

This procedure can also be extended to cases where the probability for any one concentration depends on the previous concentration.

Two main results can be obtained from the analysis just described. First, when there are high probabilities for higher concentrations in the environment in between the starvation periods, then the optimal S (S_{op}) is such that there will be enough storage materials to survive during the whole period of starvation, *even for starvation periods with very low probability*. If, however, there are high probabilities for lower concentrations then S_{op} can cover a spectrum of possibilities (see Parnas & Cohen, 1976). In general a higher probability or a longer duration of the starvation period causes starvation to be the dominant factor in determining S_{op}.

In order to understand this last remark one has to bear in mind that in nonphotosynthetic microorganisms, storage materials can increase the rate of growth at low external carbon concentrations, in addition to their role during starvation periods. However, when there are high probabilities for low external concentrations in between the starvation periods, then in general the role of the storage materials shifts from being a combination of increasing the growth rate during growth periods and survival during starvation to mainly survival.

On the basis of the above conclusions it was of interest to examine how a population of photosynthetic microorganisms would behave if the starvation period were prolonged artificially over that normal to its natural environment. Under natural conditions a prolonged period of starvation for a photoperiodic organism might be regarded as an indication that such starvation periods have become more likely. One would expect therefore that, when light is provided again following a long period of darkness, the level of S would be higher than before the darkness, at the same light intensities. Moreover the synthesis of paramylum might even occur before the synthesis of proteins when light is returned, in order for the algae to be prepared for long *unexpected* starvation periods. Results of such an experiment are shown in Figure 4.3.4. Starch synthesis started immediately after illumination began, and the level reached was 1.33 times higher than it had been before the starvation. Although not shown in the Figure, the new high level of starch was maintained for at least three more days. The beginning of protein synthesis and cell division on the other hand is delayed one day, and the new level is the same as the old level. Similar results have been obtained in two other experiments. On the whole they agree with the predictions of our model.

Similar results were found in nonphotosynthetic *Euglena* (Blum & Buetow, 1963), where, after a long starvation in the dark and following addition of acetate, the synthesis of paramylum started earlier and at a higher rate than the synthesis of proteins.

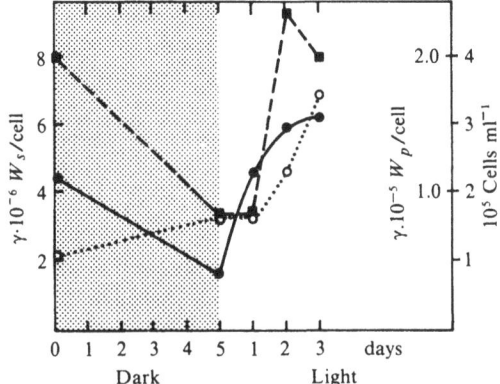

Figure 4.3.4. The synthesis of starch and protein (γ). All division is in continuous light following 5 days of starvation in the dark. Light intensity $6.46 \cdot 10^2$ lx. Preincubation for 3 days in the same continuous light. ●, starch per cell; □, protein per cell; ○, cells per ml. From Cohen & Parnas (1976).

Conclusions concerning the policy of storage material synthesis in microorganisms

There are different roles for storage materials in microorganisms. In nonphotosynthetic organisms, storage materials might regulate the growth rate when the substrate concentration is variable (Parnas & Cohen, 1976); they supply maintenance energy to increase survival during periods of starvation. In photosynthetic microorganisms, storage materials provide, during the night, both the energy for cell division and maintenance energy. The common role of storage materials for photosynthetic and nonphotosynthetic microorganisms as well as other types of populations is to enable the organism to survive during predicted and unpredicted starvation periods.

We have provided theoretical and experimental evidence for the assertion that aspects of the optimal policy are as follows.

(*a*) In an environment in which a falling substrate concentration is a signal for forthcoming starvation, synthesis of storage materials should occur at the time of the lowest substrate concentration for a period before starvation.

(*b*) When there is a signal or certainty with respect to the time of the need for storage materials, the organism should first grow and then sharply switch to prepare storage materials during the time intervals between periods of starvation.

(*c*) The optimal policy when there is a probability for starvation is almost always to store enough storage materials to allow complete survival during the starvation period, even when the probability for starvation is rather low.

(*d*) In a randomly varying environment, the occurrence of a starvation period usually increases the probability of an additional starvation period. We should therefore expect an increased production of storage materials following a starvation period.

Application of the model to metazoa

According to the hypothesis of Slobodkin (1962), the level of storage materials should be minimal in order to allow maximal number of progeny when there is no certain and known need for them. Thus, the level of storage materials should be the same in species whose life cycles are essentially the same.

By contrast, Calow & Jennings (1974, 1977) showed, as can be seen in Figure 4.3.5, that various species of platyhelminthes have different levels of storage materials. Moreover, they demonstrated that there is a direct correlation between the level (and type) of storage materials and the

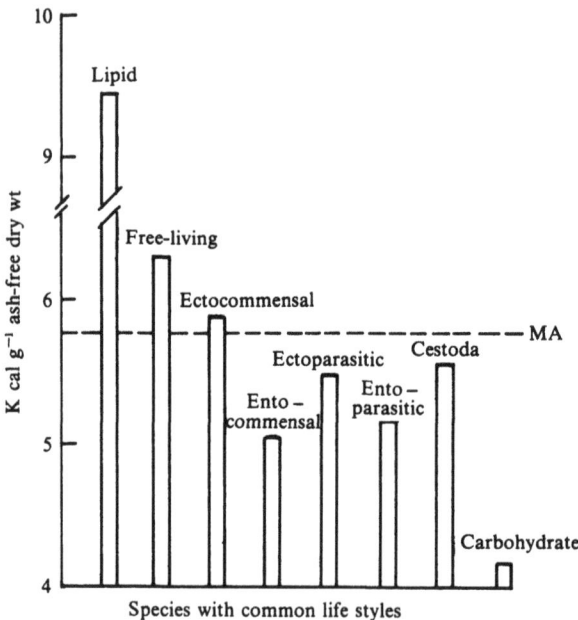

Figure 4.3.5. Mean energy values (Kcal g^{-1} ash-free dry weight) of platyhelminthes with common life styles (from Calow & Jennings, 1974). The values for lipid, carbohydrate and MA (the mean value for whole animals from other phyla) are from Cummins & Wuychek (1971).

expected starvation periods in their natural environment. In free-living platyhelminthes the level is highest, as one expects from the fact that such platyhelminthes can be faced with long and unpredictable starvation periods. The level of storage materials is lowest in entoparasites and entocommensals in which the periods of starvation are shorter, if they appear at all. This correlation between the level of storage materials and the expected appearance and length of starvation in the natural environments fits Conclusions 3 and 4 for microorganisms.

In metazoa, Calow & Jennings performed an experiment similar to the one reported here for algae, in which the duration of the starvation was artificially prolonged. They faced triclades with different feeding schedules. Those individuals that were fed every seven days stored more lipids than those which were fed every day, or once every two or three days. Such behavior cannot result from an excess hypothesis, whereby storage materials are synthesized only when surplus food is available. On the contrary, the behavior follows an optimal policy according to future demand.

Summary

The main idea expressed in this section is that the level of storage materials as well as the time of their synthesis can be correctly determined by an optimal policy. This policy takes into account the present cost and the future gain resulting from the storage of reserve materials.

The main and common role of the storage materials is to enable microorganisms as well as metazoans to survive during periods of starvation. In addition to this common role, there are other roles that differ in the different groups of living systems.

According to the optimal policy, an organism in a random environment, (with respect to the supply of food) should have a constant level of storage materials in an amount that usually covers the longest expected starvation period. In a deterministic environment, the organism should also prepare enough storage materials to cover the whole period of starvation. There is, however, one important difference between deterministic and random environments: in the former the level of storage materials is not constant. It rises sharply before the certain starvation.

The 'excess theory' also follows from the above-described model, but as an extreme case. Obviously when extra food is available, the loss function approaches zero and therefore any gain, even the smallest, will cause synthesis of storage materials. This result emerges from the optimal policy, but it cannot be the policy itself.

References

Blum, V. V. & Buetow, D. E. (1963). Biochemical changes during acetate deprivation and repletion in *Euglena*. *Exp. Cell Res.* **29**, 407–21.

Brodie, P. F. (1975). Cetacean energetics, an overview of intraspecific size variation. *Ecology* **56**, 152–61.

Calow, P. & Jennings, V. B. (1974). Calorific values in the phylum Platyhelminthes: the relationship between potential energy, mode of life and the evolution of entoparasitism. *Biol. Bull.* **147**, 81–94.

——— (1977). Optimal strategies for the metabolism of reserve materials in microbes and metazoa. *J. Theoret. Biol.* **65**, 601–3.

Cohen, D. & Parnas, H. (1976). An optimal policy for the metabolism of storage materials in unicellular algae. *J. Theoret. Biol.* **56**, 1–18.

Cook, J. R. (1961). *Euglena gracilis* in synchronous division. III. Biosynthetic rates over the life cycle. *Biol. Bull. Mar. Biol. Lab.*, Woods Hole **121**, 277–89.

——— (1963). Adaptations in growth and division in *Euglena* effected by energy supply. *J. Protozool.* **10**, 436–44.

——— (1966). Photosynthetic activity during the division cycle in synchronized *Euglena gracilis*. *Pl. Physiol.* **41**, 821–5.

Cummins, K. W. & Wuychek, J. C. (1971). Caloric equivalents for investigations in ecological energetics. *Mitt. Int. ver. Theor. Angew. Limmol.* **18**, 1–158.

Hase, E., Morimura, Y. & Tamiya, H. (1957). Some data on the growth physiology of *Chlorella* studied by the technique of synchronous culture. *Arch. Biochem. Biophys.* **69**, 149–65.

Holme, T. (1957). Continuous culture studies on lysogene in *Escherichia coli* B. *Acta Chem. Scand.* **11**, 763–75.

Leedale, G. F. (1959). Periodicity of mitosis and cell division in the Euglenineae. *Biol. Bull. Mar. Biol. Lab.*, Woods Hole **116**, 162–74.

Mayer, A. M. (1963). *A Study of the Problem of Light Intermittency and Photosynthetic Yields in Mass Cultures of Algae*. Technical Report, Jerusalem, Department of Botany, Hebrew University.

Odum, E. P., Marshall, S. G. & Marples, T. G. (1965). The caloric content of migrating birds. *Ecology* **46**, 901–4.

Palmstierna, H. (1956). Lysogene-like polyglucose in *Escherichia coli* B. during the first hours of growth. *Acta Chem. Scand.* **10**, 567–77.

Parnas, H. & Cohen, D. (1976). The optimal strategy for the metabolism of reserve materials in microorganisms. *J. Theoret. Biol.* **56**, 19–55.

Schindler, D. W., Clark, A. S. & Gray, J. R., (1971). Seasonal calorific values of freshwater zooplankton, as determined with a Phillipson bomb calorimeter modified for small samples. *J. Fish. Res. Bd Can.* **28**, 559–64.

Slobodkin, L. B. (1962). Energy in animal ecology. In *Advances in Ecological Research*, vol. 1, ed. J. B. Gragg, New York, Academic Press, pp. 69–101.

Spectorov, C. S., Slobodskaya, G. A. & Nichiporovich, A. A. (1963). In *Studies on Microalgae and Photosynthetic Bacteria*, Tokyo, Japanese Society of Plant Physiologists, The University of Tokyo Press, pp. 141–9.

Tamiya, H., Iwamura, T., Shibata, L., Hase, E. & Nihei, T. (1953). Correlation between photosynthesis and light independent metabolism in the growth of *Chlorella. Biochim. Biophys. Acta* **12**, 23–110.

Wilkinson, J. F. (1959). The problem of energy-storage compounds in bacteria. *Exp. Cell Res. Suppl.* **7**, 111–30.

4.4 ACCEPTABLE AND UNACCEPTABLE MODELS OF LIVER REGENERATION IN THE RAT

Introduction

The cells of the rat liver rarely divide in the normal adult. If, however, in an operation called a **hepatectomy**, some 65% of the liver is removed by surgical ablation of the two larger lobes, there is a great burst of mitosis in the cells of the remaining two lobes that peaks some 26 h after the operation and then declines. Within two weeks or so, these lobes grow to the size of a normal liver and the system becomes quiescent (see Bucher, 1963, for review). This phenomenon has been much studied but the mechanism responsible for regeneration has not yet been elucidated.

There are, however, two hypothetical mechanisms that, it is thought, might work. First, the liver could continually make an inhibitor of cell division (see e.g. Glinos, 1958), the concentration of which would normally be high enough to suppress mitosis. Following hepatectomy, the concentration of this inhibitor in the blood and liver remnant would drop as the molecule is assumed to have a short half-life. As a result of this decrease, mitosis would occur, the liver would grow, more inhibitor would be made and the *status ante quo* would be restored.

The second theory suggests that liver size is controlled by function rather than by size so that, after hepatectomy, the functionally inadequate liver grows until it can fulfill all its biochemical tasks (see e.g. Goss, 1964). The mechanism by which this is thought to be achieved is that some waste product builds up in the presence of an inadequately sized liver and this stimulates mitosis; as the liver increases in size so the waste product level drops and, eventually, the normal liver size is restored.

In this section, simple mathematical models that incorporate these two mechanisms are put forward and the extent to which each can explain the known experimental facts is examined. In particular an attempt is made to find functional parameters of the inhibitor and stimulator that allow the best match of the prediction of each model to the published data on regeneration. For a rather more detailed analysis of liver regeneration, see Bard, 1978 and 1979.

Experimental data

An enormous amount of work has been published on liver regeneration and this section mentions only the facts essential for the analysis. Observations have been made in four contexts which to some extent overlap. These are studies on the rates and localization of mitosis in regenerating liver, physiological experiments designed to see how regeneration is controlled, attempts to interfere with regeneration by giving drugs or other compounds to animals, and attempts to isolate stimulators or inhibitors of mitosis in the liver. The results of the last two classes may be briefly summarized: chemicals have been found that mimic to a limited extent the effects of regeneration in the intact liver (e.g. α-hexachlorocyclohexane, see Schulte-Hermann, 1977), but their mode of action is unknown. No specific inhibitors or stimulators of the hepatocyte cell cycle have, however, been isolated from the liver, although Verly, Deschamps & Pushpathadam (1971) have extracted a substance that inhibits liver cells in S-phase from proceeding to cell division. This substance has not been shown to stop cells entering S-phase, a prior step in regeneration. (For basic information on the cell division cycle, see Section 6.6.)

The most important data on regeneration are those that describe the rates at which cells divide in the regrowing liver. As a typical set of observations on this, we can consider the work of Grisham (1962), who measured mitotic indices every few hours for the three days after 65% hepatectomy. He found that after an 18 h lag the mitotic index rose and then slowly dropped (Figure 4.4.1). In the accompanying autoradiographical study he showed that it was the parenchymal cells ($\sim$90% of the liver mass) that initially took up $[^3H]$thymidine and that the other cells of the liver divided later. More recently, Fabrikant (1968) has shown that the parenchymal cells show in a different breed of rats a second peak of mitosis at around 56 h after surgery.

The mitotic response when small amounts of liver are removed is too small to measure accurately and growth has to be measured indirectly by the uptake of radioactive thymidine. The major new observation made with this technique (see e.g. Bucher, 1963) is that the peak of the uptake rises steeply as increasing amounts of liver are removed (Figure 4.4.2). There is a further interesting observation that has been made from autoradiographic studies using $[^3H]$thymidine uptake: Fabrikant (1968) and Rabes, Wirshing, Tuczek & Iseler (1976) have shown that cells near the periportal areas of the liver, where the blood enters, divide sooner than those near the hepatic veins, where blood drains away from the parenchymal cells.

Physiological studies have now shown, after considerable controversy, that there are components in the blood that play a role in regeneration. Sakai (1970) showed that, in parabiotic rats (i.e. pairs of animals with their

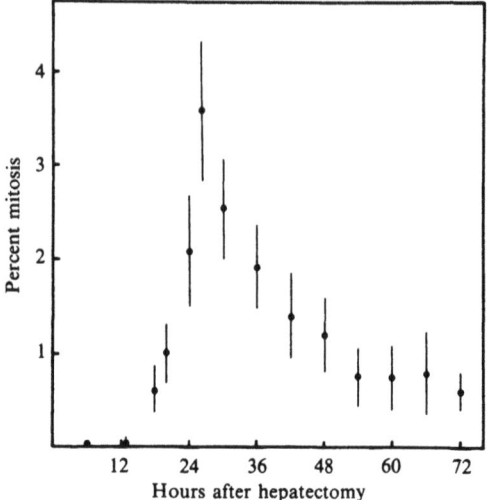

Figure 4.4.1. Mitotic indices in the regenerating livers of rats after 65% hepactectomy as measured by Grisham (1962) (from Bard, 1978).

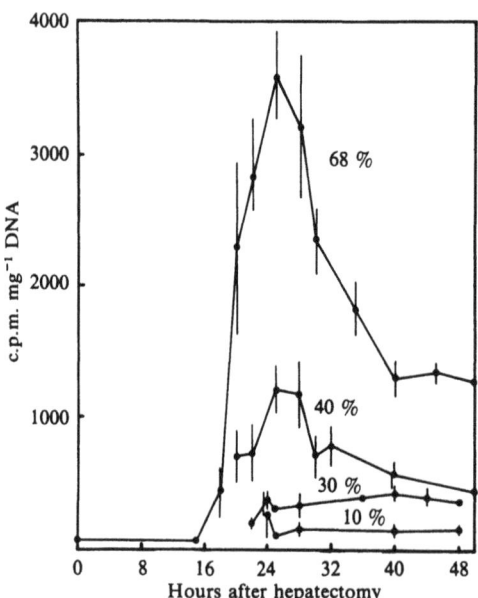

Figure 4.4.2. [³H]Thymidine incorporation rates in the livers of young rats after various amounts of hepatectomy (from Bard, 1978). The data are taken from the experiments of Bucher (1963).

blood supplies linked), where one partner has had a large partial hepatectomy, the liver of the control rat takes up considerable amounts of [³H]thymidine. This experiment proves that there are blood-borne factors in regeneration but does not show whether they are inhibitors or stimulators of mitosis. Some early, preliminary experiments by Glinos (1958) indicate that it is an inhibitor. He replaced part of a rat's plasma with saline in order to dilute any factor present: he observed a burst of mitosis, a result incompatible with a stimulator of mitosis being present.

A note on units

Below, equations will be derived for the two models which describe how the liver could grow as the concentrations of inhibitor or stimulator vary. To simplify calculations of concentration as the liver size changes, we will consider the volume v of the liver rather than its mass. For a 150 g rat, the liver weighs about 6 g (Grisham, 1962) and thus has a volume of 6 ml; the blood volume is 15 ml. In such a liver, the resting mitotic index is about 0.0004% (i.e. in sections of normal liver, some 4 cells in 10^4 are seen to be in mitosis) and mitosis itself takes about 1 h (Fabrikant, 1968). It is therefore convenient to measure time in hours, and dv/dt is thus the mitotic rate per milliliter of liver. Concentration will be measured in arbitrary units per milliliter.

The inhibitor model

The essential facts that any model must incorporate or explain are that there is a blood-borne humoral factor and that the larger the amount of liver removed the larger is the mitotic response. The essence of the inhibitor model is that the liver makes a substance of short half-life that is secreted into the blood (this is, as will be seen, a crucial point) and reduces the likelihood of a cell entering the division cycle. It is clear that the greater the amount of liver removed, the lower will be the equilibrium concentration of the inhibitor and thus the greater the mitotic response. As the liver grows and the inhibitor concentration rises, so the mitotic rate will decrease from its peak and return to the initial value. It is clear that qualitatively, at least, the inhibitor model can explain the basic data.

Before deriving the equations, it is worth making explicit some assumptions that lie behind the simple model. The time taken for the inhibitor to be transferred from liver to blood is considered to be short and, a greater assumption, the inhibitor concentrations are assumed to be the same in both liver and blood. The inhibitor itself exerts an immediate concentration–dependent effect on a cell so that the likelihood of a cell entering the mitotic cycle can be defined by a simple dose–response function. Such cells are assumed to start enlarging immediately after the decision is made [there is evidence to support this (Tongendorff, Trebin & Ruhenstroth-Bauer, 1975)] so that dv/dt represents both the immediate increase in size and the mitotic rate at the end of the cell cycle (assumed constant for all cells). Finally, it is necessary to assume that the mitotic rate never gets so high that a cell would have to divide a second time before it had completed a first mitosis; this turns out to be roughly so.

For the model, let the liver be homogeneous having a volume v and make an inhibitor at a rate p per unit volume the name and concentration of which are c and whose decay constant is q (so that the half-life is $0.69/q$). The inhibitor is secreted into the blood (volume w) and the equilibrium values of inhibitor and liver size are c_0 and v_0. The rate of liver growth is given by an unknown function $f(c)$ and the rate of cell death in the liver growth is given by r (assumed to equal the resting mitotic rate). The role of r in these equations is to define the resting equilibrium. The rate of growth is thus given by

$$dv/dt = v[f(c) - r], \tag{1}$$

and the mitotic index is $(1/v)\,dv/dt$. As inhibitor is distributed uniformly between liver and blood the rate at which its concentration grows is given by

$$dc/dt = [pv/(v + w)] - qc. \tag{2}$$

It is worth noting the role of the blood volume in these equations: if $w = 0$, then the inhibitor concentration is independent of liver size and there can be no regulation. The blood thus acts as a reservoir of inhibitor.

The condition that these equations give a stable equilibrium is that $f(c)$ be an inhibitor so that $f'(c) < 0$. (Exercise 1.) It turns out that there is a direct return to this point under physiological conditions. Using the optimal value of $f(c)$ which will be determined later, the condition that the equilibrium point is a node rather than a focus turns out to be that $q > 4c_0 r$ (Exercise 1) or that the half-life of the inhibitor is greater than 100 h or so.

Inserting those parameters that can be measured into the equations, they become

$$dv/dt = v[f(c) - 0.0004], \tag{3}$$

$$dc/dt = [pv/(15 + v)] - qc, \tag{4}$$

and it can be seen that if $f(c)$ and q, the functional parameters of the inhibitor, are inserted, then the value of p can be determined from the equilibrium conditions. We must therefore find values of $f(c)$ and q which, when substituted, will give solutions that match the data. The time between entry into cell cycle and mitosis must also be found so that the delay between hepatectomy and mitosis can be allowed for. If all this can be done, the model is shown to be at least plausible; if parameters cannot be found then the model must be rejected.

The appropriate form for the dose–response function is not immediately clear, but some constraints on it come from the data shown in Figures 4.4.2 and 4.4.5. They show first that, as inhibitor concentration drops to about one third of its value, the mitotic rate rises by two orders of magnitude; and secondly that the response is nonlinear. Various dose–response functions have been tried and the only simple one that gives solutions that match the data is as might be expected from the data,

$$f(c) = A \exp(-\alpha c),$$

where for convenience we let $\alpha = 1$ and have dimension $[\text{conc}]^{-1}$.

A computer program has been written to solve (3) and (4) when values of c_0 and q are given. Trial and error show that the optimal match of theory to data is given when $c_0 = 8.8 \pm 0.2$ and the decay constant is $0.25 \pm 0.03 \text{ h}^{-1}$ (implying a half-life of 2.9 ± 0.4 h) and the time between entering the mitotic cycle and cell division has to be 15 h (a reasonable time for the cell cycle). If the decay constant is greater, the rise in mitosis is faster, whereas with a lower decay constant the mitotic rate rises more slowly. The height of the peak is determined by $f(c)$ if the half-life is kept constant.

The match between theory and data is quite good when the optimal parameters are substituted in the equations: the mitotic index rises rapidly to a peak and then declines slowly (Figure 4.4.3). Quantitatively, the predicted peak is not as sharp as that in the data and the rate of descent rather greater. The reason for this is that $f(c)$ acts on all cells in the liver and includes cells that have already entered the cell cycle. However, given the simplicity of the model and the large number of assumptions, the fit is acceptable. Over several days (Fig. 4.4.4), the liver size increases rapidly and returns to 85% of its original size in about ten days. The inhibitor concentration likewise increases after its initial decay.

The mitotic rate prediction of the theory for ablations of less than 65% cannot be matched as, for such operations, the only data are on thymidine incorporation studies. However, if it is assumed that such incorporation is proportional to the mitotic rate, then a scaled comparison between the predicted mitotic index and the incorporation rates can be made (Figure 4.4.5). The theoretical curve matches well the data of Bucher (1963) and shows the threshold effect that is expected. The form of the curve derives, of course, from the nonlinear dose–response curve.

In summary, therefore, the inhibitor model can explain the data on liver regeneration provided that the inhibitor has a half-life of about 3 h and a negative exponential dose–response curve.

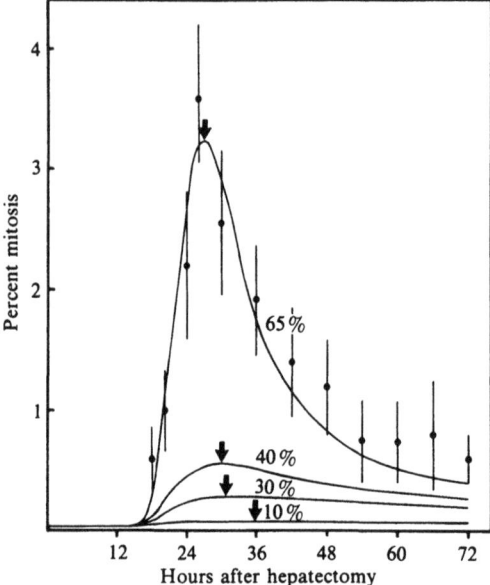

Figure 4.4.3. Theoretical mitotic indices predicted by the inhibitor theory, using a half-life of 2.9 h, a negative exponential dose–response curve and a 15 h delay between entry to the mitotic cycle and mitosis (from Bard, 1978). The data points of Grisham (1962) are matched quite well; the form of the curves is similar to the data on [^{3}H]thymidine incorporation obtained by Bucher (1963). The arrows show the times of maximal mitosis.

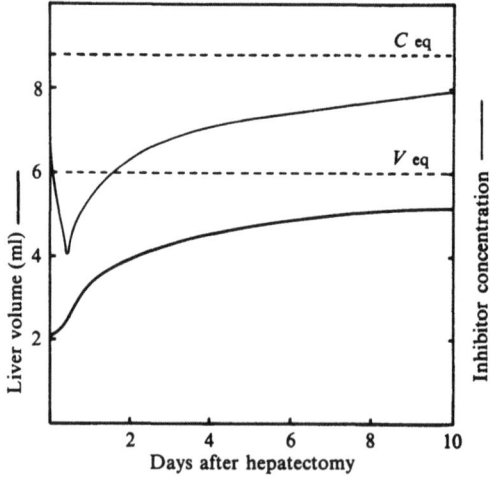

Figure 4.4.4. The regeneration of liver size and restoration of inhibitor concentration after 65% hepatectomy predicted by the inhibitor theory with optimal parameters (from Bard, 1978). C_{eq} and V_{eq} are the inhibitor concentration and liver volume, respectively, at equilibrium.

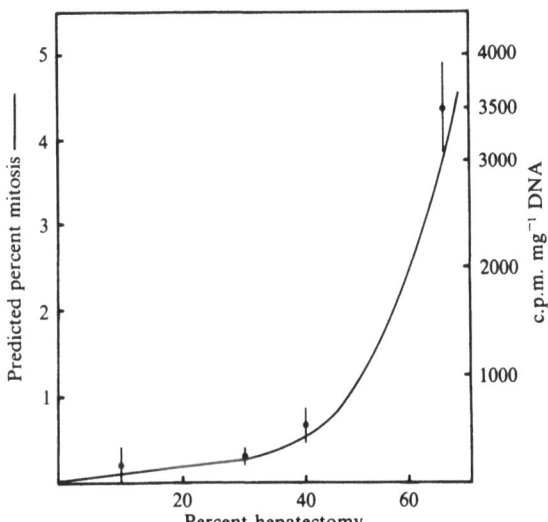

Figure 4.4.5. Theoretical peak mitotic indices as a function of degree of hepatectomy (line) are scaled to match optimally the [³H]thymidine incorporation data of Bucher (1963) (from Bard, 1978).

Function-dependent regeneration

In this approach, it is assumed that liver growth after ablation is stimulated by a waste product that would be almost completely degraded by the intact liver (Goss, 1964). The greater the concentration of waste product the greater will be the stimulus for liver cells to divide. Again, this model appears to describe the essential features of regeneration. However, a deeper analysis shows that there are some surprising aspects to this mechanism.

Again, let the liver volume be v and the blood volume be w; let a waste product c' be generated by the body at a rate s and let it be degraded by the liver at a rate u per unit volume, the waste product having the same concentration in both liver and blood. The assumptions that were discussed for the inhibitor case are assumed to hold here. If the rate at which the liver cells divide is given by the function $g(c')$, which increases with c' so that $f'(c') > 0$ then, as before,

$$dv/dt = v(g(c') - r),\tag{5}$$

and the rate at which the waste product builds up is given by

$$dc'/dt = \frac{s - uv}{v + w} = \frac{u(v_0 - v)}{v + w},\tag{6}$$

as $s = uv_0$ at equilibrium.

Although these equations have an equilibrium point at some (v_0, c_0'), this point is not stable as can be seen by linearizing these equations around (v_0, c_0'). For this, let $v = v_0 + x$ and $c' = c_0' + y$; then

$$dx/dt = 0 + v_0 g'(c')y \quad [\text{as } g(c_0') = r \text{ by definition}],$$

$$dy/dt = [-ux/(v_0 + w)] + 0.$$

Thus it can be seen (Exercise 2) that the equilibrium point is a center rather than a node or focus, and the trajectories of the system in (c, v)-space are a series of ellipses around this point. Qualitatively, this means that, if part of the liver is removed, the remainder, instead of growing back to its original size, overshoots and becomes larger than it was in order to deal with the backlog of waste product that accumulated while the liver was undersized. The now oversized liver will reduce the amount of waste product to below its original concentration so that mitosis in the liver is below the death rate and the liver eventually becomes smaller than the equilibrium size; the waste product will now build up and a cycle will start again.

The fact that (5) and (6) are nonlinear implies that the linear analysis given will not hold far away from the equilibrium point and that in all likelihood the nonlinearity will result in this point being, in fact, a weakly stable or unstable focus rather than a center. Thus the long-term solutions cannot be predicted from linearizing the equations. However, it is the initial behavior of the model which is biologically relevant and, for a qualitative description of this, the elementary analysis adequately predicts the form of the solution.

As an example of the implications of this state of affairs we can consider $g(c') = A \exp(\alpha c')$ and let $c_0' = \alpha = 1$ and let $\alpha = [\text{conc}]^{-1}$ so that $A = 0.00015$; u, which is also arbitrary, can be set equal to 1 and the other parameters maintain the values they had in the previous model. Solving the equations (Figure 4.4.6: the case where the half-life of the stimulator $= \infty$, or its decay constant $q = 0$) shows that, within a few days of a 65% hepatectomy, the liver volume rises to about 14 ml and then slowly drops as cell death occurs. In the long term, it turns out that after about three months the liver has dropped below its equilibrium size and then starts to grow again; with other dose–response functions, the detailed solution would be different but such oscillations will always occur.

As the overshoot of liver size after regeneration has not been observed in practice, it is clear that the simple model is wrong. It can be improved by suggesting that there is a second mechanism for degrading c' by, for example, giving it a decay constant q. Equation (6) becomes

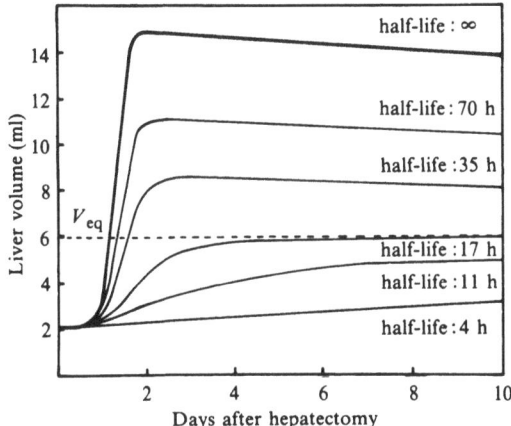

Figure 4.4.6. Theoretical growth for liver regeneration assuming that regeneration is controlled primarily by function and the consequent build up of an inhibitor. The thick line assumes that there is no secondary route for the degradation of the inhibitor (half-life: ∞); the other curves show how the liver is predicted to grow for a range of half-lives when the waste product can be degraded by a second route (from Bard, 1978).

$$\frac{dc'}{dt} = \frac{u(v_0 - v)}{v + w} - qc'. \tag{7}$$

This equation has a stable equilibrium point when $c' = 0$ but the dose-response function is so insensitive at the low values of c' which occur when $v \sim v_0$ that one can use the previous parameters to see what values of q are required to give a reasonable prediction for the system. It turns out (Figure 4.4.6) that when the decay constant is about 0.08 (half-life: 17 h) the liver returns to its original value in about eight days, a prediction not incompatible with the data. However, an examination of the mitotic indices implied by the model (Figure 4.4.7) shows that the long half-life demands a mitotic peak that not only occurs some 48 h after surgery (rather than the 26 h found in practice) but that the peak is far higher than that measured experimentally. The reason for this unexpectedly large amount of mitosis is that while the liver is small a backlog of waste product builds up.

The simple stimulator model is clearly unable to explain the experimental data. The addition of a second degradative route for the waste product improves the model but, at least in the form given here, leaves it still inadequate. The possibility that a still more sophisticated version would match the data cannot be excluded but it is probably correct to say that the model is unacceptable in the light of the current data.

346   *Case studies in kinetics*

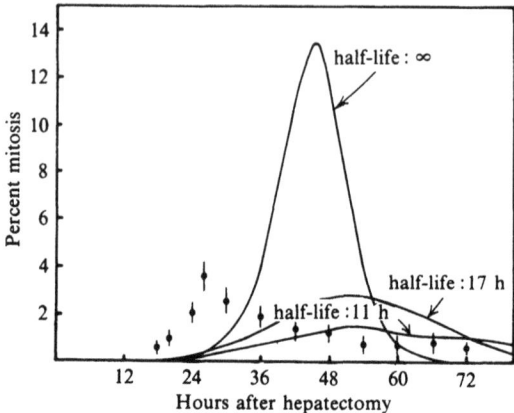

Figure 4.4.7. The mitotic indices predicted by the mitotic-stimulator theory assuming a delay of 15 h for the cell cycle (from Bard, 1978). Three half-lives for nonhepatic degradation of the waste product are shown: ∞ (no degradation); 17 h (optimal fit for growth; Figure 4.4.6); and 11 h. It can be seen that the optimal fit for growth does not match the data of Grisham (1962).

Discussion

Analysis of the two models put forward to explain liver regeneration shows clearly that a liver-synthesized inhibitor can control the system while a waste product mitotic stimulator alone cannot. The possibility that a more complex regulatory system where for example a second organ assays some property of the liver and in turn regulates liver size, cannot be excluded, but there is no reason to invoke such a model until there is compelling evidence to exclude the simple inhibitor system. Indeed this latter system is compatible with much of the published data and in accordance with the intuitions of experimentors in the field (e.g. Sakai, 1970). One fact that the model does not explain, however, is the cellular hypertrophy that occurs in the liver soon after hepatectomy and that seems to be independent of the later mitotic response (Tongendorff *et al.*, 1975). In the context of the waste product stimulator model, on the other hand, Bucher (1963) commented that there was no evidence to support the view that functional demand played a role in liver homeostasis and there has been no evidence since then to contradict her view.

There is some additional data that supports the inhibitor rather than the stimulator model of liver homeostasis. Fabrikant (1968) observed that, in the lobes remaining after hepatectomy, mitosis first occurred near arteries rather than veins. Assuming that a controlling molecule would be made in the liver and drain into such veins, the concentration would be higher there than near the arteries. Were the molecule a mitotic stimulator, one would then expect a higher mitotic rate near the vein; the fact that mitosis is higher in the periportal area around the artery is compatible with the molecule being an inhibitor.

An important aspect of any theory in biology is that it can make experimental predictions. One can use the inhibitor equations to predict the amount of mitosis expected if the dilution experiments of Glinos, (1958) were repeated in more detail using stored plasma or saline to replace the plasma of a rat. The interested reader is referred to Bard (1978), where possible experiments to investigate the half-life and the dose–response curves of an inhibitor are considered.

There are two major problems with the analysis presented here: first, the considerable number of assumptions made may not all be valid. It is, for example, unlikely that the inhibitor concentration will be the same in the liver and in the blood or that all cells take exactly the same time to complete their cycle. These and other assumptions could be allowed for by expanding the equations and the complexity of the stimulation. No useful purpose would be served by this, for the additional number of unmeasurable and arbitrary parameters that would have to be introduced would give too much freedom to the system and allow the data to be matched with an inadequate number of constraints.

The second and perhaps more interesting difficulty lies in the nature of the inhibitor–liver interaction. This is concealed in the function $f(c)$ and it is assumed that it is biochemically easy to make an interaction described by a negative exponential. This is probably so (e.g. Walter, Parker & Yčas, 1967) but the mathematical approach used here gives no clue as to the details of the interaction. It is, however, worth noting that the inhibitor is a member of the class of molecules known as 'chalones' which are tissue-specific mitotic inhibitors. Such molecules have been discussed at some length but there is in most cases a shortage of hard data on their existence in tissues where they would be expected to exist (see Forscher & Houck, 1973, for review) even though their presence has been discussed for over thirty years (see e.g. Weiss & Kavanau, 1957).

Note added in proof

It has now been possible to incorporate into the equations for the inhibitor model a count of the proportions of the regenerating liver available to enter the mitotic cycle (Bard, 1979). In this improvement cells already in the cycle are excluded and cells that have passed through the cycle and divided are counted twice. The major predictions of this model are that the best value of inhibitor half-life is about 9 h and that the mitotic index curve has a subsidiary peak when a second burst of mitosis occurs (see Grisham, 1962).

Exercises

1   Use linear theory (Appendix Section A.3) to examine the stability of the non-trivial equilibrium point (i.e. not $v = 0$, $c = 0$) for (1) and (2). Verify the relevant statements made in the text.

2   (a) Repeat Exercise 1 for (5) and (6).
      (b) Repeat Exercise 2(a), using (7) instead of (6).

References

Bard, J. B. L. (1978). A quantitative model of liver regeneration in the rat. *J. Theoret. Biol.* **73**, 509–30.
———(1979). A quantitative theory of liver regeneration in the rat. II. Matching an improved mitotic inhibitor model to the data. *J. Theoret. Biol.* **79**, 121–36.
Bucher, N. L. R. (1963). Regeneration of mammalian liver. *Int. Rev. Cytol.* **15**, 245–300.
Fabrikant, J. I. (1968). The kinetics of cellular proliferation in regenerating liver. *J. Cell Biol.* **36**, 551–65.
Forscher, B. K. & Houck, J. C. (1973), eds. *Chalones: Concepts and Current Research*, Nat. Canc. Inst. Mon. No. 38.
Glinos, A. D. (1958). The mechanism of liver growth and regeneration. In *The Chemical Basis of Development*, ed. W. D. McElroy and B. Glass, Baltimore, Johns Hopkins Press, pp. 813–39.
Goss, R. J. (1964). *Adaptive Growth*, London, Logos Press.
Grisham, J. W. (1962). A morphological study of DNA synthesis and cell proliferation in regenerating rat liver. *Cancer Res.* **22**, 842–9.
Rabes, H. M., Wirshing, R., Tuczek, H. V. & Iseler, G. (1976). Analysis of cell cycle compartments of hepatocytes after partial hepatectomy. *Cell Tiss. Kinet.* **9**, 517–32.
Sakai, A. (1970). A humoral factor triggering DNA synthesis after partial hepatectomy in the rat. *Nature, Lond.* **228**, 1186–7.
Schulte-Hermann, R. (1977). 2 stage control of cell proliferation induced in rat liver by α-hexachlorocyclohexane. *Cancer Res.* **37**, 166–71.
Tongendorff, J., Trebin, R. & Ruhenstroth-Bauer, G. (1975). Critique of the 'critical mass' hypothesis of the regeneration of liver cells, *Amer. J. Pathol.* **80**, 519–24.
Verly, W. G., Deschamps, Y. & Pushpathadam, J. (1971). The hepatic chalone I: Assay method for the hormone and purification of the rabbit liver chalone. *Can. J. Biochem.* **49**, 1376–83.
Walter, C., Parker, R. & Yčas, M. (1967). A model for binary logic in biochemical systems. *J. Theoret. Biol.* **15**, 208–17.
Weiss, P. & Kavanau, J. L. (1957). A model of growth and growth control in mathematical terms. *J. Gen. Physiol.* **41**, 1–47.

4.5 CHAOS

Simple models of biological systems have generally been expected to exhibit simple dynamic behavior. However, recently simple models have been developed which exhibit surprisingly complex behavior. For example, Mackey & Glass (1977) suggested the following model for hematopoiesis (blood cell differentiation). Let P be the concentration of mature circulating blood cells. Assuming that these cells are produced in the bone marrow and that there is a significant time delay, τ, between the initiation of cellular production in the bone marrow and the release of the mature cells in the blood, they write the following differential-delay equation to describe the changes in P:

$$\frac{dP(t)}{dt} = \left[\frac{\beta_0 \theta^n P(t-\tau)}{\theta^n + P(t-\tau)}\right] - \gamma P, \tag{1}$$

where β_0, θ, n and γ are constants. Numerical studies show that for small values of τ, P approaches a stable equilibrium value. However, if τ is increased, numerical solutions to (1) show that the blood cell population level oscillates, instead of remaining constant (see Figure 4.5.1a). Further increases in τ produce lower frequency oscillations with periods of 2, 4, 8 and 16 times the original period, as well as population changes that show no apparent periodicity, i.e. are **aperiodic** or **chaotic** (see Figure 4.5.1b). One can rigorously show that if no delay were present in (1), i.e. $\tau = 0$, or in any other first order ordinary differential equation, then the dynamics would be relatively simple. This conclusion is based on the fact that Smale (1967) proved that it is only for systems of three or more first order differential equations that chaotic behaviors appear. (In more precise technical language, Smale demonstrated that only systems with a dimension of at least three can have a new type of stable attractor, called a **strange attractor**.) In order to understand how complicated dynamics can arise without involving ourselves in the difficult analysis of systems of three or more nonlinear differential equations, we shall examine an example from population biology developed in terms of a first-order difference equation. Although this is an aside from the main mathematical

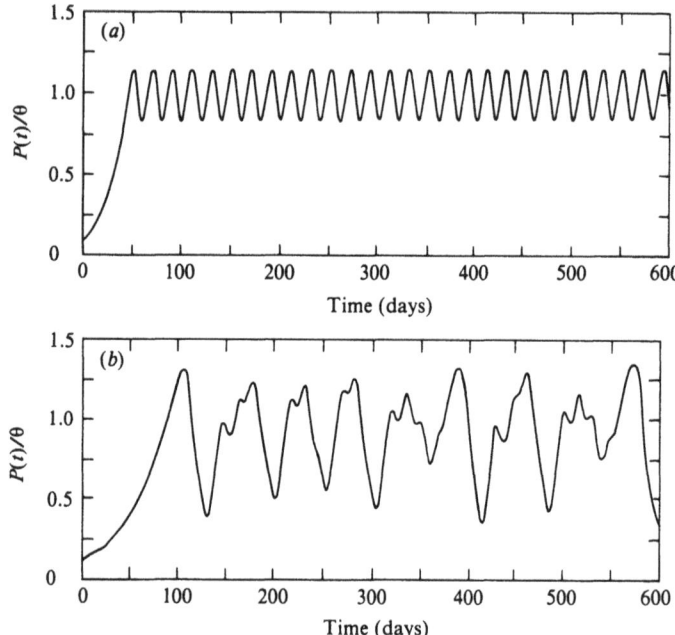

Figure 4.5.1. Numerical solutions to (1). (a) $P(0) = 0.1$, $\gamma = 0.1$ day^{-1}, $\beta_0 = 0.2$ day^{-1}, $n = 10$, and $\tau = 6$ days. The period of oscillation is 20 days. (b) Same as (a) except $\tau = 20$ days. Motion is now aperiodic. (From Mackey & Glass, 1977; copyright © 1977 by the American Association for the Advancement of Science.)

topic of ordinary differential equations in this volume, difference equations are used extensively in population biology, genetics and epidemiology (Hoppensteadt, 1975, 1976; May & Oster, 1976) as well as in numerical analysis, and hence they are of interest in their own right.

Consider a seasonally breeding population the generations of which do not overlap. Many natural populations, particularly temperate zone insects, are of this type. The way in which the population changes from generation to generation may be expressed in the general form

$$x_{t+1} = f(x_t), \qquad t = 0, 1, 2, \ldots, \tag{2}$$

where x_t is related to the size of the population in generation t. The function $f(x)$ will usually be nonlinear, that is, exhibit what an ecologist calls **density dependence**. Equation (2) is then a first-order, nonlinear difference equation.

A number of such models, employed in the ecological literature, have the property that x tends to increase from one generation to the next when it is small, but decreases when it is large. Thus $f(x)$ contains a single maximum. Also it is typical for the terms in $f(x)$ to have a common factor x, so one can express $f(x)$ as $f(x) = xg(x)$. This implies $f(0) = 0$, i.e. if a population vanishes one year it remains zero forever after. A specific example is the 'logistic' difference equation

$$N_{t+1} = rN_t[1 - (N_t/\theta)],$$

which can be written in the form

$$x_{t+1} = f(x_t) = rx_t(1 - x_t) \tag{3}$$

by the substitution $x = N/\theta$. It is this simple equation which we shall study.

For a fixed value of the parameter r, one can plot x_{t+1} versus x_t and obtain a graph of the function f as shown in Figure 4.5.2. Since only

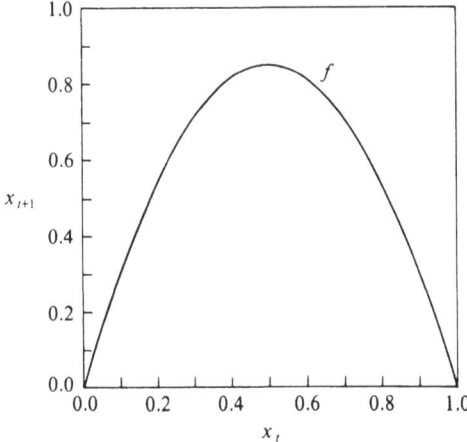

Figure 4.5.2. The graph of the function f defined by (3), for $r = 3.4$.

nonnegative populations are of interest, we restrict the values of x to the interval $0 \leqslant x \leqslant 1$. On this interval f is symmetrical about $x = \frac{1}{2}$, i.e.

$$f(\tfrac{1}{2} + y) = r(\tfrac{1}{2} + y)(\tfrac{1}{2} - y) = f(\tfrac{1}{2} - y), \qquad 0 \leqslant y \leqslant \tfrac{1}{2}.$$

Moreover f has a maximum value of $r/4$ at $x = \frac{1}{2}$, as we see below:

$$df/dx = r(1 - x) - rx = r(1 - 2x) = 0, \tag{4}$$

implying f has zero slope at $x = \frac{1}{2}$. Furthermore,

$$\left. \frac{d^2f}{dx^2} \right|_{x=\frac{1}{2}} = -2r < 0, \tag{5}$$

so $x = \frac{1}{2}$ is a maximum. From (3), $f(\frac{1}{2}) = r/4$. Since $x_{t+1} = f(x_t)$, the maximum value of $f(x)$ must be less than or equal to 1. Therefore we restrict our attention to $0 \le r \le 4$. For $0 \le x \le 1$, $x(1-x) < 1$ and consequently if $r \le 1$, x continually decreases, asymptotically approaching $x = 0$. For nontrivial dynamic behavior we require $1 < r \le 4$.

Dynamics

The dynamics of a population can easily be computed from the graph of f as shown in Figure 4.5.3a. For a given x_0 draw a line upward until it intersects the graph of f. At this point draw a horizontal line to the x_{t+1}-axis and read off the value of x_1. To obtain the next value of x we repeat this process beginning with x_1 on the x_t-axis. A simple graphical method of locating the position of x_1 on the x_t-axis is to draw the 45° line, $x_t = x_{t+1}$. Then follow the horizontal line from x_1 to the 45° line. At the intersection point draw a vertical line downward and mark the value of x_1 on the x_t-axis. To obtain x_2 draw a vertical line upward from x_1 until it intersects the graph of f. Then draw a horizontal line to the x_{t+1}-axis and read off the value of x_2. Continuing in this way the sequence $x_0, x_1, x_2, x_3, \ldots$ can be obtained. This graphical procedure can be simplified by observing that every path from the 45° line to an axis is traversed twice in opposing directions. Eliminating these steps, as shown in Fig. 4.5.3b, yields a rapid method of following trajectories. Further by adding a time axis as shown in Fig. 4.5.3c and projecting the values of x_t downward one can obtain a graph of x_t versus t.

Equilibrium points

Just as for differential equations, one defines an **equilibrium point** (or **fixed point** or **steady state**) of a difference equation as a value of x which remains constant in time. Setting $x_{t+1} = x_t = x^*$ and solving the resulting algebraic equation

$$x^* = f(x^*) \tag{6}$$

yields the possible equilibrium points of the difference equation (2). An equivalent graphical method is to find the points where the curve $f(x)$ intersects the 45° line, $x_{t+1} = x_t$. This is illustrated in Fig. 4.5.4. For (3) one must solve

$$x^* = rx^*(1-x^*).$$

As is easily seen there are two equilibrium points: the trivial solution $x^* = 0$, which we shall not deal with further, and the nontrivial solution

$$x^* = 1 - (1/r). \tag{7}$$

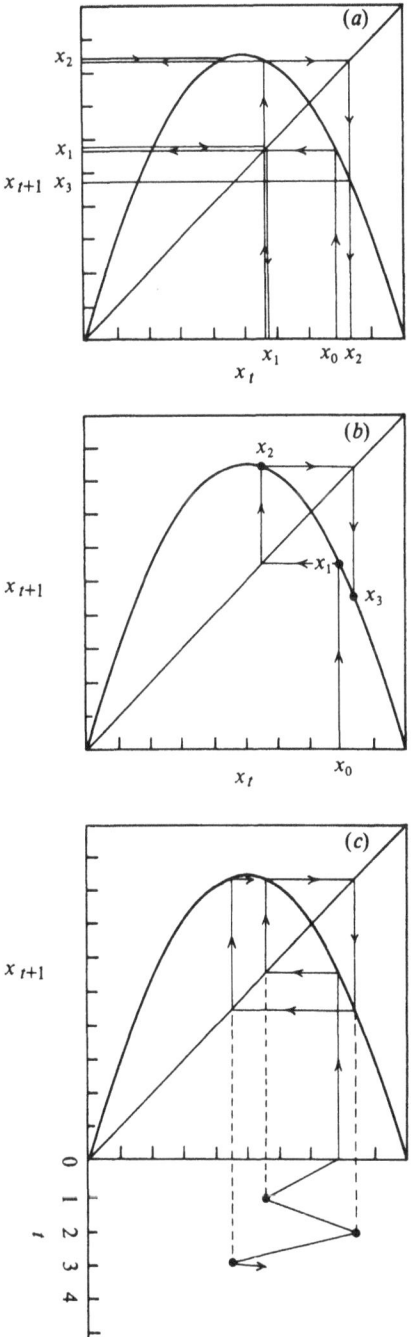

Figure 4.5.3. (a) and (b) graphical methods of determining x_t, $t = 1, 2, 3, \ldots,$ as described in the text. (c) A graphical method of generating a plot of x_t versus t.

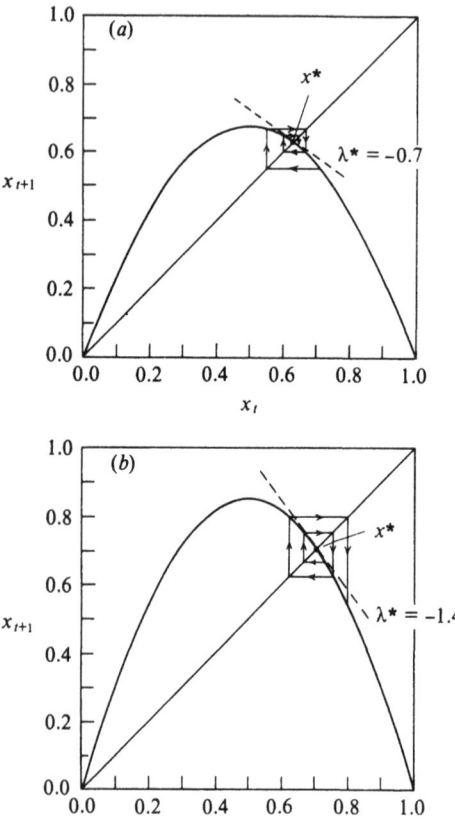

Figure 4.5.4. The graph of a function f defined by (3) with $r = 2.7$ (a) and $r = 3.4$ (b) (adapted from May, 1976). The fixed points are the places where the curve $f(x)$ intersects the $45°$ line, $x_{t+1} = x_t$. The dashed lines indicate the slope λ^* of $f(x)$ at the fixed points. In (a) the slope λ^* is between -1 and $+1$ and the fixed point is stable; for (b) the slope λ^* is less than -1, and the point is unstable.

Stability

If a trajectory that starts near the equilibrium point x^* converges to x^*, then x^* is said to be **locally stable**. For a one-dimensional system, the stability of an equilibrium point depends on the slope of f at x^*. We denote this slope by λ^*, i.e.

$$\lambda^* \equiv \frac{df}{dx}\bigg|_{x=x^*}. \tag{8}$$

As Fig. 4.5.4 illustrates, if the slope of f at x^* lies between $45°$ and $-45°$ (that is if $-1 < \lambda^* < 1$), then the equilibrium point will be locally stable, attracting all trajectories in its neighborhood. For (3), we see using (4) and (7) that

$$\lambda^* = r(1 - 2x)|_{x = x^* = 1 - (1/r)} = 2 - r. \tag{9}$$

Thus the equilibrium point x^* is stable if and only if $1 < r < 3$.

To derive this stability result in an analytical fashion, write

$$x_t = x^* + x_t',$$

where x_t' denotes a small excursion from x^*. Then using a Taylor series approximation for $f(x^* + x_t')$, (3) becomes

$$x^* + x_{t+1}' = f(x^* + x_t') \approx f(x^*) + \lambda^* x_t'. \tag{10}$$

Since $x^* = f(x^*)$, (10) becomes

$$x_{t+1}' = \lambda^* x_t'. \tag{11}$$

Consequently, if $|\lambda| > 1$, $|x_t'| \to \infty$ as $t \to \infty$, whereas, $|x_t'| \to 0$ as $t \to \infty$, if $|\lambda^*| < 1$. When $|x_t'| \to 0$, $x_t \to x^*$ and x^* is said to be locally stable.

What happens when $r > 3$? To answer this question, it is useful to examine the map which relates populations two generations apart, i.e. x_{t+2} and x_t. By iterating (2), one sees

$$x_{t+2} = f[f(x_t)]; \tag{12}$$

or introducing a simplifying notation

$$x_{t+2} = f^{(2)}(x_t). \tag{13}$$

For the function defined by (3),

$$x_{t+2} = rx_{t+1}(1 - x_{t+1}) = r^2 x_t(1 - x_t)[1 - rx_t(1 - x_t)]. \tag{14}$$

Fixing the value of r, one can plot x_{t+2} versus x_t and obtain a graph of $f^{(2)}$. Notice that just as f was symmetric about $\frac{1}{2}$, $f^{(2)}$ must be symmetric about $\frac{1}{2}$ since x_t only appears in (14) in the combination $x_t(1 - x_t)$. The maxima and minima of $f^{(2)}$ can be obtained by differentiation. Using the chain rule

$$\frac{df^{(2)}}{dx} = \frac{df}{dx}\bigg|_{x = f(x)} \frac{df}{dx}. \tag{15}$$

From (4),

$$df^{(2)}/dx = r[1 - 2f(x)]r(1 - 2x) = r^2[1 - 2rx(1 - x)][1 - 2x]. \tag{16}$$

A point p such that

$$\frac{df}{dx}\bigg|_{x=p} = 0$$

is called a *critical point* of f. From (16) we see that the critical points of $f^{(2)}$ are $x = \frac{1}{2}$ and any real solutions to the equation

$$1 - 2rx + 2rx^2 = 0, \tag{17}$$

i.e.

$$x = \frac{2r \pm \sqrt{[4r(r-2)]}}{4r}. \tag{18}$$

Thus for $r < 2$, $x = \frac{1}{2}$ is the only critical point. For $r > 2$ there are three critical points: $x = \frac{1}{2}$ and the two values of x given by (18). One can further show that for $r < 2$, $x = \frac{1}{2}$ is a maximum, whereas for $r > 2$, $x = \frac{1}{2}$ is a minimum and the two values of x given by (18) are maxima. In Figure 4.5.5, $f^{(2)}$ is plotted for various values of r.

Fixed points of period 2

Fixed points of period 2 are points invariant under two iterations of the map f. These points, written $(x^*)^{(2)}$, can be found either algebraically by solving the equation

$$(x^*)^{(2)} = f^{(2)}[(x^*)^{(2)}], \tag{19}$$

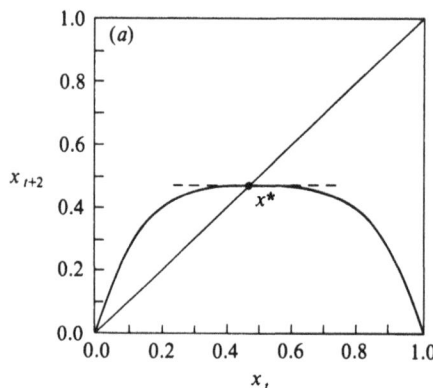

Figure 4.5.5. The relationship between x_{t+2} and x_t as described by (14) for various values of r (from May, 1976). (*a*) $r = 1.9$. For $r \leqslant 2$, $f^{(2)}$ has a single hump and only one fixed point. (*b*) $r = 2.7$. When $r > 2$, $f^{(2)}$ has two humps. The basic fixed point x^* is stable if $1 < r < 3$. (*c*) $r = 3.4$. The 45° line intersects $f^{(2)}$ in three places when $r > 3$. The basic fixed point x^* is unstable, since the slope of $f^{(2)}$ at x^* is greater than 1. The two new fixed points, $(x_1^*)^{(2)}$ and $(x_2^*)^{(2)}$, are stable for this value of r giving rise to cycles of period 2.

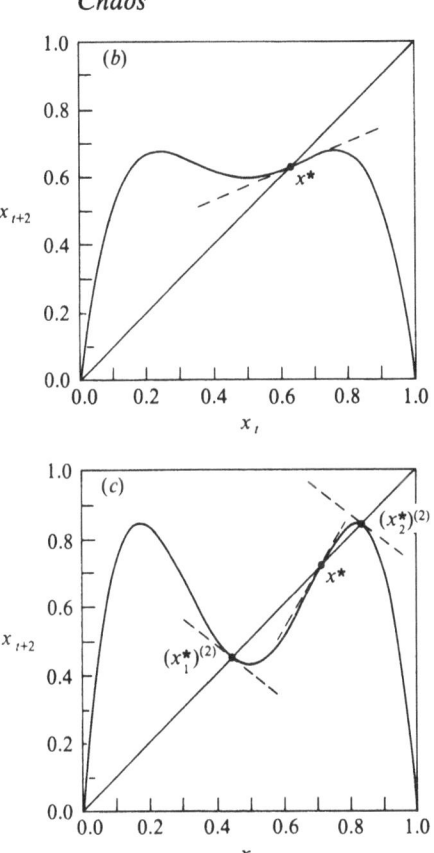

Figure 4.5.5. Continued.

or graphically from the intersection of the map $f^{(2)}(x)$ with the 45° line as shown in Fig. 4.5.5. Clearly, the equilibrium point x^* of (6) is a solution of (19). A fixed point of period 1 is a degenerate case of a fixed point of period 2.

The stability of a fixed point of period 2 is determined by the slope of $f^{(2)}(x)$ at the point $x^{*(2)}$. By the chain rule

$$\frac{d}{dx}f[f(x)]\bigg|_{x=(x^*)^{(2)}} = \frac{df}{dx}\bigg|_{x=f[(x^*)^{(2)}]}\frac{df}{dx}\bigg|_{x=(x^*)^{(2)}}. \tag{20}$$

For the degenerate fixed point $(x^*)^{(2)} = x^*$, $f(x^*) = x^*$ and thus

$$\frac{df^{(2)}}{dx}\bigg|_{x=x^*} = \left[\frac{df}{dx}\bigg|_{x=x^*}\right]^2 = (\lambda^*)^2. \tag{21}$$

This fact can now be used to see what happens when the fixed point x^* becomes unstable. If $|\lambda^*| < 1$ (i.e. $r < 3$) then x^* is stable, and from (21), the slope of $f^{(2)}(x)$ at x^* must lie between $0°$ and $45°$ as shown in Fig. 4.5.5. For slopes in this range, $f^{(2)}$ will only intersect the $45°$ line once as shown in the figure. At $|\lambda^*| = 1$, i.e. $r = 3$ in (3), $f^{(2)}$ is tangent to the $45°$ line at x^*. For $|\lambda^*| > 1$ ($r > 3$), the humps of $f^{(2)}$ become so pronounced that the slope of $f^{(2)}$ at x^* is steeper than $45°$. Thus x^* is unstable. However, $f^{(2)}$ now intersects the $45°$ line at two new points $(x_1^*)^{(2)}$ and $(x_2^*)^{(2)}$ both of which are stable.

In summary, as $f(x)$ becomes more steeply humped because of increases in r, the fixed point x^* becomes unstable. This occurs at $r = 3$. For $r > 3$ two new and initially stable fixed points of period 2 arise. In time the system then alternates between the fixed points $(x_1^*)^{(2)}$ and $(x_2^*)^{(2)}$ giving rise to periodic behavior.

The stability of the period 2 cycle depends on the slope of $f^{(2)}$ at $(x_1^*)^{(2)}$ and $(x_2^*)^{(2)}$. From (20) it is easy to see that the slope is the same at these two points. This stability-determining slope has the value $\lambda = 1$ at the birth of the 2-point cycle and then decreases through zero towards $\lambda = -1$ as the hump in $f(x)$ steepens with increasing values of r. Beyond the point $\lambda = -1$ the 2-point cycle becomes unstable and generates a stable 4-point cycle. Examining $f^{(4)}$ one can show that in this range of r values, $f^{(4)}$ has 4 humps which steepen as r increases. The slope of $f^{(4)}$ at the intersection with the $45°$ line changes from $+45°$ to $-45°$ as the kinks in $f^{(4)}$ grow, and thus the 4-point cycle goes unstable. As one continues to increase r, the same process repeats itself giving rise to a hierarchy of stable cycles with period 8, 16, 32, ... 2^n. Although this process produces an infinite sequence of cycles with periods 2^n, $n \to \infty$, the range of r values wherein any one cycle is stable progressively decreases, so that the entire process is convergent, being bounded above by some critical value, r_c, of r. For (3), $r_c = 3.5700 \ldots$ (May, 1976).

Beyond the limiting value r_c, the behavior is quite different, and we suggest that the reader simulate the difference equation (3) on a hand calculator to see this. Cycles of arbitrarily long periods appear after r_c. At first these cycles all have even periods, with x_t alternating between values above and below x^*. These cycles may be very complicated with periods of several thousand points, but they will seem rather like a noisy cycle of period 2. As r continues to increase, the first odd period cycle appears at $r = 3.6786 \ldots$ (May, 1976). Initially these cycles have very long odd periods, but as r increases cycles with smaller and smaller odd periods appear, until a 3-point cycle appears at $r = 3.8284 \ldots$. Beyond this point, cycles of all integer periods are present, i.e. changing the initial point x_0 gives rise to cycles of all lengths. But even more surprisingly, there are an uncountable number of initial points x_0 which give totally *aperiodic*

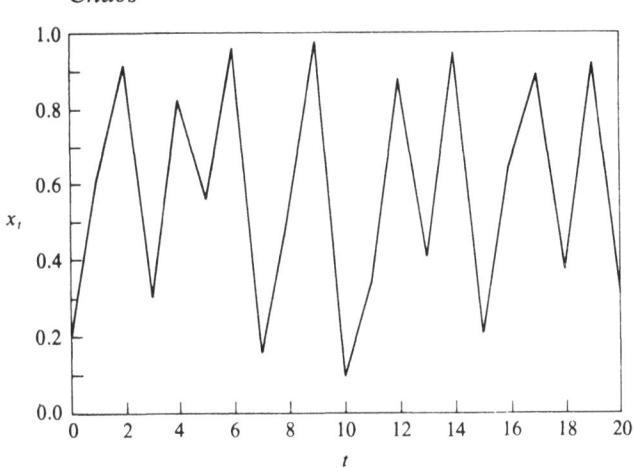

Figure 4.5.6. The solution to (3) with $r = 3.9$ and $x_0 = 0.2$ for $t = 0, 1, 2, \ldots, 20$. No value occurs twice and the pattern does not repeat itself.

trajectories; no matter how many points x_t are generated, the pattern never repeats (see Fig. 4.5.6). This situation where an infinite number of different orbits can occur has been christened 'chaotic' by Li & Yorke (1975). The original proof of this result by Li & Yorke was entitled 'Period three implies chaos.'

In the chaotic region a single trajectory has a completely well-defined and deterministic behavior. However, slight changes in the initial conditions can give rise to very different long-term behavior. If one observed a process governed by (3) with r in the chaotic regime the orbit might look indistinguishable from one generated by a stochastic process. Numerical simulations tend to confirm this (May, 1976). Other difference equations with nonanalytic functions $f(x)$, such as

$$x_{t+1} = \begin{cases} rx_t & \text{if } x_t < \frac{1}{2} \\ r(1 - x_t) & \text{if } x_t > \frac{1}{2}, \end{cases} \tag{22}$$

can be proven to have truly random behavior (of. May, 1976). For (3) one can show that for any specified parameter value there is one unique cycle that is stable, and attracts essentially all initial points (Smale & Williams, 1976). The remaining infinite number of other cycles and uncountable number of aperiodic trajectories, occur for a set of initial conditions that have measure zero. However, since any particular stable cycle is likely to occur only for an extraordinarily narrow range of r values, and because of the long time one expects to elapse before the transients associated with any given initial condition damp out, in practice the unique cycle is unlikely to be detected by observation. Thus a stochastic description of the dynamics is likely to be appropriate even though the underlying process is deterministic. This has been well appreciated in statistical mechanics for very large systems. What is new here is that *very simple* deterministic dynamics can give rise to a situation that looks stochastic. The implications of this for ecology and other areas of science are most unsettling. From a modeling point of view, it means that it may be impossible to distinguish data generated by a simple deterministic process, such as (3), from stochastic noise or experimental error in measurement. Alternatively, phenomena that look chaotic may come from simple underlying dynamic laws. Ruelle & Takens (1971) have taken this viewpoint in trying to explain turbulence in fluids.

Other examples and applications

In the chaotic regime arbitrarily close initial conditions can lead to trajectories, which after a sufficiently long time, diverge widely. Thus even with a simple model in which all parameters are determined exactly, long-term prediction is impossible. Lorenz (1963, 1964), in attempting to predict weather, noticed this phenomenon and called it the '*butterfly effect*': assuming one could describe the atmosphere exactly by a deterministic model, the fluttering of a butterfly's wings could change the initial conditions and in the chaotic regime alter the long-term predictions of the model. Besides noticing the chaotic behavior of (3), Lorenz (1963) also discovered a system of three ordinary differential equations which apparently also exhibit chaos. These equations are rather simple:

$$dx/dt = -ax + ay, \tag{23}$$

$$dy/dt = xz + bx - y, \tag{24}$$

$$dz/dt = xy - cz. \tag{25}$$

For $a = 10$, $b = 28$ and $c = \frac{1}{3}$ aperiodic behavior is observed in numerical simulations.

A simple mechanical system composed of a double pendulum with nonlinear springs, gives rise to chaotic motion. A picture of the orbits is given by Arnold & Avez (1970).

Turbulence in fluid motion has been proposed to occur via a sequence of bifurcations, similar to those we described for difference equations (see Ruelle & Takens, 1971).

Chemical reactions have recently been discovered that seem to exhibit chaotic behavior. Olsen & Degn (1977) describe this effect in the oxidation of NADH by O_2 catalyzed by horseradish peroxidase. Schmitz, Graziani & Hudson (1977) observed what appears to be chaotic states in the Belousov–Zhabotinskii reaction run in an isothermal continuous flow stirred tank reactor. Schmitz has also observed chaotic behavior in an experimental study of hydrogen oxidation on a platinum catalyst. Rössler (1977) has suggested abstract kinetic systems which exhibit chaos.

A method of discovering chaotic behavior

The existence of aperiodic behavior may be established for any first-order difference equation by utilizing a theorem of Li & Yorke (1975). Starting at the critical point, b, in Fig. 4.5.7, trace out the next two iterates of the

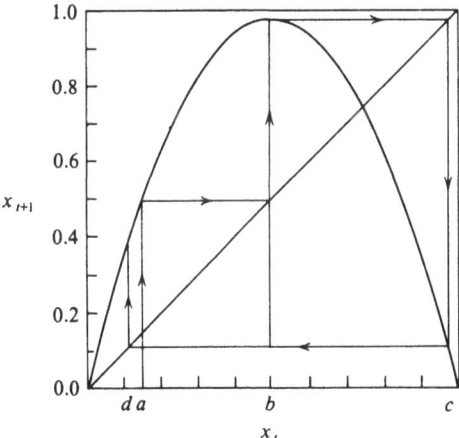

Figure 4.5.7. Method of showing the existence of chaotic solutions. If $d \leq a$ then a period-3 cycle exists and there are initial conditions which generate aperiodic trajectories.

map, $c = f(b)$ and $d = f(c)$, as well as the pre-image a, $b = f(a)$. If $a \geq d$ (i.e. if d is to the left of a) then there exists an infinite number of periodic points and the function can generate aperiodic motions.

This procedure was used experimentally by Olsen & Degn (1977). From experimental observations of the O_2 concentration in a peroxidase-catalyzed oxidation reaction of NADH (Fig. 4.5.8), they extracted the amplitudes of the oscillations. Plotting each amplitude, they obtained a transition map f, and utilized the Li & Yorke procedure to show that chaotic behavior existed. This is illustrated in Fig. 4.5.9.

Conclusion

For mathematical simplicity, the bulk of our discussion in this section has concerned a difference equation of the type found in population

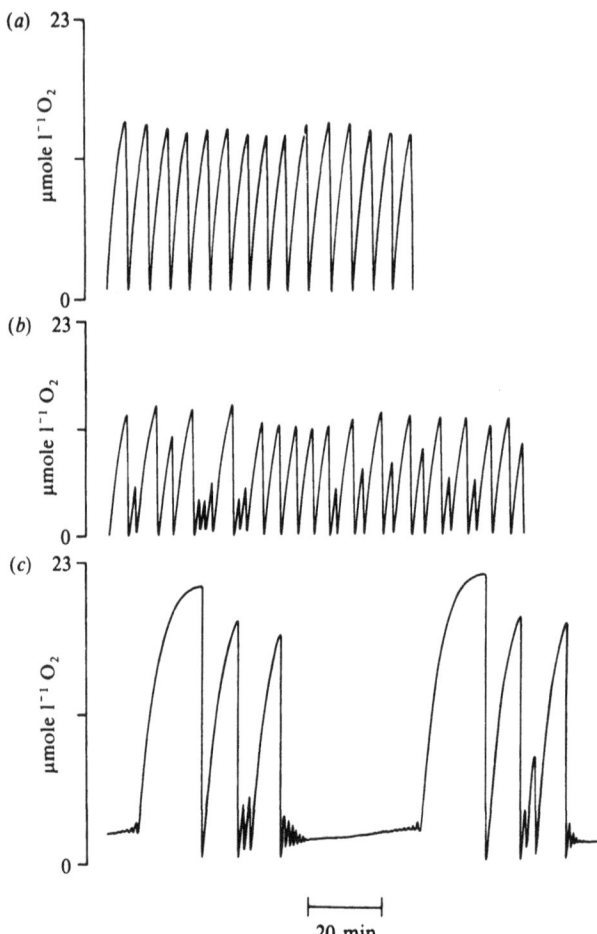

Figure 4.5.8. Oscillations in O_2 concentration in the peroxidase-catalyzed oxidation of NADH in a system open to O_2 (from Olsen & Degn, 1977). The overall peroxidase-catalyzed reaction is $2NADH + 2H^+ + O_2 \rightarrow 2NAD^+ + 2H_2O$. The concentration of peroxidase was varied in the three experiments shown; its value being (a) 0.90 μmole l^{-1}; (b) 0.55 μmole l^{-1} and (c) 0.45 μmole l^{-1}.

dynamics. We have, however, given specific examples of apparently chaotic behavior in chemical systems. Moreover, the system of (23)–(25) is of the type that governs interacting cell populations, so the chaotic solutions of these equations might be expected to find counterparts in cellular biology. In general, suitable combinations of nonlinearity, delay, and high-order systems promote chaotic behavior, so one expects more and more instances of such behavior to reveal themselves in biology.

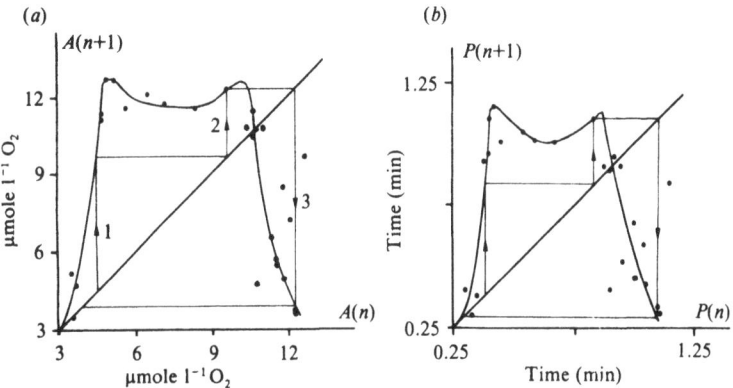

Figure 4.5.9. (*a*) The amplitudes of the oscillations in Figure 4.4.8*b* $A(n+1)$, plotted against the preceding amplitudes $A(n)$. (*b*) The oscillation periods in Figure 4.4.8*b*. $P(n+1)$ plotted against the preceding periods, $P(n)$. Trajectories with arrows were drawn and show that the transition functions allow a period-3 cycle. (From Olsen & Degn, 1977.)

References

Arnold, V. I. & Avez, A. (1970). *Ergodic Problems in Classical Mechanics*, New York, Benjamin.

Hoppensteadt, F. C. (1975). *Mathematical Theories of Populations: Genetics and Epidemics*, Philadelphia, SIAM.

—— (1976). *Mathematical Methods of Population Biology*, New York, Courant Institute of Mathematical Sciences.

Li, T.-Y. & Yorke, J. A. (1975). Period three implies chaos. *Amer. Math. Monthly* **82**, 985–92.

Lorenz, E. N. (1963). Deterministic nonperiodic flows. *J. Atmosph. Sci.* **20**, 130–41.

—— (1964). The problem of deducing the climate from the governing equations. *Tellus* **16**, 1–11.

Mackey, M. C. & Glass, L. (1977). Oscillation and chaos in physiological control systems. *Science* **197**, 287–9.

May, R. M. (1976). Some mathematical models with very complicated dynamics. *Nature, Lond.* **261**, 459–67.

May, R. M. & Oster, G. (1976). Bifurcations and dynamic complexity in simple ecological models. *Amer. Natural.* **110**, 573–99.

Olsen, L. F. & Degn, H. (1977). Chaos in an enzyme reaction. *Nature, Lond.* **267**, 177–8.

Rössler, O. E. (1977). Chaos in abstract kinetics: two prototypes. *Bull. Math. Biol.* **39**, 275–89.

Ruelle, D. & Takens, F. (1971). On the nature of turbulence. *Commun. Math. Phys.* **20**, 167–92.

Schmitz, R. A., Graziani, K. R. & Hudson, J. L. (1977). Experimental evidence of chaotic states in the Belousov–Zhabotinskii reaction. *J. Chem. Phys.* **67**, 3040–4.

Smale, S. (1967). Differentiable dynamical systems. *Bull. Amer. Math. Soc.* **73**, 747–817

Smale, S. & Williams, R. F. (1976). The qualitative analysis of a difference equation of population growth. *J. Math. Biol.* **3**, 1–4.

5 Mathematical immunology

5.1 AN INTRODUCTION TO THE IMMUNE SYSTEM

Immunology is one of the most exciting and rapidly developing areas of modern biology. The excitement is in part a consequence of the fact that the study of the immune system has provided clues to elucidating some of the general principles regulating the behavior of mammalian cells (cf. Hood, Huang & Dreyer, 1977; Hood, Campbell & Elgin, 1975; Hood & Prahl, 1971), while at the same time providing better understanding of how the body guards its integrity against infectious disease and malignant transformations. Underlying the great advances in immunology during the last decade have been the elucidation of the molecular structure of the antibody molecule by Edelman and Porter and the presence of a theoretical framework, the theory of clonal selection, from which to interpret experimental findings. Although we shall not be able to discuss most of the recent advances in immunology here, throughout this chapter we shall discuss the basic properties of the immune system with the aim of showing that both theoretical and practical questions of considerable importance in immunology cannot be answered by purely experimental techniques and require the development of mathematical models. We shall discuss some of the modern applications of mathematics to immunology, but refer the reader to Bell & Perelson (1978) for a more complete review of recent research. For additional background material on immunology we recommend modern textbooks such as those by Eisen (1974) and by Hood, Weissman & Wood (1978).

The observation that people who recover from certain diseases rarely succumb to them again – they are immune – has been made repeatedly over the centuries. The word immune derives from the latin 'immunis' meaning no money and was used to describe people who did not have to pay tax. Beginning in the Middle Ages attempts were made to induce immunity by inoculating well persons with material scraped from the skin of individuals suffering from small pox. It was not until 1890 that the basis for these immune responses was revealed by Von Behring and Kitasato to be a substance in the blood serum. The inoculated materials were called antigens, and the substances capable of neutralizing them that subsequently appeared in the serum were called antibodies.

Antigens and antibodies

According to more modern definitions, an **antigen** is any substance which is capable of inducing antibody formation and which can react specifically with these antibodies. Both properties are important because there exist substances known as **haptens** which do not stimulate the formation of antibodies, but which still react specifically with antibody. Generally haptens are small molecules with molecular weights of less than 1000 daltons. There also exist substances, such as mitogens, which can nonspecifically stimulate antibody production. Because antigens can be very large molecules or even whole cells, it is useful to distinguish between the antigen as a whole and its **antigenic determinants**, i.e. those small portions of the antigen that react chemically with antibody and consequently are responsible for the specificity of the antigen–antibody interaction.

Antibodies are proteins formed in response to an antigen. They are also called **immunoglobulins** (Ig) since they appear in the globulin fraction of serum. In man, five classes of immunoglobulin have been distinguished on the basis of electrophoretic mobility and amino acid sequence: IgG, IgM, IgE, IgA and IgD. Immunoglobulin G (IgG) constitutes about 75% of the immunoglobulin in normal serum. Its molecular structure is shown in Figure 5.1.1. The molecule has two identical heavy (H) polypeptide chains (MW $\sim 50\,000$) and two identical light (L) polypeptide chains (MW $\sim 25\,000$). Each chain is composed of a **constant region**, whose amino acid sequence is more or less the same for all immunoglobulins of a given class, and a **variable region**, whose amino acid sequence differs greatly from one antibody molecule to another, even within the same class. The great sequence diversity found among antibody variable regions is responsible for the specificity antibody exhibits in its reactions with antigen. Electron microscopic examination of purified IgG molecules show that they have a Y shape, with the angle between the arms of the Y being variable, suggestive of great flexibility in the hinge regions of the molecule (see Figure 5.1.1). This flexibility probably allows the IgG molecule simultaneously to bind two antigenic determinants. The other classes of immunoglobulin have structures similar to that of IgG, with IgM and IgA being composed of multiple IgG-like subunits (see Figure 5.4.6; also see pp. 22–8).

Lymphocytes

Antibodies are produced by a type of white blood cell, the **B lymphocyte**. These cells are morphologically indistinguishable from another class of lymphocyte, the **T lymphocytes**, which are chiefly responsible for cell-

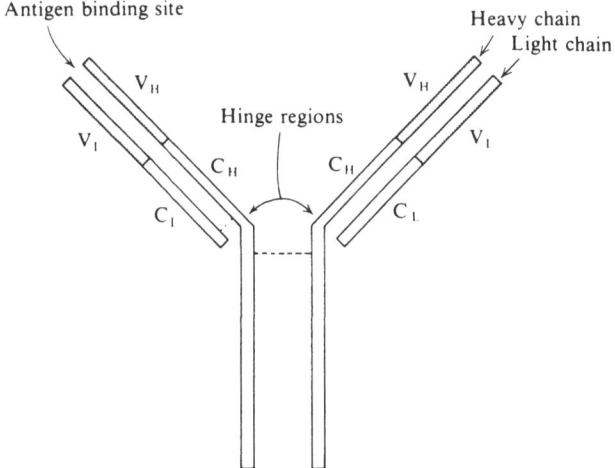

Figure 5.1.1. The molecular structure of IgG. The molecule is composed of four polypeptide chains, two identical light chains (L) and two identical heavy chains (H). Each chain contains a variable region (V_H or V_L) in which the amino acid sequence differs greatly between molecules and a constant region (C_H or C_L) where the amino acid sequence is more or less the same in all IgGs of a given class. Each molecule also contains two identical binding sites for antigen, composed of amino acids from both the heavy and light chain variable regions. The hinge regions of the heavy chains are flexible thus allowing the molecule to bind simultaneously two antigenic determinants.

mediated (i.e. nonantibody mediated) immune responses such as graft rejection and delayed hypersensitivity reactions. T cells can also interact with B cells and play either a helper or suppressive role in regulating the antibody response of the animal.

B cells have on their surfaces about 10^5 immunoglobulin molecules which act as receptors for antigen. The variable regions of all the surface immunoglobulin on a single B cell appear to be identical; thus a particular B cell can only detect a restricted class of antigens. Approximately 1% of the B cells which bind a given antigen become stimulated to divide actively and secrete antibody. Consequently, simply binding antigen is not sufficient to trigger a B cell into antibody secretion. For some antigens, known as **T-dependent antigens**, T cells or a T cell produced factor is required for B cell stimulation. For other antigens, **T-independent antigens**, T cells are not required. In general, T-independent antigens are polymeric molecules, such as pneumonococcal polysaccharide, with repeating arrays of antigenic determinants, that can be expected to bind to B cells very strongly. What the precise biochemical signal or signals are that cause B cell stimulation is not yet known.

All the antibody secreted by a stimulated B cell has the same variable region, and hence the same antigen specificity, as the cell's receptors. A single B cell therefore produces antibody of one specificity. Antigen generally stimulates many different B cells, so that the antibody detected in the serum is very heterogeneous. As an example, the hapten 4-hydroxy-5-iodo-3-nitro-phenacetyl conjugated to protein has been shown to stimulate between 100 and 200 different B cell clones within a single CBA/H mouse, and has been estimated to stimulate over 3000 different clones within the whole inbred CBA/H strain (Kreth & Williamson, 1973). With time, the heterogeneity of the immune response generally decreases and the average affinity of the antibody for the antigen generally increases. This phenomenon is known as the **maturation of the immune response** (Siskind & Benacerraf, 1969).

Stimulated B cells not only produce antibody, but also differentiate into more specialized cell types: plasma cells and memory cells. **Plasma cells** do not divide, but secrete antibody at very high rates. **Memory cells** resemble unstimulated B cells; they do not secrete antibody, but in the presence of antigen are triggered into antibody secretion and proliferation. The presence of large numbers of memory cells generated after a first contact with antigen is believed to underlie the common observation of immunity to many infectious diseases. Memory cells and plasma cells as well as the control of B cell proliferation, differentiation and antibody secretion will be discussed more fully in Section 5.4.

Clonal selection

One of the most intriguing aspects of the immune system is its ability to respond to almost any foreign substance. A simple explanation, the **instructive theory** suggested by a number of scientists, including Linus Pauling, was that antigen somehow instructed an animal to make complementary antibody. For example, Pauling (1940) hypothesized that the antigen could be used as a template in determining how the antibody molecule folded into its final three-dimensional structure. It was later found that all proteins, including antibodies, have a unique three-dimensional structure determined solely by the protein's amino acid sequence. Thus an antibody molecule could be denatured by heating, and then upon cooling in the absence of antigen, returned to its original configuration. An alternative theory, **clonal selection**, was proposed by Jerne (1955) and Burnet (1959). According to this theory an animal maintains a large repertoire of B cells with different antigen specificities. Antigen, when encountered by the animal, then selectively binds and stimulates those cells in the repertoire capable of making complementary antibody, see Figure 5.1.2. The major difference between the instructive and clonal

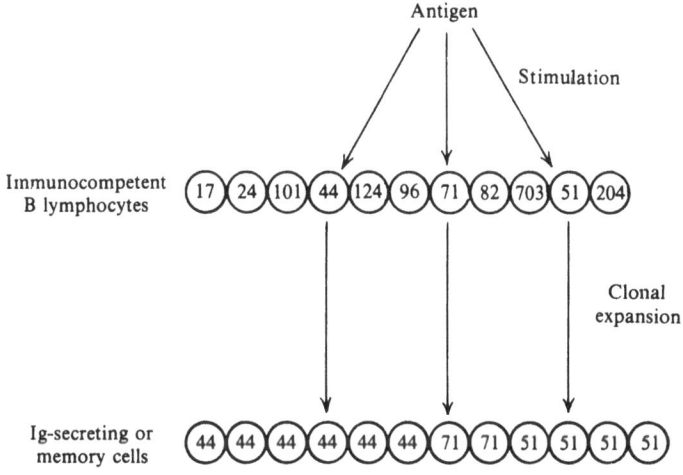

Figure 5.1.2. Schematic representation of clonal selection. Antigen stimulates a small proportion of immunocompetent B cells to proliferate into a clone of antibody secreting cells. Each clone produces a different immunoglobulin. Clones of memory cells may also be produced.

selection theories can be made clear by the following analogy. In order to serve the needs of its patrons for many copies of appropriate works an information service can operate in one of two modes. It may keep no books on hand, but on learning the patron's needs, it will assemble and copy the appropriate materials (instructive theory). Alternatively, an information service (library) may maintain a large collection of books ordered at random and hope to have in its collection some book that is reasonably appropriate to the needs of any patron (clonal selection). If it does, it need only duplicate the material.

Clearly, for clonal selection to work the library must maintain a very large diverse collection of books (B cell specificities). Exactly how large it needs to be can be estimated from a mathematical model as we show in Section 5.2. The overwhelming preponderance of experimental information supports the clonal selection theory (cf. Nossal, 1977), and it is generally accepted as a valid outline of the functioning of the immune system. For example, if an animal is injected with a radioactively labeled antigen of such high specific activity that it kills or inactivates any cell to which it binds, one finds on subsequent challenge that the animal is incapable of responding to the unlabeled form of this antigen, but its response to unrelated antigens is unaffected.

There is an interesting parallel between Darwin's theory of natural selection and the theory of clonal selection. Organisms such as bacteria survive rapid changes in their environment by maintaining a very heterogeneous population, some members of which have the genetic capability to survive in extreme environments. When the environment changes most organisms may die, but the survivors multiply rapidly and replenish the population. Infectious disease is a type of environmental change that vertebrates must cope with, but because of their limited population sizes and low birth rates they cannot employ the same survival strategy as bacteria. Evolution seems to have side-stepped this difficulty with the development of the immune system. In humans this consists of a population of 10^{12} lymphocytes which show great diversity and which are one of the most rapidly dividing cell types. Viewing these freely moving cells as a symbiotic population, clonal selection is nothing more than natural selection. Although this viewpoint is an intriguing one, it has not yet been exploited by mathematical biologists. For the most part sophisticated population dynamic and population genetic models of the immune system are lacking (see Bell, Perelson & Pimbley, 1978 for an introduction to existing models).

References

Bell, G. I. & Perelson, A. S. (1978). An historical introduction to theoretical immunology. In *Theoretical Immunology*, ed. G. I. Bell, A. S. Perelson and G. Pimbley Jr, New York, Marcel Dekker, pp. 3–41.

Bell, G. I., Perelson, A. S. & Pimbley, G., Jr (1978). *Theoretical Immunology*, New York, Marcel Dekker.

Burnet, F. M. (1959). *The Clonal Selection Theory of Acquired Immunity*, Nashville, Tennessee, Vanderbilt University Press.

Eisen, H. M. (1974). *Immunology*, 2nd edn, revised reprint, Hagerstown, Maryland, Harper and Row.

Hood, L., Campbell, J. H. & Elgin, S. C. R. (1975). The organization, expression, and evolution of antibody genes and other multigene families. *Ann. Rev. Genet.* **9**, 305–53.

Hood, L., Huang, H. V. & Dreyer, W. J. (1977). The area-code hypothesis: the immune system provides clues to understanding the genetic and molecular basis of cell recognition during development. *J. Supramolec. Struct.* **1**, 531–59.

Hood, L. & Prahl, J. (1971). The immune system: a model for differentiation in higher organisms. *Adv. Immunol.* **14**, 291–351.

Hood, L. E., Weissman, I. E. & Wood, W. B. (1978). *Immunology*, Reading, Mass., Addison-Wesley.

Jerne, N. K. (1955). The natural selection theory of antibody formation. *Proc. Nat. Acad. Sci., USA* **41**, 849–56.

Kreth, H. W. & Williamson, A. R. (1973). The extent of diversity of antihapten antibodies in inbred mice: anti-NIP (4-hydroxy-5-iodo-3-nitro-phenacetyl) antibodies in CBA/H mice. *Eur. J. Immunol.* **3**, 147–52.

Nossal, G. J. V. (1977). B-lymphocyte receptors and lymphocyte activation. In *International Cell Biology 1976–1977*, ed. B. R. Brinkley and K. R. Porter, New York, Rockefeller University Press, pp. 103–11.

Pauling, L. (1940). A theory of the structure and process of formation of antibodies. *J. Amer. Chem. Soc.* **62**, 2643–57.

Siskind, G. W. & Benacerraf, B. (1969). Cell selection by antigen in the immune response. *Adv. Immunol.* **10**, 1–50.

5.2 A MATHEMATICAL LOOK AT CLONAL SELECTION[1]

Inman (1978) has estimated that a mammal, such as a laboratory mouse, must be capable of distinguishing as foreign at least 10^{16} different haptens. In order that this many haptens may be recognized a large population of different antibody molecules is required, but not necessarily 10^{16} of them. A mouse possesses approximately $5 \cdot 10^8$ B cells and thus at most can synthesize this number of antibodies. A more realistic estimate of the number of distinct antibodies an animal can make, what immunologists call the repertoire size, is 10^6–10^7 (Jerne, 1967; Edelman, 1974). In this section we shall develop a mathematical model that helps explain how this number of different antibodies can protect us from 10^{16} or more antigens.

Antibodies recognize antigenic determinants on the basis of three dimensional shape, electric charge, dipole moment, and possibly a few other parameters. For simplicity, we shall call this set of characteristics the **generalized shape**. Suppose one can describe the generalized shape of the antigen binding site or, as it is sometimes called, the **antibody combining region** by N parameters: the width, length and depth of the combining region, its charge, etc. Then a point in an N-dimensional space which we shall call 'shape space,' S, specifies the generalized shape of an antibody with regard to its antigen binding properties. If an animal can make N_{Ab} different antibodies, then the N-dimensional shape space for that animal would contain N_{Ab} distinct points. One would expect these points to lie in some finite volume V since there is only a restricted range of widths, lengths, etc. that an antibody combining site can assume. For example, one would never find an antibody with a combining site one meter long!

Antigenic determinants are also characterized by generalized shapes which we expect to lie within V. For example a 2 nm combining site cannot be expected to recognize a determinant 10 nm long. In order to estimate the number of antibodies required to recognize any antigen, let

---

[1] The work described in this section is based upon Perelson & Oster (1979).

us assume that an antibody and antigenic determinant fit together perfectly if they have the same shape, i.e. lie at the same point in V. (This is clearly a fiction because antigenic determinants and antibodies must have complementary shapes and thus necessarily lie at different points in V. However, it is a useful fiction because it simplifies our argument and yields the same result as a more precise treatment.) If the shapes of the antibody and the antigenic determinant are not quite complementary, the antibody may still bind to the determinant but not as tightly as it would to its complementary image. To describe this we will assume each antibody binds all antigenic determinants that are within a small distance ε of it in shape space. Thus, as we illustrate in Figure 5.2.1, around each antibody there is a ball or sphere of radius ε which depicts the set of shapes that an antibody can bind.

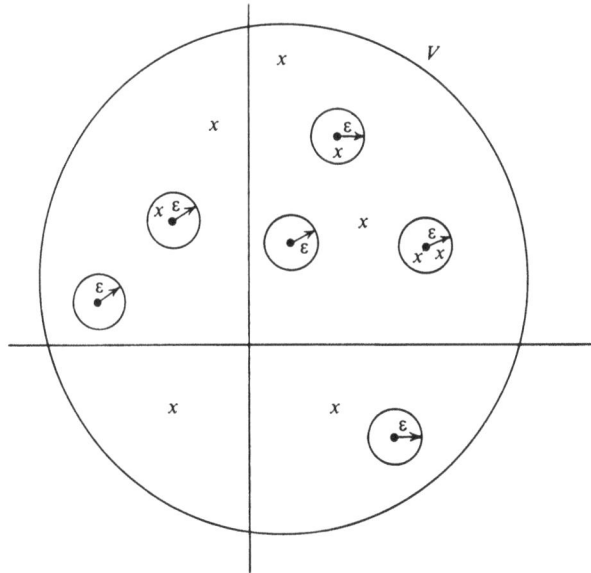

Figure 5.2.1. Pictorial representation of shape space S. Within the space there is a volume V in which antibody combining site (●) and antigenic determinant (x) shapes are located. An antibody is assumed to be able to bind all antigens within a ball of radius ε surrounding it.

To complete our estimate let us assume that V is a ball of radius R centered around a typical antibody shape and that antibody shapes are distributed randomly and uniformly throughout V. Now if the set of balls surrounding each antibody shape completely fill V, then the shape of any antigenic determinant would lie within at least one ball, and hence there would be in the repertoire an antibody which could bind any antigen the animal could conceivably encounter. We can obtain some rough conditions for this to be true as follows:

Each N-dimensional ball of radius ε takes up a volume equal to $c_N \varepsilon^N$, where c_N is a constant that depends on the dimension (e.g. $c_2 = \pi$, $c_3 = 4\pi/3$). If the repertoire size is N_{Ab}, then the volume taken up by N_{Ab} balls would be expected to be somewhat less than $N_{Ab} c_N \varepsilon^N$ since some balls may overlap. The total volume of V is $c_N R^N$. Consequently, if

$$N_{Ab} c_N \varepsilon^N \gg c_N R^N,$$

or equivalently if

$$N_{Ab} \gg (R/\varepsilon)^N, \tag{1}$$

the balls overlap considerably and must cover most of V.

It is possible to do simple probabilistic calculation to make these statements more precise (Perelson & Oster, 1979). Let p be the probability that a randomly encountered antigen shape lies within a ball of radius ε. This probability is simply the ratio of the volume of the ball to the volume of V, i.e.

$$p = \frac{c_N \varepsilon^N}{c_N R^N} \equiv \hat{\varepsilon}^N, \tag{2}$$

where $\hat{\varepsilon} \equiv \varepsilon/R$ and $0 \le \hat{\varepsilon} \le 1$. Now, the probability that the random antigen does not lie within a ball of radius ε is $1 - p$, and the probability that it does not lie within any of N_{Ab} balls of radius ε is $(1-p)^{N_{Ab}}$. Rewriting this as

$$(1-p)^{N_{Ab}} = \exp[N_{Ab} \ln(1-p)], \tag{3}$$

and noticing that the Taylor series expansion of the logarithmic term around 1 gives $\ln(1-p) \simeq -p$, if $p \ll 1$, we obtain

$$(1-p)^{N_{Ab}} = \exp(-N_{Ab}p). \tag{4}$$

Finally, the probability, P, that at least one antibody in the repertoire recognizes the antigen is given by

$$P = 1 - (1-p)^{N_{Ab}}, \tag{5}$$

and hence from (2) and (4)

$$P = 1 - \exp(-N_{Ab}\hat{\varepsilon}^N). \tag{6}$$

We can now see that our original geometrically derived estimate of the repertoire size given by (1) is quite good. If $N_{Ab} = 10(R/\varepsilon)^N$, then from (6), $P = 0.99995$ and the immune system would recognize all but 5 out of every 100 000 antigenic determinants. If N_{Ab} were somewhat larger the immune system could do much better. For $N_{Ab} = 20(R/\varepsilon)^N$ only 2 out of every billion antigens would escape detection, whereas for $N_{Ab} = 50(R/\varepsilon)^N$, $1 - P = 2 \cdot 10^{-22}$.

The parameter p or equivalently $\hat{\varepsilon}^N$ can be determined experimentally from measurements of the fraction of all B cells that respond to a random antigen. Such measurements show $\hat{\varepsilon}^N$ generally lies between 10^{-5}–10^{-4}. Thus an animal with a repertoire size of 10^6 or 10^7 operating on the basis of clonal selection would have an enormously reliable immune system. For example, with $\hat{\varepsilon}^N = 10^{-5}$ and $N_{Ab} = 10^7$, $1 - P = \exp(-100) = 3.7 \cdot 10^{-44}$ and essentially all antigens would be recognized.

It is known that invertebrates do not make antibody. Du Pasquier (1973, 1976) has examined the vertebrates to find representatives sufficiently small that their size might limit their repertoire size. He discovered that young tadpoles containing only 10^6 lymphocytes can mount what appears to be a normal immune response. Since not all lymphocytes are B cells, and because antigen specificities are likely to be repeated in the B cell population, N_{Ab} is probably near 10^5. If $\hat{\varepsilon}^N = 10^{-4}$, then $P = 0.99995$. However, if $\hat{\varepsilon}^N = 10^{-5}$, $P = 0.63$ and the tadpole would recognize only 63% of the antigens it encounters. If N_{Ab}, and hence P, were any smaller the immune system would be so unreliable that it would play almost no role in protecting the animal. It is intriguing that vertebrates with smaller and yet functioning immune systems have not been found.

References

Du Pasquier, L. (1973). Ontogeny of the immune response in cold blooded vertebrates. *Curr. Top. Microbiol. Immunol.* **61**, 37–88.

—— (1976). Phylogenesis of the vertebrate system. In *The Immune System*, ed. F. Melchers and K. Rajewsky, Berlin, Springer-Verlag, pp. 101–15.

Edelman, G. M. (1974). Origins and mechanisms of specificity in clonal selection. In *Cellular Selection and Regulation in the Immune System*, ed. G. M. Edelman, New York, Raven Press, pp. 1–38.

Inman, J. K. (1978). The antibody-combining region: Speculations on the hypothesis of general multispecificity. In *Theoretical Immunology*, ed. G. I. Bell, A. S. Perelson and G. H. Pimbley Jr, New York, Marcel Dekker, pp. 243–78.

Jerne, N. K. (1967). Summary: waiting for the end. *Cold Spring Harbor Symp. Quant. Biol.* **32**, 591–603.

Perelson, A. S. & Oster, G. F. (1979). Theoretical studies of clonal selection: minimal antibody repertoire size and reliability of self-nonself discrimination. *J. Theoret. Biol.* **81**, 645–70.

5.3 MODELS FOR THE INTERACTION OF ANTIGEN WITH CELLS

Most, if not all, cells have receptors on their surface by which they can obtain information about their environment and interact with other cells. Common examples are hormone and chemotactic receptors. Cells of the immune system (such as lymphocytes), and cells involved in allergic responses (mast cells and basophils), have immunoglobulin on their surface which allows them to detect the presence of antigen. The binding of antigen to these cells is the first step in immune and allergic responses. The goal of this section is to present a mathematical framework for analyzing such interactions and to show how the mathematical theory can be utilized to understand various experimental observations about cell behavior.

Antigens bind to B cells by chemically interacting with their immuno-globulin receptors. The antigen may have only a single chemical group or **antigenic determinant** which is recognized by the cellular receptor, in which case it is termed **monovalent**, or it may have multiple copies of the determinant, in which case it is termed **multivalent**. For example, the haptenic group dinitrophenyl (DNP) can be chemically coupled to many proteins and polysaccharides. An antigen which contains one DNP would be monovalent, whereas an antigen with 10 DNPs would be multivalent, and would be said to have **valence** 10. We shall assume throughout that all the determinants on a multivalent antigen are identical.

When a multivalent antigen binds to a B lymphocyte it can induce a spatial redistribution of the cell's immunoglobulin receptors. This can be graphically demonstrated using a fluorescent antigen. At first the B cell is seen to be uniformly fluorescent. However, after a few minutes at 37 °C the fluorescence takes on a spotted or patchy appearance. The fluorescent patches then move to one pole of the cell forming a 'cap' (Taylor, Duffus, Raff & de Petris, 1971). The first part of this process is called **patching** and the latter part **capping**. Patching and capping do not occur with mono-valent antigens and thus it is assumed that the antigen cross-links the immunoglobulin receptors on the cell surface and causes the formation of large antibody–antigen complexes. One of our goals will be the study of aggregate formation on cell surfaces.

In allergic individuals histamine is released by mast cells and basophils which have immunoglobulin E (IgE) on their surfaces. It has been shown that histamine release is only induced by multivalent antigens and not by monovalent antigens. The IgE on mast cells and basophils is not synthesized by these cells, but by lymphocytes as are all other immunoglobulins. IgE in the blood serum is then adsorbed by mast cells and basophils and used by these cells as a receptor. In an interesting experiment, Segal, Taurog & Metzger (1977) chemically cross-linked two IgE molecules and then allowed mast cells to adsorb the cross-linked IgE. The mast cells released histamine, impressively demonstrating the importance of receptor cross-linking and establishing that the minimum cross-linking unit needed for histamine release is two linked IgE molecules. Using this as a motivation we shall also study the establishment of cross-links by bivalent antigens on cell surfaces.

It is easiest to begin a study of receptor–antigen interactions by discussing the chemical kinetic theory of such reactions in solution. Later we shall confine our study to the case of receptors bound to cell surfaces. However, for the most part, this will have little effect on the theory. Because immunoglobulin molecules have two antigen binding sites, we will restrict our attention to systems in which one reactant (antibody) is bivalent and the other reactant (antigen) multivalent.

Antigen–antibody reactions: a coagulation example

Writing down the differential equations describing the growth of an antigen–antibody aggregate is surprisingly complex. In fact, it appears never to have been done in full generality. To obtain some insight into the dynamics of aggregate growth, we shall first consider a system that contains only one type of chemical species which reacts irreversibly with itself. This can then be viewed as a study of coagulation or the irreversible condensation of bifunctional molecules of type $A-(BA)_{n-1}-B$, where A and B designate groups capable of reacting with one another to form an AB bond. To relate this to the antibody problem, let the AB monomer be an undissociable complex formed between a bivalent antibody and a bivalent antigen. The A group would then be a free antibody combining site, whereas the B group would be a free antigenic determinant. In a true antibody–antigen reaction linear aggregates can also form which have only A groups or only B groups at their ends. The aggregation problem that we are studying here is therefore only a simplified model of an antigen–antibody reaction, but one which exhibits many of the mathematical difficulties seen in studying aggregation phenomena.

If $(AB)_n$ designates a molecule containing n AB units, the reactions we are studying can be represented as

$$(AB)_m + (AB)_{n-m} \xrightarrow{k} (AB)_n, \qquad m = 1, 2, \ldots, n-1, \tag{1}$$

where the second order rate constant, k, describing the reaction between an A and B group is assumed to be independent of chain length. That is, we assume k is the same for all values of n and m.

Let the concentration of $(AB)_n$ be $\bar{c}_n$. Then the rate equation for species $(AB)_n$ is given by

$$d\bar{c}_n/d\bar{t} = k \sum_{m=1}^{n-1} \bar{c}_m \bar{c}_{n-m} - 2k\bar{c}_n \sum_{m=1}^{\infty} \bar{c}_m, \tag{2}$$

where we introduce the convention that the first sum on the right side of the equation is identically zero when $n = 1$. This is the fundamental equation of coagulation, first given by Smoluchowski (1916, 1917). The first term represents the formation of species $(AB)_n$ by reactions between pairs of smaller aggregates as represented in (1). A factor of $\frac{1}{2}$ needs to be included so that all pairs are counted only once. Further, if n is even, then this sum also contains a term resulting from the reaction of $(AB)_{n/2}$ with itself. This term also needs to be multiplied by $\frac{1}{2}$ since the relative number of collisions in a mixture of molecules X, Y and Z, between X and X, X and Y, X and Z, are in the ratio $c_X^2, 2c_X c_Y, 2c_X c_Z$ (Tanford, 1961, p. 589). This factor of $\frac{1}{2}$ that should multiply the first sum is cancelled by a statistical factor of 2 that must also be included since there are 2 ways in which an AB complex can be formed in reaction (1); i.e. the A group of $(AB)_n$ can react with the B group of $(AB)_{n-m}$, or vice versa. This factor 2 remains multiplying the second sum in (2), which represents the reaction of $(AB)_n$ with all other aggregates giving rise to aggregates $(AB)_{n+1}, (AB)_{n+2}, \ldots$

We restrict our attention to the initial condition, $\bar{c}_n(0) = 0$ for all n except $n = 1$. Thus we are studying aggregates that form from the monomer AB only. Introducing the dimensionless variables

$$c_n = \bar{c}_n/\bar{c}_1(0) \tag{3a}$$

$$t = k\bar{c}_1(0)\bar{t}, \tag{3b}$$

the problem reduces to solving

$$dc_n/dt = \sum_{m=1}^{n-1} c_m c_{n-m} - 2c_n \sum_{m=1}^{\infty} c_m, \tag{4}$$

with initial condition

$$c_n(0) = \begin{cases} 1 & \text{if } n = 1 \\ 0 & \text{if } n > 1. \end{cases} \tag{5}$$

Following Ziff (1978), we solve (4) by introducing an auxiliary function, $G(z, t)$, known as a **generating function**. Define

$$G(z, t) = \sum_{n=1}^{\infty} z^n c_n(t). \tag{6}$$

Then

$$\frac{\partial G}{\partial t} = \sum_{n=1}^{\infty} z^n \frac{dc_n}{dt}$$

$$= \sum_{n=1}^{\infty} z^n \sum_{m=1}^{n-1} c_m c_{n-m} - 2 \sum_{n=1}^{\infty} z^n c_n \sum_{m=1}^{\infty} c_m. \tag{7}$$

From (6) notice that

$$G(1, t) = \sum_{n=1}^{\infty} c_n(t) \tag{8}$$

and that

$$[G(z, t)]^2 = \left(\sum_{n=1}^{\infty} z^n c_n \right) \left(\sum_{m=1}^{\infty} z^m c_m \right)$$

$$= (zc_1 + z^2 c_2 + z^3 c_3 + \ldots)(zc_1 + z^2 c_2 + z^3 c_3 + \ldots)$$

$$= z^2 c_1^2 + z^3 (c_1 c_2 + c_2 c_1) + z^4 (c_1 c_3 + c_2^2 + c_3 c_1) + \ldots$$

$$= \sum_{n=2}^{\infty} z^n \sum_{m=1}^{n-1} c_m c_{n-m}. \tag{9}$$

Using the summation convention introduced after (2), and results (8) and (9), we see that (7) can be written as

$$\frac{\partial G(z, t)}{\partial t} = [G(z, t)]^2 - 2G(z, t)G(1, t). \tag{10}$$

The initial condition for this equation, obtained using (5) and (6), is

$$G(z, 0) = \sum_{n=1}^{\infty} z^n c_n(0) = zc_1(0) = z. \tag{11}$$

To solve (10) and (11), we first set $z = 1$ and obtain

$$\frac{dG(1, t)}{dt} = -G(1, t)^2, \qquad G(1, 0) = 1. \tag{12}$$

By simple integration

$$\int_1^{G(1,t)} G(1, \tau)^{-2} \, dG(1, \tau) = - \int_0^t d\tau,$$

so that

$$[1/G(1, t)] - 1 = t$$

or

$$G(1, t) = 1/(1 + t). \tag{13}$$

To complete the solution subtract (12) from (10), giving

$$\frac{\partial G(z, t)}{\partial t} - \frac{\partial G(1, t)}{\partial t} = [G(z, t)]^2 - 2G(z, t)G(1, t) + [G(1, t)]^2. \tag{14}$$

On employing the definition $H(z, t) = G(z, t) - G(1, t)$, (14) becomes

$$\partial H/\partial t = [H(z, t)]^2 \tag{15}$$

with

$$H(z, 0) = z - 1. \tag{16}$$

The solution to (15) is

$$H = \frac{z - 1}{1 - (z - 1)t}, \tag{17}$$

and consequently,

$$G(z, t) = \frac{1}{1 + t} + \frac{z - 1}{1 - (z - 1)t}, \tag{18}$$

or

$$G(z, t) = \frac{z}{(1 + t)(1 + t - zt)}. \tag{19}$$

Letting $x = zt/(1 + t)$, we see that

$$G(x, t) = \frac{1}{t(1 + t)} \cdot \frac{x}{(1 - x)} = \frac{x(1 + x + x^2 + x^3 + \ldots)}{t(1 + t)}$$

$$= \frac{1}{t(1 + t)} \sum_{n=1}^{\infty} x^n$$

and hence

$$G(z, t) = \sum_{n=1}^{\infty} z^n \frac{t^{n-1}}{(1 + t)^{n+1}}. \tag{20}$$

Comparing (6) and (20) we finally obtain

$$c_n(t) = t^{n-1}/(1 + t)^{n+1}. \tag{21}$$

The concentration of monomer c_1 starts at unity, and monotonically decreases in amount as $(1 + t^2)^{-1}$. All other c_n, $n > 1$, begin at zero, reach a maximum at time

$$t_{n,\max} = (n - 1)/2,$$

and then approach zero, finally as t^{-2}. For all n,

$$c_{n+1}(t)/c_n(t) = t/(1 + t) < 1, \tag{22}$$

so $c_{n+1}(t) < c_n(t)$ at each instant of time. The maximum amount of each aggregate also decreases with increasing n, i.e.

$$c_n(t_{n,\max}) \geqslant c_{n+1}(t_{n+1,\max}).$$

This behavior is illustrated in Figure 5.3.1.

Before leaving this example it is useful to show that the result (21) can be obtained in another manner, which is outside the realm of traditional mathematical solutions. The method is one pioneered by the polymer chemist Paul Flory (1936) and forms the basis of our later discussion of patching. We call this approach a combinatorial method.

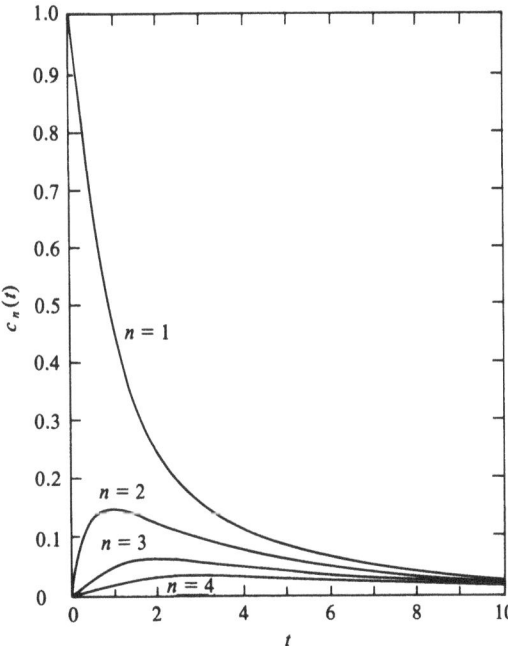

Figure 5.3.1. A plot of $c_n(t)$ versus t for various values of n. Notice that c_1 decreases monotonically, whereas for $n > 1$, c_n increases from zero through a maximum and then asymptotically returns to zero.

Let $p(t)$ be the fraction of the total number of A (or B) groups which have reacted at time t. The variable p is called the **extent of reaction** since $p(0) = 0$ and as p increases towards one it measures the progress of the reaction. Let $c_A(t)$ and $c_B(t)$ be the concentrations of unreacted A and B groups at time t. At $t = 0$ all A and B groups are in the form of monomers AB. In light of the normalization chosen in $(3a)$, $c_A(0) = c_B(0) = 1$. As the reaction progresses free A groups disappear by reacting with free B groups, thus

$$\mathrm{d}c_A/\mathrm{d}t = -c_A c_B, \tag{23}$$

where the rate constant k has been incorporated into the time scale as in $(3b)$. Since A groups react with B groups, the number of reacted A and B groups must be equal. Hence $1 - c_A = 1 - c_B$ or $c_A = c_B$ and thus (23) becomes

$$\mathrm{d}c_A/\mathrm{d}t = -c_A^2, \qquad c_A(0) = 1. \tag{24}$$

This is identical to (12) and thus

$$c_A(t) = 1/(1+t). \tag{25}$$

The extent of reaction, p, is the fraction of reacted A groups, i.e.

$$p(t) = [c_A(0) - c_A(t)]/c_A(0) = t/(1+t). \tag{26}$$

Notice that $p(t)$ is also the fraction of reacted B groups or equivalently the probability that an A group (or B group) has reacted.

Let P_n be the probability that an A group picked at random is part of an n-mer, $(AB)_n$. Thus

$$P_n = \frac{\text{number of A groups in } n\text{-mers}}{\text{total number of A groups}} \tag{27}$$

so that

$$P_n = \frac{nc_n}{c_A(0)} = nc_n. \tag{28}$$

If an A group is to be part of an n-mer, $A-(BA)_{n-1}-B$, then reactions must have occurred linking $(n-1)$ AB molecules together and no linkages must exist at the two ends. The probability that a linkage exists is p and thus $p^{n-1}(1-p)^2$ is the probability of finding an n-mer. Since an A group picked at random could be any of the n A groups in the n-mer, we see that

$$P_n = np^{n-1}(1-p)^2. \tag{29}$$

From (28) and (29) we find that

$$c_n = p^{n-1}(1-p)^2 = t^{n-1}/(1+t)^{n+1}, \tag{30}$$

which is precisely the result we obtained earlier using generating functions. The advantage of the combinatorial method, however, is that it can be used to analyze more complicated reaction mechanisms for which generating function methods either fail or are very complicated to implement.

Reactions between bivalent antibodies and multivalent antigens

When bivalent and multivalent molecules react, very large complicated aggregates can be formed. In order to visualize such aggregates it is convenient to represent them by what are called **linear graphs** (Harary, 1969). We shall assume the multivalent antigens are all f-valent ($f \geqslant 2$). To form a graph we represent each antigen by a **vertex** or **node** and each bivalent antibody by an **edge** or **branch**, as shown in Figure 5.3.2. The number of edges incident on a vertex, v, is called the **degree** of the vertex, deg (v). For all vertices in the graphs we shall deal with, $1 \leqslant \deg (v) \leqslant f$.

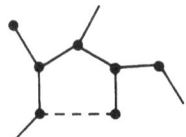

Figure 5.3.2. Graph representing an aggregate composed of f-valent ($\bullet$) and bivalent (—) units. Here $f = 3$ and the graph is a tree so long as the edge represented by the dotted line is absent from the graph.

Usually in a graph each edge is terminated at both ends by a vertex. Here, however, antibodies which are bound at only one of their combining sites are represented by edges terminated by a single vertex. Such edges will be called **external** edges, while edges that connect two distinct vertices are **internal** edges. A graph which contains no loops or cycles is called a **tree**. Here we shall restrict our attention to aggregates whose graphs are trees. This implies that all reactions are bimolecular and at each reaction the aggregate grows by addition of new units or becomes smaller by dissociation. Intramolecular rearrangements of existing aggregates create loops; such reactions will be prohibited.

Figure 5.3.3 shows a reaction between two aggregates. Notice that the reaction product is still a tree. One can view the reaction as the assembly of a tree made out of 'tinker toys'; one places a rod (external edge or free antibody site) into a hole (free antigen site) on a joint (vertex or multivalent ligand) during the forward reaction and removes the rod

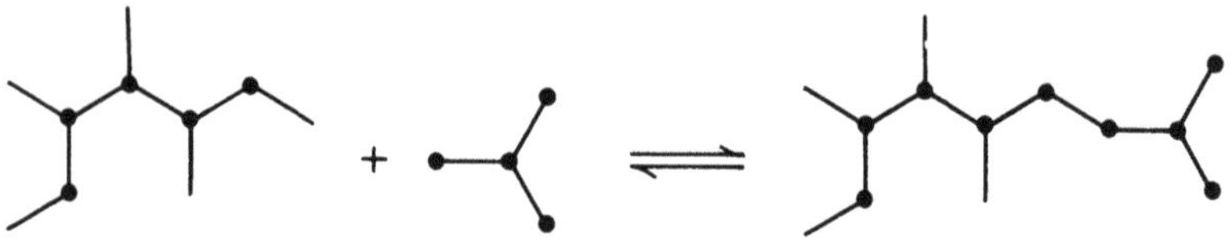

Figure 5.3.3. A reaction between two aggregates.

during the reverse reaction. If we assume that there is a single forward rate constant, k_f, describing the rate at which rods are placed into holes, and a single reverse rate constant, k_r, describing the rate at which rods are removed from holes, irrespective of the aggregate size, then we can easily write down the differential equations describing the growth of the aggregate (Perelson & DeLisi, 1975), and solve them via the combinatorial method. The assumption that there is only a single rate constant characterizing interactions between free antibody and antigen sites (rods and holes) on aggregates of all sizes is known as the **equivalent site hypothesis**.

The parameters that might be used to describe an antigen–antibody aggregate or graph are v, the number of vertices (antigens); e, the number of edges (antibodies); s, the number of external edges (free antibody sites); and l, the number of free antigen sites. For a tree any two of these four parameters can be chosen as independent. For example, if the number of antigens, v, and the number of free antibody sites, s, are chosen as independent variables then one can show (Perelson & DeLisi, 1975)

$$e = s + v - 1, \tag{31a}$$

and

$$l = v(f - 2) + 2 - s. \tag{31b}$$

Alternatively, if the number of antibodies, e, and the number of free antigen sites, l, are chosen as independent variables, then one has

$$s = [e(f - 2) + f - l]/(f - 1), \tag{32a}$$

and

$$v = (e - 1 + l)/(f - 1). \tag{32b}$$

Using v and s as independent variables, and denoting an aggregate as a (v, s)-mer, a reaction between aggregates can be written as:

$$(v_1, s_1) + (v_2, s_2) \xrightarrow{k_f} (v_1 + v_2, s_1 + s_2 - 1). \tag{33a}$$

The reverse reaction can be symbolically denoted as

$$(v, s) \xrightarrow{k_r} (v_1, s_1) + (v - v_1, s - s_1 + 1). \tag{33b}$$

The differential equation describing the growth of a (v, s)-mer has four major terms which represent: (a) formation of a (v, s)-mer by the association of two smaller aggregates, (b) the disappearance of (v, s)-mer because of the association of a (v, s)-mer with any other aggregate, (c) the disappearance of a (v, s)-mer by dissociation, and (d) the creation of a (v, s)-mer by the dissociation of a larger aggregate. When $f > 2$, explicitly writing the term corresponding to (d) is difficult, because one needs to count the number of ways a $(\tilde{v}, \tilde{s})$-mer, with $\tilde{v} \geq v + 1$, can break down into a (v, s)-mer by single bond breakage. These differential equations are very messy and we will in general avoid using them. Written below as an illustration are the equations for bivalent antigens ($f = 2$) reacting with bivalent antibodies with $c_s(v, t)$ denoting the concentration of a (v, s)-mer at time t. When $f = 2$, s may only take on the three values 0, 1, and 2 and we have

$$\frac{dc_0(v, t)}{dt} = -2k_{\mathrm{f}}S(t)c_0(v, t) + 2k_{\mathrm{f}} \sum_{v'=1}^{v-1} c_0(v', t)c_1(v - v', t)$$

$$-2(v - 1)k_{\mathrm{r}}c_0(v, t) + k_{\mathrm{r}}\left[\sum_{v'=v+1}^{\infty} 2c_0(v', t) + \sum_{v'=v}^{\infty} c_1(v', t) \right],$$

$$\frac{dc_1(v, t)}{dt} = -k_{\mathrm{f}}[L(t) + S(t)]c_1(v, t) + 4k_{\mathrm{f}} \sum_{v'=1}^{v} c_0(v', t)c_1(v - v', t)$$

$$+ k_{\mathrm{f}} \sum_{v'=1}^{v-1} c_1(v', t)c_1(v - v', t) - k_{\mathrm{r}}(2v - 1)c_1(v, t)$$

$$+ 2k_{\mathrm{r}}\left[\sum_{v'=v+1}^{\infty} c_0(v', t) + \sum_{v'=v+1}^{\infty} c_1(v', t) + \sum_{v'=v}^{\infty} c_2(v', t) \right],$$

$$\frac{dc_2(v, t)}{dt} = -2k_{\mathrm{f}}L(t)c_2(v, t) + 2k_{\mathrm{f}} \sum_{v'=1}^{v} c_1(v', t)c_2(v - v', t)$$

$$-2vk_{\mathrm{r}}c_2(v, t) + k_{\mathrm{r}} \sum_{v'=v+1}^{\infty} [c_1(v', t) + 2c_2(v', t)],$$

$$v = 0, 1, 2, \ldots, \quad (34)$$

where

$$S(t) = \sum_{v=1}^{\infty} c_1(v, t) + 2 \sum_{v=0}^{\infty} c_2(v, t), \quad (35a)$$

and

$$L(t) = 2 \sum_{v=1}^{\infty} c_0(v, t) + \sum_{v=1}^{\infty} c_1(v, t). \quad (35b)$$

Recalling the example of the coagulation of AB particles, we saw that one could determine the concentration of n-mers via Flory's combinatorial method without explicitly using the differential equations. This also is the case for antibody–antigen reactions, as was shown by Goldberg (1952). We shall now briefly indicate how the concentration of (v, s)-mers, $c_s(v, t)$ can be obtained using the combinatorial method.

Let the extent of reaction, p, be the fraction of antibody sites that are bound. If S_0 is the total concentration of antibody sites (twice the concentration of antibodies in the system) and S is the concentration of free antibody sites, then

$$p(t) \equiv [S_0 - S(t)]/S_0. \tag{36}$$

Further, let L_0 be the total concentration of antigen sites (f times the concentration of antigen in the system), let L be the concentration of free antigen sites, and let

$$p_L(t) \equiv [L_0 - L(t)]/L_0 \tag{37}$$

be the fraction of antigen sites that are bound. Since each bound antigen site is bound to an antibody site

$$L_0 - L = S_0 - S \tag{38}$$

and

$$p_L = rp, \tag{39}$$

where

$$r \equiv S_0/L_0 \tag{40}$$

is the ratio of total antibody to total antigen sites.

During an antigen–antibody reaction, free antibody sites react with free antigen sites according to the equation

$$dS/dt = -k_f SL + k_r(S_0 - S). \tag{41}$$

Solving (38) for L, we find that (41) becomes

$$dS/dt = -k_f S(L_0 - S_0 + S) + k_r(S_0 - S), \tag{42a}$$

with initial condition

$$S(0) = S_0. \tag{42b}$$

This is a so-called Riccati differential equation which can easily be solved for $S(t)$, yielding (DeLisi & Perelson, 1976)

$$S(t) = \frac{-\lambda_2(S_0 - \lambda_1)\exp(-\xi t) + \lambda_1(S_0 - \lambda_2)}{S_0[1 - \exp(-\xi t)] + \lambda_1 \exp(-\xi t) - \lambda_2}, \tag{43}$$

where

$$\xi = k_f(\lambda_1 - \lambda_2),$$

$$\lambda_1 = \frac{-[(k_r/k_f) + L_0 - S_0] + \sqrt{\eta}}{2},$$

and

$$\lambda_2 = \frac{-[(k_r/k_f) + L_0 - S_0] - \sqrt{\eta}}{2},$$

where $\eta = [(k_r/k_f) + L_0 - S_0]^2 + (4k_r S_0/k_f)$. Substituting this solution into (36) determines $p(t)$.

What remains of the calculation is to find an expression for $P(v, s)$ the probability that a free site (antibody or antigen) picked at random is part of a (v, s)-mer. We shall call the chosen free site a *root*. Once this is found, the concentration of (v, s)-mers can be easily determined, since

$$P(v, s) = \frac{\text{number of free sites on } (v, s)\text{-mers}}{\text{total number of free sites}}. \tag{44}$$

The total number of antigen sites on a (v, s)-mer containing v f-valent antigens is vf. Since bivalent antibodies must join the v antigens together, a (v, s)-mer must contain exactly $v - 1$ internal edges which occupy $2(v - 1)$ sites. Additional antigen sites may be filled with external edges, but since each of these has one free site the total number of free sites on a (v, s)-mer remains $fv - 2v + 2$. Hence

$$P(v, s) = (fv - 2v + 2)c_s(v, p)/[L(p) + S(p)], \tag{45}$$

where (36), (37) and (39) imply

$$S(p) = S_0(1 - p), \tag{46}$$

$$L(p) = L_0(1 - p_L) = L_0(1 - pr). \tag{47}$$

The probability $P(v, s)$ can also be written in the form

$$P(v, s) = \rho_L \omega_L \Omega_L + \rho_S \omega_S \Omega_S, \tag{48}$$

where the subscripts L and S label free antigen and antibody sites, respectively. ρ_L is the probability that the root is a free antigen site, ω_L the probability that the root is on a (v, s)-mer, given that the root is a free antigen site, and Ω_L is the number of ways a (v, s)-mer can be formed given that the root is a free antigen site. Analogous definitions hold for ρ_S, ω_S and Ω_S with the root being a free antibody site. By definition

$$\rho_L = L/(L + S) = L_0(1 - pr)/(L + S), \tag{49}$$

and

$$\rho_S = S/(L+S) = L_0 r(1-p)/(L+S). \tag{50}$$

The ω's are found in a manner totally analogous to that used in the case of coagulation, i.e.

$$\omega_L = (p_L p)^{v-1}[p_L(1-p)]^s (1-p_L)^{fv-2v+1-s}. \tag{51a}$$

In order to understand this equation recall that in a (v, s)-mer the v antigens are connected together by $v - 1$ antibodies. The probability that two antigens are connected is $p_L p$, since a site on the first antigen must be bound to an antibody site (this occurs with probability p_L), and the other site on the bivalent antibody must be bound to a second antigen (this occurs with probability p). Since there are $v - 1$ connections we obtain the first term in $(51a)$. The second term represents the probability that s antigen sites are attached to singly bound antibodies (external edges). The last term represents the probability that the remaining antigen sites are free. There is a maximum of $fv - 2v + 1$ free antigen sites, excluding the root in a complex containing v antigens. Since s of these sites are filled by singly bound antibodies, $fv - 2v + 1 - s$ free antigen sites remain.

Similar considerations lead to the expression

$$\omega_S = p(p_L p)^{v-1}[p_L(1-p)]^{s-1}(1-p_L)^{fv-2v+2-s}. \tag{51b}$$

Here the root is a free antibody site. Thus there is one antibody which has a free site with probability one and a bound site with probability p. This contributes the first term to $(51b)$. The remaining terms are the same as in $(51a)$ except now there are only $s - 1$ singly bound antibodies left that need to be accounted for in the last two terms. In Figure 5.3.4 we illustrate this computational procedure for a $(2, 3)$-mer with $f = 3$.

Computing Ω_L and Ω_S is a complicated combinatorial problem. Here we just state the results and refer the reader to DeLisi & Perelson (1976) or Stockmayer (1943) for details.

$$\Omega_L = \frac{(fv-v)!}{s!\,v!\,(fv-2v+1-s)!}, \tag{52a}$$

$$\Omega_S = \frac{(fv-v)!}{(s-1)!\,v!\,(fv-2v+2-s)!}. \tag{52b}$$

Combining (45)–(52) yields

$$c_s(v, p) = L_0 r^{v+s-1} p^{2v-2+s}(1-pr)^{fv-2v+2-s}(1-p)^s \frac{(fv-v)!}{(fv-2v+2-s)!\,v!\,s!}, \tag{53}$$

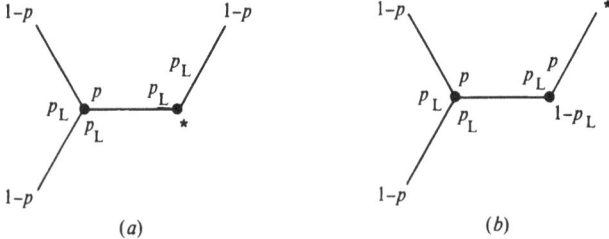

Figure 5.3.4. (a) The procedure for calculating ω_L. (b) The procedure for calculatory ω_S. In both cases the aggregate is a $(2, 3)$-mer with $f = 3$. The (*) indicates the root and is the position from which one begins assigning probabilities. In (a) the root is a free antigen site and thus the probability that the antibody drawn to the right of the root is connected to the aggregate is $p_L(1-p)$, i.e. the probability that an antigen site is bound multiplied by the probability of a free antibody site. In (b) the root is a free antibody site and thus the probability that the antibody of which it is a part is attached to the aggregate is p.

a result obtained by Stockmayer (1943) and Goldberg (1952). When $f = 2$ (53) provides the solution to (34) with p determined by (36) and (43). In this case

$$c_0(v, p) = \tfrac{1}{2}L_0 r^{v-1} p^{2v-2}(1-pr)^2, \qquad v \geq 1 \tag{54a}$$

$$c_1(v, p) = L_0 r^v p^{2v-1}(1-pr)(1-p), \qquad v \geq 1 \tag{54b}$$

$$c_2(v, p) = \tfrac{1}{2}L_0 r^{v+1} p^{2v}(1-p)^2, \qquad v \geq 0. \tag{54c}$$

Aggregate formation on lymphocyte membranes[2]

We are finally in a position to study antigen–antibody reactions on cell surfaces. One of our goals will be to develop a quantitative theory of patching on lymphocyte membranes. In this study we shall use the Flory combinatorial method. Thus our main task will be the development of a model to determine the extent of reaction on a cell surface, given certain environmental conditions such as the concentration of antigen in the solution surrounding the cell. There will also be some minor differences in computing the probabilities ω_L and ω_S and combinatorial factors Ω_L and Ω_S, primarily because all antigens on a cell surface must have at least one site bound, otherwise they would not be on the membrane.

The model that we shall use to determine the extent of reaction is a modification of a model proposed by Bell (1974) for the binding of

---

[2] This section is based upon DeLisi & Perelson, 1976.

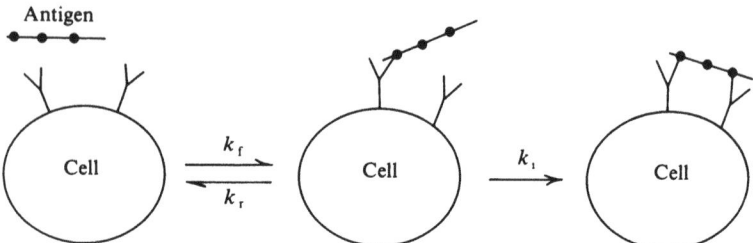

Figure 5.3.5. Model for the binding of a multivalent antigen to a cell.

multivalent antigen to cells. We assume that the interaction of antigen with a lymphocyte occurs in two steps as shown in Figure 5.3.5. During the first step the antigen binds to a single receptor with rate constant k_f. This singly bound antigen is then assumed either to dissociate with rate constant k_r or to bind irreversibly other receptor molecules, with rate constant k_i, thereby becoming multivalently bound and causing the cross-linking of two or more receptors. The assumption of irreversible binding is a very good one for experiments such as patching which last from a few minutes to, say, half an hour. For longer time scales, or for molecules which bind weakly, one might want to employ a model containing a sequence of reversible reaction steps, as we shall do in studying histamine release.

Let S_0 be the concentration of total receptor sites and S be the concentration of free receptor sites. With C and m denoting the concentrations of free and singly bound antigen, respectively, and L denoting the concentration of free antigen sites on the membrane, the equations of the model are

$$dS/dt = -k_fCS + k_rm - k_iLS, \tag{55a}$$

$$dm/dt = k_fCS - k_rm - (f-1)k_imS, \tag{55b}$$

$$dL/dt = (f-1)(k_fCS - k_rm) - k_iLS, \tag{55c}$$

where we have assumed that C, the free antigen concentration in the medium surrounding the cell, remains constant.

We shall now assume that the interaction between antigen and lymphocytes is in **quasi-equilibrium**; i.e. the antigen comes on and off the cell many times before the second irreversible step, cross-link formation, occurs successfully. Translated into mathematics these assumptions take the form

$$k_r \gg fk_iS_0, \tag{56a}$$

$$dm/dt = 0, \tag{56b}$$

and hence from (55*b*)

$$m = KCS, \tag{57}$$

where $K \equiv k_f/k_r$. (This result can be obtained in a more rigorous fashion via the theory of singular perturbation (cf. Section 5.5).) Substituting (57) into (55*a*) and (55*c*), one finds

$$dS/dt = -k_i LS = dL/dt, \tag{58}$$

and hence

$$d(S-L)/dt = 0,$$

or

$$S - L = \text{constant}. \tag{59}$$

The constant can be determined from the values of S and L at the time quasi-equilibrium is established (DeLisi & Perelson, 1976). Thus

$$dS/dt = -k_i S(S - \text{constant}). \tag{60}$$

The extent of reaction is given by the fraction of receptor sites which are bound, that is

$$p(t) = [S_0 - S(t)]/S_0. \tag{61}$$

The solution to (60) thus determines the extent of reaction. A more precise theory could be obtained by solving (55) for $S(t)$ without the quasi-equilibrium assumption, but this does not appear to be possible in closed form. DeLisi & Thakur (1977) examined numerical solutions to (55).

Patching

One problem that would arise in comparing experimental and theoretical results on patching is the definition of a patch. For example, one could define a patch as an aggregate containing more than a given number of receptors. Then if one computes the aggregate size distribution, $c_s(v, t)$, one could determine the concentration of patches. Rather than doing this complicated calculation let us assume that we are dealing with an infinite system so that infinite sized aggregates can form. It turns out that large aggregate growth occurs with increasing rapidity as the aggregation process proceeds. (A 10 000-mer can form very rapidly once 5000-mers are present, but forms very slowly if only monomers are present.) Consequently, we can estimate the time for large aggregate growth or patching by estimating the time at which it becomes possible for infinite sized aggregates to appear. It may seem surprising but this estimate can be made easily.

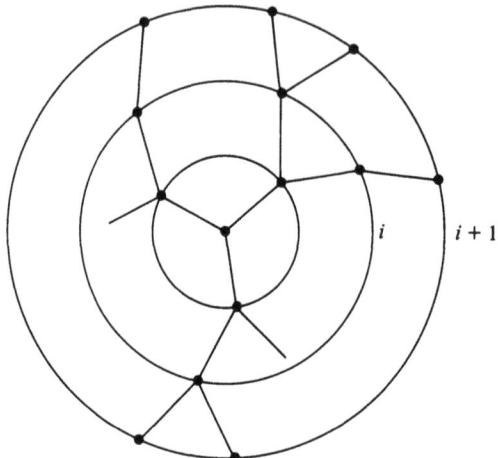

Figure 5.3.6. Graphical representation of an aggregate formed between bivalent antibodies, represented by lines, and f-valent antigens, represented by black dots. The graph is a tree and is arbitrarily centered. Each f-valent antigen on the i-th circle can be connected to at most $f-1$ antigens on the $(i+1)$-th circle. If each such connection occurs with probability q, then the expected number of antigens on the $(i+1)$-th circle is equal to the number on the i-th circle times $(f-1)\,q$. If $(f-1)\,q > 1$ there is a nonzero probability for the aggregate to be of infinite size.

Consider the graph of an aggregate shown in Figure 5.3.6. Let q be the probability that a given antigenic site is attached to an antigenic site on another antigen by an antibody connector. If each antigen has valence f, then at most $f-1$ edges can connect an antigen on circle i with antigens on circle $i+1$, because one connection leads back to circle $i-1$. Since q is the probability of each of these connections, the expected number of edges leading to the $(i+1)$-th circle is $q(f-1)N_i$, where N_i is the number of antigens on circle i. The ratio of the expected number of antigens on circle $i+1$ to those on circle i is therefore

$$(N_i + 1)/N_i = q(f-1). \tag{62}$$

It is evident from (62) that if $q(f-1) < 1$, then N_{i+1} will eventually reach zero and infinite aggregate formation will be impossible. Conversely, if

$$q \geqslant q_c \equiv 1/(f-1) \tag{63}$$

then there is a nonzero probability that an infinite-sized aggregate will appear.

Viewing aggregation processes as graphs is also done in physics in the study of phase transitions and is known as **percolation theory** (Frisch & Hammersley, 1963; Shante & Kirkpatrick, 1971; Essam, 1972). Equation (63) is the result obtained for percolation through a Bethe lattice of degree f.

The result (63) can also be obtained in a more traditional mathematical fashion. Suppose that one explicitly computes the aggregate size distribution as a function of v and p, and then examines the moments, M_j, of that distribution, i.e.

$$M_j \equiv \sum_{v=0}^{\infty} \sum_{s=0}^{fv-2v+2} v^j c_s(v, p). \tag{64}$$

One finds that the sums defining M_j diverge for $j \geqslant 2$, when $q \geqslant q_c$. Thus, q_c is the radius of convergence for these sums (DeLisi & Perelson, 1976; Stockmayer, 1943). The second moment, M_2, is proportional to the weight average molecular weight of the aggregates (Tanford, 1961). That this becomes infinite implies that some aggregates must become unbounded.

For antibody–antigen reactions in solution $q = p_L p = rp^2$. On the cell surface q is a more complicated function of p and depends on the model chosen for antigen binding (DeLisi & Perelson, 1976). However, there exists a critical value of p, which we call p_c, at which $q = 1/(f-1)$ and infinite-sized aggregates become possible. Further, since p is a function of t, one can find the critical time, t_c, at which $p(t_c) = p_c$. By identifying the time it takes a cell to patch with t_c, one can ascertain how factors such as antigen valence (f), antigen concentration (C), and receptor affinity (K) affect patching. The results of an approximate computation performed by DeLisi & Perelson (1976) are shown in Figure 5.3.7. Here a dimensionless critical patching time, $k_i S_0 t_c$, is plotted against KC for various values of the antigen valence. The curves have asymptotes at $KC = f-1$ and $KC = 1/(f-1)^2$. For values of KC less than $f-1$ or greater than $1/(f-1)^2$ large aggregates never form. This is easily understood. At low values of KC, there is insufficient antigen on the surface for large aggregates to form. At high values of KC, most receptors are attached to two singly bound antigens, leaving few receptor sites available for cross-linking. It is only at intermediate values of KC that conditions are conducive for patching. For such KC values the patching time varies, reaching a minimum near $KC = 1$. Also, one might expect, as the antigen valence increases the minimum time for patching decreases. Highly multivalent antigens form patches much more easily than low valence antigens. In fact, for $f = 2$, only linear chains are possible and p_c is reached only in the limit as $t \to \infty$.

The hypothesis that patching is a necessary first step in lymphocyte activation allows one to draw a number of very intriguing implications from Figure 5.3.7. (a) If patching must occur within a certain time after antigen exposure to be an effective stimulus, then Figure 5.3.7 shows that

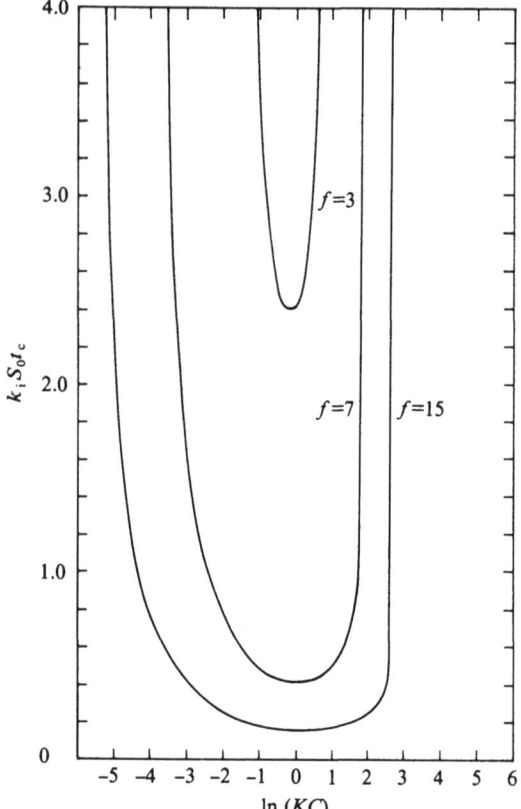

Figure 5.3.7. A dimensionless critical patching time, $k_iS_0t_c$, plotted against $\ln(KC)$ for various values of the antigen valence. The patching time approaches infinity when KC approaches either $(f-1)^{-2}$ or $(f-1)$. (From DeLisi & Perelson, 1976; © Academic Press, Inc., London.)

there may be a limiting valence, f_c, below which stimulation could not occur. By binding to an accessory cell such as a macrophage, antigen could be presented to a B cell in a highly multivalent form. It is known that polymeric antigens can stimulate B cells directly, but that other antigens require the presence of accessory cells. (*b*) Patching occurs most rapidly when $KC \approx 1$. Assuming that the events subsequent to patching in the triggering process are either very rapid or occur with little variation in rate, then cells which patch first will be triggered first. Thus, as antigen is eliminated during an immune response, C will decrease and the optimal value of K for rapid triggering will increase. The preferential triggering of cells expressing high affinity receptors and the consequent proliferation and antibody secretion by these cells would lead, over time, to an increase in the average serum antibody affinity. Such increases are observed (Siskind & Benacerraf, 1969). (*c*) Since patching can only occur over a specified affinity–concentration range, one would expect the same for triggering. Further this range should depend upon the antigen valence, but the optimal antigen concentration for stimulation should be valence independent as shown in Figure 5.3.7. Kishiomoto & Ishizaka (1975) observed both of these effects when comparing the immune response to anti-immunoglobulin and polymerized anti-immunoglobulin.

Histamine release

The cross-linking of IgE molecules on the surfaces of mast cells or basophils leads to the release of histamine by these cells. A number of experiments have shown that small amounts of cross-linking produce histamine release. Large aggregate formation (patching) is not necessary for release and in some circumstances may inhibit release. Siraganian, Hook & Levine (1975) showed that histamine release from basophils can be activated *in vitro* by simple bivalent antigens. IgE is a bivalent antibody. As we have just shown, when bivalent antigens react with bivalent antibodies only linear aggregates form and patching does not occur.

In a typical in-vitro histamine release experiment, basophils are incubated with IgE directed against a bivalent antigen such as bis-benzylpenicilloyl-1,6-diaminohexane, $(BPO)_2$. Aliquots of these passively sensitized basophils are then exposed to varying concentrations of antigen and the amounts of histamine released are measured. An aliquot of basophils is also lysed to measure their total histamine content. The data are then presented as a plot of the percentage of the total cellular histamine released as a function of antigen concentration. Figure 5.3.8. shows a typical 'histamine release curve'. It has been observed repeatedly in several laboratories that the histamine release curve for many different kinds of antigen is shaped somewhat like a gaussian or skewed gaussian distribution. Dembo & Goldstein (1978) and Dembo, Goldstein, Sobotka & Lichtenstein (1978) have suggested that histamine release in experiments with bivalent antigens is a monotonically increasing function of the equilibrium number of cross-links on the cell surface, and thus the histamine release curve should resemble a plot of the equilibrium concentration of cross-links versus antigen concentration. Using the equivalent site hypothesis, we shall derive an expression for the number of cross-links as a function of antigen concentration and demonstrate the general resemblance of the graph of this function to the histamine release curve shown in Figure 5.3.8. Dembo & Goldstein (1978) first derived these results from a more detailed model that took

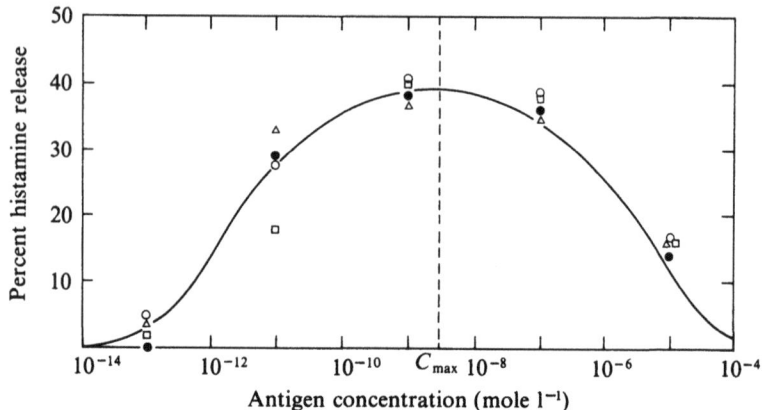

Figure 5.3.8. A histamine release curve (adapted from Dembo *et al.*, 1978, with permission). Data are plotted for four separate antigens of different length, all of the form BPO—NH—$(CH_2)_n$—NH—BPO, with $n = 3(\bullet)$, $6(\triangle)$, $9(\bigcirc)$ and $12(\square)$.

into account all possible receptor states, including cycle formation. Further details on the use of the equivalent site hypothesis in analyses of this type can be found in Perelson (1979) and Perelson & DeLisi (1980).

Using the equivalent site hypothesis, we can proceed much as before. Let C, m, and M be the concentration of free, singly bound and doubly bound antigen. We shall assume that the concentration of antigen in the medium surrounding the basophils is sufficiently large that the binding of antigen to cells does not appreciably change C, i.e., we assume C is constant. Further, we let S and S_0 be the free and total receptor site concentrations. Then, by conservation of receptor sites,

$$S_0 = S + m + 2M. \tag{65}$$

Since S, m and M are not independent we need only two differential equations to describe the dynamics of the system. We choose to solve for m and M; the relevant equations are

$$dm/dt = k_f CS - k_r m - k_i Sm + 2k'_i M, \tag{66}$$

$$dM/dt = k_i Sm - 2k'_i M, \tag{67}$$

where k_f and k_r describe the binding of antigen to the cell surface, k_i is the forward rate constant for the formation of a doubly bound antigen, and k'_i is the reverse rate constant describing the dissociation of doubly bound antigen as shown in Figure 5.3.9.

At equilibrium $dm/dt = dM/dt = 0$. Therefore

$$K \equiv k_f/k_r = m/CS, \tag{68}$$

$$K_i \equiv k_i/k'_i = 2M/mS. \tag{69}$$

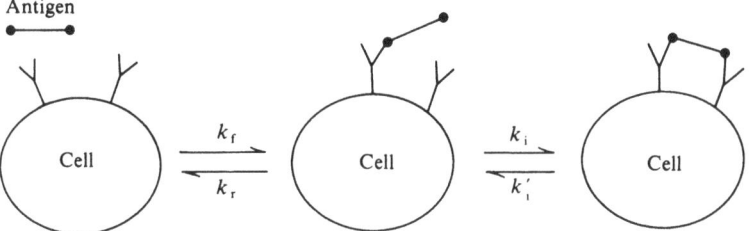

Figure 5.3.9. Model for the binding of a bivalent antigen to a cell.

Solving (65) for S, substituting into (68) and (69) and rearranging, we find that

$$m = \beta(S_0 - 2M), \tag{70}$$

where

$$\beta \equiv KC/(1 + KC) \tag{71}$$

and

$$M = mK_i(S_0 - m)/2(1 + mK_i). \tag{72}$$

Rearranging (72) one finds

$$S_0 - 2M = \frac{S_0 + m^2 K_i}{1 + mK_i}, \tag{73}$$

which on substitution into (70) yields the quadratic equation

$$m^2(1 - \beta)K_i + m - \beta S_0 = 0, \tag{74}$$

whose solution is

$$m = \frac{-1 + \sqrt{[1 + 4\beta(1 - \beta)S_0 K_i]}}{2(1 - \beta)K_i}, \qquad \beta \neq 1. \tag{75a}$$

The other solution to the quadratic equation is negative and can be ignored. When $\beta = 1$, the quadratic equation reduces to a linear one and

$$m = S_0, \qquad \beta = 1. \tag{75b}$$

Notice from (75) that

$$m = 0 \quad \text{when } \beta = 0, \tag{76}$$

as one would expect.

Substituting (75a) into (72) and performing some algebraic manipulation, we find that

$$M = \frac{1 - \sqrt{[1 - 4\beta(1 - \beta)K_i S_0]} + 2\beta(1 - \beta)K_i S_0}{4K_i\beta(1 - \beta)}. \tag{77}$$

The concentration of cross-links on the cell surface is M. Thus one would like to examine how M varies with β. First, notice that $\beta(1-\beta)$ is symmetrical around $\beta = \frac{1}{2}$. That is, if

$$g(\beta) \equiv \beta(1-\beta),$$

then

$$g(\tfrac{1}{2} + \Delta\beta) = (\tfrac{1}{2} + \Delta\beta)(\tfrac{1}{2} - \Delta\beta) = g(\tfrac{1}{2} - \Delta\beta).$$

Since M is a function of $\beta(1-\beta)$, M is also symmetrical around $\beta = \frac{1}{2}$, i.e. $M(\tfrac{1}{2} + \Delta\beta) = M(\tfrac{1}{2} - \Delta\beta)$. The histamine release curve in Figure 5.3.8 is also symmetrical, but around the maximum in the curve.

We now show that the maximum value of M occurs at $\beta = \frac{1}{2}$. Rather than differentiate (77) which involves a square root, we can differentiate (72) using the chain rule and set the derivative to zero. This yields

$$\frac{d(2M/K_i)}{d\beta} = \frac{[m'(S_0 - m) - mm'](1 + mK_i) - m'K_i m(S_0 - m)}{(1 + mK_i)^2} = 0, \tag{78}$$

where $m' = dm/d\beta$. Rearranging, we see that the m' terms cancel and (78) reduces to the quadratic equation

$$m^2 K_i + 2m - S_0 = 0, \tag{79}$$

with one positive solution

$$\bar{m} = \frac{-1 + \sqrt{(1 + S_0 K_i)}}{K_i}, \tag{80}$$

where the over-bar denotes the value of m at $dM/d\beta = 0$.
From (70)

$$\bar{\beta} = \bar{m}/(S_0 - 2\bar{M}), \tag{81}$$

which in combination with (73) yields

$$\bar{\beta} = \frac{\bar{m}(1 + \bar{m}K_i)}{S_0 + \bar{m}K_i^2}, \tag{82}$$

where the overbar again denotes evaluation at $dM/d\beta = 0$. Substitution of (80) into (82) yields the desired result

$$\bar{\beta} = \tfrac{1}{2}. \tag{83}$$

To see that this extremum is in fact a maximum note that at $\beta = 0$, $m = 0$ by (76), and that by (72), $M = 0$. Similarly at $\beta = 1$, $m = S_0$ by (75b) and from (72), $M = 0$. Since $M \geqslant 0$ and only has one extremum, $\beta = \frac{1}{2}$ corresponds to a maximum in M.

Histamine release curves are plotted as a function of C, not as a function β. Since $\beta = KC/(1+KC)$, the maximum at $\beta = \frac{1}{2}$ corresponds to $KC = 1$ or $\log KC = 0$. Let $x = \log KC$. Then from (71),

$$\beta(x) = 10^x/(1+10^x).$$

Now,

$$1 - \beta(x) = 1/(1+10^x) = 10^{-x}/(1+10^{-x}) = \beta(-x),$$

and

$$\beta(x) = 1 - \beta(-x).$$

Thus

$$\beta(x)[1 - \beta(x)] = \beta(-x)[1 - \beta(-x)],$$

and the cross-linking curves are symmetrical about the origin in x or $\log KC$, the precise type of symmetry shown in Figure 5.3.8.

Another interesting result can be derived from the fact that $\bar{\beta} = \frac{1}{2}$. From (81) one obtains

$$\tfrac{1}{2} = \bar{m}/(S_0 - 2\bar{M}),$$

or

$$\bar{m} + \bar{M} = S_0/2. \tag{84}$$

Thus at maximum cross-linking the number of bound antigens equals the number of antibody molecules on the cell surface.

For β near 0 or 1 we can obtain an approximate expression for the concentration of cross-links, M. Let $\delta = \beta(1-\beta)K_i S_0$. Then (77) can be written

$$M = S_0\left[\frac{1+2\delta - \sqrt{(1+4\delta)}}{4\delta}\right]. \tag{85}$$

Expanding $\sqrt{(1+4\delta)}$ in a Taylor series around 1, we obtain

$$\sqrt{(1+4\delta)} \cong 1 + 2\delta - 2\delta^2 + 4\delta^3 - 10\delta^4 + \ldots . \tag{86}$$

Substituting (86) into (85) yields

$$M = (S_0\delta/2)(1 - 2\delta + 5\delta^2 - \ldots). \tag{87}$$

To first order

$$M \approx S_0\delta/2 = \beta(1-\beta)K_i S_0^2/2, \tag{88}$$

and thus the concentration of cross-links is proportional to the square of the receptor concentration, as one might expect.

Effects of monovalent hapten

If the amount of histamine released by basophils depends on the number of cross-links, then procedures which reduce the number of cross-links should reduce the histamine release. One way to accomplish this reduction is to add a monovalent hapten which competes for receptor sites.

Consider basophils incubated in the presence of bivalent antigen and monovalent hapten. As before, we assume that such incubation does not significantly alter the antigen and hapten concentrations. Let H be the hapten concentration, which we assume is constant. Hapten binds to free receptors and thus

$$dH/dt = -k_h HS + k_h' H_b, \tag{89}$$

where H_b denotes the concentration of bound hapten, and k_h and k_h' are the forward and reverse rate constants for the hapten binding reaction. At equilibrium, $dH/dt = 0$ and so

$$H_b = K_h HS, \tag{90}$$

where $K_H \equiv k_h/k_h'$.

Equations (66) and (67) for the binding of antigen still apply, but now the conservation law (65) becomes

$$S_0 = S + m + 2M + H_b. \tag{91}$$

Substituting (90) into (91) and solving for S, one finds

$$S = (S_0 - m - 2M)/(1 + K_H H). \tag{92}$$

Thus the equilibrium equations for m and M, (68) and (69), become

$$m = \frac{KC}{1 + K_H H + KC}(S_0 - 2M) \tag{93}$$

and

$$M = \frac{mK_i(S_0 - m)}{2(1 + K_H H + mK_i)}. \tag{94}$$

If we let

$$K' \equiv \frac{K}{1 + K_H H} \tag{95}$$

and

$$K_i' = \frac{K_i}{1 + K_H H} \tag{96}$$

then (93) and (94) become

$$m = \frac{K'C}{1 + K'C}(S_0 - 2M), \tag{97}$$

and

$$M = \frac{mK_i'(S_0 - m)}{2(1 + mK_i')}. \tag{98}$$

The solutions to (97) and (98) are precisely the same as the solutions to (70) and (72) with K and K_i replaced by K' and K_i'. In particular, the maximum of the histamine release curve occurs at $K'C/(1 + K'C) = \frac{1}{2}$ or $K'C = 1$. Thus, if we denote the antigen concentration for maximum histamine release $C_{\max}$, we have

$$C_{\max} = 1/K' = (1 + K_H H)/K. \tag{99}$$

For $H = 0$, we have $C_{\max} = 1/K$ as before. However, if hapten is present, $C_{\max}$ increases. Consequently, if histamine release is a monotonically increasing function of cross-linking there should be a shift in the maximum of the histamine release curve to the right as H increases. Further, at high antigen concentrations, the expansion of M in (87) is valid, and to first order, using (88) and (71) with K and K_i replaced by K' and K_i',

$$M = K'CK_i'S_0^2/2(1 + K'C)^2. \tag{100}$$

For $K'C \gg 1$,

$$M \simeq K_i'S_0^2/2K'C = K_iS_0^2/2KC, \tag{101}$$

which is independent of H. Thus one predicts that, at high C, histamine release curves for all values of H, approach the $H = 0$ curve. Lastly, the model predicts that the histamine release curve in the presence of hapten always lies below the $H = 0$ curve. This is because

$$\delta' = \frac{K'CK_i'S_0}{(1 + K'C)^2} = \frac{KCK_iS_0}{(1 + K_H H + KC)^2} \leqslant \delta, \tag{102}$$

with equality holding only at $H = 0$. As seen by (85), M can be expressed as a function of δ. When $H \neq 0$, K and K_i are replaced by K' and K_i', and thus δ needs to be replaced by δ'. Differentiation of (85) leads to the conclusion $dM/d\delta > 0$ and thus M is a strictly monotone increasing function of δ, or δ' when $H \neq 0$. Because $\delta' < \delta$, $M(\delta') < M(\delta)$ when $H \neq 0$, and the histamine release curve in the presence of hapten should be below the curve measured in the absence of monovalent hapten. As shown by Dembo *et al.* (1978), experimental data verify all of these predictions. This is illustrated in Figure 5.3.10.

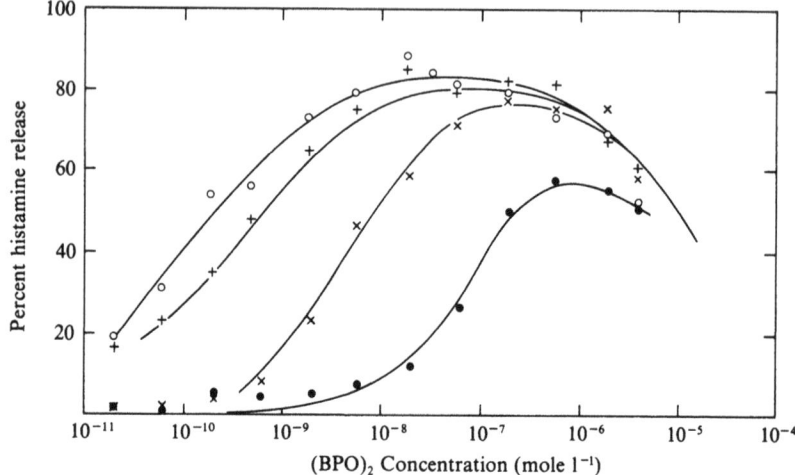

Figure 5.3.10. The effect of adding a monovalent hapten, benzylpenicilloyl formyl-L-lysine, BPO_1, on the histamine release curve for the bivalent antigen $(BPO)_2$ (from Dembo *et al.*, 1978). There was either no BPO_1 present (O), or it was present at concentrations of $10^{-8}(+)$, $10^{-7}(\times)$ and $10^{-6}(\bullet)$ mole l^{-1}.

References

Bell, G. I. (1974). Model for the binding of multivalent antigen to cells. *Nature, London,* **248**, 430–1.

DeLisi, C. and Perelson, A. (1976). The kinetics of aggregation phenomena. I. Minimal models for patch formation on lymphocyte membranes. *J. Theoret. Biol.* **62**, 159–210.

DeLisi, C. & Thakur, A. K. (1977). Antigen binding to receptors on immunocompetent cells. II. Thermodynamic and biological implications of the receptor cross-linking requirement for B cell activation. *Cell. Immunol.* **28**, 416–26.

Dembo, M. & Goldstein, B. (1978). Theory of equilibrium binding of symmetric bivalent haptens to cell surface antibody: application to histamine release from basophils. *J. Immunol.* **121**, 345–53.

Dembo, M., Goldstein, B., Sobotka, A. K. & Lichtenstein, L. (1978). Histamine release due to bivalent penicilloyl haptens: control by the number of cross-linked IgE antibodies on the basophil plasma membrane. *J. Immunol.* **121**, 354–8.

Essam, J. W. (1972). Percolation and cluster size. In *Phase Transitions and Critical Phenomena*, vol. 2, ed. C. Domb and M. S. Green, New York, Academic Press, pp. 197–270.

Flory, P. J. (1936). Molecular size distribution in linear condensation polymers. *J. Amer. Chem. Soc.* **58**, 1877–85.

Frisch, H. L. & Hammersley, J. M. (1963). Percolation processes and related topics. *SIAM J.* **11**, 894–918.

Goldberg, R. J. (1952). A theory of antibody–antigen reactions. I. Theory for reactions of multivalent antigen with bivalent and univalent antibody. *J. Amer. Chem. Soc.* **74**, 5715–25.

Harary, F. (1969). *Graph Theory*, Reading, Mass., Addison-Wesley.

Kishiomoto, T. & Ishizaka, K. (1975). Regulation of antibody response in vitro. IX. Induction of secondary anti-hapten release by anti-immunoglobulin and enhancing soluble factor. *J. Immunol.* **114**, 585–91.

Perelson, A. S. (1979). A model for the reversible binding of bivalent antigen to cells. In *Physical Chemical Aspects of Cell Surface Events in Cellular Regulation*, ed. C. DeLisi and R. Blumenthal, New York, Elsevier/North Holland, pp. 147–61.

Perelson, A. S. & DeLisi, C. (1975). A systematic and graphical method for generating the kinetic equations governing the growth of aggregates. *J. Chem. Phys.* **62**, 4053–61.

—— (1980). Receptor clustering on a cell surface. I. Theory of receptor cross-linking by ligands bearing two chemically identical functional groups. *Math. Biosci.* **48**, 71–110.

Segal, D. M., Taurog, J. D. & Metzger, H. (1977). Dimeric immunoglobulin E serves as a unit signal for mast cell degranulation. *Proc. Nat. Acad. Sci., USA* **74**, 2993–7.

Shante, V. K. & Kirkpatrick, S. (1971). An introduction to percolation theory. *Adv. Phys.* **20**, 325–57.

Siraganian, R. P., Hook, W. A. & Levine, B. B. (1975). Specific *in vitro* histamine release from basophils by bivalent haptens: evidence for activation by simple bridging of membrane bound antibody. *Immunochem.* **12**, 149–57.

Siskind, G. W. & Benacerraf, B. (1969). Cell selection by antigen in the immune response. *Adv. Immunol.* **10**, 1–50.

Smoluchowski, M. V. (1916). Drei Vorträge über Diffussion, Brownsche Molekularbewegung und Koagulation von Kolloidteilchen. *Phyzik. Z.* **17**, 585–99.

—— (1917). Versuch einer mathematischen Theorie der Koagulationskinetik kolloider Lösungen. *Z. Phys. Chem.* **92**, 129–68.

Stockmayer, W. H. (1943). Theory of molecular size distributions and gel formation in branched-chain polymers. *J. Chem. Phys.* **11**, 45–55.

Tanford, C. (1961). *Physical Chemistry of Macromolecules*, New York, Wiley.

Taylor, R. B., Duffus, W. P. H., Raff, M. C. & de Petris, S. (1971). Redistribution and pinocytosis of lymphocyte surface immunoglobulin molecules induced by anti-immunoglobulin antibody. *Nature New Biol.* **223**, 225–9.

Ziff, R. (1978). *Singular Solutions to the Coagulation Equation*. Los Alamos Technical Report LA-UR-78-851.

5.4 APPLICATION OF CONTROL THEORY TO IMMUNOLOGY[3]

Lymphocyte proliferation and differentiation

Prior to antigen stimulation most B lymphocytes are small round cells (5–8 μm diameter) with few mitochondria, no discernible polysomes or endoplasmic reticulum and a dense nucleus which almost fills the entire cell. These **small lymphocytes** generally do not divide and are thought to be resting in the G_0 period of the cell cycle. When stimulated by antigen, small lymphocytes undergo 'blast transformation' and become large round cells (up to 15 μm diameter) with abundant cytoplasm containing many mitochondria, polysomes and endoplasmic reticulum. These **large lymphocytes** secrete antibody, divide rapidly and some differentiate into plasma cells. Some large lymphocytes may revert back into small lymphocytes, where they probably function as 'memory cells.' Memory cells are nondividing cells which do not secrete antibody, but which can be stimulated into vigorous antibody production by future encounters with antigen. The formation of memory cells is one of the goals of preventive immunization programs and the presence of large populations of memory cells is believed to be responsible for our not getting many diseases a second time.

Mature plasma cells resemble secretory gland cells. Between 5 and 40% of all protein synthesized by the plasma cell is immunoglobulin. The cell is believed to survive somewhere between a few days and a few weeks and is thought not to divide. The mature plasma cell by giving up its ability to divide and specializing in antibody synthesis and secretion can produce antibody substantially faster than the large lymphocyte. (For further biological details about lymphocytes, memory cells and plasma cells see Eisen, 1974.)

These biological facts lead us to ask the following interesting question: What is the best or optimal strategy for the immune system to follow in allocating its cells among the lymphocyte, memory cell, and plasma cell

---

[3] The models described in this section are based upon the collaborative work of Perelson, Mirmirani & Oster (1976, 1978) and Perelson, Goldstein & Rocklin (1980).

populations? We shall first examine responses to *T-independent antigens*, i.e. antigens which can stimulate B cells in the absence of T cells. T-independent antigens generally do not cause the formation of memory cells and thus we can deal solely with lymphocytes and plasma cells. Later, we shall include the possibility of memory cell production.

The situation we shall consider is a typical experimental one for which data are available. A single injection of a non-replicating T-independent antigen, such as a bacterial cell wall polysaccharide, is given to a mouse. This stimulates the production of L_0 large lymphocytes which replicate and produce antibody at modest rates. These cells, however, can also differentiate into nondividing, short-lived plasma cells which secrete antibody at extremely rapid rates.

The immune system could conceivably act in many different ways. The L_0 stimulated cells could immediately differentiate into plasma cells to give rapid early antibody secretion, but at the risk of having the plasma cells die before the antigen was eliminated. Alternatively, the large lymphocytes could remain in the dividing state and enlarge the antigen reactive cell population. Once a substantial population of large lymphocytes is built up, some fraction of these cells could differentiate further into plasma cells. In a system operating in this manner, little antibody would be secreted initially, but at later times rapid and massive secretion could occur. Another possible strategy would be to have some fraction of the cells, say half, stay as large lymphocytes and the remaining fraction differentiate into plasma cells.

There are an infinite number of possible strategies, and we would like to know which is 'best,' in a sense to be defined precisely below. Questions such as this can be answered by constructing a mathematical model of the system and then using the technique of *optimal control theory* to find the best operating strategy for the system. Before attempting to do this we shall discuss the motivation for approaching biological questions from the viewpoint of optimization.

The philosophy underlying the use of optimal control theory in biology

Although one tends to think of natural selection as an optimizing process, there is no reason *a priori* that the immune system or any other biological system should behave in a globally optimal fashion. As Jacob (1977) has pointed out, evolution is a tinkering process. It is random and historically based so that small adjustments are made on existing mechanisms. Thus ample opportunity exists for evolution to get trapped in local maxima; there may be other structures with higher fitness but to reach them may require a temporary, but fatal, decrease in overall fitness. Evolution is a slow and erratic process, so there is no guarantee that even if a system is evolving in the 'right' direction it has reached its optimum. Further, an organism has to cope with many competing influences so that an improvement in one direction, may be accompanied by a sacrifice in another. Under such circumstances one might not be able to give any operational meaning to the notion of optimality and one might have to interpret optimality as some sort of best-compromise solution.

The course of evolution or even its final result can not be predicted in detail by optimization theory. For example, if one asks what is the optimal way for an immune system to eliminate antigen in minimal time, no solution exists. It would clearly be better for the immune system to utilize cells which secreted antibody faster than plasma cells, which could divide rapidly, and which were long lived, than to allocate cells between lymphocytes, plasma cells and memory cells. For reasons which we do not yet fully understand, evolution led to an immune system which utilizes many cell types.

The way we shall proceed is to build a detailed model of the immune system, including the known cell types and the known biological restrictions on their behavior. Thus the system itself will be fixed; we shall not try to predict if evolution should have arrived at the current immune system. Rather we shall ask what is the best method of operation for our model of the immune system. Although, there may not be a reason *a priori* to believe that the immune system follows an optimal allocation policy, many real biological systems when examined closely appear to perform a variety of tasks in an optimal fashion (Macevicz & Oster, 1976; Oster & Wilson, 1978; and see Section 4.3). Since the action of Darwinian evolution leads to improvement of traits which affect the reproductive success of an organism, the theoretical objections we have raised may have been partially circumvented by the immune system, which has been evolutionarily static for many tens of millions of years, and which is subject to extreme and direct selection pressure. This notwithstanding, it will only be through direct comparison of the computed optimal allocation policy with experimentally observed policies that realistic conclusions about the efficacy of the immune response can be made.

Besides their obvious usefulness in assessing the efficiency of the immune response, optimal control calculations also help one understand *why* certain events might occur. Thus, for example, if under some circumstances one observes that plasma cells are made early during the immune response, while under other circumstances they are made late in the immune response, one might be perplexed about this divergence in behavior. However, if an optimal control calculation showed that these behaviors were optimal, or just advantageous, then these behaviors would make sense to us in the light of evolution.

A simple model for the immune response to a T-independent antigen

Figure 5.4.1 depicts the simple model we shall consider for the immune response to a T-independent antigen. An experimental animal, such as a laboratory mouse, is given a single injection of a T-independent antigen. This antigen stimulates the formation of a population of large lymphocytes, L, which secrete antibody specifically directed against the antigen.

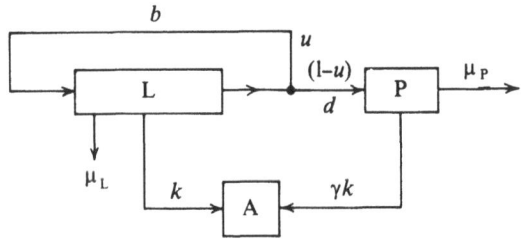

Figure 5.4.1. Model of large lymphocyte (L) proliferation, differentiation into plasma cells (P), and antibody (A) secretion (adapted from Perelson *et al.*, 1976).

At any time t, we assume a fraction $u(t)$, $0 \leq u(t) \leq 1$, of these cells proliferate with constant *per capita* birth rate b, while the remaining fraction, $1 - u(t)$, differentiate into plasma cells, P, with constant *per capita* differentiation rate d. Lymphocytes and plasma cells are assumed to die at *per capita* rates μ_L and μ_P, respectively, and to secrete antibody, A, at *per capita* rates k and γk, $\gamma > 1$, respectively. Determinations of the rate of protein synthesis of these two cell types indicate that γ can be as large as 1000, although values between 10 and 100 might be more typical. The problem we wish to consider is how should an animal apportion its stimulated cells between lymphocytes and plasma cells, so as to secrete an amount of antibody A^* sufficient to neutralize the antigen in minimal time.

The equations describing this model are

$$dA/dt = k(L + \gamma P), \tag{1}$$

$$dL/dt = bu(t)L - d[1 - u(t)]L - \mu_L L, \tag{2}$$

$$dP/dt = d[1 - u(t)]L - \mu_P P, \tag{3}$$

with initial conditions

$$A(0) = 0, \qquad L(0) = L_0, \qquad P(0) = 0. \tag{4}$$

The value of L_0 depends on the dose of antigen given to the animal, but for each antigen dose it is a constant. To ensure that the large lymphocyte population can undergo net growth, we assume

$$b_L \equiv b - \mu_L > 0. \tag{5}$$

Different choices of the function $u(t)$ produce different sets of differential equations (1)–(3), and hence produce different solution curves which reach the plane $A - A^* = 0$ at different times. This is schematically illustrated in Figure 5.4.2 where the solution curves are

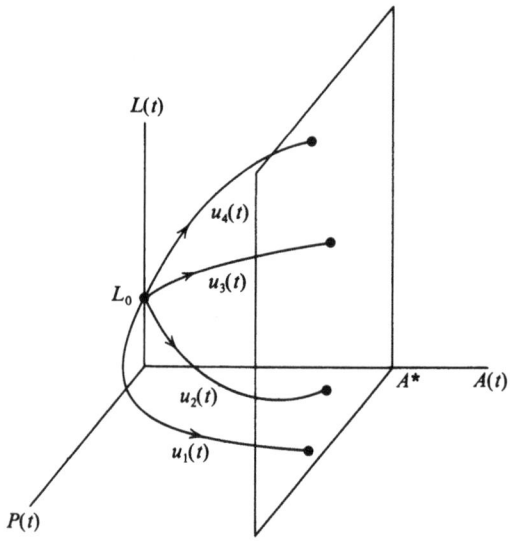

Figure 5.4.2. Schematic illustration of the minimal time control problem. Different choices of the control u, generate different trajectories eminating from the point $A = P = 0$, $L = L_0$. We wish to find the control u^* which causes a trajectory to hit the plane $A - A^* = 0$ in the shortest possible time.

parameterized by time. We shall choose the function $u(t)$ out of the class $U(t)$ of all piecewise continuous functions with $0 \le U(t) \le 1$. Roughly speaking, a piecewise continuous function is a function that can be broken up into a finite number of pieces, each of which is continuous. We have chosen to examine this class of functions because we wish to allow $u(t)$ to jump discontinuously between 0 and 1 and thus act as a proliferation or differentiation switch. The function $u(t)$ is called a **control** since changing it affects the trajectory.

Stated mathematically, the optimization problem is

$$\min_{u(t) \in U(t)} J \equiv \int_0^T dt,$$

where the final time T is determined by $A(T) = A^*$, subject to the dynamic constraints (1)–(3) and the static constraint

$$0 \le u(t) \le 1.$$

In general, the optimization criterion J can depend on all the state variables and the control, i.e.

$$J \equiv \int_0^T f_0[L(t), P(t), A(t), u(t)]\, dt.$$

However, here we shall only study the case $f_0 = 1$.

The optimization problem we have just described is known as a **minimum-time optimal control problem** and can be solved by **Pontryagin's maximum principle**. Many texts, including those by Leitmann (1966), Athans & Falb (1966) and Clark (1976), discuss this method in detail. Here we shall only present its application. To use Pontryagin's method one defines a scalar function H, called a *Hamiltonian*. For the problem at hand

$$H = \lambda_0 f_0 + \lambda_1[k(L + \gamma P)] + \lambda_2[bu - d(1 - u) - \mu_L]L$$
$$+ \lambda_3[d(1 - u)L - \mu_P P], \tag{6}$$

where

$$\lambda_0 = \text{constant} \le 0, \qquad f_0 = 1, \tag{7}$$

and λ_i, $i = 1, 2, 3$, satisfy the following set of differential equations:

$$d\lambda_1/dt = -\partial H/\partial A = 0, \tag{8}$$

$$d\lambda_2/dt = -\partial H/\partial L = -k\lambda_1 - [bu - d(1 - u) - \mu_L]\lambda_2 - d(1 - u)\lambda_3, \tag{9}$$

$$d\lambda_3/dt = -\partial H/\partial P = -\gamma k\lambda_1 + \mu_P\lambda_3, \tag{10}$$

with final conditions

$$\lambda_1(T) = 1, \qquad \lambda_2(T) = \lambda_3(T) = 0. \tag{11}$$

Notice that H is a linear combination of the right-hand sides of the differential equations (1)–(3) describing the model and the integrand of the cost functional, J, to be minimized. The additional or **adjoint** variables λ_i are somewhat akin to Lagrange multipliers in static minimization problems. The final conditions are chosen so that the vector $(\lambda_1, \lambda_2, \lambda_3)$ is normal to the target set $S \equiv \{A, L, P | A - A^* = 0\}$ at the final time $t = T$.

Pontryagin's principle now states that if $u^*(t)$ is an optimal control, then there exists a set of adjoint variables, all of which are not simultaneously zero, that satisfies (7)–(11), such that H is maximized when $u = u^*$. There is a further technical requirement, not to be discussed further here, that $H[\boldsymbol{\lambda}(t), A^*(t), L^*(t), P^*(t), u^*(t)] = 0$, $t \in (0, T^*)$, where A^*, L^*, and P^* are the solutions to (1)–(3) when $u = u^*$, T^* is the optimal final time, and $\boldsymbol{\lambda} = (\lambda_0, \lambda_1, \lambda_2, \lambda_3)$.

Pontryagin's principle only provides a set of necessary conditions for minimizing the final time. A control which satisfies these necessary conditions is called an **extremal control**, whereas a control which satisfies a set of necessary and sufficient conditions is called an **optimal control**. To find an extremal control u^* we must maximize H. Although we shall not demonstrate it here, for the problem at hand this procedure gives rise to a control which also satisfies the sufficient conditions for an optimal control (Perelson *et al.*, 1976).

Rewriting the Hamiltonian (6) as

$$H = [(b + d)\lambda_2 - d\lambda_3]Lu + \text{terms not involving } u, \tag{12}$$

and noting from (2) that L is positive if $L_0 > 0$, one immediately sees that H is maximized when u is chosen as

$$u^*(t) = \begin{cases} 1 & \text{if } \sigma(t) > 0 \\ \in (0, 1) & \text{if } \sigma(t) = 0 \text{ on a nonzero time interval,} \\ 0 & \text{if } \sigma(t) > 0 \end{cases} \tag{13}$$

where

$$\sigma(t) \equiv (b + d)\lambda_2(t) - d\lambda_3(t) \tag{14}$$

is the *switching function*.

Using the adjoint equations (8)–(10), one can exclude the possibility of $\sigma(t)$ being zero on a nonzero time interval (Perelson *et al.*, 1976). Thus the extremal control $u^*(t)$ is always equal to one of the extreme values, 0 or 1, and never equal to an intermediate value. This type of control is called **bang–bang**. In engineering systems such controllers are very economical, for instead of providing a machine that can generate any control, one only needs a device capable of existing in two states. For example, an on–off switch is less expensive than a rheostat. In biology, where bang–bang control is analogous to the all-or-none-principle, similar principles of economy may have been important in optimization by natural selection.

By integrating the adjoint equations backwards in time, one can determine if $u^*(t)$ is always 0, 1, or switches between these values. A full analysis, given by Perelson *et al.* (1976), yields the following results.

(*a*) If $(\gamma - 1)d < b$, then $u^* = 1$ for all t; i.e. lymphocytes should proliferate without differentiating into plasma cells. In assessing whether the immune system should produce lymphocytes or plasma cells, one needs to compare the antibody secreted by a lymphocyte that differentiates into a plasma cell with the amount secreted by a proliferating lymphocyte and all its progeny. Thus if a plasma cell secretes antibody 100 times as fast as a lymphocyte, in a given time period it would still be beneficial to produce only large lymphocytes if the total antibody output of one lymphocyte and all of its progeny exceeded that of the plasma cell. Realistic parameter values are $b \approx d$ and $\gamma \geq 10$. Thus lymphocytes do not proliferate fast enough for this strategy to be beneficial.

(b) If $(\gamma - 1)d > b$ then it is advantageous to produce plasma cells. The time at which this differentiation should proceed depends upon all the cellular parameters as well as the ratio $\alpha \equiv A^*/kL_0$. If α is smaller than a critical value, α_c, then the extremal strategy is $u^*(t) = 0$, for all t; i.e. immediately begin producing plasma cells. This strategy is sensible only if plasma cells during their short life times can produce an amount of antibody A^*; otherwise lymphocyte proliferation is needed. If α is larger than α_c, the optimal strategy is

$$u^*(t) = \begin{cases} 1 & 0 \leq t < t^* \\ 0 & t^* < t \leq T \end{cases}, \tag{15}$$

i.e. proliferate first then switch to plasma cell production. The parameter α_c is related to the total amount of antibody L_0 lymphocytes can secrete if they immediately begin differentiating into plasma cells without proliferating first.

When the strategy given by (15) is followed, the switching time t^* and the optimal final time T^* are

$$t^* = \frac{1}{b_L} \ln\left(\frac{A^* b_L + kL_0}{b_L kL_0 G(t^*)}\right), \tag{16}$$

and

$$T^* = t^* + \tau^*, \tag{17}$$

where

$$G(\tau^*) \equiv \left[\frac{1}{b_L} + \frac{\mu_P + \gamma d}{\mu_{Ld}\mu_P} - \frac{\mu_P - \mu_{Ld} + \gamma d}{\mu_{Ld}(\mu_P - \mu_{Ld})}\right] \exp\left(-\mu_{Ld}\tau^*\right)$$

$$+ \left(\frac{\gamma d}{\mu_P(\mu_P - \mu_{Ld})}\right) \exp\left(-\mu_P\tau^*\right), \tag{18}$$

$$b_L \equiv b - \mu_L, \, \mu_{Ld} \equiv \mu_L + d,$$

and τ^* is the solution to

$$\sigma(\tau^*) = B \exp\left(-\mu_P\tau^*\right) - C \exp\left(-\mu_{Ld}\tau^*\right) + D = 0. \tag{19}$$

Here

$$B = \frac{d\gamma k}{\mu_P}\left(1 + \frac{b+d}{\mu_P - \mu_{Ld}}\right), \tag{20}$$

$$C = \frac{(b+d)k}{\mu_P}\left[\frac{\mu_P + \gamma d}{\mu_{Ld}} + \frac{\gamma d}{\mu_P - \mu_{Ld}}\right], \tag{21}$$

$$D = \frac{k}{\mu_{Ld}}\left[b + d + \frac{\gamma d b_L}{\mu_P}\right], \tag{22}$$

and we have assumed $\mu_P \neq \mu_{Ld}$ as is the case when typical biological parameter values are employed.

Correlating typical antigen doses used to effect a noticeable immune response with values of A^*, one finds that α is typically orders of magnitude greater than α_c. *Thus the prediction of this optimal control model is that following the injection of a T-independent antigen into an experimental animal there should first be lymphocyte proliferation followed late in the immune response by lymphocyte differentiation into plasma cells.* The model also predicts the time at which the switch from $u^*(t) = 1$ to $u^*(t) = 0$ should occur.

In Figure 5.4.3 we illustrate the dynamics of an immune response following the bang–bang strategy. Notice that even though $u^*(t)$ changes discontinuously, the lymphocyte and plasma cell populations change

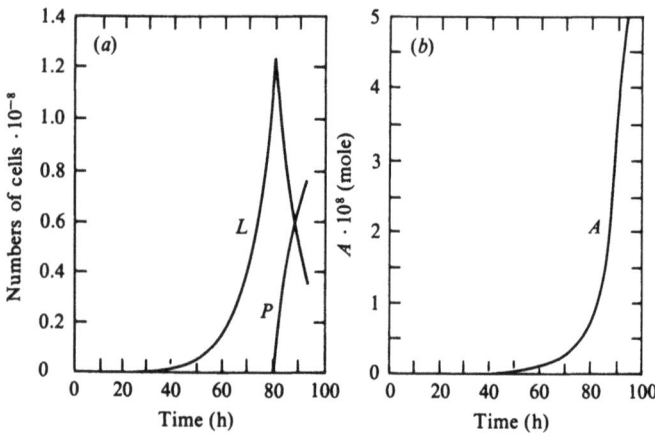

Figure 5.4.3. The dynamics of an immune response following the bang–bang strategy (adapted from Perelson *et al.*, 1976). (*a*) The number of large lymphocytes and plasma cells as a function of time. (*b*) The amount of antibody secreted as a function of time, with $b = d = 0.1\ \mathrm{h}^{-1}$, $\mu_P = 0.02\ \mathrm{h}^{-1}$, $\mu_L = 10^{-5}\ \mathrm{h}^{-1}$, $\gamma = 10$, $L_0 = 4 \cdot 10^4$, $k = 6 \cdot 10^{-18}$ mole cell^{-1}h^{-1} and $A^* = 5 \cdot 10^{-8}$ mole.

continuously. When u^* switches from 1 to 0 all lymphocytes do not become plasma cells, rather the process of differentiation begins and lymphocytes transform into plasma cells by a first-order kinetic process. Sensitivity studies show that in the range of realistic parameter values, these curves are very representative of the optimal response dynamics.

The model upon which our optimal control calculations have been based is very simple and neglects a number of important biological features. However, making the model more realistic and more compli-cated does not change the basic pattern of bang–bang control. One can directly include antigen and minimize the time to bring the antigen concentration down to a safe level. The extremal control is again bang–bang when $d(\gamma - 1) > b$. Other biological features can be included in the model: a source for additional stimulated lymphocytes, a maturation delay in which the decision to differentiate into a plasma cell is made before lymphocyte mitosis, or a time-varying antibody secretion rate, $k(t)$. In all of these cases the extremal control remains bang–bang and at most one switch occurs (Perelson *et al.*, 1976; Mirmirani, 1977). Because the bang–bang strategy remains extremal under a variety of modifications of the basic model, one is tempted to believe that this strategy may have some biological significance. In fact, this is the case.

The available biological evidence confirms the qualitative conclusion of early lymphocyte proliferation followed by late plasma cell develop-ment. Kinetic studies by Baker, Stashak, Amsbaugh & Prescott (1971*a*, *b*) have shown that, in the response to type III pneumococcal poly-saccharide, an antigen that does not produce any detectable memory, cells which slowly secrete antibody arise early in the immune response and grow exponentially, while rapid antibody secretors are detected late in the response. If one identifies the slow secretors with large lymphocytes and rapid secretors with plasma cells, these observations are in accord with the predictions of the optimal control model. Morphological studies by Russel & Diener (1970) and Zagury *et al.* (1976) have also shown that antibody-secreting plasma cells are generally absent during the beginning of the immune response and increase in numbers late in the response.

If the immune system does regulate its cell populations in a bang–bang fashion, how does it decide when to switch? For T-dependent antigens, which require the presence of both T cells and B cells for generating a response, substantial evidence has accumulated showing that antigen alone is sufficient to cause B cell proliferation, but that a factor secreted by stimulated T cells is needed to trigger B cell differentiation. What controls the release of this T cell factor is unknown. In the response to T-independent antigens, T cells are not required and other explanations must be sought. Experimental studies of B cell growth in cell culture have shown that proliferation occurs at low cell densities, whereas differentiation to plasma cells occurs at high cell densities (Melchers, Coutinho, Heinrich & Andersson, 1975). Whether density dependence is an important control parameter *in vivo* is not yet established. However, these experiments argue for the presence of closed loop control (see Chapter 3) in the B cell response.

Optimal strategy for memory cell formation

The common phenomenon of immunity to infectious disease is thought to be the result of the generation, during the first encounter with an antigen, of specialized lymphocytes, termed **memory cells**. These cells are very long lived and recirculate through the body, seeking out antigen. When antigen is encountered, they respond, as do other small lymphocytes, by transforming into large lymphocytes, proliferating and secreting antibody. The major difference in the immune response to a second encounter with antigen, is that the number of initially reactive cells is greatly increased because of the presence of memory cells, giving rise to a more rapid and vigorous antibody response than during the first encounter with antigen.

Although there is some question as to precisely which B cells become memory cells (see Perelson *et al.*, 1978) we shall adopt the model shown in Figure 5.4.4 in which memory cells, M, are generated from large

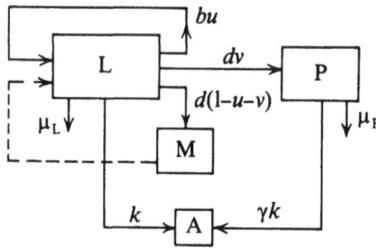

Figure 5.4.4. Model of lymphocyte (L) proliferation and differentiation into plasma cells (P) and memory cells (M) (adapted from Perelson *et al.*, 1978). The dotted line indicates that on rechallenge an animal's memory cells can become stimulated and transform into large lymphocytes.

lymphocytes. As shown in the figure, memory cells are believed not to secrete antibody, but to transform into large lymphocytes (dashed line) on subsequent encounters with antigen. The immune response generated by an animal's first encounter with antigen is called the **primary response**, whereas the response generated by the second or subsequent antigen encounters is generally called the **secondary response**.

At any instant of time, large lymphocytes now have three choices: they can remain proliferating large lymphocytes and secrete modest amounts of antibody, they can differentiate into short-lived plasma cells capable of rapidly secreting antibody, or they can differentiate into nonantibody-secreting memory cells and be held in reserve for future antigen encounters. One can again ask how the lymphocyte population should be allocated in order to provide optimal survival value to the organism. If one chooses the elimination of antigen or equivalently the production of A^* antibodies in minimal time as an optimization criterion, then in response to a single challenge with antigen no memory cells should be produced. Memory cells produce no antibody and thus diverting lymphocytes into memory cells must necessarily lengthen the response. A more realistic situation is one in which antigen is encountered many times, with probability p_i for the i-th encounter. The appropriate optimization criterion would then seem to be

$$\min J = \sum_{i=1}^{\infty} p_i \int_0^{T_i} dt, \tag{23}$$

where T_i is the time required to neutralize the antigen on the i-th encounter by secreting A_i^* antibodies. The linear weighting of the response times can be interpreted in terms of the notion of a Pareto optimum commonly used in economics. A more complete discussion of this aspect of the problem can be found in the article by Perelson *et al.* (1978).

To simplify the problem and explicitly calculate the extremal strategy for B memory cell production, we shall consider the case of two encounters with antigen, the first occurring with unit probability and the second with probability p. Although we shall not do so here, one can show that in the response to N encounters with antigen, the qualitative behavior of the strategy to the i-th antigen encounter, $i \leq N-1$, is the same as for the extremal strategy for a primary response in a model with two antigen encounters, whereas the qualitative behavior of the extremal strategy to the N-th antigen encounter is the same as in the extremal strategy for a secondary response in a two-encounter model (Perelson *et al.*, 1978).

In the two-encounter situation, the optimization criterion is

$$\min J = \int_0^{T_1} dt + p \int_0^{T_2} dt. \tag{24}$$

The dynamical equations of the model are simply generalizations of (1)–(3) and, as can be seen from Figure 5.4.4, are

$$dA/dt = k(L + \gamma P), \tag{25}$$

$$dL/dt = bu(t)L - dv(t)L - d[1 - u(t) - v(t)] - \mu_L L, \tag{26}$$

$$\mathrm{d}P/\mathrm{d}t = dv(t)L - \mu_P P, \tag{27}$$

$$\mathrm{d}M/\mathrm{d}t = d[1 - u(t) - v(t)]L. \tag{28}$$

The initial conditions for the primary response are

$$A(0) = P(0) = M(0) = 0 \quad \text{and} \quad L(0) = L_0, \tag{29}$$

whereas for the secondary response the initial conditions are

$$A(0) = P(0) = 0, \qquad M(0) = M_{20} \quad \text{and} \quad L(0) = \Lambda M(T_1) + L_{20}.$$

We have assumed that the second encounter with antigen occurs sufficiently long after the first encounter that the primary response has ceased, all secreted antibody has been catabolized and all remaining plasma cells have died. The initial number of large lymphocytes is some fraction Λ of the memory cells generated at the end of primary response, $M(T_1)$, which have survived until the second antigen encounter, plus a small number, L_{20}, of large lymphocytes generated from stimulated virgin small lymphocytes. Not all surviving memory cells become stimulated and thus $M(0) = M_{20}$, a constant.

This optimal control problem can again be solved by a judicious application of Pontryagin's principle. First, notice that during the secondary response no memory cells should be produced since antigen is surely never to be seen again. Thus the secondary response should be carried out exactly as we computed above for the response to a T-independent antigen. For A_2^* large the control is bang–bang with a single switch. From (16)–(22) with the value of L_0 replaced by $\Lambda M(T_1) + L_{20}$ an explicit formula can be written down giving T_2^* as a function of $M(T_1)$. Thus in studying the primary response it suffices to minimize

$$J' = \int_0^{T_1} \mathrm{d}t + pK(M^*), \tag{30}$$

where

$$K(M^*) = \frac{1}{b_L} \ln \left(\frac{\alpha_2 b_L + \Lambda' M^* + 1}{\Lambda' M^*} \right), \tag{31}$$

$$M^* = M(T_1), \qquad \Lambda' = \Lambda/L_{20}, \quad \text{and} \quad \alpha_2 = A_2^*/kL_{20}. \tag{32}$$

The functional J' differs from J in (24) in that constant terms appearing in the formula for T_2^* (M^*) have been deleted.

The extremal control for the primary response, explicitly computed by Perelson *et al.* (1978), is again bang–bang. However, now there is more than one switch, and the sequence of switches depends on the value of p, the probability of encountering the antigen again. For all values of p, memory cells are only produced at the end of the response. This is sensible since by our assumptions memory cells play no role in the primary response and thus the best time to make them is after one is assured of having coped with the antigen. For the biologically interesting case of large A_1^* one finds that the only biologically reasonable extremal strategy is (L, P, M); i.e. first proliferate as a large lymphocyte, then switch to plasma cell differentiation and finally make memory cells.

The dynamics of an immune response following the (L, P, M) strategy is illustrated in Figure 5.4.5. Notice the primary response takes nearly 100 h, while the secondary response takes only 30 h. Also observe that in

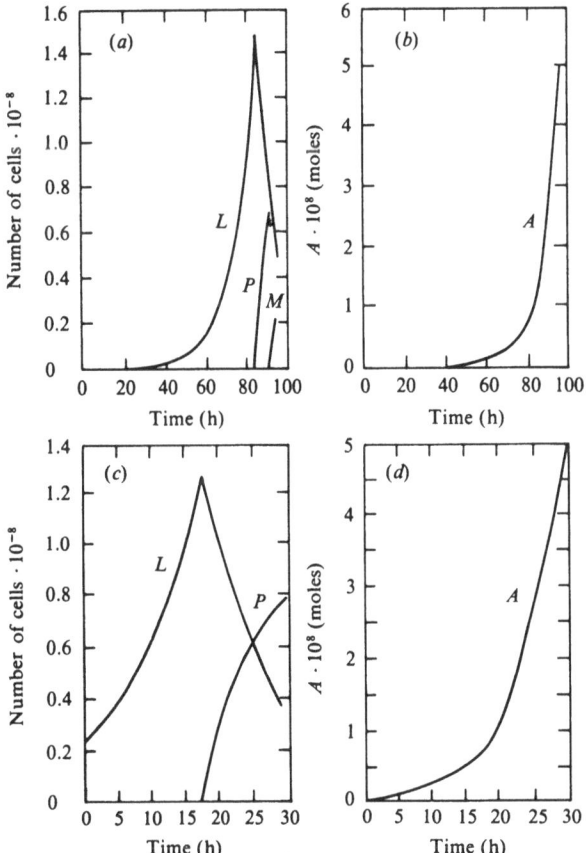

Figure 5.4.5. Dynamics of the optimal primary and secondary responses with $p = 0.1$, $k = 6.25 \cdot 10^{-18}$ mole cell^{-1} h^{-1}, $A_1^* = A_2^* = 5 \cdot 10^{-8}$ mole, $L_0 = 4 \cdot 10^4$, and $L_{20} = 4 \cdot 10^3$ (from Perelson *et al.*, 1978). (a) The number of large lymphocytes, plasma cells and memory cells produced as a function of time during the primary response. (b) The number of moles of antibody secreted as a function of time during the primary response. (c) The number of large lymphocytes and plasma cells produced during the secondary response. (d) The amount of antibody secreted during the secondary response.

the primary response the antibody concentration is nearly zero for the first 60 h, whereas in the secondary response no measurable lag occurs in the production of antibodies.

Most, but not all, biological evidence supports the conclusion of late memory cell development. For example, memory cells are believed to be found in the germinal centers of lymph nodes, transient structures whose formation is induced by antigen late in an immune response. Destruction of the germinal centers was found to leave the dynamics of antibody formation unaffected, but eliminated immunological memory (Grobler *et al.*, 1974; Humphrey, 1976). Although other explanations may be possible, the most obvious is that germinal centers are required for the production of memory cells and that such generation takes place after the formation of plasma cells. When the kinetics of memory cell formation have been studied, it has generally been found that the major increase in the number of B memory cells occurs after a peak in the number of antibody forming cells has been detected (Cunningham & Sercarz, 1971; Neiderhuber & Moller, 1973). However, one study found that antibody forming cells and memory cells appeared simultaneously (Williamson, McMichael & Zitron, 1974). Further, some of the detailed properties of memory cells with regard to the class of antibody they display on their surfaces and the class and affinity of antibody they secrete seems consistent with the notion that they are generated at the end of the primary response (Perelson *et al.*, 1978).

The IgM–IgG switch

During the primary immune response to a T-dependent antigen, two distinct classes of antibody are secreted, immunoglobulin M (IgM) and immunoglobulin G (IgG). In the serum of an animal exposed to antigen, one first notices IgM and then later in the response IgG is detected. Using the hemolytic plaque assay, described in Section 5.5, it has further been shown that a single cell secreting IgM switches to the production of IgG (Nossal, Szenberg, Ada & Austin, 1964; Bleux, Ventura & Liaupoulos, 1977). The IgM and the IgG secreted by a single cell have the same variable region and hence the same specificity for antigen. During a secondary response the majority of antibody produced is IgG, and although some IgM is made, no switch is apparent. Since each polypeptide chain in an antibody molecule is coded for by two genes, a constant region gene and variable region gene, a cell must call into play a complicated genetic mechanism to switch the constant region gene joined to a particular variable region gene. Optimization theory is one tool for understanding the evolutionary rationale for such a mechanism. In order quantitatively to examine the possibility that the IgM–IgG switch provides some advantage to an animal, we shall develop a model for the interaction of the immune system with a growing antigen and then attempt to optimize the performance of the immune system with respect to the class of antibody secreted.

An animal has available to it a number of defenses against infectious disease (cf. Mims, 1976). However, **complement dependent lysis** stands out because the constant regions of IgM and IgG play a part in the activation of this defense mechanism. Complement, which is discussed at length in Section 5.5, is a series of serum glycoproteins which bind to cells that are tagged as foreign by having antibody on their surfaces. The fixing of complement to the cell surface initiates a cascade of chemical reactions that lead to the insertion of a glycoprotein cylinder in the cell membrane through which water and ions can flow. This is believed to cause the lysis of the cell by osmotic forces.

The initiation of the complement reaction is thought to occur by the binding of C1, the first complement component, to a pair of immunoglobulin Fc regions. Since IgM has five Fc regions (see Figure 5.4.6), whereas IgG has one such region, complement fixation requires the presence of one IgM or two IgG molecules in close proximity on the cell surface. Studies by Humphrey (1967) showed that for a sheep erythrocyte a mean of approximately 800 IgG molecules would be required to attach before lysis occurred. Perelson & Wiegel (1979) established the probability, P, of two IgG molecules being separated by a distance ε or less, given that

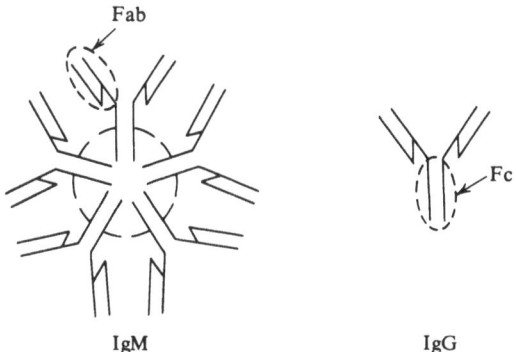

IgM   IgG

Figure 5.4.6. The structure of IgM and IgG (from Perelson, 1978). IgM contains five subunits joined by disulfide bridges (circular dashed line) and a J chain (not shown). Each subunit is similar to a single IgG. When a cell switches from IgM to IgG production, both immunoglobulins are believed to contain identical variable regions, located within the antigen binding fragments (Fab), but different crystallizable fragments (Fc). The complement binding sites are located on the Fc portions of the immunoglobulin molecule. Two such sites are thought to be required to bind the first complement component, C1.

there were a total of N IgG molecules on the surface. They found that

$$P = 1 - \exp\left[-N(N-1)/2M\right], \tag{33}$$

using the biologically reasonable assumption $N/M \ll 1$, where $M = A/\pi\varepsilon^2$ and A is the surface area of the cell. Using this theory one finds that somewhat less than 800 IgG molecules would be required to give a probability of 0.5 for the formation of an IgG pair. Nevertheless, the number of IgG molecules required to fix complement on a red cell is at least two orders of magnitude greater than the number of IgM molecules, and thus at first sight it would seem best if the immune system secreted only IgM.

Although IgM and IgG have the same specificity for antigen they do not bind to a cell with equal efficiency. Studies of the dynamics of complement-dependent red cell lysis by IgM and IgG in the hemolytic plaque assay (see Section 5.5) have indicated that IgM rapidly binds and dissociates from the cell surface with an equilibrium constant indicative of single site binding (DeLisi 1975a, b, 1977; Goldstein & Perelson, 1976), whereas IgG is known to bind bivalently to surfaces (Hornick & Karush, 1972). If IgG binds bivalently and IgM monovalently to a pathogenic organism, then the equilibrium constant for IgG binding may be as much as 10^4 times as great as that of IgM, implying that a cell in a solution containing equal amounts of IgM and IgG would be much more likely to have 800 IgG molecules on its surface than 1 IgM molecule. In such circumstances, it would seem best if the immune system secreted only IgG. However, during the initial phases of an immune response to, say, pathogenic bacteria, there simply may not be enough antibody to put 800 molecules of IgG on each pathogen's surface, and, in fact, with large infections this may take considerable time.

For example, a mouse with a total of $5 \cdot 10^8$ B lymphocytes out of which, say, 10^4 are responding, might have to cope with 10^{10} bacteria. Since a lymphocyte secretes about 10^3 antibodies per second, it would take $8 \cdot 10^5$ seconds or 222 h to secrete enough antibody to have 800 per bacterium. If one also considers that the bacteria multiply during this period, there may never be 800 antibodies per bacterium. However, with IgM it would only take 10^3 s or 0.28 h to have enough antibody to place one molecule on each bacterium. Consequently, during the initial period of a severe infection only IgM can lead to cell lysis. Thus first producing IgM and then switching to IgG may be an optimal strategy for a primary immune response. This strategy may not be necessary during a secondary response in which, say 10^6 instead of 10^4 lymphocytes are responding, for then it would only take 2.2 h to secrete enough IgG to have 800 molecules per pathogen.

This example is meant only to be illustrative of the general principles involved. The criterion of 800 molecules per pathogen is probably an overestimate for organisms as small as a typical bacterium (see (33) for the dependence on cell surface area). Also the equilibrium distribution of antibody between the solution and the cell surface has been neglected. Perelson (1978) and Perelson, Goldstein & Rocklin (1980) develop a detailed optimization model containing ten nonlinear differential equations which considers these factors, as well as the reproduction and death of pathogen, the detailed binding kinetics of IgM and IgG, the natural catabolism of free antibody, the probabilistic nature of complement-dependent killing, and the dependence of the best strategy on antigen dose and numbers of lymphocytes responding. Suffice to say that detailed analysis of the model confirms the heuristic arguments given above.

References

Athans, M. & Falb, P. L. (1966). *Optimal Control,* New York, McGraw-Hill.

Baker, P. J., Stashak, P. W., Amsbaugh, D. F. & Prescott, B. (1971*a*). Characterization of the antibody response to Type III pneumococcal polysaccharide at the cellular level. I. Dose–response studies and the effect of prior immunization on the magnitude of the antibody response. *Immunology* **20**, 469–80.

———— (1971*b*). Characterization of the antibody response to Type III pneumococcal polysaccharide at the cellular level. II. Studies on the relative rate of antibody synthesis and release by antibody producing cells. *Immunology* **20**, 481–92.

Bleux, C., Ventura, M. & Liaupoulos, P. (1977). IgM–IgG switch-over among antibody-forming cells in the mouse. *Nature, Lond.* **267**, 709–11.

Clark, C. W. (1976). *Mathematical Bioeconomics.* New York, Wiley.

Cunningham, A. J. & Sercarz, E. E. (1971). The asynchronous development of immunological memory in helper (T) and precursor (B) cell lines. *Eur. J. Immunol.* **1**, 413–21.

DeLisi, C. (1975*a*). The kinetics of hemolytic plaque formation. IV. IgM plaque inhibition. *J. Theoret. Biol.* **52**, 419–40.

———— (1975*b*). The kinetics of hemolytic plaque formation V. The influence of geometry on plaque growth. *J. Math. Biol.* **2**, 317–31.

———— (1977). Detection and analysis of recognition and selection in the immune response. *Bull. Math. Biol.* **3**, 705–19.

Eisen, H. M. (1974). *Immunology,* 2nd edn, Hagerstown, Maryland, Harper and Row.

Goldstein, B. & Perelson, A. S. (1976). The electrophoretic plaque assay – theory. *Biophys. Chem.* **4**, 349–62.

Grobler, P., Buerki, H., Cottier, H., Hess, M. W. & Stoner, R. D. (1974). Cellular bases for relative radioresistance of the antibody-forming system at advanced stages of the secondary response to tetanus toxoid in mice. *J. Immunol.* **112**, 2154–65.

Hornick, C. L. & Karush, F. (1972). Antibody affinity. III. The role of multivalency. *Immunochemistry* **9**, 325–40.

Humphrey, J. H. (1967). Haemolytic efficiency of rabbit IgG anti-Forssman antibody and its augmentation by anti-rabbit IgG. *Nature, Lond.* **216**, 1295–6.

—— (1976). The still unsolved germinal center mystery. *Adv. Exp. Med. Biol.* **66**, 711–23.

Jacob, F. (1977). Evolution and tinkering. *Science* **196**, 1161–66.

Leitmann, G. (1966). *An Introduction to Optimal Control*, New York, McGraw-Hill.

Macevicz, S. & Oster, G. (1976). Modeling social insect populations. II. Optimal reproductive strategies in annual eusocial insect colonies. *Behav. Ecol. Sociobiol.* **1**, 265–82.

Melchers, F., Coutinho, A., Heinrich, G. & Andersson, J. (1975). Continuous growth of mitogen-reactive B-lymphocytes. *Scand. J. Immunol.* **4**, 853–8.

Mims, C. A. (1976). *The Pathogenesis of Infectious Disease*, London, Academic Press.

Mirmirani, M. (1977). Optimization studies in population dynamics. Ph.D. dissertation, Berkeley, Univ. of California.

Niederhuber, J. E. & Moller, E. (1973). Antigenetic markers on mouse lymphoid cells: origin of cells mediating immunological memory. *Cell. Immunol.* **6**, 407–19.

Nossal, G. J. V., Szenberg, A., Ada, G. L. & Austin, C. M. (1964). Single cell studies on 19S antibody formation. *J. Exp. Med.* **119**, 485–502.

Oster, G. F. & Wilson, E. O. (1978). *Ecology and Evolution of Castes in Social Insects*, Princeton, New Jersey, Princeton University Press.

Perelson, A. S. (1978). The IgM–IgG switch looked at from a control theoretic viewpoint. In *Lecture Notes in Controls and Information Science*, vol. 6 *Optimization Techniques*, Proc. 8th IFIP Conference Würzburg, Sept. 1977, Part 1, ed. J. Stoer, Berlin, Springer-Verlag, pp. 431–40.

Perelson, A. S., Goldstein, B. & Rocklin, S. (1980). Optimal strategies in immunology. III. The IgM–IgG switch. (*J. Math. Biol.*, in press.)

Perelson, A. S., Mirmirani, M. & Oster, G. F. (1976). Optimal strategies in immunology. I. B-cell differentiation and proliferation. *J. Math. Biol.* **3**, 325–67.

—— (1978). Optimal strategies in immunology. II. B memory cell production. *J. Math. Biol.* **5**, 213–56.

Perelson, A. S. & Wiegel, F. W. (1979). A calculation of the number of IgG molecules required per cell to fix complement. *J. Theoret. Biol.* **79**, 317–32.

Russel, P. J. & Diener, E. (1970). The early antibody forming response to *Salmonella* antigens. A study of morphology and kinetics *in vivo* and *in vitro*. *Immunology* **19**, 651–67.

Williamson, A. R., McMichael, A. J. & Zitron, I. M. (1974). B memory cells in the propagation of stable clones of antibody forming cells. In *The Immune System, Genes, Receptors, Signals*, ed. E. E. Sercerz, A. R. Williamson and C. F. Fox, New York, Academic Press, pp. 387–406.

Zagury, D., Bernard, J., Lemieuz, S., Mazie, J. C., Avrameas, S. & Bussard, A. E. (1976). The relationship between storage and secretion of specific antibody by immune lymphoid cells: the ultrastructural localization of antiperoxidase antibodies in plaque forming cells of the rabbit popliteal lymph node. *Eur. J. Immunol.* **6**, 194–9.

5.5 THEORY OF IMMUNOASSAYS

Complement

One of the earliest applications of mathematics to immunology was in the development of theories to explain the biological action of complement. Rather than explicating only the current theory, we will take an historical perspective. This will show interesting and different types of mathematics that have been employed in understanding complement, and will also provide an instructive example of the trial and error that is often required before a satisfactory theory is attained.

Near the end of the nineteenth century a number of investigators noted the ability of fresh blood serum to kill certain microorganisms, including bacteria. This bactericidal property of serum was found to be heat labile and was thus attributed to the presence of an enzyme-like substance that helped or 'complemented' antibody in its protective function. Ehrlich and Morgenroth named this substance **complement**.

Early quantitative studies (e.g. Leschley, 1914) on the ability of complement to lyse red blood cells previously exposed to antibody showed that the percentage of cells lysed depended on the amount of complement added in a nonlinear, sigmoidal way, as shown in Figure 5.5.1. Thus small amounts of complement were without effect, but lysis began as the amount of complement increased, then increased sharply, and finally tapered off as 100% lysis was approached. Von Krough (1916) found an empirical equation that fitted the data very accurately. He let x be the quantity of complement used, y the fraction of cells lysed, and found

$$\log x = \log K + (1/n) \log [y/(1-y)], \tag{1}$$

where K and n are constants. Figure 5.5.2 shows the data in Fig. 5.5.1 fitted to the **Von Krough equation** (1); i.e. a plot of $\log x$ versus $\log [y/(1-y)]$ is a straight line with slope $1/n$. K is the value of x at 50% lysis, since at that point $y/(1-y)$ equals 1. The Von Krough equation (1) can also be expressed as

$$x = K[y/(1-y)]^{1/n},$$

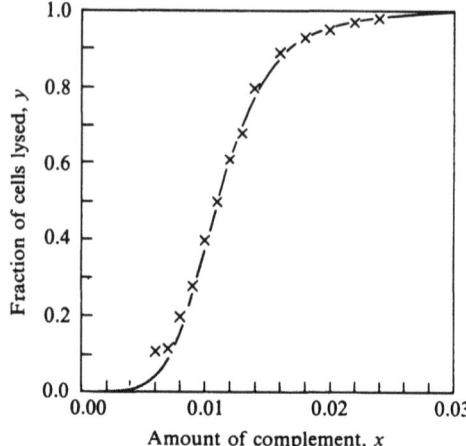

Figure 5.5.1. The fraction of red cells lysed versus the amount of complement used. The ×'s denote data from Leschley (1914). The solid curve is a theoretical one based on the Von Krough equation (1).

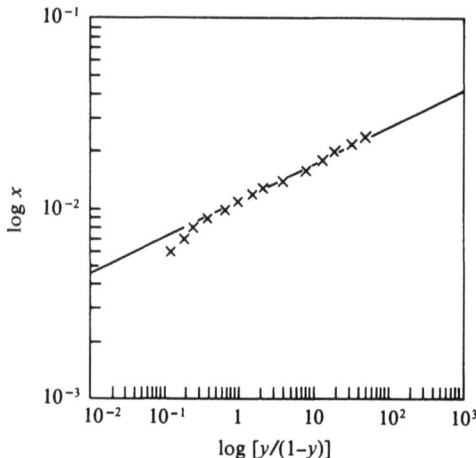

Figure 5.5.2. A plot of $\log x$ versus $\log[y/(1-y)]$. The ×'s denote the data of Figure 5.5.1. The solid line is a fit of the data using the Von Krough equation with $1/n = 0.191$ and $K = 0.011$.

or solving for y as a function of x in the form

$$y = x^n/(x^n + K^n). \tag{2}$$

This is commonly known as Hill's equation (Hill, 1910). The log–log plot suggested by Von Krough with the abscissa and ordinate interchanged, today is commonly called a **Hill plot**, and n the Hill **coefficient** (see Section 1.4).

Brooks (1919) noted that red cells may vary in their susceptibility to lysis, with some cells requiring more complement than others, and suggested that the sigmoid nature of the hemolysis curve reflected this differential susceptibility. Alberty & Baldwin (1951) showed on the basis of a mathematical model that S-shaped hemolysis curves would be expected for a suspension of identical cells. Since the 'Alberty–Baldwin theory is rather general and can be used to analyze a variety of binding problems, it is worthwhile examining it in some detail, even though it later proved not to be correct for complement dependent lysis.

Assume that each red blood cell has on its surface m identical regions at which a reaction requiring both antibody and complement can occur. Further, assume that a cell does not lyse until a critical number, r, of these regions have reacted. This type of model is known as a **cumulative damage** or **multi-hit theory**. After a given amount of complement has been added to the system, a certain fraction, p, of the total number of reactive regions in the cell suspension will have reacted. This fraction p can also be thought of as the **extent of reaction** or the probability that a region chosen at random has reacted. Because of the stochastic nature of any binding process all cells may not have the same number of reacted groups on them. The fraction of cells with i groups reacted, $F(i)$ (equivalently, the probability of a cell having i groups reacted), is given by the *binomial distribution* (Maynard Smith, 1968)

$$F(i) = \binom{m}{i} p^i (1-p)^{m-i}, \tag{3}$$

where the binomial coefficient

$$\binom{m}{i} \equiv \frac{m!}{i!(m-i)!} \tag{4}$$

is the number of different combinations of m objects taken i at a time.

The fraction of cells lysed, y, is predicted to be the fraction of cells with r or more reacted groups, i.e.

$$y = \sum_{i=r}^{m} F(i) = \sum_{i=r}^{m} \binom{m}{i} p^i (1-p)^{m-i}. \tag{5}$$

For any given m and r, y may be calculated for different values of p. Such calculations yield S-shaped curves which fit the Von Krough equation for those values of y which can be measured with some accuracy ($0.1 \leqslant y \leqslant 0.9$) (Alberty & Baldwin, 1951). If one now assumes that the extent of reaction p is directly proportional to x, the amount of complement added, these plots can be compared to experimental hemolysis curves. For values of r about 10 and for m about 50 or greater, the theoretical and experimental curves agree quite well.

One can also use the Alberty–Baldwin theory to examine the hypothesis that red cells vary in their susceptibility to lysis. Heterogeneity of red cells can be introduced into the theory by assuming that different cells require a different critical number, r, of regions to have reacted with antibody and complement. Then if $G(r)$ is the fraction of cells which require the reaction of r groups to produce lysis, the fraction of all cells lysed, y, is

$$y = G(1) \sum_{i=1}^{m} \binom{m}{i} p^i (1-p)^{m-i}$$

$$+ G(2) \sum_{i=2}^{m} \binom{m}{i} p^i (1-p)^{m-i}$$

$$+ \ldots + G(m) \sum_{i=m}^{m} \binom{m}{i} p^i (1-p)^{m-i},$$

that is

$$y = \sum_{r=1}^{m} G(r) \sum_{i=r}^{m} \binom{m}{i} p^i (1-p)^{m-i}. \tag{6}$$

Assuming that the heterogeneity $G(r)$ is a Gaussian function with the 'tails' at twice the standard deviation neglected, Alberty & Baldwin (1951) showed that the shapes of the hemolysis curves and the slopes of the Von Krough plots are not sensitive to the heterogeneity if r is about 10. However, for r much larger than 10, the introduction of heterogeneity has a marked effect and could make the shape of the lysis plot correspond with the experimental plots. Consequently, Alberty & Baldwin conclude it is not possible to distinguish from theory alone between homogeneous red cells with r about 10 or heterogeneous red cells with a much larger r.

The Alberty–Baldwin theory of cumulative damage was thought to be correct for about ten years, until Manfred Mayer and his colleagues performed a new set of hemolysis experiments in which the total number of red cells in the assay system was varied (Mayer, 1961). At the usual red cell concentration of $6.7 \cdot 10^6$ cells ml^{-1} the value of $1/n$ in the Von Krough equation is about 0.2, which corresponds to $r = 10$ in the Alberty–Baldwin theory. However, if $2 \cdot 10^9$ cells ml^{-1} are used, $1/n = 0.43$ and a value of $r = 2$ would be needed to fit the experimental data. To accommodate this shift in r the theory would need to be modified in an *ad hoc* fashion to make the lytic susceptibility of a red cell dependent upon the cell density.

Mayer (1961) formulated a new **single-hit theory** of complement action that took account of the fact that complement was not a single substance, but rather a collection of at least nine serum glycoproteins, named C1, C2, . . . , C9, which acted sequentially to produce lysis. Mayer hypothesized that the occurrence of a single, but random, critical event is all that is necessary to force a red cell on the pathway to complement dependent lysis. We now know that the critical event is the binding of the first complement component, C1, to one IgM molecule or a pair of IgG molecules in close proximity on the cell surface. This is then followed by the sequential binding of the remaining complement components and the formation of a lytic lesion. These reactions are dramatically illustrated in the review by Mayer (1973) and will not be dealt with here.

Mathematically, the one-hit theory says that 'hits' or critical events occur randomly on the cell surface. The probability of a cell having sustained k hits, $P(k)$, being given by the *Poisson distribution* (Maynard Smith, 1968)

$$P(k) = \lambda^k \exp(-\lambda)/k!, \tag{7}$$

where the parameter λ is the mean number of hits per cell. Assuming a single hit is all that is required to produce hemolysis, then all cells with one or more hits should be lysed. Thus the theory predicts that the fraction of unlysed cells is given by $P(0)$, i.e.

$$1 - y = P(0) = \exp(-\lambda) \tag{8}$$

or

$$\lambda = -\ln(1 - y) \tag{9}$$

where, as before, y is the fraction of cells lysed. The mean number of hits per cell, λ, can be equated with the mean number of lesions per cell and thus determined by electron microscopic observation. Such determinations have been performed by Humphrey & Dourmashkin (1965) and for the most part agree with the one-hit theory.

The result (8) can also be obtained without recourse to the Poisson distribution. Let p be the probability of a hit. Then $(1-p)^N$ is the probability of *no hits* with N shots. Observe

$$(1-p)^N = \exp[N \ln(1-p)].$$

For $p \ll 1$, a Taylor expansion of $\ln(1-p)$ yields

$$\ln(1-p) \approx -p,$$

and hence

$$(1-p)^N \approx \exp(-Np).$$

The mean number of hits

$$\lambda = Np, \tag{10}$$

and thus

$$P(0) \equiv (1-p)^N = \exp(-\lambda). \tag{11}$$

This result becomes precise in the limit $p \to 0$, $N \to \infty$ and Np remaining constant.

We shall now discuss another assay, designed to detect antibody secreting cells, which is based upon the complement-dependent lysis of red cells. Diffusion theory is used, and this is the subject of Chapter 7. The following discussion appears here because it is connected with immunology, but the discussion could just as well have served as an example in Chapter 7.

Hemolytic plaque assay

The humoral immune response to a single antigenic determinant is generally characterized by a great heterogeneity in the types of antibody produced. Kreth & Williamson (1973) studying the response to NIP (4-hydroxy-5-iodo-3-nitro-phenacetyl) estimated that at least 3000 distinct clones of cells were producing antibody. Further, the fraction of all B cells stimulated to produce antibody to a single antigen is generally 10^{-5}–10^{-4} (Jerne, 1974; Edelman, 1974). The small number of cells stimulated by antigen and the great heterogeneity of their products are facts that combine to make it extremely difficult to follow the population dynamics of lymphocytes during an immune response or to analyze the antibodies produced by these cells. Thus the simultaneous development by Jerne & Nordin (1963) and Ingraham (Ingraham, 1963; Ingraham & Bussard, 1964) of a means of detecting antibody forming cells (AFC) was an important milestone in immunology. The technique, known as the **hemolytic plaque assay**, has allowed much information to be obtained about the dynamics of the immune response and the properties of AFC. We shall now describe the plaque assay and develop a mathematical description of plaque growth which can be used to obtain otherwise unavailable quantitative information about the secretion of antibody by a single cell.

The experimental technique

The original assay system introduced by Jerne & Nordin (1963) was used to detect lymphoid cells producing antibody directed against red blood cell (RBC) antigens. In a typical experiment, a thin agar layer is prepared that contains nutrient medium, sheep RBCs, and spleen cells isolated from a mouse previously immunized with sheep RBCs. About 40% of the spleen cells are B lymphocytes, and thus it is expected that some cells in the agar layer are producing antibody against determinants on sheep RBCs. The layer is maintained for 1 h at 37 °C and then flooded with complement. RBCs with sufficient bound antibody are lysed by the complement, causing the formation of clear areas in the agar, called (direct) **plaques**. Each plaque has at its center an AFC which secreted the antibody responsible for initiating the complement-dependent hemolysis. Generally the secreted antibody detected in this way is IgM. To obtain lysis of a cell by IgG and complement requires about 800 bound molecules of IgG as compared to the one molecule of IgM required for IgM-dependent lysis. To increase the sensitivity of the assay so as to be able to detect AFCs secreting IgG, one adds anti-immunoglobulin (anti-Ig) serum to the layer prior to the addition of complement. Since many anti-Ig molecules can bind to each bound IgG, lysis can occur when far fewer than 800 IgG molecules are bound. Plaques produced by the addition of anti-Ig are called **indirect plaques**. In what follows we shall only discuss the theory of direct plaque formation.

To detect AFCs secreting antibodies directed against antigens other than sheep RBCs, say dinitrophenyl (DNP), one chemically couples the antigen to the RBCs placed in the agar layer. Then antibodies specific for DNP bind red cells and cause their lysis in the presence of complement. If complement is added to the layer at the beginning of the experiment rather than after 1 h, one can observe how the plaque grows in time.

Mathematical theory of direct plaque growth

As an idealized model, consider an AFC that secretes IgM antibodies at a constant rate S. Assume the AFC is at the origin of a homogeneous medium, infinite in lateral extent, and of thickness h. There are three separate cases of experimental interest: (a) the layer is very thick compared to the dimensions of a typical plaque so that the problem can be treated as if h were infinite; (b) the layer is sufficiently thin that diffusion perpendicular to the layer occurs so fast that it can be neglected on the time scale of the experiment and the problem can be treated as if it were two dimensional; (c) the layer thickness needs to be explicitly considered. The theory for case (c) is relatively complicated and we defer discussion of it until we complete the discussion of cases (a) and (b).

The antibodies released by the AFC are assumed to diffuse through the media with constant diffusion coefficient D and to bind to sites on the surfaces of RBCs. (Readers unfamiliar with the elements of diffusion theory are advised to consult Section 6.1 before proceeding.) Further, it will be assumed that even though IgM has 10 combining sites, it binds to RBCs with only one of its combining sites. This would be the case if the antigenic determinants on the RBC surface were sufficiently far apart or if the IgM molecule were very rigid as some experiments indicate (Edberg, Bronson & Van Oss, 1972).

According to these assumptions free antibody molecules diffuse, react chemically with sites on RBC surfaces, and are generated by a source at the origin. The appropriate equation for the concentration of free antibody $c(\mathbf{r}, t)$ in cases (a) and (b) above, which we shall refer to as the (infinite) three-dimensional and two-dimensional cases, are

$$\partial c/\partial t = D\nabla^2 c - k_f c(\rho_0 - \rho) + k_r \rho + S\delta(\mathbf{r}) \quad \text{(three-dimensional)}, \tag{12}$$

$$\partial c/\partial t = D\nabla^2 c - k_f c(\rho_0 - \rho) + k_r \rho + (S/h)\delta(\mathbf{r}) \quad \text{(two-dimensional)}, \tag{13}$$

with boundary condition $c(\mathbf{r}, t) \to 0$ as $|\mathbf{r}| \to \infty$ for all finite t, and initial condition $c(\mathbf{r}, 0) = 0$ for $\mathbf{r} \neq 0$. Here $\mathbf{r}$, $\delta(\mathbf{r})$, and ∇^2 are the three-dimensional or two-dimensional position vector, Dirac delta function and Laplacian as appropriate.[4] The factor h is included in the source term of (13) so that this term when integrated over any volume that contains the entire source shows that S antibodies are secreted per second, i.e.

$$\int_0^h \iint_A (S/h)\delta(x)\delta(y)\,\mathrm{d}x\,\mathrm{d}y\,\mathrm{d}z = S, \tag{14}$$

where A is any area containing the origin. The rate of IgM binding to RBCs is given by the product of the forward rate constant k_f, the free antibody concentration c, and the free site concentration, $\rho_0 - \rho$, where ρ_0 is the total concentration of antigenic sites and ρ is the concentration of sites bound with IgM. The dissociation rate of bound IgM is the product of the reverse rate constant, k_r, and the bound IgM concentration. This concentration is given by ρ since we have assumed each IgM is bound to only one site.

The concentration of bound sites, $\rho(\mathbf{r}, t)$, changes according to the equation

$$\partial \rho/\partial t = k_f c(\rho_0 - \rho) - k_r \rho, \tag{15}$$

with auxilary conditions

$$\rho(\mathbf{r}, 0) = 0 \quad \text{for all } \mathbf{r},$$

$$\rho(\mathbf{r}, t) \to 0 \quad \text{as } |\mathbf{r}| \to \infty \text{ for all finite } t. \tag{16}$$

Because bound sites are sites on RBCs they do not diffuse; their concentration changes only by chemical reaction as reflected in (15).

---

[4] In three-dimensions $\mathbf{r} = (x, y, z)$, $\delta(\mathbf{r}) = \delta(x)\delta(y)\delta(z)$ and $\nabla^2 = (\partial^2/\partial x^2) + (\partial^2/\partial y^2) + (\partial^2/\partial z^2)$, whereas in two-dimensions $\mathbf{r} = (x, y)$, $\delta(\mathbf{r}) = \delta(x)\delta(y)$ and $\nabla^2 = (\partial^2/\partial x^2) + (\partial^2/\partial y^2)$. The delta function is in essence defined by the property

$$\int_a^b \delta(x)f(x)\,\mathrm{d}x = f(0), \quad \text{if } a < 0 \quad \text{and} \quad b > 0.$$

The binding and dissociation of IgM from RBCs occurs very rapidly, so that one would expect that the concentrations of free and bound IgM to be related as if they were in equilibrium. Further, with single site binding, the free IgM concentration, which is the result of secretion by a single AFC, is never high enough for a large fraction of the RBC sites to be bound at measurable distances from the AFC (Perelson & Goldstein, 1977). Thus $\rho_0 - \rho \approx \rho_0$ and at chemical equilibrium, where $\partial\rho/\partial t = 0$, one expects from (15)

$$\rho = Kc\rho_0 \tag{17}$$

where

$$K \equiv k_f/k_r. \tag{18}$$

Further, with $\partial\rho/\partial t = 0$, (12) and (13) simplify to pure diffusion equations

$$\partial c/\partial t = D\nabla^2 c + S\delta(\mathbf{r}) \quad \text{(three-dimensional)}, \tag{19}$$

$$\partial c/\partial t = D\nabla^2 c + (S/h)\delta(\mathbf{r}) \quad \text{(two-dimensional)}. \tag{20}$$

Results (17), (19) and (20) greatly simplify further calculations. Although these equations were derived in an *ad hoc* fashion using the quasi-equilibrium assumption, $\partial\rho/\partial t = 0$, they can be derived in a more precise mathematical fashion using the theory of **singular perturbations**. We illustrate this briefly for the two-dimensional system. Perelson & Segel (1978) deal with this topic at greater length. Readers uninterested in the formal mathematical basis of the quasi-equilibrium assumption should skip to the next subsection in which the plaque equations (17), (19) and (20) are solved.

Singular perturbation analysis

Our first task is to put (12), (13), (15) and (16) into a scaled dimensionless form (see Appendix Section A.4). In this endeavor we follow the standard scaling procedure that attempts to ensure that terms preceded by a small dimensionless parameter are negligible (Lin & Segel, 1974). If we can find such a scaling with $\partial\rho/\partial t$, but not $\partial c/\partial t$, multiplied by a small parameter, say ε, then in the limit $\varepsilon \to 0$ we should obtain the quasi-equilibrium equations.

Examining values of various physical parameters (see Table 5.5.1) one finds that $\varepsilon = K\rho_0$ typically is small, $\varepsilon \ll 1$. Further, when this is the case, the time for the IgM binding reaction to come to equilibrium is of the order $1/k_r$ (Perelson & Segel, 1978). Thus to follow the establishment of equilibrium one writes a set of *short time* or 'inner equations' with $\bar{t} = t/(1/k_r) = k_r t$ as the time variable. It is known (see Section 7.2) that

Table 5.5.1. *Typical parameter values (adapted from Perelson & Segel, 1978)*

Parameter	Range of values	Typical value
k_f	$5 \cdot 10^{-16}$–$3 \cdot 10^{-15}$ cm^3 molecule^{-1} s^{-1}*	10^{-15} cm^3 molecule^{-1} s^{-1}*
k_r	1–100 s^{-1}	10 s^{-1}†
$K = k_f/k_r$	$5 \cdot 10^{-18}$–$3 \cdot 10^{-15}$ cm^3 molecule^{-1}	10^{-16} cm^3 molecule^{-1}
D_{20,H_2O}		$2 \cdot 10^{-7}$ cm^2 s^{-1}
$D_{20,gel}$	$4 \cdot 10^{-8}$ cm^2 s^{-1}–$8 \cdot 10^{-8}$ cm^2 s^{-1}	
S	10^2–$1.5 \cdot 10^3$ molecules s^{-1}	10^3 molecules s^{-1}
h	10^{-3}–$5 \cdot 10^{-1}$ cm	10^{-2} cm
e	10^3–10^6 epitopes cell^{-1}	10^5 epitopes cell^{-1}
ρ_{RBC}	10^8–10^9 cells cm^{-3}	$4 \cdot 10^8$ cells cm^{-3}
$\rho_0 = e\rho_{RBC}$	10^{11}–10^{15} epitopes cm^{-3}	$4 \cdot 10^{13}$ epitopes cm^{-3}
$\varepsilon = K\rho_0$	10^{-5}–10^{-1}	$4 \cdot 10^{-3}$
$\kappa = KS/hD$	$5 \cdot 10^{-9}$–$3 \cdot 10^{-2}$	$5 \cdot 10^{-5}$

* These values are a factor of 10 higher than those usually quoted for single site binding since here k_f refers to the rate at which an IgM molecule containing 10 independent sites binds to a single RBC determinant.
† These values correspond to antibody–hapten reactions for which one can be assured the binding is by single-site attachment.

the root mean square distance an antibody diffuses during time t is proportional to $\sqrt{(Dt)}$. (This result is sometimes called the Einstein equation.) Thus during time $1/k_r$ concentration changes will occur over lengths of order $\sqrt{(D/k_r)}$. Using this as a length scale, we choose $\bar{\nabla}^2 = \nabla^2 D/k_r$ and $\bar{\delta}(\bar{\mathbf{r}}) = \delta(\mathbf{r})D/k_r$, where $|\bar{\mathbf{r}}| = |\mathbf{r}|/\sqrt{(D/k_r)}$. Antibodies are secreted at rate S. Therefore during time $1/k_r$, S/k_r antibodies are secreted into the diffusion volume hD/k_r. A typical free antibody concentration is $(S/k_r)/(hD/k_r) = S/hD$ and hence as a dimensionless concentration we choose $\bar{c} = chD/S$. The number of bound sites varies between zero and its equilibrium value. From (15) one sees that at equilibrium $\rho = Kc\rho_0/(1 + Kc)$. Since c is typically S/hD, ρ is typically $\kappa\rho_0/(1 + \kappa)$, where $\kappa = KS/hD$. For the usual experimental parameters (see Table 5.5.1), $\kappa \ll 1$, and thus we can use $\kappa\rho_0$ as a scale for ρ. Letting $\bar{\rho} = \rho/(\kappa\rho_0)$, (13) and (15) become upon substitution

$$\partial\bar{c}/\partial\bar{t} = \bar{\nabla}^2\bar{c} - \varepsilon[\bar{c}(1 - \kappa\bar{\rho}) - \bar{\rho}] + \bar{\delta}(\bar{\mathbf{r}}), \tag{21}$$

$$\partial\bar{\rho}/\partial\bar{t} = \bar{c}(1 - \kappa\bar{\rho}) - \bar{\rho}, \tag{22}$$

with initial conditions

$$\bar{c}(\bar{\mathbf{r}}, 0) = 0, \qquad \mathbf{r} \neq 0, \tag{23}$$

$$\bar{\rho}(\bar{\mathbf{r}}, 0) = 0 \quad \text{for all } \mathbf{r}, \tag{24}$$

and boundary conditions

$$\lim_{|\mathbf{r}| \to \infty} \bar{c}(\bar{\mathbf{r}}, \bar{t}) = \lim \bar{\rho}(\bar{\mathbf{r}}, \bar{t}) = 0 \quad \text{for all finite } \bar{t}. \tag{25}$$

The 'outer' or *long time* equations are obtained by recognizing that after a short period of rapid change, the original time scale is no longer appropriate. A suitable long time variable is $t^* = \varepsilon \bar{t} = k_f \rho_0 t$. A typical time is now $1/(k_f \rho_0)$ and by the Einstein equation an appropriate length scale is now $\sqrt{(D/k_f \rho_0)}$. Using the same argument for scaling concentrations and recognizing that ρ will remain near its equilibrium value we choose as variables

$$c^* = chD/S, \qquad \rho^* = \rho/(\kappa \rho_0), \tag{26}$$

$$|\mathbf{r}^*| = \frac{|\mathbf{r}|}{\sqrt{(D/k_f \rho_0)}}, \qquad (\nabla^*)^2 = \frac{\nabla^2 D}{k_f \rho_0}, \qquad \delta^*(\mathbf{r}^*) = \frac{D\delta(\mathbf{r})}{k_f \rho_0}. \tag{27}$$

In terms of these long time variables (13) and (15) become

$$\partial c^*/\partial t^* = (\nabla^*)^2 c^* - c^*(1 - \kappa \rho^*) + \rho^* + \delta^*(\mathbf{r}^*) \tag{28}$$

and

$$\varepsilon(\partial \rho^*/\partial t^*) = c^*(1 - \kappa \rho^*) - \rho^*. \tag{29}$$

The initial conditions at $t = 0$ are no longer appropriate, but the boundary conditions remain

$$\lim_{|\mathbf{r}^*| \to \infty} c^*(\mathbf{r}^*, t^*) = \lim_{|\mathbf{r}^*| \to \infty} \rho^*(\mathbf{r}^*, t^*) = 0 \quad \text{for all finite } t^*. \tag{30}$$

To proceed one generally does a formal perturbation expansion in ε, but here we are only interested in the lowest order, $\varepsilon = 0$, terms. With $\varepsilon = 0$, (28) and (29) become

$$\partial c^*/\partial t^* = \nabla^{*2} c + \delta^*(\mathbf{r}^*) \tag{31}$$

$$\rho^* = c^*/(1 + \kappa c^*), \tag{32}$$

whereas (21) and (22) become

$$\partial \bar{c}/\partial \bar{t} = \bar{\nabla}^2 \bar{c} + \bar{\delta}(\bar{\mathbf{r}}) \tag{33}$$

and

$$\partial \bar{\rho}/\partial \bar{t} = \bar{c}(1 - \kappa \bar{\rho}) - \bar{\rho}. \tag{34}$$

Notice the remarkable feature of this solution, $\bar{c} = c^* = chD/S$, and thus the free antibody concentration obeys a pure diffusion equation that is uniformly valid for all time, i.e. (20). At long times the bound site concentration $\rho^* = c^*$, since $\kappa \ll 1$, which formally yields the heuristically derived (17).

Solution to the plaque equations

The approximate equations (17), (19) and (20) form the basis of plaque theory. The solution to the diffusion equation with a point source is well known and can be found in many books; for example, Carslaw & Jaeger (1959). For infinite two- and three-dimensional systems

$$c(r, t) = (S/4\pi Dr)\, \text{erfc}\,[(r^2/4Dt)^{\frac{1}{2}}] \quad \text{(three-dimensional),} \tag{35}$$

$$c(r, t) = (S/4\pi Dh)E_1(r^2/4Dt) \quad \text{(two-dimensional),} \tag{36}$$

where r is radial distance measured from the origin, erfc is the complementary error function and E_1 is the exponential integral of order 1. Both of these functions are defined by definite integrals

$$\text{erfc}\,(z) = \frac{2}{\sqrt{\pi}} \int_z^\infty \exp\,(-t^2)\,dt, \tag{37}$$

$$E_1(z) = \int_z^\infty t^{-1} \exp\,(-t)\,dt, \tag{38}$$

but their values are tabulated, just as is the case with sin, cos, or log (Abramowitz & Stegun, 1964).

Assuming that a RBC will lyse if N or more antibodies are bound to it, there is a well-defined **plaque radius**, r_p. Let ρ_{RBC} be the concentration of RBCs in the assay system, and assume that each RBC has on its surface e epitopes or antigenic determinants, so the total antigenic site concentration $\rho_0 = \rho_{\text{RBC}}e$. Since the numbers of bound antibodies and bound sites are equal, $N \equiv \rho(r_p, t)/\rho_{\text{RBC}}$. Using (17)

$$N = Kc(r_p, t)\rho_0/\rho_{\text{RBC}} = Kec(r_p, t). \tag{39}$$

Substitution of (35) or (36) into (39) gives an implicit equation for r_p:

$$4\pi DN/KeS = (1/r_p)\, \text{erfc}\,[(r_p^2/4Dt)^{\frac{1}{2}}] \quad \text{(three-dimensional),} \tag{40}$$

$$4\pi DhN/KeS = E_1(r_p^2/4Dt) \quad \text{(two-dimensional).} \tag{41}$$

The left side of (41) is a constant, which implies that the argument of E_1 must also be constant. Thus the theory predicts that for sufficiently thin layers $r_p^2/t = \text{constant}$. Plots of plaque area versus time should therefore be linear. Ingraham & Bussard (1964); Nossal, Bussard, Lewis

& Mazie (1970); and Nossal & Lewis (1971) among others have observed this type of plaque growth.

In three-dimensional systems, antibodies diffuse over a larger volume than in two dimensions and hence one might expect r_p^2 to grow more slowly than the first power of t. If this were the case, then $r_p^2/t \to 0$ as $t \to \infty$. Because $\mathrm{erfc}(z) \to 1$ as $z \to 0$, (40) would then predict that the plaque radius would approach a constant value

$$\lim_{t \to \infty} r_p = \frac{KeS}{4\pi DN}. \tag{42}$$

This result is indeed true and can be obtained in a rigorous fashion by setting $\partial c/\partial t = 0$ in (19), solving for the steady state free antibody concentration c, and substituting into (39).

Finite layers[5]

Any real assay system has finite thickness and consequently the infinite two- and three-dimensional results obtained above are only approximations. To see how good these approximations are one needs to solve the finite layer problem. Again by singular perturbation analysis one can reduce the problem to the analysis of a pure diffusion problem with a point source. To simplify the geometric aspects of the problem assume that the AFC (source) is at the center of a layer of thickness h. Antibodies can not pass through the boundaries of the layer and thus the equation to solve is

$$\partial c/\partial t = D\nabla^2 c + S\delta(r), \tag{43}$$

subject to the boundary conditions

$$\partial c/\partial z = 0 \quad \text{at } z = \pm l, \tag{44a}$$

and

$$\lim_{x^2+y^2 \to \infty} c(x, y, z) = 0 \quad \text{for all finite } t, \tag{44b}$$

where $\pm l = \pm h/2$ are the positions of the boundaries.

This problem can be solved by the **method of images**, a technique commonly used in electrostatics. To use the method we extend the medium to $z = \pm\infty$ and replace the boundary conditions by fictitious (image) sources placed outside the boundaries. The positions and magnitudes of the image sources are chosen in such a way that $\partial c/\partial z = 0$ at $z = \pm l$.

---

[5] This subsection is based upon Goldstein & Perelson (1977).

First, consider the problem in which there is only a single boundary at $z = l$. The actual source is at $z = 0$. If another source of equal strength is placed at $z = 2l$, then the flux outward at $z = l$ from the original source is exactly counterbalanced by the inward flux from the image source. Using the coordinate system shown in Figure 5.5.3, the concentration of free

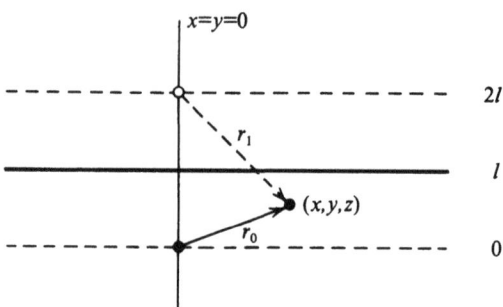

Figure 5.5.3. Actual source (●) at $(0, 0, 0)$ and image source (○) at $(0, 0, 2l)$ needed to satisfy a no-flux boundary condition at $z = l$.

antibody at position (x, y, z) resulting only from the original source is, by (36), $(S/4\pi D r_0)$ erfc $[(r_0^2/4Dt)^{\frac{1}{2}}]$, where $r_0 = (x^2 + y^2 + z^2)^{\frac{1}{2}}$. If only the fictitious source at $(0, 0, 2l)$ were present, then the free antibody concentration at (x, y, z) would be $(S/4\pi D r_1)$ erfc $[r_1^2/4Dt)^{\frac{1}{2}}]$, where $r_1 = [x^2 + y^2 + (2l - z)^2]$ is the distance from the source to the point (x, y, z). Since (43) is linear, the actual concentration at any point in the layer is the sum of the concentrations of the original and image sources, i.e.

$$c(x, y, z, t) = \frac{S}{4\pi D} \left\{ \frac{\text{erfc} \left[(r_0^2/4Dt)^{\frac{1}{2}} \right]}{r_0} + \frac{\text{erfc} \left[(r_1^2/4Dt)^{\frac{1}{2}} \right]}{r_1} \right\}. \quad (45)$$

When two boundaries are present, an infinite number of image sources are needed. Every time an image source is introduced to satisfy the boundary condition at one boundary, it causes the boundary condition to be violated at the other boundary. In Figure 5.5.4 we show the positions of the required image sources. Source -1 is required to satisfy the boundary condition at $z = -l$, however, since it contributes a flux across the $z = +l$ plane, source 2 must be placed an equal distance above (x, y, z). The presence of source 2 disturbs the flux equality at $z = -l$, requiring a source -2, and so forth. The contribution to the solution from each new source decreases quite rapidly, as the process converges to an answer. From the figure one can see that the positions of the sources are

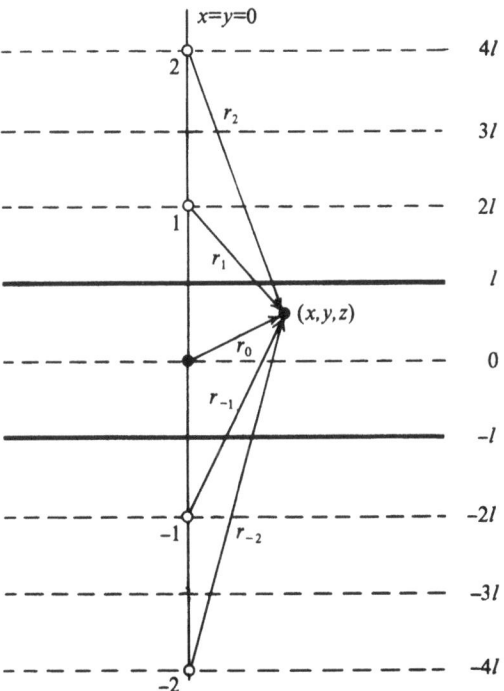

Figure 5.5.4. Actual source (●) at the origin and the image sources (○) needed to satisfy the no-flux boundary conditions at $z = \pm l$.

such that

$$r_j^2 = x^2 + y^2 + (z + 2lj)^2, \qquad j = 0, \pm 1, \pm 2, \tag{46}$$

and thus

$$c(x, y, z, t) = \frac{S}{4\pi D} \left\{ \sum_{j=-\infty}^{\infty} (r_j)^{-1} \operatorname{erfc} \left[(r_j^2/4Dt)^{\frac{1}{2}} \right] \right\}. \tag{47}$$

The $j = 0$ term corresponds to the infinite three-dimensional solution.

The sum in (47) will converge rapidly for short times when the erfc terms are small. As shown in Goldstein & Perelson (1977), one can rewrite (47) in the following form that converges rapidly at long times

$$c(R, z, t) = \frac{S}{4\pi Dh} \left[E_1(R^2/4Dt) + \sum_{j=1}^{\infty} I_j(t) \cos \left(j\pi z/l \right) \right], \tag{48}$$

where $I_j(t)$ is given by the integral

$$I_j(t) = \int_0^t (t')^{-1} \exp \left[(-R^2/4Dt') + (j^2\pi^2 Dt'/l^2) \right] dt', \tag{49}$$

and $R^2 = x^2 + y^2$. Here the first term on the right side corresponds to the two-dimensional result.

Utilizing the expansions (47) and (48) we can estimate the errors involved with using the simpler three-dimensional and two-dimensional results for various systems. This is done in detail in Goldstein & Perelson (1977) and will not be reported on here.

References

Abramowitz, M. & Stegun, I. A. (1964). *Handbook of Mathematical Functions*, Washington, D.C., National Bureau of Standards.

Alberty, R. A. & Baldwin, R. L. A. (1951). A mathematical theory of immune hemolysis. *J. Immunol.* **66**, 725–35.

Brooks, S. C. A. (1919). A theory of the mechanism of disinfection, hemolysis and similar processes. *J. Gen. Physiol.* **1**, 61–80.

Carslaw, H. S. & Jaeger, J. C. (1959). *Conduction of Heat in Solids*, 2nd edn, London, Oxford University Press.

Edberg, S. C., Bronson, P. M. & Van Oss, C. J. (1972). The valency of IgM and IgG rabbit anti-dextran antibody as a function of the size of the dextran molecule. *Immunochemistry* **9**, 273–88.

Edelman, G. M. (1974). Origins and mechanisms of specificity in clonal selection. In *Cellular Selection and Regulation in the Immune System*, ed. G. M. Edelman, New York, Raven Press, pp. 1–38.

Goldstein, B. & Perelson, A. S. (1977). The hemolytic plaque assay: theory for finite layers. *Biophys. Chem.* **7**, 15–32.

Hill, A. V. (1910). Possible effects of the aggregation of the molecules of haemoglobin on its dissociation curve. *J. Physiol.* **40**, iv–viii.

Humphrey, J. H. & Dourmashkin, R. R. (1965). Electron microscope studies of immune cell lysis. In *Complement*, Ciba Foundation Symposium, ed. G. E. W. Wolstenholme and J. Knight, London, Churchill, pp. 175–89.

Ingraham, J. S. (1963). Identification individuelle des cellules productrices d'anticorps par une réaction hémolytique locale. *C. R. Acad. Sci.* **256**, 5005–8.

Ingraham, J. S. & Bussard, A. (1964). Application of a localized hemolysin reaction for specific detection of individual antibody forming cells. *J. Exp. Med.* **119**, 667–84.

Jerne, N. K. (1974). Clonal selection in a lymphocyte network. In *Cellular Selection and Regulation in the Immune System*, ed. G. M. Edelman, New York, Raven Press, pp. 39–58.

Jerne, N. K. & Nordin, A. A. (1963). Plaque formation in agar by single antibody-producing cells. *Science* **140**, 405.

Kreth, H. W. & Williamson, A. R. (1973). The extent of diversity of anti-hapten antibodies in inbred mice: Anti-NIP (4-hydroxy-5-iodo-3-nitro-phenacetyl) antibodies in CBA/H mice. *Eur. J. Immunol.* **3**, 141–7.

Leschley, W. (1914). *Studier over Komplement*, Aarhus, Stiftsbortrykkeriet.

Lin, C. C. & Segel, L. A. (1974). *Mathematics Applied to Deterministic Problems in the Natural Sciences*, New York, Macmillan.

Mayer, M. (1961). Development of the one-hit theory of immune hemolysis. In *Immunochemical Approaches to Problems in Microbiology*, ed. M. Heidelberger and O. Plescia, New Brunswick, New Jersey, Rutgers University Press, pp. 268–79.

—— (1973). The complement system. *Sci. Amer.* **229** (5), 54–66.

Maynard Smith, J. (1968). *Mathematical Ideas in Biology*, London, Cambridge University Press.

Nossal, G. J. V., Bussard, A. S., Lewis, H. & Mazie, J. C. (1970). *In vitro* stimulation of antibody formation by peritoneal cells. I. Plaque technique of high sensitivity enabling access to the cells. *J. Exp. Med.* **131**, 894–916.

Nossal, G. J. V. & Lewis, H. (1971). Functional symmetry among daughter cells arising *in vitro* from single antibody-forming cells. *Immunology* **20**, 739–53.

Perelson, A. S. & Goldstein, B. (1977). Antigen modulation of antibody forming cells: the relationship between direct plaque size, antibody secretion rate and antibody affinity. *J. Immunol.* **118**, 1649–54.

Perelson, A. S. & Segel, L. A. (1978). A singular perturbation approach to diffusion reaction equations containing a point source, with application to the hemolytic plaque assay. *J. Math. Biol.* **6**, 75–85.

Von Krough, M. (1916). Colloidal chemistry and immunology. *J. Infect. Dis.* **19**, 452–77.

6 Some applications of partial differential equations in biology

6.1 THE GENERAL BALANCE LAW AND THE DIFFUSION EQUATION

We begin this section by deriving a general balance law that constitutes the framework of most partial differential equations used in mathematical biology, or indeed in natural science generally. (A partial differential equation is merely a differential equation that deals with functions of more than one variable, and hence contains partial derivatives.) We shall then derive a 'constitutive law' to supplement the general balance law; this leads to the diffusion equation. Later sections will treat other constitutive laws and other equations.

Our equations can be used to describe the motion of particles ranging from molecules to bacteria. Indeed, when we refer to 'particles' it is helpful to keep in mind both these examples.

The general balance law

For simplicity we shall at first confine ourselves to a situation where all variation is with respect to a single coordinate x. Thus we shall define $C(x, t)$ as the particle density or particle concentration at a point x and time t. This means that C gives, at time t, the number of particles per unit volume in a small region centered at x.

The function $C(x, t)$ is assumed to be as smooth as is necessary for the various operations, such as differentiation, that we shall perform on it. By employing such a smooth function we are idealizing the true particulate situation, but this idealization is expected to be appropriate under many conditions of interest. Certainly it is required that the distance between particles is small compared to distances over which concentrations of interest vary significantly. (See Lin & Segel, 1973, for further comments on this type of idealization.)

Consider the number of particles located within a rectangular box B bounded by unit areas in the planes $x = x_0$ and $x = x_0 + L$ (Figure 6.1.1). In words the **general balance law** states that

> the rate of change of particles in B equals the net creation of particles in B plus the net rate at which particles flow into B across the boundaries.

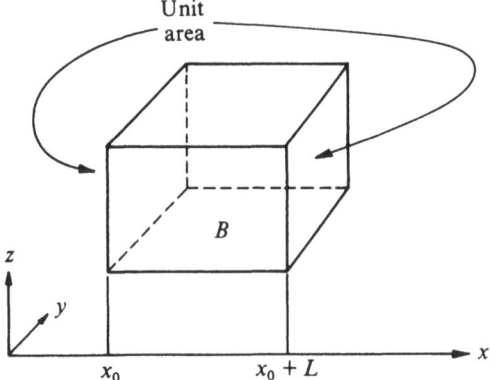

Figure 6.1.1. A rectangular domain B. The general balance law states that the net rate at which particles enter B equals the net rate at which particles are created within B. For symbols, see text.

We now make some definitions that will enable us to write the balance law in mathematical terms.

Let $Q(x, t)$ be the net creation rate, the birth rate per unit volume minus the corresponding death rate. (For chemical molecules, 'births' arise from the combination of other molecules and 'deaths' from the break-up of the given molecular species.) Also let the **flux density** or **current** $J(b, t)$ give the net rate at which particles cross a unit area in the plane $x = b$ (where b is any number), where a net crossing rate is counted as positive (negative) if a majority of particles are crossing in the direction of increasing (decreasing) x (see Figure 6.1.2).

With the aid of the above definitions, the general balance law takes the mathematical form

$$\frac{\partial}{\partial t} \int_{x_0}^{x_0+L} C(x, t)\, \mathrm{d}x = \int_{x_0}^{x_0+L} Q(x, t)\, \mathrm{d}x + J(x_0, t) - J_0(x_0 + L, t),$$

(1)

Figure 6.1.2. Particle trajectories passing through a unit area that is perpendicular to the x-axis, during a certain second of time. The flux density for this period is $4 - 2 = 2$ particles per unit area per second.

where x_0 and L are arbitrary. Note that the signs on the J terms have been taken in accordance with the necessity of tallying the rate at which particles pass *into* B. Note also that we have only kept track of the particle flow through the boundaries marked 'unit area' in Figure 6.1.1. Because of our hypothesis that there is no variation with z and y, during any time interval exactly as many particles flow into B through one of the other two pairs of faces as flow out through the opposite pair.

To put the balance law into more usable form we employ the integral mean value theorem (A.1.116) to write (1) as follows:

$$\frac{\partial}{\partial t} C(q_1, t)L = Q(q_2, t)L + J(x_0, t) - J(x_0 + L, t). \tag{2}$$

Here all we know of q_1 and q_2 is that

$$x_0 \leqslant q_1 \leqslant x_0 + L, \qquad x_0 \leqslant q \leqslant x_0 + L.$$

Equation (2) is true for arbitrary positive L. We choose to divide by L and consider the limit $L \to 0$. In this limit, the unknown points q_1 and q_2 are forced closer and closer to x_0 so that (assuming that C is continuous) we obtain the result

$$\partial C(x_0, t)/\partial t = Q(x_0, t) - [\partial J(x_0, t)/\partial x]. \tag{3}$$

Since x_0 is arbitrary we may write the general balance law as the differential equation

◆◆ $$\partial C(x, t)/\partial t = Q(x, t) - [\partial J(x, t)/\partial x]. \tag{4}$$

We remark that (1) is the more fundamental form of the balance law, for passage to (4) required us to assume that the integrands in (1) were continuous functions. There are certain applications, notably in gas dynamics, where the integrands are discontinuous. The discontinuities are then regulated by 'jump laws' or 'shock conditions' which can be derived from (1), but this is beyond our scope here. (See for example, Lin & Segel, 1973, pp. 482–5.)

Generalization of the balance law to the case of three-dimensional variation requires us to introduce K and L, the fluxes across planes of unit area placed normally to the y- and z-axes, respectively. By a straightforward extension of our reasoning in the one-dimensional case, one can see that the appropriate generalization of (4) is

$$\frac{\partial C(x, y, z, t)}{\partial t} = Q(x, y, z, t) - \left[\frac{\partial J(x, y, z, t)}{\partial x} \right.$$

$$\left. + \frac{\partial K(x, y, z, t)}{\partial y} + \frac{\partial L(x, y, z, t)}{\partial z} \right]. \tag{5}$$

Fick's law

To proceed further we shall temporarily confine our attention to inert particles that are neither created or destroyed, so that $Q = 0$. Returning to the one-dimensional case, let us attempt to formulate the simplest reasonable law that might describe the diffusion of these particles, owing to their random motion. What we must do is to relate $J(x, t)$ to the configuration of particles in the neighborhood of x. But concentrations *near x* can be obtained by means of Taylor's series if we know all partial derivatives of C *at x*. (The present discussion is for a fixed time t, so that time can essentially be ignored for the moment.) Indeed, a slight notational modification of the Taylor series (A.1.30) is

$$C(x + h, t) = C(x, t) + h \frac{\partial C(x, t)}{\partial x} + \frac{h^2}{2!} \frac{\partial^2 C(x, t)}{\partial x^2} + \frac{h^3}{3!} \frac{\partial^3 C(x, t)}{\partial x^3} + \ldots. \quad (6)$$

This shows explicitly how to calculate all concentrations sufficiently near x (h must be small enough for the series to converge) in terms of the partial derivatives at x. Because of this possibility, we can translate the reasonable assumption that 'the flux $J(x, t)$ depends on concentrations near x' into the equation

$$J(x, t) = F \left[C(x, t), \frac{\partial C(x, t)}{\partial x}, \frac{\partial^2 C(x, t)}{\partial x^2}, \ldots \right]. \quad (7)$$

To progress, consider the expected flux if the graph of C near x has the three different possible forms illustrated in Figure 6.1.3. In Case 1, the flux density J is expected to be zero, for just as many particles will cross from left to right as will cross from right to left. In Case 2, J is expected to be positive, for there are more molecules to the left of x and hence more are expected to pass from left to right than from right to left.

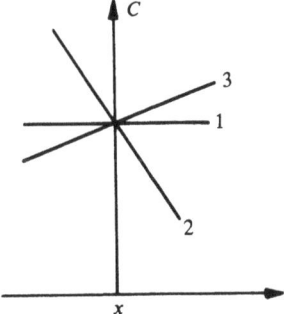

Figure 6.1.3. Three possible slopes, at point x, of a graph of concentration C versus distance.

Correspondingly, J is expected to be negative in Case 3. This discussion suggests that one might be able to obtain an adequate characterization of the flux if one deleted all but the arguments C and $\partial C/\partial x$ from the function F in (7). We have seen that eliminating $\partial C/\partial x$ as well would be too drastic a simplification; knowing only C one could not even guess the direction of flux.

If one measured J for a number of different values of $\partial C/\partial x$, for fixed C, one would expect a result such as that depicted in Figure 6.1.4. Let us

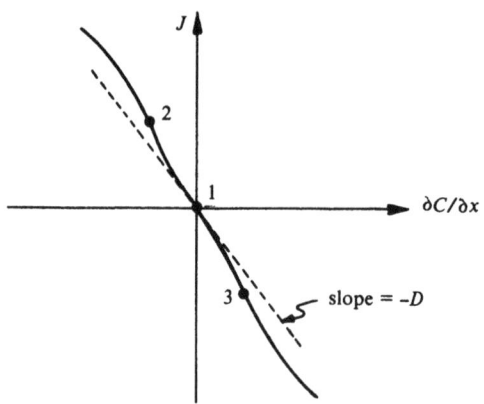

Figure 6.1.4. Typical graph of flux J as a function of the concentration gradient $\partial C/\partial x$. Heavy points correspond to the three lines of Figure 6.1.3. The dashed line provides an approximation to the graph that is accurate when $\partial C/\partial x$ is sufficiently small.

now consider a class of experiments where $\partial C/\partial x$ is sufficiently small to permit approximation of the curve in Figure 6.1.4 by a straight line through the origin with negative slope. That is

◆◆ $J = -D\, \partial C/\partial x,$ \hfill (8)

where $D = D(C)$ is a positive quantity that in general will be different for different concentrations. This simplest reasonable assumption for the dependence of flux on concentration is called **Fick's law**. If we introduce it and the assumption $Q = 0$ into the general one-dimensional balance law (4) we obtain the one-dimensional diffusion equation

$$\frac{\partial C}{\partial t} = \frac{\partial}{\partial x}\left[D(C)\,\frac{\partial C}{\partial x}\right].$$ \hfill (9)

In practice, concentration variations are usually not sufficient to produce noticeable deviations in D – i.e. D can be regarded as a constant. In this case the diffusion equation takes the more usual form

◆◆ $\quad \partial C/\partial t = D\,\partial^2 C/\partial x^2.$ $\hfill(10)$

Essentially the same argument shows that, in the three-dimensional case, the simplest reasonable assumptions for the three components of flux are

$$J = -D\,\partial C/\partial x, \qquad K = -D\,\partial C/\partial y, \qquad L = -\partial C/\partial z. \hfill(11)$$

Here the same **diffusion coefficient** D has been used in all three directions, for we assume, as is most often true, that the random motion of the molecules is equally vigorous in all directions. Combining (5) and (11) we obtain the three-dimensional diffusion equation for constant D.

$$\partial C/\partial t = D[(\partial^2 C/\partial x^2) + (\partial^2 C/\partial y^2) + (\partial^2 C/\partial z^2)]. \hfill(12)$$

We have derived (12) as the simplest reasonable overall description of random particle motion, but decades of use have shown this equation to be suitable in a very wide variety of situations (Crank, 1975).

A problem in diffusion

A concrete problem will illustrate various matters that must be taken into account in dealing with partial differential equations of the type we have just derived. Consider the following possible method for determining the diffusion constant of a substance in water. Take two tall cylinders that are closed at one end. Fill one with pure water and the other with an aqueous solution where the substance is at the uniform concentration C_0. Lay the cylinders on a table with the open ends joined together and observe the diffusion of the substance into the pure water (Figure 6.1.5).

Let the cylinders each be of length L, and orient the x-axis as shown in Figure 6.1.5. Then we must solve the diffusion equation for the concentration $C(x, t)$ in the given range of x:

$$\partial C/\partial t = D(\partial^2 C/\partial x^2), \qquad -L < x < L. \hfill(13a)$$

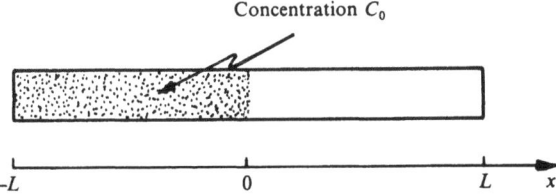

Figure 6.1.5. The initial state when a uniform mixture of particles is brought into contact with pure solvent. Comparison of observation with theory allows determination of the diffusion coefficient. For symbols, see text.

In addition, we must prescribe the situation from which the diffusion starts. This is done by means of the following *initial condition*:

$$C(x, 0) = \begin{cases} C_0, & -L < x < 0, \\ 0, & 0 < x < L. \end{cases} \tag{13b}$$

(The initial value of the concentration at the single point $x = 0$ is immaterial.) Moreover, we must write in mathematical terms the fact that no chemical will flow through the boundaries of the region under consideration, at $x = -L$ and $x = L$. Since the flux is proportional to $\partial C/\partial x$, by (8), the appropriate *boundary conditions* are

$$\partial C/\partial x = 0 \quad \text{at } x = -L \quad \text{and at } x = L.$$

The combination of (13a), initial conditions (13b) and boundary conditions (13c) is typical of problems in partial differential equations. Several more examples will be presented in this chapter. By this means the reader should acquire some feeling for appropriate problems in this field, equations in some space–time domain plus appropriate additional equations on the boundary of the domain.

The diffusion coefficient D can be estimated by matching observations with the solution to problem (13). One would expect this solution to have the qualitative behavior given by the graphs in Figure 6.1.6. This suggests

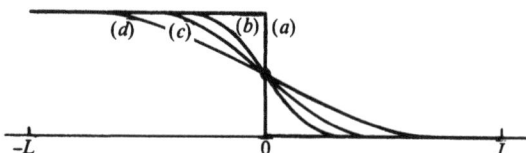

Figure 6.1.6. Curves (a), (b), (c), and (d) mark the expected qualitative behavior of the graph of concentration versus distance, given the initial condition of Figure 6.1.5. Since there are as yet imperceptible changes at $x = \pm L$, the x-domain can be regarded as infinite with little loss in accuracy.

that if the experiment is not continued too long, then there would be no harm in assuming that the tube is infinitely long. One has little experience of 'infinitely far boundaries' so that boundary conditions are not easy to guess with precision. This leads to the idealized problem

$$\partial C/\partial t = D(\partial^2 C/\partial x^2), \quad -\infty < x < \infty, \quad t > 0. \tag{14a}$$

$$C(x, 0) = \begin{cases} C_0, & -\infty < x < 0, \\ 0, & 0 < x < \infty. \end{cases} \tag{14b}$$

C and its derivatives 'not badly behaved' as $x \to \infty$, $x \to -\infty$.

$$\tag{14c}$$

It turns out that the idealized problem has the solution

$$C = C_0\left[1 - \frac{1}{\sqrt{(2\pi)}}\int_{-\infty}^{z} \exp\left(-s^2/2\right) \mathrm{d}s\right],$$

$$z \equiv x(2Dt)^{-\frac{1}{2}}, \qquad t > 0. \tag{15}$$

(The solution to the original problem, (13), is considerably more complicated.) How was (15) obtained? It is our contention that this matter need not be of major concern to the principal readership for whom this book is intended. Biologists can generally not spare the time to master the extensive theory of partial differential equations, and they can turn to theoretically oriented colleagues for help in solving well-formulated mathematical problems. Moreover, even though they do not know how a purported solution was obtained they can often verify that it *is* in fact a solution without too much difficulty. In the present case the verification proceeds as follows.

Using the chain rule (A.1.11) and the fundamental theorem of calculus (A.1.117) we see that

$$\frac{\partial C}{\partial t} = -\frac{C_0}{\sqrt{(2\pi)}}\frac{\mathrm{d}}{\mathrm{d}z}\int_{-\infty}^{z} \exp\left(-s^2/2\right)\mathrm{d}s\bigg|_{z=x(2Dt)^{-\frac{1}{2}}} \frac{\partial z}{\partial t}$$

$$= -\frac{C_0}{\sqrt{(2\pi)}}\cdot\exp\left(-z^2/2\right)\bigg|_{z=x(2Dt)^{-\frac{1}{2}}}\cdot x \cdot -\tfrac{1}{2}(2Dt)^{-\frac{3}{2}}\cdot 2D. \tag{16}$$

Similarly,

$$\frac{\partial C}{\partial x} = -\frac{C_0}{\sqrt{(2\pi)}}\frac{\mathrm{d}}{\mathrm{d}z}\left[\int_{-\infty}^{z} \exp\left(-s^2/2\right)\mathrm{d}s\right]_{z=x(2Dt)^{-\frac{1}{2}}}\frac{\partial z}{\partial x}$$

$$= -\frac{C_0}{\sqrt{(2\pi)}}\exp\left(-x^2/4Dt\right)\cdot(2Dt)^{-\frac{1}{2}}$$

$$D\left(\frac{\partial^2 C}{\partial x^2}\right) = -\frac{DC_0}{\sqrt{(2\pi)}}\cdot(2Dt)^{-\frac{1}{2}}\exp\left(-x^2/4Dt\right)\frac{\partial}{\partial x}\left[-\frac{x^2}{4Dt}\right]. \tag{17}$$

It is now easily seen that (16) and (17) are identical, proving that equation (14*a*) is indeed satisfied.

For fixed negative x, $z \to -\infty$ as $t\downarrow 0$, i.e. as t decreases to zero. Thus the integral in (15) vanishes (by (A.1.111*b*)) so that the first part of the initial condition (14*b*) is satisfied. For fixed positive x, $z \to +\infty$ as $t\downarrow 0$ so that the second part of (14*b*) will be satisfied if

$$\int_{-\infty}^{\infty} \exp\left(-s^2/2\right)\mathrm{d}s = \sqrt{(2\pi)}. \tag{18a}$$

The simplest tables of definite integrals contain (18a) or the equivalent formula (which uses the fact that the integrand is an even function, i.e. it is symmetrical about $x = 0$)

$$\int_0^\infty \exp(-s^2/2)\, ds = \sqrt{(\pi/2)}. \tag{18b}$$

As for the behavior when $|x| \to \infty$, here (for fixed positive t) $|z| \to \infty$ and Figure 6.1.7 shows that the solution behaves reasonably. We therefore have reason to feel that (15) would satisfy a more precisely formulated version of the boundary condition (14c), and we can regard (14) as having been verified.

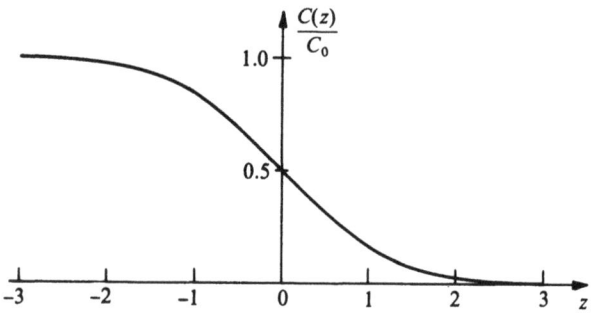

Figure 6.1.7. A graph of the function $C/C_0 = 1 - (2\pi)^{-\frac{1}{2}} \int_{-\infty}^z \exp(-s^2/2)\, ds$. The graph is drawn from points listed in Table XXVb of Burington (1955).

The basic generalization about diffusion

When the two liquids, solution and pure water, are brought together, a mixture region spreads out from the contact plane $x = 0$. Let us define as 'virtually undisturbed' a region where the concentration remains within about 5% of its initial value. From Figure 6.1.7, we see that the 'virtually undisturbed' region is defined approximately by $|z| < 2$, i.e. the effects of diffusion are almost entirely confined to the region $|x| < 3\sqrt{(Dt)}$. Other calculations show that the root mean squared distance that a particle diffuses is $(2NDt)$ when the random motion takes place in N dimensions, $N = 1, 2, 3$. (See Rubinow, 1975, pp. 207–9). The results of these calculations together with computations of first passage times (Section 6.2) can be summarized in the extremely useful generalization that the order of the magnitude of the distance L that particles diffuse in time t is given by

◆◆ $L \approx \sqrt{(Dt)}$, D the diffusion coefficient.   (19)

As one application of (19), consider the fact that a small molecule (molecular weight ≈ 100) has a diffusion constant of magnitude $10^{-5}\,\text{cm}^2\,\text{s}^{-1}$. Consequently diffusion over a distance of 1 μm (scale of procaryote cells) takes place in 10^{-3} s, but the diffusion of 10 μm (scale of eucaryote cells) takes around 10^{-1} s. Since typical reaction rates are in the millisecond range, this simple quantitative observation suggests a reason why eucaryotic cells (e.g. mammalian cells) but not procaryotic cells (e.g. bacteria) have evolved an elaborate system of organelles to concentrate various chemical activities. More will be said about this matter shortly.

Motility characterized as a diffusivity

Let us consider another application of diffusion theory in biology. Moving microorganisms are often characterized with such qualitative terms as 'very motile' or 'moderately motile.' Segel, Chet & Hennis (1977) emphasized that the effective diffusion constant, D, of a population of organisms can usefully be regarded as a measure of motility. As a simple assay for D, these authors suggested placing a capillary tube containing only water into an aqueous solution of bacteria with concentration C_0. The number of bacteria whose random motion takes them into the tube should then be counted after a certain time T. It was argued that the following problem for the bacterial density $C(x, t)$ provided an adequate model for the diffusion of bacteria into the capillary (which is taken to lie along the positive x-axis).

$$\partial C/\partial t = D(\partial^2 C/\partial x^2), \qquad 0 < x < \infty, \qquad t > 0.$$

$$C(x, 0) = 0, \qquad 0 < x < \infty. \qquad C(0, t) = C_0, \qquad t > 0. \tag{20}$$

C and its derivatives 'not badly behaved' as $x \to \infty$.

The solution of this problem is just twice the solution of (14), for Figure 6.1.7 (or symmetry considerations) shows that the function $C(x, t)$ of (14) retains the constant value $C_0/2$ at $x = 0$. If A is the cross-sectional area of the tube, the number N of bacteria that are expected to enter the tube in time T is given by $A \int_0^\infty C(x, T)\,dx$. Carrying out the integration, one obtains the following formula for D:

$$D = \pi N^2/(4C_0^2 A^2 T). \tag{21}$$

Segel *et al.* (1977) provided evidence for the validity of (21) by showing that roughly the same value of D was obtained for a given bacterial population even though various experiments used different values of T and of C_0. Typically D was $\frac{1}{4}\,\text{cm}^2\,\text{h}^{-1}$. In the ten minutes duration of a typical experiment, (19) indicates that bacteria are expected to diffuse a distance of magnitude

$$\left(\frac{1}{4}\frac{cm^2}{h}\cdot\frac{1}{4}\right)^{\frac{1}{2}}\approx\frac{1}{5}cm.$$

The tubes were over 3 cm long, so that simplifying the problem by assuming a semi-infinite tube appears well justified. An illustration of the value of a quantitative characterization of motility can be provided from Adler's (1969) results that 3000 *Escherichia coli*: K12 bacteria entered a certain capillary in 1 h while in another experiment with a different strain but essentially the same experimental conditions, 11 000 bacteria entered the tube in 90 min. Is this second strain significantly more motile than strain K12? Formula (21) gives an affirmative answer, for prolonging the time T by a factor of 1.5 is expected to increase the number N of bacteria that enter the capillary only by a factor of $(1.5)^{\frac{1}{2}}\approx 1.2$.

Exercises

1   Write out a derivation of (5).
2   (*a*) Generalize (12) to the case where D is a function of C. Make sure that your answer reduces to (9) when there is no variation with y and z.

(*b*) Show that (18*a*) implies (18*b*).

(*c*) Some tables of integrals may contain neither (18*a*) nor (18*b*), but rather

$$\int_0^\infty \exp\left(-t^2\right) dt = \sqrt{\pi}.$$

Show that this last result is equivalent to the two previously mentioned.

3   Verify (21) with the aid of the following steps.

(*a*) Use integration by parts to obtain

$$\int_0^\infty C\,dx = xC\Big|_0^\infty - \int_0^\infty x\frac{dC}{dx}\,dx.$$

(Note it can be shown that as $x\to\infty$, $xC\to 0$ because $C\to 0$ much faster than $x\to\infty$.)

(*b*) Perform the integration by making the substitution $u = x^2/4Dt$.

4   Ants secrete an alarm substance (pheromone) under suitable provocation. W. H. Bossert & E. O. Wilson [*J. Theoret. Biol.* **5** (1963), 442] used diffusion theory to make quantitative

studies of this phenomenon. The first situation they considered was one in which an ant instantly releases Q molecules of pheromone at a point on the ground (regarded as a plane). The appropriate variable to use here is ρ, the distance from the point of release. In 'ρ' coordinates (wherein the surfaces $\rho = $ constant are spheres) the diffusion equation is

$$\frac{\partial C}{\partial t} = \frac{D}{\rho^2} \cdot \frac{\partial}{\partial \rho}\left[\rho^2\left(\frac{\partial C}{\partial \rho}\right)\right]. \tag{22}$$

It is the purpose of this exercise to verify that the relevant solution is

$$C = 2Q(4\pi Dt)^{-\frac{3}{2}} \exp\left(-\rho^2/4Dt\right). \tag{23}$$

(a) Show that the expression given in (23) satisfies (22).

(b) Since the exponential approaches zero much faster than the remaining factor becomes infinite, the concentration C approaches zero as t decreases to zero – for every $\rho \neq 0$. It remains to show that exactly Q molecules are diffusing. Study of spherical coordinates shows that one must prove that

$$2\pi \int_0^\infty \rho^2 C(\rho, t)\, \mathrm{d}\rho = Q. \tag{24}$$

Show that (24) holds by noticing that the integral can be written as a multiple of

$$\int_0^\infty \rho \frac{\mathrm{d}}{\mathrm{d}\rho}\left[\exp\left(-\rho^2/4Dt\right)\right] \mathrm{d}\rho.$$

Then integrate by parts and use (18b). (This is just the first step in the Bossert–Wilson discussion. For further developments, consult the original paper, or Section 5.5 of Rubinow, 1975.)

5 Suppose that $C = C_0$ at $x = 0$, $C = C_1$ at $x = L$, and C is in a steady state (i.e. $\partial C/\partial t = 0$). Assuming (10) find $C(x)$, $0 < x < L$, and show that the flux density J is given by

$$J = -D[(C_1 - C_0)/L]. \tag{25}$$

Equation (25) is called 'the integrated form of Fick's law' or usually just **Fick's law**.

References

Adler, J. (1969). Chemoreceptors in bacteria. *Science* **166**, 1588–97.

Burington, R. (1955). *Handbook of Mathematical Tables and Formulas*. 3rd edn, reprinted with corrections, Sandusky, Ohio, Handbook Publishers, Inc.

Crank, J. (1975). *The Mathematics of Diffusion*, 2nd edn, Oxford, Clarendon Press.

Lin, C. C. & Segel, L. A. (1973). *Mathematics Applied to Deterministic Problems in the Natural Sciences*, New York, Macmillan.

Rubinow, S. (1975). *Introduction to Mathematical Biology*, New York, Wiley Interscience.

Segel, L. A., Chet, I. & Hennis, Y. (1977). A simple quantitative assay for bacterial motility. *J. Gen. Microbiol.* **98**, 329–37.

6.2 TRANSIT TIMES

When examining pictures of cells and organelles, we cannot escape from wondering why these membrane structures have their particular size and shape. Since the diffusion of metabolites and other molecules is closely involved in all cellular processes, it is reasonable to conjecture that the design of cellular structures is aimed, in large part, toward minimizing diffusion delays. To examine this idea more closely, one must acquire a better knowledge of the effects of the geometry on diffusion times for structures that commonly occur in living systems. We shall now introduce the basic theory that is necessary for such a study and present one or two examples of its use. A much fuller discussion can be found in the paper by Hardt & Cone (1980).

Consider a situation in which particles are released from a site (the source) and diffuse freely until they reach their target (the sink). In the following, we address ourselves to the question of how long it takes before the diffusing particles 'find' their target. Although we use the term 'find' to describe the fact that particles reach the sink, all that really happens is that the random movements of the particles eventually bring them to hit the target by accident. It must also be kept in mind that the particles possess no sense of their concentration: each one of them makes its own random journey irrespective of the presence of the others.

It is often helpful to regard the sink as an absorbing boundary. This is permissible because once it has reached the sink for the first time the particle is no longer of interest, so that we can remove it from consideration.

Since diffusion proceeds by random motions the particles do not all reach the target simultaneously. A characteristic time for this diffusion process, the mean of the diffusion times, is termed the **transit time**.

The transit time can be computed by the following 'thought experiment'. Release a pulse of particles from the source at time $t = 0$ and record the arrival times of particles at the sink. The transit time is the mean of the arrival times. Putting this experiment in mathematical terms, we have to solve the following diffusion problem.

$$\frac{\partial C}{\partial t} = D[(\partial^2 C/\partial x^2) + (\partial^2 C/\partial y^2) + (\partial^2 C/\partial z^2)]. \qquad (1a)$$

Initial condition: $C = 0$ everywhere except at the source.

$$(1b)$$

Boundary condition: $C = 0$ at the sink,

Flux $= 0$ at all other boundaries. $(1c)$

From the solution to this equation we can obtain the flux of particles at the sink as a function of t. This indicates the rate of arrival of the particles and permits calculation of the transit time.

The trouble with this approach, which is widely used in the literature (see for example Weiss, 1967, or Adam & Delbruck, 1968), is that it is difficult to solve the time-dependent diffusion equation in all but the simplest geometries. Hence, we must adopt a different approach if we want to be able to calculate transit time for the complex geometries commonly found in cellular structures. A hint that a simpler method of calculation may exist stems from the fact that the time-dependent approach provides too much information; it gives the distribution of diffusion times from which we have to calculate the mean.

We shall now show that the mean diffusion time can be obtained directly and more easily from the steady state behavior of the system. To this end we must consider the following problem.

$$(\partial^2 C/\partial x^2) + (\partial^2 C/\partial y^2) + (\partial^2 C/\partial z^2) = 0 \qquad (2a)$$

Boundary conditions: $C = C_0$ at the source, $(2b, c)$

$C = 0$ at the sink.

In the steady state situation described by (2) a time-independent continuous flux of particles both enters and leaves the system. The number N of particles in the system is constant. Our key observation is that this number equals the flux J into the system (which tells how many new particles enter from the source per unit time) multiplied by the average time τ that it takes to diffuse from the source to the sink. Thus the transit time τ is given in terms of the number N of particles in the system described by (2), and the flux J of particles into this system, by the simple formula

$$\tau = N/J. \qquad (3)$$

To illustrate this approach, let us first consider the simplest case of one-dimensional diffusion. We shall calculate the transit time for a plane of particles that are released a distance L from a parallel plane sink

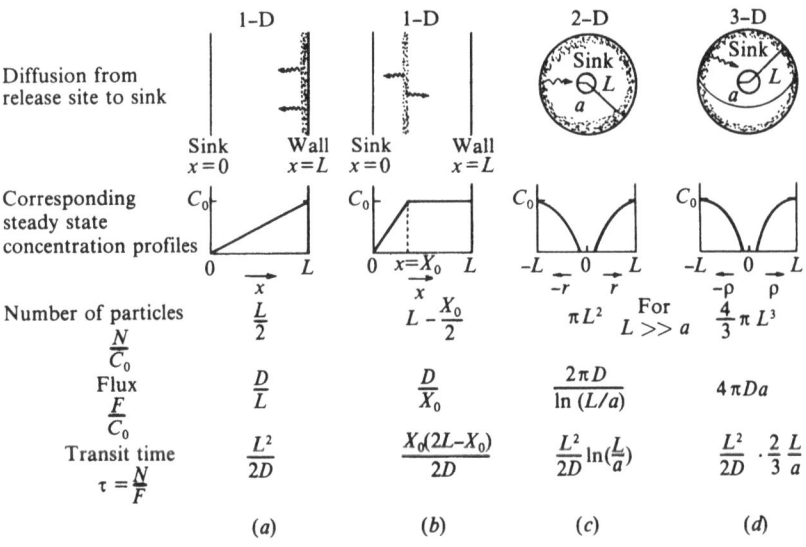

Figure 6.2.1. Transit time calculations. (*a*) One dimensional diffusion through a distance L. (*b*) One dimensional diffusion in the presence of a reflecting wall. (*c*) Two-dimensional diffusion into a sink. (*d*) Three-dimensional diffusion into a sink. For symbols, see text.

(Figure 6.2.1*a*). The appropriate particular case of (2) is

$$D\,(d^2C/dx^2) = 0; \qquad C(L) = C_0, \qquad C(0) = 0. \qquad (4a, b, c)$$

Integrating (4*a*) twice we find that $C(x) = \alpha x + \beta$. The constants α and β are determined by imposing boundary conditions (4*b*) and (4*c*), yielding $C(x) = C_0 x/L$. Recalling (6.1.8), we note that the flux J through a unit cross-sectional area of the system is given by

$$J = -D[dC(x)/dx] = -DC_0/L.$$

The number of particles in a rectangular plane of length L, with unit cross-sectional area, is

$$N = \int_0^L C(x)\,dx = C_0 L/2.$$

According to (3), then, the transit time is

$$\tau = N/J = L^2/2D.$$

Note the essential equivalence of this result with (6.1.19).

Figure 6.2.1*b* illustrates a more complicated case of one dimensional diffusion. Here the particles are released a distance X_0 from the sink, while a reflecting wall is located a distance L from the sink. The appropriate steady state diffusion problem is

$$D(\mathrm{d}^2C/\mathrm{d}x^2) = 0, \qquad \mathrm{d}C/\mathrm{d}x = 0 \quad \text{at } x = L,$$

$$C = C_0 \quad \text{at } x = X_0, \qquad C = 0 \quad \text{at } x = 0. \tag{5}$$

The required calculations are again easy, and are left as an exercise.

Figure 6.2.1.*c, d* illustrates two- and three-dimensional diffusion in the case where particles are released a distance $L - a$ from a sink of radius a, $a < L$. To formulate the appropriate diffusion equations, one must introduce, respectively, the coordinate r that measures the distance from the *line* forming the axis of the two-dimensional cylindrical sink, and the coordinate ρ that measures the distance from the *point* forming the center of the three-dimensional spherical sink. Modifications of the diffusion equation to make it appropriate to these situations (Crank, 1975, pp. 69 and 89) lead to the following problems.

For two-dimensional diffusion (with cylindrical symmetry)

$$\frac{\mathrm{d}}{\mathrm{d}r}r\left[\frac{\mathrm{d}C(r)}{\mathrm{d}r}\right] = 0,$$

$$C = C_0 \quad \text{at } r = L, \qquad C = 0 \quad \text{at } r = a. \tag{6}$$

For three-dimensional diffusion (with spherical symmetry)

$$\frac{\mathrm{d}}{\mathrm{d}\rho}\left(\rho^2 \frac{\mathrm{d}C(\rho)}{\mathrm{d}\rho}\right) = 0,$$

$$C = C_0 \quad \text{at } \rho = L, \qquad C = 0 \quad \text{at } \rho = a. \tag{7}$$

When L is large compared to a, the results for two- and three-dimensional diffusion are, respectively,

$$\tau = \frac{L^2}{2D}\ln\left(\frac{L}{a}\right), \qquad \tau = \frac{L^2}{2D} \cdot \frac{2}{3}\frac{L}{a}. \tag{8}$$

Comparing with the one-dimensional result $\tau = L^2/2D$, one sees the marked influence of dimensionality. In typical instances, it takes moderately longer for a particle to 'find' its target in two dimensions compared to one, and much longer in three dimensions. This general observation already suggests that when speed of reaction is desired, systems will evolve to confine reactants to regions in or near two-dimensional surfaces such as membranes.

To make more precise the observation about the desirability of two-dimensional diffusion in sufficiently large systems, let us consider the time it takes for a substrate molecule to find the active site of an enzyme (radius a) from a distance L. We wish to compare diffusion in three dimensions through an aqueous medium with two-dimensional diffusion within a membrane. Here the generally smaller transit time in two dimensions is countered by the fact that diffusion is slower in the more viscous membrane medium. If D_w and D_m denote the diffusion coefficients in the water and in the membrane, respectively, then by equating the transit times given in (8) we can obtain a formula for the cell dimension L above which the location of enzymes in membranes reduces diffusion delays. We obtain

$$L/a = \tfrac{3}{2}(D_w/D_m) \ln (L/a). \tag{9}$$

Substituting the realistic values $a \approx 1$ nm and $D_w/D_m \approx 100$ we obtain $L \approx 1$ μm, which corresponds to the size of bacteria. To put the matter boldly, it appears that the size of the active site of enzymes may be the principal factor that governs the absolute dimensions of cellular structures. (For further discussion of this point, see Hardt & Cone, 1980.)

Exercises

1   (*a*) Solve (5) to obtain the results of Figure 7.2(*b*).
        (*b*) Solve (6) to obtain the results of Figure 6.2(*c*).
        (*c*) Solve (7) to obtain the results of Figure 6.2(*d*).

2   (*a*) Verify (9) using the transit times for diffusion in two and three dimensions given by (8).
        (*b*) Solve (9) by iterations.

References

Adam, G. & Delbruck, M. (1968). Reduction of dimensionality in biological diffusion processes. In *Structural Chemistry and Molecular Biology*, ed. A. Rich, and N. Davidson, San Francisco, Freeman, pp. 198–215.

Crank, J. (1975). *The Mathematics of Diffusion*, 2nd edn, Oxford, Clarendon Press.

Hardt, S. & Cone, R. A. (1980). Diffusion in small structures. *J. Theoret. Biol.*, submitted for publication.

Weiss, G. H. (1967). First passage time problems in chemical physics. *Adv. Chem. Phys.* **13**, 1–18.

6.3 FACILITATED DIFFUSION

Membrane transport

The transport of substances into and out of a cell is mediated by a membrane surrounding the cell called the **plasma membrane**. Structurally, the plasma membrane consists of a double layer of lipid molecules. This contains, on both surfaces of the membrane, a mosaic of globular proteins whose function is at the present time obscure. The ability of many substances, called **permeants**, to penetrate the interior of cells has been found to be proportional to their lipid solubility. Some permeants that are notable exceptions to this observation are ions, water, urea, and certain metabolites.

The passage of ions through membranes requires an elucidation of the electrical nature of membranes, and there is insufficient experimental evidence at present to permit broad generalizations to be made. The permeation of water and urea molecules into cells has led to the concept that membranes contain **pores** through which water and other small water-soluble molecules can permeate.

Certain metabolites such as sugars and amino acids penetrate cells at a rate that is considerably faster than can be understood on the basis of their lipid solubility. Their permeation has led to the idea of **facilitated diffusion** as an important specialized membrane transport system for nonelectrolytes. Here we shall present the theory of facilitated transport, in both elementary and sophisticated mathematical formulations. The basic concept of the theory is that in the membrane reside **carrier** molecules with which a given permeant can react. Transport of a permeant molecule from the outside of the cell to the inside is accomplished by means of three steps: association with a carrier molecule at the interior surface of the membrane, transport of the permeant–carrier complex across the surface, and dissociation of the permeant molecule at the interior surface.

Widdas's formula for facilitated diffusion

We denote the carrier molecular species by C. The carrier is confined to the cell membrane of uniform thickness L in which it is able to diffuse. A ligand S in the surrounding medium, which is both extracellular and intracellular, reacts with C in a reversible manner to form a complex Y, i.e.

$$S + C \underset{k_-}{\overset{k_+}{\rightleftharpoons}} Y. \tag{1}$$

Here k_+ and k_- denote the forward and backward rate constants, respectively. We shall use subscripts e and i to denote exterior and interior sides of the membrane, respectively, and lower case letters to denote the concentrations of the quantities S, C, and Y. Thus, s_e and s_i are exterior and interior ligand concentrations. The concentrations of the ligand–carrier complex just inside the membrane and facing the exterior and interior regions are denoted by y_e and y_i.

We shall now make the quasi-steady state hypothesis (c.f. Section 1.1), and assume that s_e and s_i are changing very slowly in time. We shall assume further that the rates of association and dissociation of ligand with carrier are considerably faster than the rates of diffusion of the carrier across the width of the membrane. Thus we infer that a *condition of local reaction equilibrium* exists at any position in the membrane, but in fact we will make use of it only at either side of the membrane. Equivalently, we can state that the diffusion of carrier is the rate limiting step in the transport process. Hence, from (1), it follows that

$$0 = k_+ sc - k_- y, \tag{2}$$

where we can supply subscripts e or i to s, c and y.

Because the carrier molecules are constrained to the membrane, the number of carrier molecules in the membrane is conserved. Further, assume that the carrier molecule is very much more massive than the ligand, so that the diffusion of carrier molecules is unaffected by their state of binding. Then the concentrations of liganded and bare carrier molecules satisfy the relation

$$c + y = c_0, \tag{3}$$

where c_0 is the initial carrier concentration in the membrane, at a time when no ligand molecules are available for reaction. In (3), c and y can be evaluated at any position in the membrane. Let us eliminate c from (2) by means of (3) and introduce the dissociation constant K_d defined as

$$K_d \equiv k_-/k_+ = sc/y. \tag{4}$$

Equation (2) becomes with the aid of (3) and (4)

$$y = c_0 s/(K_d + s).   \tag{5}$$

We assume that experiments measuring currents resulting from facilitated diffusion are conducted over a sufficiently long time so that a steady state of diffusion exists during most of the time course of the experiment. Thus, we assume that the current density, considered positive in the inward direction, is given to sufficient approximation by the integrated version of Fick's Law (see Exercise 6.1.5)

$$j = (D_y/L)(y_e - y_i),   \tag{6}$$

where D_y is the diffusion constant of the ligand–carrier complex. (In the language of Section 6.1, the current density is the flux density, the number of molecules crossing a unit area per unit time.) The ligand current into the cell is identified as that associated with the ligand–carrier complex. We emphasize here that the derivation requires only that j be proportional to $(y_e - y_i)$, i.e. that the transport law across the membrane be linearly related to the concentration difference of complex. In fact, in a membrane that is only two molecules thick, it is a moot point as to whether Fick's laws of diffusion are applicable to it. Substituting (5) into (6), we obtain

$$\blacklozenge\blacklozenge \qquad j = \frac{D_y c_0}{L}\left(\frac{s_e}{K_d + s_e} - \frac{s_i}{K_d + s_i}\right).   \tag{7}$$

This is **Widdas's formula** for facilitated diffusion (Widdas, 1952, 1954). Equation (8) can also be written as

$$j = \frac{D_y c_0 K_d}{L}\frac{(s_e - s_i)}{(K_d + s_e)(K_d + s_i)}.   \tag{8}$$

Note that if s_e and s_i are both small compared to K_d, then the denominator in (8) is effectively constant, and measurements of the flux as a function of $s_e - s_i$ will be consistent with the hypothesis that the permeant is crossing the membrane by simple diffusion.

The first term on the right in (7) represents the unidirectional flux into the cell (i.e. that which obtains when $s_i = 0$). Such a flux is commonly measured with the aid of radioactive labeling of the external substrate.

Permeant accumulation in cells

In practice, a common experimental arrangement is to observe the concentration of a permeant inside a cell at more or less regular intervals, measured in minutes or even hours, following exposure of the cell to an external medium in which s_e is maintained constant. Thus, we need to know the time-dependent behavior of s_i. To obtain it, we observe that the change with time of permeant molecules inside a cell is governed by the equation

$$\frac{d}{dt}(Vs_i) = Aj \tag{9}$$

where V is the volume of the cell, Vs_i is the number of permeant molecules in the cell at any time t, A is the area of the membrane, j is the current density of ligand–carrier complex, and Aj is the current of ligand molecules entering the cell through the membrane. Hence, if the experimental conditions maintain the cell volume constant, (9) becomes with the aid of (8)

$$\frac{ds_i}{dt} = -k \cdot \frac{K_d}{K_d + s_e} \cdot \frac{s_i - s_e}{K_d + s_i}, \tag{10}$$

where

$$k \equiv ADc_0/VL.$$

Assuming $s_i(0) = 0$, we readily integrate (10) to obtain

◆◆ $$\log\left(\frac{s_e - s_i}{s_e}\right) + \frac{s_i}{(K_d + s_e)} = -k\left[\frac{K_d}{(K_d + s_e)^2}\right]t. \tag{11}$$

According to (11) $s_i \to s_e$ as $t \to \infty$. If $s_e \ll K_d$, the second term on the left hand side above can be neglected, and (11) becomes

$$s_i = s_e[1 - \exp(-\lambda t)], \tag{12}$$

where

$$\lambda = kK_d/(K_d + s_e)^2. \tag{13}$$

If simple diffusion were the mechanism of transport through the membrane, then (12) would likewise hold with λ independent of s_e (see, for example, section 5.6 in Rubinow, 1975). The difference between the functional forms of $s_i(t)$ appearing in (11) and (12) is only apparent at small times and is difficult to observe experimentally. Hence, in order to assert that facilitated diffusion is the transport mechanism, it is essential to demonstrate the dependence of λ on s_e as displayed by (13).

The principal support for the existence of the facilitated diffusion conception lies in many observations of transport of simple sugars into human erythrocytes (reviewed in Stein, 1967). Although the transport data for any one monosaccharide can be supported by the simple model presented above, there appear to be many discrepancies between observation and theory when the data are reviewed *in toto* (Lieb & Stein, 1972;

Lefevre, 1975). This has lead to a number of modifications in the carrier concept, such as the assumption that the transport mechanism is asymmetric, or that the diffusion rate for the free carrier is different from that of the permeant–carrier complex (Kotyk, 1973). Even accepting that transport of a given permeant is carrier mediated, it is by no means clear that diffusion is the mechanism of transport of the carrier from the membrane's outer surface to the inner surface. The subject of membrane research is being actively pursued at the present time and no doubt new discoveries in the near future will help clarify the theoretical conceptions of membrane transport.

Wyman's equation for facilitated diffusion

The most thoroughly documented example of facilitated diffusion is the transport of oxygen by the carriers myoglobin and hemoglobin (for a review, see Wittenberg, 1970). These two proteins, the former found in great abundance in muscle, and the latter in red cells, play important roles in the transport of oxygen to tissue in mammals. Myoglobin is a protein, approximately one-fourth the size of hemoglobin, which contains one iron atom at a binding site with a very great affinity for oxygen. Hemoglobin contains four such subunits, although the affinities are quantitatively different from that of myoglobin.

The classical physiological explanation for the function of myoglobin in muscle is to serve as an oxygen store. However, such a function now appears to be important only in diving mammals such as whales, and a more recent proposal is that myoglobin serves primarily to facilitate the transport of oxygen. This proposal has come about principally through the discovery of Scholander (1960) and of Wittenberg (1959) of the phenomenon of facilitated diffusion of oxygen by hemoglobin and myoglobin, and its subsequent intensive quantitative investigation (Wittenberg, 1966).

The basic experiment that was performed was to separate two gas chambers by a section of millipore filter paper. In one chamber oxygen was maintained at a fixed partial pressure p_e, and in the other the oxygen partial pressure was maintained constant at a lower partial pressure p_i. (In practice, p_i was made to be zero.) The filter paper was soaked in an aqueous solution of carrier molecules (either myoglobin or hemoglobin). Hence, the two gas chambers could be considered to be separated by a solution slab of carrier molecules, the filter paper merely acting as a mechanical support for the solution. By collecting the oxygen that appeared on the low pressure side of the solution (and removing it), the steady state flux of oxygen through the slab could be measured. This basic experiment was repeated for different values of the pressure difference

$p_e - p_i$. If simple translational diffusion were the only mechanism of oxygen transport through the slab, then the resultant oxygen flux would be linearly related to $p_e - p_i$, in accordance with Fick's law. That is so because of Henry's law, which states that, at a solution–air interface, the concentration of a solute in the solution side of the interface is proportional to the partial pressure of the solute constituent in the gas. Thus,

$$s_e = \sigma p_e, \qquad s_i = \sigma p_i, \tag{14}$$

where s_e and s_i are substrate concentrations on the high and low pressure sides of the solution respectively, and σ is a constant of proportionality characteristic of the solute and solvent of the solution, called the solubility coefficient.

This linear relation is actually observed when the carrier hemoglobin is poisoned by exposure to cyanide, so that it is unable to react with oxygen. The observations are illustrated by the experimental points lying along the straight line through the origin labeled MetHb in Figure 6.3.1 in which oxygen current is plotted as a function of the pressure difference p between the two sides of the slab. However, when the hemoglobin is not

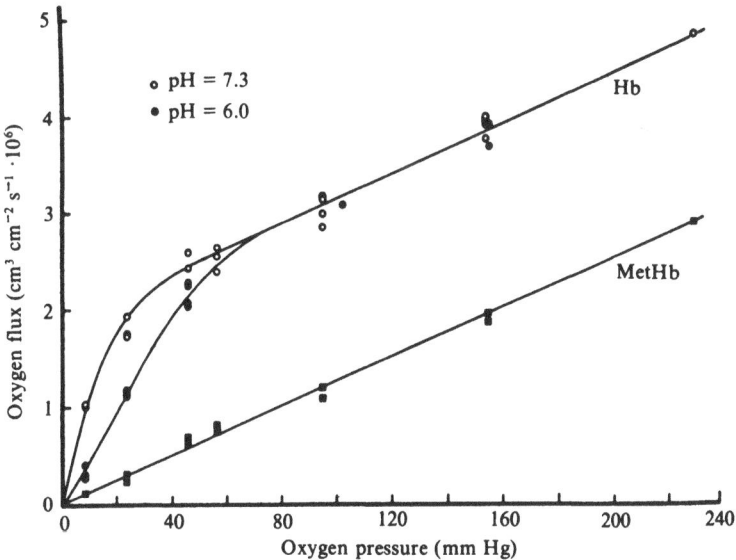

Figure 6.3.1. The curve labeled Hb represents the radioactive oxygen flux through a slab of hemoglobin solution as a function of the oxygen partial pressure, maintained equal on both sides of the slab. One side contained radioactive ^{18}O as well as ^{16}O, while the other side contained only non-radioactive ^{16}O. The lower curve labeled MetHb represents the oxygen flux through the same slab containing a solution of met-hemoglobin which is incapable of combining with oxygen. (From Hemmingsen, 1965, Fig. 14.)

so treated, there is a large increase in transported oxygen, also shown in the figure, which rapidly attains a fixed value, independent of p. The difference between the observed flux and that due to translational diffusion alone is called the **facilitated flux**.

The mathematical theory underlying these experiments was given by Wyman (1966), and leads to a single nonlinear ordinary differential equation for the concentration of oxygen in the slab. We shall present its derivation here (following Rubinow, 1975, pp. 336–337).

Let $s = s(x, t)$ denote ligand concentration, $c = c(x, t)$ denote carrier concentration, and $y = y(x, t)$ denote the concentration of ligand–carrier complex, at position x and time t. Here x measures distance in the slab of width L. The ligand and carrier are assumed to obey the reaction scheme of (1). Let D, D_c, and D_y denote the diffusion constants of ligand, carrier, and ligand–carrier complex, respectively, in the solution. Then s, c, and y are assumed to satisfy the one dimensional reaction–diffusion equations (see Section 6.1) in the region $0 < x < L$,

$$\partial s/\partial t = D(\partial^2 s/\partial x^2) - k_+sc + k_-y,$$

$$\partial c/\partial t = D_c(\partial^2 c/\partial x^2) - k_+sc + k_-y, \tag{15}$$

$$\partial y/\partial t = D_y(\partial^2 y/\partial x^2) + k_+sc - k_-y.$$

We shall impose the steady state condition and replace the left hand sides of the above equations with zeros. We shall retain the same symbols for the concentrations, but recognize that they are now functions of x alone, e.g.

$$0 = D(d^2s/dx^2) - k_+sc + k_-y, \tag{16a}$$

$$0 = D_c(d^2c/dx^2) - k_+sc + k_-y, \tag{16b}$$

$$0 = D_y(d^2y/dx^2) + k_+sc - k_-y. \tag{16c}$$

By adding (16b) and (16c) we see that

$$\frac{d^2}{dx^2}(D_cc + D_yy) = 0. \tag{17}$$

Hence, integrating twice we find that for any position x in the slab,

$$c(x) + y(x) = c_0. \tag{18}$$

Here c_0 is a constant, representing the uniform concentration of carrier in the slab in the complete absence of ligand, a quantity that is easily measured. In obtaining (18), we have utilized the condition that the carrier is much more massive than the ligand, so that $D_c = D_y$, and that the diffusive current density of both the carrier and complex solute species must vanish at the boundaries of the slab, or,

$$dc/dx = dy/dx = 0, \qquad x = 0, L. \tag{19}$$

The above relations constitute the boundary conditions that must be satisfied by c and y. Correspondingly, the boundary conditions satisfied by s are that

$$s(0) = s_e, \qquad s(L) = s_i. \tag{20}$$

Here we tacitly assume that the ligand, being a small molecule, is readily able to cross the air–solution interface.

Let us add the first and third of the equations of (16). A single integration yields the relation

$$D(ds/dx) + D_y(dy/dx) = -j, \tag{21}$$

where j is a constant, to be determined. The second term on the left represents the negative of the ligand–carrier current density at x. If we set $x = 0$ or $x = L$ in (21), we see that this quantity vanishes, so that j represents the steady state value of the current density of ligand entering or leaving the slab. Integrating (21) gives

$$Ds(x) + D_c y(x) = A - jx, \tag{22}$$

where A is a second constant of integration, to be determined.

Equations (18) and (22) determine c and y uniquely if s and the parameters A and j are known. Hence, it is only necessary to solve (16a) for s. Using the aforementioned two equations to eliminate c and y, (16a) becomes **Wyman's equation** for facilitated diffusion (Wyman, 1966)

$$\blacklozenge\blacklozenge \qquad D\frac{d^2 s}{dx^2} = k_+ c_0 s + \frac{1}{D_c}(k_+ s + k_-)(Ds - A + jx). \tag{23}$$

The required boundary conditions are, from (20) and from (19) and (21),

$$s(0) = s_e, \qquad s(L) = s_i, \qquad D(ds/dx) = -j, \qquad x = 0, L. \tag{24}$$

Wyman's equation is a nonlinear equation and it has not been found possible to solve it using analytical methods. However, Murray (1971) pointed out that for a certain range of values of the parameters appearing in (23), Wyman's equation is susceptible to approximate solution by the methods of **singular perturbation theory**. To recognize this readily, we introduce nondimensional variables and parameters into (23), whence it can be written as

$$\varepsilon (d^2 s'/dx'^2) = \beta s' + (s' + \alpha)(s' - A' + j'x'), \qquad 0 < x' < 1, \tag{25}$$

where

$$s/s_e = s', \qquad x/L = x', \qquad A/Ds_e = A', \qquad jL/Ds_e = j',$$

$$k_-/k_+s_e = \alpha, \qquad D_c c_0/Ds_e = \beta, \qquad D_c/k_+s_e L^2 = \varepsilon.$$

The usual perturbation procedure in solving a differential equation containing a small parameter is to seek an approximate solution in powers of the small parameter,

$$s'(x') = s'^{(0)}(x') + \varepsilon^{\frac{1}{2}} s'^{(1)}(x') + \varepsilon s'^{(2)}(x') + \dots \tag{26}$$

If we substitute (26) into (25) and neglect all positive powers of ε, it becomes an algebraic equation for $s'^{(0)}$. This solution will not in general satisfy the boundary conditions (24). This state of affairs is a consequence of the small parameter multiplying the highest derivative term in the differential equation, which is an indication that the special considerations of 'singular' perturbation theory have to be applied to it. It turns out that there are 'boundary layers' near $x = 0$ and $x = L$ where s' varies rapidly, and therefore derivative terms are not negligible, even if multiplied by a small parameter. By taking these boundary layers into account in an appropriate manner, singular perturbation theory shows how to correct the series solution (26) so that it gives a good approximation to the true solution even in the narrow boundary layers near $x = 0$ and $x = L$. The correction consists of terms which are not negligible in the boundary layer but are negligible everywhere else.

In applying Wyman's equation to the data of Wittenberg (1966) concerning the facilitated diffusion of oxygen by myoglobin and hemoglobin, Murray observed that $\varepsilon \sim 10^{-6}$, so that the singular perturbation criterion is easily satisfied. He found the zero order solution $s'^{(0)}$ and subsequently the first order solution $s'^{(1)}$ (Mitchell & Murray, 1973). Associated with the singular perturbation expansion (26) is the steady state current density j' expressed in the form

$$j' = j'^{(0)} + \varepsilon^{\frac{1}{2}} j'^{(1)} + \varepsilon j'^{(2)} + \dots. \tag{27}$$

It was found that $j'^{(0)}$ consists of two terms, which in terms of dimensional parameters reads

$$j^{(0)} = \frac{D(s_e - s_i)}{L} + \frac{D_c c_0 K_d (s_e - s_i)}{L(K_d + s_e)(K_d + s_i)}. \tag{28}$$

The first term on the right of (28) is the current density arising from translational diffusion of the ligand through the solution, and the second term represents the current density resulting from facilitated diffusion. We see that this latter term is identical with Widdas's formula.

Wittenberg (1966) repeated the measurement of the facilitated flux illustrated in Figure 6.3.1 for various values of the concentration of the carrier in the slab (the quantity c_0 of the theory), whether hemoglobin or myoglobin. He plotted the facilitated flux as a function of the concentration of heme (the iron containing subunit), equal to either myoglobin or $\frac{1}{4}$ hemoglobin concentration, as shown in Figure 6.3.2. The data for the

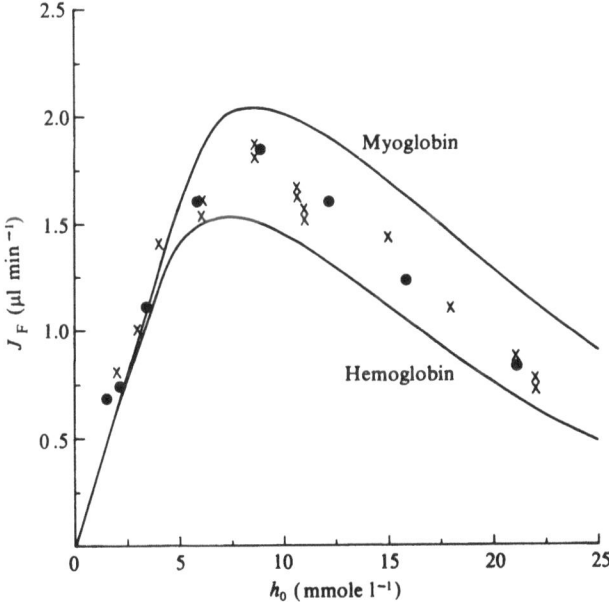

Figure 6.3.2. The facilitated flux of oxygen, J_F, through a slab of hemoglobin solution or myoglobin solution is shown as a function of heme concentration h_0 (proportional to c_0 of the theory). The solid circles (Mb) and crosses (Hb) represent the measurements of Wittenberg (1966). The solid lines are theoretical, and based on the singular perturbation solution to order ε. The quantity J_F is defined as j of the text minus the nonfacilitated flux given by the first term on the right-hand side of (28). (From Rubinow & Dembo, 1977.)

facilitated diffusional flux of oxygen by hemoglobin were found to be adequately representable by (28). (Solid line labeled Hemoglobin in Figure 6.3.2.) In making the comparison between theory and experiment, the cooperative behavior of hemoglobin is neglected and a value of K_d representing the 'average' behavior of the four binding sites is utilized. It is also necessary to determine the behavior of D and D_c as a function of c_0, the carrier concentration in the slab. This has been done and it is found that D_c in particular decreases rapidly with c_0. That explains why the facilitated flux as a function of c_0 attains a maximum and then decreases (see Figure 6.3.2). This is a triumph of the theory, because there are no adjustable parameters appearing in (28), all of them being determined by independent observations. However, it was found that $j'^{(0)}$ was not adequate to represent the measured facilitated diffusional flux of oxygen by myoglobin. In fact Wittenberg found (Figure 6.3.2) that the facilitated flux per mole of heme was approximately the same regardless of whether myoglobin or hemoglobin was utilized as the carrier, in spite of the fact that these proteins have different equilibrium constants and different diffusion constants. This observation has been dubbed 'Wittenberg's paradox'.

The resolution of this paradox was provided (Rubinow & Dembo, 1977) by calculation of the higher order terms of order $\varepsilon^{\frac{1}{2}}$ and ε appearing in (27), which are seen to be corrections to Widdas's formula. Although these terms could be expected to be negligible because of the small value of ε, and in fact are so for the case of hemoglobin, they are not negligible for the case of myoglobin. The cause for this can be ascertained, because the more mathematical derivation of the formula for the facilitated flux based on the reaction–diffusion equations (16), not only resolves the paradox, but also permits a more refined examination of the nature of the approximations made in the derivation of the solution.

Thus, the validity of the approximate solution (26) requires not only that $\varepsilon \ll 1$, but also that $\varepsilon \ll \alpha$ and $\varepsilon \ll \beta$. When these conditions are expressed in dimensional terms, they become, respectively,

$$L^2/D_c \gg 1/k_+s_e, \qquad L^2/D_c \gg 1/k_-, \qquad L^2/D \gg 1/k_+c_0. \tag{29}$$

These conditions state (see Section 6.1) that the time for diffusion of a carrier molecule across the slab must be very much greater than the reaction time for a carrier molecule to, first, associate with a ligand molecule, and secondly, dissociate from the ligand–carrier complex. In addition, the time for diffusion of a ligand molecule across the slab must be very much greater than the characteristic time for disappearance of ligand because of association. In other words, for facilitation to occur, ligand molecules must be able to jump on the carrier quickly and jump off it quickly. The parameter $1/\alpha$, which is a measure of the dissociation time, is about twenty times greater for myoglobin than for hemoglobin. This has the consequence that the coefficients $j'^{(1)}$ and $j'^{(2)}$ in (27) are very large, and their contribution, even when multiplied by $\varepsilon^{\frac{1}{2}}$ and ε, respectively, is significant in comparison with $j'^{(0)}$, sufficiently so that the data of Wittenberg for myoglobin can be understood quantitatively.

Part of the explanation of Wittenberg's paradox lies in unavoidable experimental errors in the observations of myoglobin facilitation, reported by Wittenberg, which reduced the magnitude of the facilitated flux in a systematic manner. That is why the experimental observations of myoglobin-facilitated oxygen flux lie below the theoretical curve in Figure 6.3.2. The physical explanation of the theoretical difference between myoglobin and hemoglobin facilitation, which is indeed a failure of the Widdas formula, lies in the fact that the advantage myoglobin possesses over hemoglobin in being able to diffuse more rapidly, as a consequence of its smaller molecular weight, is largely offset by its relative inability to release oxygen after association with it has occurred.

References

Hemmingsen, E. A. (1965). Accelerated transfer of oxygen through solutions of heme pigments. *Acta Physiol. Scand.* **64**, *Suppl.* **246**, 1–53.

Kotyk, A. (1973). Mechanisms of nonelectrolyte transport, *Biochim. Biophys. Acta* **300**, 183–210.

Lefevre, P. G. (1975). The present state of the carrier hypothesis. In *Current Topics in Membranes and Transport*, vol. 7, ed. F. Bronner and A. Kleinzeller, New York, Academic Press, pp. 109–215.

Lieb, W. R. & Stein, W. D. (1972). Carrier and non-carrier models for sugar transport in the human red blood cell. *Biochim. Biophys. Acta* **265**, 187–207.

Mitchell, P. J. & Murray, J. D. (1973). Facilitated diffusion: the problem of boundary conditions. *Biophysik* **9**, 177–90.

Murray, J. D. (1971). On the molecular mechanism of facilitated oxygen diffusion by haemoglobin and myoglobin. *Proc. R. Soc. Lond. B, Biol. Sci.* **178**, 95–110.

Rubinow, S. I. (1975). *Introduction to Mathematical Biology*, New York, Wiley.

Rubinow, S. I. & Dembo, M. (1977). The facilitated diffusion of oxygen by hemoglobin and myoglobin. *Biophys. J.* **18**, 29–42.

Scholander, P. F. (1960). Oxygen transport through hemoglobin solutions. *Science* **131**, 585–90.

Stein, W. D. (1967). *The Movement of Molecules across Cell Membranes*, New York, Academic Press.

Widdas, W. F. (1952). Inability of diffusion to account for placental glucose transfer in the sheep and consideration of the kinetics of a possible carrier transfer. *J. Physiol., Lond.* **118**, 23–39.

—— (1954). Facilitated transfer of hexoses across the human erythrocyte membrane. *J. Physiol., Lond.* **125**, 163–80.

Wittenberg, J. B. (1959). Oxygen transport: a new function proposal for myoglobin. *Biol. Bull., Woods Hole* **117**, 402.

—— (1966). The molecular mechanism of hemoglobin – facilitated oxygen diffusion. *J. Biol. Chem.* **241**, 104–14.

—— (1970). The myoglobin-facilitated oxygen diffusion: role of myoglobin in oxygen entry into muscle. *Physiol. Rev.* **50**, 559–636.

Wyman, J. (1966). Facilitated diffusion and the possible role of myoglobin as a transport mechanism. *J. Biol. Chem.* **241**, 115–21.

6.4 MORPHOGENETIC PATTERNS AND REACTION-DIFFUSION EQUATIONS

The main purpose of this section is to examine theoretical developments which provide support for the suggestion that spatial structures may arise in developing organisms through the combined agency of chemical reaction and diffusion. The formation of such structures is called **morphogenesis**. Our survey will not be comprehensive, but enough references will be given to afford an entry into the literature. Also see the review by Nicolis & Prigogine (1977).

Turing instabilities

Accounts of morphogenetic theory rightly begin with the outstanding work of Turing (1952). He postulated that biological form is underlain by a pre-pattern constituted by certain critical chemicals called **morphogens**. This suggestion was not novel. The main contribution of Turing's paper was to demonstrate mathematically how biologically meaningful patterns of morphogens could arise from a uniform assemblage of cells. The essence of his argument will be given in the following paragraphs.

Consider a line of idealized cells whose centers are located at the points $x = nh$, where h is the width of the cells and $n = 0, 1, 2, \ldots$. It will be assumed that only variations in x need be taken into account, although generalization to two and three dimensions presents no difficulties of principle. Two morphogens will be considered. Let their respective average concentrations in the cell centered at $x = nh$ be denoted by $C_n^{(1)}$ and $C_n^{(2)}$. These concentrations will be deemed to change by chemical reaction and by passage of chemical between adjacent cells at a rate proportional to the concentration difference between these cells. Concentrations will thus be governed by the equations

$$dC_n^{(1)}/dt = R^{(1)}(C_n^{(1)}, C_n^{(2)}) - P_1 \cdot [C_n^{(1)} - C_{n-1}^{(1)}]$$

$$+ P_1 \cdot [C_{n+1}^{(1)} - C_n^{(1)}], \tag{1a}$$

$$dC_n^{(2)}/dt = R^{(2)}(C_n^{(1)}, C_n^{(2)}) - P_2 \cdot [C_n^{(2)} - C_{n-1}^{(2)}]$$

$$+ P_2 \cdot [C_{n+1}^{(2)} - C_n^{(2)}]. \tag{1b}$$

Here $R^{(1)}(\ ,\ )$ and $R^{(2)}(\ ,\ )$ are functions that denote intercellular reaction rates, while P_1 and P_2 are constant membrane permeabilities. If cells are numbered from left to right, the first of the two permeability terms gives the rate at which the morphogen enters the n-th cell from the left, the second of these terms gives the rate at which material passes out of the n-th cell into the cell at its right. (To check that the signs are correct, we note for example that the term $-P_1[C_n^{(1)} - C_{n-1}^{(1)}]$ properly gives a negative entry rate into the n-th cell if $C_n^{(1)} > C_{n-1}^{(1)}$.)

Let us limit consideration to N cells, for example by identifying the first cell with the $(N+1)$-st (corresponding to cells arranged on a circle) or by assuming that there is no flux from the zero-th and the $(N+1)$-st cells, or that either of these cells is actually a bath held at some fixed concentration. As Turing did, one can then investigate the solutions of the resulting $2N$ equations plus appropriate auxiliary conditions. Here we shall pursue another possibility, also considered by Turing, by assuming that the cell width h is sufficiently small to permit the following Taylor approximations near $x = nh$ (compare (A.1.30)).

$$C_{n+1}^{(i)} \equiv C^{(i)}(x+h)$$

$$\approx C^{(i)}(x) + h\left(\frac{\partial C^{(i)}}{\partial x}\right) + \frac{h^2}{2}\left(\frac{\partial^2 C^{(i)}(x)}{\partial x^2}\right), \tag{2a}$$

$$C_{n-1}^{(i)} \equiv C^{(i)}(x-h)$$

$$\approx C^{(i)}(x) - h\left(\frac{C^{(i)}(x)}{\partial x}\right) + \frac{h^2}{2}\left(\frac{\partial^2 C^{(i)}(x)}{\partial x^2}\right). \tag{2b}$$

Upon substituting $(2a, b)$ into (1), one finds that (1) can be approximated by the continuous equations

$$\frac{\partial C^{(1)}}{\partial t} = R^{(1)}(C_1, C_2) + D_1\left(\frac{\partial^2 C^{(1)}}{\partial x^2}\right), \tag{3a}$$

$$\frac{\partial C^{(2)}}{\partial t} = R^{(2)}(C_1, C_2) + D_2\left(\frac{\partial^2 C^{(2)}}{\partial x^2}\right), \tag{3b}$$

where the effective diffusion constants D_i are given by

$$D_i = h^{-2}P_i, \qquad i = 1, 2.$$

These **reaction–diffusion equations** should be sufficiently accurate to describe patterns whose scale is large compared to a single cell diameter. They will also be valid in cases where the cell walls pose no appreciable barrier to diffusion. (A pair of reaction–diffusion equations has already been presented, in another context, in Section 5.5.)

Considerable attention has been paid to the possibility that patterns can be generated if there is a sink for morphogens at one end of a line of cells and a source at the other end. (See for example Lawrence, Crick & Munro, 1972.) Here we shall concentrate on the spontaneous generation of patterns from an originally homogeneous or near-homogeneous distribution of morphogen. In particular, we shall assume that $(3a, b)$ have a time-invariant and spatially uniform solution

$$C^{(1)} = C_0^{(1)}, \qquad C^{(2)} = C_0^{(2)};$$

$$C_0^{(1)} \quad \text{and} \quad C_0^{(2)} \quad \text{constants.} \tag{4}$$

Upon substitution of (4) into (3), we see that the existence of a uniform solution requires that

$$R^{(1)}(C_0^{(1)}, C_0^{(2)}) = 0, \qquad R^{(2)}(C_0^{(1)}, C_0^{(2)}) = 0. \tag{5}$$

At first it may seem difficult to envisage how a nonuniform pattern can arise from what is essentially a collection of uniform cells. Let us suppose, however, that at some time $t = 0$ the morphogen concentration is not exactly uniform. Small deviations or **perturbations** from perfect uniformity are expected to be present in any system. If we can find conditions under which such deviations become larger, then we can anticipate that some pattern of nonuniform concentrations would arise from an initial state that is virtually uniform. Later we shall see that the pattern may be much more orderly than the random perturbations that initiated it.

Analysis of the growth or decay of perturbations is considered in a branch of mathematics called **stability theory**. An illustration of the main ideas of this theory is presented in Section 6.5, in the context of slime mold aggregation. The same ideas apply here. In this section, therefore, we shall content ourselves with a general description of the results. Readers who are interested in learning the mathematical procedure are urged to read Section 6.5 and then to attempt the relevant exercises. Of course, many details can be found in the literature, including the original paper of Turing (1952). (In this chapter, we concentrate on partial differential equations. Stability theory for ordinary differential equations is considered at some length in Chapter 3 and in Appendix A.3.)

It turns out to be sufficient to consider situations wherein the perturbations have the spatial dependence of the cosine or sine. We shall consider for definiteness the case

$$C^{(i)}(x, t) = C_0^{(i)} + C^{(i)\prime}(x, t),$$

$$C^{(i)\prime} = A_i(t) \cos qx, \qquad i = 1, 2, \tag{6a, b}$$

where $2\pi/q$ is the wavelength of the perturbation. Once behavior of sinusoidal perturbations has been calculated, the theory of Fourier series allows one to determine the behavior of arbitrary perturbations by representing them as a generalized sum of sinusoidal terms.

Substitution of $(6a)$ into the governing equations (3), keeping in mind (5), 'linearizing' the resulting equations to take advantage of the fact that only small disturbances from steady state are considered and then making the assumption $(6b)$ (Exercise 1) – all this leads to ordinary differential equations for the amplitudes A_i of the form

$$dA_1/dt = a_{11}A_1 + a_{12}A_2 - D_1q^2A_1, \tag{7a}$$

$$dA_2/dt = a_{21}A_1 + a_{22}A_2 - D_2q^2A_2. \tag{7b}$$

Here, the a_{ij} are constants that can be determined from the reaction functions $R^{(1)}$ and $R^{(2)}$. These functions in turn will generally depend on several parameters involving reaction rates, temperature, etc. We shall single out one parameter, denoted by θ, for special attention. We write $a_{ij} = a_{ij}(\theta)$ to emphasize the dependence of the coefficients on this parameter. We wish to know whether the amplitude functions $A_1(t)$ and $A_2(t)$ grow or decay with time t. Precisely this kind of question arises in Chapter 3, in the context of control theory. Of the various methods considered there, the easiest to utilize is based on guessing that $(7a)$ and $(7b)$ have solutions

$$A_1(t) = \alpha_1 \exp(\sigma t), \qquad A_2(t) = \alpha_2 \exp(\sigma t), \tag{8}$$

where α_1, α_2, and σ are constants. Substitution of (8) into (7) shows that if we wish to avoid the uninteresting 'trivial case' when the constants α_1 and α_2 are both zero, then σ must satisfy a quadratic equation of the form

$$\sigma^2 + b(\theta, q)\sigma + c(\theta, q) = 0. \tag{9}$$

We have made explicit the dependence of the coefficients in (8) on the parameter combination θ and on the perturbation wavelength $2\pi/q$.

It follows at once from (9) that if $\sigma = 0$ then

$$c(\theta, q) = 0. \tag{10}$$

This equation implicitly defines a relationship between θ and q. Typically, the graph of this relationship has a minimum as is shown in Figure 6.4.1, with positive (negative) values of σ being confined to the region above (below) the curve $c = 0$. Suppose that development begins with θ at some low value θ_0. Then, as can be seen from Figure 6.4.1, σ will be negative no matter what the perturbation wavelength $2\pi/q$. The uniform state is then called **stable**. But suppose further that θ changes very slowly, as a result of various cellular processes, until it attains a value just above the critical value θ_c corresponding to the minimum of the curve $c(\theta, q) = 0$. The uniform state is then **unstable**, for there are disturbances whose amplitude would grow exponentially in time. The wavelength of these disturbances is near $2\pi/q_c$, where q_c is shown in Figure 6.4.1.

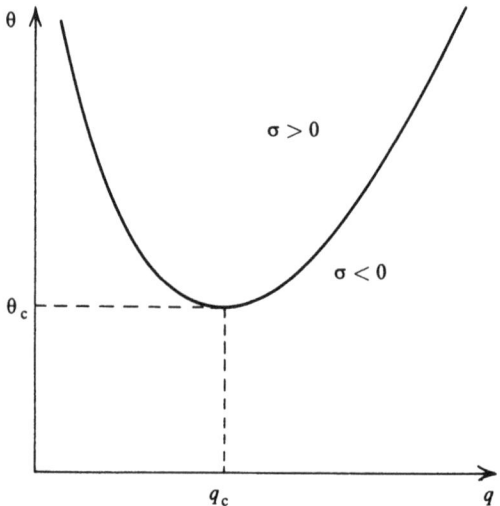

Figure 6.4.1. Domains of stability ($\sigma > 0$) and instability ($\sigma < 0$) in a parameter plane with coordinates q (where $2\pi/q$ is the wavelength of the sinusoidal perturbation) and θ (where θ is some parameter whose slow change can bring about instability). For q_c and θ_c, see text.

We have reached the essence of Turing's discovery. Reaction–diffusion equations can be such that a slow change in one or more parameters can bring about a situation where a disturbance of a certain critical wavelength $2\pi/q_c$ can begin to grow. This wavelength sets the size of the pattern that is generated by the Turing instability mechanism. The essence of a typical computation of the critical wavelength can be seen by reading Exercise 3.

The conditions for diffusive instability

We have seen that when the parameter θ is slightly above θ_c then the only growing disturbances are those with wavelength near $2\pi/q_c$. All disturbances characterized by other wavelengths decay; in particular there is a decay of disturbances of infinite wavelength, with $q = 0$. But setting $q = 0$ in (7) is equivalent to setting $D_1 = D_2 = 0$, so that solutions to (7) will decay if D_1 and D_2 are set equal to zero. In other words, the Turing instability patterns arise in situations that are stable in the absence of diffusion. Put another way, Turing **diffusive instabilities** (as they are called) are 'caused' by diffusion. This is a somewhat paradoxical result, for diffusion is normally expected to smooth concentration irregularities, not to enhance them.

A physical explanation of how diffusion can bring about instability was provided by Segel & Jackson (1972). The explanation rests on the fact that analysis shows necessary conditions for diffusive instabilities of the system (7) to be

$$a_{11}a_{22} < 0, \qquad a_{21}a_{12} < 0. \tag{11a, b}$$

For definiteness, let us satisfy these conditions by taking

$$a_{11} = -\bar{a}_{11}, \qquad a_{21} = -\bar{a}_{21};$$
$$\bar{a}_{11}, a_{12}, \bar{a}_{21}, a_{22} \text{ all positive.} \tag{12}$$

Then the disturbance equations (7), in the absence of diffusion, take the form

$$dA_1/dt = -\bar{a}_{11}A_1 + a_{12}A_2, \qquad dA_2/dt = -\bar{a}_{21}A_1 + a_{22}A_2. \tag{13}$$

We want solutions of these equations to decay (stability), and growth to occur only by addition of diffusion to the system. For stability, Exercise 2 shows that it is necessary that

$$\bar{a}_{11} > a_{22}, \qquad \bar{a}_{21}a_{12} > \bar{a}_{11}a_{22}. \tag{14a, b}$$

Note that solutions to the system (13) decay even though the perturbation A_2 by itself (with $A_1 = 0$) would grow exponentially. This is because the perturbation A_1 not only decays by itself, but also forces the A_2 perturbation to zero through the coupling terms $a_{12}A_2$ and $-\bar{a}_{21}A_1$. Because of these properties the name **destabilizer** or **activator** is given to A_2 and **stabilizer** or **inhibitor** to A_1.

We next observe that the inequality in (14) is reversed if $\bar{a}_{11}$ is so large that

$$\bar{a}_{11} > \bar{a}_{21}a_{12}/a_{22}. \tag{15}$$

Too rapid decay of the inhibitor brings about instability of the system. The reason is that the inhibitor disappears so fast that it has no time to exercise its stabilizing influence on the activator.

Now we can see how diffusion can destabilize the system. With the addition of diffusion, (13) becomes

$$dA_1/dt = -(\bar{a}_{11} + q^2 D_1)A_1 + a_{12}A_2,$$
$$dA_2/dt = -\bar{a}_{21}A_1 + (a_{22} - q^2 D_2)A_2. \tag{16}$$

Diffusion effectively stabilizes both the activator and the inhibitor, by making greater the decay rate of both these chemicals considered by themselves. If the latter effect is sufficiently greater than the former, then we can have a too rapid decay of the inhibitor by means of which the coupled system can be destabilized. It can be shown (Exercise 3a) that a necessary condition for this to happen is

$$\frac{D_2}{a_{22}} < \frac{D_1}{a_{11}}. \tag{17a}$$

To interpret (17a), let us introduce the notion of a chemical's **diffusion range**, defined as $\sqrt{(\tau D)}$ where D is the chemical's diffusion constant and τ is a time over which reaction significantly changes the concentration of the chemical. By (6.1.19) the range is an estimate of the distance that the chemical will diffuse during the time τ. We now see that (17a) can be regarded as stating that the range of the inhibitor must be less than that of the activator, for $1/a_{22}$ and $1/\bar{a}_{11}$ are the e-folding times for the uncoupled activator and inhibitor perturbations (see Exercise 4). Requirement (14a) for diffusive instability states that if interaction is neglected, the inhibitor must decay faster than the activator grows. Taken together (14a) and (17a) imply the necessary condition that

$$D_1 > D_2, \tag{17b}$$

i.e. that the inhibitor must diffuse faster than the activator. The necessary and sufficient condition for diffusive instability is provided in (25), as part of Exercise 3. Condition (25) implies both (17a) and (17b).

Recall that the necessity for a two-constituent system permitting diffusive instability to consist of an activator and an inhibitor stems from requirement (11a) that a_{11} and a_{22} have opposite signs. Suppose $A_1 \equiv 0$. If $a_{22} > 0$ then

$$dA_2/dt = a_{22}A_2,$$

showing that the activator perturbation A_2 grows if the rest of the system is held at its steady state. Such self-stimulated growth is given the name **autocatalysis**. From the point of view of thermodynamics, it is interesting to know that since Turing patterns require autocatalysis they can only occur if the system is sufficiently far from chemical equilibrium. This point has been emphasized by Prigogine and his colleagues (see Glansdorff & Prigogine, 1971). They associate the term **dissipative structure** with Turing patterns, to emphasize the need for some sort of dissipation in the system, as a condition for structure to arise from the initially near-uniform state. The dissipation need not be provided by diffusion. Though this point has not been extensively investigated, there is little doubt that the same kinds of patterns could arise if chemicals moved by a suitable form of facilitated or active transport instead of by diffusion.

The simulations of Gierer and Meinhardt

Considerable progress in demonstrating the remarkable pattern-forming capabilities of reaction–diffusion systems has been reported in a series of papers by A. Gierer and H. Meinhardt. This work has been well reviewed by Meinhardt (1978) so that we only mention a few of its major points. Thus, from here on our exposition takes on the character of a rather rapid survey.

In their formulation of the requirements for pattern forming systems, Gierer & Meinhardt (1972, 1974) argue that an activator–inhibitor interaction that is stable if concentrations are spatially uniform can become unstable to a local activator concentration if the inhibitor diffuses much faster than the activator. Although activator catalyzes the formation of its own inhibitor, the latter rapidly diffuses away and permits activator concentration to grow. There is a formal analogy to pattern formation by lateral inhibition in brain function.

To follow these arguments a little further, let us adopt the Gierer–Meinhardt notation and assume that the chemical interaction of activator a and inhibitor h is governed by the equations

$$\dot{a} = f(a, h), \qquad \dot{h} = g(a, h).$$

Gierer & Meinhardt argue that rapid diffusion will cause equilibration of inhibitor to a level determined by average activator concentration. They therefore regard inhibitor concentration $\hat{h}(a)$ as given implicitly, when h and a are near steady state values, by

$$g[a, \hat{h}(a)] = 0. \tag{18}$$

Thus although the condition $\partial f/\partial a > 0$ will provide magnification of a local activator peak, Gierer & Meinhardt assert that over the relatively large diffusion range of the inhibitor concentration there will be stability of the average activator concentration, and hence a pattern, if the following inequality holds at steady state conditions:

$$\frac{\mathrm{d}f[a, \hat{h}(a)]}{\mathrm{d}a} < 0, \qquad \text{i.e. } \frac{\partial f}{\partial a} + \frac{\partial f}{\partial h} \cdot \frac{\mathrm{d}\hat{h}}{\mathrm{d}a} < 0. \tag{19a, b}$$

But from (18)

$$\frac{\partial g}{\partial a} + \frac{\partial g}{\partial h} \cdot \frac{\mathrm{d}\hat{h}}{\mathrm{d}a} = 0. \tag{20}$$

Elimination of $\mathrm{d}\hat{h}/\mathrm{d}a$ from (19b) and (20) yields

$$\frac{\partial f}{\partial a} \cdot \frac{\partial g}{\partial h} - \frac{\partial f}{\partial h} \cdot \frac{\partial g}{\partial a} < 0.$$

Although in different notation, this inequality is fundamentally the same as (15), a necessary condition for instabilities of the type considered by Turing (1952).

We have seen how necessary conditions for Turing instabilities were arrived at by Meinhardt & Gierer by heuristic combination of the concepts of autocatalysis and lateral inhibition. We shall illustrate below how Meinhardt & Gierer used these conditions to construct various equations that exhibit a variety of biologically interesting properties in as simple a way as possible (consistent with being interpretable on a molecular basis).

The early work of Gierer & Meinhardt brought out the important point that there is no need to hypothesize that patterns arise from a uniform state. A slight existing inhomogeneity (as is found for example in the egg cells of many species) may 'fire' the system into a patterned array of chemical concentrations if conditions are similar to those permitting instabilities of the Turing type.

Turing's (1952) paper was mainly concerned with analytical stability calculations of the kind mentioned at the beginning of this section. Under conditions where such calculations predict instability, the perturbations soon grow so large that the simplifying linearization procedure ceases to be valid. Turing himself died soon after he wrote his 1952 paper, but even in that work there appeared some analytical and numerical results on the fate of perturbations when they cease to be small. These results showed that either the nonlinear terms will eventually act to limit the growth of the initially small disturbances or disturbance growth will continue until it is stopped by total depletion of one of the chemicals that is participating in the reaction.

Gierer & Meinhardt's work was all done with the aid of the computer, so their research was never hindered by the severe mathematical difficulties that arise in the analytical treatment of perturbations which are too large to permit linearization of the governing equations. Moreover, the Gierer–Meinhardt equations were designed so that perturbation growth is intrinsically limited by some form of saturation. Incidentally, the numerical analysis necessary for computer solutions of partial differential equations essentially replaces partial differential equations of the form (3) by a system of ordinary differential equations of the form (1). Thus the Gierer–Meinhardt results can be viewed as arising from the interplay of cells containing chemicals that react intracellularly and also permeate to neighboring cells.

When considering spatial inhomogeneities in concentration that can arise in activator–inhibitor systems, it is convenient to distinguish situations in which a typical dimension of the region in question is of the order of the critical wavelength $2\pi/q_c$ from other situations in which this dimension is considerably larger than the critical wavelength. In the later case, periodic or near-periodic patterns typically result from the instability of the uniform state. In the former case, however, only a 'piece' of such a pattern can be accomodated in the space available, so that a typical concentration graph will show a high concentration at one end of the region and a low concentration at the other. For example, see Figure 6.4.2a, which is concerned with activator and inhibitor concentrations $A(x, t)$ and $H(x, t)$ governed by

$$\frac{\partial A}{\partial t} = \frac{c\rho(x)A^2}{(1+\kappa A^2)H} - \mu A + D_a \frac{\partial^2 A}{\partial x^2} + \rho_0 \rho(x), \tag{21a}$$

$$\frac{\partial H}{\partial t} = c\rho'(x)A^2 - \mu A + D_h \frac{\partial^2 H}{\partial x^2} + \rho_1. \tag{21b}$$

Here $\rho(x)$ and $\rho'(x)$ are source distributions of activator and inhibitor.

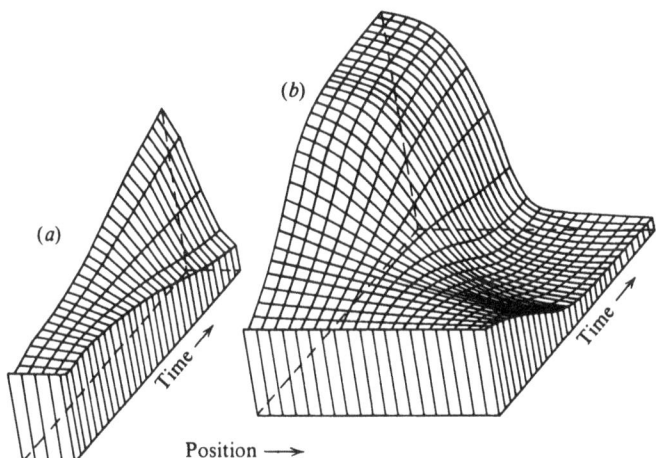

Figure 6.4.2. Activator concentrations from computer solutions of (21) (from Meinhardt, 1978, Fig. 3). (a) Development of a gradient of activator concentration from a nearly uniform state. (b) Same as (a), except for a larger region. The activator peak occupies approximately the same fraction of the region as in (a), affording an example of regulation.

The importance of the result depicted in Figure 6.4.2a lies in the fact that it shows how a gradient in chemical concentration can arise quite simply and naturally, without the need to postulate a source in the region where the chemical concentration is high and a sink where it is low. We have already mentioned that it is a classical hypothesis of embryology that cells can somehow interpret such a gradient to form various patterns.

Regulation of a developmental pattern occurs if the pattern occupies roughly the same proportion of the organism irrespective of its absolute size. Figure 6.4.2*b* shows that the activated region occupies roughly the same proportion of a piece of 'tissue' four times the size of its counterpart in Figure 6.4.2*a*. Other figures by Meinhardt (1978) show that regeneration of a correctly proportioned activator region can occur from either half of the 'tissue' of Figure 6.4.2.

Periodic structures can be generated in domains that are large compared to $2\pi/q_c$. Figure 6.4.3 shows three different arrangements of

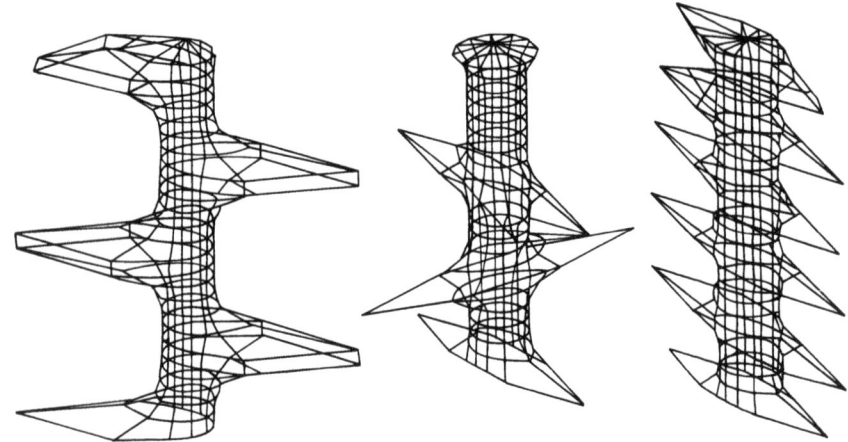

Figure 6.4.3. Various arrangements of activator peaks, with the different patterns emerging because of different relations between such factors as size, inhibitor range, and growth speed (from Meinhardt, 1978, Fig. 5).

activator peaks, resulting from different hypotheses concerning reaction and growth on a stem-like structure. If high activator concentration triggered leaf formation, such patterns could be responsible for the various leaf arrangements in plants. Meinhardt also demonstrates that simple reaction–diffusion systems can generate quite regular patterns that could be the precursors of bristles in insects or stomata in leaves.

In previous examples, it was high activator concentration that was hypothesized to trigger some form of differentiation. In the example to be discussed now (Meinhardt, 1976), it was found necessary to separate the function of reinforcing a deviation from uniformity (activation) from the function of triggering differentiation. Here, then, cell differentiation will be regarded as triggered by the switching of a substance Y from low to high concentration. Instead of an inhibitor, a similar role is played by a substance S that is depleted from an initially high value in the course of activator accumulation. Growth is simulated by periodic addition of new cells. The following equations are solved numerically, in a region that is large compared to the critical wavelength.

$$\partial A/\partial t = 0.008A^2S - 0.04A + 0.0065[(\partial^2 A/\partial x^2) + (\partial^2 A/\partial y^2)],$$
(22a)

$$\partial S/\partial t = 0.05 - 0.008A^2S - 0.25\,YS$$
$$+ 0.18[(\partial^2 S/\partial x^2) + (\partial^2 S/\partial y^2)],$$
(22b)

$$\partial Y/\partial t = 0.00032A - 0.1\,Y + Y^2/(1 + 10\,Y^2).$$
(22c)

Figure 6.4.4 depicts how a pattern of branching filaments is automatically generated as the tissue grows. Meinhardt (1976) has discussed the application of this and other filamentous patterns to the phenomenon of venation in leaves. G. J. Mitcheson (1979, unpublished) has proposed a somewhat different theory of venation, based on experiments of Sachs (1978).

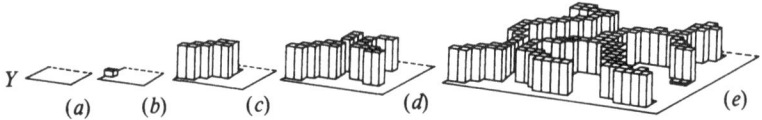

Figure 6.4.4. Computer solutions of (22). A branching pattern of differentiated cells (high 'Y' concentration) in a three-chemical system (from Meinhardt, 1976, Fig. 3).

It should be stressed that it is by no means easy to find systems with all the various properties exhibited in Figure 6.4.2–6.4.4. Nonlinear equations are such that if their structure or parameters are changed to improve the realism of one property then, more often than not, other properties turn out to be far less suitable than they were before the change. Much intuition, ingenuity, and computer time must be invested before appropriate systems can be found.

Analytical methods in nonlinear stability analysis

Although the computer is indispensible for demonstrating the full richness of complicated systems behavior, analytical investigations can often deepen understanding of the interplay of influences in nonlinear systems. The paper by Fife (1976) is a good example of progress made in the analysis of reaction–diffusion systems. By assuming that the activator diffuses much more slowly than the inhibitor, Fife is able to show something of the generality of pattern forming mechanisms in reaction–

diffusion systems. An example of a recent contribution to this line of investigation is the paper by Mimura & Nishiura (1979).

Experimental confirmation

Perhaps the most successful confrontation of experiment and reaction–diffusion theory occurs in the study of facilitated transport (see Section 6.3). Another instance of agreement between theory and experiment can be found in the study of a system where a membrane containing the immobilized enzyme papain is used to hydrolyse benzoyl-L-arginine ethyl-ester. The reaction liberates hydrogen, and is autocatalytic for a range of hydrogen concentration. Computer calculations predicted that for a certain range of parameters there would be a steep jump in hydrogen concentration which would move periodically back and forth across the membrane (Caplan, Naparstek & Zabusky, 1973). Using the theory to guide the choice of experimental conditions, Naparstek, Thomas & Caplan (1973) experimentally confirmed the existence of the oscillation. The predicted period was 18 s; this compares well with an observed period of 20 s.

The most dramatic instance of spatial pattern in a reaction–diffusion system is found in the Belousov–Zhabotinski reaction. The paper of Winfree (1974) contains an authoritative and insightful account of the main phenomena, made visible by color changes. These phenomena include concentric 'target' waves, spiral waves, and colliding waves annihilating each other. There have been a number of excellent mathematical studies of the Belousov–Zhabotinski patterns, but much remains open. It is not strictly possible to regard all this theory as confirmed by experiment, since the actual chemistry is very complicated and is thus much simplified in the various mathematical treatments. The recent monograph of Tyson (1976) provides a good introduction to the literature.

The principal biological evidence in favor of reaction–diffusion prepatterns comes from the variety of experiments in various fields that can be economically and elegantly explained by the theory. Perhaps the most impressive single collection of such explanations can be found in the survey paper of Meinhardt (1978), to which we have extensively alluded. Another recent example of the use of reaction–diffusion theory can be found in a theoretical study, by Kauffman, Shymko & Trabert (1978), of the formation of lines of clonal restriction in *Drosophila* eggs and imaginal discs.

Indisputable evidence for reaction–diffusion patterns would be the identification of participating chemicals and confirmation that the experimentally determined modes of reaction and diffusion (or other types of transport) would indeed produce the observed patterns. Specific chemicals that play key roles in development have been identified: for example, the auxins and gibberellins in plants (see Sachs, 1978) and cyclic-AMP in slime mold (Darmon, Brachet & Pereira da Silva, 1975; Gerisch, Fromm, Huesgen & Wick, 1975). In Hydra, substances have been found that promote head and tentacle formation (Schaller, 1973) and suppress bud formation (Berking, 1977). But as yet, there has been no decisive evidence that any chemical can truly be regarded as a morphogen in the sense required by the pattern formation theories. Still, one must agree with the first biologist who made a serious effort to assess the relevance of Turing's (1952) paper that 'there are many morphological and histological developments, of very different kinds, which could be explained by the theory and which, thus far, have not been satisfactorily explained by any other theory' (Wardlaw, 1955).

Exercises

1 Carry out the calculations necessary to obtain (7). In particular show that the constants a_{ij} are given by

$$a_{ij} = \frac{\partial R_i(C_1, C_2)}{\partial C_j}\bigg|_{C_1 = C_1^{(0)}, C_2 = C_2^{(0)}}.$$

The calculations require the generalization of Taylor series to two variables

$$R(C_1 + h, C_2 + k) = R(C_1, C_2) + \left[\frac{\partial R(C_1, C_2)}{\partial C_1}\right]h$$

$$+ \left[\frac{\partial R(C_1, C_2)}{\partial C_2}\right]k + \ldots,$$

where the omitted terms are negligible for our purposes here.

2a Show that both roots of the quadratic equation

$$\sigma^2 + b\sigma + c = 0,$$

are either negative or have negative real parts if and only if $b > 0$, $c > 0$.

2b Use the result of part (a) to show that (14) gives necessary and sufficient conditions for the decay of solutions to (13).

3a Show (using Exercise 2a and keeping in mind (14)) that (16) has solutions that grow with time if

$$Q(k^2) < 0, \tag{23}$$

484 *Partial differential equations*

where

$$Q(q^2) = D_1 D_2 q^4 - (D_1 a_{22} - D_2 \bar{a}_{11})q^2 + \bar{a}_{21}a_{12} - \bar{a}_{11}a_{22}.$$

Thereby deduce that (17*a*) is a necessary condition for diffusive instability.

3*b*   Demonstrate that (23) will hold if and only if $Q(q^2)$ is negative at its minimum. Show that this minimum occurs when $q^2 = q_m^2$, where

$$q_m^2 = (D_1 a_{22} - D_2 \bar{a}_{11})/2D_1 D_2. \tag{24}$$

Deduce the final form of the instability condition

$$D_1 a_{22} - D_2 \bar{a}_{11} > 2(D_1 D_2)^{\frac{1}{2}}(a_{12}\bar{a}_{21} - \bar{a}_{11}a_{22})^{\frac{1}{2}}. \tag{25}$$

4   Consider the problem $dA/dt = cA$, $A(0) = A_0$, c and A_0 constants. Show that $A/A_0 = e(e^{-1})$ at time $t = |c^{-1}|$ if c is positive (negative). This time for increase (decrease) by a factor e is called the *e*-**folding time**.

References

Berking, S. (1977). Bud formation in *Hydra* – inhibition by an endogenous morphogen. *W. Roux' Archiv.* **181**, 215–25.

Caplan, S. R., Naparstek, A. & Zabusky, N. (1973). Chemical oscillations in a membrane. *Nature, Lond.* **245**, 364–6.

Darmon, M., Brachet, P. & Pereira da Silva, L. H. (1975). Chemotactic signals induce cell differentiation in *Dictyostelium discoideum*. *Proc. Nat. Acad. Sci., USA* **72**, 3163–6.

Fife, P. C. (1976). Pattern formation in reacting and diffusing systems. *J. Chem. Phys.* **64**, 645–68.

Gerisch, G., Fromm, H., Huesgen, A. & Wick, U. (1975). Control of cell contact sites by cyclic AMP pulses in differentiating *Dictyostelium* cells. *Nature, Lond.* **255**, 547–9.

Gierer, A. & Meinhardt, H. (1972). A theory of biological pattern formation. *Kybernetik* **12**, 30–9.

——— (1974). Biological pattern formation involving lateral inhibition. *Lectures on Math. in the Life Sci.* **7**. Providence, R.I., Amer. Math. Soc., pp. 163–83.

Glansdorff, P. & Prigogine, I. (1971). *Thermodynamic Theory of Structure Stability and Fluctuations*. New York, Wiley Interscience.

Kauffman, S. A., Shymko, R. M. & Trabert, K. (1978). Control of sequential compartment formation in *Drosophilia*. *Science* **199**, 259–70.

Lawrence, P. A., Crick, F. H. C. & Munro, M. (1972). A gradient of positional information in an insect *Rhodnius*. *J. Cell Sci.* **11**, 815–22.

Meinhardt, H. (1976). Morphogenesis of lines and nets. *Differentiation* **6**, 117–23.

―――― (1978). Models for the ontogenetic development of higher organisms. *Rev. Physiol. Biochem. Pharmacol.* **80**, 47–104.

Mimura, M. & Nishiura, Y. (1979). Spatial patterns for an interaction–diffusion equation in morphogenesis. *J. Math. Biol.* **7**, 243–63.

Naparstek, A., Thomas, D. & Caplan, S. R. (1973). An experimental enzyme–membrane oscillator. *Biochem. Biophys. Acta* **323**, 643–6.

Nicolis, G. & Prigogine, I. (1977). *Self Organization in Non-Equilibrium Systems.* New York, Wiley-Interscience.

Sachs, T. (1978). Patterned differentiation in plants. *Differentiation* **11**, 65–73.

Schaller, C. H. (1973). Isolation and characterization of a low-molecular-weight substance activating head and bud formation in hydra. *J. Embryol. Exp. Morph.* **29**, 27–38.

Segel, L. A. & Jackson, J. L. (1972). Dissipative structure: an explanation and an ecological example. *J. Theoret. Biol.* **37**, 545–59.

Turing, A. (1952). The chemical basis for morphogenesis. *Phil. Trans. Roy. Soc. Lond.* B **237**, 37–72.

Tyson, J. J. (1976). *The Belousov–Zhabotinski Reaction*, Lecture Notes in Biomathematics, vol. 10, ed. S. Levin, Berlin, Springer-Verlag.

Wardlaw, C. W. (1955). Evidence relating to the diffusion–reaction theory of morphogenesis. *New Phytol.* **54**, 39–48.

Winfree, A. (1974). Rotating chemical reactions. *Scient. Amer.* **230**, 82–95.

6.5 ANALYSIS OF POPULATION CHEMOTAXIS

Chemotaxis is said to occur when an organism moves preferentially toward a relatively high concentration of some chemical (positive chemotaxis) or away from such a concentration (negative chemotaxis). This phenomenon is of interest on two levels. Firstly, the chemically directed movement of cell populations is one of the basic processes in developmental biology. Secondly, one way to approach the general problem of sensory transduction is to investigate the molecular details of how a microorganism transduces a chemical 'sensation' to an alteration of its motile behavior.

In this section we shall derive a partial differential equation for describing the chemotactic movement of large populations of organisms. This equation will be applied to an analysis of aggregation in slime mold amoebae. Wave-like population movements in chemotactic bacteria will be discussed in Section 6.7. Amoebae and bacteria are prime model systems for investigating chemotaxis at the population and molecular levels.

Aggregation of cellular slime molds

Soil is the natural habitat of the cellular slime mold amoebae. When conditions are favorable, individual amoebae emerge from their resistant spore stage and begin to engulf bacteria. As long as there is an adequate amount of their bacterial food, the amoebae move about, eat bacteria, and divide. When their food supply is exhausted, however, after an interim **interphase** period of random motion that lasts several hours, the amoebae collect together in more or less evenly spaced centers of **aggregation**. Amoebae at a given center form a multicellular slug that typically moves about for a few hours before rounding up and erecting a stalk of dead cellulose-filled cells that bears a cluster of new spores. If the wind carries the spore container to a new favorable spot, the cycle will begin again.

Of considerable importance are both the morphogenetic (form producing) movement of the aggregation process and the differentiation, wherein an initially homogeneous cell population splits into two types – stalk cells and spore cells. These processes have been studied intensively in recent years, particularly in the species *Dictyostelium discoideum*. Here we shall discuss only aggregation.

Why does aggregation occur? Evidence accumulated over a number of years that chemotaxis was involved, but only recently was the attractant identified (for *Dictyostelium* and some related species). The attractant is cyclic-AMP (cAMP), a chemical that also plays an essential role in mammalian physiology. It has been shown that the amoebae not only secrete cAMP but also an enzyme, phosphodiesterase, which converts the cAMP to another chemical (5'-AMP) that is not attractive to the cells.

The basic reference on this subject is the book of Bonner (1967). A recent review is that of Gerisch & Malchow (1976).

Formulation of equations for slime mold

We can employ the general balance equation (6.1.4) as the basis for a mathematical description of the chemotactic movement of the slime mold population. To this end, let $a(x, t)$ describe the number of amoebae per unit area. For simplicity we shall assume that the amoebic density a depends only on the time t and a single spatial coordinate x. No new principles are required to generalize our discussion to the more realistic case where a depends on two spatial coordinates.

The general balance equation (6.1.4) is

$$\partial a/\partial t = Q - \partial J/\partial x. \tag{1}$$

As before, Q is the birth rate per unit area minus the death rate per unit area. We shall take $Q = 0$, for there is little cell division in the absence of food and little cell death in the relatively short time (a few hours) over which the aggregation takes place.

We regard the cellular flux J, as being composed of two parts, one due to undirected random motion and the other to chemotaxis. We write

$$J = J_{random} + J_{chemotactic}. \tag{2}$$

Having already discussed the flux resulting from random motion, we at once adopt the simplest reasonable hypothesis

$$J_{random} = -\mu(\partial a/\partial x). \tag{3}$$

Equation (3) is essentially the same as (6.1.8), except that we prefer to use the letter μ as the effective diffusion coefficient, for 'mu' starts with the same consonant as 'motility'.

To obtain the simplest reasonable expression for the chemotactic flux, one can argue as follows. Just as proportionality to the derivative in particle density is the simplest way that the random flux can depend on the local particle concentration, so it is reasonable to conjecture that proportionality to the derivative of cAMP density is the simplest way that the chemotactic flux can depend on the local cAMP concentration. Denoting the cAMP density (i.e. the cAMP concentration) by $\rho(x, t)$, the assumption

$$J_{\text{chemotactic}} \sim \partial\rho/\partial x \tag{4}$$

is thereby suggested. For a given chemical gradient, if the local amoeba density a is doubled then the net flux would be expected to double, at least until amoebic density gets so high that the cells interfere with one another. This leads to a more precise assumption than (4), namely

$$J_{\text{chemotactic}} = \chi a (\partial\rho/\partial x). \tag{5}$$

Here the chemotactic sensitivity coefficient χ quantifies the response of a typical cell to the attractant gradient. In contrast to (3), there is no minus sign in (5), for the cells disperse from a local cell concentration but they are attracted by a local cAMP concentration.

It is reasonable to assume that both χ and μ depend on the cAMP concentration, but they can doubtless be influenced by many other chemicals as well as by the general physiological state of the cell. Thus, combining (1) (with $Q = 0$), (2), (3), and (5) we obtain the fundamental chemotaxis equation

$$\frac{\partial a}{\partial t} = -\frac{\partial}{\partial x}\left[-\mu(\rho)\frac{\partial a}{\partial x} + \chi(\rho)\frac{\partial\rho}{\partial x}\right] \tag{6}$$

which was first derived by Keller & Segel (1970). If information about the attractant concentration ρ is provided then (6) can be solved to give the resulting behavior of the chemotactic amoebae. For example, one might wish to calculate the response of a population to a point-like time varying source of cAMP, for such sources have been constricted experimentally (Robertson, Drage & Cohen, 1972). Also a partially formed aggregation center forms a source for the remaining population.

Here we shall apply (6) to demonstrate that aggregation can commence by means of an instability, in a population of entirely identical cells. The instability mechanism is now believed to form part of the explanation for aggregation in slime molds. Moreover the concept of instability is one with many applications in biology and in other fields of science.

We must supplement (6) with an equation for $\rho(x, t)$. We again start with the general balance equation (6.1.4), writing

$$\frac{\partial\rho}{\partial t} = Q_{\text{cAMP}} - \frac{\partial}{\partial x}(J_{\text{cAMP}}). \tag{7}$$

The net 'birth rate' Q_{cAMP} of the attractant equals the rate at which cAMP is secreted by the cells minus the rate at which it is destroyed by the enzyme. We write

$$Q_{cAMP} = fa - k\rho. \tag{8a}$$

Here f is the secretion rate per cell and k is generally a function of ρ. For example, k would be proportional to $(K + \rho)^{-1}$, K a constant, if the cAMP-phosphodiesterase reaction obeyed Michaelis–Menten kinetics. We shall take k to be a constant. This is accurate with Michaelis–Menten kinetics if $\rho \ll K$; in any case, we do not expect unreasonable results if we replace $k(\rho)$ by an average value.

We assume that the flux of cAMP is by simple diffusion:

$$J_{cAMP} = -D(\partial\rho/\partial x). \tag{8b}$$

Combining (7), (8a), and (8b) we obtain

$$\frac{\partial\rho}{\partial t} = fa - k\rho + D\frac{\partial^2\rho}{\partial x^2}. \tag{9}$$

Let us assume that the amoebae are confined between two walls at $x = 0$ and $x = L$. Since neither amoebae nor chemicals can pass through the walls, we must impose the 'no flux' boundary conditions

$$\partial\rho/\partial x = 0 \quad \text{and} \quad \partial a/\partial x = 0 \quad \text{at } x = 0 \text{ and } x = L. \tag{10}$$

In our analysis of (6) and (9) we shall only consider the simplest case, where the coefficients μ, χ, f, k, and D are all positive constants. Segel & Stoeckley (1972) dealt with a fuller set of equations, where in particular the details of the cAMP–phosphodiesterase interaction were considered.

Stability analysis of the uniform solution

The governing equations (6) and (9) have a uniform steady solution $a = a_0$, $\rho = \rho_0$ wherein a_0 and ρ_0 are constants. Since the derivative of a constant is zero, it follows from (6) and (9) that

$$a = a_0, \quad \rho = \rho_0 \quad \text{if } fa_0 = k\rho_0. \tag{11a, b, c}$$

The boundary conditions (10) are satisfied by our uniform solution. Note that (11c) expresses the following required condition for a uniform steady solution: given a uniform amoebic population of density a_0, a uniform cAMP level ρ must adjust itself to a value ρ_0 wherein the rate of cAMP secretion fa_0 is exactly balanced by the rate of cAMP destruction $k\rho_0$.

We shall now demonstrate that although the uniform steady solution (11) is always a mathematical possibility, there exist circumstances under which we can be sure that this solution will not be observed. These are

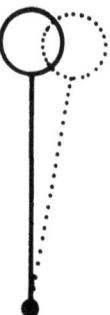

Figure 6.5.1. A pendulum in a steady state. Although mathematically possible, this state will not be observed because it is unstable. If the pendulum is slightly disturbed (dotted configuration) it will diverge from its original position.

circumstances under which a situation that starts out *almost* exactly as described in (11) will soon diverge further and further from (11). The classical example of such an unstable steady state is an inverted pendulum. (See Figure 6.5.1.) The possibility of instability has already been encountered in our discussion of reaction–diffusion theory. (Also see Appendix A.3.)

Let us be a little more precise in our formulation of the notion of 'unobservability'. We assert that any given steady solution of a set of equations which model a physical situation will not be observed unless an initial state that is *near* the given solution develops so that future states are closer and closer to the given solution. By starting our initial state near the given solution, we represent the observation that small disturbances are inevitable in any physical situation. For example, a slammed door, a footstep, or a zephyr of wind can all move the pendulum from its mathematically possible perfectly inverted state.

The basic mathematical calculations that lend precision to the remarks of the previous paragraph fall under the rubric of (linearized) **stability theory**. We present these calculations in detail here for our slime mold equations. We thus extend to partial differential equation the stability theory for ordinary differential equations that is discussed in Section 3.7. A discussion of stability theory for partial differential equations that is also brief but is somewhat more advanced and comprehensive can be found in Section 15.2 of Lin & Segel, 1973.

The first step is to introduce new variables measuring departures from the solution whose stability is under investigation. In this case, denoting the new variables by a' and ρ' we have

$$a'(x, t) = a(x, t) - a_0, \qquad \rho'(x, t) = \rho(x, t) - \rho_0. \tag{12}$$

Next, we substitute into the governing equations so that they will now be expressed in terms of the new variables. Substituting $a = a_0 + a'$ and $\rho = \rho' + \rho_0$ into (6) and (9) and remembering that the derivative of a constant is zero, we find that

$$\frac{\partial a'}{\partial t} = \mu \frac{\partial^2 a'}{\partial x^2} - \chi \left[a_0 \frac{\partial^2 \rho'}{\partial x^2} + a' \frac{\partial^2 \rho'}{\partial x^2} + \frac{\partial a'}{\partial x} \cdot \frac{\partial \rho'}{\partial x} \right]. \tag{13a}$$

$$\partial \rho'/\partial t = fa' - k\rho' + D(\partial^2 \rho'/\partial x^2). \tag{13b}$$

We now make a simplification for mathematical reasons. We begin with the fact that it is relatively easy to solve an equation wherein every term is proportional to a' or one of its derivatives or to ρ' or one of its derivatives. (The proportionality factor may in general depend on the independent variables x and t.) Such equations are called **linear**. In the present case (13b) is linear but (13a) is not, because of the terms $a'(\partial^2 \rho'/\partial x^2)$ and $(\partial a'/\partial x)(\partial \rho'/\partial x)$. As it stands then, the system (13) is nonlinear and difficult to solve. We remedy the situation by **linearizing**, that is, by deleting all nonlinear terms. This is permitted if we restrict our consideration to small disturbances a' and ρ', and treat derivatives of these quantities as being likewise small. Then we are justified in neglecting the product of two small terms (like $a'(\partial^2 \rho'/\partial x^2)$), for the magnitude of such a product is much less than that of a single small term (like $a_0(\partial^2 \rho'/\partial x^2)$).

When departure or disturbance variables like a' and ρ' (and their derivatives) are small, they are called **perturbations**. By paying the price of restricting ourselves to an examination of perturbations of our basic solution (11), we gain the advantage of having to deal with relatively simple linear equations. It may well be informative to consider larger disturbances, and indeed much modern mathematical research is being devoted to enlarging the class of disturbances that can be efficiently treated by analytical methods. Nonetheless, as our present example will illustrate, much of value can be learned from the relatively simple calculations of linear stability theory.

In the present case, the linearized equations for the perturbations a' and ρ' are

$$\frac{\partial a'}{\partial t} = \mu \frac{\partial^2 a'}{\partial x^2} - \chi a_0 \frac{\partial^2 \rho'}{\partial x^2}, \qquad \frac{\partial \rho'}{\partial t} = fa' - k\rho' + D \frac{\partial^2 \rho'}{\partial x^2}. \tag{14a, b}$$

It is relatively easy to find solutions to these equations, for all the coefficients are constants. The general theory of such equations is based on solutions of the form

$$a' = A \exp(\alpha t) \exp(\beta x), \qquad \rho' = R \exp(\alpha t) \exp(\beta x), \tag{15}$$

where it may be required to take one or more of the constants A, R, α and β to be complex. To avoid going into the complications necessitated by the appearance of complex numbers, here we merely 'guess' from the outset that special cases of (15) will suffice and assume that (14a) and (14b) possess solutions of the form

$$a' = A \cos (qx) \exp (\sigma t), \qquad \rho' = R \cos (qx) \exp (\sigma t), \qquad (16a, b)$$

where A, R, q and σ are (real) constants. These solutions of (16) may seem very special, but later we shall show that their behavior can be used to ascertain the behavior of quite general perturbations.
    Since

$$\mathrm{d}/\mathrm{d}x \, \cos (qx) = -q \sin qx$$

and

$$\sin n\pi = 0 \quad \text{for } n = 1, 2, \ldots .$$

the boundary conditions (10) will be satisfied by (16) if

$$q = q_n \quad \text{where } q_n = n\pi/L, \qquad n = 1, 2, 3, \ldots . \qquad (16c)$$

Since $\sin (-qx) = -\sin (qx)$ nothing new is added by taking $n = 1, -2, \ldots$ in (16c). The case $q = 0$ is possible, but will be given special treatment later.
    If it turns out that the constant σ is positive then the factor $\exp (\sigma t)$ grows with increasing t and the steady solution (11) is said to be **unstable** to the perturbation (16). If σ is negative then (11) is said to be **stable** to the particular perturbation (16). For the purpose of making a decision on stability, we now determine the relationship between σ and the other parameters of the problem, by requiring that expressions of the form (16) indeed provide a solution to the system of (14).
    Upon substituting (16) into (14) we see that every term contains the same factor $\sin (qx) \cos (\sigma t)$. Cancelling this common factor, we find that

$$\sigma A = -\mu q^2 A + \chi a_0 q^2 R, \qquad \sigma R = fA - kR - Dq^2 R. \qquad (17a, b)$$

Equations (17a) and (17b) are relations between the various constants of (16) that must hold if (14) is indeed to have a solution of the form of (16). (It is easy to see that if we had retained the nonlinear terms, then the equations would not have a solution of the simple form (16). See Exercise 1a.)
    We now rearrange (17a) and (17b) to bring out the fact that they can be regarded as two equations for the two unknowns A and R:

$$(\sigma + \mu q^2)A - \chi a_0 q^2 R = 0, \qquad -fA + (\sigma + k + Dq^2)R = 0.$$

$$(18a, b)$$

Assuming that $\sigma + \mu q^2 \neq 0$ we can solve (18a) for A and substitute the result into (18b). This gives

$$A = \left(\frac{\chi a_0 q^2}{\sigma + \mu q^2}\right) R, \qquad R\left[-\left(\frac{f\chi a_0 q^2}{\sigma + \mu q^2}\right) + \sigma + k + Dq^2\right] = 0.$$

$$(19a, b)$$

Equation (19b) leaves us with two possibilities. Either

$$R = 0, \quad \text{or} \quad -\frac{f\chi a_0 q^2}{\sigma + \mu q^2} + \sigma + k + Dq^2 = 0. \qquad (20a, b)$$

If $R = 0$ then $A = 0$ by (19a) and, as (15) shows, we have the **trivial solution** corresponding to a zero perturbation. The trivial solution is aptly named, for our 'discovery' that a zero perturbation is permitted is equivalent to the already determined fact that the original problem has the solution (11).

Nontrivial perturbations are those for which (20b) is satisfied. After a little algebraic rearrangement, this equation takes the form

$$\sigma^2 + b\sigma + c = 0, \quad \text{where } b = k + (\mu + D)q^2,$$

$$c = q^2[\mu(k + Dq^2) - \chi a_0 f]. \qquad (21)$$

If we denote the two roots of this quadratic by σ^+ and σ^- then

$$\sigma^+ = \tfrac{1}{2}[-b + \sqrt{(b^2 - c)}], \qquad \sigma^- = \tfrac{1}{2}[-b - \sqrt{(b^2 - c)}]. \qquad (22)$$

It can be shown that $b^2 - c > 0$ (Exercise 2) so that σ^+ and σ^- are real numbers. Since $b > 0$ we deduce that

$$\sigma^- < 0, \qquad \sigma^+ < 0 \quad \text{if and only if } c > 0. \qquad (23)$$

The role of initial conditions

We have come to the heart of the matter by deriving a necessary and sufficient condition $c > 0$ for a stable perturbation. But to see the whole problem in perspective, we must retrace our steps a bit. This will enable us to ascertain that the solutions that we have found are sufficiently general to permit satisfaction of appropriate initial conditions.

First of all, we see from (16c) that there is in fact an infinite set of q's corresponding to solutions of the form (16a) and (16b) that satisfy the imposed boundary conditions (10). Since we denote these q's by q_n, we should denote the corresponding values of b and c in (21) by b_n and c_n:

$$b_n = k + (\mu + D)q_n^2, \qquad c_n = q_n^2[\mu(k + Dq_n^2) - \chi a_0 f];$$

$$q_n \equiv n\pi/L. \qquad (24a, b)$$

From (22) we observe that for each q_n there corresponds two values of σ. Again introducing the subscript 'n' we thus rewrite (22) as

$$\sigma_n^+ = \tfrac{1}{2}[-b_n + \sqrt{(b_n^2 - c_n)}], \qquad \sigma_n^- = \tfrac{1}{2}[-b_n - \sqrt{(b_n^2 - c_n)}]. \tag{25}$$

Thus there turns out to be a whole family of solutions with the form (16). We write

$$a' = A_n^+ \cos(q_n x) \exp(\sigma_n^+ t),$$

$$\rho' = R_n^+ \cos(q_n x) \exp(\sigma_n^+ t),$$

and

$$a' = A_n^- \cos(q_n x) \exp(\sigma_n^- t),$$

$$\rho' = R_n^- \cos(q_n x) \exp(\sigma_n^- t),$$

(26)

where the former constants A and R are now replaced by A_n^+, A_n^-, R_n^+, and R_n^-, a different constant corresponding to the different possible choices of n and of σ^+ and σ^-.

We have fulfilled the requirements of (19b) by taking the expression within the square brackets to be zero, but (19a) requires that

$$A_n^+ = \left(\frac{\chi a_0 q_n^2}{\sigma_n^+ + \mu q_n^2}\right) R_n^+,$$

$$A_n^- = \left(\frac{\chi a_0 q_n^2}{\sigma_n^- + \mu q_n^2}\right) R_n^-, \qquad n = 1, 2, 3, \ldots. \tag{27}$$

We note that because the underlying equations are linear, the sum of any finite number of solutions is also a solution (Exercise 3). We conjecture that even an infinite sum of solutions is still a solution. It thus appears that we have discovered solutions to the linearized equations (14) of the form

$$a'(x, t) = \sum_{n=1}^{\infty} \cos(n\pi x/L)[A_n^+ \exp(\sigma_n^+ t) + A_n^- \exp(\sigma_n^- t)], \tag{28a}$$

$$\rho'(x, t) = \sum_{n=1}^{\infty} \cos(n\pi x/L)[R_n^+ \exp(\sigma_n^+ t) + R_n^- \exp(\sigma_n^- t)]. \tag{28b}$$

Here the constants A_n and R_n are related by (27), and the constants σ_n^+ and σ_n^- are given by (25) and (24).

Since

$$\int_0^L \cos(n\pi x/L)\, dx = \frac{L}{n\pi} \sin \frac{n\pi x}{L} \Big|_0^L = 0, \tag{29}$$

we see that

$$\int_0^L a'(x, t)\, dx = 0, \qquad \int_0^L \rho'(x, t)\, dx = 0, \tag{30a, b}$$

at least if we make the reasonable assumption that an integral of the infinite sums in (28) equals the sum of the integral of each individual term. Equations (30*a*) and (30*b*), which hold because we omitted the possibility $q = 0$ in (16*c*), assert that the perturbations in a' and ρ' are rearrangements of the existing amoebae and chemicals between $x = 0$ and $x = L$, i.e. that no new material is added. Had we included the possibility $q = 0$ we would have found that a net initial addition of cAMP to the container would decay to zero like $\exp(-kt)$ and that a net addition of amoebae would in essence merely cause an appropriate readjustment to the uniform unperturbed value a_0 (Exercise 4).

We now assert that the solutions (28) can be specialized so that they describe the development of an arbitrary initial rearrangement of the cAMP and the amoebae. Such a rearrangement would be described by the initial conditions

$$a'(x, 0) = f(x), \qquad \rho'(x, 0) = g(x). \tag{31}$$

In (31), $f(x)$ and $g(x)$ can be arbitrary functions except that they must be small enough so that our perturbation theory is accurate, they must be continuous, and they must satisfy the condition that no net new material is added, namely,

$$\int_0^L f(x)\,\mathrm{d}x = 0, \qquad \int_0^L g(x)\,\mathrm{d}x = 0. \tag{32}$$

The mathematical problem to be solved consists of the differential equations (14*a*) and (14*b*), the boundary conditions (10), and the initial conditions (31).

From (28) we deduce that the initial conditions (31) will be satisfied providing

$$\sum_{n=1}^{\infty} (A_n^+ + A_n^-) \cos(n\pi x/L) = f(x),$$

$$\sum_{n=1}^{\infty} (R_n^+ + R_n^-) \cos(n\pi x/L) = g(x). \tag{33}$$

The problem of representing functions like f and g as infinite series of cosines is part of the classical theory of Fourier series. To obtain conditions on the coefficients in (33) we multiply both sides of the equations by $\cos(m\pi x/L)$, m being a positive integer, and integrate. Since (Exercise 5)

$$\int_0^L \cos(m\pi x/L) \cos(n\pi x/L)\,\mathrm{d}x$$

$$= \begin{cases} L/2 & \text{if } m = n \\ 0 & \text{if } m \neq n \end{cases} \tag{34}$$

we obtain

$$A_n^+ + A_n^- = f_n, \qquad R_n^+ + R_n^- = g_n; \qquad n = 1, 2, 3, \ldots \qquad (35)$$

Here f_n and g_n are known constants given by the formulae

$$f_n = 2L^{-1} \int_0^L f(x) \cos(n\pi x/L)\,dx,$$

$$g_n = 2L^{-1} \int_0^L g(x) \cos(n\pi x/L)\,dx. \qquad (36)$$

Equations (35) and (27) constitute four simple conditions on the four unknown sets of constants A_n^+, A_n^-, R_n^+ and R_n^-, so the problem comprising (14a), (14b), (10) and (31) can now be regarded as solved.

Interpretation of the mathematical results

The details of the various formulae that we have derived are not important in the present context. What must be emphasized is that appropriate combinations (28) of the special solutions assumed in (16) are sufficient to determine the temporal development of general initial perturbations $a'(x, 0)$ and $\rho'(x, 0)$. Thus, the temporal behavior that we have found for the special solutions (16) in fact governs the temporal behavior of general small perturbations to the uniform state. Behavior of the solutions (16) is determined by (23), which in the present notation is

$$\sigma_n^- < 0, \qquad \sigma_n^+ < 0 \quad \text{if and only if } c_n > 0;$$

that is, by (24b), σ_n^+ and σ_n^- are both negative if and only if

$$\chi a_0 f < \mu(k + Dn^2\pi^2 L^{-2}). \qquad (37)$$

If inequality (37) is satisfied, every term in the series (28) decays exponentially. *Any* initial perturbation will die out and the original uniform state is stable. Suppose, however, that (37) is reversed for even a single n, say $n = \bar{n}$. Then the corresponding term in (28), $\exp(\sigma_{\bar{n}} t)\cos(\bar{n}\pi x/L)$, will grow with time; as a consequence the steady state will be unstable. True, the fact that $\exp(\sigma_{\bar{n}} t)$ grows with time will be irrelevant if the coefficients A_n^+ and R_n^+ are both zero, but such an unlikely coincidence can be ignored.

It is known that the chemotactic sensitivity of the amoebae, χ, increases during the interphase period of several hours that precedes the onset of aggregation, as does the rate f at which the amoebae secrete cAMP. Other parameters may change as well. What values, then, do we give to these parameters? It is helpful to imagine that periodically, say every half hour, we take a generalized 'snapshot' of the amoebae and their chemical environment. By means of such a snapshot we can ascertain (in this 'thought experiment') the values of all the parameters in the problem, and also of the instantaneous values $a_0 + f(x)$ and $\rho_0 + g(x)$ of the amoeba and cAMP concentrations. Given these parameters and initial conditions we can employ the calculations just given to determine what the behavior of the perturbations will be, say, for the next five minutes. If and only if our calculations show a growing perturbation (instability) do we predict aggregation to commence around the time of the current snapshot. Although there will be a change in the values of the parameters during the five minutes under study, and we do not take this into account, yet this change will be small so that it should not affect our broad conclusions.

According to our stability analysis, aggregative movements will commence when slow parameter changes, particularly increases in χ and f, bring about a reversal of the inequality (37). It is clear that this reversal will first occur when the right side of (37) is smallest, i.e. when $n = 1$. Thus this analysis suggests the aggregation will begin when

$$\chi a_0 f > \mu (k + D\pi^2 L^{-2}). \tag{38}$$

Moreover, it is predicted that when aggregation commences the only growing terms in the series (28) will be those corresponding to $n = 1$, i.e. the terms proportional to $\cos(\pi x/L)$. Since $A \cos(\pi x/L)$ is positive in half the region $0 \le x \le L$ (which half it is depends on the sign of A), aggregation is predicted to commence by a depletion of one half of the region $0 \le x \le L$ at the expense of the other half.

To acquire a physical understanding of the mathematical results (38) it is helpful to multiply both sides of this inequality by $(\pi/L)^2$ and then to perform a rearrangement, resulting in

$$\left[\frac{1}{f}\right]\left[\frac{(L/\pi)^2}{\chi a_0}\right] < \left[\frac{1}{k + (\pi/L)^2 D}\right]\left[\frac{(L/\pi)^2}{\mu}\right]. \tag{39}$$

Now each term within the square brackets has the dimensions of a time (compare Appendix Section A.4). In its present form, the instability condition (39) can be understood as a competition between stabilizing factors on the right of the inequality and destabilizing factors on the left. These factors are connected with the fate of perturbations proportional to $\cos(\pi x/L)$ in cAMP and amoebae concentrations.

Let us first examine the stabilizing factors. Using (6.1.19), one can conclude that $(L/\pi)^2 \mu^{-1}$ is a reasonable estimate of the time it will take the random motion of the amoebae to erase a perturbation in amoeba density proportional to $\cos(\pi x/L)$. The other factor on the right side of (39) is an estimate of the time to efface a cAMP perturbation by the combined effects of chemical decay and diffusion. The first factor on the left side of (39) gives the magnitude of the time it takes secretion by a local amoebic concentration to build up a significantly higher local cAMP concentration. The second factor on the left estimates how long it takes a local cAMP concentration to draw in a significant number of amoebae. Thus inequality (39) states that the product of two times characteristic of the enhancement of a local perturbation is less than the product of two times characteristic of the dissipation of such a perturbation. In the language of Section 3.1 one can say that instability results when (39) holds because effects that provide positive feedback to an accidental perturbation act faster than effects that provide negative feedback.

Comparison with experiment is discussed at some length by Keller & Segel (1970) and Segel & Stoeckley (1972). Here we shall only point out briefly that there are observations of increases in χ and f, and an effective decrease in k – changes that, if carried far enough, would lead to satisfaction of the instability condition (38). Condition (38) also predicts correctly the perhaps counterintuitive result that aggregation will commence later in smaller domains. (This prediction is a consequence of the fact that smaller L means a larger right-hand side in (38).) Physically, aggregation occurs later because diffusion effaces cAMP perturbations more quickly in smaller domains.

Our relatively simple stability theory seems incapable of explaining observations of the several waves of inward movement that are sometimes observed within a single aggregation territory. For this one appears to need representations of amoebae behavior that take into account such features as the autonomous cAMP secretion of some amoebae and the delayed secretion with which other amoebae reply to suitable cAMP signals. There have been computer simulations that incorporate these features into calculations of the interactions among a number of individual amoebae. But here too instability theory seems necessary to explain part of the results (Parnas & Segel, 1978). Moreover, azimuthal instabilities have been invoked to explain the 'streaming' behavior that often accompanies aggregation (Nanjundiah, 1973). In short, it now appears that the view of aggregation as an instability certainly does not suffice to explain all the major features of the phenomenon as observed in all species of cellular slime mold, but it is an indispensable part of the explanation.

Limitations of the chemotaxis equation

We close this section with some discussion of the limitations of the basic chemotactic flux equation

$$J = -\mu\,(\partial a/\partial x) + \chi a\,(\partial \rho/\partial x). \tag{40}$$

It should be clear that our discussion applies generally to equations purporting to describe population behavior.

One limitation on the use of (40) can be seen from an approach to bacterial chemotaxis as a biased random walk. This approach begins from observations that motile bacteria typically swim in fairly straight lines but after a randomly distributed length of time, these bacteria 'tumble' and then head off in a new direction. Positive chemotaxis occurs because tumbles are less frequent when the bacteria sense increasing concentrations of a chemical attractant. An appropriate limit of a probabilistic model for individual bacterial behavior leads (Segel, 1978) to a generalization of (40), namely

$$\tau(\partial J/\partial t) + J = -\mu(\partial a/\partial x) + \chi a(\partial \rho/\partial x). \tag{41}$$

Here τ is the average time between tumbles. Essentially the same equation was obtained by Segel (1977) in an analysis that focused on how the probability of a tumble changes with the percentage of a bacterium's receptor molecules which are bound with molecules of attractant.

The presence of the term $\tau(\partial J/\partial t)$ means that the flux J no longer responds instantly to changes in the environment. As discussed by Segel (1978), one can say that there is an effect of past history, or, equivalently, that the system possesses inertia. The assumption of instantaneous response is now seen to be an idealization, but one that should be justified if response time is short compared to the time over which there are significant changes in system inputs. In bacterial chemotaxis, if the average time between tumbles is short compared to the time it would take a straight-moving bacterium to experience significant changes in attractant concentration then the term $\tau \, \partial J/\partial t$ in (41) should be negligible. But such a term could be important if bacteria are confronted with a very steep attractant gradient.

A second limitation in the use of (40) stems from biological variability. Given a population of identical organisms, one can fit theoretical predictions to observation and thereby ascertain appropriate values of μ and χ. Real organisms will be characterized by a distribution of μ and χ values. For the motility μ, for example, such a distribution is characterized by a function ϕ with the property that

$$\int_{\mu_1}^{\mu_2} \phi(\mu) \, d\mu$$

gives the percentage of organisms with motility coefficients between μ_1 and μ_2. The average motility $\bar{\mu}$ is defined by

$$\bar{\mu} = \int_0^{\infty} \mu\phi(\mu) \, d\mu.$$

It is *not* generally true that the behavior of a population calculated under the assumption that each member moves with the average motility is equal to the average of all individual behaviors. True, little error will be caused by adopting the average value of coefficients like μ provided that the distribution function ϕ is nonzero only when μ is near $\bar{\mu}$. But serious complications can ensue in the contrary case, when there is considerable heterogeneity in the population. Segel & Jackson (1973) discuss this point further, in the context of bacterial chemotaxis.

Exercises

1   (*a*) Show that (13) has no nontrivial solution of the form (16).

(*b*) Show that the system (14) satisfies a suitably generalized version of (A.1.38).

2   Given *b* and *c* as defined in (21), prove that $b^2 - c > 0$. [Hint: Derive and use the result that $\alpha x^2 + \beta x + \gamma > 0$ if $\alpha > 0$, $\gamma > 0$, and $\beta^2 - 4\alpha\gamma < 0$.]

3   Let $a_1'(x, t)$, $\rho_1'(x, t)$ and $a_2'(x, t)$, $\rho_2'(x, t)$ be two pairs of solutions to (14), and let *C* be a constant.

(*a*) Show that $a_1' + a_2'$, $\rho_1' + \rho_2'$ form a solution pair.

(*b*) Show that Ca_1', $C\rho_1'$ form a solution pair.

(*c*) Show that (*a*) and (*b*) follow from Exercise 1*b*.

4   Consider perturbations of the form (16) with $q = 0$. Show that the cAMP perturbation decays but that the amoebae perturbation neither grows nor decays.

5   Verify (34). [Hint: Use a trigonometric identity.]

References

Bonner, J. T. (1967). *The Cellular Slime Molds*, 2nd edn, Princeton, N.J., Princeton University Press.

Gerisch, G. & Malchow, D. (1976). Cyclic AMP receptors and the control of cell aggregation in *Dictyostelium*. *Adv. Cyclic Nucleotide Res.* **7**, 49–68.

Keller, E. F. & Segel, L. A. (1970). The initiation of slime mold aggregation viewed as an instability. *J. Theoret. Biol.* **26**, 399–415.

Lin, C. C. & Segel, L. A. (1973). *Mathematics Applied to Deterministic Problems in the Natural Sciences*, New York, Macmillan.

Nanjundiah, V. (1973). Chemotaxis, signal relaying, and aggregation morphology. *J. Theoret. Biol.* **42**, 63–105.

Parnas, H. & Segel, L. A. (1978). A computer simulation of pulsatile aggregation in *Dictyostelium discoideum*. *J. Theoret. Biol.* **71**, 185–207.

Robertson, A., Drage, D. J. & Cohen, M. H. (1972). Control of aggregation in *Dictyostelium discoideum* by an externally applied periodic pulse of cyclic AMP. *Science* **175**, 333–5.

Segel, L. A. (1977). A theoretical study of receptor mechanisms in bacterial chemotaxis. *SIAM J. Appl. Math.* **32**, 653–65.

———— (1978). Mathematical models for cellular behavior. In *Studies in Mathematical Biology*, ed. S. A. Levin, Washington, Mathematical Association of America, pp. 156–90.

Segel, L. A. & Jackson, J. L. (1973). Theoretical analysis of chemotactic movement in bacteria. *J. Mechanochem. Cell Motil.* **2**, 25–34.

Segel, L. A. & Stoeckley, B. (1972). Instability of a layer of chemotactic cells, attractant, and degrading enzymes. *J. Theoret. Biol.* **37**, 561–85.

6.6 CELL KINETICS

What are cell kinetics

The subject of **cell kinetics** embraces the temporal evolution of cell populations, and related quantities of interest such as generation times, age distribution, and so forth. We shall first briefly review the simple theories of classical origin that describe the change in time of the total number of cells in a population. Then we shall discuss some age-structured theories of cell populations that have been introduced and investigated in recent years. Our presentation will be especially oriented towards elucidating the relationship between cell life times, which are in general variable among the members of a population, and cell growth.

Rate of growth of cell populations

Bacteria placed in a constant nutrient environment achieve, after a relatively short transient period, a **steady state of exponential growth**, also called **log phase**. In this state, the total population N in its dependence on the time t satisfies the simple ordinary differential equation

$$dN/dt = \gamma N, \qquad t > 0, \tag{1}$$

where γ is a positive constant called the **specific growth rate**, or **fractional growth rate**. Clearly, γ can be determined from measurements of $N(t)$ by means of (1), e.g. $\gamma = (dN/dt)/N$. According to the solution of (1), the population undergoes Malthusian growth (Malthus, 1798), or,

$$N = N_0 \exp(\gamma t), \tag{2}$$

where N_0 is the initial population at time zero. The single parameter γ is considered to be an invariant property of a homogeneous cell population under constant environmental conditions, and completely characterizes its growth.

Cells grown by the batch culture technique are contained in an environment of finite extent which is only fixed initially. Therefore, it cannot be considered to be constant over an extended period of time: nutrients such as oxygen are depleted, the pH changes, and effects of crowding become established, for example, as a result of toxins released by the cells and accumulating in the nutrient bath. A suggestion to account for such effects, made originally in a demographic context (Verhulst, 1838), is that (1) should be amended to read

$$\mathrm{d}N/\mathrm{d}t = \gamma N[1-(N/\bar{N})],\tag{3}$$

where $\bar{N}$ is a constant. Because of the negative sign appearing before the last term on the right, the total growth rate decreases and ultimately becomes zero as N increases. The assumption that a death rate proportional to N^2 appears in the rate equation can be made plausible by the idea that the toxic effect on *one* cell is the resultant of the toxic effect of the remaining population, or essentially N. The toxic effect on N cells is then N times N. Equation (3), with $N(0) = N_0$, has as its solution the familiar **logistic law**

$$N = \frac{N_0\bar{N}}{N_0+(\bar{N}-N_0)\exp(-\gamma t)}.\tag{4}$$

At large times, N approaches its stable equilibrium value $\bar{N}$. Thus, the growth of the population is characterized by the two parameters γ and $\bar{N}$. Experimental support for such a growth law can be found in the classical observations of Gause (1934) of the growth of *Paramecium caudatum* in a nutrient medium of fixed volume (see Figure 6.6.1).

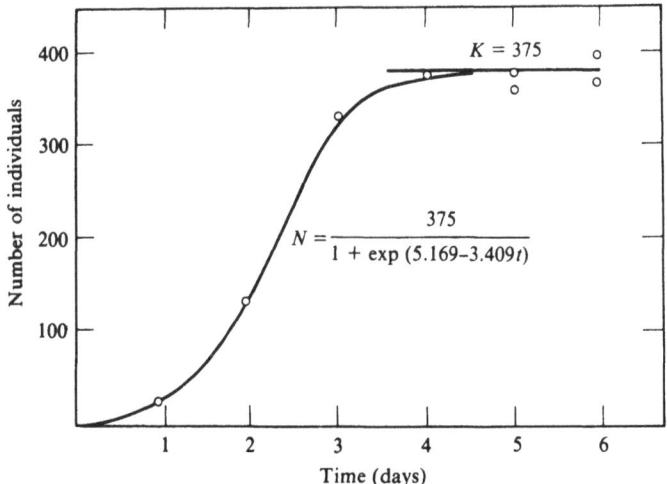

Figure 6.6.1. Data representing the growth of *Paramecium caudatum* in a nutrient medium of fixed volume is shown fitted with the logistic law (4). (From Gause, 1934.)

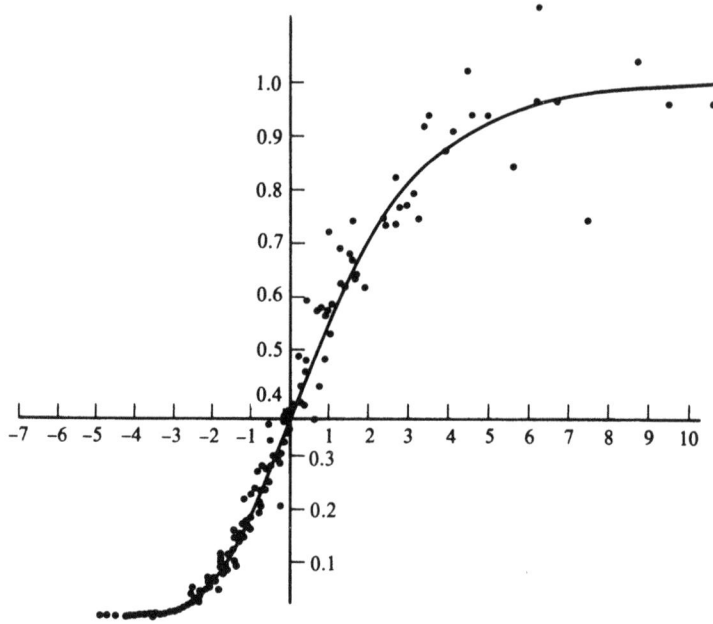

Figure 6.6.2. The growth data of tumor size (ordinate) versus time (abscissa) for 19 examples from 12 different animal tumors have been superimposed by adjusting the units of the axes for each example. The line represents a fitted 'normalized' Gompertz function, (6). (From Laird, 1965.)

It has been observed (Laird, 1965; Simpson-Herren & Lloyd, 1970) that tumor growth is well represented by a function that is very similar in appearance to the logistic law, the **Gompertz growth law** (Gompertz, 1825), as illustrated in Figure 6.6.2. This law can be derived from a generalization of (1) (Laird, Tyler, & Barton, 1965) in which the fractional growth rate γ is considered to be a time-dependent variable satisfying the differential equation

$$d\gamma/dt = -\alpha\gamma, \tag{5}$$

where α is a constant, and $\gamma(0) = \gamma_0$. Substituting the solution to the above equation, $\gamma = \gamma_0 \exp(-\alpha t)$, into (1) and integrating the resulting equation leads to the Gompertz growth law

$$N = N_0 \exp\{(\gamma_0/\alpha)[1 - \exp(-\alpha t)]\}, \tag{6}$$

where N is now the number of cells in a tumor, proportional to the tumor size. According to (6), the asymptotic value of N as t gets very large is $\bar{N} = N_0 \exp(\gamma_0/\alpha)$. Equivalently, (6) is also the solution to the differential equation, formally similar to (3),

$$dN/dt = \alpha N(\log \bar{N} - \log N). \tag{7}$$

Expressed in this form, effects of crowding are assumed to be proportional to $N \log N$.

There is no known molecular biological basis to support (7), or its equivalent representation (1) and (5), although it is widely suspected that the nonlinear terms represent a cellular interaction effect. In this connection we mention the observation that normal human fibroblast cells, capable of culture *in vitro*, have a limited potential for growth, of the order of fifty doublings (Hayflick & Moorhead, 1961), and this limitation is believed to be inherent in the cells themselves (Hayflick, 1965). Furthermore, the generation time of such cells appears to increase with generation number (Absher & Absher, 1976). These observations suggest that the decrease with time of the growth process of human fibroblasts, and perhaps other mammalian cell types, may be, at least partially, an inherent property of such cells.

The generation function

In cell kinetics, a useful parameter of interest is the **doubling time** T_d, defined as the time it takes the population to double. From (2) we see that for a population in steady exponential growth, it is determined from the fractional growth rate γ as $T_d = \ln(2/\gamma)$. It is important to keep in mind that the doubling time is not the same as the **mean generation time**, defined as the age at which a cell divides, where age is time measured from birth of a cell. An omnipresent feature of cell populations is that the generation time varies from cell to cell. In other words, there is a distribution of generation times in a cell population.

This distribution is measured as follows. Select a subpopulation of newborn cells at a given time from a cell population, i.e. all of age zero. Such a population is said to be **synchronized**. Divide the time into small intervals, and count the fraction of all the cells in the subpopulation that divide in the subsequent time interval τ to $\tau + d\tau$. Let this number equal $g(\tau) \, d\tau$. By choosing the intervals $d\tau$ to be sufficiently small, we obtain in this manner the continuous function $g(\tau)$ representing the fractional numbers of cells that divide per unit age interval. This function is called the **generation time distribution function**, or, for short, the **generation function**. For procaryotic cells and single-celled eucaryotic microorganisms, it is widely believed to be an invariant attribute, under constant growth conditions, of the given cell population, although this attribute has not been systematically investigated.

The generation time distribution function, for the protozoan *Tetrahymena geleii* is shown in Figure 6.6.3, based on the data of Prescott (1959). We see that there is a minimum generation time or age below which no cells divide, and that the generation function has the general

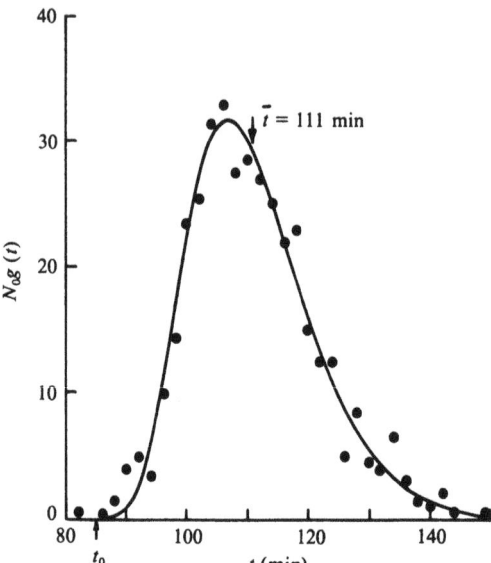

Figure 6.6.3. The solid circles represent the data of Prescott (1959) for the number of cells of the HS strain of the protozoan *Tetrahymena geleii* which divide, as a function of time measured from cell birth. The curve is a fitted γ distribution function. The number of cells investigated was $N_0 = 766$. The fitted curve assumes g is zero for $t < t_0 = 85$ min. The mean generation time is $\bar{t}$. (From Rubinow, 1968.)

appearance of a gaussian function, which is, however, skewed in the direction of long generation times. From it, we can determine the **mean generation time** T, defined as

$$T = \int_0^\infty \tau g(\tau)\, d\tau. \tag{8}$$

Note that, by definition, $g(\tau)$ is normalized to unity,

$$\int_0^\infty g(\tau)\, d\tau = 1. \tag{9}$$

What is the relation between the generation time T and the doubling time T_d? Clearly, T_d is a quantitative measure of the net fractional growth rate, or the excess increment of births over deaths in a unit time interval. Because of cell death, the generation time can be expected to be shorter than the doubling time. Suppose that there are no cell deaths. Does the doubling time equal the generation time in such circumstances? The answer is not exactly, because of the dispersion in generation times of the generation function. This can be expressed quantitatively by the variance

σ^2, defined as

$$\sigma^2 = \int_0^\infty (\tau - T)^2 g(\tau)\, d\tau. \tag{10}$$

It has been shown by a number of investigators (see for example, Painter & Marr, 1967) that the mean generation time T and the doubling time T_d are related approximately as follows,

$$T = T_d \left(1 + \frac{\log 2}{2} \cdot \frac{\sigma^2}{T_d^2}\right) + \ldots \tag{11}$$

where the ellipsis signifies a negligible correction to the formula. Note that, in these circumstances, T_d is actually less than T. The intuitive reason for this is that fast growing cells (those with shorter generation times) divide more frequently per unit time than do slow growing cells and hence have a proportionally greater influence on the doubling time.

The cell cycle

For the moment let us think about human populations, the subject of demography. We recognize that various age-dependent properties are measured and are of interest, such as the number of children borne by women of a given age interval, the age distribution of the population affected by a given disease, and so forth. Are there similar properties of cell populations that can be observed? Unfortunately, it is not feasible to attach a clock to each cell the moment it is born and keep track of its age. Nevertheless, it is possible to obtain information about meaningful events in the life history of a (large) cell population which is age-dependent.

The period of time between the birth of a cell and its disappearance as a result of division is called the **cell cycle**. Historically, and at the present time too, the only interval or **phase** of the cell cycle that is cytologically distinguishable by observation in a microscope is that during which the cell rounds up in shape, separates its duplicated DNA strands, and divides. This interval is called the **mitotic phase** of the cell cycle, or simply **mitosis**, or **M-phase**.

With the advent of radioactive labeling of cells (Taylor, Woods, & Hughes, 1957), a new phase of the cell cycle was identified. Thus, in a flash labeling experiment, cells are exposed to tritiated thymidine, a precursor of the nucleotide thymine which is found in DNA. Those cells that are in the process of duplicating their DNA will incorporate the tritiated thymidine in the duplicate DNA strand being formed. When these cells are dried and pressed against special photographic emulsions for about a week or so, the emulsions when processed contain small black spots or **grains** caused by the disintegration of the radioactive tritium contained in the thymine. Cells not in the act of duplicating their DNA do not show such grains because the thymidine is degraded within a half hour if not taken up by DNA, and does not enter the cell. Hence, exposure of cells to tritiated thymidine at a given instant of time effectively labels or identifies these cells in the phase of the cell cycle during which the DNA of the cell is being duplicated. This phase is called the **DNA-synthesis phase**, or simply **S-phase**. The S-phase of a cell is known not to overlap M-phase, so that the cell cycle divides in a natural way into four parts (Howard & Pelc, 1953):

(a) A 'gap' interval or **G_1-phase**, separating cell birth and S-phase.

(b) S-phase.

(c) A second gap interval or **G_2-phase**, separating S-phase and mitosis.

(d) M-phase.

Age-dependent properties of cell populations

A cell population is often investigated while in a steady state of exponential growth. In such a state, the population is asynchronous, i.e. cells are to be found in all stages of the cell cycle at a given time. During such growth, it is possible to measure the **mitotic index** f_M, the fraction of the cell population in mitosis, and the **labeling index** f_L, the fraction of the population in S-phase. This latter fraction is referred to as the labeling index because it is determined as the fraction of the population that is radioactively labeled following a flash labeling experiment. Following such an experiment, the fraction or percentage of cells in mitosis that are labeled can also be followed as a function of time. In this manner, a fraction-labeled mitosis curve is determined, or simply **FLM curve** (Quastler & Sherman, 1959). This usually has an oscillatory shape, which can be qualitatively understood as resulting from the entry and exit of labeled cells through M-phase.

More recently, the method of cytokinetic investigation of cell populations called **flow microfluorometry (FMF)** has been developed (Kamentsky, Melamed & Derman, 1965; Van Dilla, Trujillo, Mullaney & Coulter, 1969) which measures the amount of DNA in a cell. The measurement is accomplished by staining a suspension of cells with a fluorescent dye, making the suspension flow rapidly past a laser light beam, and detecting the scattered light from the dye molecules which are attached to the DNA. The scattered light intensity is proportional to the DNA content. Hence, the fraction of cells at a given maturation level, where maturation is measured by DNA content, can be determined.

Finally, we mention that certain chemotherapeutic drugs are phase dependent in their cytocidal action on cells. For example, arabinosyl cytosine is believed to kill only those cells that are in S-phase, when exposed to it. Hence, the action of a chemotherapeutic drug regimen in the treatment of leukemia can in principle be optimized on the basis of kinetic knowledge of the cell population being treated, its age structure, and the phase dependent nature of the cytocidal action of the chemotherapeutic agent (Rubinow & Lebowitz, 1976).

It should be clear from the above considerations that the understanding of cell kinetics has led to the necessity for age-structured theories of cell populations. In fact, such theories have been developed and exploited especially in recent years. However, it should not surprise us that this work has been anticipated to some extent by theorists in the field of demography (see for example, Keyfitz, 1968), since the measurement of age in a human population and of age-related properties do not present any particular difficulties.

The discrete compartment model of Kendall

We shall first describe one of the earliest age-dependent cell population models that was introduced, the discrete compartment model of Kendall (1948). In this model, the cell cycle is divided into K sequential compartments, cells entering the i-th compartment from the preceding $(i-1)$-th compartment, and leaving it to enter the $(i+1)$-th compartment. Let λ_i be the fractional rate of loss per unit time of cells from the i-th compartment to the $(i+1)$-th compartment, and μ_i be the fractional death rate per unit time caused by external or environmental causes. With $n_i(t)$ representing the cell number in the i-th compartment at time t, it follows that the rate of change of n_i with time is represented by the equation

$$\mathrm{d}n_i/\mathrm{d}t = \lambda_i n_{i-1} - (\lambda_i + \mu_i)n_i, \qquad i = 2, 3, \ldots, K. \tag{12}$$

The first term on the right represents cell influx from the $(i-1)$-th compartment, while the last term on the right represents cell loss from the i-th compartment.

The first compartment requires special consideration, because cells enter it only as a result of cell division occurring in the K-th compartment. For it,

$$\mathrm{d}n_1/\mathrm{d}t = 2\lambda_K n_K - (\lambda_1 + \mu_1)n_1, \tag{13}$$

where the factor 2 expresses the fact that the loss of one cell from the K-th compartment because of division leads to the birth of two cells in the first compartment.

If λ_i is a constant that is independent of i, and cell death is negligible in all compartments so that $\mu_i = 0$, (12) and (13) simplify to

$$dn_1/dt = \lambda (2n_K - n_1), \tag{14a}$$

$$dn_i/dt = \lambda (n_{i-1} - n_i), \qquad i = 2, 3, \ldots, K. \tag{14b}$$

Let us utilize these equations to represent the measurement of the generation function. Then we must neglect the term in n_K appearing in (14a), and choose as the initial condition a unit number of cells in n_1, while all other compartments are empty, e.g.

$$n_1(0) = 1, \qquad n_i(0) = 0, \qquad i = 2, 3, \ldots, K. \tag{15}$$

The solution of equations (14) is then

$$n_i(t) = ((\lambda t)^k / k!) \exp(-\lambda t), \qquad i = 1, 2, \ldots, K, \tag{16}$$

and the generation function $g(t)$, equal to $n_K(t)$, is

$$g(t) = \frac{(\lambda t)^K}{K!} \exp(-\lambda t). \tag{17}$$

This function is a 'gamma distribution', skewed in the direction of large times, and in good qualitative agreement with observed generation functions for bacteria (Powell, 1958). The Kendall model has been extended and exploited by others, notably Takahashi (1966, 1968).

The diffusion model of Stuart & Merkle

Assume there is a maturity variable μ that describes the state of maturation of a cell. Let $n(\mu, t)$ be a **maturation-time density function**, where $n(\mu, t) \, d\mu$ represents the number of cells in the maturation interval μ to $\mu + d\mu$ at time t. If we think of μ as analogous to a spatial variable, then the change with time of n obeys the balance equation, (6.1.4). The current density j of that equation is now chosen so that it represents both an orderly 'convection' through maturation states with a mean maturation velocity v_0, and a diffusive flux to take account of dispersion, some cells maturing more and some less rapidly than the mean maturation rate. Thus, we choose

$$j = v_0 n - D(\partial n / \partial \mu). \tag{18}$$

Here D represents the 'diffusion constant', which characterizes a given population. After substitution into (6.1.4), we obtain the equation of Stuart & Merkle (1965),

$$(\partial n / \partial t) + v_0(\partial n / \partial \mu) = D(\partial^2 n / \partial \mu^2), \tag{19}$$

in which cell loss caused by death has been neglected.

This theory has two noteworthy theoretical features. One is that variability in the 'effective' velocity of maturation (v_0 plus the diffusive effect) is perforce symmetric about the mean value v_0. The second is that some cells become 'younger' as time progresses because of diffusion towards smaller values of μ.

The maturity–time equation

Let us suppose that the maturation rate v is not constant, but is dependent on the maturation level μ. Further, it is plausible to assume that dispersion is a consequence of different subpopulations being associated with different maturation velocities. Then, for a given subpopulation the associated current density is just

$$j = vn. \tag{20}$$

If we substitute this into (6.1.4), and take into account the loss of cells caused by death, then we obtain perhaps the simplest equation possible for the density function n (Rubinow, 1968),

$$(\partial n/\partial t) + (\partial/\partial \mu)(vn) = -\lambda n. \tag{21}$$

Here λ is the fractional rate of loss caused by death per unit time of cells in the maturity interval μ to $\mu + d\mu$. The loss term λn could of course be included in the Stuart–Merkle equation, too.

Equation (21) holds in the maturation interval $\mu_0 < \mu < \mu_1$, where μ_0 is the maturation level of newborn cells and μ_1 is the maturation level of cells at division. Note that in both (19) and (21), the operational meaning of the variable μ has been left unspecified. It could be cell volume, DNA content, or some other physically measurable quantity that biologists may discern in the future to be a meaningful measure of cell maturity. The total population at any time $N(t)$ is obtained from n by summation:

$$N(t) = \int_{\mu_0}^{\mu_1} n(\mu, t) \, d\mu. \tag{22}$$

The utilization of additional internal variables to describe the state of a cell population has been proposed by Oldfield (1966) and others (Fredrickson, Ramkrishna & Tsuchiya, 1967), and (21) can be considered to be a special case of their equations.

The age–time equation of M'Kendrick

Suppose the maturity variable μ is replaced by age a. How is (21) to be modified? In this case the maturation velocity $v = d\mu/dt$ becomes unity, because age and time are measured by the same clocks. However, there is no longer a maximum age of a cell, so that the domain of applicability of the fundamental equation for the age–time density function $n(a, t)$ is $0 < a < \infty$. Furthermore, cells disappear at various ages because of the mitotic process, as well as cell death. Hence, the equation obeyed by $n(a, t)$ is **M'Kendrick's equation**

$$(\partial n/\partial t) + (\partial n/\partial a) = -\lambda n. \tag{23}$$

This equation was discovered by M'Kendrick (1926) in connection with the theory of epidemics. However, its application to cell populations dates from its subsequent rediscovery by Scherbaum & Rasch (1957) and von Foerster (1959). (Equation (23) is sometimes unjustly called von Foerster's equation.)

In (23) it is desirable to express the loss term λ as

$$\lambda = \lambda_d + \lambda_m \tag{24}$$

to represent the fact that there are two causes of loss, death and mitosis. The function λ_m is presumed to depend on age alone, although λ_d can in general depend on the time as well as age. Hence, in spite of the formal resemblance, (23) is by no means a special case of (21), in which λ represents the loss caused by death alone.

Further differences become apparent when we express the boundary condition satisfied by $n(a, t)$, which is

$$n(0, t) = 2 \int_0^\infty \lambda_m(a) n(a, t) \, da. \tag{25}$$

The quantity $n(0, t)$ is just the cellular **birth rate**. Note that, on the right hand side above, $n(a, t) \, da$ is the number of cells in the age interval a to $a + da$ at time t. Multiplication by $\lambda_m(a)$ gives the fraction of these cells that are dividing. Summation over all ages (integration) yields the total number of cells dividing at time t. The factor 2 appears because division of one cell produces two cells of age zero.

When (25) is supplemented by an initial condition

$$n(a, 0) = \phi(a), \tag{26}$$

where $\phi(a)$ is a given function that describes the age distribution of the population at time zero, then we are assured that M'Kendrick's equation possesses a unique solution. The total population at any time is

$$N(t) = \int_0^\infty n(a, t) \, da. \tag{27}$$

Solution to M'Kendrick's equation by generations

We shall now indicate the solution to M'Kendrick's equation by means of the generation expansion (Rubinow, 1968). Thus, let

$$n(a, t) = \sum_{j=1}^{\infty} n_j(a, t), \tag{28}$$

where $n_j(a, t)$ is the age–time density function representing the population in the j-th generation. The function n_1, which represents the initial cohort of cells whose age distribution at $t = 0$ is $\phi(a)$,

$$n_1(a, 0) = \phi(a). \tag{29}$$

How does this initial cohort of cells evolve? To answer this question, let us backtrack for the moment and seek the formal solution to M'Kendrick's equation without regard to initial or boundary conditions. It is easy to verify that for the special case in which $\lambda = 0$, (23) is satisfied by

$$n(a, t) = \chi(a - t), \tag{30}$$

where χ is an arbitrary function of its argument. This solution expresses the fact that the density function, which at $t = 0$ is $\chi(a)$, when plotted as a function of age, simply moves to the right with unit velocity as the time increases. (Such 'traveling wave' solutions frequently occur in biology and physics, see Section 6.7.) In the case where λ is not equal to zero and is a function of age alone we must take into account that the amplitude function χ decreases because of cell loss. We state without derivation that the solution (30) must be modified to read (Trucco, 1965)

$$n(a, t) = \chi(a - t) \exp\left[-\int_{\psi(a-t)}^{a} \lambda(\xi) \, d\xi\right], \tag{31}$$

where ψ is, like χ, an arbitrary function of its argument.

Using this result, we see that by choosing $\chi(a) = \phi(a)$, and $\psi(a - t) = a - t$, the function $n_1(a, t)$, which satisfies (29) at $t = 0$, is given as

$$n_1(a, t) = \phi(a - t) \exp\left[\int_{a-t}^{a} \lambda(\xi) \, d\xi\right], \qquad 0 < t < a. \tag{32}$$

Note that $n_1(a, t)$ for $a < t$ is zero, because at time t, the youngest cell of the first generation is of age t.

In general (Trucco, 1965),

$$n(a, t) = n(0, t - a) \exp\left[-\int_{0}^{a} \lambda(\xi) \, d\xi\right], \qquad t > a. \tag{33}$$

This equation is also a special case of the general solution. It states that the number of cells in the age interval a to $a + da$ at time t equals the number of cells that were of age zero at time a previously, multiplied by the fraction of these cells that have survived to age a (the exponential factor).

The second generation arises from cells of the first generation in accordance with the birth condition (25), thus,

$$n_2(0, t) = 2 \int_0^\infty \lambda_m(\tau) n_1(\tau, t) \, d\tau. \tag{34}$$

Let us substitute this expression into (33) with subscript 2 attached. Then

$$n_2(a, t) = 2 \int_0^\infty \lambda_m(\tau) n_1(\tau, t - a) \, d\tau \exp\left[-\int_0^a \lambda(\xi) \, d\xi \right]. \tag{35}$$

According to (32), $n_1(\tau, t - a)$ is not zero only when $0 < t - a < \tau$. Furthermore, $n_2(a, t)$ is zero for $a > t$, because the oldest possible age of a cell in the second generation at time t is $a = t$. Hence, (35) becomes

$$n_2(a, t) = 2 \int_{t-a}^\infty \lambda_m(\tau) n_1(\tau, t - a) \, d\tau \exp\left[-\int_0^a \lambda(\xi) \, d\xi \right], \qquad t > a, \tag{36}$$

with $n_2 = 0$ for $t < a$.

Equations (32) and (36), representing the first two generation density functions, are illustrated schematically in Figure 6.6.4. In a manner similar to the derivation of $n_2(a, t)$, $n_3(a, t)$ can be found from n_2, n_4 from n_3, and so forth, although we shall not present these equations here (see Rubinow, 1978). The generation expansion can be expected to be useful to represent experiments that extend over a relatively short time, i.e. over a time span in which only several generations are born. This solution is complementary to the long-term solution, which we shall also derive. However, before doing so, we shall first investigate how the function $\lambda_m(a)$ is experimentally determined.

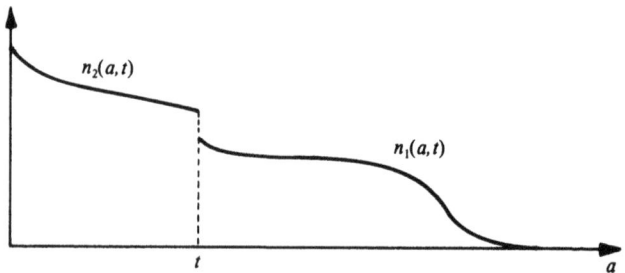

Figure 6.6.4. Schematic illustration of the first two generation functions, $n_1(a, t)$ and $n_2(a, t)$ of the generation expansion, plotted as functions of age at a given time t. Note that these two generation functions never overlap.

Experimental determination of the function λ_m

Let us try to represent the experiment by which the generation function is measured. In this experiment there is no cell death, so that $\lambda_d = 0$, and the initial age distribution function representing N_0 cells of age zero is

$$\phi(a) = N_0 \, \delta(a), \tag{37}$$

where $\delta(a)$ is the Dirac delta function. We recall that

$$\delta(a) = \begin{cases} 0, & a \neq 0 \\ \infty, & a = 0 \end{cases}, \quad \text{and} \quad \int_{-A}^{B} \delta(a) f(a) \, da = f(0), \tag{38}$$

where $f(a)$ is an arbitrary function, and the limits of integration include $a = 0$ as an interior point. The evolution in time of this cohort of cells is given by (32) as

$$n_1(a, t) = N_0 \delta(a - t) \exp\left[-\int_{a-t}^{a} \lambda_m(\xi) \, d\xi\right]. \tag{39}$$

The generation function is obtained as the fractional disappearance rate of these cells, or,

$$g(\tau) = (1/N_0) \int_0^\infty \lambda_m(a) n_1(a, \tau) \, da, \tag{40}$$

where τ denotes generation time.

Because all cells must ultimately divide,

$$\int_0^\infty \int_0^\infty \lambda_m(a) n_1(a, \tau) \, da \, d\tau = N_0, \tag{41}$$

which according to (40), is equivalent to the statement that $g(\tau)$ is normalized to unity – see (9). Substitution of (39) into (40) and using (38) yields an expression for $g(\tau)$ explicitly in terms of $\lambda_m(\tau)$,

$$g(\tau) = \lambda_m(\tau) \exp\left[-\int_0^\tau \lambda_m(\xi) \, d\xi\right]. \tag{42}$$

By integrating this equation between the limits 0 and τ and utilizing (9), we find that

$$\int_\tau^\infty g(\xi) \, d\xi = \exp\left[-\int_0^\tau \lambda_m(\xi) \, d\xi\right]. \tag{43}$$

From the ratio of (42) and (43), we find $\lambda_m(\tau)$ explicitly in terms of $g(\tau)$ as

$$\lambda_m(\tau) = g(\tau) \Big/ \int_\tau^\infty g(\xi) \, d\xi. \tag{44}$$

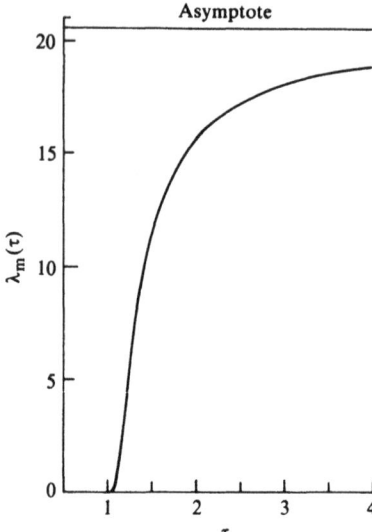

Figure 6.6.5. The function $\lambda_m(\tau)$, based on the generation function $g(\tau)$ of Figure 6.6.3, is shown, calculated from (44). The unit of time τ is t_0 of Figure 6.6.3, and the ordinate unit is $1/t_0$. (From Rubinow, 1968.)

Figure 6.6.5 shows $\lambda_m(\tau)$ calculated from the function $g(\tau)$ displayed in Figure 6.6.3.

Long-time solution of M'Kendrick's equation

Let us seek a solution of the form

$$n(a, t) = \exp(\gamma t)\bar{n}(a). \tag{45}$$

When (45) is satisfied for a positive constant value of γ, the population is said to be in a **steady state of exponential growth**. Such a solution can be expected to hold true (if at all) after a long time, when the influence of the initial age distribution of the population has been obliterated. Therefore, we shall not require of it that it satisfies the initial condition (26), and shall keep in mind that the validity of (44) is limited to 'sufficiently large' times. Substituting (45) into (23), we find that $\bar{n}(a)$ satisfies the ordinary differential equation

$$d\bar{n}/da = -[\gamma + \lambda(a)]\bar{n}. \tag{46}$$

We have assumed here that λ is a function of age alone. Its solution is readily given as

$$\bar{n} = n_0 \exp\left[-\gamma a - \int_0^a \lambda(\xi)\,d\xi\right], \tag{47}$$

where n_0 is a constant.

By normalizing $\bar{n}$ to unity, we obtain the **steady state age frequency function** or **stable age distribution** of the population, $\tilde{n}(a) = n(a)/\int_0^\infty \bar{n}(a)\,da$, which, with the use of (9) becomes

$$\tilde{n}(a) = \frac{\exp\left[-\gamma a - \int_0^a \lambda_d(\xi)\,d\xi\right]\int_a^\infty g(\tau)\,d\tau}{\int_0^\infty \exp\left[-\gamma a - \int_0^a \lambda_d(\xi)\,d\xi\right]\int_a^\infty g(\tau)\,d\tau\,da}. \tag{48}$$

When λ_d is a constant, a widely quoted result is that $\tilde{n}(a)$ is an exponentially decreasing function of a. However, we see from (48) that such a result is not strictly true because of modulation by the integral function, which depends on $g(\tau)$, appearing in the numerator above. It is only when there is no dispersion in generation times, i.e. when $g(a)$ is (infinitely) sharply peaked – $g(\tau) = \delta(\tau - T)$, and λ_d is constant – that $\tilde{n}(a)$ is a decreasing exponential in the interval $0 \le a \le T$:

$$n(a) = (\gamma + \lambda_d)\exp\left[-(\gamma + \lambda_d)a\right]/(1 - \exp\left[-(\gamma + \lambda_d)T\right]). \tag{49}$$

If (45) is to be a satisfactory solution, it must satisfy the birth condition (25). Substitution of (45) and (47) into (25) and using (42), yields the requirement

$$1 = 2\int_0^\infty \exp\left[-\lambda a - \int_0^a \lambda_d(\xi)\,d\xi\right]g(a)\,da. \tag{50}$$

This equation relates in an intimate manner two invariants of the cell population γ and $g(a)$. It is an equation of consistency they must satisfy, that cell kineticists have failed to exploit. Note that if λ_d is too large, the integral on the right-hand side will be smaller than one-half when $\gamma = 0$, and (50) will be satisfied only if γ is negative. The population is then in a **steady state of exponential decay**.

Generation times of mothers and daughters

In view of the observed variability of generation times of cells, the relationship between the generation times of mother and daughter cells was one that was investigated very early by cell kineticists. Powell (1958) found only a weak positive association of generation times between parents and daughters, by examination of individual generation times. Hughes (1955), however, claimed that the growth rate (equivalent to the generation times) of *E. coli* cells was inherited, fast (slow) growing cell subpopulations giving birth to fast (slow) growing subpopulations. The results of many other direct observations of individual mother–daughter generation times (see, for example, Powell, 1958; Prescott, 1959;

Kubitschek, 1962, 1966; Harvey, 1972) have failed to corroborate Hughes's results, only a small positive correlation being found, at best. An explanation of this discrepancy is to be found in the investigation of bacterial generation times by Powell & Errington (1963). Although these investigators again found only a weak positive correlation between the generation times of mothers and daughters, as in the earlier work of Powell, they found a surprisingly strong correlation between the generation times of cousins and second cousins. It is almost needless to add that the molecular mechanism of transferral of information about the generation time is completely unknown.

Which theory is correct?

An indirect test of the various cell population theories that we have described was provided by some very precise observations of Prescott (1959) of the individual growth in time of fifty newborn cells of the protozoan *Tetrahymena geleii* HS. He followed in time the colonies originating from these cells for several generations. By combining these observations as if they had a common time origin, the evolution in time of a single 'large' population could be very accurately obtained. The resulting growth curve for this population is shown by the circles in Figure 6.6.6. The generation for these cells was obtained by separate observations of 766 cells, as shown in Figure 6.6.3.

The age–time formalism of M'Kendrick was applied to these data (Rubinow, 1968) by utilizing λ_m, shown in Figure 6.6.5 as inferred from the generation function (Figure 6.6.3), choosing $\phi(a)$ as in (37) with $N_0 = 50$, calculating the first few generation density functions as previously indicated; and by summation according to (27) and (28), calculating $N(t)$ with which to compare the observations. We emphasize that the age–time formalism of M'Kendrick assumes no correlation between the generation times of mothers and daughters, because each cell is assumed to be subject to the same probabilistic law of division, as expressed by the function λ_m, and this property of the cell is assumed to be inborn, and unaffected by the particular generation time of its parent. Hence, the age–time model can be characterized as the 'zero memory' model.

Alternatively, the maturity time model can be applied to these data by assuming that a cell possesses 'perfect memory' of the generation time of its parent. Thus, the total population is thought of as being composed of cells of various generation times, the fraction possessing generation times between τ and $\tau + d\tau$ being $g(\tau) \, d\tau$. These cells are assumed to mature with constant velocity and to divide at age τ and give birth exclusively to cells of generation time τ – hence the appellation 'perfect memory'. The

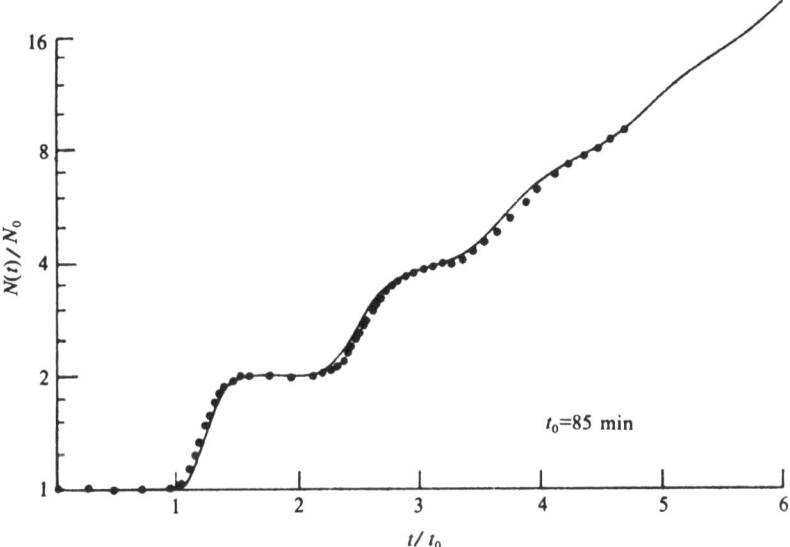

Figure 6.6.6. The solid circles represent the growth in time of a population of N_0 (= 50) newborn cells at time zero (Prescott, 1959). The line represents the prediction of the 'perfect memory' model, based on the maturity–time equation (21). (From Rubinow, 1968.)

total population is then

$$N(t) = \int_{\mu_0}^{\mu_1} \int_0^\infty n(\mu, t, \gamma) \, d\tau \, d\mu, \tag{51}$$

where $n(\mu, t, \tau)$ satisfies (23) with $v = 1/\tau$ subject to the initial condition

$$n(\mu, 0, \tau) = N_0 \delta(\mu - \mu_0) g(\tau). \tag{52}$$

As a third alternative, the predictions of the diffusion model, which had already been calculated by Stuart & Merkle (1965) utilizing the mean and variance of the generation function to determine the parameters v_0 and D, could be compared with Prescott's data.

The result of the comparison of $N(t)$ as determined for all three theories with the observations of Prescott was that only one of them could be said to be in reasonably good agreement with the observation, and that was the 'perfect memory' model. The theoretical prediction of this model is shown as the solid line in Figure 6.6.6. The disagreement of the predictions of the M'Kendrick equation and the Stuart–Merkle equation with Prescott's results represent the most serious, albeit only, indictment of these theories that has been made.

The agreement of the 'perfect memory' model with experience contains within it the seeds of a paradox: perfect memory cannot be the true relationship between the generation times of mother and daughter cells, because there is a natural dominance of fast growing cells over slow growing cells. If the maturation speed of the fastest growing cell persisted for all future generations, the generation function would be a delta function $\delta(\tau - \tau_{min})$, where τ_{min} is the shortest generation time observed, in contradiction with observation. Hence, the only consistent conclusion with the above investigation is that in a real cell population, there is persistence of memory of parental generation time over a time scale of several generations. However, this memory gradually fades because of random variation of the maturation velocity, on a time scale that is large compared to a single generation.

Lebowitz & Rubinow (1974) have proposed a theoretical resolution of this paradox. They suggest how the 'perfect memory' model can be modified so that generation functions of the form shown in Figure 6.6.3 remain an invariant property of the cell population. To accomplish this, they introduced a 'transition probability function' $K(\tau, \tau')$ which describes the probability of a cell of generation time τ' giving birth to a cell of generation time τ. Although progress has been made, many additional investigations of a quantitative nature are needed before a satisfactory answer to the question of which theory is correct is found.

We hope we have demonstrated by this brief survey that many fundamental aspects of cell population behavior are not yet understood by biologists, and that the mathematical theory of this behavior plays a natural role in its elucidation.

Exercises

1   Solve (4) by the method of separation of variables.
2   Derive (16).

References

Absher, P. M. & Absher, R. G. (1976). Clonal variation and aging of diploid fibroblasts: cinematographic studies of cell pedigrees. *Exp. Cell Res.* **103**, 247–55.

Fredrickson, A. G., Ramkrishna, D. & Tsuchiya, H. M. (1967). Statistics and dynamics of procaryotic cell populations. *Math. Biosci.* **1**, 327–74.

Gause, G. F. (1934). *The Struggle for Existence*, Baltimore, Williams & Wilkins.

Gompertz, B. J. (1825). On the nature of the function expressive of the law of human mortality and on a new mode of determining the value of life contingencies. *Phil. Trans. Roy. Soc.* **115**, 513–85.

Harvey, J. D. (1972). Parameters of the generation time distribution of *Escherichia coli* B/r. *J. Gen. Microbiol.* **70**, 109–14.

Hayflick, L. (1965). The limited *in vitro* lifetime of human diploid cell strains. *Exp. Cell Res.* **37**, 614–36.

Hayflick, L. & Moorhead, P. S. (1961). The serial cultivation of human diploid cell strains. *Exp. Cell Res.* **25**, 585–621.

Howard, A. & Pelc, S. R. (1953). Synthesis of desoxyribonucleic acid in normal and irradiated cells and its relation to chromosome breakage. *Heredity, Suppl.* **6**, 261–73.

Hughes, W. H. (1955). The inheritance of differences in growth rate in *Escherichia coli. J. Gen. Microbiol.* **12**, 265–8.

Kamentsky, L. A., Melamed, M. R. & Derman, H. (1965). Spectrophotometer: new instrument for ultrarapid cell analysis. *Science* **150**, 630–1.

Kendall, D. G. (1948). On the role of variable generation time in the development of a stochastic birth process. *Biometrika* **35**, 316–30.

Keyfitz, N. (1968). *Introduction to the Mathematics of Population*, Reading, Mass.: Addison-Wesley.

Kubitschek, H. E. (1962). Normal distribution of cell generation rate. *Exp. Cell Res.* **26**, 439–50.

—— (1966). Generation times: ancestral dependence and dependence upon cell size. *Exp. Cell Res.* **43**, 30–8.

Laird, A. K. (1965). Dynamics of tumour growth. Comparison of growth rate and extrapolation of growth curve to one cell. *Brit. J. Cancer* **19**, 278–91.

Laird, A. K., Tyler, S. A. & Barton, A. D. (1965). Dynamics of normal growth. *Growth* **29**, 233–48.

Lebowitz, J. L. & Rubinow, S. I. (1974). A theory for the age and generation time distribution of a microbial population. *J. Math. Biol.* **1**, 17–36.

Malthus, T. R. (1798). *An Essay on the Principle of Population*, 1st edn, printed for J. Johnson in St Paul's Churchyard, London, Chapter 2, p. 21.

M'Kendrick, A. G. (1926). Application of mathematics to medical problems. *Proc. Edinburgh Math. Soc.* **44**, 98–130.

Oldfield, D. G. (1966). A continuity equation for cell populations. *Bull. Math. Biophys.* **28**, 545–54.

Painter, P. R. & Marr, A. G. (1967). Inequality of mean interdivision time and doubling time. *J. Gen. Microbiol.* **48**, 155–9.

Powell, E. O. (1958). An outline of the pattern of bacterial generation times. *J. Gen. Microbiol.* **18**, 382–417.

Powell, E. O. & Errington, F. P. Jr (1963). Generation times of individual bacteria: some corroborative measurements. *J. Gen. Microbiol.* **31**, 315–27.

Prescott, D. M. (1959). Variations in the individual generation times of *Tetrahymena geleii* HS. *Exp. Cell Res.* **16**, 279–84.

Quastler, H. & Sherman, F. G. (1959). Cell population kinetics in the intestinal epithelium of the mouse. *Exp. Cell Res.* **17**, 420–38.

Rubinow, S. I. (1968). A maturity-time representation for cell populations. *Biophys. J.* **8**, 1055–73.

—— (1975). *Introduction to Mathematical Biology*. New York, Wiley.

—— (1978). Age-structured equations in the theory of cell populations. In *A Study in Mathematical Biology*, ed. S. Levin, Washington, D.C., Math. Assoc. of America.

Rubinow, S. I. & Lebowitz, J. L. (1976). A mathematical model of the chemotherapeutic treatment of acute myeloblastic leukemia. *Biophys. J.* **16**, 1257–71.

Scherbaum, O. & Rasch, G. (1957). Cell size distribution and single cell growth in *Tetrahymena pyriformis* GL. *Acta Pathol. Microbiol. Scand.* **41**, 161–82.

Simpson-Herren, L. & Lloyd, H. H. (1970). Kinetic parameters and growth curves for experimental tumour systems. *Cancer Chemother. Rep.* **54**, 143–74.

Stuart, R. N. & Merkle, T. C. (1965). Calculation of treatment schedules for cancer chemotherapy, Part II. *University of California, E.O. Lawrence Radiation Laboratory Report*, UCRL-14505, Univ. of Calif., Berkeley, California.

Takahashi, M. (1966). Theoretical basis for cell cycle analysis. I. *J. Theoret. Biol.* **13**, 202–11.

——— (1968). Theoretical basis for cell cycle analysis. II. *J. Theoret. Biol.* **18**, 195–209.

Taylor, J. H., Woods, P. S. & Hughes, W. L. (1957). The organization and duplication of chromosomes as revealed by autoradiographic studies using tritium-labeled thymidine. *Proc. Nat. Acad. Sci., USA* **43**, 122–8.

Trucco, E. (1965). Mathematical models for cellular systems. The von Foerster equation. *Bull. Math. Biophys.* **27**, 285–304; 449–71.

Van Dilla, M. A., Trujillo, T. T., Mullaney, P. F. & Coulter, J. R. (1969). Cell microfluorometry: a method for rapid fluorescent measurement. *Science* **163**, 1213–14.

von Foerster, H. (1959). Some remarks on changing populations. In *The Kinetics of Cellular Proliferation*, ed. F. Stohlman, Jr, New York, Grune & Stratton, pp. 382–407.

Verhulst, P. F. (1838). *Notice sur la loi que la population suit dans son accroissement.* Correspondence mathématique et physique publiée par A. Quételet, Brussels, vol. 10, pp. 112–21.

6.7 BIOLOGICAL WAVES

Introduction

Many biological phenomena exhibit, as their most apparent feature, a coherent pattern or waveform that moves in space. Examples to be considered here include depolarization waves propagating along nerve axons; coherent swarms of motile microorganisms advancing steadily through their environment toward a fresh supply of diffusing nutrient which they consume and seek chemotactically; chemical concentration waves carried by fluid buffer flow and diffusion through a separation column; propagation by random motility of logistically reproducing organisms through a one dimensional universe.

In this section, we explain a method of modeling such phenomena. We first derive the mathematical representation of propagating waveforms; then we discuss the general continuum scalar balance law as a tool for constructing mathematical models of phenomena such as those just mentioned. The general balance laws take the generally intractable form of nonlinear partial differential equations. We seek solutions to these equations in the form of propagating waves. This transforms our problem to an often tractable one, involving ordinary differential equations. The wave transformation, our central mathematical theme, is of great power. Indeed, it is one of the very few known methods for dealing with nonlinear partial differential equations.

More than mathematical methodology is involved, however. The balance law, (6), which we will use as the framework of our models, almost always generates systems of (nonlinear) partial differential equations (**PDEs**) of so-called **parabolic type**. The prototypical parabolic PDE is the 'heat' or 'diffusion' equation,

$$\partial T/\partial t = D(\partial^2 T/\partial x^2),$$

for the temperature T at position x and time t. This equation models the process by which an initially nonuniform temperature distribution evolves towards uniformity. (For a derivation of the heat equation, see

Section 6.1.) More complicated PDE's are called parabolic if they can be regarded (in a sense we will not define exactly) as an embellished heat equation.

There are no physically significant traveling wave solutions to the diffusion equation. But a great deal of recent mathematical research interest is focused upon the ability of nonlinear parabolic PDE systems to conduct traveling wave solutions, and upon the stability of these solutions. An exact traveling wave solution to a parabolic PDE system represents a coherent pattern that maintains its identity in the face of diffusion processes that are commonly regarded as dissipative. In many cases, a propagating pattern not only prevails against diffusive degradation but also owes its very existence to diffusion.

The foremost characteristic a biological entity must have is an ability to maintain its own 'pattern', for a time, against the unrelenting thermodynamic gnawing that diffusive/dissipative mechanisms inflict. Evolution would obviously reward any biological entity that not only copes with such forces, but also enslaves them. As just mentioned, this, in a general sense, is what is accomplished by exact wave solutions of parabolic PDE's and this is one reason for the interest in them.

To deal with the systems of ordinary differential equations that we will obtain (by transforming the PDE's), we rely heavily on the use of material presented in Appendix A.3, especially that on phase plane analysis. Thus the reader should be familiar with the content of Appendix A.3 before studying this section.

We consider a very simple example to fix ideas and to illustrate most of the concepts needed for the more interesting examples that follow. The warm-up exercise consists of a balance law, (14), for a *single* quantity (population density of motile breeding creatures). We begin with a consideration of whether any of the traveling wave solutions we find are *stable* solutions of the governing partial differential equation. This is included to make the reader aware of the stability issue that remains to be resolved after a wave solution is discovered, and to illustrate the mathematical nature of the difficulties of all such stability analyses. The difficulties are so severe that only rarely are they overcome analytically. Instead, most evidence for or against stability is obtained by numerical simulation, using the partial differential equations. In any case, analytical resolution of the stability issue is beyond the scope of this book so, after outlining the nature of the problem (for the warm-up exercise only), we abandon it, unsolved, and make no further attempt at stability analyses in examples that follow.

We generalize what we learn from this first exercise to cases involving several interacting quantities (concentrations of various chemicals, creatures, ions, etc. . . . .), and then present a sequence of biologically interesting examples at increasing levels of difficulty. We build towards a brief discussion of the justly famous Hodgkin–Huxley equations characterizing nerve axon membrane as an excitable medium along which various types of electrical depolarization waves can propagate – one of the true 'success stories' in mathematical biology.

The mathematical representation of propagating waves

At the outset, we restrict our attention to situations in which smooth functions of position and time suffice to represent the objects of interest. For example, if our problem concerns a population of microorganisms, we will let $u(x, t)$ represent the number of organisms per unit volume near position x at time t. We make no attempt to represent the position, activity, or any other property, of any particular single organism. We employ a **continuum model**, a terminology that designates a point of view that is sufficiently distant from the phenomenon being modeled so that the *microscopic* details, and the 'particles' being modeled, cannot be resolved. We concern ourselves only with the **macroscopic**, averaged, 'herd' behavior of the microscopic participants. This approach has two virtues. First, it discards at the outset the myriad details that we often have no interest in knowing even if we could discover them, and, of course, saves the labor that discovery would cost. Secondly, it makes possible a mathematical formulation upon which a very powerful body of mathematical theory can be brought to bear.

Before stating any mathematical model or problem, let us see what *form* the answer to the problems before us must assume in order to represent propagating patterns. Suppose $u(x, t)$ represents the variation with position x and time t of the concentration of something involved in a propagating wave phenomenon. Figure 6.7.1 illustrates a possible case. It shows two snapshots, one of $c(x, t)$ at an instant $t = 0$, and the other at a later time $t = t_1 > 0$. Suppose the curve illustrated at $t = 0$ moves steadily (at constant speed) down the x-axis, so that the curve illustrated at $t = t_1$ is geometrically congruent to the curve at $t = 0$. In Figure 6.7.1 $x_0 = 0$ and x_1 locate the positions of maximum concentration at times $t = 0$ and $t = t_1$. The speed of propagation of this wave is, thus,

$$c = (x_1 - x_0)/(t_1 - 0) = x_1/t_1. \tag{1}$$

If we view this waveform from a coordinate system that moves at speed c, then, in that moving coordinate system, the wave shape will not change with time. Let z measure distance, parallel to the x-axis, from the center of this moving coordinate system. Let $U(z)$ denote the shape of the waveform as seen from this system. U does not change with time. What is

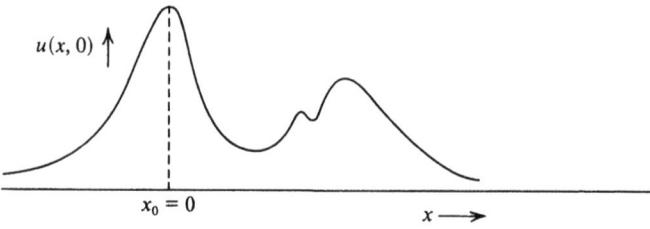

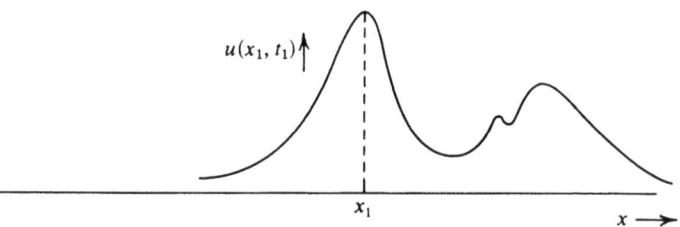

Figure 6.7.1. An arbitrary waveform propagating along the x-axis at constant speed.

the relationship between $U(z)$ and $u(x, t)$? From our definitions above, we must have

$$U(z) = u(x, 0). \tag{2}$$

This is because, at $t = 0$, $x = 0$ and $z = 0$ locate the same position. From Figure 6.7.1, we must have

$$U(0) = u(x_1, t_1),$$

and, more generally,

$$U(z) = u(x_1 + z, t_1), \tag{3}$$

because, at time $t = t_1$, $z = 0$ and $x = x_1$ locate the same position. Now, from (1), $x_1 = ct_1$, so that (3) becomes

$$U(z) = u(ct_1 + z, t_1).$$

There is nothing special about time t_1, i.e. the preceding equation holds for all time t:

$$U(z) = u(ct + z, t). \tag{4}$$

Let us identify $x = ct + z$. Thus, $z = x - ct$, and (4) becomes

$$u(x, t) = U(z) = U(x - ct). \tag{5}$$

This gives the relationship we sought, and proves that $u(x, t)$ *represents a* *fixed waveform propagating along the x axis at constant speed, c, if and only if, for some function $U(z)$, (5) holds. $z = x - ct$ is called a* **wave variable**.

The general balance law as the basic modeling tool

In this section, we investigate some problems whose answers can be expressed in the form (5). The general scheme is to construct models using the general scalar continuum balance law derived in Section 6.1 (see (6.1.4)). This is the equation

$$\partial u(x, t)/\partial t = -(\partial J/\partial x) + Q, \tag{6}$$

where J represents the **flux density** of u, and Q represents the net creation (or 'birth–death') rate density of u. To model a particular phenomenon, we must make particular choices for the way J and Q depend upon the field $c(x, t)$, etc. The resulting recipes for J and Q are called **constitutive equations**. When these are substituted into (6), we usually obtain a formidable (often intractable) PDE for $u(x, t)$. For the very special case when u moves only by diffusion, and Q is zero, we obtain the *linear* PDE (6.1.10). This is easy to solve. Slightly more generally, if

$$J = v_0 u - D(\partial u/\partial x), \tag{7}$$

and

$$Q = a_0 u(x, t), \tag{8}$$

where a_0, and v_0 are constants then (6) becomes

$$\partial u/\partial t = -v_0(\partial u/\partial x) + D(\partial^2 u/\partial x^2) + a_0 u. \tag{9}$$

Equation (9) is a **linear**[1] PDE (see Appendix A.1) and as such is mathematically tractable. Its general solution can be assembled as a linear superposition of special solutions, in a method called **separation of variables**, described, for example, in Boyce & DiPrima (1977). An example appears in Section 6.3.

The constitutive equation (7) models a situation in which u moves by random diffusion at the same time that it is swept down the x-axis at constant speed v_0 by some mechanism, a fluid flow, for example. The total flux density is a superposition of the **diffusive flux**, $-D(\partial u/\partial x)$, and the **convective flux**, $v_0 u$.

---

[1] Equation (9) is called *linear* because if $\hat{u}(x, t)$ and $\tilde{u}(x, t)$ are any two solutions of (9), then, for any two constants, b_1 and b_2, the *linear* superposition, $u = b_1\hat{u} + b_2\tilde{u}$ is also a solution of (9). The concepts of linearity and superposition for ODE's are reviewed in Appendix Section A.1.

The constitutive equation (8) models a situation in which the local 'birth rate' of u is proportional to the local concentration, u, as might be appropriate (with (7)) to model a population of motile bacteria swept along in a fluid stream, during the logarithmic phase of its reproduction dynamics.

We will not concern ourselves here with linear equations of form (9), because (9) cannot exhibit realistic ($u > 0$) propagating wave solutions (of form (5)) except in the special and uninteresting case when $D = 0$ and $a_0 = 0$.[2] Later, the proof of this claim will be made an exercise. We will concentrate, instead, upon (6) (or upon systems of several such equations) when J and/or Q depend nonlinearly on u and $\partial u/\partial x$.

A warm-up example

We illustrate, by a simple example, how seeking an exact traveling wave solution to (6) changes an intractable PDE into a tractable ordinary differential equation (**ODE**). This is the central mathematical concept of this section.

Consider a population distribution, $u^*(x^*, t^*)$ of motile creatures whose random motility results, as explained in Section 6.1, in a diffusive flux density

$$J = -D(\partial u^*/\partial x^*),$$

and whose net birth–death rate density, locally, is

$$Q = Au^*(B - u^*). \tag{9}$$

(We shall dispense with the asterisks shortly.) The quantity Q is the reproduction rate density specified by the well known logistic equation. For u^* small, the population first grows exponentially, then approaches the equilibrium density B. If u^* exceeds the carrying capacity of the environment, B, the net birth–death rate becomes negative.

Equation (6) becomes

$$\frac{\partial u^*}{\partial t^*} = D[\partial^2 u^*/\partial(x^*)^2] + Au^*(B - u^*). \tag{10}$$

We simplify (10) by introducing dimensionless variables. Express the dimensioned variables u^*, x^*, and t^* in terms of dimensionless variables,

---

[2] Then, as may be verified by direct differentiation, $u(x, t) = U(x - v_0t)$ is a solution where U is **any** differentiable function. This says that, in the context sketched above, *any* distribution of nonreproducing, nondiffusing bacteria is swept along, undistorted, at speed v_0 by a fluid stream moving at speed v_0.

u, x, t, and dimensioned scales $\bar{u}$, L, and T as follows.

$$x^* = Lx,$$
$$t^* = Tt,$$

(11)

and

$$u^*(x^*, t^*) = \bar{u}u(x, t).$$

Substituting (11) into (10), we obtain,

$$\frac{\partial u}{\partial t} = \left(\frac{DT}{L^2}\right)\frac{\partial^2 u}{\partial x^2} + [A\bar{u}T]u\left[\left(\frac{B}{\bar{u}}\right) - u\right].$$

(12)

The scales L, T, and $\bar{u}$ are at our disposal, to choose to put (12) in the simplest possible form. We select

$$\bar{u} = B,$$
$$T = 1/A\bar{u} = 1/AB,$$

(13)

and

$$L^2 = DT = D/AB.$$

These are the choices that give each dimensionless combination of parameters in (12) the value unity. Thus, (12) becomes the dimensionless equation

$$\partial u/\partial t = (\partial^2 u/\partial x^2) + u(1 - u).$$

(14)

Equation (14) is an instance of a class of PDE's formulated by Fisher (1937) to model gene diffusion in population genetics. This class of equations, and the traveling wave solutions it can exhibit, has been studied extensively by mathematicians. See, for example Kolmogoroff, Petrovsky & Piscounoff (1937), or, for a modern treatment of the entire class of PDE's to which (14) belongs, Aronson & Weinberger (1974).

We wish to know if the nonlinear PDE (14) has exact traveling wave solutions, corresponding to a population distribution of fixed shape propagating through a one-dimensional universe. To find out, we assume the existence of such a solution, moving to the right at (dimensionless) speed c. Thus, using (5), we assume

$$u(x, t) = U(z),$$

where $z = x - ct$. Using the chain rule (see Equation A.1.1.11), we compute

$$\frac{\partial u}{\partial x} = \frac{dU}{dz} \cdot \frac{\partial z}{\partial x} = \frac{dU}{dz} \cdot 1,$$

(15)

and

$$\frac{\partial u}{\partial t} = \frac{dU}{dz} \cdot \frac{\partial z}{\partial t} = -c\left(\frac{dU}{dz}\right). \tag{16}$$

Substituting (15) and (16) into (14), we obtain from the PDE, the ordinary differential equation (ODE)

$$-c\left(\frac{dU}{dz}\right) = \frac{d^2U}{dz^2} + U(1-U),$$

or

$$d^2U/dz^2 = -c\, dU/dz - U(1-U). \tag{17}$$

All possible solutions to this nonlinear ODE can be easily characterized using phase plane analysis. This is discussed thoroughly in Appendix A.3, where (17) is used as a principal example. In particular if in (A.3.4), t is replaced by z and u replaced by U, and a and ω are set to zero, (17) results.

If we regard U as measuring the density (concentration) of a population, then U must be non-negative and bounded (noninfinite). We seek, therefore, solutions of (17) that are non-negative and bounded for *all* values of z, $-\infty < z < +\infty$. The necessity for z to assume all real values is implicit in (5). In Appendix A.3, it is shown that, for $c < 2$, the phase portrait for (17) looks qualitatively as shown in Figure A.3.13, and that, therefore, no positive bounded solutions of (17) exist. Further, it is shown that, for every $c \geq 2$, (17) has a unique positive bounded solution. For $c \geq 2$, all phase portraits for (17) look qualitatively as pictured in Figure A.3.14, and the corresponding solution function $U(z)$ looks as shown in Figure A.3.15, reproduced below as Figure 6.7.2. Our analysis has shown that, theoretically, the wave front of population density shown in Figure 6.7.2 can propagate to the right at speed c. For each $c > 2$ there is a unique wave front that can do this. These are the only traveling wave solutions ($u > 0$, bounded) of (14).

The PDE–stability issue

The above chain of reasoning illustrates how the mathematical device of seeking an exact traveling wave solution to a PDE, reduces the PDE to an ODE, and how the resulting ODE often can be dealt with easily, leading to a knowledge of all possible propagating wave solutions. It illustrates, further, the issue remaining to be resolved. For each dimensionless speed, $c \geq 2$, there is a single exact propagating wave solution; which of them, if any, is a *stable* solution of the original PDE? This is, in general, a very difficult question to answer analytically. Numerical simu-

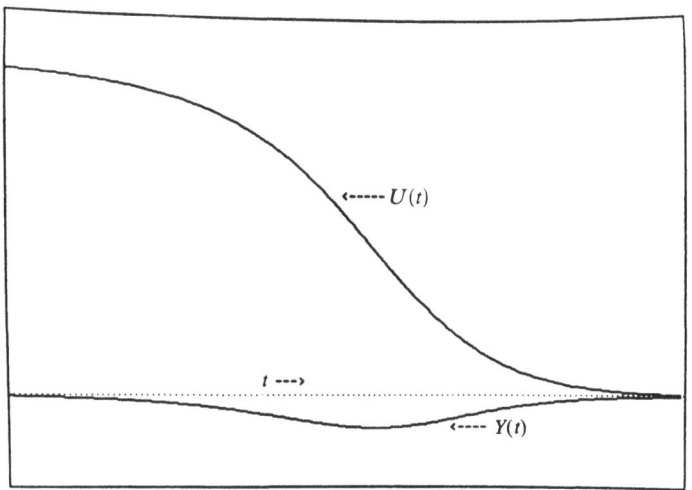

Figure 6.7.2. A propagating waveform of population density $U(z)$, for logistically-breeding, randomly-motile organisms. z measures the wave variable, and $Y(z)$ is dU/dz. $U(z)$ is the only bounded solution of (17) when $c = 2.5$.

lation (using the full PDE) can often give a practical resolution, because small numerical errors continually perturb the solution and only stable PDE solutions can survive. We will not resolve any such issue here. By omitting the resolution of this problem, we accurately represent the state of completeness to which *most* mathematical investigations of propagating wave phenomena have been brought.

In fact, the stability issue has been resolved for the warm-up example, (14). The wave propagating at speed $c = 2$ (the least propagation speed possible) is a stable solution of the PDE, and no other wave solutions are stable. This was established by Kolmogoroff *et al.* (1937) (see also Aronson & Weinberger, 1974).

Before leaving this topic, we illustrate *why* this PDE-stability issue is a difficult one. (Readers wishing to ignore this discussion may continue with the next subsection.) We pick a value for c that is 2 or greater, and ride along in a coordinate system, moving at speed c. Let

$$u(x, t) = v(x - ct, t) = v(z, t) \tag{18}$$

be the solution of (15), seen from this moving coordinate system. Again, using the chain rule, (15) becomes

$$-c(\partial v/\partial z) + (\partial v/\partial t) = (\partial^2 u/\partial z^2) + v^2(1 - v). \tag{19}$$

We want to study the question of whether a small perturbation in our exact traveling wave solution $U(z)$ can grow with time. We consider perturbations that are initially small in amplitude *compared, locally, to* $U(z)$, which, itself, gets small as $z \to \infty$. One way to do this is to express $v(z, t)$ in (19) as

$$v(z, t) = U(z)[1 + w(z, t)] \qquad (20)$$

where, at $t = 0$, $|w(z, 0)| \ll 1$.

We leave as an exercise the substitution of (20) into (19), followed by the use of the fact that $U(z)$ satisfies (17) to cancel several terms in the resulting equation. When terms that are of quadratic or higher order in small values of w are neglected, we obtain the linear PDE with variable coefficients for w

$$\partial w / \partial t = (\partial^2 w / \partial z^2) + \beta(z)(\partial w / \partial z) + \alpha(z) w,$$

where

$$\beta(z) = c + (2/U)(dU/dz), \qquad (21)$$

and

$$\alpha(z) = U(z)(2 - 3U(z)) + (c/U)(dU/dz).$$

To solve the linear equation (21) for w, we could try the **separation of variables technique**. Namely, we could seek $w(z, t)$ in the form

$$w(z, t) = T(t) Y(z). \qquad (22)$$

Substituting (22) into (21), then dividing by $T(t) Y(z)$, we obtain

$$\frac{1}{T} \cdot \frac{dT(t)}{dt} = \frac{1}{Y} \frac{d^2 Y(z)}{dz^2} + \frac{\beta(z)}{Y(z)} \frac{dY}{dz} + \alpha(z). \qquad (23)$$

The right-hand side of (23) is a function of z (only) while the left-hand side is a function of t only. The only possibility is that both sides are equal to the same constant, λ, since z and t vary independently. This means

$$\frac{1}{T} \cdot \frac{dT}{dt} = \lambda = \frac{1}{Y(z)} \frac{d^2 Y}{dz^2} + \frac{\beta(z)}{Y(z)} \frac{dY(z)}{dz} + \alpha(z), \qquad (24)$$

or

$$dT/dt = \lambda T, \qquad (25)$$

and

$$\frac{d^2 Y}{dz^2} + \beta(z) \frac{dY}{dz} + [\alpha(z) - \lambda] Y = 0. \qquad (26)$$

Equation (25) has the solution $T(t) = \exp(\lambda t)$. In general λ is a complex number: $\lambda = \lambda_r + i\lambda_i$. As is shown in Appendix A.2, $|\exp(\lambda t)| = \exp(\lambda_r t)$.

Thus if $\lambda_r > 0$ the magnitude of $T(t)$ grows exponentially large as t increases. Thus, if for any λ with a positive real part there is a solution, $Y_\lambda(z)$, of (26) for which $|Y_\lambda(z)|$ is bounded for all z, then we have found an initially infinitesimal perturbation (using (25), (22), and Y_λ),

$$v(z, t) = U(z)[1 + w(z, t)] = U(z)[1 + \varepsilon \exp (\lambda t) Y_\lambda(z)] \qquad (27)$$

where $|\varepsilon|$ is made arbitrarily small, that grows unboundedly large as t increases (according to linear theory). In this case, we would conclude that $U(z)$ is an unstable solution of the PDE.

If, on the other hand, with the real part of λ positive, no such bounded solution of (26) exists, we have a strong hint that $U(z)$ is a stable solution of the PDE. This is not conclusive because of our neglect of nonlinear terms, but generally linear stability theory correctly predicts situations where solutions are stable to small perturbations.

The general difficulty in this stability analysis is that we usually have no formula for the exact wave solution, $U(z)$, that we wish to perturb. We know only the general shape of U. Even with exact formulae for $\alpha(z)$ and $\beta(z)$, solving the ODE (26) is not a trivial matter. (Note, (26) is not autonomous, so that phase plane analysis is not an option.) Equation (26) is made even more difficult because, not knowing $U(z)$, we have no formulae for the variable coefficients $\alpha(z)$ and $\beta(z)$, each of which is determined by $U(z)$, appearing in (21) and (26). About the most we can hope for in the general case is qualitative information about the behavior of $\alpha(z)$ and $\beta(z)$. In this example problem, it is easy to show that, as $z \to \infty$, $[1/U(z)] \cdot dU/dz \to -c$, and that, as $z \to -\infty$, $1/U(z) \cdot dU/dz \to 0$. Thus, for large positive values of z, $\beta(z) \sim -c$ and $\alpha(z) \sim -c^2$. For large negative values of z, $\beta(z) \sim +c$ and $\alpha(z) \sim -1$ (since $U(z) \to 1$ as $z \to -\infty$). With this information, using the analysis of the linear ODE (26), when β and α are constants given in Appendix A.3 (see Equation A.3.15 *et seq.*), we could establish the behavior of solutions of (26) by 'patching together' known constant-coefficient solutions at $z \to -\infty$ and $z \to +\infty$. This is not simple, however. One complication, for example, is that an infinitesimal perturbation consisting of a phase shift of the entire exact traveling wave pattern to the right or left will not decay, and will appear to be unstable.

The discussion above is intended to indicate the kinds of mathematical difficulties that invariably arise when an attempt is made to analyze the stability of an exact propagating wave solution to a nonlinear PDE, when only the existence and general nature of that wave solution are known. Having seen how one sample stability analysis is begun, we will not begin another because the closing arguments of all such analyses require much deeper mathematical theory than is assumed to be known by the reader.

A final important remark about stability is in order. As discussed in Appendix A.3, whether or not $U(z)$ is a stable solution of the ODE (17) has *absolutely no bearing* upon the issue of whether $u(x, t) = U(x - ct)$ is a stable solution of the PDE (14).

Generalization to the several-variable case

Few models of interest involve only one concentration (or population density) field. More generally, several quantities, distributed spatially, interact and evolve with time. Each quantity will have its own balance law. We must therefore consider systems of several PDE's. Thus, the *answer* to the problem will consist of n functions, $u_1(x, t)$, $u_2(x, t)$, ..., $u_n(x, t)$, where, for $i = 1, \ldots, n$, $u_i(x, t)$ is the concentration (or density) of the i-th quantity of interest near position x at time t.[3] For each u_i we have the general balance law,

$$\frac{\partial u_i}{\partial t} = -\frac{\partial J_i}{\partial x} + Q_i, \qquad (i = 1, \ldots, n), \tag{28}$$

in which each flux density, J_i, and each creation rate density, Q_i, depend, in general on all of the fields $u_1, \ldots, u_n$. To model a particular phenomenon, constitutive equations specifying each of these dependencies must be hypothesized. For most applications, constitutive equations of the following form suffice:

$$Q_i = Q_i(u_1, u_2, \ldots, u_n, x, t), \tag{29}$$

and

$$J_i = J_i\left(u_1, \ldots, u_n, \frac{\partial u_1}{\partial x}, \ldots, \frac{\partial u_n}{\partial x}, x, t\right). \tag{30}$$

$$i = 1, \ldots, n$$

If, to fix ideas, the u_i represent concentrations of interacting creatures and food supplies, then (29) specifies how the (spatially) local interactions between creatures and foods augment or deplete each item locally. Equations (30) specify how each u_i-quantity moves about. This movement or flux may be passive, caused by *convection* (flow of the environment that sweeps some of the u_i with it) and/or by random motion of the individual 'particles' collectively measured by u_i (in this view, Brownian motion of particles and aimless motility of living creatures amount to the same thing). If only these transport mechanisms are involved, then each of the equations of (30) will have the same form as (7),

$$J_i = v_i u_i - D_i(\partial u_i/\partial x), \tag{31}$$

---

[3] To be more general we should consider x to be a vector, with two or three components, locating position in two or three dimensional space, and we should use the balance law given in (6.1.5). For simplicity, we consider only the one-dimensional case. It turns out, however, that we do not lose many important concepts by restricting our attention to this case.

where v_i and D_i are, respectively, the convection velocity and diffusivity appropriate to u_i. D_i and/or v_i might depend upon u_i, x, t, or even upon the other $u_j(j \neq i)$.

It may happen that the motion (flux) of u_i comes about because u_i 'senses' variations in its local environment and moves purposefully toward (respectively, away from) conditions favorable or attractive (respectively, unfavorable or odious) to u_i. See, for example, Section 6.3, where population chemotaxis of slime mold amoebae is modeled. By adopting (30), we make the assumption that local variations in environment are sensed by the u_i only through the first spatial gradients, $\partial u_i / \partial x$, of the various concentrations (see the derivation of (6.1.8) in Section 6.1). In such a situation, J_i may depend not just on $\partial u_i / \partial x$, but on the gradients $\partial u_j / \partial x$ ($j \neq i$) of other variables as well. This is why all gradients appear in each of the equations of (30).

We emphasize the point that all of the creativity of building a mathematical model of a particular biological (or any other) phenomenon, using the general framework presented here, is concentrated in the act of discovering appropriate constitutive equations in the form of (29) and (30).

Once we have (29) and (30), we substitute them into (28), to get a system of n second-order PDE's for the n functions $u_1, \ldots, u_n$. Next (provided we expect traveling wave phenomena), we seek a solution to (28) in which each u_i has the form of (5). That is, assuming that all concentration fields propagate *at the same speed*, c, we try

$$u_i(x, t) = U_i(z), \tag{32}$$

where $z = x - ct$ is the wave variable. Using the chain rule, every occurrence of $\partial / \partial x$ in (28), and (30) is replaced by

$$\frac{\partial}{\partial x} = \frac{\partial z}{\partial x} \cdot \frac{\mathrm{d}}{\mathrm{d}z} = \frac{\mathrm{d}}{\mathrm{d}z}, \tag{33}$$

and every $\partial / \partial t$ is replaced by

$$\frac{\partial}{\partial t} = \frac{\partial z}{\partial t} \cdot \frac{\mathrm{d}}{\mathrm{d}z} = -c\left(\frac{\mathrm{d}}{\mathrm{d}z}\right). \tag{34}$$

Conditions that must prevail to make an exact traveling wave solution possible

The reader should inspect the form of (28), with (29) and (30) inserted, to convince himself that *only when the Q_i and J_i have no explicit dependence upon x or t does our scheme convert the PDE system (28) into an ODE system.* That is, only when the phenomenon unfolds in a 'universe' that is spatially homogeneous and time invariant can we expect exact traveling wave solutions. No part of the universe can be different from any other part. In particular, our scheme cannot tolerate *boundary conditions* imposed at some particular x-location ((32) alone precludes this). Thus, in particular, our traveling wave solution scheme can succeed only for phenomena that occur in a universe that wraps back on itself to form a circle (a spatially periodic universe in which $x = L$ locates the same physical point as $x = 0$), or in a universe in which the boundaries are so distant that, for practical purposes, conditions imposed there can be ignored. If, for example, we are considering a nerve action potential model, we must consider an action potential spike far from the nerve soma where it was 'triggered', and far from the end-plate where it will be quenched upon arrival. The other possible (artificial) context would be an in-vitro experiment in which one end of an axon is electrically wired to the other end to make possible the unending propagation of the spike round and round a toroidal-shaped axon. The following exercise is suggested to bring home the important points made in this paragraph.

Exercise Let $n = 1$, $Q(u, x, t) = \sin^2(kt)u(x, t)(1 - u(x, t))$, and $J = -\partial u/\partial x$. Use these choices in (28) or (6), to generate an equation that resembles (14), but contains a $\sin^2(kt)$ term. This might be a model of a diffusing population whose reproduction rate is modulated diurnally or seasonally. Then, make the assumption (32). It will be found that there is no way to eliminate the explicit dependence of Q upon t. That is, the assumption (32) cannot convert the PDE (28) to an ODE. The explicit appearance of t or x in any of the equations of (29) or (30), or any boundary condition, similarly ruins the traveling wave solution scheme that we are advancing here.

Explicit t- and x-dependence was included in (29) and (30) because this often occurs in a realistic model. Only when it does not can we proceed. Supposing Q_i and J_i do not depend on x or t, substitution of (29) and (30) into (28), followed by the use of (33) and (34), yields a system of ODE's,

$$-c\left(\frac{dU_i}{dz}\right) = -\frac{d}{dz} J_i\left(U_1, \ldots, U_n, \frac{dU_1}{dz}, \ldots, \frac{dU_n}{dz}\right)$$
$$+ Q_i(U_1, \ldots, U_n), \qquad \text{for } i = 1, \ldots, n. \qquad (35)$$

This system is *autonomous*, and can be put into the standard form of (A.3.10). We will do this in the examples that follow.

If a solution to (35) exists, such that each $U_i(z)$ is bounded[4] for all

[4] We require that there be a constant M such that, for all values of z, and for all $i = 1, \ldots, n$, $|U_i(z)| < M$, because infinite concentrations, population densities, voltages, etc., are physically untenable.

values of z, $-\infty < z < +\infty$, then that ODE solution corresponds to an exact traveling wave solution of the PDE system (28). It may or may not be a stable solution. The ensemble of solution functions $U_1(z), \ldots, U_n(z)$ determines an ensemble of fixed waveforms that propagate along the x axis, all at constant speed c. Any U_i representing population density, chemical concentration, etc., cannot become negative at any z or we reject the solution as unrealistic. We now present a sequence of examples.

Chemical wave fronts in a separation column

This example involves a laboratory device used to separate an aqueous mixture of proteins into constituent fractions. It consists of a long vertical tube packed with small beads. An aqueous buffer flows slowly down through the column at a steady speed.

A sample of protein, injected at the top, is carried down through the beads. The buffer does not react with the beads and the beads do not move. The material of which the beads are made, or a material coating the beads, is chosen so that some of the proteins to be separated will have an affinity for the beads – so that protein molecules adsorb, or otherwise bind, temporarily to the beads.[5] Different proteins have different affinities for the beads and hence attach to, and detach from, the beads at different rates. Those proteins with the highest affinity for the beads (or that take longest to detach, for whatever reason) stay longest in the column and come out last.

We will make a model of the simplest interesting case of a single protein moving through the column. This will characterize the dynamics of a chemical wave front that advances down the column – how its shape and speed are related to the buffer speed, properties of the beads, diffusivity of the protein, upstream concentration of the protein, etc. Of course, the main reason for considering only a single protein is the simplicity of the equations thus obtained; this will allow us to present the main mathematical concepts without getting lost in complicated calculations. The model obviously can be generalized to characterize n different proteins.

Let $u_1(x, t)$ be the concentration of protein. We shall suppose that there is a limited number of binding or attachment sites on the beads at which protein molecules can attach. Let B denote the uniform density of these sites in the column. Let $u_2(x, t)$ be the density of protein molecules attached to the beads. (All of these densities will be measured per unit

---

[5] Another setup amounts to the same thing in essence. The beads are porous with such small pores that protein molecules squeeze through only with difficulty, and in single file, with some proteins taking longer to get through than others.

volume of column.) We need a balance law for both u_1 and u_2 (that is, in (28), (29), and (30), $n = 2$). The flux, J_2, of u_2 is zero because u_2 represents particles bound to beads that do not move.

The net creation rate density of u_2 is

$$Q_2 = k_1(B - u_2)u_1 - k_2 u_2. \tag{36}$$

This consists of production (binding) of u_2 (the k_1 term of (36)), expressing the assumption that binding is proportional to the product of the free site density, $B - u_2$, with the density of unbound protein, u_1. The rate at which attached proteins detach is proportional to the density, u_2, of attached molecules. Thus the balance law for u_2 is

$$\partial u_2/\partial t = -(\partial J_2/\partial x) + Q_2 = k_1(B - u_2)u_1 - k_2 u_2. \tag{37}$$

The net creation rate density of unbound protein molecules in the fluid, Q_1, is exactly the negative of Q_2, because for each u_1 particle 'destroyed' a u_2 particle is created, and vice versa. We take the flux density of u_1 to be of the form of (7),

$$J_1 = -D(\partial u_1/\partial x) + vu, \tag{38}$$

where D is the effective diffusivity of u_1 molecules in the fluid–bead column and v is the speed of the buffer through the column. Thus, the balance law for u_1 is

$$\frac{\partial u_1}{\partial t} = -\frac{\partial J_1}{\partial x} + Q_1$$

$$= -\frac{\partial}{\partial x}\left[-D\left(\frac{\partial u_1}{\partial x}\right) + vu_1 \right] - [k_1(B - u_2)u_1 - k_2 u_2]. \tag{39}$$

When we use the traveling wave solution assumption (32), we obtain from (37) and (39) a special case of the ODE system (35),

$$-c\left(\frac{dU_1}{dz}\right) = -\frac{d}{dz}\left[-D\left(\frac{dU_1}{dz}\right) + vU_1 \right] - [k_1(B - U_2)U_1 - k_2 U_2],$$

and

$$\tag{40}$$

$$-c(dU_2/dz) = [k_1(B - U_2)U_1 - k_2 U_2].$$

We now examine the possible solutions of this ODE system to see if it has any solutions for which $U_1(z) = u_1(x, t)$ and $U_2(z) = u_2(x, t)$ are non-negative and bounded for all values of the wave variable $z = x - ct$. If such solutions exist they represent the chemical wave fronts or concentration pulses that could propagate (at speed c) down the column, provided the column is long enough to have a region in its interior far enough from its upstream and downstream terminations that boundary conditions there are not 'felt' in the interior.

*Analysis of all possible bounded solutions of the ODE system
(40)*

If we add the equations of (40) together, the production rate density
terms must cancel. We obtain

$$-c\left[\frac{\mathrm{d}}{\mathrm{d}z}(U_1+U_2)\right]=-\frac{\mathrm{d}}{\mathrm{d}z}\left[-D\left(\frac{\mathrm{d}U_1}{\mathrm{d}z}\right)+vU_1\right].$$

Integrating both sides, we have

$$-c(U_1+U_2)=+D(\mathrm{d}U_1/\mathrm{d}z)-vU_1+E, \tag{41}$$

where E is a constant of integration. We will look for a solution in which,
at $z\to+\infty$ (far downstream), both U_1 and U_2 vanish (because sufficiently
far downstream, neither u_1 nor u_2 has arrived). Thus, evaluating (41) as
$z\to+\infty$ where U_1, U_2, and $\mathrm{d}U_1/\mathrm{d}z$ all vanish, we find $E=0$. In this
circumstance, system (40) is equivalent to (41) with $E=0$, and the second
of the equations of (40). Thus we need only deal with the following system
of two autonomous ODE's:

$$\mathrm{d}U_1/\mathrm{d}z = [(v-c)/D]U_1-(c/D)U_2,$$
$$\mathrm{d}U_2/\mathrm{d}z = -(k_1/c)(B-U_2)(U_1)+(k_2/c)U_2 \tag{42}$$

Phase plane analysis, described in detail in Appendix A.3, will tell us
all we need to know about (42). First, we introduce dimensionless
variables to reduce the parameter count. We must choose scales for z, U_1,
and U_2. Since we will not use our scaling as the basis for arguing that one
or another term is negligible, *any* scaling that makes the equations
dimensionless will suffice, but the following choices, made with hindsight,
simplify later interpretation.

For a length scale we use $L=D/v$, and let

$$z = L\tau = (D/v)\tau.$$

It will turn out that the traveling wave solution we obtain will have
$U_1(z)\to$ constant, as $z\to-\infty$. Let this upstream concentration of protein
be P_0, and let

$$U_1(z) = P_0 Y(\tau) = R_0 BY(\tau),$$

where B is the previously defined bead binding site density. We scale U_2,
which represents the concentration of bead-bound protein, by B also,

$$U_2(z) = BX(\tau).$$

τ, Y, and X are our new dimensionless variables. When the above scaled variables are substituted into (42), we obtain

$$dX/d\tau = (S/\gamma)[\sigma X - R_0(1-X)Y] \equiv F(X, Y),$$

and   (43)

$$\frac{dY}{d\tau} = -(\gamma/R_0)X + (1-\gamma)Y \equiv G(X, Y),$$

where

$$\gamma = c/v, \qquad R_0 = P_0/B,$$

σ is the ratio of effective binding constants,

$$\sigma = k_2/(k_1 B),$$

and

$$S = k_1 BD/v^2.$$

We now construct a phase portrait for (43). The X-nullcline (the locus of (X, Y) points making $F(X, Y) = 0$) is

$$Y = \frac{\sigma}{R_0} \cdot \frac{X}{1-X}. \tag{44}$$

The Y nullcline is the straight line

$$Y = \left[\frac{\gamma}{R_0(1-\gamma)} \right] X. \tag{45}$$

The steady state points for (43) are the points where these loci intersect, namely,

$$(X, Y) = (0, 0),$$

and   (46)

$$(X_0, Y_0) = [1 - \sigma(1-\gamma)/\gamma]\left[1, \frac{\gamma}{R_0(1-\gamma)} \right].$$

We will consider only parameter cases that make the steady state point (X_0, Y_0) lie in the first quadrant and argue later that no other case is of interest. Thus, to make $X_0 > 0$, we require

$$\sigma(1-\gamma)/\gamma < 1. \tag{47}$$

From (46) and (47), we have $Y_0 > 0$ provided

$$\gamma < 1. \tag{48}$$

γ is the ratio of wave speed to buffer speed, $\gamma = c/v$. Together, (47) and (48) bound this speed ratio as follows:

$$(1+\sigma)^{-1} < \gamma = c/v < 1. \tag{49}$$

We leave as an exercise the linearization of the ODE system (43) in the neighborhood of the two steady state points specified in (46). The linearized version of (43) will be found always to have two real positive eigenvalues at $(X, Y) = (X_0, Y_0)$, and real eigenvalues of opposite signs at $(0, 0)$. See Appendix A.3 for a description of the procedure and a demonstration that this means that the steady state point at $(0, 0)$ is a saddle point, while (X_0, Y_0) is an unstable node.

In Figure 6.7.3, the nullclines, steady state points, and strategically located flow arrows (at $X = 0$, where $dY/dX = -\gamma(1-\gamma)/R_0 S$) and on $Y = 0$, where $dY/dX = -\gamma^2/\sigma R_0 S$) for the ODE system (43) are shown. We have used the parameter values $S = 0.5$, $R_0 = 3.0$, $\sigma = 0.25$, $\gamma = 1.31$ in Figure 6.7.3, but any positive values of these parameters, such that (49) holds, yield the same qualitative picture.

From the material given in Appendix A.3, we know (rigorously) that, because of the nature of the flow shown in Figure 6.7.3, the only possibility for a solution to (43) with $X(\tau)$ and $Y(\tau)$ positive and bounded for all τ is a (heteroclinic) phase plane trajectory that tends to one steady state point as $\tau \to -\infty$, and to another as $\tau \to +\infty$. We now prove there is exactly one such trajectory.

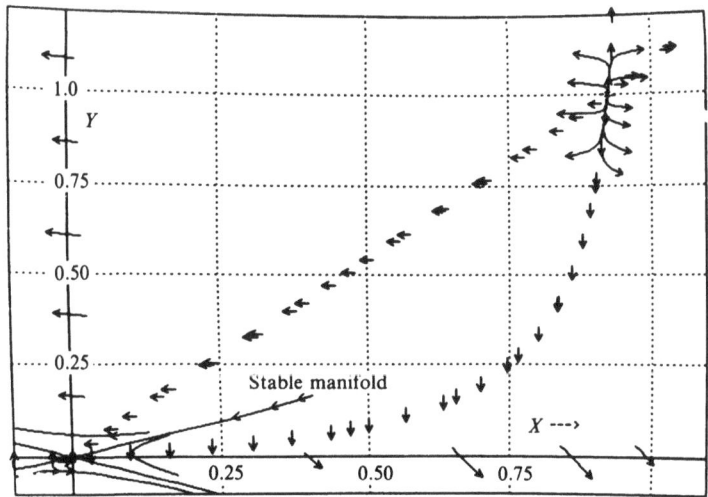

Figure 6.7.3. Preliminary phase portrait for the ODE system (43), showing the nullclines and the character of the steady state points. The parameter values are $S = 0.5$, $R_0 = 3.0$, $\sigma = 0.25$, and $\gamma = 1.31$.

There is precisely one trajectory in the first quadrant tending to the saddle point $(0, 0)$ as $|\tau| \to \infty$. This is the first quadrant branch of the stable manifold at $(0, 0)$ that lies trapped inside the crescent between the two nullclines (see Figure 6.7.3). This trajectory, traced backward in τ *must* come from the unstable node at (X_0, Y_0) (as $\tau \to -\infty$) for, because of the direction of the flow at the nullclines, it cannot enter the crescent from any point other than from that node. This ends the existence and uniqueness proof.

We want $\lim Y(\tau) = 1$, as $\tau \to -\infty$, so that $P_0 = R_0 B$ is the upstream protein concentration defined in (44). Thus we want Y_0, given in (46), to be 1. The parameter $\gamma = c/v$ is still at our disposal. Only for one value of this parameter, γ_0, will $Y_0 = 1$. This specifies the unique (dimensionless) wave speed γ_0, in terms of σ, R_0, and P_0:

$$Y_0 = [1 - \sigma(1 - \gamma_0)/\gamma_0][\gamma_0/R_0(1 - \gamma_0)] = 1,$$

or   (50)

$$\gamma_0 = c_0/v_0 = 1 \bigg/ \left[1 + \frac{1}{R_0 + \sigma}\right].$$

When γ_0, given in (50), satisfies (49), we have a legitimate exact traveling wave front solution (corresponding to the unique heteroclinic trajectory connecting the node to the saddle point in Figures 6.7.4 and 6.7.5 to the PDE system (37) and (39)). Figure 6.7.4 shows a complete phase portrait for one such case ($S = 0.5$, $R_0 = 3.0$, $\sigma = 0.25$, and, from (50), $\gamma = \gamma_0 = 1.31$). Figure 6.7.5 shows the $X(\tau)$ (bound protein) and $Y(\tau)$ (free protein) dimensionless concentration fields as functions of dimensionless wave variable

$$\tau = (x - ct)/L = (v/D)(x - ct).$$

We now note a point that is important for those who wish to study these and similar problems by numerical methods. The phase portrait in Figure 6.7.4 indicates the only feasible way to compute $X(\tau)$ and $Y(\tau)$ numerically. Namely, integrate backward in τ, starting at initial values $X(0)$, $Y(0)$ lying on the stable manifold at $(0, 0)$ (i.e. on the 'stable' eigenvector of (43) linearized near $(0, 0)$. For τ running backwards (numerically) the unstable node at (X_0, Y_0) becomes stable, so there is no difficulty approaching it numerically. If, in contrast, (43) is integrated with τ increasing, the saddle point is hard to approach because it is an unstable steady state point regardless of whether τ increases or decreases. Figure 6.7.5 was generated numerically in this fashion. In Figure 6.7.5, τ varies from 0 to 25.

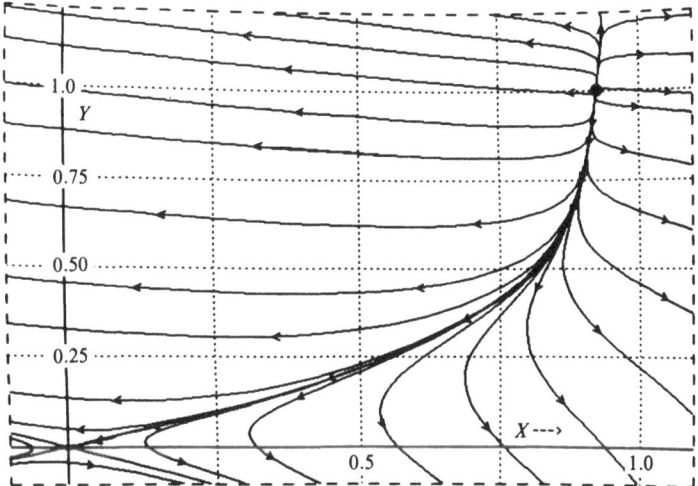

Figure 6.7.4. Finished phase portrait for the ODE system (43), when $S = 0.5$, $R_0 = 3.0$, $\sigma = 0.25$, and $\gamma = 1.31$. The single heteroclinic trajectory connecting the stable node in the upper right corner to the saddle point in the lower left corner corresponds to the unique bounded propagating wave solution to the PDE system (37) and (39). $X(\tau)$ and $Y(\tau)$ are graphed along this trajectory in Figure 6.7.5.

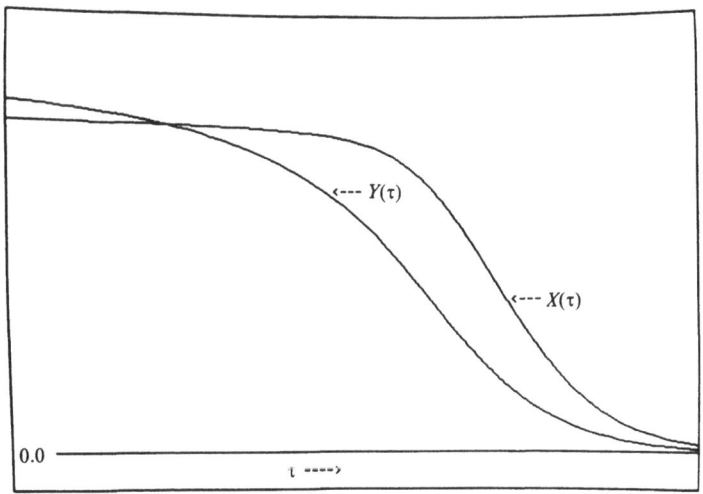

Figure 6.7.5. Propagating waveforms of dimensionless concentrations of free protein $Y(\tau)$ and of bead-bound protein $X(\tau)$, for the separation column problem. τ is the dimensionless wave variable, values ranging from 0 to 25. These waveforms correspond to $X(t)$ and $Y(t)$ traced along the unique heteroclinic trajectory in Figure 6.7.4. The parameter values are $S = 0.5$, $R_0 = 3.0$, $\sigma = 0.25$, and $\gamma = 1.31$.

We have established the existence of the kind of wave front shown in Figure 6.7.5 propagating at speed

$$c = v\gamma_0 = v\left[1 + \frac{B}{P_0 + (k_2/k_1)}\right]^{-1} < v. \tag{51}$$

This solution is possible only when γ, given in (50), satisfies (49), that is, when

$$(1+\sigma)^{-1} < [1 + (R_0 + \sigma)^{-1}]^{-1},$$

or

$$\sigma > (R_0 + \sigma)^{-1},$$

or

$$R_0 > \sigma^{-1} - \sigma. \tag{52}$$

In terms of the original dimensioned variables, (52) is

$$P_0 > \left\{\left(\frac{k_2}{k_1 B}\right)^{-1} - \frac{k_2}{k_1 B}\right\} B. \tag{53}$$

The interpretation is as follows. If the column is set up so that the ratio of detachment rate constant, k_2, to maximum attachment rate constant $k_1 B$, is greater than unity ($\sigma > 1$), then (53) is always satisfied for any upstream concentration of free protein P_0, and a wave front of the type shown in Figure 6.7.5 can propagate at the speed given in (51). According to (51), the propagation speed varies between $v/[1 + (1/\sigma)]$ and v, with the upper speed limit v approached as $P_0 \to \infty$.

If, on the other hand $\sigma = (k_2/k_1 B) < 1$ (beads have high affinity for protein and/or abundant binding sites), then (53) enforces a lower bound on the upstream protein concentration to permit traveling wave front solutions.

How can the wave front, shown in Figure 6.7.5 propagate forever without being flattened by the equilibrating effect of diffusion? The physical mechanism is apparent. Those protein molecules that, by diffusion, run out ahead of the front find an abundance of free binding sites. They therefore spend most of their time bound, waiting until the wave front catches up. Increased binding to beads at the leading edge of the wave front has a steepening effect, that, at the wave speed given by (51), just counterbalances the broadening effect of diffusion.

Is it possible for an isolated pulse solution (for which $X(\tau)$ and $Y(\tau) \to 0$ for $\tau \to -\infty$, as well as for $\tau \to +\infty$) to propagate down the column? The reasoning above, applied to the trailing edge of such a pulse rules out the possibility. At the trailing edge, the diffusive and bead-binding effects, instead of canceling, both work to flatten the pulse. Thus no pulse could propagate. Indeed, such a pulse, if it existed, would correspond to a homoclinic trajectory in the phase plane beginning (as $\tau \to -\infty$) and ending (as $\tau \to +\infty$) at the steady state point $(0, 0)$. This is clearly impossible in Figure 6.7.4. It is a worthwhile exercise in phase plane analysis to prove that, even if we make different choices of dimensionless wave speed, γ, than imposed by (47), we cannot obtain a homoclinic trajectory on which $X(\tau)$ and $Y(\tau)$ stay everywhere bounded and positive. Once this exercise is completed, we will have cataloged all possible realistic traveling wave solutions of the PDE system (37) and (40). The exact shape of the wave front (Figure 6.7.5 is an example) for a given set of parameters (v, D, k_1, k_2, B, P_0) can be ascertained numerically (with ease, on a programmable hand calculator, for example) as described above.

There are only two things remaining to be said about this example. One is a reminder that we have learned nothing about whether the wave fronts we found are stable solutions to the PDE system. The other is an exhortation that the reader think more carefully about the mathematical *method* illustrated in the example we have just presented than about the separation-column context that packaged it. A single protein species filtering through a column may not much excite the reader's imagination. However, the general phenomenon of a moving stream carrying a diffusing substance (or population . . .) through a medium with which that substance interacts (by binding, etc. . . .) occurs in many biological contexts. The above example is intended primarily to provide instruction on how to model such phenomena, and on how to cope with the PDE's.

Exercise. Using the material presented so far, the reader should be able to prove the claim made above about the limited ability of the linear equation (9) to exhibit bounded traveling wave solutions. Namely, show that it does only when $c_0 = v_0 =$ the imposed convective speed, and either $D = 0$, $a_0 = 0$, or $a_0/D > 0$.

Traveling bands of chemotactic microorganisms

Population chemotaxis of aggregating slime mold amoebae is discussed in Section 6.3. We investigate a similar model, put forward by Keller & Scgcl (1971), of the coherent moving bands or swarms that motile, chemotactic bacteria can exhibit.

Bacterial chemotaxis is a phenomenon that is currently under vigorous investigation. (See the reviews of Adler (1975), Berg (1975) and Koshland (1974), and see Holz & Chen (1978, 1979).) The basic intuitive idea is the following. A one-dimensional universe (a long capillary tube) is filled with a fluid medium in which a nutrient substance, s, is dissolved. Bacteria are inoculated into one end of the tube. They are observed to consume the nutrient in their neighborhood and to form a band that moves steadily up the tube.

By propelling itself about at random, but 'remembering' (in a primitive way) how the concentration, s, varies with time along his trajectory, a single bacterium can modulate or prejudice its meanderings so as to seek, in a statistical sense, nearby regions with higher nutrient levels. This ability of individual bacteria apparently endows a continuum herd of many to behave as if it can purposefully sense the local spatial gradient of s and move up it from regions where s is smaller (because the bacteria have consumed s there) into regions where s is larger. In this way, evidently, chemotactic competence is a mechanism responsible for the choreography of a macroscopically coherent performance by a swarm of individually 'stupid' performers, none of whom knows the global plan of the dance. The performance, as explained in the reviews cited above, consists simply of a pulse waveform of bacterial population density (called a band) that sweeps through the capillary tube at constant speed, and with an unchanging shape, consuming all the nutrient as it advances. This performance is of general interest, because it may be a simple prototype of the much more complicated 'flows' that cell ensembles must generate during embryogenesis.

Is there any need to make a mathematical model? If this is the first time the reader has learned of the phenomenon sketched above, he may believe the prose to be utterly convincing, and to have no need of support or verification by mathematical analysis. After all, if the bacteria consume the local nutrient supply, and can sense abundant food nearby down the tube, and have the ability to move toward it, then what else can happen? Of course the swarm of bacteria will advance down the tube. If the mathematical analysis turned out merely to confirm such a first intuitive impression, then a reader with no interest in the mathematics *per se* could properly regard that analysis as superfluous.

The mathematical analysis that follows leads, however, to a very different conclusion, namely that *only* certain special chemotactic response algorithms will lead to band propagation maintaining coherence over 'long' (compared to band width) distances. The mathematical quest, therefore, is to catalog *all* of the macroscopic (or phenomenological) behavior repertoires of chemotactic microorganisms that lead to propagating bands.

Constitutive equations for chemotaxis

Let $b(x, t)$ be the population density of bacteria, and $s(x, t)$ be the nutrient density, at position X and time t. Keller & Segel (1971) hypothesized the constitutive relation for the bacterial flux density

$$J_b = -\mu(\partial b/\partial x) + b\chi(\partial s/\partial x). \tag{54}$$

The first term of (54) represents the random (diffusive) component of the flux while the second term models the chemotactic component. See Section 6.3 for a discussion of this equation. In (54), we assume that, in general, $\mu = \mu(s)$, and $\chi = \chi(s)$. That is, as $s \to 0$, bacteria starve, and may seek nutrient more fervently than when s is large, so, we might expect $\chi(s)$ to increase as s decreases. Starving bacteria may thrash about randomly (μ measures the intensity of random motility) either more or less actively than well fed ones. We restrict our model to cases when the bacteria neither die nor reproduce during the experiment (reproduction can be prevented chemically).

With the creation rate density of bacteria equal to zero, and using (54), the balance law (28) for $b(x, t)$ is

$$\frac{\partial b}{\partial t} = -\frac{\partial}{\partial x}\left[-\mu(s)\frac{\partial b}{\partial x} + b\chi(s)\frac{\partial s}{\partial x}\right]. \tag{55}$$

The nutrient diffuses passively, and is eaten by the bacteria. Its balance law is

$$\partial s/\partial t = D(\partial^2 s/\partial x^2) - k(s)b, \tag{56}$$

where D is the molecular diffusivity of the food and $k(s)$ is the rate of nutrient consumption per bacterium. In general, we expect $k(s)$ to fall to zero as $s \to 0$ because no consumption is possible when no food is present. Keller & Segel (1971) took both μ and k to be constants. (A constant value of k should be interpreted to mean that, so long as there is any nutrient available, each bacterium maintains a constant consumption rate. If the food ever really runs out completely ($s = 0$), then k is to be replaced by zero.) We will consider only the case when the diffusive flux of nutrient, s, is negligible. This is both for simplicity, and because, in most applications here, both the *random motility* (measured by μ) and the *chemotactic sensitivity* (measured by $b\chi(s)$) are so much larger than the nutrient diffusivity, D, that, for practical purposes we can regard the nutrient as motionless while the bacteria sweep through it. Mathematically sophisticated readers will recognize, however, that some care is needed when dropping the $D(\partial^2 s/\partial x^2)$ (highest order) term in (56). In cases where dissolved oxygen is the limiting substrate, D in (56) is not negligible.

We seek a traveling wave solution of (55) and (56) by assuming the answer has the form of (32). Thus, we assume

$$b(x, t) = B(z), \quad \text{and} \quad s(x, t) = S(z), \tag{57}$$

where $z = x - ct$. Substituted into (55) and (56), (57) yields another instance of (35), namely

$$-c\left(\frac{dB}{dz}\right) = -\frac{d}{dz}\left\{-\mu\,\frac{dB}{dz} + b\chi\,\frac{dS}{dz}\right\}, \tag{58}$$

and

$$-c(dS/dz) = -kB. \tag{59}$$

We can integrate (58) once. We seek a solution in which, 'far ahead of the band', (i.e. at $z \to \infty$) there are no bacteria, i.e. $B \to 0$, $dB/dt \to 0$ (see Figure 6.7.10); evaluating at $z \to \infty$, we therefore see that the constant of integration, obtained when (58) is integrated, must vanish. Thus, we have

$$-cB = \mu(dB/dz) - b\chi(dS/dz). \tag{60}$$

We eliminate dS/dz in (60) using (59), to obtain the second order, nonlinear, autonomous ODE system,

$$\begin{aligned} dS/dz &= [k(S)/c]B, \\ dB/dz &= [B/c\mu(S)][k(S)\chi(S)B - c^2]. \end{aligned} \tag{61}$$

We can analyze the behavior of solutions to (61) using the phase plane analysis technique discussed in Appendix A.3. We are interested in solutions $(S(z), B(z))$ to (61) that are non-negative and bounded for *all* values of the wave variable z. In fact, since we are interested in isolated propagating swarms of bacteria, we are interested in solutions for which $B(z) \to 0$ both as $z \to \infty$ and as $z \to -\infty$. (See, for example, Figure 6.7.10.)

The first of the equations of (61) means that as $z \to -\infty$, $S(z)$ must decrease monotonically. $S(z)$ cannot fall below zero, so the only possibility is that

$$\lim_{z \to -\infty} S(z) = s_c = \text{constant}.$$

We do not know what this (presumably 'small') unconsumed nutrient concentration left in the wake of a band might be. Whatever it may be, we lose no generality in assuming $s_c = 0$ because, if $s_c > 0$, we can introduce a new dependent variable $s' = s - s_c$, and new coefficient functions $k'(s') = k(s' + s_c)$, $\mu'(s') = \mu(s' + s_c)$, $\chi'(s') = \chi(s' + s_c)$. The transformed versions of (61) will look just like (61), with k, s, μ, and χ all primed. We will have, then, $\lim s'(z) = 0$, as z tends to $-\infty$. Therefore, we assume

$$\lim_{z \to -\infty} S(z) = 0. \tag{62}$$

We wish to discover what phenomenological coefficient functions, $k(s)$, $\chi(s)$, and $\mu(s)$ make such solutions possible. Keller & Segel (1971) proved that for the special cases $\mu = \mu_0 = $ constant and $k = k_0 = $ constant, then such band solutions exist *only* if the function $\chi(s)$ goes to infinity as $s \to 0$ as fast as, or faster than, $1/s$. For the choice $\chi(s) = \delta/s$, they showed that (61) has a solution of the kind sought, and, remarkably, found an exact analytical formula for it. We postpone an interpretation of the need for, and meaning of, this singular chemotactic sensitivity function $\chi(s)$.

Keller & Odell (1975) considered arbitrary functions $\mu(s)$, $k(s)$, and $\chi(s)$, and found both necessary and sufficient conditions they must satisfy in order that (61) have bounded non-negative solutions with $\lim B(z) = 0$, as $|z|$ tend to ∞. Here, to make the analysis easier, but still to retain some generality, we consider a class of representative functions for k, μ, and χ, namely,

$$k(s) = k_0 s^\alpha, \qquad \mu(s) = \mu_0 s^r, \qquad \chi(s) = \delta_0/s^p, \tag{63}$$

where k_0, μ_0, and δ_0 are positive constants, and α, r, and p are constants. It turns out (see Keller & Odell, 1975) that only the behavior of $k(s)$, $\mu(s)$, and $\chi(s)$ near $s \to 0^+$ is important as regards the existence of traveling bands. For various choices of α, r, and p, (63) represents cases of virtually all conceivable kinds of behavior near $s \to 0$. System (63) cannot realistically represent k, μ, or χ as s becomes large, but, as is shown by Keller & Odell (1975), any kind of large-s variation can be 'spliced' into the behavior of small-s variation given in (63) without affecting the outcome of the following analysis.

Using (63), our ODE system (61), becomes

$$\begin{aligned}
dS/dz &= (k_0/c)S^\alpha B, \\
dB/dz &= (1/c\mu_0)S^{-r}[(k_0\delta_0)S^{\alpha-p}B - c^2]B.
\end{aligned} \tag{64}$$

A necessary condition on the nutrient consumption rate, $k(s) = k_0 s^\alpha$

Let A be the cross-sectional area of the capillary tube, and let N be the number of bacteria in the band. By definition,

$$N = A \int_{-\infty}^{\infty} B(z)\, dz. \tag{65}$$

Using the first of the equations of (64), we have, from (65),

$$N = A\left(\frac{c}{k_0}\right) \int_{z=-\infty}^{z=\infty} S^{-\alpha}\left(\frac{dS}{dz}\right) dz = \frac{cA}{k_0} \int_{S=0}^{S=S(\infty)=S_0} S^{-\alpha}\, dS$$

$$= \left[\frac{c}{k_0(1-\alpha)}\right] S^{1-\alpha}\Big|_{S=0}^{S=S_0}. \tag{66}$$

From (66) we see that the band will have a *finite* number of bacteria in it only if,

$$\alpha < 1. \tag{67a}$$

In this case, using (66), we have the following relationship determining band speed as a function of N, k_0, A, α, and the ambient nutrient concentration $S_0 = \lim S(z)$, as z tends to ∞.

$$c = k_0(1-\alpha)N/S_0^{1-\alpha}. \tag{67b}$$

Equation (67b) means that band speed increases as the number of bacteria in the band increases, or as the rate at which they eat (measured by k_0) increases, or as the ambient food concentration, S_0, decreases. Any of the above effects will shorten the time taken by a band to consume the nutrient in its neighborhood, and thus increase the speed at which the band must go to avoid starvation.

Steady state points of (64)

$(S, B) = (S_0, 0)$ is a steady state point of (64) for all positive values of S_0. That is, *every* point on the positive S-axis is a steady state point, and, moreover, there are no other steady state points.

Taking the S-axis horizontal, and the B-axis vertical, the B-nullcline (where trajectories are horizontal) is the curve,

$$(k_0\delta_0 S^{\alpha-p}B - c^2) = 0,$$

or

$$B = (c^2/k_0\delta_0)S^{p-\alpha}. \tag{68}$$

Dividing the second of the equations of (63) by the first, we obtain an equation giving the slope of the trajectories at the point (S, B),

$$dB/dS = (1/k_0\mu_0)S^{-(r+\alpha)}[(k_0\delta_0)S^{\alpha-p}B - c^2]. \tag{69}$$

From (69), we see that, on the S-axis ($B = 0$), the trajectories have the negative slope,

$$\left.\frac{dB}{dS}\right|_{B=0} = -c^2/(k_0\mu_0 S^{r+\alpha}). \tag{70}$$

We now construct phase portraits for (64) for various different choices of α, r, and p to see which choices permit bounded positive solutions with the band property, $\lim B(z) = 0$, as $|z|$ tends to ∞. For every case, there will be trajectories that approach steady state points (any point on the S-axis) $(S_0, 0)$ as $z \to +\infty$. We seek such solutions for which $B(z) \to 0$ as z runs toward $-\infty$.

Case I, $p - \alpha < 0$.

In this case the B-nullcline, given in (68) tends to $+\infty$ as $S \to 0^+$. Figure 6.7.6 shows the situation. Just using (69) and the location of the B-nullcline, we see (without bothering with any detailed analysis of where the trajectories go as $z \to -\infty$), that $B(z)$ cannot become zero as z gets small. We therefore reject this case.

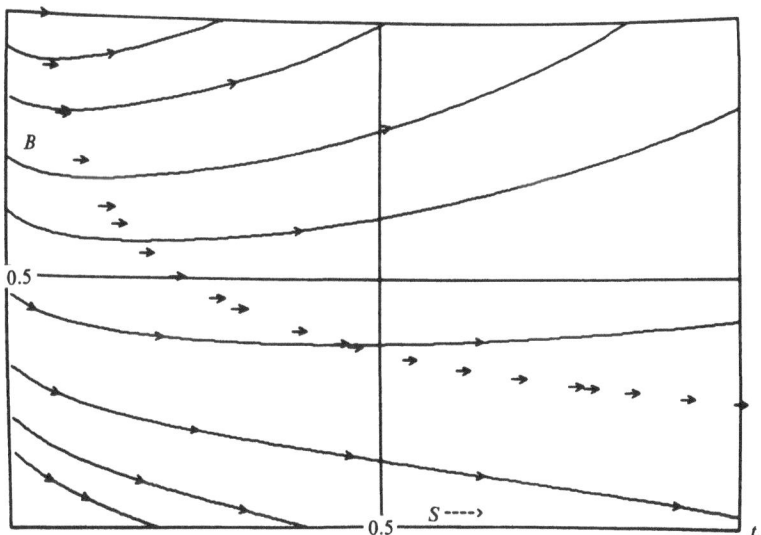

Figure 6.7.6. Generic phase portrait for the chemotactic band ODE system (64), when $p - \alpha < 0$, in which the B-nullcline tends to ∞ as s tends to 0. Here no trajectories exist that begin and end with $B \to 0$; hence no proper band solutions exist.

Case II, $p - \alpha = 0$.

In this case, the B-nullcline is horizontal and intersects the B-axis at a finite nonzero height. Figure 6.7.7 shows the situation. Pick any trajectory that approaches any steady state point $(S_0, 0)$ on the positive S-axis as $z \to +\infty$. Since this trajectory, followed backward in z, can become horizontal only as it approaches the B-nullcline (which is the horizontal line $B = c^2/k_0\delta_0$), it can never approach $B = 0$ as it is followed backward. Thus, this case, also, is rejected without bothering with any detailed analysis of the trajectories as $z \to -\infty$.

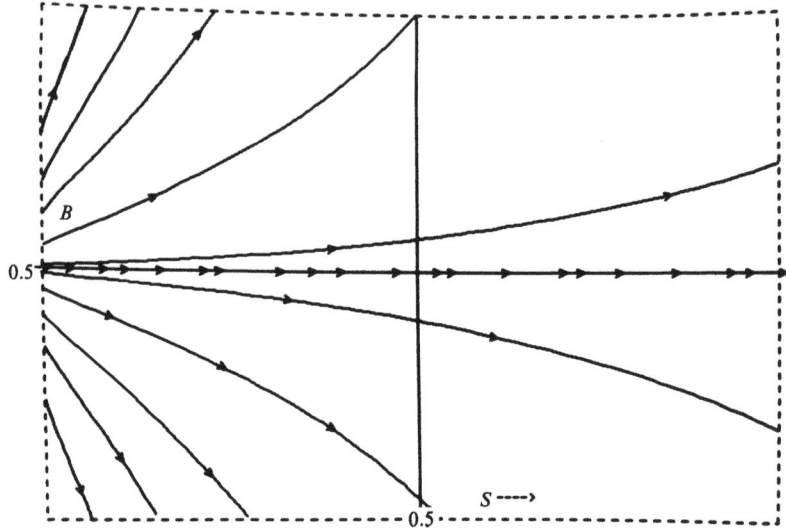

Figure 6.7.7. Phase portrait for the chemotactic band ODE system (64) when $p - \alpha = 0$. No trajectories exist that begin and end with $B \to 0$.

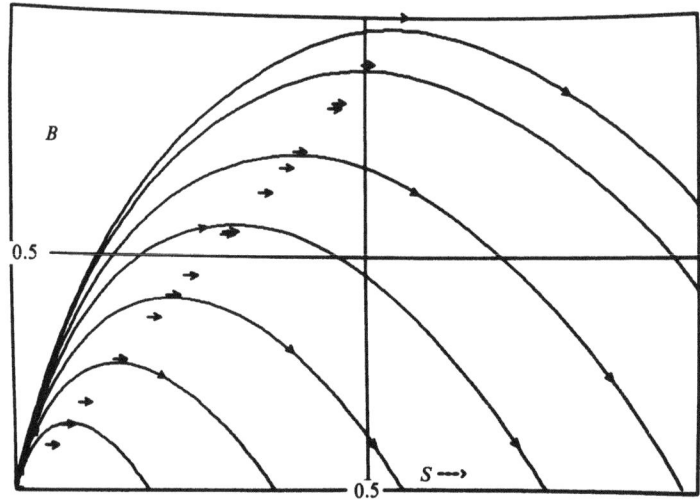

Figure 6.7.8. Phase portrait for the chemotactic band ODE system (64) for the case $k_0 = 1.0$, $k(s) = 1.0$, $\mu(s) = 1.0$, $\chi(s) = 1.25/s_0$.

This is generic for the cases when $p - \alpha = 0$. There is a one-parameter family of trajectories that begins at $(\hat{S}, B) \to (0, 0)$ and end with $B \to 0$; each corresponds to a properly behaved band solution of the sort illustrated in Figure 6.7.10.

Case III, $p - \alpha > 0$.

Here there are three subcases, $0 < p - \alpha < 1$, $p - \alpha = 1$, and $p - \alpha > 1$. In each, the B-nullcline, given by (68), intersects the origin $(0, 0)$, and rises as S increases. We illustrate the $p - \alpha = 1$ case in Figure 6.7.8, in which the B nullcline is a straight line. For Figure 6.7.8 the parameter values are $k_0 = 1.0$, $\alpha = 0$, $\mu_0 = 1.0$, $r = 0$, $\delta_0 = 1.25$, $p = 1$, $c = 1.5$. The other two subcases have the same general look, with the B-nullcline concave, either upward or downward accordingly as $p - \alpha$ is greater or less than 1. In any of these subcases, pick any trajectory, in the first quadrant, where it crosses the B-nullcline. Traced forward (z increasing), it must slant down to approach some steady state point $(S_0, 0)$ on the S-axis as $z \to \infty$. Tracing it backwards (z decreasing), B decreases, and *there is a chance* that, as $z \to -\infty$, the point $(0, 0)$ is approached. In this case, $p > \alpha$, then, it appears worthwhile to analyze what happens as $z \to -\infty$. So far, we have found

$$p > \alpha, \tag{71}$$

and (67), to be *necessary* conditions for traveling bands.

Analysis of conditions at the rear of the band (near $S = 0$)

We will prove the following:

Proposition.   If (71) holds, and

$$r + p > 1, \tag{72}$$

then, as $z \to -\infty$, $S \to 0$ and $B \to 0$. Thus, conditions (67), (71) and (72) together are sufficient to guarantee the existence of a well-behaved traveling band containing finitely many bacteria.

We prove this by assuming it is false and deriving a contradiction, that is, we assume B does not approach 0 as $S \to 0$. If B does not approach 0 as $S \to 0$, but stays positive, then, in (69), the $S^{\alpha - p}B$ term overwhelms the $-c^2$, as $S \to 0$. This is because, given (71), $S^{\alpha - p} \to +\infty$ as $S \to 0$. Then, near $S \to 0$, (69) becomes

$$dB/dS \approx (\delta_0/\mu_0)S^{-r-\alpha+\alpha-p}B,$$

or

$$\int dB/B \approx \int (\delta_0/\mu_0)S^{-r-p}\, dS. \tag{73}$$

Both sides of (73) can be integrated at once, yielding

$$\ln B = (\delta_0/\mu_0)(1 - r - p)^{-1}S^{1-r-p} + R,$$

where R is a constant of integration. The above equation is the same as

$$B = Q \exp \left[(\delta_0/\mu_0)(1 - r - p)^{-1} S^{1-r-p} \right]. \tag{74}$$

Now, if $r + p > 1$ (our hypothesis), then $(\delta_0/\mu_0)(1 - r - p)^{-1} < 0$, and, with $1 - r - p < 0$, the expression in square brackets in (74) tends to $-\infty$ as $S \to 0$. Since $\exp(-\infty) = 0$, we have $B \to 0$ as $S \to 0$, contradicting what we assumed about B. The contradiction proves our result.

Exercise. Show that (72) (assuming the necessary condition (71) holds) is *necessary* to have $B \to 0$ as $S \to 0$. The solution may be found in the article by Keller & Odell (1975).

With the exercise completed, we have, principally by phase plane analysis, the following result. *With $k(s)$, $\mu(s)$, and $\chi(s)$ given by (63), the ODE system (61) has a positive bounded solution $B(z)$, such that $\lim B(z) = 0$, as $|z|$ tends to ∞, and hence the PDE system (55) and (56) has a solitary pulse traveling band solution (with $\int_{-\infty}^{\infty} b(x, t) \, dz$ finite) if and only if*

$$\alpha < 1, \qquad p > \alpha, \quad \text{and} \quad p + r > 1. \tag{67a), (71), (72}$$

Notice that, as long as they are positive, it does not matter what k_0, μ_0, and δ_0 are (of course the 'shape' of the band is affected by k_0, μ_0, δ_0, as well as by α, p, and r).

Interpretation of results

The condition (67a) means that, however intuitively appealing the $\alpha = 1$ choice may be, making $k(s) = k_0 s$ near $s = 0$, it will not permit exact traveling band solutions. Only if $k(s)$ vanishes abruptly (with infinite slope) at $s = 0$, $k(s) = k_0\sqrt{s}$ for example, do we get bands. Equation (67a) was required to keep the number in the band finite. The consumption of nutrient is implemented by enzyme transport presumably. If the usual Michaelis–Menten kinetics are assumed, $k(s) \sim k_1 s/(k_2 + k_3 s)$, then, near $s = 0$, we have $k(s) \propto s$, violating (67).

$k(s)$ *must* vanish as $s \to 0$ (no food means no consumption) for biological realism. Thus, we need $0 < \alpha < 1$, and thus (71) requires $p > \alpha > 0$. This means, to permit exact traveling band solutions, the chemotactic sensitivity, $\chi(s) = \delta_0/s^p$, must become infinite as $s \to 0$.

The reason for this necessity is intuitively clear. The bands we seek must look something like the solutions shown in Figure 6.7.10. The effect of the random motility (diffusion with diffusivity μ) acts to advance the leading edge of the band forward, but equally, to push the trailing slope of the band backward. This diffusive flattening (broadening) of the trailing edge of the band must be counterbalanced by the sharpening effect there of chemotaxis. The strength of chemotaxis is, by (54), $b\chi(s)(\partial s/\partial x)$. In order to balance the diffusive flux $-\mu(\partial b/\partial x)$ at the trailing edge caused by *finite gradient of b there*, $\chi(s)$ must go to infinity as s (and $\partial s/\partial x$, and b) $\to 0$ in order that the product $b\chi(s)(\partial s/\partial x)$ stay finite. It is this delicate balance at the trailing edge that determines the conditions for traveling bands. In contrast, since the chemotactic *and* diffusive contributions to the bacterial flux at the leading edge both work to advance the leading edge of the band, *any* reasonable recipes for $\mu(s)$, $k(s)$, and $\chi(s)$ will allow the front to propagate.

To carry the above reasoning a bit further, the interpretation of (72) is the following. The larger r is, the faster the random motility coefficient,

$$\mu(s) = \mu_0 s^r,$$

vanishes as $s \to 0$ at the trailing edge. And, the larger r is, the smaller p need be (i.e. the 'less singular' $\chi(s) = \delta_0/s^p$ need be) to satisfy (72). Since $\chi(s)$ needs to get large as $s \to 0$ precisely to balance random motility, the trade-off suggested by (72) makes sense. If $\alpha \simeq 0$, then $\chi(s) = \delta_0/s^p$ need be only 'very slightly singular' ($p > \alpha \sim 0$ to satisfy (71)) provided the random motility $\mu(s) = \mu_0 s^r$ vanishes fast enough ($r \sim 1$ to satisfy (72)) as $s \to 0$.

On what the bands look like

For any particular set of parameters, μ_0, k_0, δ_0, p, r, α, in (63), the shape of the exact traveling wave solution, $S(z)$ and $B(z)$, can be calculated numerically. Figure 6.7.9 for example, shows a quantitatively accurate phase portrait for the case,

$$k_0 = 1.0, \qquad \alpha = 0.25,$$
$$\mu_0 = 1.0, \qquad r = 0.50,$$
$$\delta_0 = 0.50, \qquad p = 1.00.$$

Figure 6.7.10 shows $S(z)$ and $B(z)$ for this case when $S_0 = 1.0$, and $N = 1.33$, so that $c = 1$; z varies from 0 to 20, in the figure.

For isolated special choices, analytical solutions may be obtained. See, for example, Keller & Segel (1971) for explicit solutions, both analytic formulae and curves drawn, for the cases $p = 1$, $r = \alpha = 0$, $\delta_0/\mu_0 > 1$. For the cases $p = 1$, $r = \alpha = 0$ and $0 < \delta_0/\mu_0 < 1$, see Odell & Keller (1976). For these cases, the following properties of the band shapes are known. For $\bar{\delta} = \delta_0/\mu_0 > 2$, the leading edge of the band is steeper than the trailing edge; for $\bar{\delta} < 2$, the trailing edge is steeper.

See Holz & Chen (1978, 1979) for an analysis of band profiles by laser techniques. Agreement with the theory is good on many, but not all,

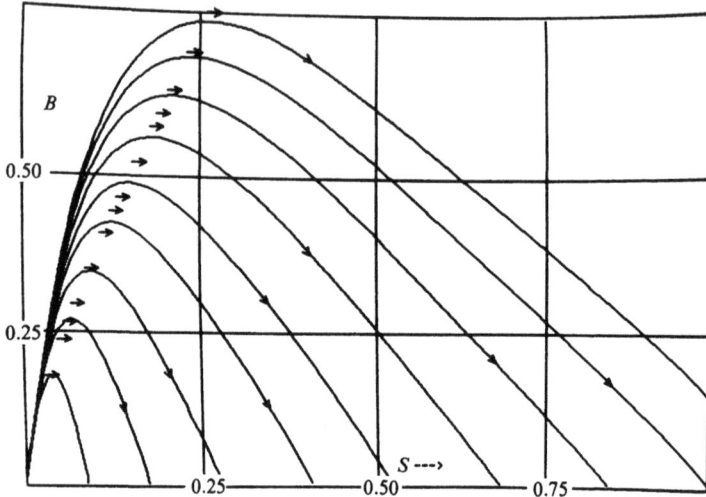

Figure 6.7.9. Phase portrait for the case $K(s) = s^{\frac{1}{4}}$, $\mu(s) = s^{\frac{1}{2}}$, $\chi(s) = 0.5/s$.

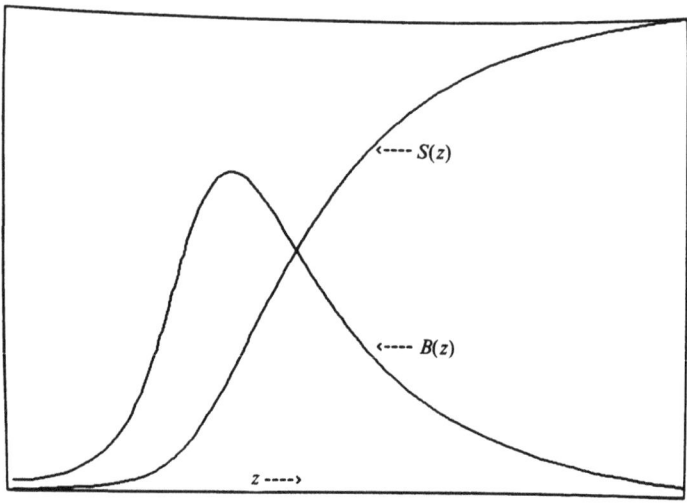

Figure 6.7.10. $S(z)$ and $B(z)$ along one trajectory from the phase portrait of Figure 6.7.9. This is an exact propagating band solution to the chemotaxis PDE system (84). The coefficient functions are $k(s) = s^{\frac{1}{4}}$, $\mu(s) = s^{\frac{1}{2}}$, and $\chi(s) = 0.5/s$, and, for the band graphed, $S_0 = 1$ and $N = 1.33$, so that $c = 1$. The wave variable ranges from 0 to 25.

points. In particular it is found that the bands slowly change their shape, owing to a gradual leakage of bacteria from the rear of the band. This can be associated with a deviation from conditions (67a), (71), and (72), when substrate concentrations are very low. Indeed, Novick (1977) has shown that, if conditions for an exact traveling wave are violated when s becomes very small, then a 'near' traveling wave can be anticipated, with a slow attenuation caused by leakage from the rear.

If a sequence of bands is considered, each with the same number of bacteria, then smaller values of $\bar{\delta}$ (that is, weaker chemotactic sensitivity relative to random motility) correspond to shorter taller 'bell-shaped' functions $B(z)$. The maximum height of the band is

$$b_{max} = \sup_{all\ z} (B(z)) = \frac{k_0 N^2}{S_0 \mu_0} \left\{ \begin{array}{ll} \bar{\delta}^{[\bar{\delta}/(1-\bar{\delta})]}, & if\ \bar{\delta} \neq 1 \\ \exp(-1), & if\ \bar{\delta} = 1 \end{array} \right\}. \tag{75}$$

The function of $\bar{\delta}$ in the curly brackets is monotone decreasing, tending to zero as $\bar{\delta} \to \infty$, and approaching 1 as $\bar{\delta} \to 0$. If the ambient nutrient concentration, δ_0, the consumption rate constant, k_0, the random motility coefficient, μ_0, and the number of bacteria in the band, N, are all held fixed, then as δ_0 (measuring the strength of chemotactic sensitivity) gets smaller (so $\bar{\delta} = \delta_0/\mu_0$ gets smaller) then this band gets taller and hence narrower! This may seem counterintuitive. The following line of reasoning explains it. If k_0, N, and S_0 are fixed, then, by (68), c is fixed also, regardless of the value of $\bar{\delta}$. As δ_0 gets smaller (hence chemotactic sensitivity decreases), the gradient in s must get steeper so that the band can track this gradient at the fixed speed $c = k_0 N/S_0$ using its diminished chemotactic sensitivity. Only a narrower (taller) band can create a sharper gradient in s. We expect this qualitative result (known rigorously from Keller & Segel (1971) and Odell & Keller (1976) for the $\alpha = r = 0$, $p = 1$ case) to hold generally.

We now undertake the explanation of a subtle point. In the special case $r = \alpha = 0$, $p = 1$, and $\delta_0/\mu_0 < 1$, an exact solution of (61) or (64) is

$$S(z) = \left\{ \begin{array}{ll} S_0[1 - \exp(-cz/\mu_0)]^{\mu_0/(\mu_0 - \delta_0)}, & if\ z \geq 0 \\ 0, & if\ z < 0 \end{array} \right\} \tag{76}$$

and

$$B(z) = \left\{ \begin{array}{l} \dfrac{c^2 S_0}{k(\mu - \delta)} \cdot \exp(-cz/\mu_0) \\ \quad \times [1 - \exp(-cz/\mu_0)]^{\delta_0/(\mu_0 - \delta_0)}, \quad if\ z \geq 0 \\ 0, \quad if\ z < 0. \end{array} \right\} \tag{77}$$

These are derived and sketched in Odell & Keller (1976) and may be verified by differentiation and substitution into (64). These are 'strange' solutions[6] because they consist of two 'pieces', spliced together at $z = 0$. If

---

[6] They were overlooked by Keller & Segel (1971).

$\delta_0/\mu_0 < \frac{1}{2}$, then, at $z = 0^+$, where $B(z) = 0$ and $S(z) = 0$, we have

$$\left.\frac{dB}{dz}\right|_{z=0^+} = +\infty. \tag{78}$$

This means (for any $\delta_0/\mu_0 < 1$) that the steady state point $(S, B) = (0, 0)$ is hit, not as $z \to -\infty$, but at a *finite* value of z (namely at $z = 0$). Then the solution to (64) 'sits' steadily at $(0, 0)$ as $z \to -\infty$. The solution function, $B(z)$, specified by (77) is continuous for all z, but has an infinitely sharp 'kink' in it (when $\delta_0/\mu_0 < \frac{1}{2}$) at $z = 0$. This means that $b(x, t) = B(x - ct) = B(z)$ given in (77) is a 'suspicious'[7] solution of the PDE system (55) and (56), because the partial derivative, $\partial^2 b/\partial x^2$, that (56) seems to demand cannot be computed at the kink point. The subtle, and very important point this confronts us with is the following. The general scalar PDE balance law, (6.1.4), was derived from a fundamental balance statement (6.1.1) *expressed in integral form* using the assumption that the flux, J, was a smooth differentiable function of x and t. Sometimes the smoothness assumption is not valid (as in this present case). Then, one must use (6.1.1), and not the PDE (6.1.4). It is claimed that (76) and (77) satisfy this integral form of the balance law.

The preceding invocation of the primacy of the integral form of the balance law notwithstanding, there is a temptation to regard the 'kinky' solutions, (76) and (77), as bizarre and physically meaningless; this seems not to be the case. Preliminary numerical evidence[8] indicates that, under these 'kink' conditions, an inoculum of bacteria at one end of the capillary tube at $t = 0$ forms itself into a traveling band that is, evidently, a stable solution to the PDE, whose shape evolves toward that of the 'kink' solution, (76) and (77), as t increases.

Numerical simulations by Scribner, Segel & Rogers (1974) indicate, also, that the exact traveling wave solutions we have discussed here are stable solutions of the governing PDE's.

Taking a lesson from the 'kink' solutions given in (76) and (77) that were obtained for the special $r = \alpha = 0$ case, we must interpret all statements above that

$$\lim_{z \to -\infty} S(z) = 0 \quad \text{and/or} \quad \lim_{z \to -\infty} B(z) = 0,$$

as meaning either that the limiting value, 0, is approached (without being actually reached) asymptotically, as $z \to -\infty$, or, alternatively, meaning that the zero values of $S(z)$ and $B(z)$ are hit at a finite value of z. As is

---

[7] The proper jargon is **weak solution**.

[8] The author of this section expresses thanks to J. Hyman, at the Los Alamos Scientific Laboratory, for his impressive 'while-you-wait' solution of some initial value problems for the nonlinear PDE system (55) and (56) (with $\mu(s) = \mu_0$, $\chi(s) = \delta_0/s$, and $k(s) \le k_0$, $\delta_0/\mu_0 < 1$) using a general purpose 'Method of Lines' numerical PDE solver package he developed.

explained in Appendix A.3, phase portraits give no information on how rapidly or slowly trajectories are traversed. For $r = \alpha = 0$, $p = 1$, $\delta_0/\mu_0 >$ 1, the point $(S, B) = (0, 0)$ is approached smoothly, asymptotically as $z \to -\infty$. For the same case, but with $\delta_0/\mu_0 < 1$, it is hit abruptly at finite z. We must anticipate instances of 'kink' solutions resembling (76) and (77) not just for the $\alpha = r = 0$, $p = 1$ case but generally. Our necessary and sufficient conditions in Keller & Odell (1975), and the conditions (76), (71), and (72) above, make no statement about the smoothness of the solutions.

With (63), (67a), (71), and (72) as sufficient conditions for exact traveling bands, and the slightly more general necessary and sufficient conditions in Keller & Odell (1975), we have cataloged all possible phenomenological behavior that can lead to exact propagating bands. The band-forming competence of a strain of microorganism waxes as the functions $k(s)$, $\mu(s)$, and $\chi(s)$ are tuned (over evolutionary time) nearer to fulfilment of these constraints.

Nerve action potential conduction

We turn now for a brief look at one of the great successes in mathematical biology: the Hodgkin–Huxley theory characterizing the ion permeability properties of nerve membrane. We make no attempt at anything like a complete exposition of this theory and the vast experimental and mathematical literature surrounding it. Instead, our aim is to illustrate the way the same mathematical structure (in this case, parabolic PDE's, derived as continuum balance laws, that conduct exact traveling wave solutions) underlies diverse phenomena.

We shall assume the reader is familiar in a general way with the basic physiology of nerves. From each neuron emanates a long cylindrical tube called an axon, along whose outer membrane (the axolemma) signals are sent in the form of propagating electrical depolarization waves. The axon membrane is now known to contain a variety of 'pumps' and 'gates', consisting of small protein islands dispersed in the bilipid layer membrane. The pumps slowly and continually pump sodium ions out across the membrane to establish a 'resting potential' (voltage drop) across the membrane. Some gates modulate the membrane's permeability (conductance) to sodium ions; others modulate its permeability to potassium ions (see Keynes, 1979, and Hille, 1978).

The membrane passes other ions, but its active modulation of sodium and potassium conductances are the most important mechanisms for action potential conduction.

By an elegant sequence of experiments on squid axon, Hodgkin, Huxley, and Katz (see Hodgkin & Huxley, 1952, and their papers with

Katz cited in the bibliography therein) obtained a quantitative description of how, in a typical piece of nerve membrane, the selective permeability to sodium and potassium changes with time in response to changes in the transmembrane potential. During these ('space-clamped') experiments, no waves propagated along the axon; a wire down its axis forced spatial homogeneity.

Armed with the resulting local data fit (which we will reproduce below), they derived the PDE governing the variation of transmembrane potential as a function of both axial position along the axon and time. Using the strategy expounded in this section, they sought and found (numerically, using a mechanical calculator!) an exact traveling wave solution to the PDE. Their triumph was that the propagating waveform reproduces the transmembrane voltage waves recorded on oscilloscopes when propagating action potential spikes race past recording electrodes inserted through axon membranes.

The basic structure of the model

We make a one-dimensional model with x representing axial displacement and t representing time. Let $\Gamma(x, t)$ be the charge density inside the axon (per unit length of axon). Following (6) we write the balance law for charge

$$\partial\Gamma/\partial t = -(\partial J/\partial x) + Q, \tag{79}$$

where J is the net axial ion current (flux of charge), and Q is the creation rate density (per unit length of axon) of interior charge.

Let V denote the displacement of transmembrane voltage from its resting value. If C is the capacitance per unit area of membrane and a is the radius of the circular cylindrical axon, then $2\pi aC$ is the capacitance per unit length of axon and

$$\Gamma = 2\pi aCV. \tag{80}$$

To be precise, since V is a displacement potential, Γ must be the displacement (from the resting value) of interior charge density.

Assuming that the interior of the axon consists of uniform cytoplasm (electrolyte) with specific resistivity, R, and that the axon is bounded on the outside by an infinite conducting fluid, we have, by Ohm's law,

$$J = -(\pi a^2/R)(\partial V/\partial x). \tag{81}$$

In this one-dimensional model we think of the $(+)$ ionic current through the membrane from exterior to interior as 'creating' interior charge. Thus

$$Q = -2\pi aI_i, \tag{82}$$

where I_i is the net electrical transmembrane current density (per unit area of membrane) resulting from net outward flux of positive ions.

R, a, and C are assumed constants. Thus, substituting (80), (81), and (82) into (79), we obtain

$$\partial V/\partial t = +(a/2RC)(\partial^2 V/\partial x^2) - (1/C)I_i. \tag{83}$$

The major contribution made by Hodgkin, Huxley and Katz was to quantify the current density, I_i. They represented I_i as the sum of three distinct currents, I_{Na} (of sodium), I_K (of potassium), and I_L (of all other ions; 'L' is for leakage). Furthermore, to characterize I_K, they were forced to introduce an **internal variable**,[9] n, and to characterize I_{Na}, two more internal variables, m and h, were required.

The data fit they obtained is the following:

$$I_i = \bar{g}_{Na}m^3 h(V - V_{Na}) + \bar{g}_K n^4 (V - V_K) + \bar{g}_L(V - V_L), \tag{84}$$

where $\bar{g}_{Na}$, $\bar{g}_K$, and $\bar{g}_L$ are constant nominal conductances, while V_{Na}, V_K, and V_L are constant equilibrium potentials, for the three ion species. The behavior of each of the internal variables, m, h, and n is determined by a differential equation for each of them:

$$dn/dt = \alpha_n(V)(1-n) - \beta_n(V)n,$$
$$dm/dt = \alpha_m(V)(1-m) - \beta_m(V)m, \tag{85}$$
$$dh/dt = \alpha_h(V)(1-h) - \beta_h(V)h.$$

α_n, α_m, α_h, β_n, β_m, β_h are functions of V specified as follows:

$$\alpha_n(V) = 0.01(V + 10) \exp\left[1 - \left(\frac{V+10}{10}\right)\right],$$

$$\alpha_m(V) = 0.1(V + 25) \exp\left[1 - \left(\frac{V+25}{10}\right)\right],$$

$$\alpha_h(V) = 0.07 \exp(V/20),$$

$$\beta_n(V) = 0.125 \exp(V/80), \tag{86}$$

$$\beta_m(V) = 4.0 \exp(V/18),$$

$$\beta_h(V) = \exp\left[-1 - \left(\frac{V+30}{10}\right)\right].$$

V is measured in millivolts. The values of the other parameters can be found in Hodgkin & Huxley (1952).

---

[9] An internal variable is a mathematical device useful for representing an extensive measurable quantity (such as I_K) whose detailed dynamics are not clearly understood; the variable may or may not turn out to correspond to a physically identifiable mechanism.

Mathematically, we regard (85) as a set of balance laws for quantities m, h, and n that are produced and destroyed locally, but have no axial flux. This makes sense if we think of m, h, and n as attached to protein gate assemblies that do not move in the axolemma.

Equations (84), (85), and (86) fit the experimental squid data in the sense that, when $\partial V/\partial x$ is set equal to zero in (83) (to model the spatially homogeneous conditions under which the data were recorded), then the system of four first-order autonomous ODE's consisting of

$$\mathrm{d}V/\mathrm{d}t = (-1/C)I_i, \tag{87}$$

with I_i given by (84), and (85) and (86), yield solutions (especially $V(t)$) that correspond with quantitative accuracy to measured data. Refer to Hodgkin & Huxley (1952) for the fascinating story of how the various components of (84), (85), and (86) were separately determined.

Do traveling wave solutions exist?

Hodgkin & Huxley naturally looked for wave solutions to the PDE system consisting of (83), (84), and (85), because nerve axons obviously conduct traveling waves. To exhibit the dependent variables, with equal emphasis, we convert the above system into the standard form (35). We look for a solution as a waveform propagating at speed c. Thus, assume

$$
\begin{aligned}
V(x, t) &= U_1(z), \\
m(x, t) &= U_2(z), \\
h(x, t) &= U_3(z), \\
n(x, t) &= U_4(z),
\end{aligned}
\tag{88}
$$

where $z = x - ct$ is the wave variable. We obtain the following fifth-order autonomous system of ODE's:

$$
-c\left(\frac{\mathrm{d}U_1}{\mathrm{d}z}\right) = +\left(\frac{a}{2RG}\right)\frac{\mathrm{d}^2 U_1}{\mathrm{d}z^2} + \frac{1}{C}\left[\bar{g}_{\mathrm{Na}}U_2^3 U_3(U_1 - V_{\mathrm{Na}})\right.
$$

$$
\left. + \bar{g}_{\mathrm{K}}U_4^4(U_1 - V_{\mathrm{K}}) + \bar{g}_{\mathrm{L}}(U_1 - V_{\mathrm{L}})\right], \tag{89}
$$

$$
-c\frac{\mathrm{d}U_2}{\mathrm{d}z} = \alpha_m(U_1)(1 - U_2) - \beta_m(U_1)U_2,
$$

$$
-c\frac{\mathrm{d}U_3}{\mathrm{d}z} = \alpha_h(U_1)(1 - U_3) - \beta_h(U_1)U_3, \tag{90}
$$

$$
-c\frac{\mathrm{d}U_4}{\mathrm{d}z} = \alpha_n(U_1)(1 - U_4) - \beta_n(U_1)U_4.
$$

The functions α_n, α_m, α_h, β_n, β_m, β_h are specified in (86).

The fifth-order ODE system (89) and (90), is of the class discussed in Appendix A.3. It is a very complicated system, however, and we will not even begin an analysis of it. The interested reader might begin a study of it with Hastings (1975). We make only a few general remarks.

If all the constant parameters are assigned values, then the search for a solitary, single-pulse solution consists of adjusting the wave speed, c, until a (homoclinic) solution trajectory is found to exist in which $V(x, t) = U_1(z) \rightarrow U_0$ both as $z \rightarrow +\infty$ and as $z \rightarrow -\infty$. For some parameter sets, numerical evidence indicates that no such c exists. For other sets, two such values of c exist, the larger giving a short wavelength fast pulse (that is now known to be a stable solution to the PDE's), and the smaller giving a long wavelength, slow pulse (now conjectured to be unstable as a solution to the PDE's (see Rinzel (1978) and Evans (1975)). Hodgkin and Huxley found both pulses. It is the fast one that corresponds beautifully to the observed solitary action potential pulse.

Action potentials are initiated by a depolarization event where the axon joins the neural soma by a depolarization event there. If the constant parameters are adjusted so as to make a single isolated pulse solution possible, then a given trigger attempt at the soma will, if it is 'strong' enough, initiate a propagating pulse that, as it moves down the axon, quickly assumes the shape predicted by the faster ODE single pulse (homoclinic) solution. Trigger events that are too weak set off a $V(x, t)$ that attenuates without propagating very far. (See Aronson & Weinberger (1974) for a demonstration of this using the Fitzhugh–Nagumo equation discussed below.)

In addition to single-pulse solutions, a rich continuous spectrum of other solutions to (89) and (90) exists. For a set of speeds, c, (less than the fast single-pulse speed) (89) and (90) have periodic (limit cycle) solutions. These correspond to infinitely repeating propagating wave trains. Here, the propagation speeds (of the periodic solutions that turn out to be stable solutions to the PDE's) vary with the axial spacing of the pulses in such a way that more closely spaced pulses propagate slower than more widely spaced ones. That is, the axon is a dispersive wave carrier. This fact may have biologically important consequences because the dispersion effect tends to scramble messages encoded into the firing frequency when those messages move on a very long axon.

In addition to periodic (limit cycle) solutions, (89) and (90) apparently possess a spectrum of (homoclinic $V(z) \rightarrow U_0$ as $|z| \rightarrow \infty$) solution trajectories that have multiple V-peaks (see Carpenter, 1977).

The Fitzhugh–Nagumo equations and other caricatures of the Hodgkin–Huxley equations

The reader who wishes to *begin* a study of the mathematical theory of nerve conduction should not begin with the Hodgkin–Huxley equations. They are too difficult and are complicated by mathematical details that, while needed for quantitative data fitting, are not crucial to the nature of the basic property of the axolemma as an excitable medium. Fortunately, an essentially complete caricature of the Hodgkin–Huxley equations is available.

The mechanism that makes wave propagation possible is that I_i in (83) be so constituted that a small increase of V brings about an influx of current, I_i, that, raising V further, autocatalyzes an accelerating increase of (depolarization) V.

Fitzhugh (1969) and Nagumo, Arimoto, & Yoshizawa (1962) introduced a caricature of the Hodgkin–Huxley equations, obtained by replacing (84) (in dimensionless form) by

$$I_i = f(V) + w, \tag{91}$$

where f is the cubic function

$$f(V) = V(a - V)(1 - V), \tag{92}$$

in which $0 \le a \le \frac{1}{2}$, and where w is governed by the linear ODE

$$\partial w / \partial t = bV \qquad (b > 0). \tag{93}$$

Thus, in (91), (92), and (93) (now called the Fitzhugh–Nagumo equations), there is only one internal variable, w, replacing the three internal variables m, h, and n, in the Hodgkin–Huxley equations. $f(V)$ described in (92) has the essential autocatalytic property just mentioned. The ODE system that results when an exact propagating wave solution for (91) and (93) is sought is only third order. With $V(x, t) = U_1(z)$, and $w(x, t) = U_2(z)$, where $z = x - ct$, we obtain

$$-c(dU_1/dz) = \frac{d^2 U_1}{dz^2} - f(U_1) - U_2,$$
$$-c(dU_2/dz) = bU_1. \tag{94}$$

The reduction in order notwithstanding, qualitatively the solutions of (94) exhibit the same types of behavior described above for the Hodgkin–Huxley equations (see Rinzel, 1978).

The system (94) is still a strongly nonlinear ODE. McKean (1970) suggested a caricature of the Fitzhugh–Nagumo caricature in which the cubic function, f, in (92), whose graph is as shown in Figure 6.7.11, is

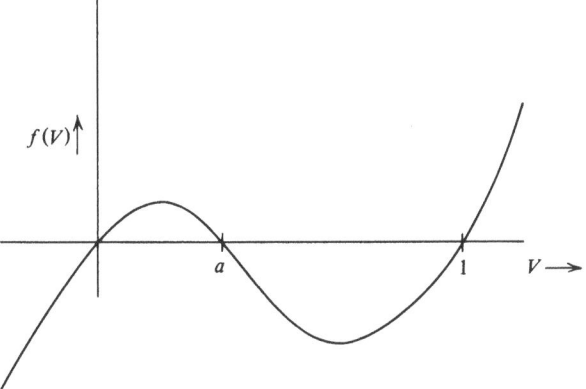

Figure 6.7.11. The cubic permeability function $f(V)$, given in (92), for the Fitzhugh–Nagumo equations.

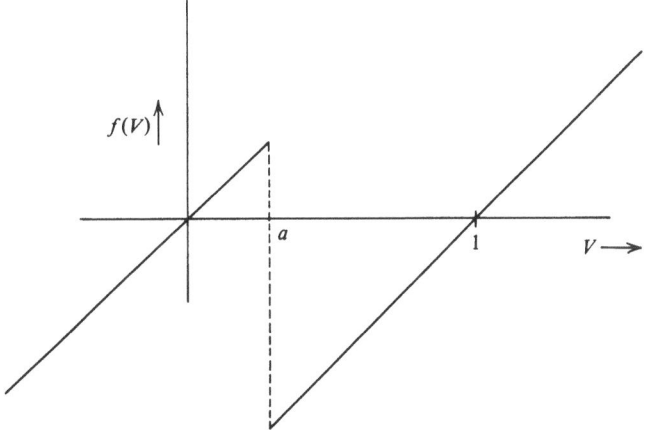

Figure 6.7.12. A graph of $f(V)$ as given in (95). This is the piecewise linear caricature of the cubic $f(V)$ in Figure 6.7.11 suggested by McKean (1970).

replaced by the piecewise linear (but discontinuous) function

$$f(V) = \begin{cases} V, & \text{if } V < a \\ V - 1, & \text{if } V \geqslant a \end{cases} \tag{95}$$

whose graph is shown in Figure 6.7.12.

The virtue of this is that (94) becomes a piecewise *linear* ODE system. Rinzel & Keller (1973) exploited this, finding, with (95), explicit formulae for almost all solutions of (94). Their analysis is accessible to students with little mathematical sophistication because solutions are proved to exist by being exhibited rather than by the much more abstract procedures needed when $f(V)$ is nonlinear. For this reason, the recommended starting point of a study of the mathematical theory of nerve conduction is Rinzel & Keller, 1973, or Rinzel, 1978.

Exercise. Using the phase plane analysis techniques given in Appendix A.3, investigate the second-order autonomous ODE system resulting when $\partial U_1/\partial z$ is set equal to zero in (94), and $f(V)$ is specified in (92), to prove that a ('space-clamped') spatially homogeneous model axon cannot exhibit limit-cycle oscillations.

References

Adler, J. (1975). Chemotaxis in bacteria. *Ann. Rev. Biochem.* **44**, 341–56.

Aronson, D. G. & Weinberger, H. F. (1974). *Partial Differential Equations and Related Topics*, ed. A. Dold and B. Eckmann, Lecture Notes in Mathematics No. 446, Berlin, Springer-Verlag.

Berg, H. (1975). Chemotaxis in bacteria. *Ann. Rev. Bioeng.* **4**, 119–36.

Boyce, W. E. & R. C. DiPrima (1977). *Elementary Differential Equations and Boundary Value Problems*, 3rd edn, New York, Wiley.

Carpenter, G. (1977). 'Periodic solutions of nerve impulse equations'. *J. Math. Anal. Appl.* **58**, 152–73.

Evans, J. W. (1975). 'Nerve axon equations. IV. The stable and unstable impulse', *Indiana Univ. Math. Jl*, **24**, 1169–90.

Fisher, R. A. (1937). *The advance of advantageous genes. Ann. Eugenics* **7**, 355–69.

Fitzhugh, R. (1969). Mathematical models of excitation and propagation in nerve. In *Biological Engineering*, ed. H. P. Schwan, New York, McGraw-Hill, Inc. 1–85.

Hastings, S. P. (1975). 'Some mathematical problems from neurobiology' *Am. Math. Monthly* **82**, 881–94.

Hille, B. (1978). Ionic channels in excitable membranes: current problems in biophysical approaches. *Biophys. J.* **22**, 283–94.

Hodgkin, A. L. & Huxley, A. F. (1952). A quantitative description of membrane current and its application to conduction and excitation in nerve. *J. Physiol.* **117**, 500–44.

Holz, M. & Chen, S-H. (1978). Quasi-elastic light scattering from migrating chemotactic bands of *Escherichia coli. Biophys. J.* **23**, 15–31.

—— (1979). Spatio-temporal structure of migrating chemotactic bands of *E. coli.* I. Traveling band profile. *Biophys. J.* **26**, 243–61.

Keller, E. F. & Odell, G. M. (1975). Necessary and sufficient conditions for chemotactic bands. *Math. Biosci.* **27**, 309–17.

Keller, E. F. & Segel, L. A. (1971). Traveling bands of chemotactic bacteria: a theoretical analysis. *J. Theoret. Biol.* **30**, 235–48.

Keynes, R. D. (1979). Ion channels in the nerve-cell membrane. *Scient. Am.* **240**(3), 126–35.

Kolmogoroff, A., Petrovsky, I. & Piscounoff, N. (1937). Etude de l'equation de la diffusion avec croissance de la matière et son application à un problème biologique. *Bull. Univ. Moskou, Ser. Internat., Sec.* A, **1**(6), 1–25.

Koshland, D. (1974). Chemotaxis as a model for sensory systems. *FEBS Lett.* **40**, *Suppl.*, 53–9.

McKean, H. P. (1970). Nagumo's equation. *Adv. Math.* **4**, 209–23.

Nagumo, J., Arimoto, S. & Yoshizawa, S. (1962). An active pulse transmission line simulating nerve axon. *Proc. IRE.* **50**, 2061–70.

Novick, A. (1977). Slowly varying bands of chemotactic bacteria. Masters Thesis, Department of Applied Mathematics, The Weizmann Institute of Science, Rehovot, Israel.

Odell, G. M. & Keller, E. F. (1976). Traveling bands of chemotactic bacteria revisited. *J. Theoret. Biol.* **56**, 243–7.

Rinzel, J. (1978). 'Integration and propagation of neuroelectric signals', in *MAA Studies in Mathematics*, vol. 15 *Studies in Mathematical Biology*, Part I *Cellular Behavior and the Development of Pattern*, ed. S. A. Levin, The Mathematical Association of America, pp. 1–66.

Rinzel, J. & Keller, J. B. (1973). Traveling wave solutions of a nerve conduction equation. *Biophys. J.* **13**, 1313–37.

Scribner, T. L., Segel, L. A. & Rogers, E. H. (1974). A numerical study of the formation and propagation of traveling bands of chemotactic bacteria. *J. Theoret. Biol.* **46**, 189–219.

7 Visual fixation and tracking in flies[1]

7.1 INTRODUCTION

What does it mean to understand a complex system?

Despite their complexity, all nervous systems appear to obey similar general principles at the level of cellular structures and mechanisms. On the one hand recognition of these general principles, especially in simple nervous systems, established neuroanatomy and electrophysiology as disciplines fundamental for neurobiology and a large body of knowledge has been accumulated concerning neuronal morphology and physiology at the level of one or a few cells. On the other hand, very little quantitative work has been done at the level of the function of an entire nervous system, or even parts of it, in order to explain its complex information-processing capabilities.

One reason for this is the enormous number of nerve cells even in very simple nervous systems or functional parts of it. If the circuitry is small (few cells) its information processing properties are trivial in almost all cases. If, however, the information processing properties of a nervous system or a part of it are nontrivial then generally many neurons will participate in its operation and *all* of them would have to be identified histologically and characterized functionally through electrophysiological recordings and other experiments (synapse characterization etc.).

This problem can be compared to that of thermodynamics. It is practically impossible to write the equations of motion of every single molecule of a gas in a cylinder and know the initial conditions at a certain instant. Even if this could be done and we could calculate the path of each molecule at each instant this enormous information would be of little practical help.

Thermodynamics provides a *phenomenological* description for the physics of a gas in a cylinder which enables us to understand its properties and predict its behavior. The phenomenological theory of thermo-dynamics has been very useful; for instance, nobody would try to use the microscopic, 'molecular' approach to calculate the volume of a gas for a

---

[1] Mathematical terms marked with a dagger † are explained in an appendix (Section 7.6).

given pressure and temperature. What we can see from this example is that there are systems which can be understood at (at least) two different levels: macroscopic phenomenological and microscopic. These are largely independent of each other although they can be connected by the theory of statistical mechanics.

In order to show what we mean by 'understanding' a complex system, e.g. that part of the brain of a fly which is responsible for visually guided orientation, another analogy (of Harmon, 1970) seems quite useful.

You are given a desert island and on it a high-speed digital computer, complete with terminal equipment, that is always in good repair. Your task is to 'understand' that system. You may import any number of people of any discipline, except that they must know nothing *a priori* about computer theory, structure, or operation.

You may obtain 'understanding' in three different ways. First, one of your teams must find out how to operate the computer; they must discern its gross input/output functions. They will 'understand' the computer when they are able to write a complete operation manual for the system.

Another group must seek understanding at a quite different level. They are required to find out, for instance, what a shift register is and how an adder works. This group will convince you of their 'understanding' when they are able to design a similar machine – not identical, but using the same principles of subsystem organization, operation, and inter-action.

A still different level of 'understanding' must be provided by the third group. It is required that they find out how the individual components operate. They must, for example, discern how a transistor works, eluci-date the properties of electronic tubes, and find out about magnetic storage. They will have reached their 'understanding' when they are able to produce complete equivalent-circuit diagrams of all components, both active and passive.

The three levels at which an 'understanding' is finally achieved can be called

(a) the level of hardware (transistors, tubes etc.);

(b) the level of algorithms (adders, memory); and

(c) the level of overall computation or logical organization of the system (operation manual).

Each level is somewhat independent of the other two. For instance, at the hardware level it is irrelevant for the description whether a transistor is used in an adder or in an and–not gate and whether the computer language is Fortran IV. At the level of the algorithms, one can see no difference between a magnetic or a flip–flop memory on the one hand and calculating or recognizing a stereo figure on the other. Finally, the system user may not care at all which algorithms are used for a fast

Fourier transform subroutine. What is important is that each level has its own validity in the description of a computer according to the problem we face and that each level can be developed independently from the other levels.

What we can learn from this analogy is that, for any complex system having nontrivial information processing capabilities, at least three levels of description should be distinguished. The exact definition of these depends on operational requirements. Coming back to our problem of 'understanding' the nervous system, we may define the levels as follows.

(*a*) *Basic component analysis*: This is the analysis of how the hardware works. In neurophysiology it includes the biophysical analysis of receptors, neurons, synapses, membranes, transmitters, etc.

(*b*) *Algorithms and functional properties*: This includes understanding of how, for instance, edge enhancement can be computed. It can be done using an algorithm based on a Fast Fourier Transform or lateral inhibition. Other examples could be understanding which algorithms may successfully compute movement of visual patterns or the stereoscopic perception of an object.

(*c*) *Theory of the overall computation*: Two questions at this level are 'what could be the basic computational organization for the control of the visual orientation behavior of the fly?' and 'what are the computations our visual system has to carry out in order to solve problems such as how to recognize a person or keep our image of the world stable while we walk?'

In general, one may say that the computation is determined by the problem a system has to solve, while the algorithms are determined by the computation and the available hardware.

It should be noted that the levels of description are only loosely related. There should be no confusion about the level at which a problem arises. This is important because the theoretical methods corresponding to one level of understanding cannot, in general, be used to answer questions arising at another.

Let us look more closely at our assertion that the distinction between the three levels of understanding is also reflected in the various theoretical methods which can be used to analyze a complex system. The classification is only indicative and there are several exceptions.

(*a*) Differential equations, chemical kinetics and noise analysis are typical theoretical methods for investigating biophysical problems involved in basic component analysis. Examples are the theory of Hodgkin & Huxley (1952), noise analysis of the axon channels (Verveen & De Felice, 1974), noise analysis of the chemical synapses (Katz & Miledi, 1973) and the Hodgkin photoreceptor model (Baylor, Hodgkin & Lamb, 1974*a*, *b*; Baylor & Hodgkin, 1974).

(*b*) Functional methods, linear and nonlinear system theory, estimation and identification methods are used in the study of algorithms and functional properties. Examples are lateral inhibition in man (Mach, 1914) and in the *Limulus* eye (Ratliff, Knight, Dodge & Hartline, 1974), movement and position detection in the fly (for a review see Poggio & Reichardt, 1976), pupillary system in humans (Stark & Baker, 1959; Stark & Sherman, 1957). Various other biological control systems have been analyzed with the same tools, for example eye movements (Yarbus, 1967) and insulin regulation (Calwill & Soeldner, 1969).

(*c*) There are as yet no clear-cut methods in the theory of the overall computation. Even the computational nature of many information processing questions is not well recognized. Behavioral data can be analyzed in various ways in order to obtain an overall understanding of a system. Ethology may be very useful in defining with precision the problems that a nervous system has to solve (see Section 7.3). At this level, computer simulation experiments are beginning to play an increasingly important role in bringing concepts and ideas into sharper focus.

In this section we look at the orientation behavior of the fly, which will be analyzed in detail in the subsequent sections at level (*c*). The flight behavior of houseflies requires elaborate flight control systems. These insects perceive motion relative to the environment and thereby stabilize their flight course: they locate and fly towards prominent objects; they track moving objects and chase other flies; they discriminate or prefer some specific visual patterns. The analysis of one of these control systems (surveyed by Reichardt & Poggio, 1976) represents an example of a theory at the computational level. The overall computation is defined and well accessible to experimentation, since it involves a complete input–output transduction, from the optical input to the behavioral motor-output. As we will see, fixation of a pattern or tracking of an object by a fly can be described independently of both the specific algorithm actually used by the nervous system and the specific cellular components from which the neuronal circuitry is built.

Flight orientation behavior: analysis on the computational level

In free flight, a fly can move in three rotational and three translational degrees of freedom. In addition, a flexible neck enables the fly to move its head independently. The behavior described here is restricted to one rotational and one translational degree of freedom, namely rotation around the vertical axis and translation in that axis. In most experiments the head of an experimental animal is fixed to the thorax by a wax bridge, thus blocking the neck movements.

Object in the environment

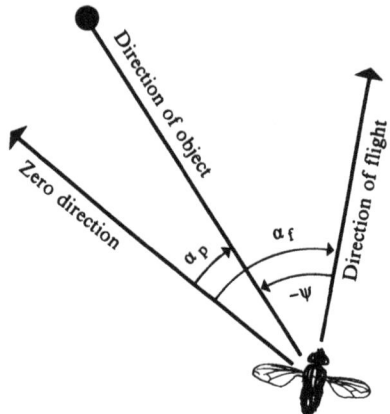

Figure 7.1.1. Angular coordinate system describing the fly's rotational degree of freedom around the vertical axis (from Reichardt & Poggio, 1976). $\alpha_p(t)$ designates the angle between an arbitrary zero direction and the direction of an object; $\alpha_f(t)$ the angle between the zero direction and the fly's direction of flight; $\psi(t) = \alpha_p(t) - \alpha_f(t)$ is the error angle between the fly's direction of flight and the object. If the fly's head is fixed to the thorax $\psi(t)$ represents the location of the object on the retina of the fly.

Figure 7.1.1 shows a coordinate system describing the angular position of the fly relative to an object in the horizontal plane in free flight. α_f designates the flight direction measured from an arbitrary zero direction, α_p denotes the angular position of an object. The angle

$$\psi = \alpha_p - \alpha_f \tag{1}$$

is the error angle, which defines the position of the object in the coordinate system of the fly. When the head is fixed to the thorax, $\psi(t)$ also represents the location of the image of the object on the retina of the fly at a particular instant t. In contrast to humans, the fly has a visual field covering almost its entire surround. When $\alpha_f = \alpha_p$ then $\psi = 0$: the fly's long axis (i.e. flight direction) points towards the object. When a 'conspicuous' object is present in the visual field of the fly it tends to turn towards this object.

The free flight situation of Figure 7.1.1 can be summarized by the signal flow diagram of Figure 7.1.2a. The 'fly' has been split into two 'boxes' – one describing the relation between visual input and torque output, the other representing the flight dynamics. A rotation of the fly around its vertical axis depends on the torque $F(t)$ generated by the wings.

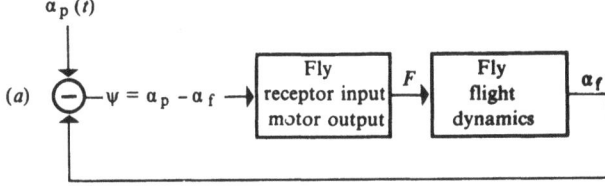

(a)

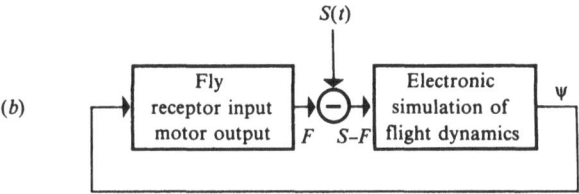

(b)

Figure 7.1.2. (a) Signal flow diagram referring to the free flight situation of Figure 7.1.1 (from Reichardt & Poggio, 1976). The fly's transduction of the visual input into angular displacement can be split into two 'boxes'. The visual input/torque output 'black-box' is followed by a 'box' representing the flight dynamics, transducing torque into angular displacement. (b) Signal flow diagram of the flight simulation device. The flight dynamics is simulated by analog electronics. $S(t) = \Theta \ddot{\alpha}_p(t) + k \dot{\alpha}_p(t)$ (see (4)) simulates the object motion.

It turns out that the free flight dynamics of rotation can be approximated by

$$\Theta \ddot{\alpha}_f(t) + k \dot{\alpha}_f(t) = -F(t), \tag{2}$$

where $\Theta = 1.5 \cdot 10^{-3}$ g cm^2 is the moment of inertia around the vertical axis of a fly, and $k = 0.18$ g cm^2 s^{-1} is an aerodynamic friction constant. The equation says that a torque $F(t)$ determines an angular acceleration $\ddot{\alpha}_f$ and velocity $\dot{\alpha}_f$. However, if the power spectrum of the flight torque of a fixed flying fly is determined, it turns out that it only contains frequency components up to $f_1 = 10$ Hz. A lower limit for a characteristic time constant of the system is therefore $\tau_1 = 100$ ms ($\tau_1 = 1/f$). Since $\Theta/k = 8$ ms is small compared to τ_1, the angular velocity is essentially proportional to the flight torque. In other words, after a step change in $F(t)$ the new stationary state $- \dot{\alpha}_f \propto F$ – is almost immediately reached (8 ms is the time constant). $F(t_0)$, the instantaneous torque of the fly, is assumed to depend on the error angle function $\psi(t)$ for $t \leq t_0$. Taking this into account and using (2) we can rewrite (1) as

$$\Theta \ddot{\psi}(t) + k \dot{\psi}(t) = -F\{\psi, t\} + S(t) \tag{3}$$

with

$$S(t) = \Theta \ddot{\alpha}_p(t) + k\dot{\alpha}_p(t). \tag{4}$$

If $\alpha_p(t) = 0$, and $\dot{\alpha}_p(t) = 0$, i.e. the object does not move with respect to the environment and its angular position is defined as zero, then $S(t) = 0$ and $\psi(t) = -\alpha_f(t)$. Equation (3) describes the free flight situation in the coordinate system of the fly. The equivalent 'closed-loop' signal flow diagram is shown in Figure 7.1.2b.

A method to simulate the free flight conditions represented by (3) is schematized in Figure 7.1.3 and described in detail in Reichardt (1973).

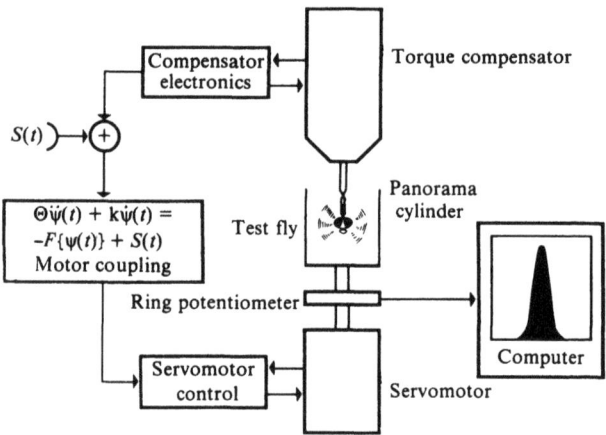

Figure 7.1.3. Simplified scheme of the experimental set-up (closed loop system) (from Reichardt & Poggio, 1976). A fly, suspended from the torque compensator, controls the velocity of a cylindrical 'panorama' by its own torque signal. The transfer properties of the compensator, the motor coupling block and the servomotor approximate free flight dynamics, (7.1.2). The instantaneous position of the panorama is transduced by a ring-potentiometer for further data processing. For more details, see text.

In this, the flying fly is fixed to a torque compensator and the environment can be moved according to (3). Under closed loop conditions the fly controls the movement of the panorama whose error angle ψ is the characteristic variable directly observed. A reconstruction of the trajectory $\alpha_f(t)$ can be calculated from a knowledge of $\psi(t)$ and $\alpha_p(t)$. The variable observed under open loop conditions, in which the panorama is moved by external signals only, is the flight torque measured at the output of the torque compensator.

A similar set-up was used to investigate the height orientation behavior of flies (Wehrhahn & Reichardt, 1975). Corresponding to (3), a second-order differential equation in the vertical coordinate z (height) was found to relate the lift of the fly to its vertical displacement. The equation reads

$$m\ddot{z}(t) + k_z \dot{z}(t) = -P\{\vartheta(z), t\} + L_0,$$

$$\vartheta(z) = \arctan(z/d).$$

(5)

Here m is the mass and k_z is a friction constant associated with this degree of freedom. The functional[†] $P\{\vartheta(z), t\} + L_0$ represents the fly's lift response to the object's vertical displacement z, at a distance d. The coordinate ϑ is the angular error with respect to the equator of the fly's eye.

The algorithms computing motion and position

The phenomenological theory described in Section 7.2 outlines the basic logical organization of the visually guided flight control system of the fly. We will see that it requires the neural network and the flight motor system to perform two main computations on the visual input. One component extracts movement information; the other provides position information.

It is clear, although sometimes overlooked, that (angular) velocity or (angular) position of an object or a pattern does not exist explicitly in the light inputs to the array of photoreceptors. Each receptor 'sees' a time-dependent light function, but none directly sees velocity or position information. The fly's nervous system extracts both from the image or, if one prefers, from the image flow. The questions are 'how'? 'which algorithms are implemented in the fly's visual system'? and 'how do the signals from the photoreceptors interact in order to compute position and velocity'? Thus, the question is at level (*b*). It is a question about the algorithms underlying these two computations or, in an essentially equivalent way, about the interactions or the systems that implement the algorithms. In this sense one can speak indifferently of an algorithm or a system (or interactions). Understanding at this level implies the capabilities (*a*) of predicting the system behavior and (*b*) of building a machine or writing a program that performs these computations as the fly does.

In some cases, it is easy to devise algorithms for a given computation and plan corresponding experiments to check whether the algorithms are actually used. But in other cases it may be not as easy. For this reason Poggio & Reichardt (1973, 1976) have developed and used a rather general approach that amounts to a kind of classification scheme for simple algorithms or systems.

The starting point of the approach is to consider an algorithm as an operation on an input to yield a corresponding output. In formal terms an algorithm can be thought of as a mapping (operator) between a space of input signals and a space of output signals. An operator can simply be seen as a law that relates an output function to each input function. In general, the operators of interest here are nonlinear: movement detection, for instance, cannot be performed by a linear system. Nonlinearity, however, is too general. Thus, one has to consider a specific class of nonlinear systems. The important point is that the input–output transduction of interest (in the fly's case) are 'smooth' and 'continuous'. Heuristically, there seems to be no 'decision' or discontinuity involved in the fly's movement or position computations.

It is important to mention here that the analysis at level (b) had its starting point in the analysis at level (a). We will see this in the course of going through the phenomenological theory in the next section.

7.2 THE PHENOMENOLOGICAL THEORY

In this section a few key experiments will be described. Their results lead to a phenomenological equation that quantitatively describes the fly's orientation behavior.

Spontaneous behavior

The first experimental finding concerns flight behavior without visual stimuli.

In a homogeneously illuminated environment a fly flies in all directions α_f with equal probability, assuming it is given sufficient time. That is, in a panorama without prominent patterns the fly spontaneously 'searches around', apparently in a random manner. To be more precise, when a contrastless panorama is used in the experimental set-up of Figure 7.1.3, the fly's torque during quiet flight turns out to be satisfactorily described as a stationary[†], zero mean, stochastic[†] signal $N(t)$, with a gaussian density distribution[†] and an exponential autocorrelation[†]

$$C_{NN}(\tau) = A \exp(-\gamma\tau). \tag{1}$$

See Figure 7.2.1a, b.

Visually induced behavior: open loop experiments

Visually guided behavior can be induced whenever a contrasted 'panorama' moves relative to the fly's retina. For instance, when a vertical stripe is oscillated around the position $\psi = 30°$, a test fly generates a positive torque. When the stripe is oscillated around $\psi = -30°$ the fly will generate a negative torque. This means that in either case the fly tends to turn towards the stripe. An example is given in Figure 7.2.2.

We can now ask whether it is necessary to oscillate the stripe in order to induce a torque response or whether a stationary stripe would give the same response. We find that, when we measure the open loop torque response towards a stationary stripe, and the head of the fly is fixed to the

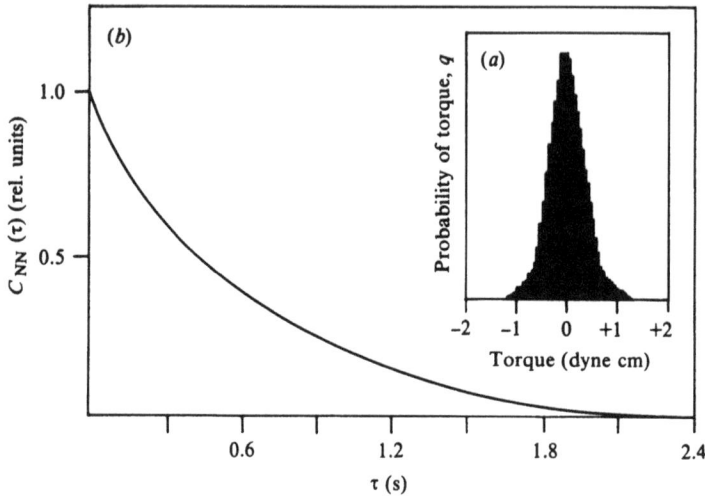

Figure 7.2.1. The inset (*a*) shows a histogram of the torque fluctuation generated by a fly in a uniformly illuminated environment (open loop experiment), during quiet, undisturbed flight. As shown by statistical tests, the histogram is well approximated by a gaussian distribution under these conditions. The random torque process leads, through the flight dynamics, to a flat asymptotic probability distribution of α_t. (*b*) The normalized autocorrelation of the open loop torque fluctuation. The autocorrelation is well described by the exponential $C_{NN}(\tau) = A \exp(-\gamma\tau)$ with $\gamma = 1.9$ (s^{-1}) and $\sqrt{A} = 0.3$ dyne cm ($\sqrt{A}$ is given by the standard deviation of the histogram in (*a*)). (From Reichardt & Poggio, 1976.)

thorax, then no torque different from zero is induced, whether the stripe is at the position $\psi = +30°$ or $-30°$ (Figure 7.2.3*a*). Apparently, a stabilized retinal image is not perceived by the fly. If, however, the head is free to move, clear response to the resting stripe can be observed. An experiment illustrating this is shown in Figure 7.2.3(*b*). It seems that by moving its head the fly itself provides the light flux changes on its retina necessary to perceive the stripe.

The experiments just mentioned represent a control for the findings illustrated in Figure 7.2.2. In addition, they are important in the quest to understand the algorithm underlying position detection. Since there is no response to a stabilized retinal image, the algorithm of position detection must be nonlinear. For a detailed description of these problems, which belong to level (*b*), see the paper by Poggio & Reichardt (1976).

The phenomenological equation

The observations just described suggest that the fly's torque underlying its orientation behavior arises from two main components. These are (1)

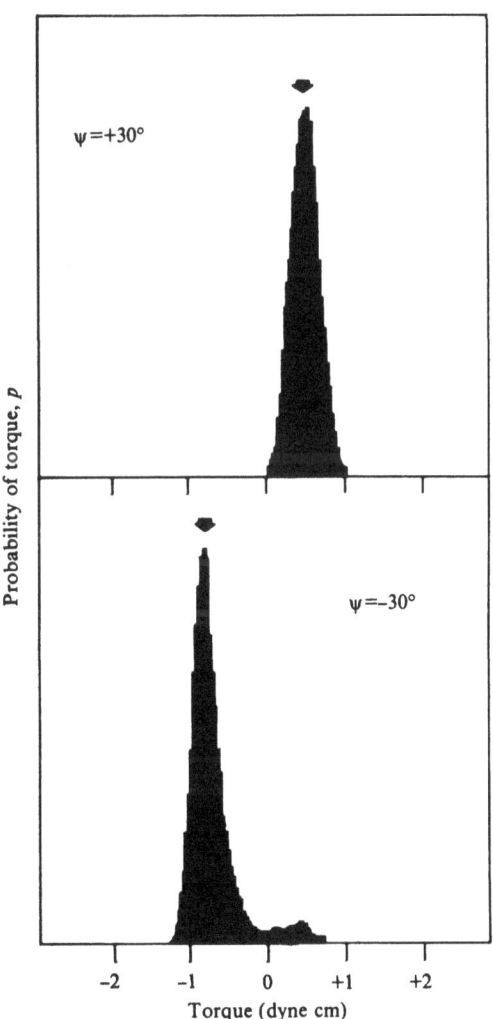

Figure 7.2.2. Histogram of the flight torque induced by a vertical black stripe of 90° length and 5° width oscillated around $\psi = +30°$ and $\psi = -30°$. Amplitudes of oscillation ±10°. The average brightness of the background was $1.75 \cdot 10^3$ cd m^{-2}. The head of the fly was fixed to the thorax. The arrows indicate the mean values of the torque histograms.

a stationary, gaussian random process $N(t)$, essentially independent of visual input, and (2) a visually induced response $R\{\psi, t\}$ regarded as a functional of ψ, the error angle of the object on the retina of the fly. Each of the two components could, in principle, be switched on by the stimulus specifically evoking it, while the other component is switched off. In all experiments, however, where the induced response $R\{\psi, t\}$ is measured,

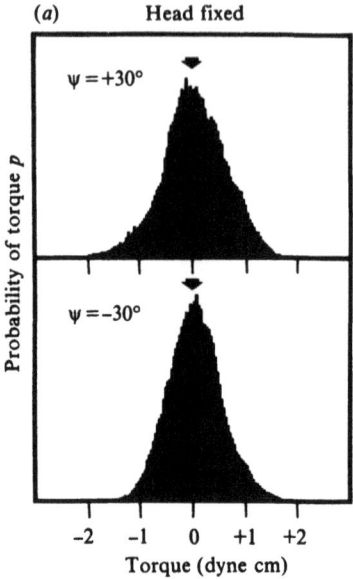

(*a*)   Head fixed

$\psi = +30°$

$\psi = -30°$

-2   -1   0   +1   +2

Torque (dyne cm)

Probability of torque *p*

Figure 7.2.3. Histogram of the flight torque induced by a black vertical stripe, resting at $\psi = +30°$ and $\psi = -30°$. Other parameters as in Figure 7.2.2. (*a*) The fly's head was fixed to the thorax by means of a wax bridge. No difference in the torque histograms is observed. (*b*) The fixed fly was able to move its head freely. The difference between the torque responses is significant.

the noise component is present too. The simplest assumption consistent with a first-order approximation of the experimental results discussed above is that $N(t)$ and $R\{\psi, t\}$ are additive in the nervous system. According to this assumption, the fly's torque can be written as

$$F\{\psi, t\} = R\{\psi, t\} + N(t), \tag{2a}$$

and (7.2.3) becomes

$$\Theta\ddot{\psi}(t) + k\dot{\psi}(t) = N(t) - R\{\psi, t\} + S(t), \tag{2b}$$

since the sign of the zero-mean random signal $N(t)$ does not matter.

Two further experimental findings support the hypothesis contained in (2*a*); one will be described in Section 7.3. The second comes from an electrophysiological study of Heide (1975), who examined the optomotor response of the fly *Musca domestica* at the neural output that controls the activity of the flight muscles. The physiological data led to the formation of a diagram of information flow from the central nervous system (CNS) to the flight muscles. Visually induced information and spontaneous yaw-turn neural commands seem to be *separate*, independent inputs, *additive* at the level of the muscle motoneurons or earlier. This indicates

(b)   Head free

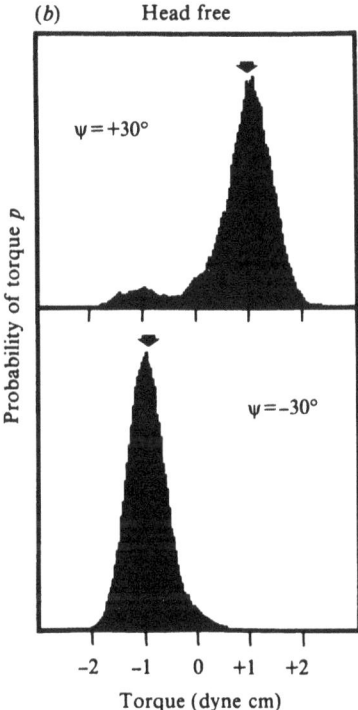

Probability of torque p

$\psi = +30°$

$\psi = -30°$

-2   -1   0   +1   +2
Torque (dyne cm)

Figure 7.2.3. Continued.

that the hypothesis given in $(2a)$ may be more than a convenient phenomenological description and that separate neural structures possibly correspond in the CNS to the two terms of $(2a)$.

A critical role in the formulation of the phenomenological theory is played by the assumption that the visually induced torque $R\{\psi, t\}$, at the instant t, is given by a functional of the past history of the object position. Examples of parameters are light intensity, the pattern itself, pattern contrast and perhaps color and other quantities. The following three points are important.

(i) The visually induced response is independent of the visual motor loop being open or closed.

(ii) The visually induced response of the fly seems time invariant, under our experimental conditions. This implies that R does not explicitly depend on time: $R\{\psi, t\}(t) = R\{\psi\}(t)$.

(iii) The value of R at time t is much more sensitive to changes of ψ in the near past than to changes at instants far away from t. Because of the flight dynamics specified in (7.1.3) the error angle ψ 'follows' the response R with a time constant of about 8 ms. Therefore under normal (natural) coupling conditions, and when the object is not moving too fast, the error angle ψ is a 'smoothed' version of R (more precisely a low pass filtered version of R).

With these assumptions, the value of the functional $R\{\psi\}$ at t should be given approximately by a function R of the values of ψ and its first n derivatives at t:

$$R\{\psi\}(t) \sim R^*(\psi(t), \dot{\psi}(t), \ddot{\psi}(t), \ldots). \tag{3}$$

'Retardation theorems', which justify this expectation and make it mathematically precise do in fact exist (Coleman, 1971, theorem 7). From (3), a Taylor expansion of R around $\psi = 0$ up to the first-order terms gives

$$R(\psi)(t) \sim D(\psi(t)) + r(\psi(t))\dot{\psi}(t). \tag{4}$$

According to the retardation theorem this equation provides a first-order approximation of the visually induced response under normal conditions. However, it is clear that approximation (4) may not hold when ψ is not smooth enough compared to R. Nevertheless, the validity of this first-order approximation will be demonstrated, for two applications, in Section 7.3.

The functions $D(\psi)$ and $r(\psi)\dot{\psi}$ depend on the particular pattern and on other parameters. Their meaning becomes quite clear from consideration of their symmetry properties. $r(\psi)\dot{\psi}$ changes sign for velocity inversion and is thus a direction-sensitive term. On the other hand, $D(\psi)$ depends only on the position ψ and is thus a position-dependent direction-insensitive term. In the special case of a narrow vertical black stripe these functions, denoted as $D^*(\psi)$ and $r^*(\psi)\dot{\psi}$, have been measured with different methods, under the condition of being 'smooth' enough. Figure 7.2.4 shows the functions $D^*(\psi)$ and $r^*(\psi)\dot{\psi}$ associated with a black vertical stripe of 5° width and 22.5° length. The instantaneous open loop response $R\{\psi\}$ to the object rotating at small constant speed around the fly was measured for each ψ. Under the assumption that (4) holds, the two components $D^*(\psi)$ and $r^*(\psi)\dot{\psi}$ were identified through their respective symmetry properties in $\dot{\psi}$. Figure 7.2.5 shows the $D^*(\psi)$ associated with a vertical black stripe measured with head free. As can be seen by comparing Figures 7.2.4 and 7.2.5, the function $D^*(\psi)$ in this situation does not differ very much from that measured with the head fixed.

The importance of the functions $D^*(\psi)$ and $r^*(\psi)\dot{\psi}$ associated with a single stripe derives from the fact that any small contrasted object will elicit a similar response, apart from scaling factors. It can be shown that the $D^*(\psi)$ and $r^*(\psi)\dot{\psi}$ associated with an arbitrary pattern can be well approximated from the functions of Figure 7.2.4 and a simple superposition rule. In this sense the knowledge of $D^*(\psi)$ and $r^*(\psi)\dot{\psi}$ associated

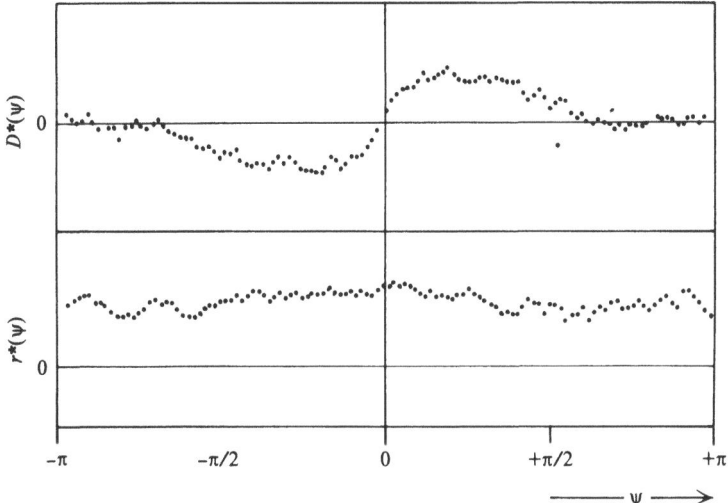

Figure 7.2.4. The functions $D^*(\psi)$ and $r^*(\psi)\dot{\psi}$, in relative units, associated with the response of the fly (see (6)) to a narrow vertical black stripe segment (5° wide, 22.5° long) (from Poggio & Reichardt, 1976). The stripe segment was rotated with a constant angular speed ($8°\,s^{-1}$) and the measured (open loop) torque was decomposed into the direction-insensitive component $D^*(\psi)$ and into the direction-sensitive one $r^*(\psi)\dot{\psi}$. The figure represents the average of five experiments, each one lasting about 6 min.

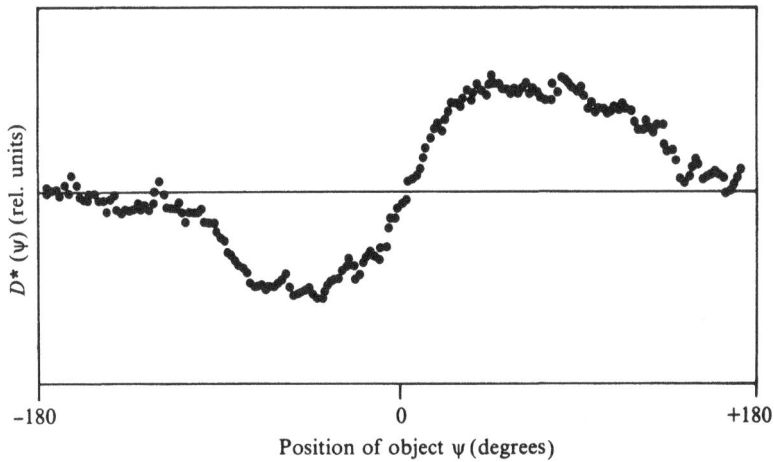

Figure 7.2.5. The function $D^*(\psi)$ for a narrow vertical black stripe. The heads of the experimental animals were free. Other parameters as in Fig. 7.2.2.

with a single stripe is basic to the phenomenological description of orientation behavior (for details see Reichardt & Poggio, 1975).

$D^*(\psi)$ turns out to be an antisymmetric function, independent, to a first approximation, of the speed $\dot{\psi}$ – at least for $2° \, s^{-1} \leqslant \dot{\psi} \leqslant 50° \, s^{-1}$. The coefficient $r^*(\psi)$ turns out to be a constant r^*. Therefore the term $(r^*(\psi)\dot{\psi} = r^*\dot{\psi})$ is direction sensitive, velocity dependent, and position independent; the other term $(D^*(\psi))$ is direction insensitive, velocity independent, and position dependent. To sum up, for a dark vertical stripe under the conditions just mentioned the visually induced response of the fly, R_{St}, takes the form

$$R_{St}\{\psi\}(t) \approx D^*(\psi(t)) + r^*\dot{\psi}(t). \tag{5}$$

We now see that the visual control system that the fly uses performs two basic operations. One transduces position or angular error into torque through the 'attractiveness' term $D(\psi)$. The other converts velocity into torque, through the term $r(\psi)\dot{\psi}$. Thus, the first computation extracts *position* information $[D(\psi)]$, the second gives *movement* information $[r(\psi)\dot{\psi}]$.

In a series of papers that are reviewed by Poggio & Reichardt (1976), an attempt is made to explain *how* these computations are performed in the visual system of the fly. The analysis takes place at the level of the algorithms, the level (b) mentioned in the introduction. It turns out that both algorithms, the one computing position, and the other computing movement, must be nonlinear. It is noteworthy that the theoretical considerations and the experiments leading to this conclusion had as their starting point the phenomenological theory that we have been describing.

The central thesis of the phenomenological theory is that the computation of position and movement (see (4)), together with the fluctuation term $N(t)$, is basic to the orientation behavior. We shall see that (4) can in fact account for fixation and tracking and even spontaneous pattern preference behavior.

Inserting (4) into (7.1.4) we find that the equation in the error angle reads, for a black vertical stripe,

$$\Theta\ddot{\psi}(t) + k\dot{\psi}(t) + r^*\dot{\psi}(t - \varepsilon) + D^*(\psi(t - \varepsilon)) = N(t) + S(t), \tag{6}$$

where ψ is chosen to be between zero and 2π. The 'dead time' or lag ε between visual input and motor responses of a fly is in the order of 20 ms, depending on the average luminance of the panorama (Reichardt, 1978). It can be neglected in almost all cases considered here. The closed loop behavior can be predicted from the open loop responses by (6). Whether the artificial closed loop conditions described in Section 7.1 are indeed representative for the free flight situation will be discussed in Section 7.3.

An equation similar to (6) can be derived in an essentially equivalent way for the height orientation behavior, whose dynamics are described by (7.1.5). The equation for a horizontally oriented dark stripe is

$$m\ddot{z} + k_z r_\vartheta^*(\vartheta)\dot{\vartheta}(z) + L^*(\vartheta(z)) = N_\vartheta(t) + S_\vartheta(t)$$

$$\vartheta = \arctan (z/d).$$

(7)

The functions $r_\vartheta^*(\vartheta)\dot{\vartheta}$ and $L^*(\vartheta)$ are shown in Figure 7.2.6. $L^*(\vartheta)$ represents the position-sensitive and $r^*(\vartheta)\dot{\vartheta}$ the direction-sensitive lift responses. $N_\vartheta(t)$ is a stochastic gaussian lift fluctuation: its quantitative characterization has been given by Wehrhahn & Reichardt (1975).

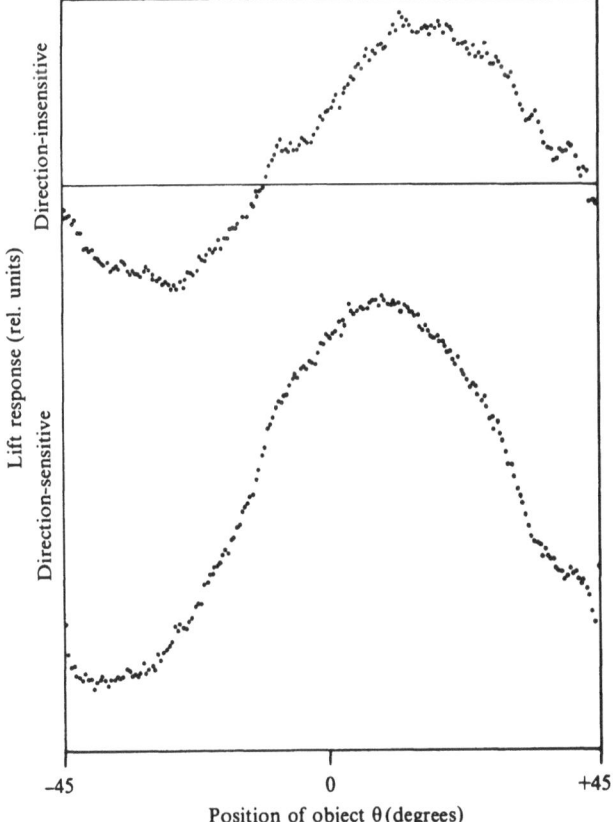

Figure 7.2.6. The functions $L^*(\vartheta)$ and $r_\vartheta^*(\vartheta)\dot{\vartheta}$ in relative units associated with the lift response of the fly (see (7)) to a horizontal black stripe 5° wide and 45° long (from Wehrhahn, 1978a). The stripe was moved with constant speed (8° s^{-1}) and the measured open loop lift was decomposed into the direction-insensitive component $L^*(\vartheta)$ and into the direction-sensitive one $r_\vartheta^*(\vartheta)\dot{\vartheta}$. The figure represents the average of five experiments.

The structure of (6) is well known. It is a nonlinear stochastic differential equation called in physics the Langevin equation. Its mathematical analysis and biological interpretation is given elsewhere (Poggio & Reichardt, 1973). A simple linear approach, which accounts for the fixation and tracking of a single object, will be presented in the next section.

7.3 LINEAR THEORY

The shape of the term $D^*(\psi)$ implies that (7.2.6) can be linearized if the fluctuations of ψ are restricted to the linear range of $D^*(\psi)$ $(-30° \leqslant \psi \leqslant 30°)$. A simple test for the validity of the linearization consists of a tracking experiment whose result has to satisfy a condition that we will now derive.

Let $\phi(\omega)$ be the power spectrum[†] of $\psi(t)$ and $\tilde{N}(\omega)$ the power spectrum of $N(t)$ (ω is the angular frequency). If $G(\omega)$ is the transfer function[†] (see Section 3.2) of the closed loop flow diagram in Figure 7.1.2(b) and the system is linear then, as in (7.6.8),

$$\phi(\omega) = |G(\omega)|^2 \cdot \tilde{N}(\omega). \tag{1}$$

Since the fluctuation $N(t)$ is gaussian and the system is linear, the error angle $\psi(t)$ is also gaussian. We now take a gaussian fluctuation $N_i(t)$ of root mean square amplitude n which is statistically independent[†] of the fluctuation $N(t)$ generated by the fly, and add it into the closed loop system at the output of the torque compensator. If the power spectrum of $N_i(t)$ is $n^2\tilde{N}_i(\omega)$, (1) becomes

$$\phi(\omega) = |G(\omega)|^2[\tilde{N}(\omega) + n^2\tilde{N}_2(\omega)]. \tag{2}$$

Integration over ω yields

$$(1/2\pi)\int_{-\infty}^{\infty} \phi(\omega)\, d\omega = (1/2\pi)\int_{-\infty}^{\infty} |G(\omega)|^2\tilde{N}(\omega)\, d\omega$$

$$+ n^2 \int_{-\infty}^{\infty} |G(\omega)|^2\tilde{N}_i(\omega)\, d\omega. \tag{3}$$

Let σ^2 be the standard deviation of the gaussian distribution of the stripe position, so that (see (7.6.5))

$$\sigma^2 = (1/2\pi)\int_{-\infty}^{\infty} \phi(\omega)\, d\omega. \tag{4}$$

Moreover let σ_0^2 and σ_i^2 be the standard deviations of the fluctuation generated by the fly and the artificial fluctuation, respectively. Thus,

$$\sigma_0^2 = (1/2\pi) \int_{-\infty}^{\infty} |G(\omega)|^2 \tilde{N}(\omega)\, d\omega, \tag{5}$$

$$\sigma_i^2 = (1/2\pi) \int_{-\infty}^{\infty} |G(\omega)|^2 \tilde{N}_i(\omega)\, d\omega. \tag{6}$$

By virtue of (4), (5), and (6), (3) becomes

$$\sigma^2 = \sigma_0^2 + n^2 \sigma_i^2, \tag{7}$$

or

$$n\sigma_i = \sqrt{(\sigma^2 - \sigma_0^2)}. \tag{8}$$

Let us recapitulate. We made the following assumptions.

(a) The fly's torque can be decomposed into a stochastic visual input term $N(t)$ independent from ψ, and a visually induced response $R\{\psi\}$.

(b) The response $R\{\psi\}$ depends only on the error angle ψ.

(c) Equation (7.2.6) can be linearized for $-30° \leqslant \psi \leqslant 30°$.

(d) $N(t)$ is approximated by a gaussian process.

We then showed by elementary theoretical considerations that the additivity relation (7) should hold between σ_0, σ_i and σ. This prediction can be checked experimentally. Indeed, Figure 7.3.1 shows the results of a series of closed loop experiments, where a fluctuation $N(t)$ of different root mean square amplitudes n was added to the torque signal and the half width of the resulting distribution was determined. Figure 7.3.1 shows that the expectation is very precisely confirmed by the experimental data.

Fixation

In the range $-30° \leqslant \psi \leqslant 30°$ (7.2.6) takes the linear form

$$\Theta\ddot{\psi} + k\dot{\psi} + r^*\dot{\psi} + \beta\psi = N(t) + S(t), \tag{9}$$

where the coefficient β represents the slope of the $D^*(\psi)$ characteristics around $\psi = 0$. The standard correlation method (see Wax, 1954; Martin, 1968) can be used to provide a solution for (9). In this context to 'solve' (9) means to characterize the gaussian process $\psi(t)$.

Since $N(t)$ is a gaussian random process, completely defined by its autocorrelation, $\psi(t)$ must also be a random process. The transfer function associated with (9) (setting $N(t) = S(t) = 0$) turns out to be given

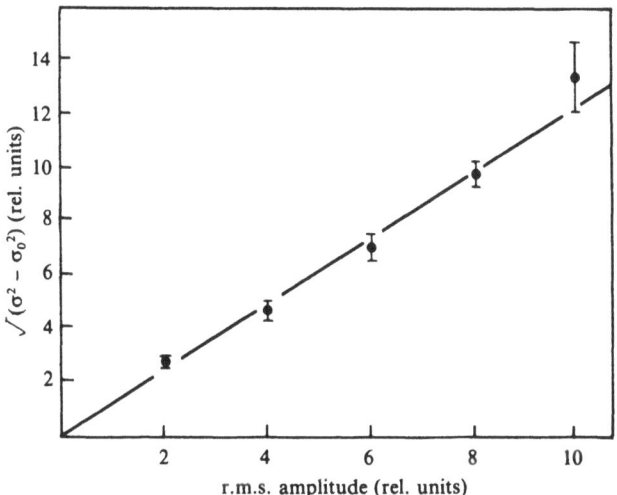

Figure 7.3.1. Experimental relationship between the standard deviation σ of a fixation histogram and the power (in arbitrary units) of an artificial gaussian noise $N_i(\psi)$ injected additively into the closed loop (from Reichardt & Poggio, 1976). σ_0 represents the standard deviation of the fixation histogram without artificial noise. The points are averages taken from five flies. The artificial gaussian noise used in this experiment had a flat spectrum up to 15 Hz. The equivalent free flight situation corresponds to tracking by the fly of an object moving with an angular speed $\dot{\alpha}_p(t)$ proportional to the artificial noise: $\dot{\alpha}_p(t) = [N_i(t)]/k$. The ordinate of the figure indicates how the root mean square (r.m.s.) of the 'error angle' ψ_{tr} increases with increasing r.m.s. of the object's velocity.

by

$$G(\omega) = 1 \Big/ \left[i\omega^2 + \left(\frac{k+r}{\Theta}\right) i\omega + \frac{\beta}{\Theta} \right]. \tag{10}$$

Since the autocorrelation of $N(t)$ is given by (6.3.1), its power spectrum is

$$\tilde{N}(\omega) = 2A\gamma/[\gamma^2 - (i\omega)^2]. \tag{11}$$

If we consider $N(t)$ as the input function and $\psi(t)$ as the output function of the system of which (10) represents the transfer function, the power spectrum of $\psi(t)$ is given by the expression

$$\phi(\omega) = \frac{\tilde{N}(\omega)/\Theta^2}{|i\omega^2 + i\omega(k + r/\Theta) + \beta/\Theta|^2}. \tag{12}$$

Integration of (12) gives the standard deviation of the asymptotic ($t \to \infty$) gaussian probability distribution

$$\sigma^2 = \frac{A}{\beta(k+r^*)} \frac{k + r^* + \gamma\Theta}{\beta + (k+r^*)\gamma + \gamma^2\Theta}. \tag{13}$$

We use the fact that $N(t)$ is gaussian and $\phi(\omega)$ is linear to obtain the equation

$$p(\psi) = (1/2\pi\sigma) \exp{(-\psi^2/2\sigma^2)}. \tag{14}$$

Equations (12) and (14) show the role played by the various parameters. The dependence on the fluctuation power is not surprising. The parameter β (the slope of the linearized $D^*(\psi)$) plays the main role. While an increase in either β or r^* leads to a 'better' stationary fixation (smaller σ), $\beta > 0$ is a necessary and sufficient condition for fixation if the condition $(k + r) > 0$ holds. Fixation cannot take place without position-sensitive information. However, speed sensitive feedback can 'improve' stationary fixation. The linear equation (9) quantitatively predicts a stationary distribution of the process $\psi(t)$ for low couplings (Θ/k up to $16 \cdot 10^{-3}$ s).

The following parameters have been determined experimentally: $A = 0.09$ [dyn^2 cm^2]; $p = 2$ [g cm^2 s^{-2}]; $k = 0.37$ [g s^{-1} cm^2]; $r^* \approx 0$ [g s^{-1} cm^2]; $\gamma = 1.9$ [s^{-1}]; $\Theta = 1.5 \cdot 10^{-3}$ g cm^2.

Calculation of σ from (14) leads to values which are found in closed loop experiments using the scheme shown in Fig. 7.1.2. This result indicates that the assumptions leading to (9) are correct. A very important point is that dynamic solutions in the expectation[†] $\langle\psi(t)\rangle$, given by (9), agree with experimental results. A good example is provided by the chasing behavior of *Fannia*, described by Land & Collett (1974), to which we now turn.

Chasing behavior of free flying houseflies *Fannia canicularis*

Two housefly species are frequently found, the common housefly (*Musca domestica*) and the lesser housefly (*Fannia canicularis*). The important difference between the two species for us is that male *F. canicularis* tend to congregate around prominent objects such as lampshades, making horizontal patrolling flights near them for long periods. Males engage in chasing females or, more commonly, other males.

Films were made by positioning a camera directly beneath a lampshade frequented by *F. canicularis*, and pointing upwards. Inspection of chasing flights from the side and below shows that most maneuvers take place in the horizontal plane.

The films were analyzed frame by frame and the courses of flies in the horizontal plane were plotted. Figure 7.3.2 shows a complete chase. Inspection of this figure shows that one animal (open circles) is being chased and is taking continuous evasive action, and that the other animal (closed circles) is trying to follow the first as closely as possible. In the chase the leading animal makes six quite distinct maneuvers: a sharp left

Figure 7.3.2. Flight paths of chasing (●) and leading (○) flies during chase (from Land & Collett, 1974). Points at 20 ms intervals. Corresponding instants on the two paths numbered at 200 ms intervals.

turn after 1, a right turn after 2, a 180° right turn between 3 and 4, a left turn at 4, a right turn immediately thereafter, and finally an extraordinary right-hand loop which leaves the fly almost on its former course. It is apparently this final maneuver that causes the pursuer to loose visual contact. In contrast to this almost random behavior, the pursuing fly seems to behave in a much more comprehensible manner. It follows each maneuver of the leading fly quite accurately, even during the final loop. It gives the impression of trying to keep up its course pointing in the direction of the leading fly, as though it were attempting to catch up with it.

The flight path of the pursuing fly can be described by (9). For the description of the chasing behavior, the noise term $N(t)$ can be neglected, since it is small compared to $S(t) = \dot{\alpha}_p(t)$. In addition, the acceleration terms of (9) can be neglected (see 7.1.2). Taking into account the delay time ε between stimulus and motor response we see that (9) becomes

$$(r^*/k)\dot{\psi}(t-\varepsilon) + (\beta/k)\psi(t-\varepsilon) = \dot{\alpha}_f(t), \tag{15}$$

which corresponds to the equation derived by Land & Collett (1974). Here α_f describes the flight path of the leading fly in the coordinate system that was introduced in Figure 7.1.1. The pursuing fly controls its forward velocity as a linear function of its distance to the leading fly (C. Wehrhahn and T. Poggio, unpublished). The qualitative data supporting this for the chase of Figure 7.3.2 are given in the inset of Figure 7.3.3. This relation was used in addition to (15) to simulate the flight path of the pursuing fly, given that of the leading fly. The simulation is shown in Figure 7.3.3. The agreement with the actual path of the chasing fly, shown in Figure 7.3.2, is rather good.

Several noteworthy points emerge from the good approximation of the experimental data by the linear theory.

(*a*) The values of Θ and k assumed for the simulation of the free flight dynamics used in the experimental set-up of Figure 7.1.3 agree very well with the values inferred by Land & Collett (1974).

(*b*) The linear approximation of the theory as expressed by (9) describes rather well the *dynamic* features of the flight path of a chasing fly.

(*c*) Two computations, one transducing position, the other velocity of the target, determine the tracking behavior of the fly.

In summary, these results provide critical evidence that the phenomenological theory which describes the artificial closed loop situations of Figure 7.1.3 applies equally well to the free flight conditions.

Recently male and female houseflies *Musca domestica* were found to chase other flies in free flight. But female chases are brief and poorly controlled compared to male chases. By the simultaneous filming of flies from above and from the side, female flies were shown to use the lower frontal part of their field of view for tracking other flies. Male flies use the upper frontal part of the their field of view for that purpose (Wehrhahn, 1979). In addition, while chasing other flies, male flies (but not females), are capable of controlling their forward velocity to a degree roughly proportional to the distance to their target (see Fig. 7.3.3). Recently sex-specific neurons have been discovered in the optic lobes of male flies. Their receptive field is situated in the dorsal and frontal part of the field of view, Hausen & Strausfeld (1979). These neurons are not found in female flies. Their association with male-specific chasing behavior is up to now purely speculative, but very suggestive. The tracking system described earlier in this chapter and used by female and male flies is different from the male-specific system outlined in the second part of this chapter, as suggested earlier by Collett & Land (1975). The angular tracking of both systems can be described by (9), although each has a different physiological location.

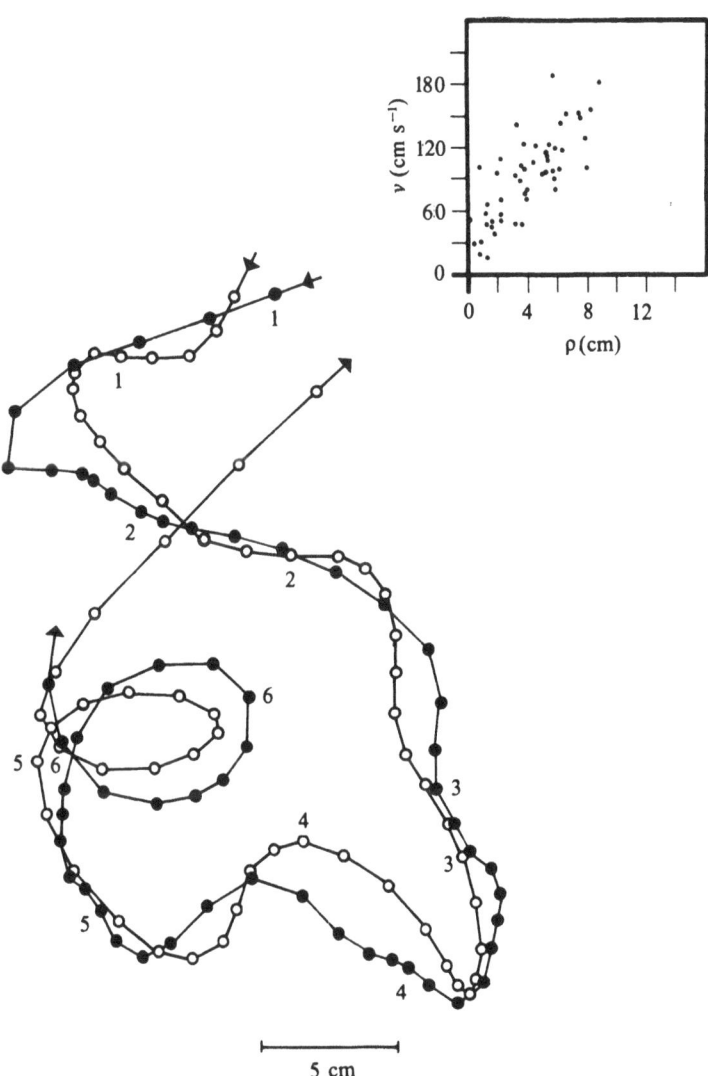

Figure 7.3.3. Attempt to simulate the chase shown in Figure 7.3.2. The course of the leading fly is taken from Figure 7.3.2 and the course of the chasing fly is plotted according to (15). The parameter values are: $\alpha/k = 20\,\mathrm{s}^{-1}$, $r^*/k = 0.7\,\mathrm{s}^{-1}$, $\varepsilon = 30\,\mathrm{ms}$. The forward velocity v of the pursuing fly was determined by the linear relation between its distance ρ to the leading fly, suggested by the inset figure. v had to be delayed by $100\,\mathrm{ms}$ with respect to ρ.

7.4 SOME ASPECTS OF THE NONLINEAR THEORY

For the considerations of this section we define a potential function $U(\psi)$ through the relation

$$D(\psi) = \partial U(\psi)/\partial \psi. \tag{1}$$

As was the case with $D(\psi)$ the potential $U(\psi)$ repeats itself over every interval of the length 2π.

$$U(\psi) = U(\psi + n \cdot 2\pi), \qquad n = 1, 2, 3, \ldots.$$

Figure 7.4.1 shows the potential $U^*(\psi)$ computed through (1) from the function $D^*(\psi)$ that represents the position response to a vertical black stripe. The potential $U^*(\psi)$ has its minimum at $\psi = 0$, the point of stable fixation, and increases from there to $\psi = \pm \pi$.

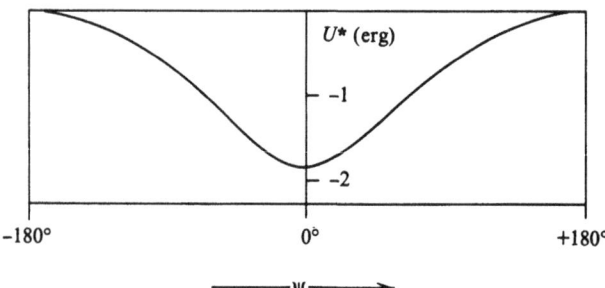

Figure 7.4.1. The potential $U^*(\psi)$ associated with the attractivity function $D^*(\psi)$ to a vertical black stripe 5° wide and 90° long (from Reichardt, 1973). Other parameters as in Figure 7.2.2. The potential is the average of 111 experiments.

Using (1) we can rewrite (7.2.6) for a vertical black stripe in the form

$$\Theta\ddot{\psi} + (k + r^*)\dot{\psi} + \partial U^*(\psi)/\partial\psi = N(t) + S(t). \tag{2}$$

As has been mentioned, equations of this type, where $N(t)$ is a 'physical' (smooth) noise process, are known in physics as Langevin equations. To solve the stochastic equation (2) one must characterize the random process $\psi(t)$ completely. The solution is carried out by transforming (2) into what is known as a Fokker–Planck equation. This is particularly easy if it is assumed that

(a) since Θ is very small the term $\Theta\ddot{\psi}$ can be neglected in (2); and that

(b) $N(t)$ is 'white' noise[†].

In the case of stable fixation then $S(t) = 0$ holds in addition. The derivation of the solution is treated in detail by Poggio & Reichardt (1973) and Reichardt & Poggio (1975). The asymptotic $(t \rightarrow \infty)$ probability distribution in ψ is found to be

$$p(\psi) = C \exp\left(-U^*(\psi) \cdot (k + r^*)/c\right), \tag{3}$$

where C is a normalization constant and c the spectral density of the white noise. Equation (3) relates the potential $U^*(\psi)$, associated with the panorama consisting of a vertical black stripe, to the stationary error angle distribution $p(\psi)$. It is important to note that relation (3) can be generalized to arbitrary patterns if the corresponding potential is known. For simplicity we discuss the case of $U^*(\psi)$ for a vertical black stripe. By analogy with the Brownian motion of the particle in a potential well, the stripe is attracted by $U^*(\psi)$ and has its stable position at $\psi = 0$. The noise process $N(t)$ leads to a gaussian probability distribution of ψ around $\psi = 0$. The friction term $(k + r^*)\dot{\psi}$ *improves* fixation: The greater $(k + r^*)$ the smaller the half width of the probability distribution of ψ.

In the tracking case described above, $S(t) \neq 0$. If $N(t)$ is negligible compared to $S(t)$, and $S(t)$ is a known function, then (2) becomes an ordinary differential equation in $\psi(t)$ that can be solved with standard methods, as we have already seen.

Tracking of a target moving at constant angular speed $\dot{\alpha}_p(t) = $ constant can be described by defining a new noncyclic potential

$$\tilde{U}(\psi) = k\dot{\alpha}_p\psi - U^*(\psi). \tag{4}$$

The associated Fokker–Planck equation can be determined as before except that $S(t) = k\dot{\alpha}_p = $ constant. It is found that

$$p(\psi) \sim \exp\left\{\left(\frac{k + r^*}{c}\right)\tilde{U}(\psi)\right\}\int_{\psi - 2\pi}^{\psi} \exp\left\{\left(\frac{k + r}{c}\right)\tilde{U}(\psi')\right\} d\psi'. \tag{5}$$

Equation (5) can be interpreted as the probability distribution of a Brownian particle in the noncyclic potential $U^*(\psi)$ shown in Figure 7.4.2. Tracking of a moving target at constant angular speed $\dot{\alpha}_p$ is then formally equivalent to the motion of a particle in a wavy inclined surface.

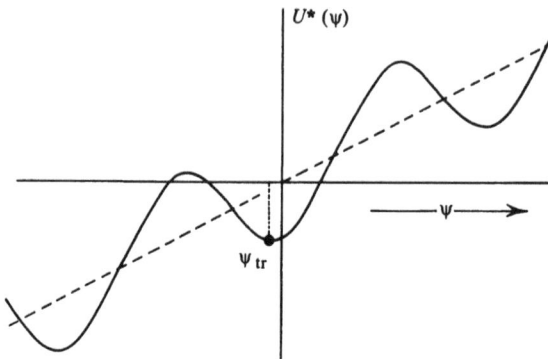

Figure 7.4.2. Tracking of a target moving at constant angular speed $\dot{\alpha}_p(t)$ is formally equivalent (through (2)) to the motion of a particle in a wavy inclined surface under the influence of a fluctuation $N(t)$ and some friction (from Reichardt & Poggio, 1976). ψ_{tr} is the mean angular lag with which the fly tracks the stripe.

In the absence of the fluctuations represented by the spontaneous torque process $N(t)$, the stripe comes to rest at the bottom of one of the wells. Tracking is impossible if there are no 'bottoms' (relative minima) because of the excessive slope of the surface (the target's speed is too high). The fluctuation process dislodges the stripe. Its average effect is to cause the stripe to slide down the plane. Sliding (loss of target) occurs more rapidly the greater the slope (speed of the target). The shallower the wells (attractiveness of the target), the stronger the effect of the random function $N(t)$ (fluctuations superimposed on the deterministic track). Figure 7.4.3 shows a closed loop experiment carried out in the scheme described in Figure 7.1.3, where the target has a constant angular speed $k(t)$. In Figure 7.4.4 a simulation using (2) with the same constant angular speed is shown.

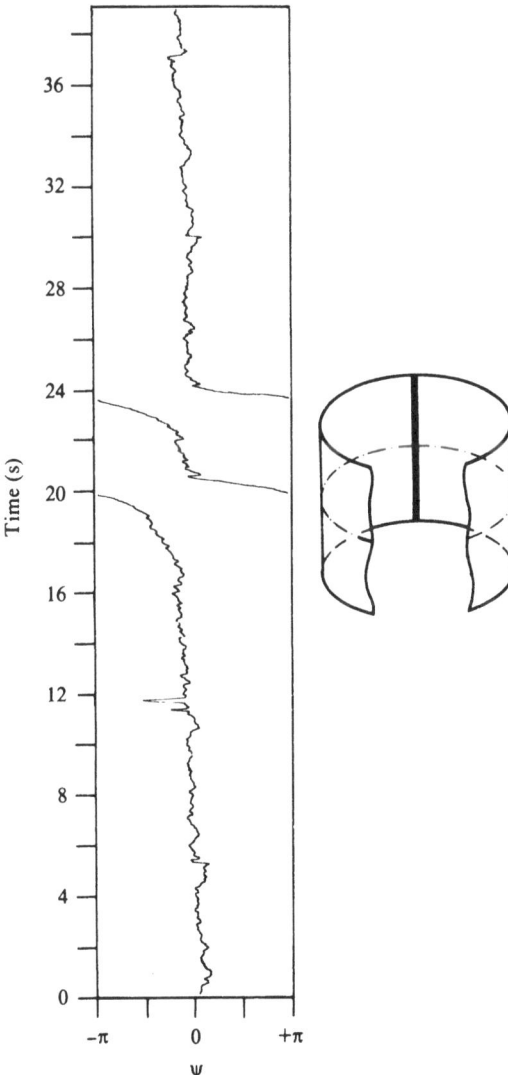

Figure 7.4.3. A typical trajectory of the error angle $\psi(t)$, modulus 2π, during tracking of a black vertical stripe moving at fast constant angular speed $\dot\alpha_p$ (W. Reichardt & T. Poggio, unpublished). The fly lags more and more behind the target, suddenly makes what appears to be a sort of nystagmus (a fast turn against the direction of target motion), again locks on the target and fixates for a while.

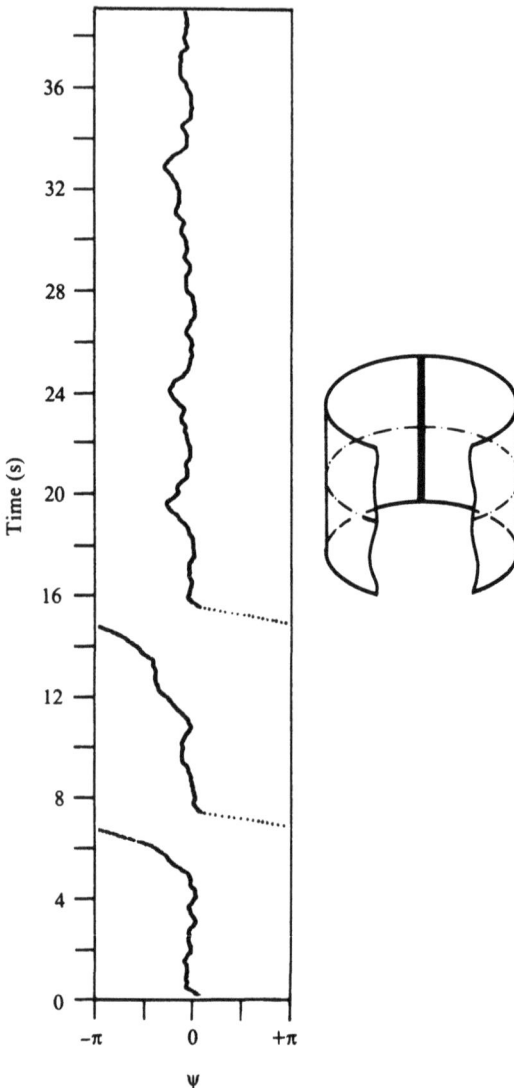

Figure 7.4.4. Simulation of a tracking experiment as shown in Figure 7.4.3 (W. Reichardt and T. Poggio, unpublished). Equation (2) was solved in an analog device.

7.5 DISCUSSION

Perhaps the most significant aspect of the phenomenological theory is the demonstration that (closed loop) orientation behavior can be quantitatively predicted from knowledge of the open loop response of the fly. The phenomenological equation (7.2.6), which takes into account the dynamics of flight, links (open loop) information processing in the visual system with the natural orientation behavior.

The critical role of this point for an understanding of the visual control system of the fly is clear. It means that one can consider the computations performed on the visual input by the nervous system as independent from the motor loop being 'open' or 'closed'. In other words, knowledge of open loop information processing, as measured in behavioral (or electrophysiological) experiments, can predict, through the phenomenological theory, the (closed loop) 'natural' orientation behavior. Furthermore, it suggests that an analysis of the computations performed under open loop conditions is in fact completely sufficient for an understanding of the behavior. This represents an essential prerequisite for the approach at level (b).

A critical assumption leads to the separation of the fly's response into a visually induced component *and* into a 'noise' term $N(t)$, essentially independent from visual input.

The arguments of Section 7.3 imply that the visually induced response can be further decomposed into a position-dependent and a velocity-dependent component. Many experimental data support the validity of these conceptual decompositions and indicate that the visual control system of the fly rests on two main computations, one converting position information into torque $D^*(\psi)$, the other converting velocity into torque $r^*\dot{\psi}$. An analogous decomposition holds for the lift response underlying height orientation (Wehrhahn & Reichardt, 1975).

An important question here concerns the physiological implications of the separation of the fly response into the three parts: $N(t)$, $r^*\dot{\psi}$, $D^*(\psi)$. In a strict sense the separation is a first-order conceptual approximation and cannot, by itself, imply any corresponding physiological separation.

However, it may provide suggestions about the structural organization of the underlying nervous network. The independence of the term $N(t)$ from visual input suggests the existence of a 'noise' channel, separate from the channels carrying position and movement information to the motor output. This simple conjecture is supported by recent electrophysiological evidence, part of which we have already mentioned (see Section 7.2). Moreover, fibers in the visual ganglia show a noise-free, visually induced activity. On the other hand, comparable recordings from the cerival connective, which links the brain with the thoracic ganglion, show the presence of spontaneous activity in addition to the visually induced activity (K. Hausen and R. Hengstenberg, personal communication). It is quite satisfactory that, in the case of $N(t)$, the conceptual separation, suggested by the phenomenological theory, seems to reflect an actual physiological separation.

Recently experimental evidence has been given for the existence of a two-dimensional potential $U(\psi, \vartheta)$ which is a generalization of the potential $U(\psi)$ defined by (7.4.1). The position response $D(\psi)$ and $L(\vartheta)$ defined by (7.2.6) and (7.2.7) were shown to depend on each other in the sense that they can be derived from a two-dimensional potential $U(\psi, \vartheta)$. The neural correlate underlying position detection therefore could be the same for both $D(\psi)$ and $L(\vartheta)$ (Wehrhahn, 1978a, b).

The spatial properties of the $D(\psi)$, $L(\vartheta)$ computation compared to the $r^*\dot{\psi}$ and $r^*_{\vartheta}(\vartheta)\dot{\vartheta}$ computation may represent an important key for anatomical and electrophysiological identifications of the underlying nervous structures. Determination of these identifications, a problem at level (a), is a prerequisite for an understanding of the visual ganglia. (The lobula plate, the smaller of the two third-order 'ganglia', seems a typical specialization of flying dipterans, perhaps serving their navigatory requirements: other insect species, e.g. locusts, do not have a lobula plate.) These questions are now in the range of electrophysiological techniques and preliminary answers should soon be forthcoming.

One question on the theoretical side of level (a) arises here. How are the algorithms implemented in the neuronal hardware? The nonlinearity of the interactions requires mechanisms at the cellular level which should work like a multiplier. This means that two (presynaptic) cells should converge onto one (postsynaptic) cell and that the signal in the postsynaptic cell is the product of the two signals in the presynaptic cells.

On the basis of an extension of neuronal cable theory (Rall, 1970), Torre & Poggio (1978) proposed a specific synaptic interaction (a presynaptic pair and an output cell) as the implementation of the algorithm of movement detection.

The intention in briefly outlining levels (b) and (a) was to show the close connection between the three levels described in the introduction.

A prerequisite for an understanding of a neural system is an analysis at levels (*c*) and (*b*). The result of this can then be applied and an understanding at the cellular level (*a*) can be achieved.

Epilogue

After this manuscript had gone to press new experiments revealed that positional information on the time scales important for free flight tracking is mediated through clockwise motion for the right eye and counterclockwise motion for the left eye. An equation describing the tracking behaviour of female flies is

$$(*) \qquad \theta\ddot{\psi} + k\dot{\psi} = N(t) + \begin{cases} R_1(\psi, |\dot{\psi}|) & \text{for } \dot{\psi} > 0 \\ R_2(\psi, |\dot{\psi}|) & \text{for } \dot{\psi} < 0 \\ 0 & \text{for } \dot{\psi} = 0, \end{cases}$$

where $R_1(\psi, |\dot{\psi}|) = -R_2(-\psi, |\dot{\psi}|)$ (Wehrhahn and Hausen, unpublished).

The term $R_1(\psi, |\dot{\psi}|)$ describes the reaction of the fly to clockwise motion. This reaction is relatively small and constant for $-180° \leq \psi < -10°$, increases for $-10° \leq \psi \leq 20°$ to a maximum comparable to that of the $D(\psi)$ and for $20° < \psi \leq 180°$ falls to the small value of $\varphi = -180°$.

The physiological counterpart of this reaction may be found in the H-cells of the lobula plate, which is part of the third optical ganglion in flies. The H-cells sensitive to motion in a clockwise direction can be found in the right lobula plate. Qualitatively they reveal positional response characteristics corresponding to $R_1(\psi, |\dot{\psi}|)$ in (*). The H-cells of the left lobula plate are sensitive to counterclockwise motion and show a positional response function like $R_2(\psi, |\dot{\psi}|)$. The exact details of the functions R_1 and R_2 are at present under investigation (Hausen and Wehrhahn, unpublished).

Equation (*) is logically equivalent to (7.2.6) and for all purely phenomenological purposes (7.2.6) should be used. The two equations (*) and (7.2.6) are mechanistically different since (*) employs only movement detection as a basic neural mechanism. Position comes into play as a spatial parameter of movement detection.

7.6 APPENDIX

The **autocorrelation** function of a random function $f(t)$ is defined as

$$\phi_{11}(\tau) = \lim_{T \to \infty} \frac{1}{2T} \int_{-T}^{T} f(t)f(t-\tau) \, dt. \tag{1}$$

Its value at the origin is related to the **mean square value** σ^2 of the random function by

$$\sigma^2 = \phi_{11}(0) = \lim_{T \to \infty} \frac{1}{2T} \int_{-T}^{T} f^2(t) \, dt. \tag{2}$$

This value is also the maximum value of the autocorrelation function:

$$\phi_{11}(0) \geqslant \phi_{11}(\tau). \tag{3}$$

The autocorrelation function measures how much a random function is dependent upon its past. In the case of Figure 7.2.1, ϕ_{11} decays exponentially with a time constant $\tau = 1.9$ s^{-1}, which means that after about 0.5 s the stochastic signal $N(t)$ is almost independent of its past.

Using the inverse of the Fourier transform given by (6), (8) means that the autocorrelation $\phi_{11}(\tau)$ can be expressed in terms of $S(\omega)$:

$$\phi_{11}(\tau) = \frac{1}{2\pi} \int_{-\infty}^{\infty} S(\omega) \exp(j\omega t) \, d\omega. \tag{4}$$

With $\tau = 0$, the above yields

$$\frac{1}{2\pi} \int_{-\infty}^{\infty} S(\omega) \, d\omega = \phi_{11}(0) = \sigma^2. \tag{5}$$

The **Fourier transform** of a function $f(t)$ is given by

$$\tilde{F}(\omega) = \int_{-\infty}^{\infty} f(t') \exp(i\omega t') \, dt'. \tag{6}$$

Similarly to the Laplace transform discussed in Chapter 3, the following rule holds: Let $f(t)$ be an input function to a linear system and $g(t)$ the corresponding output function. Then the **transfer function** $G(\omega)$ of the linear system is given by

$$G(\omega) = \frac{\tilde{g}(\omega)}{\tilde{f}(\omega)}, \tag{7}$$

where $\tilde{f}$ and $\tilde{g}$ are the Fourier transforms of f and g.

The **power spectrum** or **spectral density** $S(\omega)$ of a function $f(t)$ is the Fourier transform of its autocorrelation function:

$$S(\omega) = \int_{-\infty}^{\infty} \phi_{11}(\tau) \exp(i\omega\tau) \, d\tau. \tag{8}$$

It can be proven (see for example Papoulis, 1962, p. 246) that

$$S(\omega) = \lim_{T \to \infty} \frac{1}{2T} \left| \int_{-\infty}^{\infty} f(t) \exp(-i\omega t) \, dt \right|^2 \geq 0. \tag{9}$$

Thus given the power spectrum $\tilde{N}(\omega)$ of an input function $N(t)$ to a linear system and the power spectrum $\phi(\omega)$ of the output function $\psi(t)$, $\tilde{N}(\omega)$ and $\phi(\omega)$ are related to the transfer function $G(\omega)$ through

$$|G(\omega)| = \phi(\omega)/\tilde{N}(\omega). \tag{10}$$

If $N(t)$ is 'white noise' in the sense defined below and $S(\omega)$ the corresponding power spectrum, then

$$\frac{1}{2\pi} \int_{-\infty}^{\infty} S(\omega) \, d\omega = \sigma^2. \tag{11}$$

The **distribution function** $F(x)$ of a function $f(t)$ is the probability that $f(t) \leq x$. Thus the probability that $x \leq f(t) \leq x + h$ is $F(x+h) - F(x)$, which is approximately $hF'(x)$ if h is small.

The **expectation** $\langle x \rangle$ of a random number x with a distribution function F is an average defined precisely as

$$\langle x \rangle = \int_{-\infty}^{\infty} xF'(x) \, dx. \tag{12}$$

A **functional** is a real valued function which has a function as an argument. Example: $\phi\{f\} = \int_a^b f(t) \exp(-t) \, dt$.

The **normal (or gaussian) distribution** describes a random number having the following distribution function

$$F(x) = (1/2\pi\sigma) \int_{-\infty}^{x} ds \exp(-\tfrac{1}{2}(s-\xi)^2/\sigma^2), \tag{13}$$

where ξ is the mean of the distribution and σ is the standard deviation.

The derivative F' is called the **normal (gaussian) density**

$$F'(x) = (1/2\pi\sigma) \exp\{-\tfrac{1}{2}(x - \xi)^2/\sigma^2\}. \tag{14}$$

A **random function of time or stochastic process** is a function of time depending on chance; for example the value $f(t)$ of f at time t is a **random number** which can be described by a distribution function.

A **stationary process** f is a stochastic process whose distribution of values $f(t)$ is the same at any time t. (Refer to the definition of 'distribution' to understand this correctly.)

The two stochastic processes f and g are **statistically independent** if, for any two functionals ϕ_1 and ϕ_2,

$$\langle \phi_1(f) \rangle \cdot \langle \phi_2(g) \rangle = \langle \phi_1(f) \cdot \phi_2(\delta) \rangle. \tag{15}$$

White noise has a power spectrum which is positive and constant for all frequencies. For a physical system a random process is effectively 'white' if its power spectrum is positive and nearly constant for those frequencies where the transfer function of the system does not vanish.

References

Baylor, D. A. & Hodgkin, A. L. (1974). Changes in time scale and sensitivity in turtle photoreceptors. *J. Physiol.* **242**, 729–58.

Baylor, D. A., Hodgkin, A. L. & Lamb, T. D. (1974*a*). The electrical response of turtle cones to flashes and steps of light. *J. Physiol.* **242**, 685–727.

—— (1974*b*). Reconstruction of the electrical responses of turtle cones to flashes and steps of light. *J. Physiol.* **242**, 759–91.

Calwill, G. F. Jr & Soeldner, J. S. (1969). Glucose homeostasis: a brief review. In *Mathematical Biosciences*, Suppl. 1 *Hormonal Control System*.

Coleman, B. D. (1971). On retardation theorems. *Arch. Ration. Mech. Anal.* **43**, 1–23.

Collett, T. S. & Land, M. F. (1975). Visual control of flight behaviour in the hoverfly, *Syritta pipiens L. J. Comp. Physiol.* **99**, 1–66.

Harmon, L. D. (1970). Neural subsystems: an interpretive summary. In *The Neurosciences Second Study Program*, ed. G. C. Quarton, T. Melnechuk and G. Adelman, New York, Rockefeller University Press, pp. 486–93.

Hausen, K. & Strausfeld, N. J. (1979). Sexually dimorphic interneuron arrangements in the fly visual system. *Proc. Roy. Soc. Lond. B*, in press.

Heide, G. (1975). Properties of a motor output system involved in the optomotor response in flies. *Biol. Cybernet.* **21**, 99–112.

Hodgkin, A. L. & Huxley, A. F. (1952). A quantitative description of membrane current and its application to conduction and excitation in nerve. *J. Physiol.* **117**, 500–44.

Katz, B. & Miledi, R. (1973). The characteristics of 'end-plate noise' produced by different depolarizing drugs. *J. Physiol.* **230**, 707–17.

Land, M. F. & Collett, T. S. (1974). Chasing behaviour of houseflies (*Fannia canicularis*). A description and analysis. *J. Comp. Physiol.* **89**, 331–57.

Mach, E. (1914). *The Analysis of Sensations and the Relation of the Physical to the Psychical*, trans. C. M. Williams, rev. S. Waterlow, Chicago and London, The Open Court Publishing Company.

Martin, P. (1968). *Measurements and Correlation Functions*, New York, London, Paris, Gordon and Breach.

Papoulis, A. (1962). *The Fourier Integral and its Applications*, New York, McGraw-Hill.

Poggio, T. & Reichardt, W. (1973). A theory of the pattern induced flight orientation of the fly *Musca domestica. Biol. Cybernet.* **12**, 185–203.

—— (1976). Visual control of orientation behaviour in the fly. II. Towards the underlying neural interaction. *Quart. Rev. Biophys.* **9**, 377–438.

Rall, W. (1970). Dendritic neuron theory and dendrodendritic synapses in a simple cortical system. In *The Neurosciences Second Study Program*, ed. G. C. Quarton, T. Melnechuk and G. Adelman, New York, Rockefeller University Press, pp. 552–65.

Ratliff, F., Knight, B. W., Dodge, F. A. & Hartline, H. K. (1974). Fourier analysis of dynamics of excitation and inhibitions in the eye of *Limulus*: amplitude, phase and distance. *Vision Res.* **14**, 1155–68.

Reichardt, W. (1973). Musterinduzierte Flugorientierung. *Naturwiss.* **60**, 122.

—— (1978). Functional characterization of neural interactions through an analysis of behaviour. In *The Neurosciences, Fourth Study Program*, ed. F. O. Schmitt and F. G. Worden, Cambridge, MIT Press, pp. 81–103.

Reichardt, W. & Poggio, T. (1975). A theory of the pattern induced flight orientation of the fly *Musca domestica* II. *Biol. Cybernet.* **18**, 69–80.

Reichardt, W. & Poggio, T. (1976). Visual control of orientation behaviour in the fly. I. A quantitative analysis. *Quart. Rev. Biophys.* **9**, 311–75.

Stark, L. & Baker, F. (1959). Stability and oscillations in a neurological servomechanism. *J. Neurophysiol.* **22**, 156–64.

Stark, L. & Sherman, P. M. (1957). A servoanalytical study of consensual pupil reflex to light. *J. Neurophysiol.* **20**, 17–26.

Torre, V. & Poggio, T. (1978). A synaptic mechanism possibly underlying directional selectivity to motion. *Proc. Roy. Soc. B* **202**, 409–16.

Verveen, A. & De Felice, L. (1974). Membrane noise. *Prog. Biophys. Biol.* **28**, 189.

Wax, N. (1954). *Selected Papers on Noise and Stochastic Processes*, New York, Dover Publ.

Wehrhahn, C. (1978*a*). Flight torque and lift responses of the housefly (*Musca domestica*) to a single stripe moving in different parts of the visual field. *Biol. Cybernet.* **29**, 237–47.

—— (1978*b*). The angular orientation of the movement detectors acting on the flight lift response in flies. *Biol. Cybernet.* **31**, 169–73.

—— (1979). Sex-specific differences in the chasing behaviour of free flying flies (*Musca*). *Biol. Cybernet.* **32**, 239–41.

Wehrhahn, C. & Reichardt, W. (1975). Visually induced height orientation of the fly *Musca domestica. Biol. Cybernet.* **20**, 37–50.

Yarbus, A. L. (1967). *Eye Movements and Vision*, New York, Plenum Press.

Appendix: Mathematical topics

A.1 A CALCULUS REFRESHER

In this section we present the principal mathematical prerequisites for the main material of this book. Although much of the presentation is selfcontained, it is undoubtedly too terse to be followed by someone with no previous experience of the calculus. Yet it should serve well as a review for those whose mathematics has become rusty through disuse. As has been mentioned in the Introduction, major formulae have been indicated by a symbol. Anyone who is familiar with nearly all of these formulae has adequate mathematical preparation for present purposes. Other prospective readers might well begin with a book like the paperback *Quick Calculus* by D. Kleppner and N. Ramsay (Wiley, 1965).

There are some novelties in our approach. These will be mentioned in the following annotated list of topics.

Functions, limits, derivatives.

Chain rule, implicit differentiation.

Inverse functions.

Power series, Taylor formula with remainder.

Introduction to differential equations, linearity.

[In our presentation of calculus, we place major emphasis on the background necessary to contend with differential equations.]

Systems of differential equations.

[Systems are introduced via a brief explanation of elementary biochemical kinetics that is intended primarily for readers who are mathematicians interested in learning more about applications in biology.]

The exponential function.

[The function exp is introduced as the solution to the equation of unrestricted population growth. The basic form of this equation is achieved by an application of dimensional analysis.]

The logarithm.

The spring-mass system.

[The famous forced second-order ordinary differential equation of elementary physics is introduced here, because it is central to Chapter 7 and because a dimensionless version of the equation for the undamped, unforced case forms the basis of the next topic.]

Sines and cosines.

[These functions are defined as appropriate solutions of the differential equation $(d^2x/dt^2) + x = 0$. Derivation of their properties provides a first example of the phase plane methods to be introduced in Appendix Section A.3. Both this derivation and discussion of the exponential function serve as introductory examples of a central mathematical theme of this book – deduction of qualitative properties of the solutions to differential equations.]

The integral: definition and rules for manipulation.

[The integral mean value theorem, the fundamental theorem of calculus, integration by parts.]

Functions of several variables: Partial derivatives, chain rule, Taylor series, multiple integrals.

The reader should be warned that some of the formulae that we give may not be correct if certain conditions are violated. We have omitted these conditions here. 'Usually' there is nothing to worry about, but any standard calculus text can be consulted if doubts arise.

Function

y is a **function** of x for x between a and b, i.e.

$$y = f(x), \qquad a \leq x \leq b, \tag{1}$$

if for every x in the given interval there is a method for determining precisely one number y.

Example. The **linear function** $y = mx + b$ is defined for all x. Its graph is a straight line with slope m and y-intercept b (Figure A.1.1).

Limit

The limit of $f(x)$ as x approaches a is L, i.e.

$$\lim_{x \to a} f(x) = L, \tag{2}$$

if $f(x)$ can be made arbitrarily close to L by taking x within a certain distance of a. What happens when $x = a$ is irrelevant.

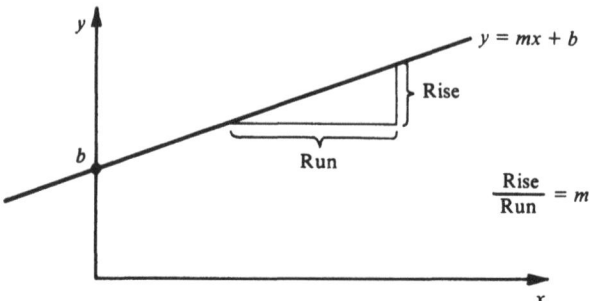

Figure A.1.1. Graph of straight line with equation $y = mx + b$.

Derivatives

The **derivative** of $f(x)$ at the point a is defined as follows.

$$\blacklozenge\blacklozenge \quad \left.\frac{df(x)}{dx}\right|_{x=a} \equiv \lim_{h \to 0} \frac{f(a+h)-f(a)}{h}. \tag{3}$$

(The symbol $\equiv$ is used when, as here, the expression on the left is defined by the expression on the right.) As shown in Figure A.1.2 a geometric interpretation of the derivatives is the slope of the tangent line to $f(x)$ at $x = a$.

If df/dx is positive (negative), then the slope of f is positive (negative) and f is increasing (decreasing). If d^2f/dx^2 is positive (negative) then f is concave upward (downward). If $df/dx = 0$ at a point, then the function at that point has either a maximum, a minimum, or a point of inflection (see Figure A.1.3).

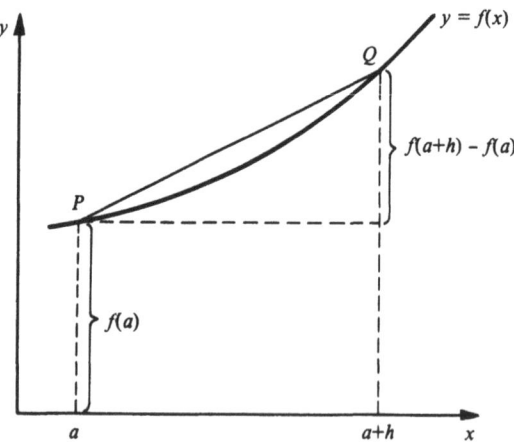

Figure A.1.2. The secant line PQ has the slope $[f(a+h)-f(a)]/h$. As $h \to 0$, $Q \to P$. From (3), the derivative thus gives the slope of the limiting secant line at P, i.e. the tangent line.

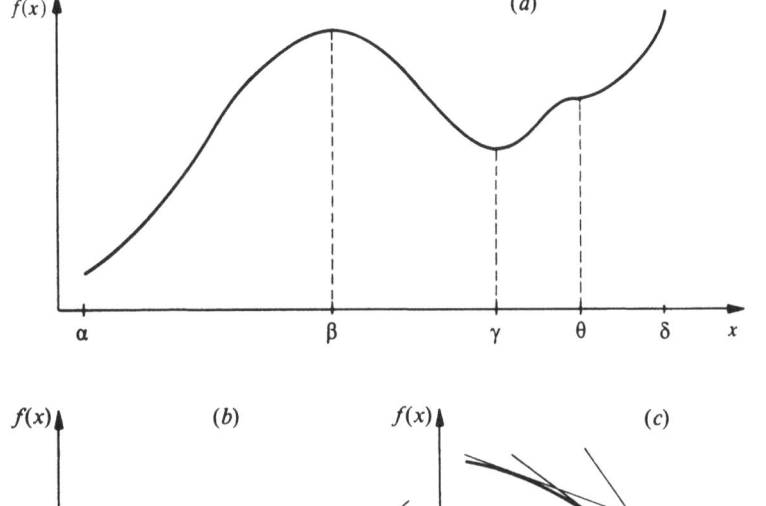

Figure A.1.3. (a) For $\alpha < x < \beta$ and $\gamma < x < \delta$, $df/dx > 0$, f increasing. For $\beta < x < \gamma$, $df/dx < 0$, f decreasing. $df/dx = 0$ at $x = \beta$ (maximum), $x = \gamma$ (minimum) and $x = \theta$ (inflection point). (b) $d^2f/dx^2 > 0$, successive tangent lines have increasing slope, f concave upward. (c) $d^2f/dx^2 < 0$, successive tangent lines have decreasing slope, f concave downward.

The second derivative is the derivative of the first derivative, etc. That is

$$\frac{d^2f(x)}{dx^2} = \frac{d}{dx}\left[\frac{df(x)}{dx}\right], \quad \frac{d^3f(x)}{dx^3} = \frac{d}{dx}\left[\frac{d^2f(x)}{dx^2}\right], \dots$$

Rules for manipulating derivatives

The derivative of a constant multiple of f is that constant times the derivative of f.

◆◆ $$\frac{d}{dx}[Cf(x)] = \frac{C\,df(x)}{dx}.$$ (4)

The derivative of the sum (difference) of two functions f and g is the sum (difference) of their derivatives

$$\blacklozenge\blacklozenge \qquad \frac{d}{dx}[f(x)+g(x)] = \frac{df(x)}{dx} + \frac{dg(x)}{dx}, \tag{5a}$$

$$\blacklozenge\blacklozenge \qquad \frac{d}{dx}[f(x)-g(x)] = \frac{df(x)}{dx} - \frac{dg(x)}{dx}. \tag{5b}$$

Derivatives of products and quotients are given by the following formulae

$$\blacklozenge\blacklozenge \qquad \frac{d}{dx}[f(x)g(x)] = f(x)\left[\frac{dg(x)}{dx}\right] + \left[\frac{df(x)}{dx}\right]g(x), \tag{6a}$$

$$\blacklozenge\blacklozenge \qquad \frac{d}{dx}\left[\frac{f(x)}{g(x)}\right] = \frac{g(x)\left[\dfrac{df(x)}{dx}\right] - f(x)\left[\dfrac{dg(x)}{dx}\right]}{[g(x)]^2}. \tag{6b}$$

All the above rules follow from the definition (3). For example

$$\frac{d}{dx}[f(x)g(x)]\bigg|_{x=a} = \lim_{h\to 0} \frac{f(a+h)g(a+h)-f(a)g(a)}{h}$$

$$= \lim_{h\to 0} \frac{f(a+h)[g(a+h)-g(a)]+g(a)[f(a+h)-f(a)]}{h}$$

$$= \lim_{h\to 0} f(a+h)\left[\frac{g(a+h)-g(a)}{h}\right]$$

$$+ \lim_{h\to 0} g(a)\left[\frac{f(a+h)-f(a)}{h}\right]$$

$$= f(a)\frac{dg(x)}{dx}\bigg|_{x=a} + g(a)\frac{df(x)}{dx}\bigg|_{x=a}.$$

Since the equality of the first and last expressions holds for any a, we have demonstrated the product formula (6a).

In the above proof, we have used the result that the limit of the sum of two quantities is the sum of their limits. We have also made use of the equality

$$\lim_{h\to 0} f(a+h) = f(a). \tag{7}$$

This equality is not true for the functions shown in Figure A.1.4. But it is true for functions that are defined at $x = a$ and that do not have 'jumps' there. Such functions are called **continuous** at $x = a$. Thus our proof of the product formula (6a) certainly requires that $f(x)$ be continuous at $x = a$. As mentioned above, in this very brief treatment we shall generally omit such 'technical' requirements that are 'usually' true.

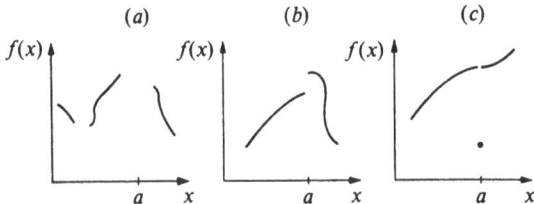

Figure A.1.4. Graphs of functions that are not continuous at $x = a$. (*a*) A function that is not defined at $x = a$. (*b*) A function where a single limit does not exist as $x \to a$. (*c*) A function where $\lim f(x)$, as $x \to a$, exists, but does not equal $f(a)$.

The derivative of the power function

It can be shown that

◆◆ $\qquad \dfrac{d}{dx}(x^b) = bx^{b-1}, \qquad b$ a constant. $\qquad\qquad$ (8)

We shall provide a proof for this, when b is a positive integer n, using the **method of induction**. This method requires that we show that the result is true for $n = 1$, and that if the result is true for $n = N$ then it is true for $n = N + 1$. (Then the result holds for $n = 1$, $n = 2$, $n = 3$, etc.)

When $n = 1$ we have

$$\frac{d}{dx}(x) = \lim_{h \to 0} \frac{(x+h)-x}{h} = \lim_{h \to 0} \frac{h}{h} = \lim_{h \to 0} 1 = 1. \qquad (9)$$

Supposing the result correct for $n = N$ we see (using (6*a*) and (9)) that

$$\frac{d}{dx}(x^{N+1}) = \frac{d}{dx}[x \cdot x^N] = x(Nx^{N-1}) + x^N = (N+1)x^N. \qquad (10)$$

The chain rule

Suppose f is a function of x and that x is in turn a function of t. Then the following formula allows efficient computation of the derivative with respect to t.

◆◆ $\qquad \dfrac{d}{dt}f[x(t)] = \dfrac{df(x)}{dx}\bigg|_{x=x(t)} \dfrac{dx(t)}{dt}. \qquad\qquad$ (11)

Formula (11) follows at once from the fact that its left side can be written in the form

$$\lim_{h \to 0} \frac{f[x(t+h)] - f[x(t)]}{x(t+h) - x(t)} \cdot \frac{x(t+h) - x(t)}{h}.$$

Example.   Suppose that $f(x) = x^3$ and $x(t) = t^2$. On the one hand

$$f[x(t)] = [t^2]^3 = t^6, \qquad \frac{d}{dt}(t^6) = 6t^5. \tag{12}$$

But the chain rule gives the same result:

$$\frac{d}{dt}[x^3] \text{ when } x = t^2$$

is

$$\frac{d}{dx}(x^3)\Big|_{x=t^2} \frac{d(t^2)}{dt} = 3x^2\Big|_{x=t^2} \cdot 2t = 3t^4 \cdot 2t = 6t^5. \tag{13}$$

Implicit differentiation

The equation

$$x^2 + y^2 = R^2, \qquad R \text{ a positive constant,} \tag{14}$$

describes a circle of radius R centered at the origin of an x–y plane (Figure A.1.5). Suppose that one wished to find the slope at some point (x_0, y_0) in the first quadrant. One could solve for y as a function of x and differentiate. Using (8) and (11) we find that

$$y = \sqrt{(R^2 - x^2)},$$

$$\frac{dy}{dx} = \frac{(-2x)}{2\sqrt{(R^2 - x^2)}} = -\frac{x}{\sqrt{(R^2 - x^2)}} = -\frac{x}{y},$$

$$\frac{dy}{dx}\Big|_{(x_0, y_0)} = -\frac{x_0}{y_0}. \tag{15}$$

Alternatively one could regard (14) as a statement that the sum of the function x^2 and the function $[y(x)]^2$ is constant. The derivative of a constant is zero, so that

$$\frac{d}{dx}\{x^2 + [y(x)]^2\} = 0, \qquad 2x + 2y\frac{dy}{dx} = 0,$$

$$\frac{dy}{dx} = -\frac{x}{y}, \qquad \frac{dy}{dx}\Big|_{(x_0, y_0)} = -\frac{x_0}{y_0}. \tag{16}$$

This is the same answer as obtained previously. The latter procedure is an example of **implicit differentiation**; 'implicit' in that we obtained a formula for dy/dx without first explicitly solving for y in terms of x.

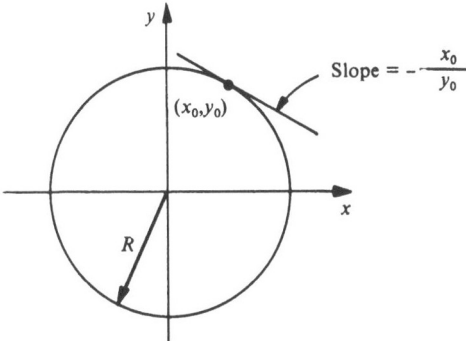

Figure A.1.5. Graph of the circle $x^2 + y^2 = R^2$, with a tangent line at a point (x_0, y_0).

Note that the calculation of the implicit differentiation formula (16) actually gives a result that is valid for all quadrants. By contrast, the derivation of (15) must be revised for the third and fourth quadrants ($y < 0$); one must begin with $y = -\sqrt{(R^2 - x^2)}$. Of course the same result as (16) is ultimately obtained.

Inverse functions

If y is an increasing function of x in a certain interval, or a decreasing function, for every x in that interval there corresponds exactly one y and vice versa (Figure A.1.6). Thus we can either regard x as a function of y or

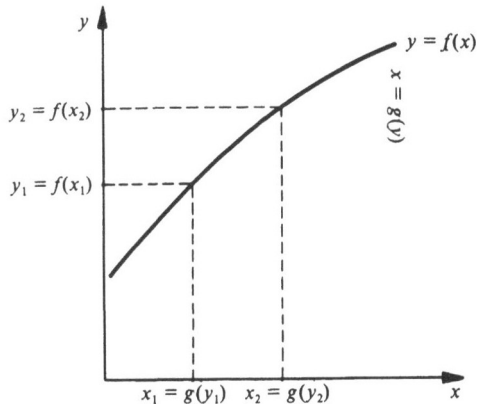

Figure A.1.6. Since $y = f(x)$ is an increasing function, we can also regard x as a function of y. To see a graph of this inverse function, $x = g(y)$, one merely needs to turn the page sideways.

y as a function of *x*, i.e.

$$y = f(x) \quad \text{or} \quad x = g(y). \tag{17}$$

We emphasize that the inverse functions *f* and *g* exist when to every *x* there corresponds one *and only one* *y* and vice versa. This one-to-one correspondence can be symbolically indicated as follows:

$$x \underset{g}{\overset{f}{\rightleftharpoons}} y.$$

As a consequence we have the relations

$$x = g[f(x)], \qquad y = f[g(y)]. \tag{18}$$

The first of these, for example, states the obvious fact that it is *x* that corresponds to *y* that corresponds to *x*.

Let *y* = *b* correspond to *x* = *a*:

$$b = f(a) \quad \text{or} \quad a = g(b).$$

Then the derivatives of *f* and *g* at the point (a, b) are reciprocals of one another. For, since $x \to a$ if and only if $y \to b$, we can write

$$\frac{df(x)}{dx}\bigg|_{x=a} = \lim_{x \to a} \frac{f(x) - f(a)}{x - a} = \lim_{y \to b} \frac{y - b}{g(y) - g(b)} = \frac{1}{[dg(y)/dy]|_{y=b}}.$$

This result can be written in abbreviated form as

$$\blacklozenge\blacklozenge \qquad \frac{dy}{dx} = \left(\frac{dx}{dy}\right)^{-1}. \tag{19}$$

Infinite series

We write

$$A = a_0 + a_1 + a_2 + a_3 + \ldots, \quad \text{or equivalently } A = \sum_{i=0}^{\infty} a_i, \tag{20}$$

if

$$\lim_{N \to \infty} |A - (a_0 + a_1 + \ldots + a_{N-1} + a_N)| = 0.$$

That is, the series (20) converges to *A* if the first *N* terms of the series come closer and closer to *A* as *N* gets larger and larger.

We write

$$f(x) = r_0(x) + r_1(x) + r_2(x) + \ldots, \quad \text{i.e.} \quad f(x) = \sum_{i=0}^{\infty} r_i(x);$$

$$a \leqslant x \leqslant b; \tag{21}$$

if our previous definition concerning infinite series holds for each point x between a and b.

Power series

Suppose we know that

$$f(x) = a_0 + a_1(x-c) + a_2(x-c)^2 + a_3(x-c)^2 + \ldots \tag{22}$$

Then, substituting $x = c$ into both sides, we see that

$$f(c) = a_0. \tag{23}$$

Furthermore, differentiating both sides of (22) we find that

$$\frac{df(x)}{dx} = a_1 + 2a_2(x-c) + 3a_3(x-c)^2 + \ldots, \tag{24}$$

so that

$$\frac{df(x)}{dx}\bigg|_{x=c} = a_1. \tag{25}$$

Similarly

$$\frac{d^2f(x)}{dx^2}\bigg|_{x=c} = 2a_2,$$

$$\frac{d^3f(x)}{dx^3}\bigg|_{x=c} = 3 \cdot 2 \cdot a_3, \ldots, \frac{d^nf(x)}{dx^n}\bigg|_{x=c} = n!a_n. \tag{26}$$

Here $n!$ ('en factorial') is defined by

$$n! \equiv n \cdot (n-1) \cdot (n-2) \cdot \ldots \cdot 3 \cdot 2 \cdot 1. \tag{27}$$

It turns out that the series representation (22) is valid in some interval centered on $x = c$, i.e. for

$$|x - c| < B, \quad \text{or equivalently } -B < x - c < B, \quad \text{for some } B \geq 0. \tag{28}$$

Certainly the representation is valid for $B = 0$, for then it is just the trivial equality

$$f(c) = a_0 + 0 + 0 + \ldots. \tag{29}$$

A necessary condition for the representation to hold is that it make sense: this means that $f(x)$ must have derivatives of all orders. For a wide class of functions, then, we have the **Taylor series representation**

◆◆ $$f(x) = \sum_{n=0}^{\infty} a_n(x-c)^n; \quad a_n = \frac{1}{n!}\frac{d^nf(x)}{dx^n}\bigg|_{x=c};$$

$$|x - c| < B, \quad B \geq 0. \tag{30}$$

A refinement of this representation is the **Taylor formula with remainder**

$$f(x) = \sum_{n=0}^{N} a_n (x-c)^n + R_{N+1}; \qquad |x-c| < B. \tag{31}$$

Here the remainder R_{N+1} after $N+1$ terms is given by

$$R_{N+1} = \frac{1}{(N+1)!} \frac{d^{N+1} f(x)}{dx^{N+1}}\bigg|_{x=q}, \qquad \text{for some } q \text{ such that } |q-c| < B. \tag{32}$$

Since the exact location of the point q is unknown, the remainder cannot generally be evaluated exactly. Nonetheless, one can often assert that R_{N+1} is less than some small quantity, and thereby obtain a good approximation to f by an N-term series. Such approximations are used in evaluating functions by computers, among other applications.

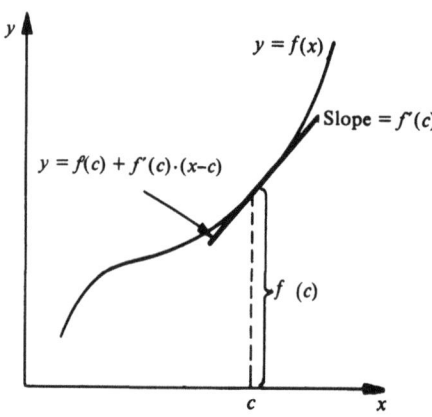

Figure A.1.7. The heavy line (slope $= f'(c)$) provides a straight-line approximation to $f(x)$ in the neighborhood of the point $x = c$.

Particularly important is the **straight line or linear approximation** when $N = 1$ (Figure A.1.7).

$$f(x) \approx f(c) + (x-c) f'(c), \qquad |x-c| < B. \tag{33}$$

NB: The symbol "$\approx$" means 'approximately equal to'. The notation $f'(c)$ is defined by

$$f'(c) \equiv \frac{df(x)}{dx}\bigg|_{x=c}. \tag{34}$$

Differential equations

Differential equations are simply equations involving derivatives. The equation

$$dx/dt = f(x, t) \qquad (35a)$$

for $x(t)$ is called **first order** because only the first derivative appears. Similarly the equation

$$d^2x/dt^2 = h(x, dx/dt, t). \qquad (36a)$$

is called **second order**.

Examples. Equation $(35a)$ arises in biology when f gives the growth rate of the population x. Equation $(36a)$ arises in physics when h gives the force per unit mass (which is equal to the acceleration, according to Newton). We expect a unique solution to $(35a)$, provided that we state the population level x_0 at the start of the experiment (time t_0). Similarly, $(36a)$ should have a unique solution provided that the initial position x_0 and the initial velocity v_0 are given. It is not a surprise, therefore, that generally there exist unique solutions to $(35a)$ and $(36a)$ supplemented by the **initial conditions**

$$x(t_0) = x_0 \qquad (35b)$$

and

$$x(t_0) = x_0, \qquad \frac{dx(t)}{dt}\bigg|_{t=t_0} = v_0, \qquad (36b)$$

respectively.

It is important to realize that a purported solution to a differential equation can be checked, even though one cannot follow the reasoning that led to the solution.

Example. Verify that

$$t^2(d^2x/dt^2) - 2t(dx/dt) + 2x = 0$$

has the solutions

$$x = t \quad \text{and} \quad x = t^2.$$

Solution:

$$t^2 \cdot \frac{d^2}{dt^2}(t) - 2t \cdot \frac{d}{dt}(t) + 2(t) = 0 - 2t + 2t = 0.$$

$$t^2 \cdot \frac{d^2}{dt^2}(t^2) - 2t\frac{d}{dt}(t^2) + 2(t^2) = 2t^2 - 4t^2 + 2t^2 = 0.$$

Linear differential equations

The special equation of n-th order, with *given* coefficients $a_n(t)$, $a_{n-1}(t), \ldots,$

$$a_n(t)\frac{d^n x}{dt^n}+a_{n-1}(t)\frac{d^{n-1}x}{dt^{n-1}}+\ldots+a_2(t)\frac{d^2 x}{dt^2}+a_1(t)\frac{dx}{dt}+a_0(t)x = 0 \qquad (37)$$

is called **linear**. If the equation is abbreviated by $L[x]=0$ then it is a simple application of the rules (4) and (5) for manipulating derivatives to verify that

◆◆ $L[\alpha+\beta]=L[\alpha]+L[\beta], \qquad L[C\alpha]=CL[\alpha].$ \qquad (38a, b)

Here $\alpha(t)$ and $\beta(t)$ are arbitrary functions, while C (and D in (39) below) are arbitrary constants. These two basic **linearity properties** can also be summed up in the following single equation, whch is equivalent to (38a) and (38b):

◆◆ $L[C\alpha+D\beta]=CL[\alpha]+DL[\beta].$ \qquad (39)

   The importance of linearity lies in the following fact. If $\alpha(t)$ and $\beta(t)$ are solutions of (37), that is if

$$L[\alpha]=0 \quad \text{and} \quad L[\beta]=0,$$

then

$$L[C\alpha]=0 \quad \text{and} \quad L[\alpha+\beta]=0,$$

for by linearity

$$L[\alpha]=CL[\alpha]=0 \quad \text{and} \quad L[\alpha+\beta]=L[\alpha]+L[\beta]=0+0=0.$$

Thus, for a linear equation there holds the **superposition principle** which asserts that a constant multiple of a solution is a solution and the sum of two solutions is a solution. The superposition principle does not in general hold for nonlinear equations. (For example it is easily verified that the equation

$$(dx/dt)^2-4x = 0$$

has one solution $x = t^2$. But the function Ct^2 is a solution only when $C = 1$.) The superposition principle lies at the core of the theory of linear differential equations. This theory is rather complete, in contrast to the theory of nonlinear equations – a subject for much current mathematical research.

Law of mass action: systems of differential equations

Biochemistry provides many applications of differential equations. Perhaps the most important single example concerns an enzyme (concentration $e(t)$ at time t) that combines with its substrate (the chemical on which the enzyme 'works' – concentration $s(t)$) to form an enzyme–substrate complex (concentration $c(t)$). The complex may break apart to yield the original enzyme and substrate molecules. On the other hand the enzyme may 'work' on the substrate so that when the complex breaks apart the substrate is transformed, by the breaking of a chemical bond for example, into the original enzyme and an altered product (concentration $p(t)$). In chemical shorthand this process is described as follows.

$$e + s \underset{k_{-1}}{\overset{k_{+1}}{\rightleftharpoons}} c \xrightarrow{k_{+2}} p + e. \tag{40}$$

The k_i are called **rate constants**. To understand their role note that there is a certain probability per unit time that enzyme and substrate molecules will collide. At low concentrations, all other things being equal, this probability is proportional both to e and to c. (If the amount of enzyme is doubled, for example, we expect twice as many collisions per unit time.) Only a certain fraction of collisions will 'successfully' result in the formation of a complex. By definition, the proportionality factor that gives the correct rate of successful enzyme–substrate collisions is k_1. The constants k_{-1} and k_{+2} have a similar interpretation. This view of chemical reaction is associated with the term **law of mass action**. It results in the **system of first-order ordinary differential equations**

$$de/dt = -k_{+1}es + k_{-1}c + k_{+2}c, \tag{41a}$$

$$ds/dt = -k_{+1}es + k_{-1}c, \tag{41b}$$

$$dc/dt = k_{+1}es - k_{-1}c - k_{+2}c, \tag{41c}$$

$$dp/dt = k_{+2}c. \tag{41d}$$

In order to specify the situation fully we must adjoin initial conditions such as

$$\text{at } t = 0; \quad e = e_0, \quad s = s_0, \quad c = c_0, \quad p = p_0. \tag{41e}$$

Here e_0, s_0, c_0, and p_0 are given constants. Note that the presence of the product of two unknown functions in the 'es' terms means that the equations are nonlinear. Such nonlinearities are entirely typical of chemical kinetics. Their presence means, mathematically, that the elementary theory of differential equations is of little help. Phenomenologically it means that a variety of interesting behaviors can be expected from systems that are influenced by chemical reactions. Essentially all biological systems are of this type; it is a central purpose of this book to exhibit, analyze, or predict some of the interesting behavior of which these systems are capable.

Exponential growth

The number of bacteria in a flask, plotted as a function of time, might look like Figure A.1.8. If we smooth out the tiny steps in the graph we get a

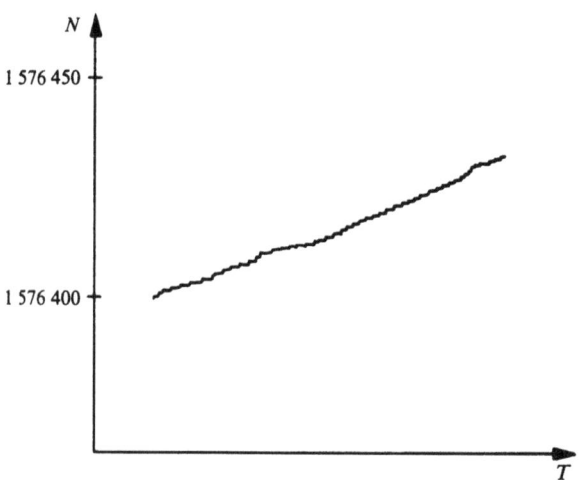

Figure A.1.8. Possible plot of the number N of bacteria in a flask as a function of the time T.

new graph that differs only imperceptibly from the true graph, though it does have the unrealistic property that there can be a fractional number of bacteria. Accepting a fractional number of bacteria as an inessential error, we may postulate the following as the simplest reasonable equation for $N(T)$, the number of bacteria at time T:

$$dN/dT = rN. \tag{42}$$

Here r is a constant called the **net birth rate**; it gives the average number of bacteria 'born' from a given bacterium, per unit time, minus the average number that have died. To expect a unique solution to the problem of determining the number of bacteria, we must add the information that the number of bacteria at some initial time T_0 is known:

$$N = N_0 \quad \text{at time } T = T_0; \quad N_0 \text{ and } T_0 \text{ constants.} \tag{43}$$

Since there are average of r net births per bacterium per unit time, $1/r$ is the average time per net birth (the **doubling time**). It seems reasonable to measure time in this intrinsic unit. Furthermore, it seems sensible to set a 'problem oriented' clock to read 'zero' at the time T_0 when the experiment is started. Consequently we define a new time t by

$$t = \frac{T - T_0}{1/r} = r(T - T_0). \tag{44}$$

It also appears sensible to measure N in multiples of the initial inocculum size N_0. We therefore define a new dependent variable x by

$$x = N/N_0. \tag{45}$$

Using the chain rule (11), we find that our problem becomes

$$dx/dt = x, \qquad x = 1, \quad \text{when } t = 0. \tag{46a, b}$$

Essence of proof.

$$\frac{dN}{dT} = \frac{d(xN_0)}{dT} = N_0\left(\frac{dx}{dT}\right),$$

$$\frac{dx}{dT} = \frac{dx}{dt} \cdot \frac{dt}{dT}, \qquad \frac{dt}{dT} = \frac{d}{dT}[r(T - T_0)] = r\frac{d}{dT}(T - T_0) = r.$$

Thus

$$dN/dT = rN \quad \text{implies } N_0 r(dx/dt) = rxN_0, \quad \text{i.e. } dx/dt = x. \tag{47}$$

The exponential function

The problem posed by (42) and (43), and of course by the special case (46), describes a reproduction process that at any instant t depends only on the population at that instant. If the population at any single instant is known, one would expect that the population is determined at all other times. That is, one would expect that (42) and (43), and (46), have unique solutions. That this is indeed the case is proved in standard books on the theory of differential equations.

The unique solution of (46) is called the **exponential function** and is written $x = \exp(t)$. Thus

♦♦ $$\frac{d}{dt}\exp(t) = \exp(t), \qquad \exp(0) = 1. \tag{48a, b}$$

It follows from (44) and (45) that the solution of (42) and (43) is

$$N = N_0 \exp[r(T - T_0)]. \tag{49}$$

Useful formulae involving the exponential function can be obtained if we note from (49) that the solution of the problem

$$dN/dT = N, \qquad N(T_0) = 1,$$

is $N = \exp(T - T_0)$. But $\exp(T - T_0)$ has the value $\exp(-T_0)$ when $T = 0$. Thus $\exp(T - T_0)$ is also the solution of the problem

$$dN/dT = N, \qquad N(0) = \exp(-T_0). \tag{50}$$

On the other hand, specialization of (49) shows that the function $\exp(-T_0)\exp(T)$ also satisfies (50). Since this problem has a unique solution, it must be that

$$\exp(-T_0)\exp(T) = \exp(T - T_0). \tag{51}$$

Relation (51) holds for arbitrary T and T_0. In particular if $T = T_0$, (51) implies

$$\exp(-T_0)\exp(T_0) = 1, \tag{52a}$$

i.e.

◆◆ $$\exp(-T_0) = \frac{1}{\exp(T_0)}, \tag{52b}$$

where we have employed (48b). Moreover if we make the substitutions $-T_0 = s$, $T = t$, (51) takes on a more symmetrical appearance:

◆◆ $$\exp(s)\exp(t) = \exp(s + t). \tag{53}$$

The function $\exp(t)$ is often written in the form e^t. To see why, we first note that if $t = s$, then (53) yields

$$\exp(2s) = [\exp(s)]^2. \tag{54}$$

If $t = 2s$, (53) and (54) imply

$$\exp(3s) = \exp(s)\exp(2s) = [\exp(s)]^3,$$

and generally

$$\exp(Ms) = [\exp(s)]^M, \tag{55}$$

where M is an positive integer. Indeed, (55) holds for an arbitrary constant M. This is easily demonstrated by showing that both sides of (55) satisfy

$$dy/ds = My, \qquad y(0) = 1.$$

The number $\exp(1)$ is given a special symbol, e:

◆◆ $$e \equiv \exp(1). \tag{56}$$

From (55), with $s = 1$, we see that

◆◆ $$\exp(M) = e^M \tag{57}$$

so that $\exp(M)$ can be regarded as 'e to the M-th power'.

Let us now examine the qualitative behavior of the function $x = \exp(t)$. Since the derivative of $\exp(t)$ is positive at $t = 0$, this function starts to increase from $x(0) = 1$ as t increases from $t = 0$. Suppose that $\exp(t)$ continually increased until an instant $t = \gamma$. Certainly

$$x(\gamma) > x(0) = 1. \tag{58}$$

Assume in addition that $x(t)$ decreased for a time when $t > \gamma$. As shown in Figure A.1.3a, there would then be a maximum at γ, implying

$$\left.\frac{dx(t)}{dt}\right|_{t=\gamma} = 0. \tag{59}$$

But the differential equation requires $dx/dt = x$, so that (58) contradicts (59). Thus $\exp(t)$ can never decrease as t increases from $t = 0$. In particular, t can certainly never become negative for positive t. It follows from (52b) that

$$\exp(t) > 0 \quad \text{for all } t. \tag{60}$$

Since (48) implies that

$$\frac{d^n(\exp(t))}{dt^n} = \exp t \quad \text{for all positive integers } n, \tag{61}$$

we can deduce that

$$\frac{d^n(\exp(t))}{dt^n} > 0 \quad \text{for all positive integers } n. \tag{62}$$

In particular (62) implies that $\exp(t)$ continually increases and is concave upward. Moreover, the facts that $\exp(t)$ increases and that $\exp(0) = 1$ imply that

$$\exp(1) \equiv e > 1.$$

Together with (57) this shows that $\exp(M)$ can be made arbitrarily large by taking M large enough. From (52b) it follows that $\exp(-M)$ can be made arbitrarily close to zero by taking M large enough. These results can be written in the following way:

$$\lim_{t\to\infty} \exp(t) = \infty, \qquad \lim_{t\to\infty} \exp(-t) = 0. \tag{63}$$

All these qualitative findings are illustrated in the graph of Figure A.1.9.

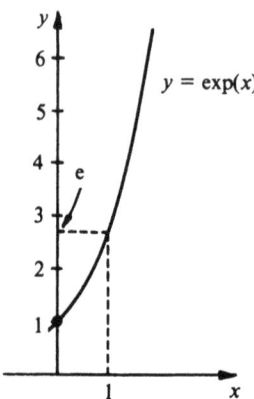

Figure A.1.9. Graph of the exponential function.

Arbitrarily precise values for $x(t) = \exp(t)$ can be obtained from its Taylor series. To calculate this series we note from (61) and (48b) that all derivatives of $\exp(t)$ have the value unity at $t = 0$. Therefore, in this case the Taylor series expansion (30) with $c = 0$ is

◆◆ $$\exp(t) = 1 + t + \frac{t^2}{2!} + \frac{t^3}{3!} + \ldots + \frac{t^n}{n!} + \ldots . \qquad (64)$$

It turns out that this series converges for all t. As one application, we note that

$$\exp(1) = e = 1 + 1 + \tfrac{1}{2} + \tfrac{1}{6} + \tfrac{1}{24} + \tfrac{1}{120} + \tfrac{1}{720} + \ldots = 2.718 \ldots . \qquad (65)$$

The logarithm

Since $\exp(t)$ increases with t, if $x = \exp(t)$ then for every positive x_0 there corresponds precisely one t_0 such that $x_0 = \exp(t_0)$. We can regard this correspondence as giving t as a function of x. The corresponding function is called the **natural logarithm**, written 'ln'. Thus

◆◆   for $x > 0$, $t = \ln x$ if and only if $x = \exp(t)$. $\qquad (66)$

Thus, by (18),

◆◆ $t = \ln[\exp(t)]$ and $x = \exp(\ln x)$. $\qquad (67a, b)$

Moreover, setting $t = 1$ in (66) we find, using (56), that

◆◆ $\ln e = 1$. $\qquad (68)$

By (19) and (48a)

$$dt/dx = 1/(dx/dt) = 1/x. \qquad (69)$$

Also by (48b), $\ln 1 = 0$. Combining this and (68), we can say that $\ln x$ is prescribed by the differential equation and initial condition

$$df(x)/dx = 1/x, \qquad f(1) = 0. \tag{70}$$

That is,

◆◆ $\qquad \dfrac{d}{dx}(\ln x) = \dfrac{1}{x}, \qquad \ln 1 = 0.$ (71a, b)

The graph of $\ln x$ can now be constructed qualitatively, using the facts that its first derivative is the positive function x^{-1} (function always increasing, but more and more slowly), and that its second derivative is the negative function $-x^{-2}$ (concave downward) (see Figure A.1.10).

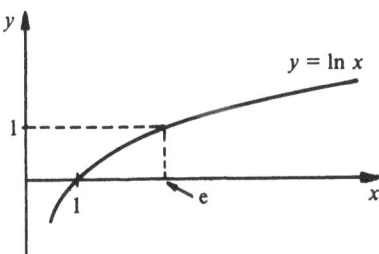

Figure A.1.10. Graph of the ln function. (As one would anticipate from Figure A.1.6, this graph is the same as that of Figure A.1.9 viewed sideways.)

Let us write (53) in the form

$$xy = \exp(\ln x + \ln y), \tag{72}$$

where we have introduced x and y by

$$s = \ln x, \qquad t = \ln y \quad \text{so that} \quad x = \exp(s), \qquad y = \exp(t).$$

Taking the natural logarithm of both sides of (72) and using (67a), we find the characteristic feature of the logarithm

◆◆ $\qquad \ln(xy) = \ln x + \ln y.$ (73)

If $y = x$, we deduce that $\ln x^2 = 2\ln x$. If $y = 2x$ it follows that $\ln(x^3) = \ln x + \ln x^2 = 3\ln x$, etc. Indeed, since both quantities satisfy $dy/dx = ay$, $y(1) = 0$,

◆◆ $\qquad \ln x^a = a \ln x$ (74)

for any real number a. For the special case $x = 10$, this formula gives

$$\ln 10^a = a \ln 10. \tag{75}$$

Comparison with the classical relation for common logarithms

$$\log 10^a = a$$

requires that

$$\log 10^a = \ln 10^a / \ln 10. \tag{76}$$

Since a is arbitrary, this is equivalent to

$$\log x = \ln x / \ln 10. \tag{77}$$

To pass from the natural logarithm to the common logarithm, then, one need only divide by $\ln 10 = 2.302 \ldots$ From (70) and (77) it follows that

$$\frac{d(\log x)}{dx} = \frac{1}{\ln 10} \cdot \frac{1}{x}. \tag{78}$$

The awkwardness of this formula compared to (70) is the reason why only the natural logarithm is generally employed in advanced mathematical discussions.

The spring-mass system

Consider a mass m suspended from a spring. If the mass is displaced a distance x upward from its equilibrium position its subsequent position $x(\tau)$ at the time τ will be governed by Newton's law

$$m \, d^2x/d\tau^2 = \text{spring force} + \text{damping force}. \tag{79}$$

(Gravity only enters in establishing the equilibrium location, not in determining departures from it.) By Hooke's law (which is essentially an application of the straight line approximation (33)), the spring force is proportional to the displacement from equilibrium. Observation suggests that the damping force (resulting from friction with the air) is proportional to the velocity. This gives

$$m \, d^2x/d\tau^2 = -kx - a \, dx/d\tau, \tag{80}$$

where k and a are constants. The signs in (80) are correct, for the spring force acts to decrease x if x is positive and to increase it if x is negative. Also, the damping force is downward if the mass is moving upward ($dx/d\tau > 0$) and upward if the mass is moving downward.

If, in addition to the forces mentioned above, there is an external force $F(\tau)$ acting on the mass, for example by means of a time-varying magnetic field, then we have

$$m \, (d^2x/d\tau^2) + a \, (dx/d\tau) + kx = F. \tag{81}$$

This second-order ordinary differential equation is called **inhomogeneous** because of the presence of F. By definition, an inhomogeneous equation is one that does not have the identically zero function as a solution.

It is well known that an equation with the form of (81) arises in important contexts other than mechanics, for example in the study of the basic interaction of resistance, inductance and capacitance in electric circuits. But the same equation plays an important role in biology too, as is seen in Chapter 7.

Here we shall be concerned only with a spring that is undamped ($a = 0$) and unforced ($F = 0$). Then application of the chain rule (11) shows that the change of variable

$$t = \frac{\tau}{(m/k)^{\frac{1}{2}}} \tag{82}$$

brings the equation into the standard form, without parameters,

$$d^2x/dt^2 + x = 0. \tag{83}$$

The chance of variable (82) results from dimensional analysis (Appendix Section A.4). We reason in this particular case as follows. The spring constant k is a force per unit length. As force equals mass times acceleration, the dimensions of k must be

$$(\text{mass})[\text{length}/(\text{time})^2]/\text{length} = (\text{mass})/(\text{time})^2.$$

Since m is a mass, the dimensions of $(m/k)^{\frac{1}{2}}$ must be time. It would appear advantageous to measure time in units of this intrinsic time $(m/k)^{\frac{1}{2}}$ and indeed such a change of variable, in (82), did lead to a simplification of the problem.

A second-order equation requires two initial conditions. Thus a full statement of the problem under consideration is

$$d^2x/dt^2 + x = 0, \qquad x(t_0) = x_0, \qquad \left.\frac{dx(t)}{dt}\right|_{t=t_0} = v_0, \tag{84}$$

where x_0 and v_0 are given constants.

It proves advantageous to introduce v as an abbreviation for the velocity dx/dt. With this, the original second-order ordinary differential equation is equivalent to the following system of two first-order ordinary differential equations:

$$dx/dt = v, \qquad dv/dt = -x. \tag{85a, b}$$

If we multiply (85a) by x, multiply (85b) by v, and add the resulting equations we find that

$$x\frac{dx}{dt} + v\frac{dv}{dt} = xv - xv = 0, \quad \text{i.e.} \quad \frac{d}{dt}(\tfrac{1}{2}x^2 + \tfrac{1}{2}v^2) = 0, \tag{86}$$

$$x^2 + v^2 = \text{constant}.$$

The constant can be determined by the value of v and x at the initial time:

$$x^2 + v^2 = x_0^2 + v_0^2. \tag{87}$$

The physical meaning of (87) is that the sum of the kinetic and potential energies remains constant, equal to its initial value. The equation takes on even more meaning if we introduce a coordinate system where x is the abscissa and v is the ordinate. Then for various values of the constant on the right-hand side, the curves of (87) form a family of concentric circles.

The **state** of the spring-mass system is given at any time t by the position $x(t)$ and the velocity $v(t)$. Evidently the state point (x, v) moves on circular orbits in the abstract x–v plane, or **phase plane** as it is called. (Use of the phase plane is discussed at some length in Appendix Section A.3.)

What is the speed of travel along the circular orbit? We have seen in (16) that the slope of the tangent to the orbit at an arbitrary point (x_1, v_1) – using present notation – is $-(x_1/v_1)$. But this slope is equal to the ratio of the 'vertical' velocity component (in the v-direction) to the 'horizontal' velocity component (in the x-direction). And the latter quantity is just v_1. Therefore

$$\text{'vertical' velocity component at } (x_1, v_1) = v_1\left(-\frac{x_1}{v_1}\right) = -x_1. \tag{88}$$

Knowing both 'horizontal' and 'vertical' velocity components we can find the speed along the orbit by Pythagoras's theorem:

$$\text{speed along orbit at } (x_1, v_1) = \sqrt{[x_1^2 + v_1^2]}. \tag{89}$$

But for any point x on a given orbit, $x^2 + v^2$ is the constant $x_0^2 + v_0^2$. Thus the phase point proceeds around the circle at this *constant* speed. Furthermore, from (88) we see that the 'vertical' velocity component is positive if and only if x is negative. This means that the phase point circles the origin in the clockwise direction (see Figure A.1.11).

Another important consequence of (87) is that (84) has a unique solution. To see this, suppose that $x_1(t)$ and $x_2(t)$ are two solutions of (84). Let $y(t) \equiv x_1(t) - x_2(t)$. Then

$$(d^2 y/dt^2) + y = 0, \qquad y(0) = 0, \qquad (dy/dt)(0) = 0.$$

By (87) we deduce that $y^2 + (dy/dt)^2 = 0$. The only way that the two non-negative numbers can add to zero is if each separately equals zero. In particular $y = 0$, i.e. $x_1(t) = x_2(t)$ as we wished to prove.

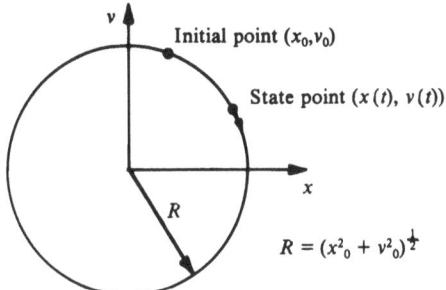

Figure A.1.11. A typical circular orbit which according to (87), is traversed by the solutions to (85). The state-point (x, v) moves clockwise around this circle at the uniform circumferential speed $(x_0^2 + v_0^2)^{\frac{1}{2}}$.

The functions sin and cos

We now define the function $\sin t$ as the unique solution to the following special case of (84):

$$(d^2x/dt^2) + x = 0, \qquad x(0) = 0, \qquad (dx/dt)(0) = 1. \qquad (90a, b, c)$$

The function $\cos t$ is defined as the unique solution of another special case of (84)

$$(d^2x/dt^2) + x = 0; \qquad x(0) = 1, \qquad (dx/dt)(0) = 0. \qquad (91a, b, c)$$

For both the sin and cos functions, (87) implies that

$$x^2 + (dx/dt)^2 = 1. \tag{92}$$

Thus for both functions x^2 has a maximum value of unity, when $dx/dt = 0$; i.e.

$$\max |x| = 1, \quad \text{when } dx/dt = 0. \tag{93}$$

Moreover the state point for both functions moves around the unit circle in the x–v plane at the speed $\{[x(0)]^2 + [v(0)]^2\}^{\frac{1}{2}} = 1$. Thus both functions repeat themselves in a time of duration 2π, the time to traverse the circumference of the unit circle at unit speed. In other words sin and cos have a period of 2π:

$$\blacklozenge\blacklozenge \qquad \sin(t + 2\pi) = \sin t, \qquad \cos(t + 2\pi) = \cos t. \tag{94}$$

As depicted in Figure A.1.12, the only difference between the trajectories of the sin and cos functions in the phase plane is that the phase point of the former begins its trajectory a quarter-circle behind. Thus

$$\blacklozenge\blacklozenge \qquad \cos(t - (\pi/2)) = \sin t, \qquad \sin(t + (\pi/2)) = \cos t. \tag{95}$$

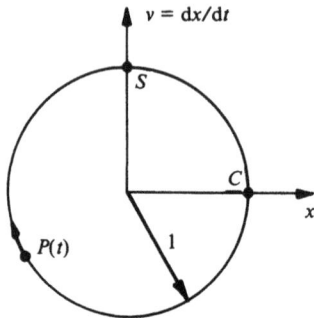

Figure A.1.12. At any time *t*, the value *x* and the derivative d*x*/d*t* of the sin and cos functions are given by the coordinates of a point *P(t)* that moves clockwise around the unit circle, at unit circumferential speed. For sin (cos), point *P* begins at *t* = 0 at the point marked *S(C)*.

The information found in (90), (91), (93), and (94), plus the fact that the behavior of *x*(*t*) in subsequent quarter-periods can obviously be found by appropriate reflections of its behavior in the first quarter-period – this is sufficient to give the qualitative behavior of the graphs of the sin and cos functions (Figure A.1.13). An arbitrarily accurate graph can be obtained from the series formulae (101) and (102) given below.

If *x*(*t*) satisfies the equations and boundary conditions of (90) then d*x*/d*t* satisfies the equations and boundary conditions of (91), for

$$\frac{d^2}{dt^2}\left(\frac{dx}{dt}\right)+\left(\frac{dx}{dt}\right)=\frac{d}{dt}\left(\frac{d^2x}{dt^2}+x\right)=0 \tag{96a}$$

by (90*a*),

$$\frac{dx}{dt}(0)=1 \tag{96b}$$

by (90*c*),

$$\frac{d^2x}{dt^2}(0)=-x(0)=0 \tag{96c}$$

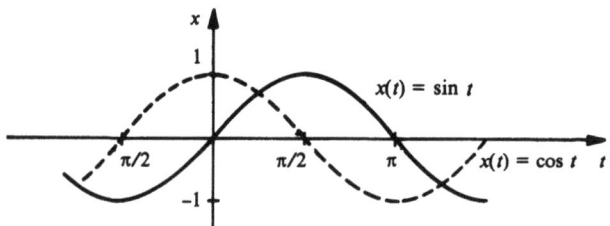

Figure A.1.13. Graphs of the sin and cos functions.

by (90*a*) and (90*b*). Since (91) has a unique solution, cos *t*, it follows that $dx/dt = \cos t$, where $x(t)$ is the unique solution of (90), sin *t*. Thus

◆◆ $\quad \dfrac{d}{dt}(\sin t) = \cos t.$ $\hfill (97)$

Similarly

◆◆ $\quad \dfrac{d}{dt}(\cos t) = -\sin t.$ $\hfill (98)$

From (92) when $x = \sin t$, and from (97), we deduce further that

◆◆ $\quad (\sin t)^2 + (\cos t)^2 = 1.$ $\hfill (99)$

Using the differential equation (90*a*) or (91*a*) to obtain the formulae

$$d^2x/dt^2 = -x, \qquad d^3x/dt^3 = -dx/dt,$$
$$d^4x/dt^4 = -d^2x/dt^2 = x, \qquad d^5x/dt^5 = dx/dt, \ldots \qquad (100)$$

we can compute all the derivatives of x at the origin from the initial conditions (90*b*, *c*) or (91*b*, *c*). Substitution into the Taylor series formula (30) with $c = 0$ yields

◆◆ $\quad \sin t = t - \dfrac{t^3}{3!} + \dfrac{t^5}{5!} - \dfrac{t^7}{7!} + \ldots,$ $\hfill (101)$

◆◆ $\quad \cos t = 1 - \dfrac{t^2}{2!} + \dfrac{t^4}{4!} - \dfrac{t^6}{6!} + \ldots.$ $\hfill (102)$

These series can be shown to converge for all *t*.

A famous identity linking sin and cos follows from the fact that $\sin(t - \alpha)$, α a constant, satisfies

$$(d^2x/dt^2) + x = 0, \qquad x(\alpha) = 0, \qquad (dx/dt)(\alpha) = 1. \qquad (103a, b, c)$$

But the general theory of differential equations states that *any solution to a second-order linear differential equation can be written in the form* $Af(t) + Bg(t)$ *where A and B are constants and f and g are two 'different' special solutions to the given equation.* By 'different' we mean that

$$(f\,dg/dt) - (g\,df/dt) \neq 0 \qquad (104)$$

except perhaps at isolated points. We thus know that any solution of (103*a*) can be written in the form

$$x(t) = A \sin t + B \cos t. \qquad (105)$$

From (103*b*) and (103*c*) we see that

$$A \sin \alpha + B \cos \alpha = 0, \qquad A \cos \alpha - B \sin \alpha = 1. \qquad (106)$$

Solving these two equations we find that

$$A = \cos \alpha, \qquad B = -\sin \alpha. \qquad (107)$$

Since (103) has a unique solution it follows from (105), (106), and (107) that

◆◆ $\sin (t - \alpha) = \sin t \cos \alpha - \cos t \sin \alpha.$ (108)

Other 'trigonometric identities' can be derived similarly.

We have defined the exp, sin, and cos functions as solutions to certain simple linear differential equations with constant coefficients. It can be shown that *any* such equations can be solved by a suitable combination of these three functions, plus polynomials in certain relatively rare cases.

Definition of the integral

Consider the problem of finding the area under the curve $y = f(x)$ between $x = a$ and $x = b$; $f > 0$, $a < b$. This area lies between the sums

$$\sum_{i=1}^{N} [f_i^{(m)} \Delta x] \quad \text{and} \quad \sum_{i=1}^{N} [f_i^{(M)} \Delta x]; \qquad \Delta x \equiv \frac{b-a}{N}. \qquad (109)$$

Here $f_i^{(m)}$ $[f_i^{(M)}]$ is the smallest (largest) value of $f(x)$ for x between $a + (i-1)(\Delta x)$ and $a + i(\Delta x)$. Moreover if Δx is small the area seems well approximated by the sum

$$\sum_{i=1}^{N} [f(x_i^*) \Delta x], \qquad (110)$$

where x_i^* is any point between $a + (i-1)\Delta x$ and $a + i(\Delta x)$ (see Figure A.1.14). It can be shown that if f is continuous for $a \leq x \leq b$ then all the sums approach a common limit as $N \to \infty$. This limit is called 'the integral of f between a and b' and is written $\int_a^b f(x) \, dx$.

The integral has many other interpretations, in addition to that as an area. For example if f, the concentration of a chemical, depends on a single coordinate x then the integral $\int_a^b f(x) \, dx$ gives the total number of molecules between $x = a$ and $x = b$.

Some rules for manipulating integrals

By definition,

◆◆ $$\int_a^a f(x) \, dx = 0, \qquad \int_b^a f(x) \, dx = -\int_a^b f(x) \, dx. \qquad (111a, b)$$

If $a < b < c$ then the interpretation of the integral as an area makes

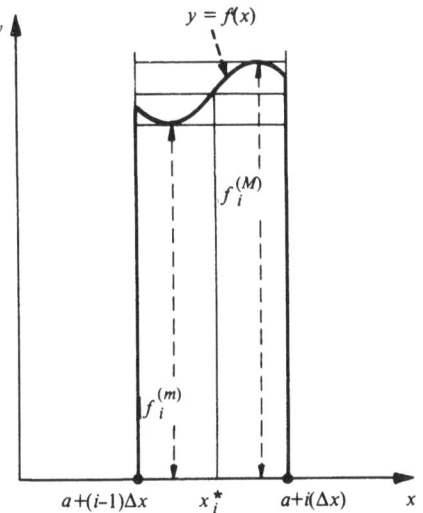

Figure A.1.14. The area bounded by the heavy line is less than $f_i^{(M)} \Delta x$, greater than $f_i^{(m)} \Delta x$, and approximately equal to $f(x_i^*) \Delta x$.

obvious the rule

$$\blacklozenge\blacklozenge \qquad \int_a^c f(x) \, dx = \int_a^b f(x) \, dx + \int_b^c f(x) \, dx. \qquad (112)$$

It follows almost directly from the definition of the integral that

$$\blacklozenge\blacklozenge \qquad \int_a^b K f(x) \, dx = K \int_a^b f(x) \, dx, \qquad K \text{ a constant.} \qquad (113)$$

It can be shown that if $dt(u)/du > 0$ then the chain rule (11) implies that

$$\int_a^b f(x) \, dt = \int_A^B f[t(u)] \frac{dt(u)}{du} \, du \qquad (114)$$

where $a = t(A)$, $b = t(B)$.

The integral mean value theorem

The area interpretation of the integral makes clear the assertion

$$m(b-a) \leqslant \int_a^b f(x) \, dx \leqslant M(b-a).$$

Here m and M are the largest and smallest values of $f(x)$ for $a \leqslant x \leqslant b$. Thus if we define I by

$$I \equiv \frac{1}{b-a} \int_a^b f(x) \, dx \qquad (115)$$

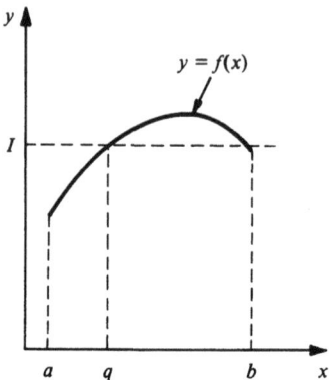

Figure A.1.15. Illustration of the fact that if f is continuous for $a \leqslant x \leqslant b$ and I is a number between the largest and smallest values of f in the interval, then there is (at least) one number q in the interval such that $I = f(q)$.

then I is between m and M. But if f is continuous, there is at least one number q between a and b such that $I = f(q)$ (Figure A.1.15). That is

◆◆ $$\int_a^b f(x)\,\mathrm{d}x = f(q) \cdot (b-a), \qquad \text{for some } q \text{ such that } a \leqslant q \leqslant b.$$

(116)

The fundamental theorem of calculus (This shows the relation between the derivative and the integral; up to now these have been unrelated concepts.)

That differentiation 'undoes' integration is shown by the result

◆◆ $$\frac{\mathrm{d}}{\mathrm{d}x} \int_a^x f(t)\,\mathrm{d}t = f(x), \qquad a = \text{any constant,}$$

(117)

which holds when f is continuous.

Proof. By the definition of the derivative

$$\frac{\mathrm{d}}{\mathrm{d}x} \int_a^x f(t)\,\mathrm{d}t = \lim_{h \to 0} \frac{1}{h} \left[\int_a^{x+h} f(t)\,\mathrm{d}t - \int_a^x f(t)\,\mathrm{d}t \right].$$

(118)

Using (112) and (116), we see that the quantity in the square brackets can be written as

$$\int_x^{x+h} f(t)\,\mathrm{d}t = f(q) \cdot h \quad \text{where } x \leqslant q \leqslant x + h.$$

(119)

The result follows after one divides by h and takes the limit.

To see that integration also (in a sense) 'undoes' differentiation consider the differential equation and initial condition

$$\frac{dy(x)}{dx} = f(x), \qquad y(a) = A. \tag{120}$$

Using (117) and (111a), we can verify that the solution is

$$y(x) = \int_a^x f(t)\, dt + A. \tag{121}$$

Employing (120), we can rewrite the preceding line as

$$y(x) = \int_a^x \frac{dy(t)}{dt}\, dt + y(a)$$

or

◆◆ $$\int_a^x \frac{dy(t)}{dt}\, dt = y(x) - y(a), \tag{122}$$

which is the desired result.

The result (122) shows that every differentiation formula implies a corresponding integration formula. For example, we have seen that

$$\frac{d(t^3)}{dt} = 3t^2 \quad \text{or} \quad \frac{d(\sin t)}{dt} = \cos t. \tag{123}$$

Substituting these into (122) we find that

$$\int_a^x 3t^2\, dt = x^3 - a^3, \qquad \int_a^x \cos t\, dt = \sin x - \sin a. \tag{124a, b}$$

The notation

$$f(x)\big|_a^b \equiv f(b) - f(a) \tag{125}$$

is often employed. With this the results of (124) are written

$$\int_a^x 3t^2\, dt = t^3\Big|_a^x, \qquad \int_a^x \sin t\, dt = \cos t\Big|_a^x, \tag{126a, b}$$

and one says that t^3 and $\cos t$ are **indefinite integrals** of $3t^2$ and $\sin t$. Equations (126a) and (126b) are said to provide the required **definite integral** between x and a by means of evaluating the appropriate indefinite integral between the limits x and a.

Integration by parts

If the substitution $y(t) = f(t)g(t)$ is made in (122) there results the remarkably useful formula

$$\blacklozenge\blacklozenge \qquad \int_a^x f(t)\,\frac{dg(t)}{dt}\,dt = f(t)g(t)\Big|_a^x - \int_a^x \frac{df(t)}{dt}\,g(t)\,dt. \tag{127}$$

Functions of several variables: partial derivatives

If the concentration of a chemical, C, is a function of three spatial coordinates and time, one often wishes to know, for example, the rate at which C changes with time at a fixed point (x, y, z), or the rate at which C changes with x while y, z, and t are held constant. These changes are given by partial derivatives, defined as follows:

$$\frac{\partial C(x, y, z, t)}{\partial t} \equiv \lim_{h \to 0} \frac{C(x, y, z, t+h) - C(x, y, z, t)}{h},$$

$$\frac{\partial C(x, y, z, t)}{\partial x} \equiv \lim_{h \to 0} \frac{C(x+h, y, z, t) - C(x, y, z, t)}{h}. \tag{128}$$

Examples. Partial derivatives are computed exactly like ordinary derivatives once one recognizes that the 'other' variables are constants and thus can be taken outside the differentiation operator. Thus

$$\frac{\partial(x^2 y^3 z^4 \sin t)}{\partial t} = x^2 y^3 z^4 \cos t, \qquad \frac{\partial(x^2 y^3 z^4 \sin t)}{\partial x} = 2xy^3 z^4 \sin t.$$

Note that if $f(x, y) = x^2 \sin y$ (for example)

$$\frac{\partial f}{\partial x} \equiv f_x = 2x \sin y, \qquad \frac{\partial f}{\partial y} \equiv f_y = x^2 \cos y,$$

$$\frac{\partial}{\partial y}\frac{\partial f}{\partial x} \equiv \frac{\partial^2 f}{\partial y\,\partial x} \equiv f_{yx} = \frac{\partial}{\partial y}(2x \sin y) = 2x \cos y,$$

$$\frac{\partial}{\partial x}\frac{\partial f}{\partial y} \equiv \frac{\partial^2 f}{\partial x\,\partial y} \equiv f_{xy} = \frac{\partial}{\partial x}(x^2 \cos y) = 2x \cos y,$$

so that

$$\blacklozenge\blacklozenge \qquad \frac{\partial^2 f}{\partial x\,\partial y} = \frac{\partial^2 f}{\partial y\,\partial x} \quad \text{or} \quad f_{xy} = f_{yx}. \tag{129}$$

Equation (129) in fact holds generally, when the partial derivatives involved are continuous.

Chain rule

The generalization of (11) to a function of two variables is

◆◆ $$\frac{d}{dt}f[x(t),y(t)] = \frac{\partial f(x,y)}{\partial x}\bigg|_{\substack{x=x(t)\\y=y(t)}} \frac{dx(t)}{dt} + \frac{\partial f(x,y)}{\partial y}\bigg|_{\substack{x=x(t)\\y=y(t)}} \frac{dy(t)}{dt}. \tag{130}$$

Taylor series

Extension of the Taylor series (30) to functions $f(x,y)$ of two variables can be made by considering the line segment

$$x = c + t\alpha, \qquad y = d + t\beta, \qquad 0 \leq t \leq 1, \tag{131}$$

joining (c,d) and a nearby point $(c+\alpha, d+\beta)$. Let

$$F(t) \equiv f(c+t\alpha, d+t\beta).$$

By (30)

$$F(t) = F(0) + t\frac{dF}{dt}\bigg|_{t=0} + \frac{t^2}{2!}\frac{d^2F}{dt^2}\bigg|_{t=0} + \dots. \tag{132}$$

Using the chain rule (130) – employing the notation f_x for $\partial f/\partial x$, etc. – we find

$$\frac{dF}{dt} = f_x\bigg|\frac{dx}{dt} + f_y\bigg|\frac{dy}{dt} = \alpha f_x| + \beta f_y|.$$

Here the vertical line means that (131) must be used to substitute for x and y in terms of t, after the partial differentiations have been carried out.
Applying the chain rule once again we find

$$\frac{d^2F}{dt^2} = \frac{\partial}{\partial x}[\alpha f_x| + \beta f_y|]\frac{dx}{dt} + \frac{\partial}{\partial y}[\alpha f_x| + \beta f_y|]\frac{dy}{dt}$$

$$= [\alpha f_{xx}| + \beta f_{xy}|]\alpha + [\alpha f_{yx}| + \beta f_{yy}|]\beta$$

$$= \alpha^2 f_{xx}| + 2\alpha\beta f_{xy}| + \beta^2 f_{yy}|,$$

where we have employed (129). By substituting into (132) and judiciously using the relations

$$t = \frac{x-c}{\alpha}, \qquad t = \frac{y-d}{\beta},$$

we arrive at the final formula

$$f(x, y) = f(c, d) + (x - c)f_x(c, d) + (y - d)f_y(c, d)$$

◆◆
$$+ \frac{1}{2!} [(x - c)^2 f_{xx}(c, d) + 2(x - c)(y - d)f_{xy}(c, d)$$

$$+ (y - d)^2 f_{yy}(c, d)] + \ldots \tag{133}$$

The same ideas can be used to extend Taylor's formula to an arbitrary number of variables. As in the single variable case, error estimates can be derived under suitable conditions.

Used often in the text is the linear approximation

◆◆
$$f(x, y) \approx f(c, d) + (x - c)f_x(c, d) + (y - d)f_y(c, d), \tag{134}$$

which can be shown to have the geometrical significance of approximating the surface $z = f(x, y)$ near $x = c$, $y = d$, by the plane that is tangent to the surface at (c, d).

Multiple integrals

Given the concentration $C(x, y, z)$, what is the number of molecules at time t inside the box B given by

$$a_1 \leqslant x \leqslant b_1, \qquad a_2 \leqslant y \leqslant b_2, \qquad a_3 \leqslant z \leqslant b_3?$$

The answer is found by a process analogous to the single variable case. The box is subdivided into small boxes of dimensions Δx by Δy by Δz, the volume of each box is multiplied by the concentration C at some interior point, and the individual concentrations are summed. The most common notation for this integral is

$$\iiint_B C(x, y, z)\, dx\, dy\, dz.$$

It can be shown that the correct final result can be calculated by first (say) integrating the x contribution for a given fixed y and z, then integrating with respect to y keeping z constant, and finally integrating with respect to z. That is, the determination of the triple integral is reduced to a three-fold calculation of a single integral.

Example.   If $a_1 = 1$, $b_1 = 2$, $a_2 = 3$, $b_2 = 5$, $a_3 = 6$, $b_3 = 8$ then

$$\iiint_B xy^2 \sin z\, dz = \int_6^8 \int_3^5 \left[\int_1^2 xy^2 \sin x\, dx \right] dy\, dz$$

$$= \int_6^8 \int_3^5 \left[\frac{2^2}{2} - \frac{1^2}{2} \right] y^2 \sin z \, dy \, dz$$

$$= \int_6^8 \frac{3}{2} \left(\frac{5^3}{3} - \frac{3^3}{3} \right) \sin z \, dz = 49 \, (\cos 6 - \cos 8).$$

Treatment of regions whose shape is not rectangular requires a somewhat more advanced consideration.

The integral mean value theorem remains valid in several dimensions. For example if C is a continuous function of (x, y, z) then

$$\iiint_B C(x, y, z) \, dx \, dy \, dz = C(p, q, r) \cdot \text{Volume of } B. \qquad (135)$$

Here (p, q, r) is some point in B. An interpretation is that the number of molecules in the box equals the volume of the box times a typical concentration at some point inside the box.

Partial fraction expansion

A function $f(s)$ is a **rational function** if it can be expressed as the ratio of two polynomials. It is a **proper** rational function if the order of the numerator is less than the order of the denominator. A partial fraction expansion provides a way of expressing a proper rational function in a form that is particularly convenient when taking inverse Laplace transforms.

Consider the proper rational function

$$f(s) = \frac{s^m + a_{m-1}s^{m-1} + \ldots + a_1 s + a_0}{s^n + b_{n-1}s^{n-1} + \ldots + b_1 s + b_0}, \qquad n > m.$$

Let $\beta_1, \ldots \beta_n$ be the roots of the equation

$$s^n + b_{n-1}s^{n-1} + \ldots + b_1 s + b_0 = 0.$$

$f(s)$ can be rewritten as

$$f(s) = \frac{s^m + a_{m-1}s^{m-1} + \ldots + a_1 s + a_0}{(s-\beta_1)(s-\beta_2)\ldots(s-\beta_n)}.$$

If all of the β's are unequal, it can be shown (Levinson & Redheffer, 1970), that there are constants $C_1, \ldots C_n$ such that for all s.

$$f(s) = \frac{C_1}{s-\beta_1} + \frac{C_2}{s-\beta_2} + \ldots + \frac{C_n}{s-\beta_n}.$$

This is made clearer by considering an example.

$$f(s) = \frac{s^2 + 6s + 3}{s^3 - 6s^2 + 11s - 6} = \frac{s^2 + 6s + 3}{(s-1)(s-2)(s-3)} = \frac{C_1}{s-1} + \frac{C_2}{s-2} + \frac{C_3}{s-3}.$$

The question is now one of finding the C's. To do this multiply the equation by the denominator.

$$s^2 + 6s + 3 = C_1(s-2)(s-3) + C_2(s-1)(s-3) + C_3(s-1)(s-2).$$

This relation must hold for any value of s and in particular it must be true when $s = 1, 2$ and 3.

$s = 1$ implies $\quad 10 = C_1(1-2)(1-3), \quad 5 = C_1.$

$s = 2$ implies $\quad 19 = C_2(2-1)(2-3), \quad -19 = C_2.$

$s = 3$ implies $\quad 30 = C_3(3-1)(3-2), \quad 15 = C_3.$

So we have found that

$$f(s) = \frac{s^2 + 6s + 3}{s^3 - 6s^2 + 11s - 6} = \frac{5}{s-1} - \frac{19}{s-2} + \frac{15}{s-3}.$$

The truth of this relation is easily checked by adding the three fractions on the right-hand side.

The procedure can be generalized to all cases where the β's are different, but it does not cope with the case where a root is repeated. Suppose β_1 has multiplicity j (i.e. β_1 is a j-repeated root of $s^n + b_{n-1}s^{n-1} + \ldots b_0 = 0$). Then

$$f(s) = \frac{s^m + a_{m-1}s^{m-1} + \ldots + a_1 s + a_0}{(s-\beta_1)^j (s-\beta_2) \ldots (s-\beta_l)}.$$

It can now be shown that there are C's such that

$$f(s) = \frac{C_1}{s-\beta_1} + \frac{C_2}{(s-\beta_1)^2} + \ldots + \frac{C_j}{(s-\beta_1)^j}$$

$$+ \frac{C_{j+1}}{(s-\beta_2)} + \ldots + \frac{C_n}{(s-\beta_l)}.$$

There are slight problems associated with evaluating the C's. Consider the example

$$f(s) = \frac{s+3}{s^3 - 4s^2 + 5s - 2} = \frac{s+3}{(s-1)^2(s-2)}$$

$$= \frac{C_1}{s-1} + \frac{C_2}{(s-1)^2} + \frac{C_3}{s-2}.$$

Again we begin by multiplying by the denominator.

$$s + 3 = C_1(s-1)(s-2) + C_2(s-2) + C_3(s-1)^2.$$

Considering the equation at $s = 1$ and $s = 2$ gives values for C_2 and C_3.

$s = 1$ implies $\quad 4 = C_2(1-2), \quad -4 = C_2.$

$s = 2$ implies $\quad 5 = C_3(2-1)^2, \quad 5 = C_3.$

This still leaves C_1 undetermined. However, C_2 and C_3 are now known and the expression must hold for all s, so any convenient s can be chosen to produce a third equation giving C_1. For example, consider the case $s = 0$. This yields

$$3 = C_1(-1)(-2) + (-4)(-2) + 5(-1)^2, \quad -5 = C_1.$$

We have shown that

$$\frac{s+3}{s^3 - 4s^2 + 5s - 2} = -\frac{5}{s-1} - \frac{4}{(s-1)^2} + \frac{5}{(s-2)}.$$

Because of their central role in linear control theory and other applications, partial fraction expansions have received a substantial measure of attention. It is possible to provide a general formula for the expansion coefficients, but its derivation requires an understanding of complex variable theory.

Complex numbers

A complex number can be regarded as an ordered pair of real numbers, (x, y), subject to the following addition and multiplication rules:

$$(x_1, y_1) + (x_2, y_2) = (x_1 + x_2, y_1 + y_2),$$

$$(x_1, y_1)(x_2, y_2) = (x_1 x_2 - y_1 y_2, x_1 y_2 + y_1 x_2).$$

The motivation for this abstract definition of a complex number becomes clear after considering the representation

$$(x, y) = x + iy,$$

where i is defined as $i = \sqrt{(-1)}$. Proceeding formally, addition is now

$$x_1 + iy_1 + x_2 + iy_2 = (x_1 + x_2) + i(y_1 + y_2),$$

and multiplication is

$$(x_1 + iy_1)(x_2 + iy_2) = x_1 x_2 + ix_1 y_2 + iy_1 x_2 + i^2 y_1 y_2,$$

but $i^2 = -1$ so that

$$(x_1 + iy_1)(x_2 + iy_2) = (x_1 x_2 - y_1 y_2) + i(x_1 y_2 + y_1 x_2).$$

In the ordered pair (x, y), x is referred to as the **real part** and y is the **imaginary part**. If $z = x + iy$, one writes $x = \mathrm{Re}\, z$, $y = \mathrm{Im}\, z$. For every complex number (x, y) its **complex conjugate**, denoted by $(x, y)^*$, is defined as $(x, -y)$. Some authors use the overbar $\bar{z}$ to denote the complex conjugate of z. That is,

$$(x + iy)^* = x - iy \quad \text{or} \quad \bar{z} = x - iy.$$

From the definition of complex numbers it follows that

$$[(x_1, y_1) + (x_2, y_2)]^* = (x_1, y_1)^* + (x_2, y_2)^*.$$

This can be seen by performing the indicated operations:

$$[x_1 + iy_1 + x_2 + iy_2]^* = \{(x_1 + x_2) + i(y_1 + y_2)\}^*$$
$$= (x_1 + x_2) - i(y_1 + y_2).$$
$$(x_1 + iy_1)^* + (x_2 + iy_2)^* = (x_1 - iy_1) + (x_2 - iy_2)$$
$$= (x_1 + x_2) - i(y_1 + y_2).$$

Similarly,

$$[(x_1, y_1)(x_2, y_2)]^* = (x_1, y_1)^*(x_2, y_2)^*.$$

Thus, if z_1 and z_2 are two complex numbers

$$(z_1 + z_2)^* = z_1^* + z_2^*, \qquad (z_1 z_2)^* = z_1^* z_2^*. \tag{1a, b}$$

Since complex numbers have two components it is natural to represent them as points on a two-dimensional plane. It is conventional to let the real part be plotted on the horizontal axis and the imaginary part on the vertical axis. See Figure A.2.1.

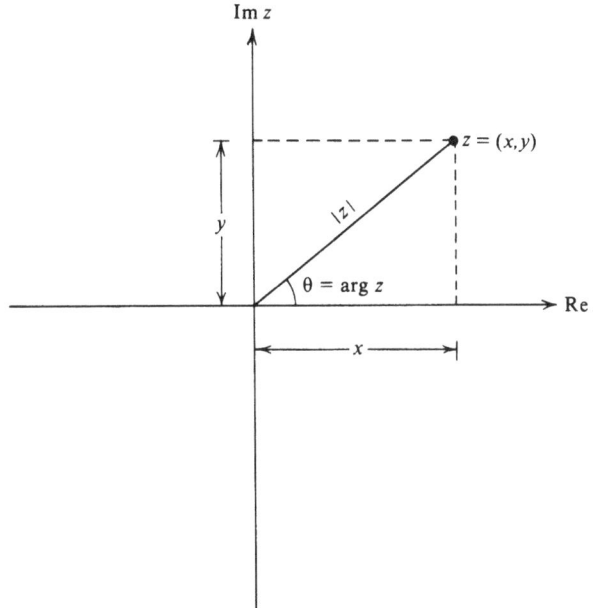

Figure A.2.1. Representation of the complex number $z = x + iy$ in the complex plane.

The distance from the origin to the point is its **magnitude** or **modulus**. By the Pythagorean theorem it is $(x^2+y^2)^{\frac{1}{2}}$, and it is denoted by $|z| = |(x, y)|$. The angle θ is called the **argument** of z and is denoted arg z.

$z(t)$ is a **complex function** of the real variable t if there are two real functions $z_1(t)$ and $z_2(t)$ such that $z(t) = z_1(t) + iz_2(t)$. One can also define complex functions f of a complex variable $z \equiv x + iy$, such as $f(z) = z^2 = z \cdot z$. Most of the standard formulae of differential calculus can be extended more or less straightforwardly to functions of a complex variable. This gives for example, the definition

$$\left.\frac{df(z)}{dz}\right|_{z=a} = \lim_{h \to 0} \frac{f(a+h) - f(a)}{h}, \tag{2}$$

and the formula

$$\frac{d}{dz}(z^n) = nz^{n-1}, \quad n \text{ a constant.} \tag{3}$$

Complex functions exp (z), sin z, and cos z are defined by substituting z for t in the appropriate infinite series (A.1.64), (A.1.101) and (A.1.102). Various relationships such as (A.1.48), (A.1.52), (A.1.53), (A.1.66), (A.1.71a), (A.1.73), (A.1.97–A.1.99), and (A.1.108) remain true for the complex functions.

A famous result that is used in the text is found by considering the exponent of the purely imaginary number iy. We have, using (A.1.64), (A.1.101), and (A.1.102).

$$\exp{(\mathrm{i}y)} = 1 + (\mathrm{i}y) + \frac{(\mathrm{i}y)^2}{2!} + \frac{(\mathrm{i}y)^3}{3!} + \dots$$

$$= 1 - \frac{y^2}{2!} + \dots + \mathrm{i}\left(y - \frac{y^3}{3!} + \dots\right).$$

This gives **de Moivre's theorem**

◆◆ $$\exp{(\mathrm{i}y)} = \cos y + \mathrm{i} \sin y. \tag{4}$$

Note that

$$|\exp{(\mathrm{i}y)}| = \cos^2 y + \sin^2 y = 1. \tag{5}$$

Using de Moivre's theorem, we can write

$$z = x + \mathrm{i}y = |z| \cos \theta + \mathrm{i}|z| \sin \theta = |z| \exp{(\mathrm{i}\theta)}.$$

Thus if

$$|z| = r, \qquad \arg z = \theta,$$

we can write any complex number in the **polar form**

$$z = r \exp(i\theta). \tag{6}$$

Using the law of exponents (53)

$$z_1 z_2 = |z_1||z_2| \exp(i\theta_1) \exp(i\theta_2) = |z_1||z_2| \exp[i(\theta_1 + \theta_2)].$$

This exhibits $z_1 z_2$ in polar form. Thus

◆◆ $\quad |z_1 z_2| = |z_1||z_2|, \qquad \arg(z_1 z_2) = \theta_1 + \theta_2 = \arg z_1 + \arg z_2. \quad (7a, b)$

As an application of the various rules that have been presented, consider the following (a result that is used in stability theory):

$$|\exp(z)| = |\exp(x + iy)| = |[\exp(x)][\exp(iy)]| = |\exp(x)||\exp(iy)| = \exp(x). \tag{8}$$

Determinants

Associated with every square array of numbers or **square matrix** A there is a number called its **determinant** denoted as $\det A$ or $|A|$. It is possible to construct an abstract definition of $\det A$ which is important in the development of sophisticated linear algebras. Here we restrict ourselves to an operational definition that makes possible the evaluation of the determinant for any square matrix (see Lipschiz, 1968).

For the case of a 2×2 matrix the definition is particularly simple:

$$\det \begin{pmatrix} a_{11} & a_{12} \\ a_{21} & a_{22} \end{pmatrix} = a_{11}a_{22} - a_{12}a_{21}.$$

For example

$$\det \begin{pmatrix} 1 & 2 \\ 3 & 4 \end{pmatrix} = 4 - 6 = -2. \tag{9}$$

For the case of a 3×3 matrix, the determinant is expressed as a sum of 2×2 determinants:

$$\det \begin{pmatrix} a_{11} & a_{12} & a_{13} \\ a_{21} & a_{22} & a_{23} \\ a_{31} & a_{32} & a_{33} \end{pmatrix}$$

$$= (-1)^{1+1} a_{11} \begin{vmatrix} a_{22} & a_{23} \\ a_{32} & a_{33} \end{vmatrix} + (-1)^{1+2} a_{12} \begin{vmatrix} a_{21} & a_{23} \\ a_{31} & a_{33} \end{vmatrix} + (-1)^{1+3} a_{13} \begin{vmatrix} a_{21} & a_{22} \\ a_{31} & a_{32} \end{vmatrix}$$

$$= a_{11}(a_{22}a_{33} - a_{23}a_{32}) - a_{12}(a_{21}a_{33} - a_{23}a_{31}) + a_{13}(a_{21}a_{32} - a_{22}a_{31}). \tag{10}$$

The 3×3 determinant has been expanded along the first row. In fact it can be shown that the same value for the determinant results if the expansion is made along any row or column according to the rules given here. The coefficients of the first row a_{11}, a_{12}, a_{13} are multiplied by (-1) raised to some power. This exponent is the sum of the row index and the column index of the 'a' coefficient, so that for a_{11} the exponent is $1 + 1$ and for a_{13} is $1 + 3$. In the general case, a_{ij} is multiplied by $(-1)^{i+j}$ where i is the row index and j is the column index.

The determinant multiplying a_{11} is the determinant of the matrix formed by deleting the first row and the first column of the original 3×3 matrix:

$$\begin{pmatrix} a_{22} & a_{23} \\ a_{32} & a_{33} \end{pmatrix} = \begin{vmatrix} a_{11} & a_{12} & a_{13} \\ a_{21} & a_{22} & a_{23} \\ a_{31} & a_{32} & a_{33} \end{vmatrix}.$$

Following the same pattern a_{12} is multiplied by the determinant of the matrix found by deleting the first row and the second column:

$$\begin{pmatrix} a_{21} & a_{23} \\ a_{31} & a_{33} \end{pmatrix} = \begin{vmatrix} a_{11} & a_{12} & a_{13} \\ a_{21} & a_{22} & a_{23} \\ a_{31} & a_{32} & a_{33} \end{vmatrix}.$$

Also

$$\begin{pmatrix} a_{21} & a_{22} \\ a_{31} & a_{32} \end{pmatrix} = \begin{vmatrix} a_{11} & a_{12} & a_{13} \\ a_{21} & a_{22} & a_{23} \\ a_{31} & a_{32} & a_{33} \end{vmatrix}.$$

The resulting 2×2 determinants are evaluated according to the previously stated rule.

The procedure can be applied to a matrix of any dimension. The $n \times n$ determinant is expanded along any row or column to produce a sum of $(n-1) \times (n-1)$ determinants. The reduction continues on until a sum of 2×2 determinants is obtained. As an example consider the 4×4 matrix expanded along the second column.

$$\begin{vmatrix} a_{11} & a_{12} & a_{13} & a_{14} \\ a_{21} & a_{22} & a_{23} & a_{24} \\ a_{31} & a_{32} & a_{33} & a_{34} \\ a_{41} & a_{42} & a_{43} & a_{44} \end{vmatrix} = -a_{12} \begin{vmatrix} a_{21} & a_{23} & a_{24} \\ a_{31} & a_{33} & a_{34} \\ a_{41} & a_{43} & a_{44} \end{vmatrix} + a_{22} \begin{vmatrix} a_{11} & a_{13} & a_{14} \\ a_{31} & a_{33} & a_{34} \\ a_{41} & a_{43} & a_{44} \end{vmatrix}$$

$$- a_{32} \begin{vmatrix} a_{11} & a_{13} & a_{14} \\ a_{21} & a_{23} & a_{24} \\ a_{41} & a_{43} & a_{44} \end{vmatrix} + a_{42} \begin{vmatrix} a_{11} & a_{13} & a_{14} \\ a_{21} & a_{23} & a_{24} \\ a_{31} & a_{33} & a_{34} \end{vmatrix}.$$

Each 3×3 determinant is reexpressed as the sum of three 2×2 determinants, which are evaluated by the first rule.

Determinants play an important role in the development of linear algebra and in particular in the solution of simultaneous equations represented in matrix form.

$$\begin{pmatrix} b_1 \\ b_2 \\ b_3 \end{pmatrix} = \begin{pmatrix} a_{11} & a_{12} & a_{13} \\ a_{21} & a_{22} & a_{23} \\ a_{31} & a_{32} & a_{33} \end{pmatrix} \begin{pmatrix} x \\ y \\ z \end{pmatrix}.$$

It can be shown that systems of this form (and also for other dimensions) will have a unique solution if the determinant of the a_{ij} matrix is not zero.

If this determinant is zero, the system either has no solution or an infinite number of solutions, depending on the values of certain other determinants. For example the equations

$$1 = x + y + z$$
$$2 = x + y$$
$$0 = -x - y + z$$

are inconsistent and have no solution, since adding the first and third equations gives

$$1 = 2z$$

while adding the second and third gives

$$2 = z.$$

This is confirmed by evaluating the determinant of the coefficient matrix (along the first column)

$$\begin{vmatrix} 1 & 1 & 1 \\ 1 & 1 & 0 \\ -1 & -1 & 1 \end{vmatrix} = +1 \begin{vmatrix} 1 & 0 \\ -1 & 1 \end{vmatrix} - 1 \begin{vmatrix} 1 & 1 \\ -1 & 1 \end{vmatrix} + (-1) \begin{vmatrix} 1 & 1 \\ 1 & 0 \end{vmatrix}$$

$$1 \cdot 1 - 1 \cdot 2 - 1 \cdot (-1) - 0.$$

However the system

$$3 = x + y$$
$$5 = y + z$$
$$4 = x + z$$

has the unique solution $x = 1$, $y = 2$, $z = 3$ and

$$\begin{vmatrix} 1 & 1 & 0 \\ 0 & 1 & 1 \\ 1 & 0 & 1 \end{vmatrix} = 2 \neq 0.$$

References

Levinson, N. & Redheffer, R. M. (1970). *Complex Variables*, San Francisco, Holden Day.

Lipschiz, S. (1968). *Schaum Outline of Theory and Problems of Linear Algebra*, New York, Schaum Outline Series, McGraw-Hill.

A.3 QUALITATIVE THEORY OF SYSTEMS OF ORDINARY DIFFERENTIAL EQUATIONS, INCLUDING PHASE PLANE ANALYSIS AND THE USE OF THE HOPF BIFURCATION THEOREM

This lengthy appendix is intended as a selfcontained module of instruction on the behavior of systems of ordinary differential equations (ODE's). Since so many phenomena that evolve continuously in time are best modeled mathematically by ODE systems, it is essential for students of theoretical biology to have a clear understanding of what a differential equation means, how it determines solutions, and how one goes about deducing the qualitative behavior that a particular ODE system can exhibit.

This appendix aims to convey these concepts without overwhelming the reader with either jargon, proof, or detailed calculation. It should be accessible to the reader who knows only basic calculus and algebra. We package these concepts in a geometric metaphor that makes possible the visualization of solutions to an ODE system as concrete geometrical objects. Intuitive geometrical reasoning (that can be made rigorous) then suffices to determine a great deal about the 'shape' and behavior of these objects, as influenced by the values of various parameters of the phenomenon being modeled. Scanning the figures of this appendix will give the reader an idea of what these 'objects' look like, and will give him or her the correct impression that the message of this appendix is carried in a pictorial medium.

To *use* differential equations for mathematical modeling, it is not necessary to know how to write down formulae for their solutions (or approximations thereunto). In most cases, no such formulae exist. In many cases for which an exact or approximate (asymptotic) analytical solution can be discovered, the solution formula is so complicated that it discloses nothing about the nature of the solution until a (geometrical) graph of it is drawn. We aim to generate the graphs, at least their general shapes, directly, without bothering with formulae for them.

Excellent computer program packages (such as CSMP; see the Introduction to this book) that numerically generate solutions to initial value problems (**IVP**'s) for ODE systems are widely available. These packages sometimes fail to produce accurate solutions when confronted with certain pathological ODE systems (see Appendix A.5), but numerical analysts have recently tuned ODE solver codes to the point that they fail less frequently than analytical techniques. Further, these programs are very easy to use, and can be employed effectively by persons who neither know nor care what makes the programs work, and who may also be ignorant of analytical procedures for determining formulae for solutions. The use of such programs usually suffices to fix the exact quantitative details of a solution to an ODE problem. Several quantitatively accurate solutions, produced numerically, surrounded by the kind of qualitative overview this appendix will explain, usually provide complete information about the content of an ODE model.

If a computer program can painlessly generate quantitatively accurate solutions to ODE problems, why should one bother to master the kind of qualitative theory this appendix presents? The answer is simple. Most ODE models of biological phenomena (or of anything else) contain a number of parameters each of which can assume a range of values. For each definite set of parameter values, the ODE system has not one, but an infinitude of solutions, most of which are superfluous. Initial conditions (**IC**'s) or other constraints select one (or some) of these solutions as the relevant one(s). To characterize the behavior of an ODE model numerically, one has to pick a particular sequence of definite parameter settings and IC's and generate numerical solutions, one for each set of settings. A blind numerical exploration of the way the solutions depend on the settings of just two or three parameters can involve a magnitude of computation and display that is forbidding either to pay for, or, even were it free, to read and understand. Often the region in parameter space within which the parameter values must lie to make the model behave realistically is so small or curiously shaped that even a thundering numerical shotgun blast will miss it. The kind of qualitative reasoning presented in this appendix can determine if there is a needle in the haystack and, if so, exactly where to locate it (numerically or analytically).

We begin by describing ODE systems and by reducing them all to a standard form, (10). We outline the problems of whether solutions exist and are unique, confining attention to initial value problems (IVP's, corresponding to models that predict the future (or past), given the present). Then we construct our geometrical metaphor, trajectories in phase space. We briefly sketch the nature of *linear* ODE systems, for which a complete theory is known. Next we explain phase plane analysis in which geometry, graphical sketching, and local magnifying glass views via linearization, collaborate in a simple way to determine (rigorously) most information about the behavior of a class of second-order systems. We discuss the use of the Hopf bifurcation theorem to determine when and how the tuning of parameter values can bring into existence a time-periodic (oscillatory) solution of an ODE system that 'previously' had none. This is an example of a central theme of this book: a gradual adjustment of conditions or parameters can cause, all of a sudden, a dramatically new kind of behavior of a model. Two bifurcation examples are worked through. One is a two-variable predator–prey problem and the other involves the three-variable cAMP oscillation/relay model derived in Section 4.1. The dominant theme of this book is the construction of ODE models of biological phenomena. This appendix analyzes, but does not derive, such models. It concludes with remarks about some bifurcations not characterized by the Hopf theorem, and a list of ODE text references.

Definitions

The **solution** to an ODE system is one or several (dependent variable) functions of a single independent variable. We will use t (for time, usually) to denote the independent variable. A **differential equation** is a relationship between t, and the dependent variable, and its derivatives with respect to t. For example,

$$\mathrm{d}u/\mathrm{d}t = u^2(1-u) \tag{1}$$

is a single **first-order** (because only first derivatives of u are involved) ODE system for $u(t)$.[1] Any function $u(t)$ satisfying (1) is a solution. Equation (1) gives the recipe by which the present value of $u(t)$ determines the slope (time rate of change) of the function $u(t)$. Figure A.3.1 portrays this recipe graphically. The arrows in Figure A.3.1 are tangent at the position (t, u) in the t–u plane to the solution curve, $u(t)$, passing through the point (t, u) to which the arrows are attached. The slope of the arrow at (t, u) is given by the right-hand side of (1), and so in this case depends only on u, not t. To construct a particular solution to (1), we start at a particular initial value, $u(0)$, and 'draw' a smooth curve that lies everywhere tangent to these tangent vectors. Figure A.3.1 shows several of these solution curves. Roughly speaking, a numerical ODE solver program starts at the chosen initial condition $u(0)$ and takes very small steps from one point to the next, moving in the direction specified by the tangent vectors (whose slopes are given by the recipe $u^2(t)[1-u(t)]$).

Think of (1) as a model, in dimensionless form, of the population u (of breeding females) at time t of creatures whose reproduction is governed by a law that says each female produces $u(1-u)$ female offspring per unit

---

[1] (1) happens to be an Abel equation and is discussed on p. 24 of Kamke, 1971. Whether or not an exact solution can be written down does not concern us here.

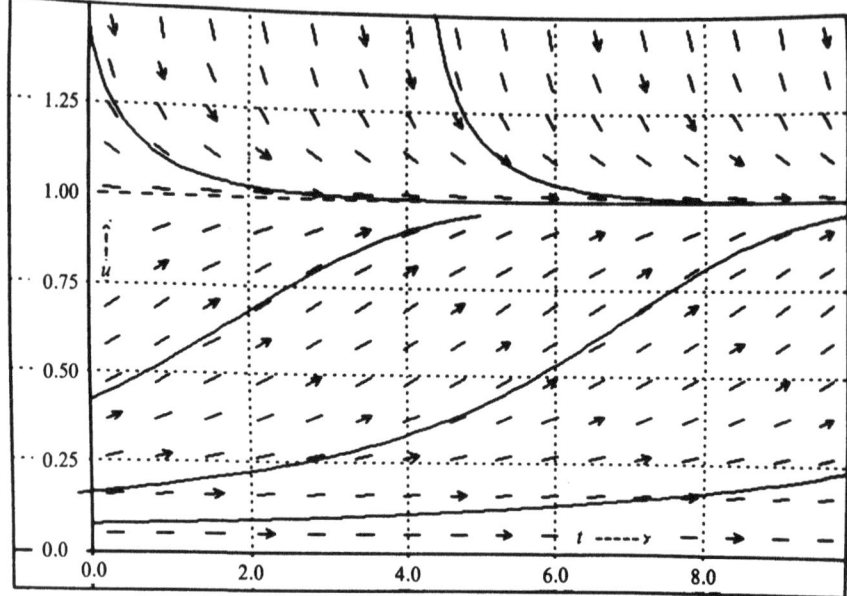

Figure A.3.1. A graphical method of constructing solutions to first-order ODE's of the form $dU/dt = f(U)$. In this example (1) is used.

time, that survive to breed in turn. (The case of a constant reproduction rate was considered in Appendix A.1. See (A.1.42).) This reproduction rate per female might fall to zero as $u \to 0$ because of mate-finding difficulties in sparse populations, and might go negative as u exceeds 1 because of overcrowding or nutrient shortage. Figure A.3.1 tells us, at a glance, all there is to know about how the initial population, $u(0)$, evolves. Namely, if $u(0) > 0$, $u(t)$ tends in a monotone fashion to 1, as t increases. For obvious reasons. $u(t) \equiv 1$ is called a **stable steady state** (or **equilibrium**) **solution** of (1). If $0 < u(0) < 1$, $u(t)$ is a sigmoid curve approaching this stable equilibrium value, 1, as $t \to \infty$ (and approaching 0 as $t \to -\infty$). If some perturbation displaces $u(t)$ away from $u = 1$, the solution returns to $u = 1$ asymptotically. The picture tells the complete story, and it can be sketched, free-hand, in a minute.

The reader who can succeed at the following exercise, generalizing what we did for (1), will glimpse the great power of this kind of geometrical, pictorial reasoning. Consider this class of first-order nonlinear ODE's

$$du/dt = f(u). \tag{2}$$

Show that, no matter what the smooth function $f(u)$ may be, no solution $u(t)$ can oscillate in t. More precisely, $u(t)$ must be a monotone increasing or a monotone decreasing function of t. Thus as $t \to +\infty$, $u(t)$ can do only one of three things: (i) $u(t) \to k$ where k is a finite constant; (ii) $u(t) \to +\infty$; (iii) $u(t) \to -\infty$.

A canonical form for ODE systems

Equation (1) is a first-order ODE. Next, we consider the general **m-th order ODE** (so called because $d^m u / dt^m$ is the highest-order derivative appearing),

$$d^m u / dt^m = h(u, du/dt, \ldots, d^{m-1}u/dt^{m-1}, t). \tag{3}$$

To fix ideas, here is a particular example, in which $m = 2$:

$$d^2 u / dt^2 = h(u, du/dt, t) = -u(1 - u) - c(du/dt) + a \sin(wt). \tag{4}$$

The constants c, a, and w in (4) are called **parameters**.

We now convert (3) into a system of m first-order ODE's by a simple transformation. We define m functions of t: $y_1(t), \ldots, y_m(t)$ as

$$y_1(t) = u(t),$$
$$y_2(t) = du/dt,$$
$$y_3(t) = d^2 u / dt^2,$$
$$\vdots$$
$$y_m(t) = d^{m-1}u/dt^{m-1}. \tag{5}$$

With these names we can express (3) in the form

$$dy_1/dt = y_2,$$
$$dy_2/dt = y_3,$$
$$\vdots$$
$$dy_{m-1}/dt = y_m,$$
$$dy_m/dt = h(y_1, y_2, \ldots, y_m, t). \tag{6}$$

The first $m - 1$ of (6) are tautologies, and the last is precisely (3) with the arguments of h renamed.

Our example (4) is rendered as follows,

$m = 2$, $y_1(t) = u(t)$, $y_2(t) = du/dt$,

$$dy_1/dt = y_2,$$
$$dy_2/dt = h(u, du/dt, t) = h(y_1, y_2, t) = -y_1(1 - y_1) - cy_2 + a \sin(wt). \tag{7}$$

Often systems of m first-order ODE's arise in their own right in a natural way, not just from the above kind of transformation of a single m-th-order ODE. The general system of m first-order ODE's looks like this:

$$dy_1/dt = f_1(y_1, \ldots, y_m, t),$$

$$dy_2/dt = f_2(y_1, \ldots, y_m, t),$$

$$\vdots$$

$$dy_m/dt = f_m(y_1, \ldots, y_m, t). \tag{8}$$

We think of this system giving rise to a generalization of the diagram in Figure A.3.1 into $m + 1$ dimensional space, in which the u-axis of Figure A.3.1 is replaced by m-space, each of whose dimensions represents one of the y_i. One initial condition singled out a particular solution curve in Figure A.3.1. To single out a particular solution curve now we need m initial conditions $y_1(t_0) = c_1$, $y_2(t_0) = c_2$, $\ldots$, $y_m(t_0) = c_m$, where t_0 is an initial instant. Most numerical packages for solving ODE problems are constructed to operate on them in the form of (8).

We need to study ODE's only in the form of (8) because, as just demonstrated, (8) embraces, as a special case, equations of order m.

If the functions f_i explicitly involve t (the way (7) does when a and w are nonzero), then (8) is called a **nonautonomous** system. Since the $y_i(t)$ change as t varies, all the f_i vary indirectly with t. However, it often happens that no f_i depends *explicitly* on the independent variable t. In this case the system (8) is called **autonomous**, the terminology indicating that the system clocks out its future evolution from its initial state without influence of the actual value of the initial time t. A more formal way to say this is that an autonomous system is not affected if t is replaced by $t + T_0$, where T_0 is any constant.

In this appendix we aim to deduce how solutions to an ODE system behave by examining the nature of the functions $f_1, \ldots, f_m$. It is easiest, conceptually, to do this if the f_i do not wander with time, t, but instead change only when the y_i change. Thus we want to confine attention to autonomous systems. In fact, this restriction is no restriction at all, by virtue of the following simple trick. If (8) is nonautonomous, then we invent a new 'dependent' variable, $y_{m+1}(t)$, whose differential equation we define to be

$$dy_{m+1}/dt = 1 \equiv f_{m+1}(y_1, \ldots, y_{m+1}),$$

and whose initial condition we always take to be $y_{m+1}(t_0) = t_0$. The reason for these choices is that the unique solution for $y_{m+1}(t)$ is $y_{m+1}(t) = t$. Thus, wherever t appears in the f_i in (8), we can replace it by y_{m+1} to obtain

$$dy_1/dt = f_1(y_1, \ldots, y_m, y_{m+1}), \qquad\qquad y_1(t_0) = c_1,$$

$$\vdots \qquad\qquad\qquad\qquad\qquad\qquad \vdots$$

$$dy_m/dt = f_m(y_1, \ldots, y_m, y_{m+1}), \qquad\qquad y_m(t_0) = c_m, \qquad\qquad (9)$$

$$dy_{m+1}/dt = f_{m+1}(y_1, \ldots, y_m, y_{m+1}) \equiv 1, \qquad y_{m+1}(t_0) = c_{m+1} \equiv t_0.$$

The system (9) is exactly equivalent to (8), but (9) is an autonomous system. We have now arrived at a **canonical** or standard form for ODE systems. It is the autonomous system (10), below, of n first-order ODE's. If our original system (8) was autonomous, we get (10) by deleting t from the argument list of each f_i, and taking $n = m$. If (8) was nonautonomous, we generate (9), and thence (10), by setting $n = m + 1$, and $c_n = t_0$. Since (9) is autonomous, the actual value of t_0 (in $y_i(t_0) = c_i$) is immaterial, so, for convenience, we will take $t_0 = 0$.

$$\left. \begin{array}{ll} dy_1/dt = f_1(y_1, \ldots, y_n) & y_1(0) = c_1 \\ dy_2/dt = f_2(y_1, \ldots, y_n) & y_2(0) = c_2 \\ \quad\vdots & \quad\vdots \\ dy_n/dt = f_n(y_1, \ldots, y_n) & y_n(0) = c_n \end{array} \right\} \; 0 \leq t \leq T. \qquad (10)$$

What we have demonstrated is that a nonautonomous system of order m is equivalent to an autonomous system of order $n = m + 1$.

Reconsider our example (7). When w, $a \neq 0$, so (7) is non-autonomous, we would recast (7) in the mold of (10) as

$$dy_1/dt = f_1(y_1, y_2, y_3) = y_2,$$

$$dy_2/dt = f_2(y_1, y_2, y_3) = -y_1(1 - y_1) - cy_2 + a \sin(wy_3),$$

$$dy_3/dt = f_3(y_1, y_2, y_3) = 1,$$

$$y_1(0) = c_1, \; y_2(0) = c_2, \; y_3(0) = t_0. \qquad\qquad (11)$$

when a or $w = 0$, the last term of the second equation vanishes, and the third equation can be discarded.

The issue of existence and uniqueness of solutions

The general quest of this appendix is to make sense of the system (10) by exploring the nature of the solutions it can exhibit. Faced with a particular set of equations in the form of (10), a mathematician would want first to determine the answer to this delicate question: given a set of initial conditions $c_1, c_2, \ldots, c_n$, how many solutions of the IVP (10) exist? The answer might be none, one, or several, and hinges upon the nature of the f_i, i.e. in part, upon how smooth they are. We want to sidestep this issue, a complete resolution of which requires many hard theorems.

It often happens that each f_i is continuously differentiable with respect to each y_i. That is, for each $1 \leqslant i, j \leqslant n$, $\partial f_i / \partial y_j$ is a continuous function of all the y_i's. In this fortunate circumstance, it is guaranteed that, when a set of initial conditions, $c_1, \ldots, c_n$, is prescribed, then, for small enough values of T, (10) has precisely one solution. Much weaker conditions on the f_i can guarantee the same outcome, but these are technicalities that need not concern us here.

Here is a simple example to show how a perfectly reasonable-looking IVP can fail to have any solution.

$$n = 1,$$

$$\mathrm{d}y_1/\mathrm{d}t = f_1(y_1) = y_1^2; \qquad y_1(0) = 1; \quad 0 \leqslant t \leqslant T. \tag{12}$$

$y_1(t) = 1/(A - t)$ is a solution of this nonlinear ODE for any constant A, as may be verified by differentiating it. With $A = 1$ chosen to fit the initial condition, we see the unique solution to the IVP (12) is $y_1(t) = 1/(1 - t)$, so long as this makes sense. All is well provided $T < 1$. If $T > 1$, however, our solution has a singularity as t crosses $t = 1$. That is, $y_1(t)$ does not exist at $t = 1$, and so, if $T > 1$, the IVP (12) has no solution at all. Evidently trouble can occur even when the functions f_i are infinitely smooth.

To illustrate how a system can exhibit two (or more) distinct solutions to the same IVP, consider another simple example,

$$\left.\begin{array}{l} \mathrm{d}y_1/\mathrm{d}t = f_1(y_1) = y_1^{\frac{1}{3}}, \\ y_1(0) = 0, \end{array}\right\} \quad 0 \leqslant t < \infty. \tag{13}$$

$y_1(t) = 0$ for all t certainly solves (13). So also does $y_1(t) = t^2/4$ for $t \geqslant 0$. We have (at least) two distinct solutions to the same IVP. The guideline above does not apply because $\mathrm{d}f_1/\mathrm{d}y_1$ fails to exist for $y_1 = 0$.

These simple examples of pathologies warn that the existence and uniqueness issue we now sidestep is something one ignores at substantial peril. We take the risk because it is unreasonable to expect the reader to have much interest in the intricacies of how a system of ODE's can generate bad behavior until he sees many examples of their good behavior.

Trusting that the ODE in (10) has a unique solution for each set of initial conditions, $(c_1, \ldots, c_n)$, we now proceed to study these solutions.

A geometric metaphor

Each solution of the system in (10) is a set of n functions $y_1(t), \ldots, y_n(t)$, each of which changes as t varies. We now think of a particular solution as a curve or **trajectory** winding through n-dimensional Euclidean space, R^n. The points on this trajectory have, at the instant t, the coordinates $(y_1(t), \ldots, y_n(t))$ in R^n. As t changes these coordinates vary and the trajectory unfolds in R^n. The entire space R^n is filled up with myriad interwoven trajectories, each one a solution of *some* initial value problems. In fact, every point on a given trajectory can serve as initial conditions for that trajectory. That is, by re-defining $t = 0$, every point visited by a given trajectory can be taken as the initial conditions singling out that trajectory among the infinitude of all solution trajectories of the ODE system.

In the preceding paragraph, we imagined that we had before us n functions of t, $y_1(t), \ldots, y_n(t)$, that solved the ODE system in (10), and that, using them, we drew an n-dimensional graph to represent the solution as a curve in R^n. This geometrical way of *representing* an already known solution can be used, inverted, to understand how one *constructs* a solution in the first place. Figure A.3.2 depicts a single solution trajectory to a hypothetical third-order ($n = 3$) ODE system of the form of (10). The arrows on the trajectory indicate the direction in which the point $[y_1(t), y_2(t), y_3(t)]$ moves as t increases. Attached to the point $[y_1(t), y_2(t), y_3(t)]$ is a vector whose components are $\{f_1[y_1(t), y_2(t), y_3(t)], \quad f_2[y_1(t), y_2(t), y_3(t)], \quad f_3[y_1(t), y_2(t), y_3(t)]\}$. Because, according to (10), $f_i[y_1(t), y_2(t), y_3(t)]$ is precisely the time rate of change of $y_i(t)$, this vector is tangent to the trajectory at the point $[y_1(t), y_2(t), y_3(t)]$. This is true for every value of t.

This means we should think of the functions $f_i(y_1, y_2, y_3)$ as endowing every point in (y_1, y_2, y_3)-space with a vector, whose components are the values of the f_i there, which is tangent to the unique trajectory through that point. (We are assuming that the f_i are smooth functions.) The length of this vector specifies the speed at which the solution passes through the point to which the vector is attached. We say that the $f_i(y_1, y_2, y_3)$ specify a **flow** on (y_1, y_2, y_3) space, and we call this space the **phase space** for the ODE system in (10). The task of finding solutions to the ODE system amounts to constructing curves in this **phase space** that are everywhere tangent to the flow vectors. These curves are called, among other things, **trajectories, solution curves, integral curves of the flow**.

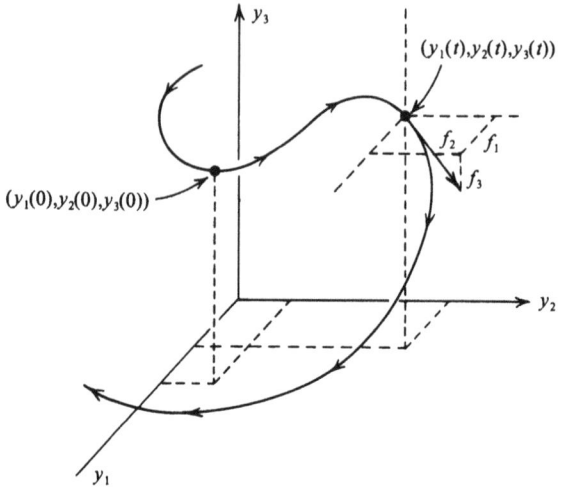

Figure A.3.2. A single typical trajectory of a third-order autonomous ODE system winding through its three-dimensional phase space. The vector construction shows how the components of the ODE's right-hand side determine the direction and speed of the trajectories at each point.

The task of solving a particular IVP in the form of (10) consists of finding, among the infinitude of trajectories, the unique integral curve of the flow, $(f_1, \ldots, f_n)$, that passes through the point whose coordinates are the specified initial values $(c_1, \ldots, c_n)$. It is this task that computer programs accomplish with ease. Roughly speaking, a numerical algorithm for solving the IVP (10), begins at the initial point $(c_1, \ldots, c_n)$ and then, taking discretely small 'steps', marches along the integral curve through that point, taking its direction from the flow, $(f_1, \ldots, f_n)$ at each new point visited.

The reason we went to some trouble above to transform a non-autonomous system to the form of (10) is that only for an autonomous system do these flow vectors hold still (constant with respect to t) during the construction process. In a nonautonomous system where the f_i's depend on t explicitly, the flow vectors swing around as t varies, ruining the geometric metaphor we are trying to build.

Conceiving of solutions to the ODE system of (10) as integral curves of the flow field imposed by the f_i's on an n-dimensional phase space, as just illustrated for $n = 3$, extends to any finite value of n (and even to infinite n) with no difficulty except that sketches such as Figure A.3.2 become impossible.

The quest to understand the qualitative behavior of the ODE system of (10) boils down, in concept, to the production of a composite portrait of all the integral curves of the flow $(f_1, \ldots, f_n)$. This is called a **phase**

portrait, and tells us everything interesting there is to know about all possible solutions of the ODE system. Later in this appendix, step-by-step instructions for the construction of a phase portrait for any second-order system ($n = 2$) will be given. For a while, though, we allow n to have any value.

Periodic solutions

Very often in biological applications, ODE systems are used to model time-periodic phenomena such as circadian rhythms, oscillatory chemical signaling, periodic passage of electrical action potential waves along nerve axons, etc. It is important to realize that a solution that is periodic (i.e. for some $\hat{T} > 0$, $y_i(t + \hat{T}) = y_i(t)$ for all t) corresponds to a solution trajectory in phase space that winds back on itself to form a closed loop. In the $n = 2$ case such a periodic orbit (loop) must be topologically identical to a circle. Part of the subject of this appendix is an effort to determine when such closed trajectories exist.

Steady state points

In our phase space, is it possible that two different trajectories intersect at a point? Suppose this happens at a point $\mathbf{p} = (p_1, \ldots, p_n)$.[2] At this point, as at all others, f has exactly one (vector) value, $\mathbf{f}(\mathbf{p})$, and this unique vector at $\mathbf{p}$ must be tangent to *all* trajectories that hit the point $\mathbf{p}$. One possibility is that, at a point $\mathbf{p}$ where two trajectories intersect, $\mathbf{f}(\mathbf{p}) = \mathbf{0}$ (which means $f_i(p_1, \ldots, p_n) = 0$, for $i = 1, \ldots, n$).

We will call such a point a **steady state point**[3] because, if $\mathbf{f}(\mathbf{p}) = 0$, then our ODE system,

$$dy/dt = \mathbf{f}(\mathbf{y}),$$

has $\mathbf{y}(t) = \mathbf{p}$ (or $y_1(t) = p_1$, $y_2(t) = p_2$, $\ldots$, $y_n(t) = p_n$) for all time t, as one steady (time-invariant) solution.

At any such point, several or a multitude of solution trajectories can intersect, and the angles they make with each other in this intersection are arbitrary.

If two or more distinct trajectories intersect at a point $\mathbf{p}$ at which $\mathbf{f}(\mathbf{p})$ is not $\mathbf{0}$, then all the intersecting trajectories would have to be tangent at the point $\mathbf{p}$ (since they all share the same nonzero tangent vector, $\mathbf{f}(\mathbf{p})$, there).

---

[2] To streamline notation, we will now start to use bold characters to denote n-vectors. For example $\mathbf{y}(t)$ denotes the point with coordinates $(y_1(t), \ldots, y_n(t))$. $\mathbf{f}(\mathbf{y})$ denotes the flow field whose components are $(f_1(\mathbf{y}), \ldots, f_n(\mathbf{y}))$. The ODE system in (10) becomes simply $dy/dt = \mathbf{f}(\mathbf{y})$.

[3] Other names for steady state points include **'critical point'**, **'equilibrium point'**, **'singular point'**.

If this were to happen, then two trajectories that touch, tangent, at $\mathbf{p}$ where $\mathbf{f}(\mathbf{p}) \neq \mathbf{0}$ would represent two distinct solutions to the same IVP

$$\mathrm{d}\mathbf{y}/\mathrm{d}t = \mathbf{f}(\mathbf{y}), \qquad \mathbf{y}(0) = \mathbf{p}.$$

It was declared above that this nonuniqueness of solutions to initial value problems could not occur in cases where the f_i were continuously differentiable functions of the y_i's. It is far from obvious why smoothness of the f_i precludes the possibility of trajectories intersecting at a point $\mathbf{p}$ where $\mathbf{f}(\mathbf{p}) \neq \mathbf{0}$, and cannot be made obvious by drawing pictures. Readers interested in why this intersection of trajectories at $\mathbf{p}$, where $\mathbf{f}(\mathbf{p}) \neq \mathbf{0}$, cannot ever occur when the f_i are smooth should attend the proof of the Picard–Lindelof theorem in any text on ODE's.

For a continuously differentiable flow $\mathbf{f}(\mathbf{y})$, *trajectories can intersect in phase space only at steady state points.* For this reason, the first step in analyzing the behavior of a particular set of ODE's, $\mathrm{d}\mathbf{y}/\mathrm{d}t = \mathbf{f}(\mathbf{y})$, is to find all of the steady state points. This is a problem in algebra: Find all $\mathbf{p}$ that make $\mathbf{f}(\mathbf{p}) = \mathbf{0}$. This may be an easy problem for some flows, $\mathbf{f}(\mathbf{y})$, and may be extremely difficult (or impossible without turning to numerical algorithms) for other flows.

Consider our example in (11). $f_3(y_1, y_2, y_3) = 1$ is never zero, so the system in (11) has no steady state points. This is generally true of all systems obtained from originally nonautonomous ODE systems by the transformation that produced the system in (9).

Suppose, however, $a = 0$ in (11). Then (11) collapses to

$$\mathrm{d}y_1/\mathrm{d}t = f_1(y_1, y_2) = y_2,$$

$$\mathrm{d}y_2/\mathrm{d}t = f_2(y_1, y_2) = -y_1(1 - y_1) - cy_2. \tag{14}$$

To find the steady state points, we seek $(p_1, p_2) = \mathbf{p}$ such that $f_1(p_1, p_2) = p_2 = 0$ and $f_2(p_1, p_2) = -p_1(1 - p_1) - cp_2 = 0$. This requires $p_2 = 0$ and $p_1 = 1$ or $p_1 = 0$. There are exactly two steady state points, $(0, 0)$ and $(1, 0)$.

A typical elementary course on ODE's would consist, in large part, of a study of certain special ODE's whose solutions can be obtained in analytical form. There are very few differential equations with this property. We will make no attempt to give a representative sampling of them. A compendium of known solutions by Kamkë (1971) should be scanned by any scientist using a differential equation to see if, by chance, his particular ODE is one of the few whose solutions are known. (For the special value, $c = 5/\sqrt{6}$, our example (14) occurs as Kamkë's equation 6.23 with solutions in terms of exponentials and Weierstrass P-functions).

The behavior of linear, constant-coefficient, ODE systems

For direct use in this book, we sketch the features of the most important class of ODE's whose exact solutions are known, namely **homogeneous linear systems with constant coefficients,** which have the following form

$$dy_1/dt = f_1(y_1, \ldots, y_n) = a_{11}y_1 + a_{12}y_2 + \ldots + a_{1n}y_n,$$

$$dy_2/dt = f_2(y_1, \ldots, y_n) = a_{21}y_1 + a_{22}y_2 + \ldots + a_{2n}y_n,$$

$$\vdots$$

$$dy_n/dt = f_n(y_1, \ldots, y_n) = a_{n1}y_1 + a_{n2}y_2 + \ldots + a_{nn}y_n. \tag{15}$$

Here the a_{ij}'s are constants. This system is called *linear* because, if $\tilde{\mathbf{y}}(t) = (\tilde{y}_1(t), \ldots, y_n(t))$ and $\hat{\mathbf{y}}(t) = (\hat{y}_1(t), \ldots, \hat{y}_n(t))$ are *any* two solutions of (15), then so also is any linear combination of $\tilde{\mathbf{y}}$ and $\hat{\mathbf{y}}$. That is, for any two constants α and β, $z(t) = \alpha\tilde{\mathbf{y}}(t) + \beta\hat{\mathbf{y}}(t)$ is also a solution. It turns out that this property allows *all* possible solutions of (15) to be constructed as linear combinations of exactly n simple solutions of (15) that we find as follows.

A multiple of the exponential function $y(t) = \exp(\lambda t)$ is the only function whose t-derivative is proportional to itself. We should therefore be able to find some constants $(e_1, \ldots, e_n)$ and an exponent λ so that

$$\begin{pmatrix} y_1(t) \\ \vdots \\ y_n(t) \end{pmatrix} = \begin{pmatrix} e_1 \\ \vdots \\ e_n \end{pmatrix} \exp(\lambda t) \tag{16}$$

is a solution of (15). Equation (16) substituted into (15) results in this system of homogeneous linear algebraic equations:

$$(a_{11} - \lambda)e_1 + a_{12}e_2 + \ldots + a_{1n}e_n = 0,$$

$$a_{21}e_1 + (a_{22} - \lambda)e_2 + \ldots + a_{2n}e_n = 0,$$

$$\vdots$$

$$a_{n1}e_1 + a_{n2}e_2 + \ldots + (a_{nn} - \lambda)e_n = 0. \tag{17}$$

Refer to any text on linear algebra for the reason why (17) has a nonzero solution, $(e_1, \ldots, e_n) \neq (0, 0, \ldots, 0)$, if and only if the determinant of the coefficients in (17) is zero. That is, we must have

$$\det \begin{vmatrix} a_{11} - \lambda & a_{12} & a_{13} & a_{1n} \\ a_{21} & a_{22} - \lambda & a_{23} & a_{2n} \\ \vdots & & & \\ a_{n1} & a_{n2} & \ldots & a_{nn} - \lambda \end{vmatrix} = 0. \tag{18}$$

(18) expands to become a polynomial of degree n in λ:

$$\lambda^n + I_{n-1}\lambda^{n-1} + I_{n-2}\lambda^{n-2} + \ldots + I_1\lambda + I_0 = 0, \tag{19}$$

where the I's are complicated nonlinear functions of the (a_{ij}), so called **scalar invariants** of the (a_{ij}) matrix.

When λ is any of the n possible zeros of this so called **characteristic polynomial**, (19), of the system (15), then, associated with that λ, there will be a set of constants $(e_1, \ldots, e_n) \neq (0, 0, \ldots, 0)$ solving (17). Such a λ is called an **eigenvalue** (or **characteristic exponent**) of the system (17) and the associated $(e_1, \ldots, e_n)$ is called an **eigenvector** (or **characteristic vector**) of (17). The special solution of (15) of the form (16) is called an **eigenfunction** of the ODE system (15). If the characteristic polynomial (19) has n *distinct* eigenvalues $(\lambda_1, \ldots, \lambda_n)$, then, associated with them, we have a set of n eigenfunctions possessing the following valuable property. *All* possible solutions of (15) can be constructed as a linear combination of these n eigenfunctions. That is, as the constants $(b_1, \ldots, b_n)$ are adjusted, the linear combination

$$\begin{pmatrix} y_1(t) \\ \vdots \\ y_n(t) \end{pmatrix} = b_1 \begin{pmatrix} e_1^1 \\ e_2^1 \\ \vdots \\ e_n^1 \end{pmatrix} \exp(\lambda_1 t) + b_2 \begin{pmatrix} e_1^2 \\ \vdots \\ e_n^2 \end{pmatrix} \exp(\lambda_2 t) + \ldots + b_n \begin{pmatrix} e_1^n \\ \vdots \\ e_n^n \end{pmatrix} \exp(\lambda_n t)$$

$$\tag{20}$$

generates all possible solutions of the ODE system (15). To solve an IVP, the constants $(b_1, \ldots, b_n)$ are adjusted to fit the initial conditions. Some complications result when the λ's are not distinct; refer to any text on ODE's, for example to that by Boyce & DiPrima (1977).

In this process it often happens that the exponents, λ_i, are complex valued (and so occur in complex conjugate pairs). Now, each $y_i(t)$ must be real valued, so, in this case, the b_i's and e_i^j's are complex so that, in the linear combination (20), all complex quantities cancel each other. To make sense of this, we use de Moivre's theorem (see Appendix A.2)

$$\exp(i\theta) = \cos\theta + i\sin\theta, \tag{21}$$

where $i = \sqrt{(-1)}$. Thus if $\lambda_k = \alpha_k + i\beta_k$ is a typical complex eigenvalue, then

$$\exp(\lambda_k t) = \exp(\alpha_k + i\beta_k)t = \exp(\alpha_k t)[\exp(i\beta_k t)]$$

$$= \exp(\alpha_k t)\{\cos\beta_k t + i\sin\beta_k t\}. \tag{22}$$

The part of (22) in curly brackets merely oscillates in t, keeping a constant modulus of unity. Thus, the real part, α_k, of the eigenvalue λ_k, determines the amplitude (envelope) of $\exp(\lambda_k t)$ to be $\exp(\alpha_k t)$. This means that if the real part of the eigenvalue λ_k is positive (negative), then the associated eigenfunction $\begin{pmatrix} e_1^k \\ \vdots \\ e_n^k \end{pmatrix}$ exp $(\lambda_k t)$ grows exponentially large (small) with time.

The last remark is the basis of the most important conclusion from our consideration of linear ODE systems. Often we do not care *exactly* what the solution to an ODE system looks like; we seek only qualitative information. In practice it is very tedious to compute the e_j^i's and b_i's above. Often a knowledge of the algebraic signs of the real parts of the eigenvalues tells us all we need to know. For example, if all eigenvalues have negative real parts, then we know from (20) and the preceding paragraph, that *all* solutions to all IVP's with (15) decay exponentially to zero as $t \to +\infty$. In contrast, if just one eigenvalue, λ_k, has a positive real part, then the solutions to *most* IVP's with (15) ($b_k = 0$ is an unlikely circumstance in 20) will explode exponentially in time. That is to say, the signs of the eigenvalues' real parts are all-important, and finding them is a problem in algebra, not differential equations. Below we will use this feature of linear systems to great advantage. Also, see Chapter 3, where useful algebraic theorems about eigenvalues are enunciated.

We remark that the linear ODE system (15) has precisely one steady state point: $(0, 0, \ldots, 0) = \mathbf{0}$, provided $\lambda = 0$ is not an eigenvalue of (17).

Second-order systems; phase plane analysis

We now restrict our attention to second-order ($n = 2$) autonomous ODE systems. This will allow us to develop **phase plane analysis**, a very powerful process by which essentially complete qualitative information about all solutions of the ODE can be deduced by graphical methods (i.e. by sketching freehand pictures) coupled with the use of second-order linear systems just discussed and judicious use of some theorems that will be described. We want to emphasize that this powerful method of studying second-order ODE's can be easily mastered by students knowing only calculus. Usually, no laborious calculations are involved. A good reference is Andronov, Leontovich, Gordon & Maier, 1973b.

The ODE system at hand is (10) with $n = 2$. Since we have only two variables, y_1 and y_2, it is convenient to call them $X(t) = y_1(t)$ and $Y(t) = y_2(t)$. We will call $f_1(y_1, y_2) = F(X, Y)$ and $f_2(y_1, y_2) = G(X, Y)$. We have collapsed (10) to

$$dX/dt = F(X, Y), \qquad dY/dt = G(X, Y). \tag{23}$$

Example. To fix ideas, we will work out a simple example as this phase plane analysis procedure is described. We will use the system in

(14); with the names changed, as above, we have

$$dX/dt = F(X, Y) = Y,$$

$$dY/dt = G(X, Y) = -X(1-X) - cY. \tag{24}$$

We shall consider only $c > 0$. Note that (24) is not a linear ODE system.

Now we want to produce the kind of phase portrait sketch described above for arbitrary n, and suggested in Figure A.3.2, but collapsed into the X-Y plane. The phase space becomes a **phase plane**, and, because $n = 2$ now, it becomes possible to represent faithfully, in a single flat picture, the entire set of integral curves of the flow specified by the functions $F(X, Y)$, $G(X, Y)$.

We review how F and G specify a flow. Pick any (every) point in the X-Y plane. Attached to that point, draw the vector whose X-Y components are $[F(X, Y), G(X, Y)]$. Figure A.3.3 indicates the construction at several points using the example system (24).

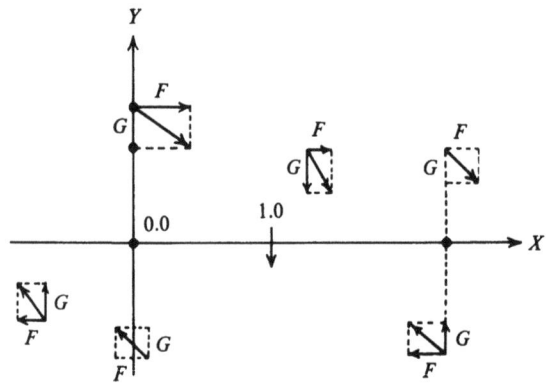

Figure A.3.3. Construction of a few of the tangent vectors for the flow specified by (24).

If we fill up the X-Y phase plane with these flow arrows, then if we construct (sketch) the curves that, at each point, run tangential to the flow arrows, we will produce all the integral curves of the flow. That is, we will depict all the solution (trajectories) (23) can exhibit.

The two spots in Figure A.3.3, at $(0, 0)$ and $(1, 0)$, locate the two steady state points we found for our example ODE (24). In the phase plane analysis procedure, it is necessary to find (at least locate qualitatively in the proper quadrant) all of the system's steady state points, for, as we have pointed out, *these are the only points at which trajectories intersect.*

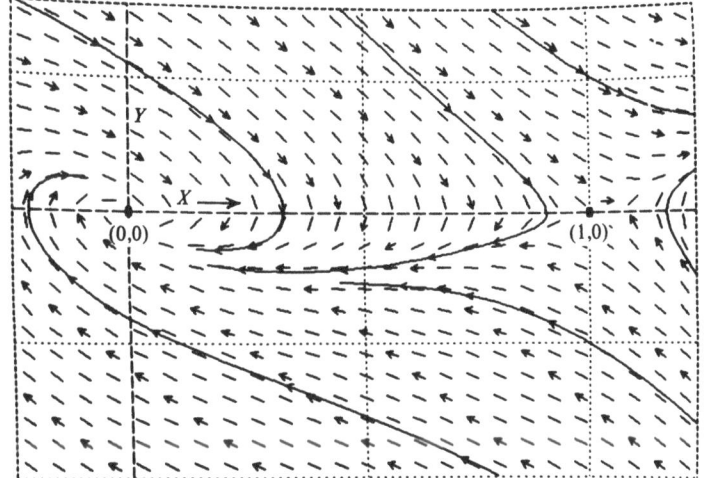

Figure A.3.4. Tangent vectors for the flow specified by the ODE (24) densely sprinkled in the phase plane. In (24), $c = 1$. Several trajectories are partially sketched in; these run everywhere tangential to the flow vectors. Linear analysis will determine the details of the flow near the two steady state points: $(0, 0)$ and $(1, 0)$.

Figure A.3.4, drawn on a computer graphics screen, shows the $X-Y$ phase plane densely sprinkled with flow arrow segments, with a few trajectory segments drawn in to indicate what we wish to accomplish. Hand-drawing this dense an array of arrows is, of course, out of the question – and unnecessary. For Figure A.3.3, A.3.4 and A.3.5, we use the value $c = 1$.

Nullclines (Step *1* of phase portrait instructions)

Here begin the step-by-step instructions for constructing a phase portrait. We need only a few of the arrows, strategically located, to characterize the flow. In particular, find the locus of points, called the X-**nullcline**, at which $F(X, Y) = 0$. This locus, determined by $F(X, Y) = 0$, comprises all points in the $X-Y$ plane at which the trajectories can be vertical (because only the Y-components of the flow arrows can be nonzero).

In our example problem, (24), this X-nullcline locus happens to be the horizontal line $Y = 0$.

Next, find the Y-nullcline locus, namely the set of points satisfying $G(X, Y) = 0$. This is the locus of all points at which the trajectories can be horizontal (because only the X-component of the flow arrows can be nonzero). In our example, the Y-nullcline locus is the parabola $Y = -X(1 - X)/c$.

These two nullcline loci intersect at, and only at, the steady state points. (This fact can be used to locate the steady state points.)

Sketch horizontal arrows along the Y-nullcline, and vertical arrows along the X-nullcline. A look at the ODE (23) will reveal which way the arrows point. In our example (24), on the Y-nullcline, the arrows point to the left when $Y < 0$ (because $dX/dt < 0$, so X is decreasing), and to the right when $Y > 0$. As you move along a nullcline, the arrows cannot reverse direction except as a steady state point is crossed, for the obvious reason that a smooth function (F or G) cannot change sign without assuming an intermediate zero value.

With the nullclines in place, the steady state points marked, and with just a few flow arrows sketched elsewhere, a general idea of the flow emerges. The slope of the flow arrows is $dY/dX = F(X, Y)/G(X, Y)$. Locate the arrows at points where their slopes are easy to compute. In our example (24), $dY/dX = -c$ at $X = 0$ and at $X = 1$, so those two loci are convenient places to draw flow arrows. Figure A.3.5 shows the state of the phase portrait at this juncture. A few trajectories have been partially drawn in. The general clockwise swirl around the steady state point at

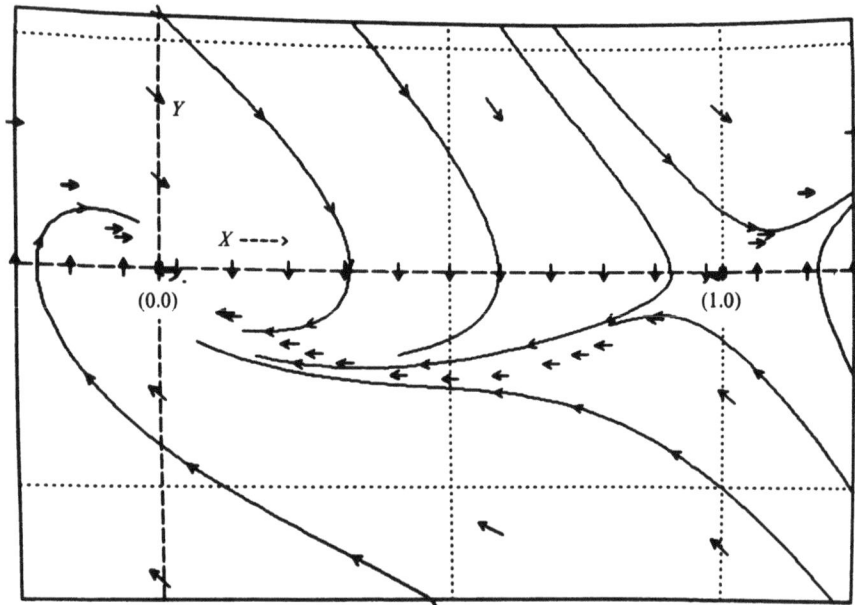

Figure A.3.5. The nullcline loci, and other loci where the flow is easy to analyze (at $X = 0$ and $X = 1$), marked by flow arrows for the ODE system (24) when $c = 1$. From this limited set of flow vectors, the general lie of the trajectories is apparent; there is no need to sketch the dense array of flow vectors shown in Figure A.3.4.

(0, 0) is apparent. The detailed pattern of trajectories in the immediate neighborhood of the steady state points is the crucial information needed to finish the phase portrait correctly. This is usually the case: we need magnified views of details at the steady state points. Linear theory gives us this. The next subsection explains how.

Linearization near the steady state points (Step 2 of the phase portrait instructions)

We wish to know what happens for X near X_0 and Y near Y_0, where (X_0, Y_0) is any steady state point of the system (23). Thus, letting x and y be infinitesimal displacements from the steady state point, we substitute $X = X_0 + x$ and $Y = Y_0 + y$ into (23). As in (A.1.133), expand both F and G as Taylor series near $x = y = 0$. Neglecting relatively small quadratics in x and y and even smaller terms of higher order (this can be rigorously justified in most cases), we obtain this **linearized** version of the first equation of (23)

$$\frac{d}{dt}(X_0 + x) = F(X_0 + x, Y_0 + y)$$

$$= F(X_0, Y_0) + \frac{\partial F}{\partial X}(X_0, Y_0)x + \frac{\partial F}{\partial Y}(X_0, Y_0)y.$$

We obtain a similar expansion of the second equation. Since, by definition of $X_0, Y_0, F(X_0, Y_0) = G(X_0, Y_0) = 0$, these expansions reduce to

$$dx/dt = a_{11}x + a_{12}y,$$

$$dy/dt = a_{21}x + a_{22}y, \tag{25}$$

where

$$a_{11} = \frac{\partial F}{\partial X}(X_0, Y_0), \qquad a_{12} = \frac{\partial F}{\partial Y}(X_0, Y_0),$$

$$a_{21} = \frac{\partial G}{\partial X}(X_0, Y_0), \qquad a_{22} = \frac{\partial G}{\partial Y}(X_0, Y_0).$$

Equation (25) is a linear, constant coefficient ODE system we have already seen. We find its eigenvalues.

$$\det \begin{vmatrix} a_{11} - \lambda & a_{12} \\ a_{21} & a_{22} - \lambda \end{vmatrix} = 0$$

or

$$\lambda^2 - (a_{11} + a_{22})\lambda + (a_{11}a_{22} - a_{12}a_{21}) = 0.$$

Let $\beta = a_{11} + a_{22}$ and $\gamma = a_{11}a_{22} - a_{12}a_{21}$. The eigenvalues, obtained by the quadratic formula, are

$$\lambda_1 = \tfrac{1}{2}(\beta + [\beta^2 - 4\gamma]^{\frac{1}{2}}),$$

and

$$\lambda_2 = \tfrac{1}{2}(\beta - [\beta^2 - 4\gamma]^{\frac{1}{2}}). \tag{26}$$

The eigenvector associated with λ_i has components

$$\begin{pmatrix} e_1^i \\ e_2^i \end{pmatrix} = (1 + s_i^2)^{-\frac{1}{2}} \begin{pmatrix} 1 \\ s_i \end{pmatrix}$$

where $s_i = (\lambda_i - a_{11})/a_{12}$, provided $a_{12} \neq 0$. We will not consider the rare instance when $\lambda_1 = \lambda_2$. We will then show that (25) can exhibit only six qualitatively different kinds of behavior and produce the corresponding generic phase portraits. The importance of this feature of the linear system (25) is that, except for rare instances, the magnified view we seek of trajectories near a steady state point for the nonlinear ODE system (23) will look like one of these six cases. We need only identify which one. The exceptions occur when two eigenvalues are identical, and especially when they are both zero, in which case our linearization discloses nothing.

A catalog of behavior possible for a second-order linear system

We consider separately the case for positive and negative values of $\beta^2 - 4\gamma$.

When $\beta^2 - 4\gamma > 0$, both λ_1 and λ_2 are real with $\lambda_2 < \lambda_1$. Then the general solution to (25) is

$$y(t) = b_1 \begin{pmatrix} e_1^1 \\ e_2^1 \end{pmatrix} \exp(\lambda_1 t) + b_2 \begin{pmatrix} e_1^2 \\ e_2^2 \end{pmatrix} \exp(\lambda_2 t) \tag{27}$$

where

$$\begin{pmatrix} e_1^1 \\ e_2^1 \end{pmatrix} \quad \text{and} \quad \begin{pmatrix} e_1^2 \\ e_2^2 \end{pmatrix} \quad \text{are eigenvectors of the matrix} \quad \begin{pmatrix} a_{11} & a_{12} \\ a_{21} & a_{22} \end{pmatrix}.$$

Saddle points. If $\gamma < 0$, then, regardless of the sign of β, λ_1 and λ_2 have opposite signs. This means the eigenfunction associated with the positive eigenvalue grows exponentially in time, while the eigenfunction associated with the negative eigenvalue decays exponentially to zero. In this case,

$$\gamma < 0,$$

the steady state point $(0, 0)$ of (25) is called a **saddle point**.

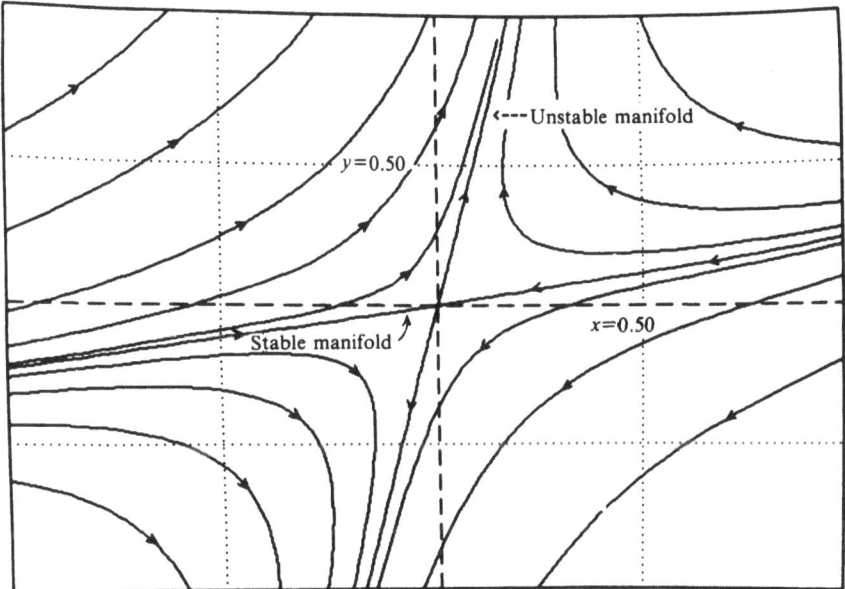

Figure A.3.6. A typical *saddle point*: one of the few generic types of steady state points the constant-coefficient linear ODE (25) can exhibit. Different values for the a_{ij} in (25) would cause different orientations of the stable and unstable manifolds. (Here $a_{11} = -1.0$, $a_{12} = 0.35$, $a_{21} = -0.45$, $a_{22} = 1.0$.) Figures A.3.7 to A.3.9 depict other generic types.

Figure A.3.6 shows the phase portrait of a typical saddle point (for the coefficient values $a_{11} = -1$, $a_{12} = 0.35$, $a_{21} = -0.45$, $a_{22} = +1$). Four trajectories intersect at a saddle point. Two form the so called **stable manifold** of the saddle point. These are the trajectories traced out by the eigenfunction associated with the negative eigenvalue. Suppose $\lambda_2 < 0$. Then these trajectories would be generated by

$$+\begin{pmatrix} e_1^2 \\ e_2^2 \end{pmatrix} \exp(\lambda_2 t) \quad \text{and} \quad -\begin{pmatrix} e_1^2 \\ e_2^2 \end{pmatrix} \exp(\lambda_2 t),$$

and so coincide with the eigenvectors

$$\pm\begin{pmatrix} e_1^2 \\ e_2^2 \end{pmatrix}.$$

These two stable manifold trajectories approach $(0, 0)$ asymptotically as $t \to +\infty$. The **unstable manifold** comprises the two trajectories that approach $(0, 0)$ as $t \to -\infty$, and are generated by the eigenfunction associated with the positive eigenvalue.

Nodes. If, with $\beta^2 - 4\gamma > 0$, $\gamma < 0$, then both eigenvalues have the same sign, the sign of β. The steady state point is then called a **node**. If $\beta > 0$, all nonzero solutions to (25) grow exponentially with t, so $(0, 0)$ is called an **unstable node**. The word 'unstable' is used because, in this case, the steady state solution, $x(t) = y(t) = 0$, is unstable to small perturbations. That is, if the initial conditions place $(x(0), y(0))$ near $(0, 0)$, the solution will run away from $(0, 0)$; all solutions diverge from $(0, 0)$. Note that saddle points, too, are unstable in the same sense except that two isolated solutions (the stable manifold) do approach $(0, 0)$ as $t \to +\infty$. Infinitesimal perturbations off this isolated manifold, however, result in eventual divergence from $(0, 0)$.

If $\beta < 0$ and $\beta^2 - 4\gamma > 0$, then both eigenvalues are negative. All solutions to (25) then decay exponentially to $(0, 0)$. In this case, $(0, 0)$ is called a **stable node**, naming a situation in which the steady state solution, $x(t) = y(t) = 0$, is stable to small perturbations; all trajectories lead to $(0, 0)$.

Clearly an unstable node looks like a stable node with the flow arrows reversed. In both cases, infinitely many trajectories intersect at $(0, 0)$, emanating from $(0, 0)$ for an unstable node, coalescing at $(0, 0)$ for a stable node.

When $\beta^2 - 4\gamma \neq 0$, one eigenvalue has a larger absolute value than the other. Thus one eigenfunction grows (or decays) exponentially faster than the other. Hence, in the case of a stable node, all trajectories (except two) come into the node tangent to the eigenvector associated with the least negative eigenvalue, λ_1. Figure A.3.7 depicts a stable node (for the coefficients $a_{11} = -1.0$, $a_{12} = 0.5$, $a_{21} = 0.5$, $a_{22} = -1.0$).

Spirals. When $\beta^2 - 4\gamma < 0$, then the eigenvalues occur as a complex conjugate pair,

$$\lambda_{1,2} = \beta/2 \pm i(4\gamma - \beta^2)^{\frac{1}{2}}/2,$$

and the general solution (27), using de Moivre's theorem, (21), becomes

$$y(t) = \exp(\beta t/2)[b_1' \sin(\omega t) + b_2' \cos(\omega t)] \tag{28}$$

where $\omega = \frac{1}{2}(4\gamma - \beta^2)^{\frac{1}{2}}$.

This solution oscillates with an amplitude envelope that grows or decays as does $\exp(\beta t/2)$. Unless $\beta = 0$, the solution trajectories are spirals. If $\beta < 0$, the trajectories spiral in to the steady state point, and $(0, 0)$ is called a **stable spiral** (or stable focus). If $\beta > 0$, $(0, 0)$ is called an **unstable spiral** (or focus) because trajectories spiral out from it. Figure A.3.8 shows an unstable spiral (for the coefficients $a_{11} = 0.5$, $a_{12} = -1.0$, $a_{21} = 1.0$, $a_{22} = -0.25$).

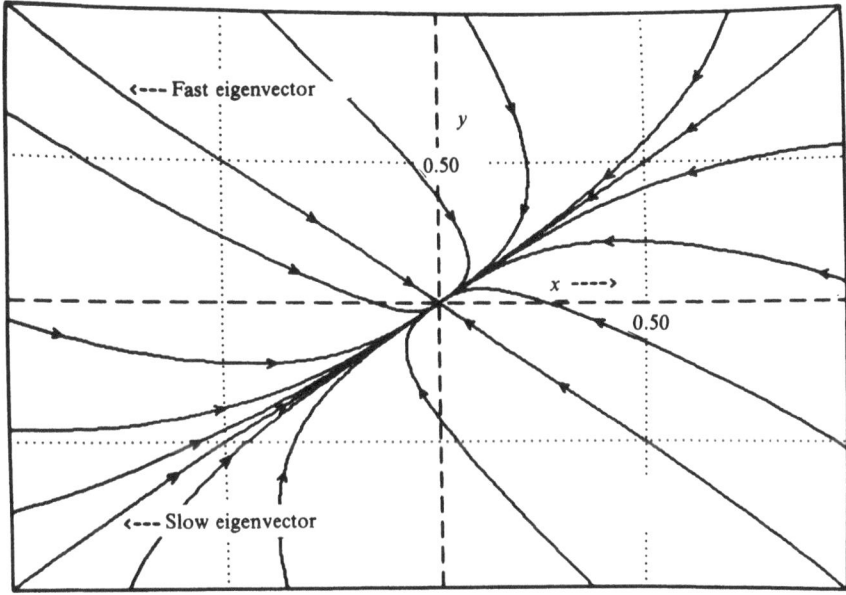

Figure A.3.7. A typical *stable node* steady state point for the ODE (25). To generate an unstable node's phase portrait, reverse all arrowheads.

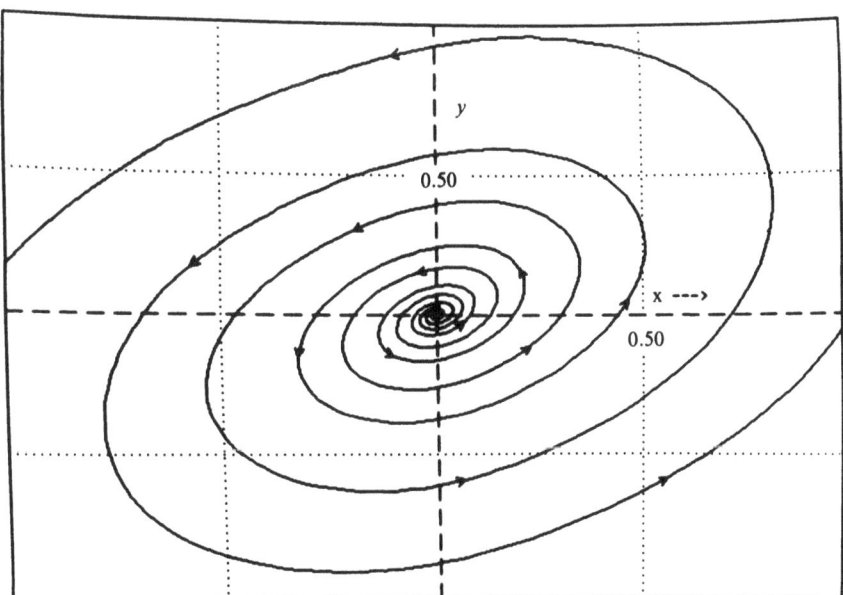

Figure A.3.8. A typical *unstable spiral* steady state point for the ODE (25). To generate a stable spiral's phase portrait, reverse all arrowheads.

Centers.   The isolated situation when $\beta = 0$, $\gamma > 0$, so that the eigenvalues are pure imaginary $\lambda_{1,2} = \pm i\sqrt{(+\gamma)}$, and the general solution of (25) is the undamped sinusoidal oscillation,

$$y(t) = b_1' \sin\left[(+\gamma)^{\frac{1}{2}}t\right] + b_2' \cos\left[(+\gamma)^{\frac{1}{2}}t\right], \tag{29}$$

is a case which turns out to be very important. For any choice of the integration constants b_1', b_2', (29) is a periodic function of t. The solution trajectories are all concentric ellipses centered at $(0, 0)$. The steady state point is called a **center**. It is termed **neutrally stable**. Solutions neither run away from nor approach $(0, 0)$ as $t \to +\infty$. Figure A.3.9 shows a linear center (for the coefficients $a_{11} = 0.5$, $a_{12} = -1.0$, $a_{21} = 2.0$, $a_{22} = -0.5$). Centers play a central role in the bifurcation theory we will discuss shortly.

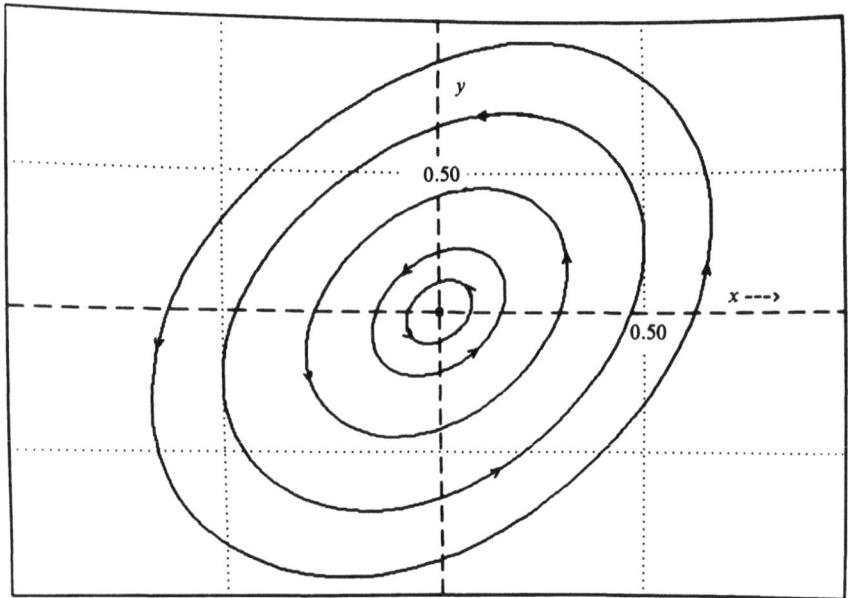

Figure A.3.9. A typical *center* steady state point for the ODE (25). The orbits are ellipses. The orientations of their major and minor axes depend upon the a_{ij} in (25).

Figure A.3.10 summarizes the kinds of behavior a linear system can exhibit. Given a linear system (25), and the attendant constants β and γ of (26), locate the point (β, γ) on Figure A.3.10 to read off what kind of steady state point $(0, 0)$ corresponds to that system.

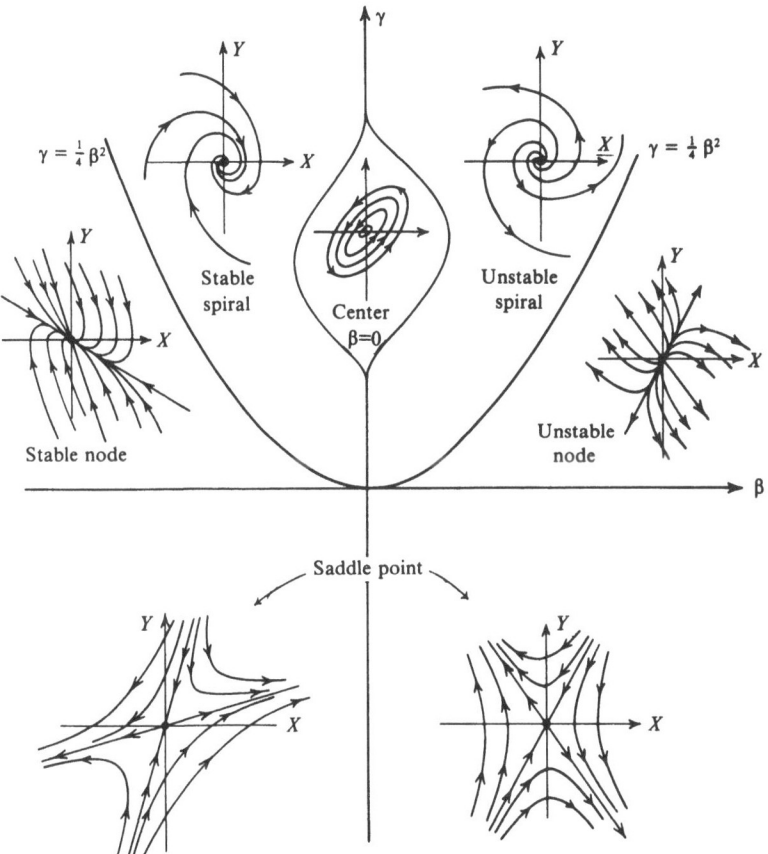

Figure A.3.10. Diagram summarizing how the values of the a_{ij} in the linear constant-coefficient ODE (25) determine the type of steady state point at $(X, Y) = (0, 0)$. The axes of this diagram are $\beta = a_{11} + a_{12}$ and $\gamma = a_{11}a_{12} - a_{12}a_{21}$. Faced with a particular instance of (25), compute β and γ only accurately enough to locate (β, γ) in one of the six different regions in the diagram.

Step 2 of phase portrait instructions: linearization

As the second step of our phase portrait construction, we linearize the flow in the neighborhood of each of its steady state points. That is, we compute the partial derivatives a_{11}, a_{12}, a_{21}, a_{22} in the linear system (25), where (X_0, Y_0) is a typical steady state point. From these, compute β and γ and ascertain, from Figure A.3.10, the type of steady state point involved.

For our example problem (24), we find, for $(X_0, Y_0) = (0, 0)$

$$\begin{pmatrix} a_{11} & a_{12} \\ a_{21} & a_{22} \end{pmatrix} = \begin{pmatrix} 0 & 1 \\ -1 & -c \end{pmatrix},$$

so $\beta = -c$ and $\gamma = 1$. β is negative, and γ is positive, so, using Figure A.3.10, $(0, 0)$ is a stable node if $\beta^2 - 4\gamma = c^2 - 4 > 0$, and is a stable spiral if $c^2 - 4 < 0$.

At $(X_0, Y_0) = (1, 0)$, we find

$$\begin{pmatrix} a_{11} & a_{12} \\ a_{21} & a_{22} \end{pmatrix} = \begin{pmatrix} 0 & 1 \\ 1 & -c \end{pmatrix},$$

so $\beta = -c$ and $\gamma = -1$. With β and γ both negative, Figure A.3.10 specifies that the steady state point $(1, 0)$ is always a saddle point. Several other examples of linearization occur in the main body of the text.

Phase portrait instructions (Step 3)

Supposing we have located each steady state point and identified its type by linearization, we can frequently finish a phase portrait. Only at steady state points can trajectories intersect. Trajectories emanate from unstable spirals, nodes and from saddle points (spirals and nodes send out infinitely many; saddle points send out two). Trajectories end only at stable nodes, spirals and saddle points (with the same counting as above). If the linearization of step 2 declares a steady state point of the nonlinear ODE system (23) to be (locally) a node, spiral, or saddle point, then that declaration can be trusted. Nonlinearities in (23) cause the (linear system) trajectories of Figures A.3.6–A.3.10 to appear warped and deformed as they are locally embedded, with $(x, y) = (0, 0)$ centered at $(X, Y) = (X_0, Y_0)$, in the nonlinear phase portrait, but the local *topology* of the flow is the same in the linear and nonlinear systems.

In our example, $(X_0, Y_0) = (1, 0)$ is a saddle point. Its unstable manifold will curve in the (X, Y) space (whereas, in a linear system, the unstable manifold is always a straight infinite line).

The claims above do not apply when linearization declares (X_0, Y_0) to be a center. The topology of a center – namely a family of concentric closed curves – is so delicate that even the smoothest nonlinear deformations can break the closure of some of the closed loops. Shortly, under the heading of 'Hopf bifurcations', we will investigate what can happen when linear theory predicts a center.

Assuming no steady state point is a center according to linear theory, step 3 in the construction of a phase portrait consists of fitting together the pieces already assembled. We have the nullclines marked, and we know what the flow must look like in each steady state point's immediate vicinity. We know that trajectories intersect, begin or end, only at steady state points. We claim that the following proposition (which can be proved rigorously) is intuitively obvious.

Proposition. Let $[X(t), Y(t)]$ be a solution trajectory of (23). As $t \to +\infty$ (or as $t \to -\infty$), only one of four things can happen.

(i) $|X(t)|$ and/or $|Y(t)| \to +\infty$.

(ii) $(X(t), Y(t)) \to (X_0, Y_0)$ where (X_0, Y_0) is a steady state point.

(iii) The trajectory is, itself, a closed loop trajectory corresponding to a t-periodic solution of (23), or else the trajectory $(X(t), Y(t))$ winds around in a bounded (finite size) spiral that approaches (but never quite arrives at) a closed loop trajectory that corresponds to a t-periodic solution of (23) called a **limit cycle**. This behavior will be discussed later in conjunction with bifurcation theory, and is illustrated in Figures A.3.17, A.3.21, and A.3.22.

(iv) As in case (iii), the trajectory winds around in a bounded spiral, approaching, in this case, a closed trajectory path made up of one or several segments, each of which consists of a solution trajectory that connects one steady state point to another. Two examples of this behavior are shown in Figure A.3.11, in which the paths are made up of solution trajectories each of which is both the unstable manifold of a saddle point and also the stable manifold of a saddle point. Figure A.3.11 shows a possible phase portrait with four steady state points A, B, C, and D. A trajectory winds out from the unstable spiral point B to approach a closed trajectory path with two segments, the top one consisting of a trajectory that is, at the same time, a branch of the unstable manifold out of A and a branch of the stable manifold into C, the bottom one being similar to the top one but with the names A and C interchanged. These segments, connecting distinct steady state points, are called **heteroclinic** trajectories.

The trajectory winding out from the unstable spiral D in Figure A.3.11 approaches a trajectory that starts as a branch of the unstable manifold out of the saddle point C, and ends as a branch of the stable manifold *at the same* saddle point C. This trajectory (approached by the spiral as $t \to +\infty$) is called a **homoclinic** trajectory.

It is important to realize that *homoclinic* and *heteroclinic* trajectories mentioned above (in general, a heteroclinic trajectory is one that approaches one steady state point as $t \to +\infty$, and approaches another steady state point as $t \to -\infty$, these steady state points being of *any* type) represent solutions to (23) that are *bounded* (finite) for all t. Closed trajectory paths consisting of a homoclinic trajectory or a collection of heteroclinic trajectories *do not correspond* to t-periodic solutions of (23) because it takes infinitely much t to go from one steady state point to another along these paths.

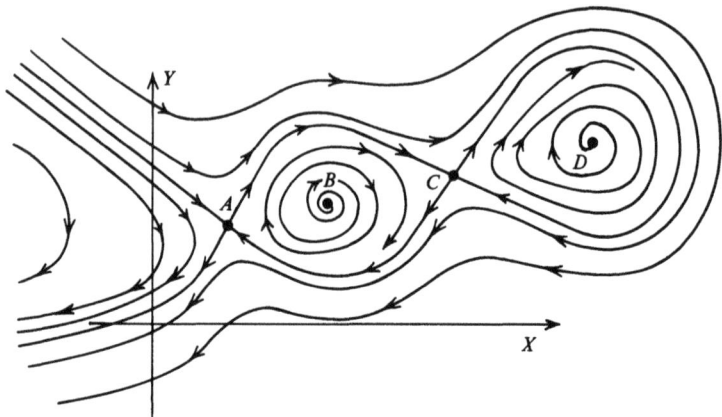

Figure A.3.11. A topologically possible phase portrait with four steady state points *A, B, C, D*. The closed loop to the right of the saddle point, *C*, formed by the coincidence of the unstable and stable manifolds at the saddle point *C* is a *homoclinic trajectory*. The upper branch of the unstable manifold out of *A*, joined to the upper branch of the stable manifold into *C*, constitutes a *heteroclinic trajectory*. It takes infinitely much time to traverse either of these special trajectories.

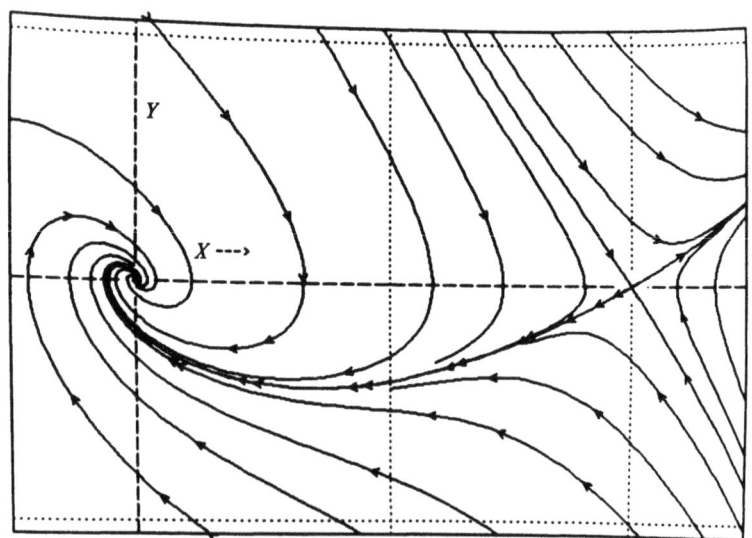

Figure A.3.12. The finished phase portrait for (24), with $c = 1$. This is the portrait begun in Figures A.3.4 and A.3.5, completed by linearizing (24) near its two steady state points to find, using the diagram in Figure A.3.10, that $(0, 0)$ is a stable spiral and $(1, 0)$ is a saddle point. The lower branch of the unstable manifold out of $(1, 0)$ that winds down into $(0, 0)$ is the unique bounded trajectory for this ODE system.

The constraints imposed by the above proposition usually leave only one possible way to fit the locally known pieces of a phase portrait together globally in a smooth way. This fact will become clear to the reader after he has worked through a number of different phase portrait constructions himself. If the constraints mentioned leave ambiguities, numerical solutions for one or two IVP's can usually clarify which of several possibilities occurs.

Figure A.3.12 is Figure A.3.5 completed: a phase portrait of the example system (24) for the case when $c = 1$, so that $c^2 - 4 < 0$ and $(0, 0)$ is a stable spiral.

Figure A.3.13 shows the phase portrait of (24) for the other qualitatively different behavior (24) can exhibit, for $c = 2.5$, so that $c^2 - 4 > 0$ and $(0, 0)$ is a stable node. Figure A.3.14 is a view of the stable node steady state point at $(0, 0)$, magnified two times relative to Figure A.3.13. Further magnification would further reduce the apparent nonlinear warping of the eigenfunctions at $(0, 0)$. In the next section, we will 'prove' that the phase portrait in Figure A.3.13 shows what the flow really does. That is, we will show how to *construct*, by hand-sketching, a qualitatively accurate phase portrait, like Figure A.3.13, without any assistance from the computer graphics facility that produced the figures.

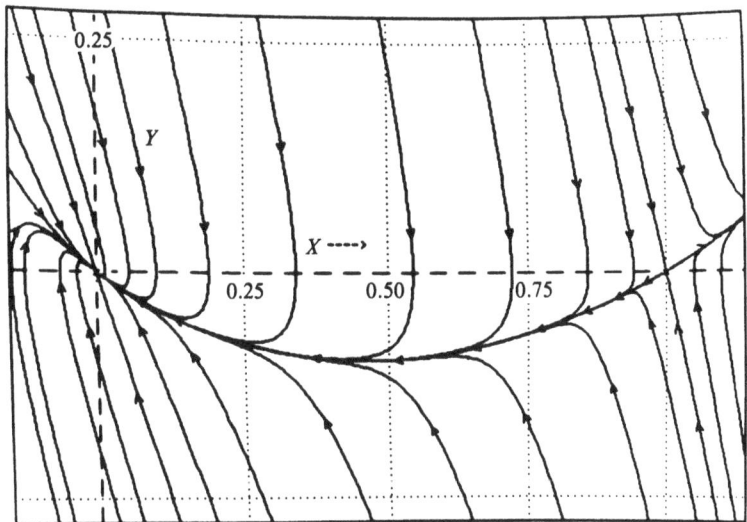

Figure A.3.13. The finished phase portrait for the ODE (24), when $c = 2.5$. this is generic for (24) when $c \geq 2.0$. Here, the unique bounded trajectory of the system is the heteroclinic trajectory consisting of the lower branch of the unstable manifold out of the saddle point, $(1, 0)$, that approaches the stable node at $(0, 0)$ without winding round it, i.e. this bounded trajectory lies entirely in the $X \geq 0$ half plane. $X(t)$ and $Y(t)$, traced along this bounded trajectory appear in Figure A.3.15, and a magnification of the region surrounding $(0, 0)$ appears in Figure A.3.14.

Interpretations of a phase portrait

We are now in a position to discuss the kind of information that can be deduced from a phase portrait, using Figure A.3.12 as an example. In Chapter 6 it is explained how the ODE system (24) results from seeking an exact traveling wave solution, with dimensionless propagation speed c, to the nonlinear parabolic partial differential equation (in dimensionless form)

$$\partial u/\partial \bar{t} = \partial^2 u/\partial x^2 + u(1-u). \tag{30}$$

This represents a population of organisms that breeds logistically and moves randomly in a one-dimensional universe. Equation (24) is obtained by seeking $u(x, \bar{t})$ in the traveling waveform $u(x, \bar{t}) = X(t)$ where t represents not time but wave variable $t = x - c\bar{t}$. In a solution to (24), $X(t)$ represents a possible population wave shape that propagates along the x-axis at speed c.

This discussion of the origin of (24) has nothing to do with the construction of the phase portrait for (24). Rather, it explains why we are interested only in certain kinds of solutions of (24). Usually one is interested only in solutions with some special characteristic. Here, we are interested only in solutions for which $X(t)$ is always non-negative and X is bounded (finite) for all t ($t = x - c\bar{t}$ must cover the domain $-\infty < t < \infty$). Since X represents a population density, it cannot be negative, and infinite population densities are untenable. Thus, only solution trajectories of (24), lying wholly in the $X \geq 0$ half plane, that never run off to $x = +\infty$ are of interest. The only possibilities, therefore, are trajectories described in cases (ii), (iii), and (iv) in the above proposition.

This proposition (that can be proved rigorously) is offered as intuitively evident. *If the functions F and G in (23) are smooth, then any closed loop trajectory (limit cycle) (as described in case (iii) above) must surround at least one steady state point.* A good metaphor for this concept is a head of hair with a general swirl to the direction field assumed by the hairs tangent to the scalp. What we assert is the guaranteed existence of a 'cowlick', namely a tuft of hair somewhere in the 'middle' of the swirl that sticks straight out from the scalp, refusing to be combed flat, because it cannot both lie tangent to the scalp and, because of its swirl, point in all directions at once. The option of the flow field $(F(X, Y), G(X, Y))$ (the hair) sticking straight out from the X–Y plane (the scalp) is not open, so the only possibility is a steady state point (a bald spot) somewhere in the interior of a swirl. A rigorous proof of this claim is not trivial and, though amusing, would delay us too long.

This last proposition immediately rules out closed trajectories in the positive X half plane in either of the phase portraits (Figures A.3.12 and A.3.13) for (24), because such a loop would have to surround $(1, 0)$, the only steady state point in $X > 0$. By inspecting the flows in Figures A.3.12 and A.3.13 (or in the preliminary version of A.3.12 given in Figure A.3.5), this is topologically impossible. Cases (i), (iii), and (iv) are ruled out.

Thus, a positive, bounded X trajectory must, at one or both ends, hit the steady state saddle point at $(1, 0)$. Only four trajectories do this: the two branches of the stable and unstable manifolds at $(1, 0)$. The branches of the stable and unstable manifolds extending to the right of $(1, 0)$ are ruled out, as they run to $X \to +\infty$. The stable manifold approaching $(1, 0)$ from the left is ruled out for it must come from the $X < 0$, $Y > 0$ quadrant. Only the branch of the unstable manifold leaving $(1, 0)$ toward $(0, 0)$ remains a candidate.

In the phase portrait in Figure A.3.12 representing cases where $c < 2$, so that $(0, 0)$ is a stable spiral, it is obvious that this unstable manifold, out of $(1, 0)$ down and to the left, must cross the Y-axis into the $X < 0$, $Y < 0$ quadrant. Thus, no positive X solution bounded for all t is possible when $c < 2$.

For the other cases, $c > 2$, so that $(0, 0)$ is a stable node; it *is* possible that the unstable manifold out of $(1, 0)$, down and to the left, runs into the stable node at $(0, 0)$ without crossing the X axis. With c fixed at a value greater than 2, this could happen (and definitely does happen in Figure A.3.13) and would represent the *unique* X-positive bounded traveling wave solution to (30) propagating at speed c. It would be unique because there is exactly one branch of the unstable manifold out of $(1, 0)$ to form the $t \to -\infty$ section of this solution.

Figure A.3.13, produced numerically on a computer graphics facility, shows what does happen to the unstable manifold from $(1, 0)$ when $c > 2$, namely it does run into $(0, 0)$ without X going negative. Thus we have a (single) heteroclinic trajectory in the positive X half plane. In Figure A.3.13, $c = 2.5$.

We now *prove* that the unstable manifold out of $(1, 0)$ to the left does not cross the Y-axis. This will finish the proof of the existence and uniqueness of an exact traveling wave solution to the PDE (30), at propagation speed c when $c > 2$. We know already that none exists when $c < 2$.

The detailed considerations we now make, depending on the particular nature of the example ODE system (24), illustrate the kind of final analysis that most phase portraits require for their completion. At this level, each ODE system must be treated individually and in detail. The phase portrait sketching done up to this point suggests the crucial questions needing answers, and suggests a geometrical way to answer them, but usually a formal proof, including some calculations, is needed at the end.

We now prove one feature of one particular trajectory to illustrate the closing game. We will leave other features unproved. For example, is there a value of $c < 2$ for which the unstable manifold out of $(1, 0)$ to the left, swings around $(0, 0)$ to come back to $(1, 0)$ as the stable manifold from the left? This is certainly topologically possible, and does happen for some ODE systems. As an exercise, the reader is invited to prove geometrically that, for c sufficiently greater than zero to make $(1 - c)(1 - 2c) < c^4$, this cannot happen in our example flow.

Our proof begins with the linearizations of (24) at $(0, 0)$ and at $(1, 0)$. At $(0, 0)$, $\beta = -c$ and $\gamma = 1$, so the eigenvalues are $\lambda_1 = [-c + (c^2 - 4)^{\frac{1}{2}}]/2$ and $\lambda_2 = [-c - (c^2 - 4)^{\frac{1}{2}}]/2$. The eigenvector associated with the eigenvalue λ has components (e_1, e_2) determined by

$$\begin{pmatrix} 0 - \lambda & 1 \\ -1 & -c - \lambda \end{pmatrix} \begin{pmatrix} e_1 \\ e_2 \end{pmatrix} = \begin{pmatrix} 0 \\ 0 \end{pmatrix}, \quad \text{i.e.} \begin{cases} -\lambda e_1 + e_2 = 0, \\ -e_1 - (c + \lambda)e_2 = 0. \end{cases}$$

Thus, $e_2 = \lambda e_1$, meaning that the slope of this eigenvector at $(0, 0)$ is λ, for this particular example. The eigenvector associated with the least negative eigenvalue, λ_1, (marking the 'slow eigenfunction', so called because its decay to 0 is slowest) is tangent to all trajectories approaching $(0, 0)$ except for two of them that lie on the eigenvector associated with the most negative (fastest decaying) eigenvalue, λ_2. (This was explained under Nodes, p. 670 above.)

The Y-nullcline, given by $Y = -X(1 - X)/c$, has a slope of $-1/c$ at $(0, 0)$. We claim this slope is less negative than the slope, λ_1, of the least negative eigenvalue (when $c > 2$). That is, we claim,

$$\lambda_1 = \tfrac{1}{2}[-c + (c^2 - 4)^{\frac{1}{2}}] < -1/c.$$

This is equivalent to

$$(c^2 - 4)^{\frac{1}{2}} < c - (2/c).$$

Since, with $c > 2$, the right-hand side is positive, the latter inequality holds if and only if,

$$c^2 - 4 < c^2 - 4 + (4/c^2).$$

Which, of course, is true: $0 < 4/c^2$.

Similarly, we can show that the slope of the unstable manifold at $(1, 0)$ is $\tfrac{1}{2}[-c + (c^2 + 4)^{\frac{1}{2}}]$, and that this has a lesser slope than has the Y nullcline at $(1, 0)$ (whose slope is $1/c$).

These considerations locate the eigenvectors and Y-nullcline at $(0, 0)$ as shown in Figure A.3.14. Similarly, we deduce that the unstable manifold out of $(1, 0)$ to the left is trapped, near $(1, 0)$, in the crescent formed by the $Y = 0$ X-nullcline and the $Y = -X(1 - X)/c$ Y-nullcline. It escapes this crescent by crossing the Y-nullcline, and, as it does, it is

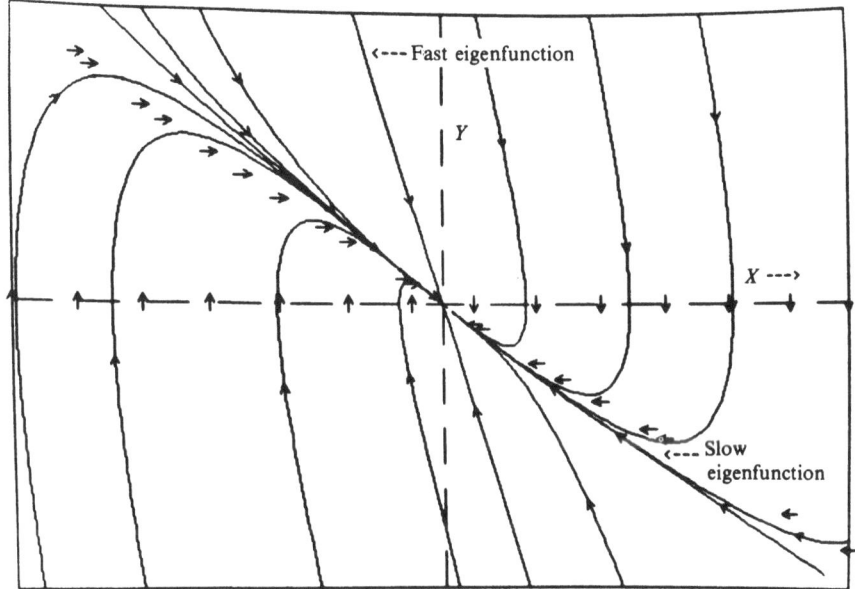

Figure A.3.14. A magnification (×2 relative to Figure A.3.13) of the stable node at $(0, 0)$ for the ODE (24) when $c = 2.5$. See text for the proof that the fast and slow eigenfunction trajectories lie as pictured.

trapped to the right of both eigenvectors into $(0, 0)$ from below. Since it cannot cross either of these eigenfunction trajectories (or any other trajectory), it must run into $(0, 0)$ tangent to the eigenvector associated with λ_1 at $(0, 0)$, never crossing the Y-axis. This ends the proof. It was necessary in order to preclude the possibility of both eigenvectors at $(0, 0)$ having slopes less negative than the Y-nullcline at $(0, 0)$, permitting the unstable manifold out of $(1, 0)$ to swing to the left of the node eigenvectors at $(0, 0)$ across the Y-axis, thence to approach $(0, 0)$ from the $X < 0$, $Y > 0$ quadrant.

Let us take stock of what we have accomplished as regards the example problem (24). The first thing to note is that, once one fully understands the phase portrait construction procedure, what we did with the ODE (24) involves no hard work and can be done in perhaps fifteen minutes. With this trivial labor we know all interesting aspects of the behavior of (24) (and hence of *exact* traveling wave solutions to 30).

We know that for every $c > 2$, there is precisely one realistic traveling wave solution, and, for $c < 2$, none. We have not *solved* (24); we have not produced an analytical representation of solutions. For most ODE systems this is impossible anyway. However, we know exactly where to look for the traveling wave solution.

We do not know, from the phase portrait, the speed at which a point moves along any particular trajectory. (Phase portraits never tell us this.) That is, we do not know the details of how rapidly $X(t)$ and $Y(t)$ vary with t. To find this information we can compute $X(t)$, $Y(t)$ numerically; the phase portrait tells us exactly how to do this. Namely, assign initial conditions that locate X and Y very close to the $(1, 0)$ steady state point *on the unstable manifold*, and, using some general numerical ODE solver, integrate forward in time to a sufficiently large value of time such that $X(t)$, $Y(t)$ approach $(0, 0)$ closely. These initial conditions would be $X(0) = 1 - \varepsilon$, $Y(0) = \frac{1}{2}[-c + (c^2 + 4)^{\frac{1}{2}}]\varepsilon$ for $\varepsilon = 10^{-6}$ say.

Without this knowledge of where to start, and in what t-direction to integrate, a numerical search for the single trajectory of (24) corresponding to an exact positive-bounded X traveling wave solution of (30) would have no realistic chance of success.

Alternatively, to find formulae for $X(t)$ and $Y(t)$, we could try to patch together the local representations of the appropriate eigenfunctions of (24) linearized at $(0, 0)$ and at $(1, 0)$.

Figure A.3.15 shows a plot, generated numerically as described above, of X and Y versus t. Remember t originated as a wave variable, $t = x - c\tilde{t}$, so Figure A.3.15 represents the shape of a wave propagating to the right at dimensionless speed $c = 2.5$. In Figure A.3.15, t varies from 0 to 25.

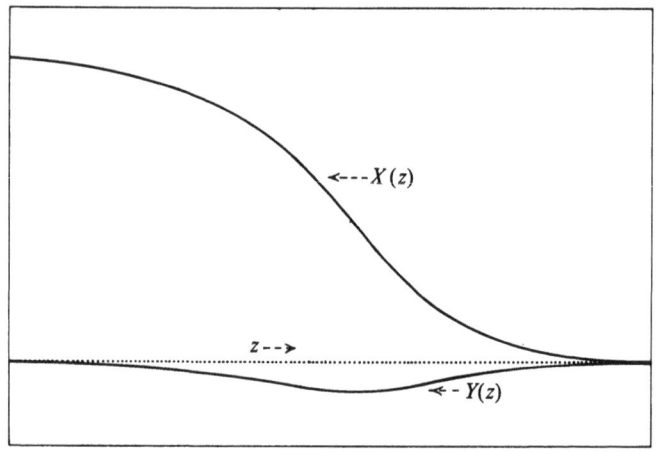

Figure A.3.15. Graphs of $X(t)$ and $Y(t)$ along the unique heteroclinic trajectory of the ODE (24) when $c = 2.5$ (see Figure A.3.13). Since the ODE (24) was generated by seeking an exact traveling wave solution, propagating at speed $c = 2.5$, to the PDE (30), $X(t)$ and $Y(t)$ graphed here represent the unique bounded propagating wave solution, of speed $c = 2.5$, to (30). In this context t is the wave variable (see text), ranging from 0 to 25.

We now abandon our example system (24). The power of phase plane analysis will become apparent only after several examples have been worked through. Section 6.7 gives several. The remainder of this appendix gives two more, as we turn now to bifurcation theory to find out how a prediction by linear theory that a steady state point is a center can be a harbinger of the birth of true periodic solution of a nonlinear ODE system.

Bifurcation theory

We will now explain one way in which the tuning of a parameter can bring into existence a closed loop trajectory in phase space, corresponding to a t-periodic solution, of an ODE system. To be more precise, we display explicitly the dependence of the ODE flow upon a parameter, μ, and rewrite the general n-th order autonomous ODE system (10) in the following (vector) form

$$d\mathbf{y}/dt = \mathbf{f}(\mathbf{y}; \mu). \tag{31}$$

Recall that the bold characters stand for n-vectors, so $\mathbf{y}(t) = (y_1(t), \ldots, y_n(t))$, and $\mathbf{f}$ has n components, the i-th one being $f_i(y_1, \ldots, y_n; \mu)$.

In our previous treatment of the ODE system (14) or (24), c was a parameter. We considered all positive values of c, and found that for $c < 2$, the phase portrait for (24) was qualitatively different than the $c > 2$ case. In the following sections we will investigate a different kind of qualitative change in a phase portrait that can be brought about by a parameter alteration. Roughly speaking, we want to characterize the ways that, for μ on one side of a critical value μ_0, (31) has no closed loop trajectories in a certain neighborhood singled out for inspection, but, for μ at, or on the other side of μ_0, a closed loop trajectory does exist in that neighborhood.

When this happens, we say that a **bifurcation occurs as** μ crosses μ_0, and we call μ_0 a **bifurcation value** of the **bifurcation parameter** μ.

Usually an ODE system has a number of parameters involved in its coefficients. Potentially, any one of these, or any combination, might serve as μ in (31). Not infrequently, several different kinds of bifurcations, giving rise to **periodic orbits** (another name for closed loop trajectories), can occur in the same ODE system as its various parameters are tuned. To catalog the full variety of possible bifurcations is complicated. (See, for example, Takens, 1973.) We adopt, here, the understanding that all but one of the parameters have fixed (but arbitrary, hence 'adjustable') values, while a single parameter, μ, is selected for tuning. Having said that, it is important to say, also, that in real applications it will never be obvious initially just which parameter, or combination thereof, to select as μ. Various possibilities must be tried, as will be explained below.

Bifurcation to periodic solutions in a linear system

To begin, we give the simplest possible example of the bifurcation of a periodic orbit. Consider a special, $n = 2$, linear case of (31)

$$dy_1/dt = f_1(y_1, y_2; \mu) = \mu y_1 - y_2,$$

$$dy^2/dt = f_2(y_1, y_2; \mu) = y_1 + \mu y_2. \tag{32}$$

In terms of previous notation $\beta = a_{11} + a_{22} = 2\mu$ and $\gamma = a_{11}a_{22} - a_{21}a_{12} = \mu^2 + 1$ (thus $\beta^2 - 4\gamma = -4 < 0$). Referring to Figure A.3.10, we see that, since $\gamma > 0$ for all μ, the steady state point $(0, 0)$ of (32) is a stable spiral when $\mu < 0$, a center (with an infinitude of closed concentric elliptical orbits surrounding $(0, 0)$) when $\mu = 0$, and an unstable spiral when $\mu > 0$. This fits our above description of a bifurcation, $\mu_0 = 0$ is the bifurcation value of μ.

Can a linear system of the form (25) exhibit any other kind of bifurcation in which a closed trajectory appears? Figure A.3.10 supplies the answer. Regardless of how μ is embedded in the (a_{ij}) coefficients in (25) (assuming the a_{ij}, and hence β and γ, depend smoothly on μ), varying μ can do nothing more than smoothly translate the point (β, γ) in the β–γ plane. Only for $\gamma > 0$ and $\beta = 0$ does the phase portrait of (25) exhibit closed trajectories. The bifurcation in the example above occurred as a change in μ caused the point (β, γ) to cross the γ-axis where $\gamma > 0$. The only other kind of bifurcation possible is one in which the point (β, γ) moves onto the positive γ-axis as it crosses $(\beta, \gamma) = (0, 0)$. This kind of bifurcation is pathological, because, just at bifurcation with $\beta = 0$ and $\gamma = 0$, *every* point on some straight line is a steady state point, and, in the discussion below, this kind of bifurcation will be excluded. With this qualification, we have seen as our first example of a bifurcation, the only ('legitimate') bifurcation to a periodic orbit that a linear second-order ODE system can exhibit. We turn next to nonlinear systems.

A nonlinear bifurcation example to fix ideas

Before describing a general bifurcation theorem, we set up a simple nonlinear example by which we can illustrate general abstract ideas as they are presented. Consider two interacting populations, one of exploiters, one of victims. $E^*(t^*)$ and $V^*(t^*)$ represent, in dimensioned form, the population densities at time t^* of breeding female exploiters and victims, respectively. In dimensioned form, our differential equations are these (asterisks denote dimensioned quantities):

$$dV^*/dt^* = b^* V^{*2}(V_0^* - V^*) - a^* E^* V^*,$$

and

$$dE^*/dt^* = c^*(V^* - V_c^*)E^*. \tag{33}$$

We motivate (33) briefly. The second equation declares the birth rate of female exploiters (that survive to breed in their turn) to increase linearly as V^* increases, and to be negative if the victim population V^* is less than a critical victim population V_c^*. That is, the victim population must exceed a threshold if the exploiters are to prosper. When there are no exploiters, $E^* = 0$, the first equation of (33) declares the birth rate per female (of females that survive to breed) of victims to be proportional to $V^*(V_0^* - V)$. We have already exhibited how this equation behaves (when $E^* = 0$) in Figure A.3.1, because, if we define $t^* = (b^* V_0^{*2})^{-2} t$ and $V^*(t^*) = V_0^* u(t)$ (see below), then the equation above for V^* is identical to (1).

The second term in the above equation for V^* represents the rate at which victims are consumed by exploiters. It models a situation in which encounters between exploiters and victims occur at random, and a fraction, a^*, of these encounters end in 'exploitation'.

Equation (33) contains five parameters, b^*, V_0^*, a^*, V_c^*, c^*. We scale the variables in (33) to make everything dimensionless, and, in the process, reduce this too-large parameter count. We seek dimensioned characteristic scales $\bar{T}$, $\bar{V}$, and $\bar{E}$, for t^*, V^*, and E^* respectively, with $\bar{T}$, $\bar{V}$, $\bar{E}$ chosen to yield maximum simplification. That is, let

$$t^* = \bar{T}t,$$

$$V^*(t^*) = \bar{V}X(t),$$

$$E^*(t^*) = \bar{E}Y(t), \tag{34}$$

where t, X, and Y are dimensionless. Substituting (34) into (33) we obtain,

$$dX/dt = (\bar{T}b^* \bar{V}^2)X^2[(V_0^*/\bar{V}) - X] - (\bar{T}a^*\bar{E})XY,$$

and

$$dY/dt = (\bar{T}c^* \bar{V})[X - (V_c^*/\bar{V})]Y. \tag{35}$$

The parameter combinations in parentheses are dimensionless. $\bar{T}$, $\bar{E}$, and $\bar{V}$ are at our disposal to simplify (35). There is no unique best way to select them. Here, we choose

$$\bar{V} = V_0^*, \qquad \bar{T} = (b^*\bar{V}^2)^{-1} = (b^*V_0^{*2})^{-1},$$
$$\bar{E} = (a^*\bar{T})^{-1} = (a^*b^*V_0^{*2})^{-1}. \tag{36}$$

Then, (35) becomes

$$dX/dt = X[X(1-X) - Y] \equiv F(X, Y, \mu),$$

and

$$dY/dt = k(X - 1/\mu)Y \equiv G(X, Y, \mu), \tag{37}$$

in which the parameter count has been reduced from five to two. We see that only the dimensionless combinations of original parameters,

$$k = c^*/b^*V_0^*$$

and

$$\mu = V_0^*/V_c^*, \tag{38}$$

are important in (33). We assume $k > 0$, $\mu > 0$, and we are interested only in solutions to (37) that keep X and Y positive.

The nullclines. The ODE system (37) is the kind of system on which we can use phase plane analysis. We begin construction of a phase portrait for (37). As step 1, we determine the nullclines. The Y-nullcline locus, that makes $G(X, Y, \mu) = 0$, consists of the vertical line

$$X = 1/\mu$$

and the X-axis, $Y = 0$. The X nullcline locus consists of the Y-axis, $X = 0$, and the parabola $Y = X(1-X)$. The X and Y nullcline loci intersect at three points, the steady state points

$$(X_0, Y_0) = (0, 0), (1, 0), \quad \text{and} \quad [\mu^{-1}, \mu^{-1}(1 - \mu^{-1})].$$

As the second step, we linearize F and G near the three steady state points. We leave the computation of the partial derivatives $a_{11} = \partial F/\partial X$, $a_{12} = \partial F/\partial Y$, $a_{21} = \partial G/\partial X$, $a_{22} = \partial G/\partial Y$, all evaluated at (X_0, Y_0) as an exercise. The results are as follows.

$$\text{At } (X_0, Y_0) = (0, 0), \qquad \begin{pmatrix} a_{11} & a_{12} \\ a_{21} & a_{22} \end{pmatrix} = \begin{pmatrix} 0 & 0 \\ 0 & -k/\mu \end{pmatrix}.$$

Thus $\beta = a_{11} + a_{22} = -k/\mu$ and $\gamma = 0$. From Figure A.3.10, we see that (β, γ) lies on the line delineating stable nodes and saddle points. We will not explore further the detailed geometry of this steady state point because these details do not influence the general outcome in this case, and because sketching flow arrows near $(0, 0)$ will tell us all we need to know.

$$\text{At } (X_0, Y_0) = (1, 0), \quad \begin{pmatrix} a_{11} & a_{12} \\ a_{21} & a_{22} \end{pmatrix} = \begin{pmatrix} -1 & -1 \\ 0 & k(1-\mu^{-1}) \end{pmatrix}.$$

Thus $\beta = -1 + k(1 - \mu^{-1})$ and $\gamma = -k(1 - \mu^{-1})$. At this stage we must consider various values of μ. We claim that, if $\mu < 1$, then for any positive initial values, $X(0)$, $Y(0)$, $\lim X(t) = 1$, as $t \to \infty$, and $\lim Y(t) = 0$, as $t \to \infty$. That is, $\mu < 1$ means the Y population goes toward extinction. The proof of this is left as an exercise. It can be done by analyzing (37) directly, or by sketching a phase portrait for the $\mu < 1$ case. The latter method is suggested for phase portrait practice. We do not discuss the $\mu < 1$ case further.

If $\mu > 1$, then $\gamma < 0$, and Figure A.3.10 tells us that $(1, 0)$ is a saddle point.

When $\mu > 1$, the third steady state point $(\hat{X}_0, \hat{Y}_0) = [\mu^{-1}, \mu^{-1}(1 - \mu^{-1})]$ lies in the $X > 0$, $Y > 0$ quadrant. At this point, we have

$$\begin{pmatrix} a_{11} & a_{12} \\ a_{21} & a_{22} \end{pmatrix} = \begin{pmatrix} \mu^{-1}(1 - 2\mu^{-1}) & -\mu^{-1} \\ k\mu^{-1}(1 - \mu^{-1}) & 0 \end{pmatrix}. \tag{39}$$

Thus, $\beta = \mu^{-1}(1 - 2\mu^{-1})$ and $\gamma = k\mu^{-2}(1 - \mu^{-1}) > 0$.

We see that if $\mu = 2$, then $\beta = 0$ while $\gamma = (k/8) > 0$, so that, from Figure A.3.10, $(\hat{X}_0, \hat{Y}_0) = [\mu^{-1}, \mu^{-1}(1 - \mu^{-1})] = (\frac{1}{2}, \frac{1}{4})$ is a center according to linear theory. It was mentioned above, and will be made precise below, that this occurrence is the harbinger of a periodic (closed trajectory) orbit.

As long as $k > 0$, and $\mu > 1$, γ is seen to be positive. Thus, from Figure A.3.10, we see that our steady state point is either stable or unstable according to whether β is negative or positive. That is, when $\mu < 2$, making β negative, $(\hat{X}_0, \hat{Y}_0)$ is either a stable spiral or a stable node. On the other hand when $\mu > 2$, making β positive, $(\hat{X}_0, \hat{Y}_0)$ is an unstable spiral or an unstable node.

In both cases, $\mu > 2$, $\mu < 2$, we have a node if $\beta^2 - 4\gamma > 0$, and a spiral if $\beta^2 - 4\gamma < 0$. Using the above values of β and γ, we see that $k > (1 - 2\mu^{-1})^2 / 4(1 - \mu^{-1})$ means $(\hat{X}_0, \hat{Y}_0)$ is a spiral; otherwise (X_0, Y_0) is a node. Table A.3.1 summarizes.

If we think of $k > 0$ as fixed, then for $\mu < 2$, but μ 'close' to 2, $(\hat{X}_0, \hat{Y}_0)$ must be a stable spiral. For $\mu > 2$, but 'close' to 2, $(\hat{X}_0, \hat{Y}_0)$ is an unstable spiral. Only when μ is adjusted 'far' from 2 can $(\hat{X}_0, \hat{Y}_0)$ be a node; how 'far' depends on the value of k. If $k > \frac{1}{4}$ then, for all $\mu > 2$, $(\hat{X}_0, \hat{Y}_0)$ is an unstable spiral.

Table A.3.1. *Typing of steady state point* $(\hat{X}_0, \hat{Y}_0) =$ $(\mu^{-1}, \mu^{-1}(1 - \mu^{-1}))$ *according to linear theory*

	$k > (1 - 2\mu^{-1})^2/4(1 - \mu^{-1})$	$k < (1 - 2\mu^{-1})^2/4(1 - \mu^{-1})$
$\mu < 2$	Stable spiral	Stable node
$\mu = 2$	Center	Impossible since $k > 0$
$\mu > 2$	Unstable spiral	Unstable node

We have a number of cases to sort out. The Hopf bifurcation theorem will do the sorting for us. We will describe this valuable theorem below. To motivate the student to master this (abstract) theorem, we will try to construct a reliable phase portrait for the ODE system (37) *without* using the Hopf theorem. This effort will fail, but will pinpoint, geometrically, exactly the issue the Hopf theorem resolves.

A phase portrait attempt without using the Hopf theorem.   Let us select values $k = 1$, $\mu = 1.8$. We have found the following features of the phase portrait of (37).

(i) $(1, 0)$ is a saddle point.

(ii) $(\hat{X}_0, \hat{Y}_0) = (\mu^{-1}, \mu^{-1}(1 - \mu^{-1})) = (0.556, 0.247)$ is a stable spiral.

(iii) The X-nullcline (where trajectories are vertical) comprises the Y-axis and the parabola $Y = X(1 - X)$.

(iv) The Y-nullcline (where trajectories are horizontal) comprises the X-axis and the vertical line $X = \mu^{-1} = 0.556$.

We choose strategic locations for other flow arrows. When $X = 1$, $F(X, Y) = -XY$, so the slope of the flow arrows on $X = 1$ is

$$dY/dX|_{X=1} = G/F|_{X=1} = -k[1 - (1/\mu)] = -0.44.$$

In Figure A.3.16, this information has been used to sketch in nullclines, the flow arrows on $X = 1$, an inward bound spiral at $(\hat{X}_0, \hat{Y}_0)$, and the unstable manifold out of the saddle point at $(1, 0)$. The stable manifold of $(1, 0)$ is the X-axis.

Aside.   Figure A.3.16 includes some details of the flow near the steady state point at $(0, 0)$, the one on which our linearization gave an ambiguous decision. We see that the flow near $(0, 0)$, for $X < 0$, is what we would expect at a stable node. The flow near $(0, 0)$, for $X > 0$, is what we would expect at a saddle point. Since (β, γ) for $(0, 0)$ lay on the boundary, in Figure A.3.10, separating saddle points and stable nodes, the appearance of the flow near $(0, 0)$ in Figure A.3.16 makes sense. The parts of the phase portrait where $X < 0$ or $Y < 0$ are not relevant in this population dynamics context, but are shown for general interest.

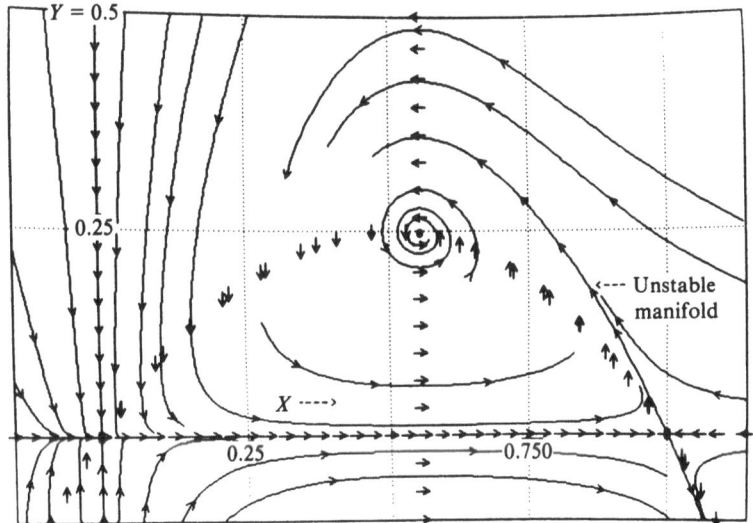

Figure A.3.16. Graphical depiction of the information in items (i) to (iv) above. This is the beginning of a phase portrait for the ODE (37) when $k = 1$ and $\mu = 1.8$ (generic for the cases when $1 < \mu < 2$ and $k > (1 - 2\mu^{-1})^2/4(1 - \mu^{-1})$). The parts of the flow shown in this figure leave in doubt the fate of the upper branch of the unstable manifold out of the saddle point at $(1, 0)$.

Does the unstable manifold rising out of $(1, 0)$ spiral down into the steady state (stable spiral) point $(\hat{X}_0, \hat{Y}_0)$? Or, possibly, is there a closed trajectory surrounding $(\hat{X}_0, \hat{Y}_0)$ that this unstable manifold approaches asymptotically as $t \to +\infty$? Figure A.3.17 shows a topologically possible phase portrait illustrating this latter possibility. In it, the inward spiraling trajectory that approaches $(\hat{X}_0, \hat{Y}_0)$ as $t \to +\infty$ approaches (but never actually reaches) the closed trajectory surrounding $(\hat{X}_0, \hat{Y}_0)$ as $t \to -\infty$.

Figure A.3.18 shows another (topological) possibility in which the flow has no periodic orbits; the unstable manifold rising out of $(1, 0)$ spirals in toward $(\hat{X}_0, \hat{Y}_0)$, approaching it (but never actually arriving) as $t \to +\infty$.

Sketching will not decide between these two possibilities. We are trying to decide, in this example, whether a closed periodic orbit (as appears in Figure A.3.17) does or does not exist. We turn now to the Hopf theorem for help in making this decision.

The Hopf bifurcation theorem

The Hopf theorem makes precise a concept we will first describe intuitively, with attention restricted to the $n = 2$ case. Consider a second-order

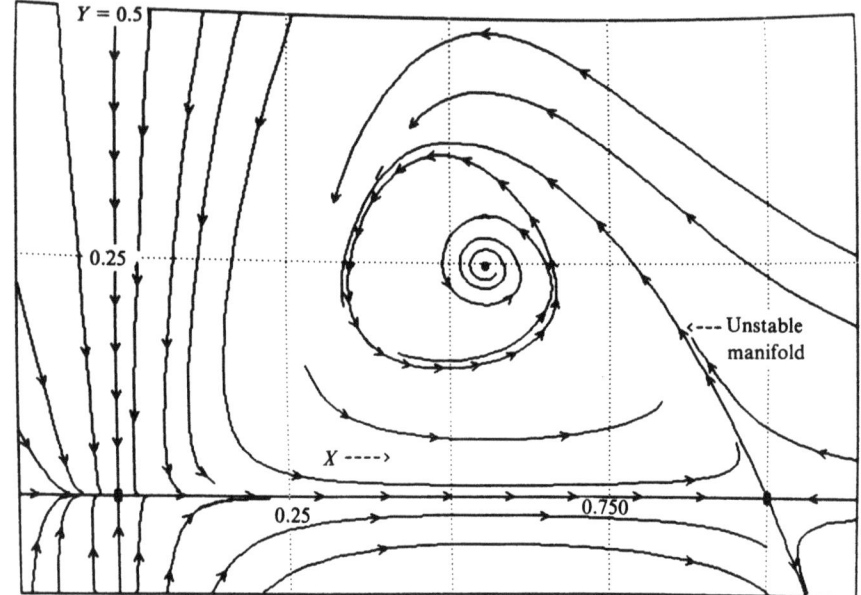

Figure A.3.17. A topologically possible completion of the phase portrait begun in Figure A.3.16. Here, there is a single closed orbit surrounding the stable spiral, and the unstable manifold out of (1, 0) spirals down toward this orbit.

autonomous ODE system of the sort (31) whose smooth flow $\mathbf{f}(\mathbf{y}; \mu)$ depends smoothly on the parameter μ.

Let $\mathbf{E}(\mu) = [E_1(\mu), E_2(\mu)]$ be the coordinates of an isolated steady state point (for different values of the parameter, μ, this steady state point will have different locations). Let μ vary over some interval $I = [a, b]$. Suppose that, for $\mu = b$, the steady state point, $\mathbf{E}(b)$, is an unstable spiral according to linear theory and that, for $\mu = a$, $\mathbf{E}(a)$ is a stable spiral according to linear theory. That is, as μ is continuously tuned from a to b, the steady state point, $E(\mu)$, changes from a stable to an unstable spiral. The Hopf theorem asserts that, for second-order ODE systems, when the flow $\mathbf{f}(\mathbf{y}, \mu)$ satisfies certain conditions, as μ is tuned from a to b the phase portrait (in the vicinity of the steady state point $\mathbf{E}(\mu)$) must change from a stable to an unstable spiral in *just one of three* different possible metamorphoses. The three possible paths of intermediate states are shown in Figure A.3.19.

In the upper path, as μ is increased from a to b, a closed orbit appears, with infinitesimal diameter, as some critical (bifurcation) value of μ is passed. Just as this occurs, the steady state point changes type from a stable to an unstable spiral. The closed orbit, surrounding the steady state

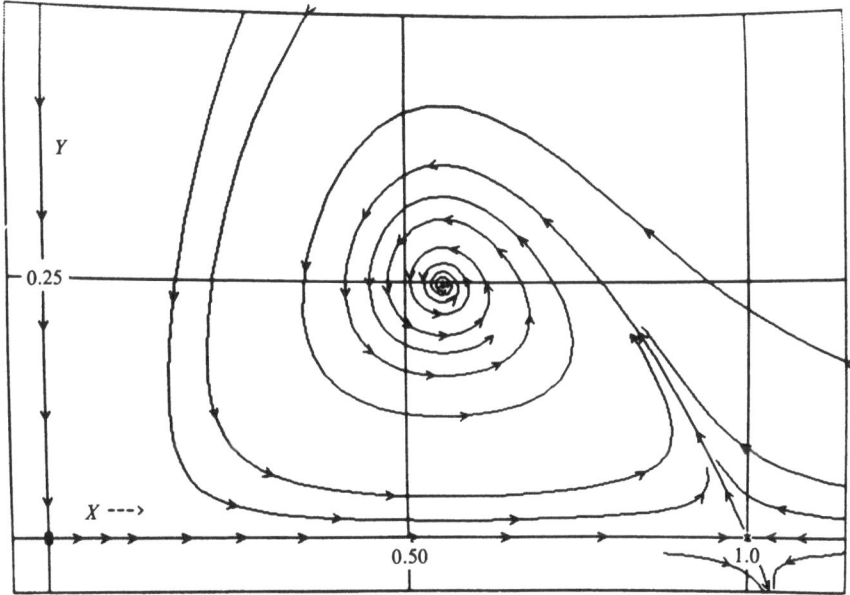

Figure A.3.18. Another topologically possible completion of the phase portrait begun in Figure A.3.16. Here, no closed orbits exist; the unstable manifold out of $(1, 0)$ spirals into the stable spiral steady state point. The Hopf theorem will resolve which, if either, of Figures A.3.17 and A.3.18 is correct for the ODE at hand: (37).

point $E(\mu)$, grows in diameter as μ increases. In this case, the trajectories spiraling away from the newly unstable steady state point asymptotically approach the closed orbit surrounding the steady state point as $t \to +\infty$. The trajectories on the outside, spiraling in toward what used to be (for $\mu = a$) a stable spiral steady state point, never get there; instead they approach the newly born closed orbit as $t \to +\infty$. A closed orbit that arises along this upper path corresponds to a t-periodic trajectory that is stable to small perturbations away from itself, because all trajectories in its neighborhood approach it as $t \to +\infty$. Such a closed orbit is called a **stable limit cycle**. Along this upper path in Figure A.3.19, at any given setting of μ, there is, at most, a single isolated closed orbit in the neighborhood of the steady state point.

Along the middle path of metamorphosis in Figure A.3.19, as μ is increased beyond a, the inward spiral to the steady state point becomes less tightly wound. At a single intermediate value, $\mu = \mu_0$, there appears an infinitude of concentric closed orbits surrounding the steady state point. As soon as μ exceeds this critical setting, all closed orbits disappear and $E(\mu)$ becomes an unstable spiral that keeps spiraling outward. As we saw above, *linear* ODE systems bifurcate to periodic trajectories in this

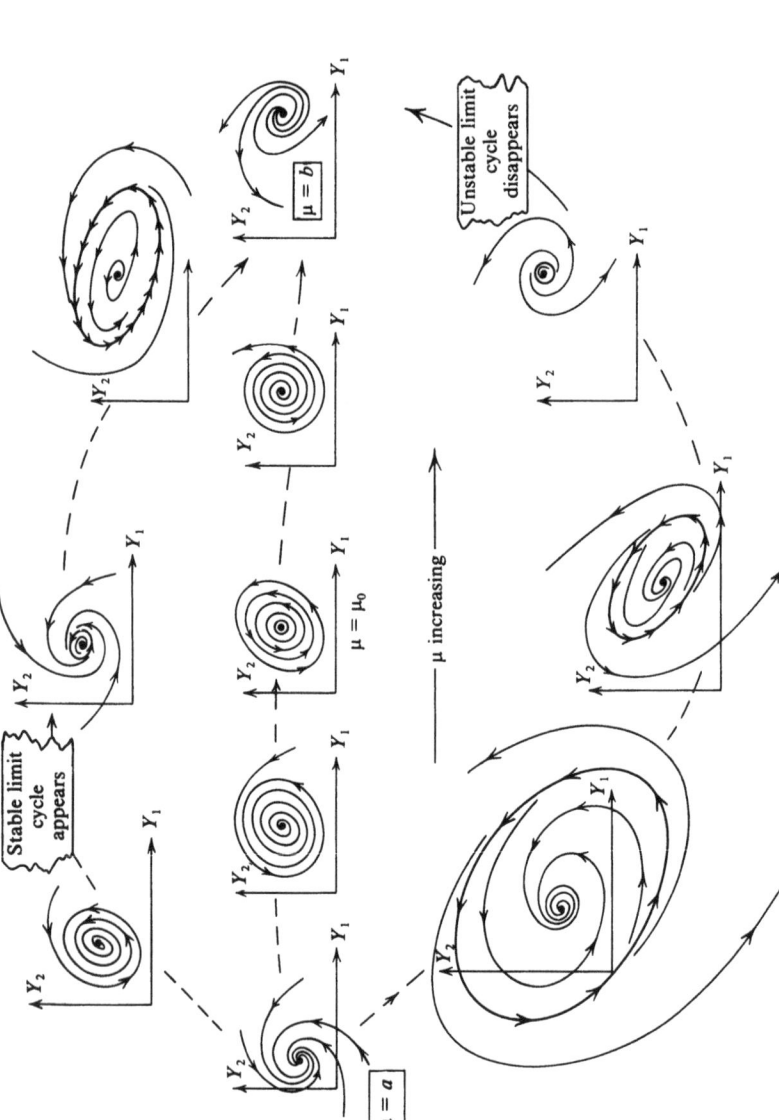

Figure A.3.19 The three (and only three) possible metamorphoses of a stable spiral into an unstable spiral as the parameter μ is increased, from a value a below to a value b above, the isolated value μ_0 at which the linearized ODE system has pure imaginary characteristic exponents, when the hypotheses of the Hopf bifurcation theorem are satisfied. Each dashed line strings together several snapshots of the phase portrait along one of the three metamorphosis paths. The way closed orbits appear or disappear is of interest.

way only, and as we explained, closed orbits generated this way are neither stable nor unstable, but neutrally stable.

In the lower path in Figure A.3.19, the diameter of a closed orbit surrounding the steady state point shrinks as μ is increased from a toward b, until, at some critical value, μ_0, $a < \mu_0 < b$, it shrinks to zero diameter causing disappearance of the closed orbit. On this path, before the closed orbit disappears, the trajectories inside it spiral in, approaching the steady state point as $t \to +\infty$, and spiral out to approach the closed orbit as $t \to -\infty$. On the outside of the closed orbit, trajectories spiral out away from it as $t \to +\infty$. For a given setting of μ, there is at most one closed orbit near the steady state point. A periodic solution, corresponding to a closed orbit along the bottom path in Figure A.3.19, is *unstable* to small perturbations. That is, even though the closed orbit corresponds to an exact t-periodic solution, the smallest perturbation away from that orbit will cause eventually large displacements away from it, because all trajectories near the closed orbit diverge from it as $t \to \infty$. These closed orbits are called **unstable limit cycles**.

The three paths by which a stable spiral can metamorphose into an unstable spiral, as depicted in Figure A.3.19, should seem intuitively reasonable ones and certainly topologically possible ones. If you study Figure A.3.19 and use some imagination you will think of other paths from the $\mu = a$ portrait to the $\mu = b$ portrait. For example, why could it not happen that, along the upper path, *two* distinct concentric closed orbits appear surrounding the steady state point and both expand in diameter as μ increases? For certain flows (that are Lipschitz continuous, but not analytical, see Chaffee, 1968) this does, in fact, occur. Another example is the topology seen in Figure A.3.17, a closed trajectory 'stable from the outside' but 'unstable from the inside'. The 'conditions' on the flow given in the Hopf theorem, however, preclude all possible paths except the three shown in Figure A.3.19. Some of the excluded bifurcations are discussed in the final subsection of this Appendix.

Are unstable limit cycles irrelevant?

If t in our ODE system really stands for time that always increases in the phenomenon being modeled, then the only limit cycle trajectories of interest are stable ones. Although an unstable limit cycle corresponds to an exact periodic solution of the ODE system, this will never be 'observed' in a 'real' context because infinitesimal random perturbations, always present, would displace the solution off the closed trajectory, whence it would be carried far away by the diverging flow.

Often, however, the independent variable, t, stands for something besides real time. For example, in our warm-up exercise (24), t could stand for wave variable in a quest for exact traveling wave solutions to the PDE (30). When this is the case, it is crucial to understand that there is absolutely no correlation between the stability of a solution trajectory to the ODE system and the stability of the corresponding exact traveling wave solution as a solution to an initial value problem for the original PDE. To see that this must be the case, consider that, if an ODE solution trajectory is stable to small perturbations for t increasing, then it will be unstable for t decreasing. But 'running t backwards' corresponds merely to reversing the sign of the wave speed, or to moving to the left along the spatial axis instead of to the right. Such rearrangements of the coordinate system in which one *describes* an exact traveling wave solution cannot conceivably affect the stability of the solution described as a solution to the PDE. For this reason, unstable limit cycles are often sought just as ardently as are stable ones; both can correspond to spatially and temporarily periodic exact traveling wave solutions of a PDE.

We now give an informal statement of the Hopf theorem, followed by an interpretation. More complete treatments can be found in a monograph by Marsden & McCracken (1976) or in the article by Cronin (1977). The former reference gives a thorough treatment of all facets of the Hopf theorem and many applications; it is written on an abstract level. The latter collects many important theorems relevant in the study of t-periodic solutions of ODE systems, both autonomous and non-autonomous. See also Andronov *et al.* 1973a, for a complete treatment of the $n = 2$ case. Note that the Hopf theorem applies for n-th order systems for arbitrarily large n.

Theorem (E. Hopf).   Consider the n-th order ODE system

$$\left.\begin{aligned} dy_1/dt &= f_1(y_1, \ldots, y_n; \mu), \\ dy_2/dt &= f_2(y_1, \ldots, y_n; \mu), \\ &\vdots \\ dy_n/dt &= f_n(y_1, \ldots, y_n; \mu), \end{aligned}\right\} \quad d\mathbf{y}/dt = \mathbf{f}(\mathbf{y}; \mu), \quad (40)$$

in which the f_i are sufficiently smooth functions of the y_j and μ (four times continuously differentiable will suffice; in the classical Hopf theorem, the f_i were assumed to be *analytical* functions of $\mathbf{y}$ and μ). Note that (40) is *autonomous* in that no f_i depends on t explicitly.

As μ varies in some interval I, let $\mathbf{E}(\mu) = [E_1(\mu), \ldots, E_n(\mu)]$ be the coordinates of an isolated[4] steady state point of (40), i.e. for all $\mu \in I$, $\mathbf{f}(\mathbf{E}(\mu); \mu) = \mathbf{0}$. (It can and must be proved that the $E_i(\mu)$ are smooth functions of μ.)

[4] That is, there must be a neighborhood of phase space containing $\mathbf{E}(\mu)$, but containing no other steady state point.

For each $\mu \in I$, let $\lambda_1(\mu)$, $\lambda_2(\mu)$, ..., $\lambda_n(\mu)$ be the characteristic exponents of the flow f linearized in the neighborhood of the steady state point $\mathbf{E}(\mu)$. (It can and must be proved that these can be arranged so that each λ_i is a smooth function of μ.) The linearized system referred to is the ODE system (15) in which the coefficients, $a_{ij} = \partial f_i/\partial y_j(\mathbf{E}(\mu); \mu)$. Refer to our previous discussion of (15) for the meaning of the characteristic exponents $\lambda_i(\mu)$.

Suppose that, for $\mu \in I$, all but the first two characteristic exponents have negative real parts. That is, suppose that, for all $\mu \in I$,

$$\mathrm{Re}\ (\lambda_3(\mu)) < 0, \ldots, \mathrm{Re}\ (\lambda_n(\mu)) < 0. \tag{41}$$

Note that, when $n = 2$, this hypothesis is automatically satisfied. Note also that, if $\lambda_3, \ldots, \lambda_n$ all have *positive* real parts, t can be reversed to obtain (41).

Suppose, further, that $\lambda_1(\mu)$ and $\lambda_2(\mu)$ form a complex conjugate pair of characteristic exponents for $\mu \in I$. That is:

$$\lambda_1(\mu) = r(\mu) + i\omega(\mu),$$

$$\lambda_2(\mu) = r(\mu) - i\omega(\mu),$$

where r and ω are real.

Suppose finally that there is an isolated value $\mu_0 \in I$ such that $r(\mu_0) = 0$, so that at $\mu = \mu_0$, $\lambda_1(\mu_0)$ and $\lambda_2(\mu_0)$ are a pure imaginary complex conjugate pair, and suppose

$$\omega(\mu_0) \neq 0,$$

and

$$(\mathrm{d}r/\mathrm{d}\mu)(\mu_0) > 0. \tag{42}$$

(This last condition means that $r(\mu) > 0$ for $\mu > \mu_0$, near μ_0, while $r(\mu) < 0$ for $\mu < \mu_0$.)

Then one and only one of the following three things occurs.

Case I. For $\mu = \mu_0$, the steady state solution $\mathbf{E}(\mu_0)$ really is a center. That is, for a single value, $\mu = \mu_0$, there is an infinite collection of concentric closed orbits 'surrounding' the steady state point $\mathbf{E}(\mu_0)$. In this case, for $\mu \neq \mu_0$, but μ near μ_0, there are no periodic orbits 'near' $\mathbf{E}(\mu)$.

Case II. There is a number $b > \mu_0$, so that for any value of μ, $\mu_0 < \mu < b$, there is one (and just one) closed orbit 'near' the steady state point $\mathbf{E}(\mu)$, 'surrounding' it. This **one parameter family of closed orbits** bifurcates out of the steady state point, $\mathbf{E}$, in the sense that, as $\mu \to \mu_0$, the 'diameter' of the closed orbit varies with $|\mu - \mu_0|^{\frac{1}{2}}$. In this case, for $\mu \leqslant \mu_0$, $\mu \in I$, there is no closed orbit 'near' $\mathbf{E}(\mu)$.

Case III. There is a number $a < \mu_0$ so that for any value of μ, $a < \mu < \mu_0$, there is one (and just one) closed orbit 'near' the steady state point $E(\mu)$, 'surrounding' it. This one parameter family of closed orbits bifurcates out of the steady state point, E, in the same way as described above. For $\mu \geq \mu_0$, there is no closed orbit 'near' $E(\mu)$.

Moreover, if, for $\mu = \mu_0$, the steady state point $E(\mu_0)$ (i.e. a center by linear theory) is a **stable steady state solution** (which means all trajectories in the phase space 'near' $E(\mu_0)$ lead asymptotically to $E(\mu_0)$ as $t \to +\infty$), then case II above obtains and each orbit in the one parameter family of closed orbits corresponds to a *stable* limit cycle. Thus case II corresponds to the upper 'path' in Figure A.3.19. Case II bifurcations are called **supercritical**.

If, in contrast, for $\mu = \mu_0$, the steady state point $E(\mu_0)$ is unstable (i.e. some trajectories starting arbitrarily close to $E(\mu_0)$ lead away from $E(\mu_0)$, then case III above obtains. For each $a < \mu < \mu_0$, the single closed orbit that exists near $E(\mu)$ is an unstable limit cycle. Thus case III corresponds to the bottom 'path' in Figure A.3.19. Case III bifurcations are called **subcritical**.

Case I corresponds to the middle 'path' in Figure A.3.19.

Discussion of geometrical aspects of the Hopf theorem

Below we will explain how to do an explicit calculation that can, in most cases, determine whether case, I, II, or III occurs. We pause now to consider the crucial geometrical concept that makes the Hopf theorem work for cases in which $n > 2$.

The hypothesis (41) that $\mathrm{Re}\,[\lambda_3(\mu)] < 0, \ldots, \mathrm{Re}\,[\lambda_n(\mu)] < 0$ for all $\mu \in I$, but that $\lambda_1(\mu_0)$, $\lambda_2(\mu_0)$ form a nonzero pure imaginary complex conjugate pair has the following consequence. Linearization of the flow near $E(\mu_0)$, and finding the characteristic exponents (the $\lambda_i(\mu_0)$) there, yields an approximation of the solution trajectories, $y_i(t)$, near $E(\mu_0)$, given by (20). If $\lambda_3, \ldots, \lambda_n$ all have negative real parts, then the contributions made to each $y_i(t)$ by the eigenfunctions associated with $\lambda_3, \ldots, \lambda_n$, all decay exponentially fast to zero as t increases. While this decay occurs, the contributions to each $y_i(t)$ by the two (sinusoidally oscillating but non-decaying) eigenfunctions associated with the pure imaginary λ_1 and λ_2 survive. That is, all but the first two terms of (20) effectively 'disappear' as t increases, no matter where, near $E(\mu_0)$, the initial conditions locate the starting point of a trajectory.

Conceived geometrically, each pair of complex eigenfunctions in (20) determines a two dimensional surface (a plane, or two-dimensional vector subspace of R^n) in phase space intersecting $E(\mu_0)$. Each real eigenfunction determines a one-dimensional subspace (a straight line) of phase space, intersecting $E(\mu_0)$. These subspaces are those generated (spanned) by the

$$\begin{pmatrix} e_1^j \\ \vdots \\ e_n^j \end{pmatrix}$$

eigenvectors in (20). The claim that *all* solutions to the linearized ODE system (15) are generated by adjustment of the constants b_i in (20) says that these one- and two-dimensional 'eigen' subspaces just described 'unite' by linear combination to fill the entire n-dimensional phase space. The decay of all but the first two eigenfunctions means, geometrically, that, according to our linear approximation, given any initial point 'near' $\mathbf{E}(\mu_0)$, the trajectory through that point either leads (as $t \to +\infty$) directly to the steady state point $\mathbf{E}(\mu_0)$ or else collapses onto (toward) the two-dimensional surface, through $\mathbf{E}(\mu_0)$, specified by (spanned by) the two (sine–cosine) eigenfunctions associated with λ_1, λ_2. Roughly speaking then, the Hopf theorem hypotheses mean that, according to linear approximation, all the interesting (nondecaying) action occurs on a two dimensional (planar) surface containing $\mathbf{E}(\mu_0)$.

In a modern proof of the Hopf theorem, it is established rigorously that, for the full nonlinear ODE system, there really is such a two-dimensional surface in the phase space, containing $\mathbf{E}(\mu_0)$, playing the above role. This surface is called the **center manifold**. This center manifold exists and behaves as described for all μ near μ_0, not just for $\mu = \mu_0$. It is not, in general, a plane, but locally, it 'looks' like a portion of a two-dimensional plane smoothly warped by the nonlinearities. The global shape of this surface remains in doubt. For different flows, $\mathbf{f}(\mathbf{y}; \mu)$, it might extend infinitely like a plane, or wrap back on itself into a cylindrical shape, or close completely as does a sphere. For use in the Hopf theorem we need only know that, locally, it looks like a smoothly warped plane. The center manifold is (by formal definition) an **invariant manifold** of the flow, that is, any trajectory that 'starts' on the center manifold stays on it forever as t tends to both $+\infty$ and $-\infty$. Because of the Hopf theorem's hypotheses, for the (linear theory) reason sketched above, all trajectories beginning anywhere near $\mathbf{E}(\mu)$ either fall into $\mathbf{E}(\mu)$ or approach asymptotically the center manifold as $t \to +\infty$.

Once this is established, the only topological possibilities (in the arbitrary n case) for the generation (bifurcation) of a closed orbit out of a steady state point are exactly those possibilities that exist in the $n = 2$ dimensional case, and these possibilities are few. The smoothness hypothesis on the $\mathbf{f}_i(\mathbf{y}; \mu)$, and the hypothesis (42) that $(d/d\mu)\{\mathrm{Re}\,[\lambda_1(\mu_0)]\} \neq 0$, and $\mathrm{Im}\,[\lambda, (\mu_0)] \neq 0$, further limit these possibilities to just the three cases I, II, and III.

If we think of the closed orbits lying on the two-dimensional center manifold, which they do, it makes sense now to describe them as 'surrounding' $\mathbf{E}(\mu)$, and in case I, as 'concentric'. Because of the center manifold, Hopf bifurcations are essentially two dimensional phenomena that take place on a two dimensional slice of n-dimensional phase space. The concept and theorem can be extended to the $n = +\infty$ case for use in characterizing bifurcations of a partial differential equation system (refer to Marsden & McCracken, 1976, sections 8 and 9).

Above, the terminology 'near' is used repeatedly, as for example, in 'there is no closed orbit "near" $\mathbf{E}(\mu)$'. The Hopf theorem yields *local* results. In a more formal statement of it, all claims would be confined to an appropriate 'sufficiently small neighborhood of $\mathbf{E}(\mu)$'. A description is given of what goes on inside it, and no claims are made about what happens elsewhere in phase space, near other steady state points, for example.

An important qualitative feature of the theorem is its result in cases II and III that, as μ grows away from μ_0, the 'size' or 'diameter' of the limit cycle grows as $|\mu - \mu_0|^{\frac{1}{2}}$. The birth of a closed trajectory is thus abrupt because the rate of growth of its diameter with respect to a change in μ is infinite at $\mu = \mu_0$.

What becomes of a limit cycle as μ is adjusted 'far' from the bifurcation value, μ_0? The Hopf theorem gives no hint. There are several possibilities. It could expand toward infinite diameter. It could expand, then, with further changes of μ, contract and eventually merge into a steady state point. This merger could be a Hopf bifurcation also. It could expand to merge asymptotically with a **homoclinic trajectory**. This can happen in the $n = 2$ or arbitrary n cases. We will not try here to pursue this issue.

The condition that $dr(\mu_0)/d\mu > 0$ (where $r(\mu) = \mathrm{Re}\,[\lambda_1(\mu)] = \mathrm{Re}\,[\lambda_2(\mu)]$), in our hypothesis for the Hopf theorem, is often given as $dr(\mu_0)/d\mu \neq 0$. We give it as an inequality to simplify the theorem's conclusions. In applications, the user must identify a suitable bifurcation parameter and, if it turns out that the above inequality faces in the wrong direction, replace μ by $\mu_0 - \mu$, or $1/\mu$, etc., to aim the inequality in the right direction. This is the reason for naming $\mu = V_0^*/V_c^*$ in (38), leading to the appearance of μ^{-1} in many of the equations that followed (38).

The stability calculation

If case I in the Hopf theorem's conclusions does not occur, then which of cases II or III occurs hinges upon whether the steady state point $E(\mu_0)$ is stable or unstable at the critical value of the parameter μ. Calculational recipes have been derived to decide this issue. These are based upon Taylor series expansions of the $f_i(y_1, \ldots, y_n; \mu_0)$ near $y_i = E_i(\mu_0)$ in which terms up to third order are kept. If a certain combination of the coefficients in these expansions (the partial derivatives of the f_i) has a negative value (positive value), then $\mathbf{E}(\mu_0)$ is stable (unstable) and case II (case III) obtains. If this combination of parameters has the value zero, then no conclusion can be drawn. Instead higher-order terms must be kept in the Taylor series expansion, and a much more complicated combination of coefficients must be evaluated, with the decision again hinging on the algebraic sign of the outcome. Again, a zero outcome means still higher-order terms must be kept, *ad infinitum*.

Given a Taylor series expansion of the f_i, kept to third order, the appropriate combination of coefficients for the recipe, mentioned above and given below, can be derived by trying to construct a so-called 'Lyapunov function' for $\mathbf{f}(\mathbf{y}, \mu)$ at $\mathbf{E}(\mu_0)$. We make no attempt to do this, but only give the resulting recipe which can be used effectively whether or not its derivation is understood.

We give the recipe for the $n = 2$ case only. It is much more complicated for higher-order systems and may be found in Marsden & McCracken (1976, sections 4 and 4A, especially p. 133). The recipe below is the one just cited, with notation specialized to the $n = 2$ case. As will be seen, this recipe requires the 'user' to transform the y_i variables (linearly) so as to bring the linearized version of $\mathbf{f}(\mathbf{y}, \mu)$ near $\mathbf{E}(\mu_0)$ into a standard 'canonical' format. Alternative recipes exist that do not require this 'preconditioning'. (See, for example, Andronov *et al.*, 1973a, for the $n = 2$ case only.) The one we give is chosen because it is a special case of a recipe that works for all n.

The stability calculation recipe for the $n = 2$ case

With $n = 2$, we rename variables so our ODE system appears as in (37):

$$dX/dt = F(X, Y; \mu),$$
$$dY/dt = G(X, Y; \mu). \tag{37}$$

By hypothesis, when $\mu = \mu_0$, (37) linearized near $\mathbf{E}(\mu_0)$ gives rise to a coefficient matrix

$$\begin{pmatrix} a_{11} & a_{12} \\ a_{21} & a_{22} \end{pmatrix} = \begin{pmatrix} \partial F/\partial X & \partial F/\partial Y \\ \partial G/\partial X & \partial G/\partial Y \end{pmatrix}\Bigg|_{\substack{X=E_1(\mu_0) \\ Y=E_2(\mu_0) \\ \mu=\mu_0}}, \tag{43}$$

that has pure imaginary eigenvalues $\pm i\omega(\mu_0)$. This coefficient matrix must be in the canonical form

$$\begin{pmatrix} 0 & |\omega(\mu_0)| \\ -|\omega(\mu_0)| & 0 \end{pmatrix}. \tag{44}$$

If it is not in the form (44), then new variables $\tilde{X} = \alpha_{11} X + \alpha_{12} Y$, $\tilde{Y} = \alpha_{21} X + \alpha_{22} Y$ must be introduced so that the new ODE system

$$d\tilde{X}/dt = \tilde{F}(\tilde{X}, \tilde{Y}; \mu),$$

$$d\tilde{Y}/dt = \tilde{G}(\tilde{X}, \tilde{Y}; \mu),$$

has a coefficient matrix of its linearized version in the form (44), when $\mu = \mu_0$. We will illustrate below by example how this is done. Notice (44) is the coefficient matrix of a linear system (25) whose solution trajectories are perfect concentric circles in the phase plane.

Compute the following partial derivatives, all to be evaluated for $(\tilde{X}, \tilde{Y}, \mu) = [\tilde{E}_1(\mu_0), \tilde{E}_2(\mu_0), \mu_0]$:

$$A_{20} = \partial^2 \tilde{F}/\partial \tilde{X}^2, \qquad B_{20} = \partial^2 \tilde{G}/\partial \tilde{X}^2,$$

$$A_{11} = \partial^2 \tilde{F}/\partial \tilde{X} \, \partial \tilde{Y}, \qquad B_{11} = \partial^2 \tilde{G}/\partial \tilde{X} \, \partial \tilde{Y},$$

$$A_{02} = \partial^2 \tilde{F}/\partial \tilde{Y}^2, \qquad B_{02} = \partial^2 \tilde{G}/\partial \tilde{Y}^2,$$

$$A_{30} = \partial^3 \tilde{F}/\partial \tilde{X}^3, \qquad B_{03} = \partial^3 \tilde{G}/\partial \tilde{Y}^3,$$

$$A_{12} = \partial^3 \tilde{F}/\partial \tilde{X} \, \partial \tilde{Y}^2, \qquad B_{21} = \partial^2 G/\partial \tilde{X}^2 \, \partial \tilde{Y}. \qquad (45)$$

Now compute

$$D = [A_{30} + A_{12} + B_{21} + B_{03}]/|\omega(\mu_0)|$$

$$+ [-A_{11}(A_{20} + A_{02}) + B_{11}(B_{20} + B_{02}) + A_{20}B_{20} - A_{02}B_{02}]/\omega(\mu_0)^2. \qquad (46)$$

If $D < 0$, then $\mathbf{E}(\mu_0)$ is a stable steady state point and case II above obtains. If $D > 0$, then $E(\mu_0)$ is an unstable steady state point and case III above obtains. If $D = 0$, no conclusion can be drawn, as explained above; higher partial derivatives of F and G are needed.

Conclusion of the $n = 2$ exploiter–victim bifurcation example:
interpretations

To illustrate how the Hopf theorem is *used*, we complete the exploiter-victim example begun above. First, we verify that the hypotheses of the Hopf theorem are satisfied by the ODE system (37) of our exploiter-victim example. It is in the form of (40). We are interested in its steady state point

$$[E_1(\mu), E_2(\mu)] = [\mu^{-1}, \mu^{-1}(1 - \mu^{-1})].$$

Equation (37) linearized near this steady state point has the coefficient matrix (39). $\mu_0 = 2$ is the bifurcation value of μ. At $\mu = \mu_0 = 2$, β (of Figure A.3.10) $= 0$, and the characteristic exponents are pure imaginary.

The flow, $\mathbf{f}(\mathbf{y}; \mu)$, in (37) is infinitely many times continuously differentiable with respect to $y_1 = X$, $y_2 = Y$, and μ so long as μ is kept away from $\mu = 0$. Since $n = 2$, the Hopf theorem's hypothesis (41) is automatically satisfied.

From (39) we have the characteristic exponents as

$$\lambda_1, \lambda_2 = \{\mu^{-1}(1 - 2\mu^{-1}) \pm [\mu^{-2}(1 - 2\mu^{-1})^2 - 4k\mu^{-2}(1 - \mu^{-1})]^{\frac{1}{2}}\}/2$$

when $\mu = \mu_0 = 2$, and (by continuity) for μ 'near' $\mu_0 = 2$, the quantity in square brackets is negative. Thus the real part of this complex conjugate pair of eigenvalues is

$$r(\mu) = \mu^{-1}(1 - 2\mu^{-1})/2,$$

and the imaginary part is

$$\omega(\mu) = \mu^{-1}[4k(1 - \mu^{-1}) - (1 - 2\mu^{-1})^2]^{\frac{1}{2}}.$$

Moreover,

$$dr/d\mu = \mu^{-2}(\tfrac{1}{2} - 2\mu^{-1}).$$

At $\mu = \mu_0 = 2$, $dr/d\mu = \tfrac{1}{8} > 0$ and $\omega(\mu_0) = (k/8)^{\frac{1}{2}} \neq 0$. Thus all of the Hopf theorem's hypotheses are satisfied. We know that one and only one of cases I, II, or III obtains.

Without going further, we have just ruled out the possibility of the kind of (stable from the outside, unstable from the inside) closed orbit shown in Figure A.3.17 bifurcating out from the steady state point because this is not one of the possibilities in cases I, II, or III. We have not ruled out cases I or II. We proceed to the stability calculation described above.

At the bifurcation value $\mu = \mu_0 = 2$, the linear approximation coefficient matrix (39) has the form

$$\begin{pmatrix} a_{11} & a_{12} \\ a_{21} & a_{22} \end{pmatrix} = \begin{pmatrix} \partial F/\partial X & \partial F/\partial Y \\ \partial G/\partial X & \partial G/\partial Y \end{pmatrix}\Bigg|_{\mathbf{E}(\mu_0)} = \begin{pmatrix} 0 & \tfrac{1}{2} \\ k/4 & 0 \end{pmatrix}. \tag{47}$$

This is not in the canonical form (44), so we must change variables. Let $\omega_0 \equiv \omega(\mu_0) = (k/8)^{\frac{1}{2}}$. We find numbers α_{11} and α_{22} so that the transformation

$$\tilde{X} = \alpha_{11}X, \qquad \tilde{Y} = \alpha_{22}Y, \tag{48}$$

generates that canonical form. Then (37) becomes

$$d\tilde{X}/dt = \tilde{X}[\tilde{X}(1 - \tilde{X}/\alpha_{11})/\alpha_{11} - (\tilde{Y}/\alpha_{22})] = \tilde{F}(\tilde{X}, \tilde{Y}),$$

and

$$d\tilde{Y}/dt = k[(\tilde{X}/\alpha_{11})(-\mu^{-1})]\tilde{Y} = \tilde{G}(\tilde{X}, \tilde{Y}). \tag{49}$$

The new linear coefficient matrix at the 'new' steady state point, $\tilde{\mathbf{E}}(\mu_0) =$

$(\tilde{X}, \tilde{Y}) = (\alpha_{11}/\mu_0, (\alpha_{22}/\mu_0)(1 - \mu_0^{-1})) = (\alpha_{11}/2, \alpha_{22}/4)$, is

$$\begin{pmatrix} a_{11} & a_{12} \\ a_{21} & a_{22} \end{pmatrix} = \begin{pmatrix} \partial\tilde{F}/\partial\tilde{X} & \partial\tilde{F}/\partial\tilde{Y} \\ \partial\tilde{G}/\partial\tilde{X} & \partial\tilde{G}/\partial\tilde{Y} \end{pmatrix}\Bigg|_{\tilde{E}(\mu_0)} = \begin{pmatrix} 0 & -\alpha_{11}/2\alpha_{22} \\ k\alpha_{22}/4\alpha_{11} & 0 \end{pmatrix}.$$

We pick α_{11} and α_{22} to put this matrix in the canonical form (44). Thus, we need

$$k\alpha_{22}/4\alpha_{11} = \alpha_{11}/2\alpha_{22} \quad \text{and} \quad k\alpha_{22}/4\alpha_{11} < 0,$$

or

$$(\alpha_{11}/\alpha_{22})^2 = k/2 = 4\omega_0^2,$$

or

$$\alpha_{11}/\alpha_{22} = -2|\omega_0| = -2(k/8)^{\frac{1}{2}}.$$

We select $\alpha_{11} = 1$ and $\alpha_{22} = -1/(2|\omega_0|) = (2/k)^{\frac{1}{2}}$. The 'new' ODE system is

$$d\tilde{X}/dt = \tilde{F}(\tilde{X}, \tilde{Y}) = \tilde{X}\{\tilde{X}(1 - \tilde{X}) + 2|\omega_0|\tilde{Y}\},$$

and

$$d\tilde{Y}/dt = \tilde{G}(\tilde{X}, \tilde{Y}) = k(\tilde{X} - \mu^{-1})\tilde{Y}. \tag{50}$$

The coefficient matrix for the linearization of (50) near $(\tilde{X}, \tilde{Y}) = (\tilde{E}_1(\mu_0),$ $\tilde{E}_2(\mu_0)) = (Y_2, -1/8|\omega_0|)$ will have, by derivation of $\tilde{X} = \alpha_{11}X + \alpha_{12}Y$, $\tilde{Y} = \alpha_{21}X + \alpha_{22}Y$, the canonical form (44). It is always possible to find constants α_{11}, α_{12}, α_{21}, α_{22} to accomplish this.

Using (5) we compute the following (the verification of which is left as an exercise):

$A_{20} = -1$, $B_{20} = 0$,

$A_{11} = 2|\omega_0|$, $B_{11} = k$,

$A_{02} = 0$, $B_{02} = 0$,   and $D = (-6/|\omega_0|) + (2|\omega_0|/\omega_0^2) = -4/|\omega_0|$.

$A_{30} = -6$, $B_{03} = 0$,

$A_{12} = 0$, $B_{21} = 0$.

Thus, $D < 0$ and, in our example, only case II occurs. This means, for $\mu = \mu_0 = 2$, the steady state point (we switch back to X and Y, $\hat{X}$ and $\hat{Y}$ having served their purpose), $\mathbf{E}(\mu_0) = (\frac{1}{2}, \frac{1}{4})$, is a stable (spiral) point; there is no closed orbit (near $(\frac{1}{2}, \frac{1}{2})$) when $\mu = \mu_0 = 2$. And, there are no closed orbits surrounding $\mathbf{E}(\mu) = [\mu^{-1}, \mu^{-1}(1 - \mu^{-1})]$ for μ less than, but near, $\mu_0 = 2$, at least there is none that contracts to this (moving) point as $\mu \to 2$, $\mu < 2$. Thus Figure A.3.18, not Figure A.3.17, is the correct phase portrait for $\mu = 1.8$, and, qualitatively, for each $\mu < 2$.

In contrast, for each value of μ greater than, but close to $2 = \mu_0$, there is one (stable limit cycle) closed orbit surrounding $\mathbf{E}(\mu)$. Armed with this information we can confidently sketch in phase portraits for the ODE system (37) when $\mu > 2$. We use the previously considered information about nullclines, etc. Into the general flow surrounding (X_0, Y_0), we fit a closed orbit whose diameter gets larger as μ increases. Trajectories outside of it spiral down to it. Those inside, emanating from the unstable spiral steady state point $[\mu^{-1}, \mu^{-1}(1 - \mu^{-1})]$, spiral out toward it. The result is shown in Figure A.3.20, for which $\mu = 2.05$, and in Figure A.3.21, for which $\mu = 2.15$. In both cases, $k = 1$. Evidently the size (diameter) of the limit cycle increases dramatically as μ increases past 2 (it varies as $|\mu - 2|^{\frac{1}{2}}$).

A general intuitive consideration of the flow in Figures A.3.20 and A.3.21 should convince the reader that, as $\mu \to +\infty$, the limit cycle must grow asymptotically so that its left 'side' approaches the Y-axis, its lower side approaches the X-axis, and its right and top sides approach the unstable manifold rising out of $(1, 0)$. The steady state point enclosed in the limit cycle moves down the X-nullcline (the parabola, $Y = X(1 - X)$) toward $(0, 0)$.

Figure A.3.22 shows $X(t)$ and $Y(t)$ as they vary with t when X and Y, solving (37), traverse the limit cycle in Figure A.3.21. This illustrates how

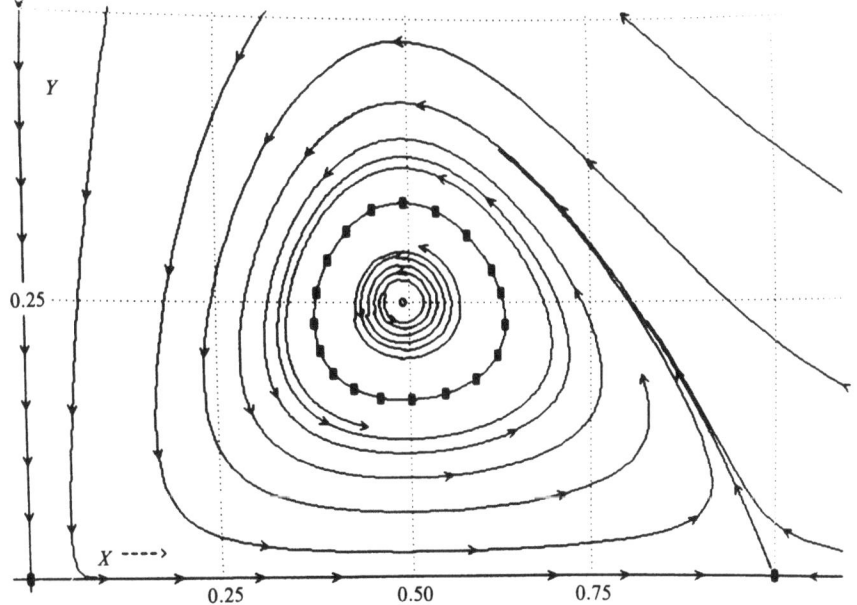

Figure A.3.20. The completed phase portrait for the ODE (37) when $k = 1$ and $\mu = 2.05$ (the bifurcation value of $\mu_0 = 2.0$). There is a single closed orbit. Its diameter expands when μ is increased further past the bifurcation value, as is shown in Figure A.3.21.

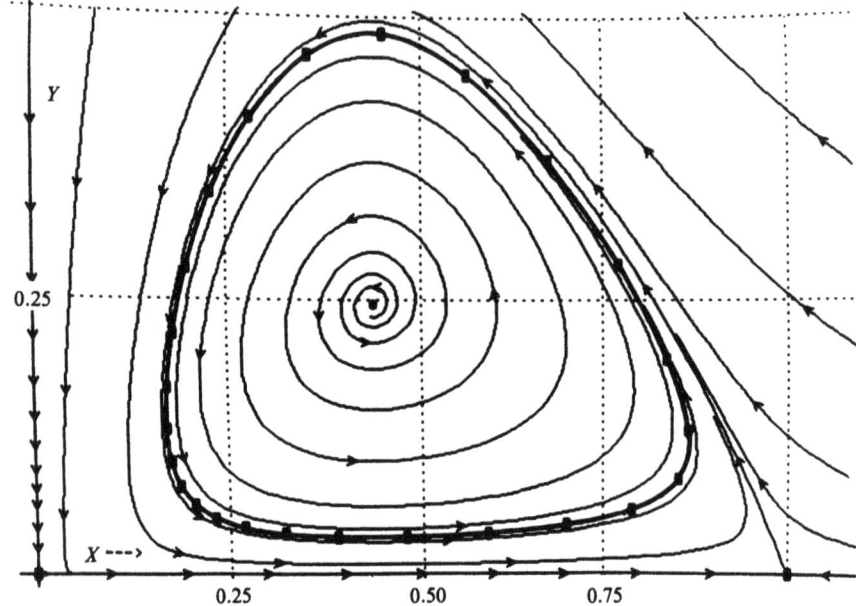

Figure A.3.21. The phase portrait for (37) when $k = 1$ and $\mu = 2.15$. The unique closed orbit corresponds to infinitely repeating time-periodic oscillations in the population densities of exploiters, Y, and victims, X. This oscillation is graphed in Figure A.3.22.

a closed trajectory in phase space corresponds to t-periodic solutions of the ODE system.

Let us briefly reconsider the exploiter–victim problem we began as ODE system (33). Our investigation has proved that the crucial parameter for this system is the dimensionless combination $\mu = V_0^* / V_c^*$. If this has a value smaller than 2, then, regardless of the values of the other parameters, the state one would observe in exploiter–victim populations left alone for a long time would be the stable steady state solution $V^*(t) = V_c^*$ and $E^*(t) = (b^*/a^*) V_c^* (V_0^* - V_c^*)$. No oscillations would be observed.

In contrast, if $V_0^* > 2 V_c^*$, the steady state coexistence point is unstable to small perturbations, and only the periodic population oscillations (stable limit cycles) represented in Figure A.3.22 would be seen.

Consider now a situation in which V_c^* is fixed, in which the parameters b^*, a^*, c^* (in (33)) may change very gradually with time, and V_0^*, initially less than $2V_c^*$, very gradually increases. Until V_0^* evolves to exceed $2V_c^*$, the exploiter–victim population will sit at the stable coexistence steady state point (which very slowly moves in phase space as the parameters very slowly change). But, just as V_0^* exceeds $2V_c^*$, a qualitatively new

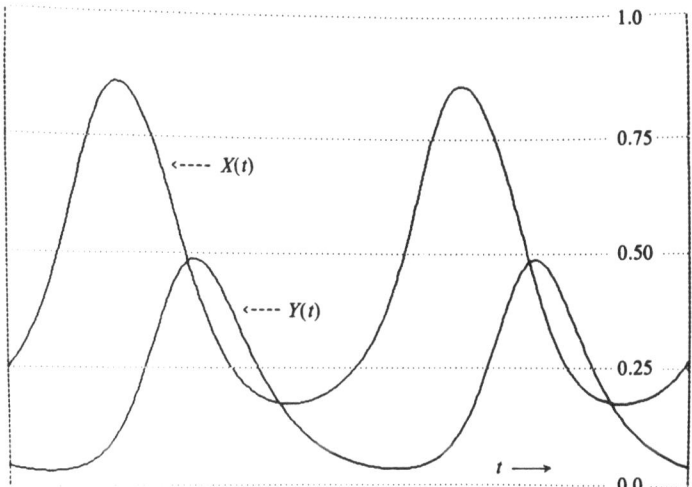

Figure A.3.22. Victim, $X(t)$, and exploiter, $Y(t)$, population densities, as determined by the ODE system (37) when $k = 1$ and $\mu = 2.15$, graphed as functions of time t as the unique closed orbit in Figure A.3.21 is traversed twice.

kind of behavior will suddenly appear. Both populations will oscillate, with an oscillation amplitude that increases sharply as V_0^* increases. The period of this oscillation will be approximately determined by $\omega(\mu) = [4k(1 - \mu^{-1}) - (1 - 2\mu^{-1})^2]^{\frac{1}{2}}/\mu$ where $\mu = V_0^*/V_c^*$ and $k = c^*/b^*V_0^*$ as determined by (38). The period will be $\sim 2\pi/\omega(\mu)$, and this allegory will be a valid one if none of the five parameters in (33) changes significantly over this time period, but varies more slowly. This allegory, backed by the Hopf theorem and our phase plane analysis, contains one of the mathematical structures by which a gradual ('slow' timescale) evolution of parameters can bring about suddenly a wholly new qualitative behavior (limit cycles) as seen on a 'fast' time scale. This allegory can be simulated numerically as follows. To the second-order ODE system (37) append another ODE that makes μ increase 'slowly'. For example

$$dX/dt = F(X, Y, \mu) = X[X(1 - X) - Y],$$

$$dY/dt = G(X, Y, \mu) = k(X - \mu^{-1})Y,$$

$$d\mu/dt = H(X, Y, \mu) = \varepsilon, \qquad \varepsilon \ll 1. \tag{51}$$

Use initial conditions

$$X(0) = \text{anything positive},$$

$$Y(0) = \text{anything positive},$$

and

$$\mu(0) = 1.8 \text{ (or anything less than 2).}$$

Pick values for k ($k = 1$, say) and ε ($\varepsilon = 0.005$, say), and use a numerical ODE integration package to solve the IVP. This is suggested as an exercise for those with access to computing machinery. The results should be striking. You will have to terminate the simulation before $\mu(t)$ becomes too large (10 say) or your ODE integration routine will crash because, as μ gets large, the ODE system (51) gets 'stiff'.

An example of a Hopf bifurcation in a three-dimensional phase space illustrating the possibility of any of several parameters playing the role of 'the' bifurcation parameter

As an example of a Hopf bifurcation in a system of dimension greater than two, where phase plane reasoning does not carry the day (but the Hopf theorem does), we investigate some details of the cyclic AMP (cAMP) oscillation model proposed by Goldbeter & Segel (1977) and described in Section 4.1. This example is rather complicated, so that some readers may wish to skip to the concluding paragraph of this appendix.

In the model of Goldbeter & Segel (1977), $\alpha(t)$, $\beta(t)$, and $\gamma(t)$ represent normalized concentrations of intracellular ATP (α), intracellular cAMP (β), and extracellular cAMP (γ). These are the governing differential equations:

$$d\alpha/dt = v - \sigma\phi,$$

$$d\beta/dt = q\sigma\phi - k_t\beta,$$

$$d\gamma/dt = (k_t\beta/h) - (k\gamma), \tag{52}$$

with

$$\phi(\alpha, \gamma) = \alpha(1+\alpha)(1+\gamma)^2/[L + (1+\alpha)^2(1+\gamma)^2]. \tag{53}$$

This third-order autonomous system of ordinary differential equations involves seven constant parameters v, σ, q, k_t, h, k, L. For what point set in the seven-dimensional $(v, \sigma, q, k_t, h, k, L)$-space does the system (52) exhibit spontaneous oscillations? We will consider this question in detail here, and seek an answer in the form of a characterization of the boundary of the oscillation point set. Note that the regions just outside the boundary of this domain are the locations (parameter settings) to inspect in the search for 'relay' behavior, some instances of which were found and exhibited by the numerical analysis reported in Section 4.1.

Our strategy consists of an attempt to recast the system (52) into a form permitting application of the Hopf bifurcation theorem. If this turns out to be possible, we succeed in understanding the mathematical mechanism that generates bifurcations from stable quiescent behavior of the system to limit cycle oscillations, and, more importantly, we characterize the parameter tunings that delineate oscillatory and quiescent domains in parameter space. Otherwise, we learn that oscillations, if they exist in the system at hand, are generated by some other mechanism. In fact, we will find that Hopf bifurcations produce the oscillatory behavior of the cAMP in the problem at hand.

Before proceeding, let us consider the question of whether the analytical work facing us is worthwhile. In general, numerical calculations are preferable when either (i) they will get to the answer faster or (ii) the analytical answer is foreseen to be so complicated that numerical evaluation of the formula, and computer-drawn graphs of it, will be required to understand what the answer means.

As is pointed out in Chapter 3, a parameter space of dimension as high as 7^5 is far too vast to search randomly or by inspecting each of ten values, say, for each of the seven parameters (10^7 sets of parameter values). We need to know almost precisely where to look for the bifurcation boundary in 7-dimensional space in order to have a good chance of finding it. The analytical work ahead will fill this need; this is its justification. (The intuition of Goldbeter & Segel (1977) did guide them to *part* of the bifurcation boundary – a most useful achievement. Our aim is to characterize the whole boundary. This is not merely a mathematical exercise, for it will permit closer comparison of theory with experiment.)

The tactics we shall employ (which are of general applicability) are as follows.

(a) Find the steady state points the system admits.

(b) Linearize the differential equations in the neighborhood of each steady state point.

(c) Find constraints upon the several parameters that force there to be one pair of purely imaginary characteristic exponents of the linear system of (b) while the other characteristic exponents have (for example) negative real parts. These constraints are precisely the mathematical characterization of 'surfaces' across which bifurcations *could* occur.

(d) Resolve details about whether the Hopf theorem's hypothesis (42) holds, and which side of the bifurcation locus, if any, corresponds to stable limit cycles. The 'right' way to do this is by implementing the

---

[5] While α, β, γ, L, q, and h are already dimensionless in the problem at hand, t^{-1}, v, σ, k, and k_t are all measured in inverse time. It turns out that, by making these quantities dimensionless, we can collapse the number of essential parameters from seven to five. We will not do this here because a 5-dimensional parameter space is still forbiddingly vast to search, and because the new dimensionless names would steal from the reader any mastery of this problem's nomenclature obtained by reading Section 4.1.

n-dimensional version of the stability calculation given above. The practical way is to do it numerically.

We note that, as long as all variables and parameters are positive, the ODE system (52) has a flow that is infinitely differentiable with respect to all variables and parameters, so the smoothness-of-flow hypothesis of the Hopf theorem is satisfied.

Steady state points

The possible steady state values of α, β, γ are those that make the right-hand sides of the differential equations (52) vanish. This requires

$$\beta_0 = qv/k_t, \qquad \gamma_0 = qv/kh, \tag{54}$$

and

$$\phi(\alpha_0, \gamma_0) = \phi(\alpha_0, qv/kh) = s, \tag{55}$$

where $s = v/\sigma$.

At this point we postpone the details of how α_0 (determined by equation (55)) depends upon the parameters v, σ, q, k_t, h, k, and L, noting only that the functional form of ϕ makes (55) a quadratic equation for α_0. Thus, for a given set of the parameters, there exist at most two steady state points (one for each zero of the quadratic). For some parameter sets, (55) will yield complex α_0, showing that no steady state points exist. For those parameter sets for which (55) yields real-valued α_0's, one of the α_0's will be negative, hence meaningless as a chemical concentration, so that for these parameter sets there will be a unique steady state point.

We denote departures from equilibrium as

$$x(t) = \alpha - \alpha_0, \qquad y(t) = \beta - \beta_0, \qquad z(t) = \gamma - \gamma_0.$$

ϕ, as a function of the new x, y, z variables, needs a new name: $\Phi(x, z) = \phi(\alpha, \gamma) = \phi(x + \alpha_0, z + \gamma_0)$. Equation (55) becomes

$$\Phi(0, 0) = s = v/\sigma. \tag{56}$$

Linearization

We linearize the system (52) in the neighborhood of $(x, y, z) = (0, 0, 0)$. The result is the following set of three linear, first order, differential equations with constant coefficients.

$$\mathrm{d}x/\mathrm{d}t = -Ax - Bz,$$

$$\mathrm{d}y/\mathrm{d}t = qAx - k_t y + qBz,$$

$$\mathrm{d}z/\mathrm{d}t = (k_t/h)y - kz. \tag{57}$$

Here $A = \sigma \, \partial\phi/\partial\alpha \, (\alpha_0, \gamma_0) = \sigma \, \partial\Phi/\partial x \, (0, 0)$, and

$$B = \sigma \frac{\partial\phi}{\partial\gamma}(\alpha_0, \gamma_0) = \sigma \frac{\partial\Phi}{\partial z}(0, 0).$$

The important linearization procedure has been discussed above. To afford further confidence in this technique, we now provide the details of linearizing the equation

$$d\alpha/dt = v - \sigma\phi(\alpha, \gamma).$$

Written with our new names, this equation is

$$d(x + \alpha_0)/dt = v - \sigma\phi(x + \alpha_0, z + \gamma_0),$$

or

$$dx/dt = v - \sigma\Phi(x, z).$$

Taylor series expansion of Φ near $(x, z) = (0, 0)$ yields

$$dx/dt = v - \sigma\left[\Phi(0, 0) + \frac{\partial\Phi}{\partial x}(0, 0) \cdot x + \frac{\partial\phi}{\partial z}(0, 0) \cdot z \right.$$

$$\left. + \text{higher order terms we neglect} \right].$$

According to (56), $v - \sigma\Phi(0, 0) = 0$, so that the linearized equation is

$$dx/dt = -\left[\sigma \frac{\partial\Phi}{\partial x}(0, 0) \right]x - \left[\sigma \frac{\partial\Phi}{\partial z}(0, 0) \right]z. \tag{58}$$

With abbreviations A and B introduced for the expressions in the square brackets, whose calculation is somewhat lengthy, (58) is the first of the equations of (57). We expect the approximation (58) to be valid when $|x|$ and $|z|$ are sufficiently close to zero.

Characteristic exponents

The equations of (57) are a special case of the general, linear, constant-coefficient, homogeneous, ODE system (15) discussed above. The general solution is the superposition of exponential eigenfunctions (20) above, where the characteristic exponents, $\lambda_1, \lambda_2, \lambda_3$, are determined by

$$\det \begin{vmatrix} -A - \lambda & 0 & -B \\ qA & -k_t - \lambda & qB \\ 0 & k_t/h & -k - \lambda \end{vmatrix} = 0,$$

or

$$(-A - \lambda)[(-k_t - \lambda)(-k - \lambda) - (k_t/h)qB] - (BqAk_t/h) = 0.$$

This is a third degree polynomial in λ, namely:

$$p(\lambda) = \lambda^3 + [A + k + k_t]\lambda^2 + [Ak + Ak_t + kk_t - (qBk_t/h)]\lambda + Akk_t = 0.$$

(59)

The three zeros of (59), the unique three values of λ (called *characteristic exponents*) that make (20) solve (15), could be written down explicitly using the handbook formula for roots of a cubic equation, but the resulting formulae would be both complicated and uninformative. In fact, we do not want to know what the roots of this cubic are; we want only to know when the set of three zeros comprises one real zero and one pure imaginary conjugate pair. That is, we insist that $p(\lambda)$ in (59) be (with r and η real):

$$p(\lambda) = (\lambda + r)(\lambda + i\eta)(\lambda - i\eta) = 0,$$

or, expanding,

$$p(\lambda) = \lambda^3 + r\lambda^2 + \eta^2\lambda + r\eta^2 = 0.$$

(60)

Polynomials (59) and (60) are identical when and only when

$$r = A + k + k_t,$$

$$\eta^2 = Ak + Ak_t + kk_t - (qBk_t/h) > 0,$$

(61)

$$r\eta^2 = Akk_t.$$

The first and third of these equations yield an expression for

$$\eta^2 = Akk_t/r = Akk_t/(A + k + k_t),$$

that must be the same as the second of the above equations. Thus we see that necessary and sufficient conditions for there to be two pure imaginary characteristic exponents, and one real one are

$$Akk_t/(A + k + k_t) = Ak + Ak_t + kk_t - (qk_tB/h) = \eta^2 > 0.$$

(62)

All of the seven parameters in this problem v, σ, q, k_t, h, k, and L are inherently positive. But, to know whether the inequality part of (62) can hold, we must first learn something about the signs of A and B. To compute

$$A = \sigma \frac{\partial \phi}{\partial \alpha}(\alpha_0, \gamma_0),$$

we rewrite ϕ, given in (53), as $\phi(\alpha, \gamma) = \alpha(1 + \alpha)/(\Gamma + (1 + \alpha)^2)$, where $\Gamma \equiv L/(1 + \gamma)^2 > 0$. Let γ be any fixed positive value. It is clear that ϕ, as a function of α, looks as pictured in Figure A.3.23 and that $\partial \phi/\partial \alpha > 0$, so that $A > 0$. An explicit formula for A is

$$A = \sigma\{1 + 2\alpha_0 - 2\alpha_0(1 + \alpha_0)^2/[\Gamma + (1 + \alpha_0)^2]\}/[\Gamma + (1 + \alpha_0)^2].$$

(63)

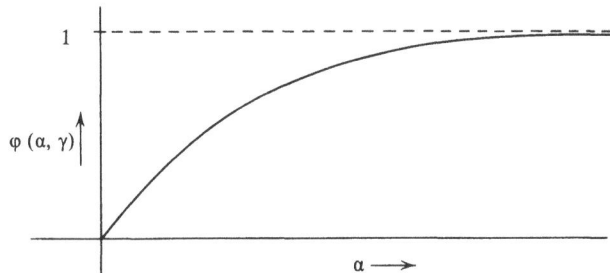

Figure A.3.23. The shape of the intracellular cAMP production rate, ϕ, as a function of intracellular ATP, α (the conversion enzyme's substrate), for some (any) fixed value of extracellular cAMP concentration, γ (the conversion enzyme's regulator).

Since α_0 must be positive to be physically meaningful, a lower bound on A is obtained by setting positive Γ to zero in the denominator of the fraction preceded by the minus sign. This gives

$$A \geq \sigma(1 + 2\alpha_0 - 2\alpha_0)/[\Gamma + (1 + \alpha_0)^2],$$

that is

$$A \geq \sigma/[\Gamma + (1 + \alpha_0)^2]. \tag{64}$$

This lower bound on A goes to zero as α_0 becomes infinitely large.

The guaranteed positivity of A insures that, if the equality part of (62) can be met, then the inequality part, $\eta^2 > 0$, holds also. Further, with A positive, it follows from the first of the equations of (61) that r must be positive. Thus, whenever we have two pure imaginary characteristic exponents in the system at hand, the other characteristic exponent is negative. This, in turn, keeps alive the hope that the system may bifurcate to *stable* limit cycle solutions.

We now face this relatively clean problem in nonlinear algebra:

Find values of parameters v, σ, q, k_t, h, k, and L that satisfy

$$\blacklozenge\blacklozenge \qquad \frac{Akk_t}{A + k + k_t} = Ak + Ak_t + kk_t - (qk_tB/h), \tag{62}$$

where

$$\blacklozenge\blacklozenge \qquad A = \sigma\frac{\partial\phi}{\partial\alpha}(\alpha_0, \gamma_0), \qquad B = \sigma\frac{\partial\phi}{\partial\gamma}(\alpha_0, \gamma_0),$$

with

$$\blacklozenge\blacklozenge \qquad \phi(\alpha, \gamma) \equiv \alpha(1 + \alpha)(1 + \gamma)^2/[L + (1 + \alpha)^2(1 + \gamma)^2] \tag{53}$$

such that $\gamma_0 = qv/kh$ and α_0 is a real positive solution of (the quadratic in α_0)

◆◆ $\phi(\alpha_0, \gamma_0) = s = v/\sigma$.   (55)

The bifurcation locus

If Hopf bifurcations occur in this problem, they occur only in the neighborhood of parameter sets that solve the problem indicated above.

We want to know approximately what the point set satisfying the above problem looks like geometrically, as a surface, we presume, in $(v, \sigma, q, k_t, h, k, L)$-space. One approach toward this goal proceeds as follows. Solve the quadratic (in α_0) equation (55) for α_0. Accept only positive values of v, σ, q, h, k, L (k_t is not involved here) that make at least one root, α_0, positive, and accept only positive α_0. Then write out exact expressions for A and B and hence transform (62) into a single nonlinear algebraic equation in $(v, \sigma, q, k_t, h, k, L)$. Since neither α_0, nor γ_0, nor A, nor B involves k_t, this expanded version of (62) is easily seen to be a quadratic equation for k_t, whose solution can be recorded explicitly.

The point set we seek comprises those values of v, σ, q, h, k, and L that give this quadratic in k_t real positive zeros (roots). The full calculations would, after much work, yield long expressions of little obvious significance. We will thus begin these calculations, but not finish them, turning instead to an asymptotic (approximate) conclusion that yields both a close approximation to the solution, and simple easily understood formulae for it.

The quadratic equation (55), expanded, is

$$(1-s)\alpha_0^2 + (1-2s)\alpha_0 - s(1+\Gamma) = 0,$$

where $\Gamma = L/[1 + (qv/kh)^2]$. The solutions of this quadratic are

$$\alpha_0 = \{-1 + 2s \pm [1 + 4s(1-s)\Gamma]^{\frac{1}{2}}\}/[2(1-s)].   (65)$$

For a fixed value of Γ, we piece together a plot of α_0 as a function of s. We see that the two values of α_0 are identical when $1 + 4s(1-s)L = 0$, that is, when $s = \frac{1}{2}[1 - (1+\Gamma^{-1})^{\frac{1}{2}}]$ or $s = \frac{1}{2}[1 + (1+\Gamma^{-1})^{\frac{1}{2}}]$.

Local analysis of (65), for example, determination of local maxima and minima and points of inflection, confirms that Figures A.3.24 and A.3.25 give qualitatively correct graphs of α_0 as determined by (65) for the two distinguishable cases $\Gamma < 1$ (in Figure A.3.24) and $\Gamma > 1$ (in Figure A.3.25). We see that only the largest root in (65) can be positive and *is* positive when and only when

$$0 < s = v/\sigma < 1.   (66)$$

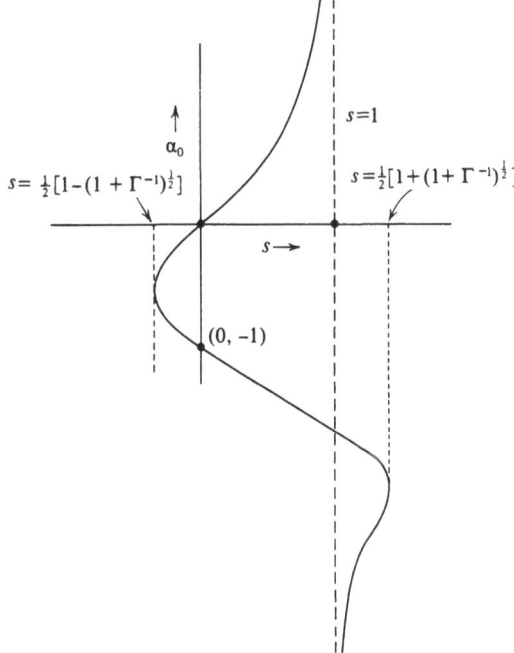

Figures A.3.24 and A.3.25. The steady state concentration of intracellular ATP, α_0, as a function of the scaled constant supply rate of ATP, $s = v/\sigma$. For cases when $\Gamma < 1$, see Figure A.3.24; the situation when $\Gamma > 1$ is shown in Figure A.3.25. Only the $\alpha_0 > 0$ portions of these graphs are physically meaningful.

Condition (66) is necessary and sufficient for existence of a meaningful steady state solution. We see that, as $s \to 1$, the equilibrium value of intracellular ATP, α_0, becomes unbounded. For any set of positive parameters $v, \sigma, q, k_t, h, k,$ and L, satisfying (66) there is a unique meaningful steady state point.

Straightforward algebra yields these expressions for A and B:

$$A = v[(1 + 2\alpha_0)/(1 + \alpha_0) - 2s]/\alpha_0, \tag{67}$$

and

$$B = 2v[1 - s(1 + \alpha_0)/\alpha_0]/(1 + \gamma_0) = 2vsL/\alpha_0(1 + \alpha_0)(1 + \gamma_0)^3. \tag{68}$$

Verification of (67) and (68) is left as an exercise in calculus. We know already that A is positive, by (64), and (68) shows that B is also positive. Goldbeter & Segel (1977) used identical values for the decay constants, k and k_t, in their numerical analysis. To simplify the algebra, we now make the same restriction: $k_t = k$. Equation (62) becomes

$$Ak = (A + 2k)(2A + k - qB/h), \tag{69}$$

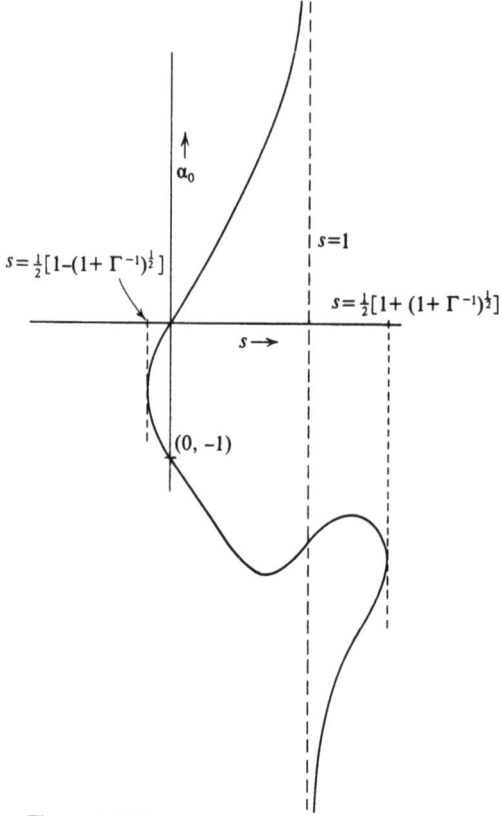

Figure A.3.25.

which we regard as a quadratic equation for k (even though A and B depend upon k). Thus, solving the quadratic, (69) is equivalent to

$$k = \tfrac{1}{2}\hat{H}(1 - (2A/\hat{H}) + [-(2A/H)^{\frac{1}{2}}]),^6 \tag{70}$$

where $\hat{H} = qB/h = QB$, where $Q = q/h$.

The following problem in nonlinear algebra remains to be solved.

Let $Q = q/h$, $s = v/\sigma$, v, and L be fixed and positive with $0 < s < 1$. Let

◆◆ $j = k/v$ and let $\delta(j) = 1 + Q/j$. \tag{71}

Let

◆◆ $p(j) = [\delta^2(j) + 4s(1-s)L]^{\frac{1}{2}}$,

◆◆ $\alpha_0(j) = (2s - 1 + p(j)/\delta(j))/2(1-s)$. \tag{72}

---

[6] Careful analysis reveals that only the largest root is meaningful.

Let

◆◆ $H(j) = 2QsL/\alpha_0(j)[1 + \alpha_0(j)]\delta^3(j)$

$\equiv 2Q(1 - s(1 + \alpha_0)/\alpha_0)/\delta(j).$ (73)

Let

◆◆ $D(j) = \delta^2(j)p(j)/QsL.$ (74)

Let

◆◆ $F(j) = \frac{1}{2}H(j)\{1 - D(j) + [1 - D(j)]^{\frac{1}{2}}\}.$ (75)

Find all real positive values of k satisfying

◆◆ $j = F(j).$ (76)

Equation (72) is the positive root of (65) (the only positive α_0). Equation (73) is the expression for $\hat{H}/v = QB/v$ obtained from (68). Equation (74) is a simplified expression for $D = 2A/H$ using (67) in which (72) is substituted for α_0. $F(j)$ in equation (76) is the right-hand side of (70) using (74) for $2A/H$, and the equation to be solved for j, (76), is just (70).

Those acquainted with numerical procedures for finding zeros (i.e. roots) of nonlinear functions will see that, for any fixed set of parameters $(Q, s, v, \text{ and } L)$, it is a simple numerical task to find values of j that solve $j - F(j) = 0$; simple, that is, if one has decent estimates of solution values for j with which to begin an iterative search, and knowledge of how many solutions exist.

The next several pages consist of informal functional analysis, concluding with asymptotic estimates based on certain parameter size restrictions. These considerations produce, first, a proof that there can exist at most two j values satisfying the boxed problem (for a given Q, s, L set), and, secondly, *quite simple* formulae for asymptotically estimating the two corresponding values of k (when they exist) in terms of q, h, v, σ, L. These are (93) and (94) below giving $k_1(q, h, v, \sigma, L)$ and $k_2(q, h, v, \sigma, L)$. As these 'fixed' parameters (q, h, v, σ, L) are varied through parameter space, $k = k_1(q, L, v, \sigma, L)$ and $k = k_2(q, h, v, \sigma, L)$ sweep out the two 'halves' of the surface in parameter space across which Hopf bifurcations could occur.

Using (93) and (94), with numerical verification and refinement, Figures A.3.27 and A.3.28 have been constructed. These depict the bifurcation surface in (σ, v, k)-space for the fixed values of $q = 10^2$, $L = 10^6$, and $h = 10$ used by Goldbeter and Segel in their numerical analysis.

Readers unconcerned with the details of *how* Figures A.3.27 and A.3.28 are constructed are urged to ignore those details, and to concentrate, instead, on the *interpretations* of Figures A.3.27 and A.3.28.

Solution of the problem (71)–(76). $\delta(j)$ is a monotone decreasing function of j that tends to ∞ as $j \to 0$, so that $\alpha_0(j)$ is a monotone increasing function of j bounded from above. Thus, from the second expression for $H(j)$ in (73), $H(j)$ is seen to be a monotone increasing function of j, bounded from above, $D(j)$ is a monotone decreasing function of j. These features make $F(j)$ a monotone strictly increasing function of j, bounded from above by some least upper bound F_∞.

Let j_0 denote the value of j (if such exists) that makes $F(j) = 0$. $D(j_0) = 1$ determines j_0 if it exists. The least value attained by $D(j)$ occurs as $j \to +\infty$, where $\delta = 1$, and is $[1 + 4s(1-s)L]^{\frac{1}{2}}/QsL$. $D(j) \to +\infty$ monotonically as $j \to 0$, so that a necessary and sufficient condition for j_0 to exist is

$$[1 + 4s(1-s)L]^{\frac{1}{2}}/QsL < 1, \tag{77}$$

and this condition guarantees uniqueness of j_0 as well. Suppose this condition is met. Then straightforward calculus shows that dF/dj is infinite at $j = j_0$. These considerations prove that the graph of F in Figure A.3.26 is qualitatively correct.

We seek values of j that solve $j = F(j)$, i.e. we want values of j (if any exist) where the 45° line in Figure A.3.26 intersects the function $F(j)$.

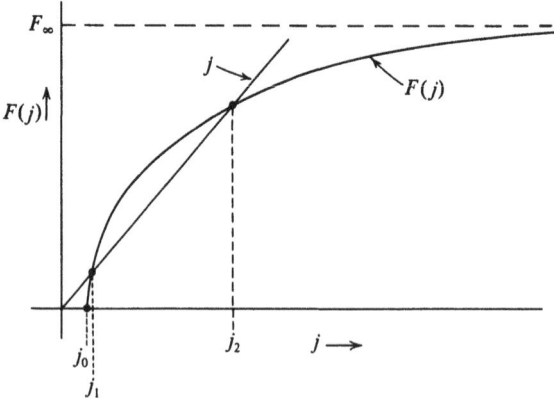

Figure A.3.26. The shape of the function $F(j)$, given by (75). To solve the nonlinear equation (76), it is necessary to find the intersection(s) of the 45° line with the curve $F(j)$.

We see from Figure A.3.26 that, for given (v, σ, q, h, L), there will be no such points, or one, or two. (The case depicted in Figure A.3.26 shows two, j_1 and j_2.)

From Figure A.3.26, we also learn that a (numerically implemented) fixed point iteration, in which we compute successive approximations $j_2^{(1)}, j_2^{(2)}, \ldots, j_2^{(n)}$, etc., using the formula

$$j_2^{(n+1)} = F(j_2^{(n)}), \tag{78}$$

starting with a reasonable first guess $j_2^{(0)}$, will surely converge to as close an approximation of j_2 as may be desired if j_2 exists. This is because $|dF/dj| < 1$ at $j = j_2$, a necessary and sufficient condition for convergence of the fixed point iteration defined in (78). Further, an attempt to find j_1 by fixed point iteration will certainly fail because, from Figure A.3.26 $|dF/dj| > 1$ at $j = j_1$. Refer to 'fixed point iteration' in any text on numerical analysis for a discussion of these matters.

Asymptotics

In order to proceed, we simplify matters by looking only at parameter sets that fulfil these constraints:

$$L \gg Q \equiv q/h \gg 1. \tag{79}$$

Goldbeter & Segel (1977) used $q/h = 10$ and $L = 10^6$ in their numerical analysis.

We now exhibit, informally, a selfconsistent approximation scheme, based on the assumption that

$$1 < \delta^2(j) \ll 4s(1-s)L. \tag{80}$$

With this assumption, (71) becomes

$$p \sim 2(s(1-s)L)^{\frac{1}{2}}. \tag{81}$$

Equations (80) and (81) together make $p/\delta \gg 1$, so that (72) becomes

$$\alpha_0 \sim p/2(1-s)\delta \gg 1. \tag{82}$$

Consequently, $(1+\alpha_0)/\alpha_0 \sim 1$, with which (73) becomes

$$H \sim 2Q(1-s)/\delta. \tag{83}$$

We know from the analysis above that $0 < D < 1$. Accordingly, we use $1 - (1/2D)$ as an approximation for $(1-D)^{\frac{1}{2}}$. (This errs by only 25% when D is as large as 0.75.) (75) now becomes approximately

$$F = \tfrac{1}{2}H(2 - \tfrac{3}{2}D). \tag{84}$$

Substituting (83) and (74) into (84), we obtain from (76) the following approximate equation determining j:

$$j\delta(j) = 2Q(1-s)[2-(\tfrac{3}{2}\delta^2 p/QsL)]. \tag{85}$$

Replacing $\delta(j)$ by its definition, and p by (81), we obtain from (85),

$$j \cong Q(1-2s) - 3(1-s)^{\frac{3}{2}}(1+Q/j)^2/(sL)^{\frac{1}{2}}. \tag{86}$$

One zero of this cubic equation for j is obviously negative, hence meaningless. We seek approximate estimates of the other two zeros when they are real.

We restrict attention to s values sufficiently removed from 0 that $1/sL \ll 1$. In this case, the last term of (86) is negligible unless $Q/j \gg 1$, so we replace $1+(Q/j)$ by Q/j in (86). When Q/j is not large, Q/j is a poor approximation of $1+(Q/j)$, but then the entire term is negligible anyway. We have

$$j - Q(1-2s) + \{3Q^2(1-s)^{\frac{3}{2}}/(sL)^{\frac{1}{2}}\}(1/j^2) \cong 0. \tag{87}$$

Two distinct positive j solutions suggest themselves. First, if $j \ll 1$, then the first term is negligible compared to the other two. A balance of the latter two terms yields

$$j = j_1(s; Q, L) \cong (3Q)^{\frac{1}{3}}[(1-s)^3/sL]^{\frac{1}{6}}/[1-2s]^{\frac{1}{3}}. \tag{88}$$

Second, if $j > 1$, the last term of (87) is negligible, leaving simply

$$j = j_2(s; Q, L) \cong Q(1-2s). \tag{89}$$

Let us check the selfconsistency of our approximation scheme, i.e. let us use the results (88) and (89) to check the validity of the assumptions that produced them. For a careful discussion of what can be concluded from a consistency check, see Lin & Segel (1974), Section 6.1. Equation (88) can hold only so long as $j_1 \ll 1$, or, using (88)

$$s(1-2s)^2 \gg 9Q^2/L. \tag{90}$$

While it remains valid, (88) gives the smaller value of j, hence the larger value of δ. Thus, (80), an assumption made to obtain (88), corresponds to the constraint

$$Q^2/j_1^2 = Q(1-2s)/3[(1-s)^3/sL]^{\frac{1}{3}} \ll 4s(1-s)L. \tag{91}$$

This *is* satisfied for $s \in (0, \tfrac{1}{2})$, when

$$s \gg Q^2/144L,$$

a condition implied by (90). Equation (89) can hold only when $j_2 \gg 1$, or

$$1-2s \gg 1/Q. \tag{92}$$

Given (79), the s values satisfying (90) and (92) fill most of the interval $(0, \frac{1}{2})$, excluding small zones at both ends. Figure A.3.27 depicts the two curves $j = j_1(s; Q, L)$ and $j = j_2(s; Q, L)$ on the interval $(0, \frac{1}{2})$. These are the dashed-line curves. For the case shown, $L = 10^6$ and $Q = 10$ as in Goldbeter & Segel (1977). The solid curve in the same figure is the 'true' locus in the (s, j) plane of points satisfying the problem (71–6). The true solutions were obtained numerically, using the estimates provided by (88) and (89) to provide initial guesses at j solution values to initiate standard root-finding routines.

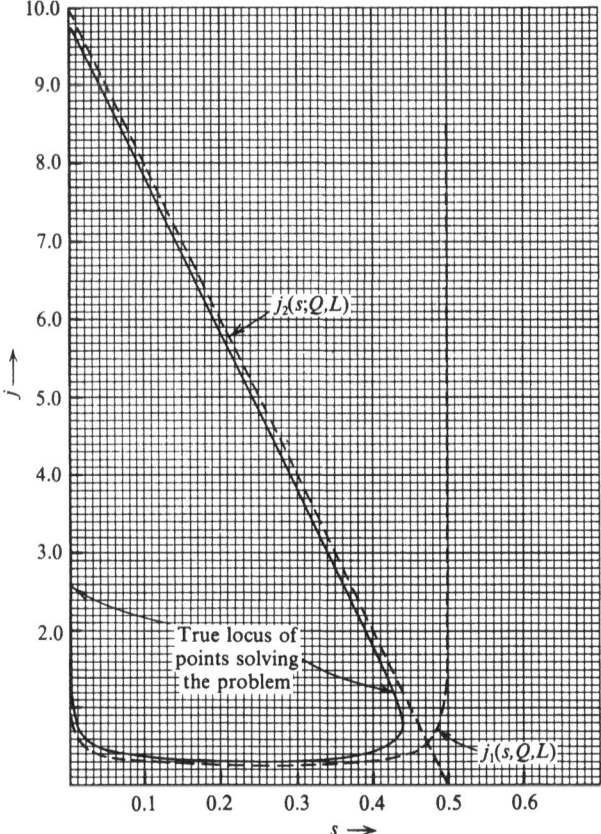

Figure A.3.27. A complete and quantitatively accurate dimensionless representation of the potential bifurcation locus in parameter space (with $L = 10^6$ and $q/h = 10$ fixed) for the Goldbeter–Segel cAMP relay/oscillator ODE system (52). The solid line (obtained numerically) shows the true locus. The dashed lines show estimates made by our asymptotic analysis. The axes measure $j = k/v$ and $s = v/\sigma$.

As can be seen in Figure A.3.27, our informal asymptotic estimates, (88) and (89) for the upper and lower halves of the (s, j) solution curve, are quite good except for strips near the edges of the $(0, 0.5)$ s interval excluded by constraints (90) and (92).

Figure A.3.27 in fact fully represents the potential bifurcation surface in (σ, k, v)-space when the parameters L and $Q = q/h$ are held fixed. We redraw Figure A.3.27 now, in terms of the parameters k, v, and σ to produce a representation of the true shape of the box in (k, v, σ)-space schematized in Figure 3 of Goldbeter & Segel (1977). Returning to the original variable names, (88) and (89) become

$$k = K_1(q, h, v, \sigma, L) = v(2q/h)^{\frac{1}{2}}[1 - v/\sigma)^3/(vL/\sigma)]^{\frac{1}{4}}/(1 - 2v/\sigma)^{\frac{1}{2}}, \quad (93)$$

and

$$k = K_2(q, h, v, \sigma, L) = \{q(1 - 2v/\sigma)/h\}v. \quad (94)$$

For fixed values of q, h, σ, and L, the translation of Figure A.3.27 into the k–v plane produces the curves shown in Figure A.3.28 ($q/h = 10$, $L = 10^6$, $\sigma = 1.2$).

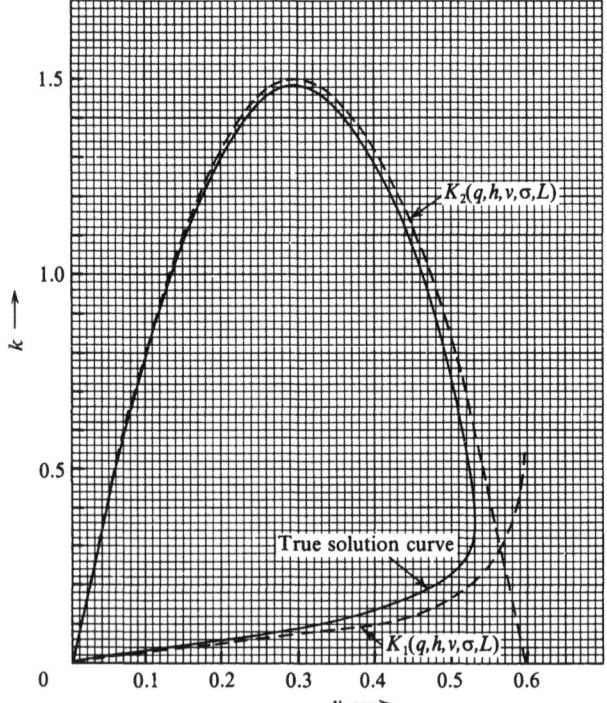

Figure A.3.28. A representation in the v–k plane of the bifurcation locus in Figure A.3.27, for the fixed parameter values $q/h = 10$, $L = 10^6$, and $\sigma = 1.2$.

Finally, Figure A.3.29 shows the true shape of the 'box' in parameter space schematized in Figure 3 of Goldbeter & Segel (1977). Each constant-σ cross-section of this surface is derived from Figure A.3.27 and is a scaled copy of Figure A.3.28. One such cross-section is depicted in Figure 4.1.17. More precisely, Figure A.3.29 shows an inverted cone-shaped surface in (k, v, σ)-space (whose shape would change as q/h and L were changed; here $q/h = 10$ and $L = 10^6$, and also $k = k_t$ is assumed) across which Hopf bifurcations *could* occur. According to our analysis, this cone expands upward infinitely.

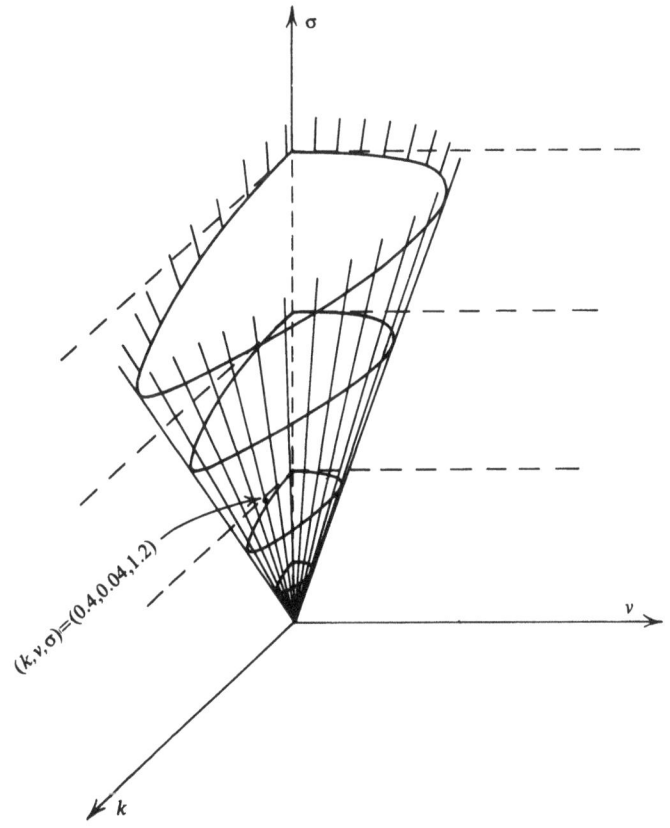

Figure A.3.29. The true shape of the bifurcation locus 'box' in parameter space schematized in Figure 3 of Goldbeter & Segel (1977). Each constant-σ cross-section of this surface resembles the curve in Figure A.3.28 (which is the $\sigma = 1.2$ cross-section).

The point $(k, v, \sigma) = (0.4, 0.04, 1.2)$ locates a parameter setting found (numerically) by Goldbeter & Segel to yield 'relay behavior.' Directly across the bifurcation surface shown, they found autonomous oscillatory behavior.

In passing, we extol the virtues of using dimensionless variables (see Appendix A.4). They are efficient. Figure A.3.29 is complex and only qualitatively accurate. Figure A.3.27, using the dimensionless variables $j = k/v$ and $s = v/\sigma$, contains all the information Figure A.3.29 represents, and is quantitatively accurate into the bargain.

Concluding remarks on the cAMP oscillator/relay ODE system

We now discuss what we have learned about the cAMP oscillation/relay ODE system, (52), and what requires further investigation. We set out to apply the Hopf bifurcation and were brought up hard against the nontrivial algebraic problem of determining the $(v, \sigma, k, k_t, q, h, L)$-parameter locus at which (52), linearized near its only meaningful steady state point, has a pair of purely imaginary characteristic exponents. We solved this problem, at least for the $k = k_t$ and $L \gg q/h \gg 1$ zones of parameter space, so we now know exactly where to look in parameter space for oscillatory behavior of (52). The Hopf theorem tells us the following. If we select any combination of parameters of this problem, calling that combination μ (or μ^{-1}), and choosing it so that tuning μ will carry us across the cone-shaped surface in Figure A.3.29 (in a direction not tangent to that surface so that hypothesis (42) of the theorem is satisfied) then, so long as the purely imaginary pair of characteristic exponents is not zero (hypothesis (42) again) at the crossing point, (52) must exhibit closed orbits in its (α, β, γ) phase space in one of the three ways prescribed in the conclusions of the Hopf theorem.

We do not know which case occurs. Possibly all three do, each near a different 'part' of the cone-shaped surface in Figure A.3.29. To resolve these issues analytically would be quite difficult. First, of course, we would have to pick a parameter combination to call μ. Then we would have to derive formulae for the real and imaginary parts of the pair of characteristic exponents that become purely imaginary as the cone-shaped surface is crossed, and to verify analytically hypothesis (42). Finally, and this would be expected to generate an impenetrable algebraic thicket, we would have to carry out the three dimensional version of the stability calculation cited above. It is questionable whether the calculations involved would be worth the labor they would require.

An exercise in numerical experimentation

Numerical integration of the ODE system (52) presents an attractive alternative to the analytical stability calculation. We know exactly where to look in parameter space for oscillatory behavior. The needed numerical exploration is left as an exercise for interested students with access to computing machinery. The exploration could be done as follows. Pick two points, P_1 and P_2, lying close together, one on each side of the cone-shaped potential bifurcation surface shown in Figure A.3.29. (Actually only the quantitatively accurate Figure A.3.27 will suit this purpose.) At each parameter setting, P_1 and P_2, compute the characteristic exponents by solving (numerically) the cubic equation (59). If, with P_1 and P_2 'very close together', there is a complex conjugate pair of characteristic exponents with a small negative real part at one point, P_i, and a small positive real part at the other point, and a clearly nonzero imaginary part at both P_1 and P_2, then, almost certainly a Hopf bifurcation occurs as the parameters are tuned to move from P_1 to P_2 across the cone-shaped potential bifurcation surface.

To go further, solve an IVP for the system (52), using initial conditions $[\alpha(0), \beta(0), \gamma(0)]$ near the steady state point determined by (54) and (55), and parameter values determined by whichever of P_1 and P_2 gave a complex conjugate pair of characteristic exponents with a positive real part. This IVP may yield a solution that tends to t-periodic behavior (as illustrated in Figure 4.1.13). If not then reverse time in (52) (that is, multiply the right-hand sides of (52) by -1), and solve another IVP, this time using, to determine the parameters, whichever of P_1, P_2 gave a complex conjugate pair of characteristic exponents with a negative real part. Approach to a t-periodic solution with t running backwards would indicate bifurcation to an unstable limit cycle that, of course, would never be observed experimentally. If neither of these two IVP's tends to a t-periodic solution, then either the parameter settings, P_1 and P_2, are too far from the bifurcation surface, or there is, at one of the parameter setting points, P_1 or P_2, a limit cycle that is unstable for both time directions or the numerical ODE solver failed, or case I of the Hopf theorem occurs. With luck, and skill with numerical ODE solvers, one can discover where in parameter space (52) has periodic solutions, and whether those limit cycles are stable.

In Figure A.3.29, we indicate the location of the point $(k, v, \sigma) = (0.4, 0.04, 1.2)$, with $k_t = k$, $L = 10^6$, $q = 10^2$, and $h = 10$, found by Goldbeter & Segel (1977) to yield (nonoscillatory) relay behavior of (52). This point lies outside the cone-shaped potential bifurcation surface. Directly across this surface, inside it (for $v = 0.10$, with all other parameters having the above values), oscillatory behavior was found numerically (corresponding to a stable limit cycle). Near the point indicated in Figure A.3.29, case II of the Hopf theorem evidently occurs. The bifurcation (and relay) behavior elsewhere near the cone-shaped surface awaits your discovery.

This ends our second bifurcation example. We conclude this appendix with a brief sketch of some types of bifurcations not characterized by the Hopf theorem.

Bifurcations not characterized by the Hopf theorem

The hypothesis on smoothness of the flow, and (42): $\omega(\mu_0) \neq 0$, $dr/d\mu\,(\mu_0) > 0$ are not nit-picking mathematical details. To the contrary, violation of any of them permits other kinds of bifurcations of limit cycles out of steady state points. The Hopf theorem is useful because it, through its hypotheses, characterizes the kinds of bifurcation that appear frequently in applications. In this section we sketch other possibilities.[7] This material can safely be omitted by readers who do not possess much mathematical sophistication.

To study the various kinds of bifurcations that can occur, it is useful to consider only circular limit cycles. Since, by smooth nonlinear warping, any phase space can be deformed to make any particular limit cycle in it circular, we lose no generality. We restrict attention to the $n = 2$ case, and introduce polar coordinates

$$X(t) = r(t) \cos [\theta(t)],$$
$$Y(t) = r(t) \sin [\theta(t)]. \tag{95}$$

To consider only limit cycles that are circles in the X–Y plane centered at $(0, 0)$ we further restrict attention to flows that have this form:

$$dr/dt = rf(r, \mu),$$
$$d\theta/dt = 1,$$

where f is a smooth function of r. For this flow $(X, Y) = (0, 0)$ is a steady state point. If $r_0(\mu)$ is any root of the function $f[f(r_0, \mu) = 0]$, then the circle $r(t) = r_0(\mu)$ is a limit cycle of the flow (95).

We differentiate (95), to obtain

$$dX/dt = (dr/dt) \cos \theta - r \sin \theta(d\theta/dt) = rf(r, \mu) \cos \theta - r \sin \theta,$$
$$dX/dt = f(r, \mu)X - Y,$$

and similarly, $dy/dt = X - f(r, \mu)Y$.

Since $r = (X^2 + Y^2)^{\frac{1}{2}}$, the flow, in Cartesian coordinates, is

$$dX/dt = f[(X^2 + Y^2)^{\frac{1}{2}}, \mu]X - Y = F(X, Y; \mu),$$
$$dY/dt = X + f[(X^2 + Y^2)^{\frac{1}{2}}, \mu]Y = G(X, Y; \mu). \tag{96}$$

[7] Section 3A of Marsden & McCracken (1976) is a good reference for the material below.

Now we are ready to illustrate various kinds of bifurcations. For the Hopf theorem to apply, the function $f(r, \mu)$ must be such as to make F and G four times continuously differentiable functions of X, Y, μ. Some care is necessary to insure this, as the Cartesian–polar coordinate transformation has a singularity at $r = 0$.

A prototypical illustration of case II of the Hopf theorem's conclusions arises if we set

$$f(r, \mu) = \mu - r^2. \tag{97}$$

Case III is illustrated if we set

$$f(r, \mu) = -(1 - \mu^{-1} - r^2). \tag{98}$$

For (97), the bifurcation occurs at $\mu = 0$, and for (98), the bifurcation occurs at $\mu = 1$. Verification of these claims is left as an exercise.

Consider the case

$$f(r, \mu) = (\mu - r^2)^2. \tag{99}$$

Then, given any positive μ, $r(t) = r_0(\mu) = \mu^{\frac{1}{2}}$ is the radius of a circular limit cycle. Note that for all r, except $r = 0$ and $r = r_0(\mu) = \mu^{\frac{1}{2}}$, $dr/dt = rf(r, \mu) > 0$. Clearly, when $\mu < 0$, $(X, Y) = (0, 0)$ is an unstable spiral steady state point. Just as μ is tuned greater than zero, a limit cycle bifurcates out of $(0, 0)$. Trajectories inside it spiral out toward it, and trajectories outside of it spiral further out away from it (as would occur in Figure A.3.17, were t reversed). For this choice of $f(r, \mu)$, the flow (96) is analytical (infinitely smooth). The bifurcation it generates is ruled out by the Hopf theorem, and it is left as an exercise to discover which hypothesis of the Hopf theorem is violated.

Consider the analytical case

$$f(r, \mu) = (\mu - r^2)(2\mu - r^2). \tag{100}$$

As μ is tuned to be greater than zero, two distinct limit cycles, with radii $r = \mu^{\frac{1}{2}}$, $r = (2\mu)^{\frac{1}{2}}$, bifurcate from the steady state point at $(X, Y) = (0, 0)$. The inner one is stable from inside and outside, while the outer one is stable from the outside but unstable from the inside. The Hopf theorem does not allow this kind of bifurcation either and, again, the discovery of the violated hypothesis is left as an exercise. This example illustrates a class of bifurcations characterized by Chaffee's theorem (see Chaffee, 1968).

Obviously, by inventing complicated functions $f(r, \mu)$, with many real roots, each collapsing to zero as $\mu \to 0$, arbitrarily complicated bifurcations of arbitrarily many limit cycles out of a steady state point can be concocted. It is important to understand that the Hopf theorem we have studied explains only one class of these. For practical application purposes, however, the Hopf theorem frequently proves adequate, and, for the most part, the bifurcations excluded by it can be regarded as bizarre.

Conclusion: a warning

To misinterpret this appendix as a comprehensive summary of known and practically useful results in the theory of ODE's would be to mistake the flashy jacket of a serious book for the contents of the volume. The jacket serves its purpose if it entices those who see it to read the book it wraps. A good book jacket should merely caricature, with a kernel of truth, one or two interesting facets of the contents inside. Interpret this appendix as such a jacket. It could wrap selected parts of the following texts on ODE's. Lefschetz (1957); Hurewicz (1958); LaSalle & Lefschetz (1961); Hartman (1964); Hale (1969); Andronov *et al.* (1973*a, b*); Arnold (1973); Marsden & McCracken (1976); Boyce & DiPrima (1977).

References

Andronov, A. A., Leontovich, E. A., Gordon, I. I. & Maier, A. G. (1973*a*). *Theory of Bifurcations of Dynamic Systems on a Plane*, New York, Halsted Press.

——— (1973*b*). *Qualitative Theory of Second Order Dynamic Systems*, New York, Halsted Press.

Arnold, V. I. (1973). *Ordinary Differential Equations*, Cambridge, Mass., MIT Press.

Boyce, W. E. & DiPrima, R. C. (1977). *Elementary Differential Equations and Boundary Value Problems*, 3rd edn, New York, Wiley. [See especially chapter 9 for phase plane analysis.]

Chaffee, N. (1968). The bifurcation of one or more closed orbits from an equilibrium point of an autonomous differential system. *J. Different. Equat.* **4**, 661–79.

Cronin, A. J. (1977). Some mathematics of biological oscillations. *SIAM Rev.* **19**(1), 100–38.

Goldbeter, A. & Segel, L. A. (1977). Unified mechanism for relay and oscillation of cyclic AMP in *Dictyostelium discoideum*. *Proc. Nat. Acad. Sci. USA* **74**, 1543–7.

Hale, J. K. (1969). *Ordinary Differential Equations*, New York, Wiley-Interscience.

Hartman, P. (1964). *Ordinary Differential Equations*, New York, Wiley.

Hurewicz, W. (1958). *Lectures on Ordinary Differential Equations*, Cambridge, Mass., MIT Press.

Kamkë, E. (1971). *Differentialgleichungen Lösungsmethoden und Lösungen*, New York, Chelsea Publishing Co. [Reprint of the 1948 version of the third edition.]

LaSalle, J. P. & Lefschetz, S. (1961). *Stability by Lyapunov's Direct Method with Applications*, New York, Academic Press.

Lefschetz, S. (1957). *Differential Equations: Geometric Theory*, New York, Wiley-Interscience.

Lin, C. C. & Segel, L. A. (1974). *Mathematics Applied To Deterministic Problems in the Natural Sciences*, New York, Macmillan.

Marsden, J. & McCracken, M. (1976). *The Hopf Bifurcation and its Applications*, New York, Springer-Verlag.

Takens, F. (1973). Unfolding of certain singularities of vector fields: generalized Hopf bifurcations. *J. Different. Equat.* **14**, 476–93.

A.4 DIMENSIONAL ANALYSIS

We present here, largely through simple examples, the main ideas of dimensional analysis.

An elementary example

Suppose that it is known that the initial velocity V_0 of a chemical reaction depends solely on the initial substrate concentration S_0, the maximum velocity V and the value K_m of S_0 at which $V_0 = 0.5\ V$ (cf. Section 1.1). In mathematical language, V_0 is a function of S_0, V, and K_m, i.e.

$$V_0 = f(S_0, V, K_m). \tag{1}$$

It would appear to require a whole book to graph experimental or theoretical results concerning V_0. In such a volume, a single page would contain graphs of V_0 versus S_0 for several values of K_m and a fixed value of V. Similar graphs would appear on different pages, for different values of V.

Further thought indicates that the situation is much simpler. One begins by recognizing that the equation

$$V_0 = K_m + VS_0,$$

for example, is not a possible relation among the various quantities. The reason is that V_0 and V are both concentrations divided by time, while K_m and S_0 are both concentrations. If time units were switched from seconds to minutes, both V_0 and V would be multiplied by 60, but K_m and S_0 would be unaltered. Thus (1) could not hold in both systems of units. But physical laws cannot depend on what standard of reference a scientist chooses to use!

To avoid this difficulty, we note that V_0/V is a **dimensionless** reaction velocity, for this ratio has the *same value whatever units are used.* Similarly, S_0/K_m is a dimensionless initial concentration and there are no other dimensionless groups except combinations of V_0/V and S_0/K_m. If we define

$$y = V_0/V \quad \text{and} \quad x = S_0/K_m, \tag{2}$$

then, it must be that $y = f(x)$. For example, the Michaelis–Menten formula

$$V_0 = VS_0/(K_m + S_0) \tag{1.1.10}$$

becomes

$$y = x/(1 + x)$$

when the dimensionless variables (2) are employed.

Any form of relationship among the variables except $y = f(x)$ would bring us into the type of difficulty that was elucidated in the previous paragraph. Moreover, recognition that the data can be represented by the *curve $y = f(x)$* will save the effort of recording a book full of data.

Another example will illustrate a more comprehensive approach to dimensional analysis that is generally used in more complicated problems. The steps to be used in this approach are italicized.

Procedure for dimensional analysis of a functional relation

Consider the conversion of substrate of concentration S into product of concentration P by means of an enzyme of concentration E. The chemical reaction is symbolized by

$$E + S \underset{k_{-1}}{\overset{k_{+1}}{\rightleftharpoons}} C \xrightarrow{k_{+2}} E + P, \tag{3}$$

where C represents the concentration of the enzyme–substrate complex, and the k's are rate constants. (See Section 1.1.) Let the initial concentrations of S, E, C and P be S_0, E_0, 0, and 0 respectively. Then the substrate concentration, S, depends on the time t and the five parameters $S_0, E_0, k_{+1}, k_{-1},$ and k_{+2}:

$$S = f(t, S_0, E_0, k_{+1}, k_{-1}, k_{+2}). \tag{4}$$

Our first step is to *ascertain the dimensions of all the quantities* that appear on the right-hand side of (4). Normally, the fundamental units of mass (M), length (L), time (T) are used, plus temperature and electric charge if relevant. Here t of course has the dimensions of time; we write, using conventional square brackets,

$$[t] = T.$$

Both S_0 and E_0 are concentrations, numbers of molecules per unit

volume. Volume has dimensions length^3 so that we write
$$[S_0] = L^{-3}, \qquad [E_0] = L^{-3}.$$

The rate constant k_{-1} gives the proportionality between C and the temporal rate of change of S (cf. (1.1.3)). Thus
$$[k_{-1}] = \left[\frac{dC/dt}{S}\right] = \left[\frac{C/t}{S}\right] = \left[\frac{1}{t}\right] = T^{-1}.$$

Similarly $[k_{+2}] = T^{-1}$.

The rate constant k_{+1} gives the proportionality between dS/dt and SE. Thus
$$[k_{+1}] = \left[\frac{dS/dt}{SE}\right] = \left[\frac{1}{tE}\right] = T^{-1}L^3.$$

Dimensionless combinations of parameters must have the form of products of powers to achieve the required cancellation of M, L, and T. To find systematically all the dimensionless combinations of the six quantities on the right side of (4) we thus consider the combination
$$Q \equiv t^{a_1} S_0^{a_2} E_0^{a_3} k_{+1}^{a_4} k_{-1}^{a_5} k_{+2}^{a_6} \tag{5}$$

where the a's are constants. Inserting the dimensions we have found, we obtain
$$[Q] = T^{a_1} L^{-3a_2} L^{-3a_3} [T^{-1}L^3]^{a_4} T^{-a_5} T^{-a_6} = T^{a_1 - a_4 - a_5 - a_6} L^{-3(a_2 + a_3 - a_4)}. \tag{6}$$

To enforce the requirement that the product of powers Q is to be dimensionless we must have
$$a_1 - a_4 - a_5 - a_6 = 0, \qquad a_2 + a_3 - a_4 = 0. \tag{7}$$

Starting to *eliminate variables*, we see that
$$a_1 = a_4 + a_5 + a_6, \qquad a_2 = -a_3 + a_4.$$

Thus if α, β, γ, and δ are arbitrary constants, the general solution to (7) is
$$a_3 = \alpha, \qquad a_4 = \beta, \qquad a_5 = \gamma, \qquad a_6 = \delta,$$
$$a_1 = \beta + \gamma + \delta, \qquad a_2 = -\alpha + \beta. \tag{8}$$

We now substitute the most general possible form of the parameters into Q. Here this requires substituting (8) into (5), yielding
$$Q = t^{\beta + \gamma + \delta} S_0^{-\alpha + \beta} E_0^{\alpha} k_{+1}^{\beta} k_{-1}^{\gamma} k_{+2}^{\delta} = (E_0/S_0)^{\alpha} (k_{+1}S_0 t)^{\beta} (k_{-1}t)^{\gamma} (k_{+2}t)^{\delta}.$$

Taking all but one of the arbitrary parameters to be zero and the remaining one to be unity, in all possible ways, we see that the dimensionless parameters must be some combination of a certain definite set of dimensionless parameters. Here, this set is
$$(E_0/S_0), \ k_{+1}S_0 t, \ k_{-1}t, \ k_{+2}t. \tag{9}$$

We now *form a dimensionless version of the quantity in which we are interested*, for example S/S_0. We know that *this must be a function of dimensionless quantities*, which leads to the result

$$S/S_0 = F[(E_0/S_0), k_{+1}S_0t, k_{-1}t, k_{+2}t]. \tag{10}$$

Note that only four dimensionless parameters are involved in (10), compared to six parameters in the original relationship (4). Such a reduction in the number of parameters makes the use of dimensionless variables valuable in almost all quantitative investigations, whether experimental, numerical, or analytical.

It may be convenient (or more biochemically meaningful) to express (10) in terms of other dimensionless parameters, formed by combining the parameters of (9). For example, one may use the dimensionless parameters

$$\varepsilon \equiv \frac{E_0}{S_0}, \qquad \kappa \equiv \frac{k_{-1}t + k_{+2}t}{k_{+1}S_0t} = \frac{k_{-1} + k_{+2}}{k_{+1}S_0}, \qquad \lambda \equiv \frac{k_2t}{k_1S_0t} = \frac{k_2}{k_1S_0}, \tag{11a}$$

and the dimensionless time

$$\tau \equiv (k_{+1}S_0t)(E_0/S_0) = k_{+1}E_0t. \tag{11b}$$

With this, for some function G, (10) is replaced by

$$S/S_0 = G(\tau, \varepsilon, \kappa, \lambda). \tag{12}$$

The dimensionless parameters are different, but there are still four of them. This invariance in the number of dimensionless parameters is a general phenomenon.

Nondimensionalizing differential equations

If the equations governing a situation are known, the benefits of non-dimensionalization can be obtained by dividing each variable by an appropriate combination of parameters, to make the variables dimensionless. For example the chemical process (3) is governed by the equations for E, S, C, and P that are given in (1.1.3) and by the initial conditions of (1.1.4):

$$dE/dt = -k_{+1}ES + (k_{-1} + k_{+2})C,$$

$$dS/dt = -k_{+1}ES + k_{-1}C,$$

$$dC/dt = k_{+1}ES - (k_{-1} + k_{+2})C, \tag{13}$$

$$dP/dt = k_{+2}C,$$

$$E(0) = E_0, \qquad S(0) = S_0, \qquad C(0) = 0, \qquad P(0) = 0.$$

Let us introduce the dimensionless concentrations

$$e = E/E_0, \qquad s = S/S_0, \qquad c = C/E_0, \qquad p = P/S_0, \qquad (14a)$$

and the dimensionless time

$$\tau = k_{+1}E_0 t. \qquad (14b)$$

To change the derivative terms to the new variables we must use the chain rule (A.1.11); for example

$$\frac{\mathrm{d}E}{\mathrm{d}t} = \frac{\mathrm{d}(E_0 e)}{\mathrm{d}t} = E_0\frac{\mathrm{d}e}{\mathrm{d}t} = E_0\frac{\mathrm{d}e}{\mathrm{d}\tau}\cdot\frac{\mathrm{d}\tau}{\mathrm{d}t} = k_{+1}E_0^2\frac{\mathrm{d}e}{\mathrm{d}\tau}.$$

As is generally the case, the correct answer can be obtained by substitution followed by a 'translation' of constants past the derivative:

$$\frac{\mathrm{d}E}{\mathrm{d}t} = \frac{\mathrm{d}(E_0 e)}{\mathrm{d}(\tau/k_{+1}E_0)} = \frac{E_0}{(1/k_{+1}E_0)}\frac{\mathrm{d}e}{\mathrm{d}\tau} = k_{+1}E_0^2\frac{\mathrm{d}e}{\mathrm{d}\tau}.$$

Introducing $(14a, b)$ into (13) we thus obtain the dimensionless equations

$$\varepsilon\,(\mathrm{d}e/\mathrm{d}\tau) = -es + \kappa c,$$

$$\mathrm{d}s/\mathrm{d}\tau = -es + (\kappa - \lambda)c,$$

$$\varepsilon\,(\mathrm{d}c/\mathrm{d}\tau) = es - \kappa c,$$

$$\mathrm{d}p/\mathrm{d}\tau = \lambda c;$$

$$e(0) = 1, \qquad s(0) = 1, \qquad c(0) = 0, \qquad p(0) = 0.$$

The only parameters that appear are ε, κ, and λ – in contrast with the five parameters k_{+1}, k_{-1}, k_{+2}, E_0 and S_0 of (13). Thus the solution can depend only on ε, κ, λ and the dimensionless time τ. In particular s must have this dependence; since $s = S/S_0$ we have derived (12).

The use of dimensionless variables to simplify equations is illustrated several times in this text. More comprehensive treatments of dimensional analysis can be found for example in the books of Bridgman (1931), Huntley (1967), or Lin & Segel (1974).

References

Bridgman, P. W. (1931). *Dimensional Analysis*, rev. edn, New Haven, Conn., Yale University Press. (Paperback edn, 1963.)

Huntley, H. E. (1967). *Dimensional Analysis*, New York, Dover Publications, Inc.

Lin, C. C. & Segel, L. A. (1974). *Mathematics Applied to Deterministic Problems in the Natural Sciences*, New York, Macmillan.

A.5 A BRIEF INTRODUCTION TO THE NUMERICAL SOLUTION OF ORDINARY DIFFERENTIAL EQUATIONS

We shall present here some basic concepts of the theory of ordinary differential equations and the numerical solution of initial value problems. This is a problem because computers do not solve equations but carry out operations such as adding, subtracting, multiplying and dividing. The major errors encountered by approximating the ordinary differential equations and by solving the approximating equations on a computer will be demonstrated. It is hoped that the reader will end up with an idea about the problems involved, and the errors which may creep up on him or her as he or she uses some standard numerical routine. This should facilitate a more intelligent use of available routines.

Equations and systems

We shall be dealing with the solution of the equation

$$y' = \mathrm{d}y/\mathrm{d}x = f(x, y), \tag{1}$$

with an initial condition at $x = 0$, say,

$$y(0) = y_0. \tag{2}$$

The function $y = y(x)$ in (1) could also represent a vector of unknowns, in which case f is also a vector.

Given a system of first-order equations one can transform it into a single equation of higher order, and vice versa. This procedure is not unique. Consider for example the system of equations

$$u' = 998u + 1998v, \tag{3a}$$

$$v' = -999u - 1999v, \tag{3b}$$

with initial conditions

$$u(0) = 1, \qquad v(0) = 0. \tag{4}$$

Differentiating $(3a)$ and substituting $(3b)$ for v' and $(3a)$ for v, we end up with one second-order equation for u

$$u'' + 1001u' + 1000u = 0,$$
$$u(0) = 1, \qquad u'(0) = 998. \tag{5}$$

One can then solve for u from (5) and then for v from (3b). Conversely, given the problem (5), one may proceed to obtain a first-order system of equations via

$$v' = w$$
$$w' = -1000u - 1001w, \tag{6}$$
$$u(0) = 1, \qquad w(0) = 998.$$

We shall concentrate from now on mostly on a single first-order ordinary differential equation. The generalization to systems is standard and may be found in all textbooks on the subject.

It is important to realize that (1) has infinitely many solutions if we do not supply the additional initial condition (2). Given the problem (1) and (2), a unique solution exists under suitable conditions on the function $f(x, y)$ (see e.g. conditions on f in the convergence theorem below).

Stable–unstable equation

Again consider (1), and suppose we work out the solution for two initial values $y_0^{(1)}$ and $y_0^{(2)}$. Intuitively we expect that if $y_0^{(1)}$ and $y_0^{(2)}$ are close then $y^{(1)}(x)$ and $y^{(2)}(x)$ should be close for finite x. This is part of the 'well-posedness' of the problem. Even so, the deviation between the solutions might grow or not grow with increasing x. If the difference between the solutions grows we term the equation **unstable**. If not it is **stable**. Consider, for example, the equations

$$y' = y, \tag{7}$$

and

$$y' = -y; \tag{8}$$

see Figure A.5.1. Equation (7) is unstable, and equation (8) is stable.

Numerical approximation

The methods which we shall talk about go under the heading of **difference methods**. We approximate the solution of (1) by obtaining values y_i at a discrete set of points x_i (the **nodes**), such that

$$x_0 = a, \qquad x_{i+1} = x_i + \Delta x_i, \tag{9}$$

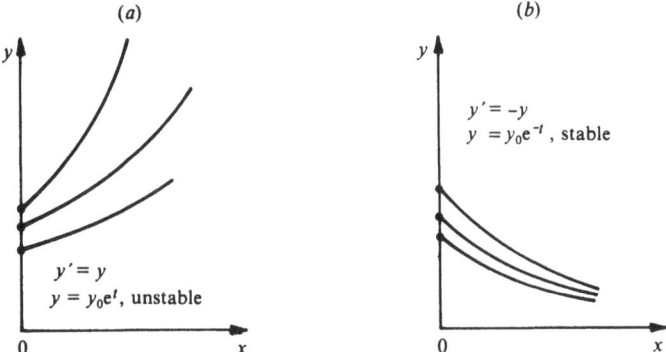

Figure A.5.1. (*a*) Unstable equation, solutions diverge. (*b*) Stable equation, solutions converge.

where $x = a$ is the left-hand side of the interval we deal with, and Δx_i are the step sizes; see Figure A.5.2. The ideal situation would be if we could take the values y_i to be $y(x_i)$, the exact solution at the nodes and, for values in between, use some interpolation method. This of course is impossible in that we do not know the solution, and in general, we cannot find it. We must find a way to obtain the solution approximately. This is done by solving a different, but related, problem. We shall treat some of the approximating problems later, but first we shall consider errors.

The above strategy introduces two types of errors called **generated errors**.

(1) One error results from the difference between the solution of the differential equation and the problem we actually pose and try to solve. Loosely speaking this is termed **truncation error**.

(2) A second error arises from the fact that numbers, in general, cannot be represented exactly on the computer, in the process involved, because of the finite accuracy of the computer words. This is termed **round-off error**.

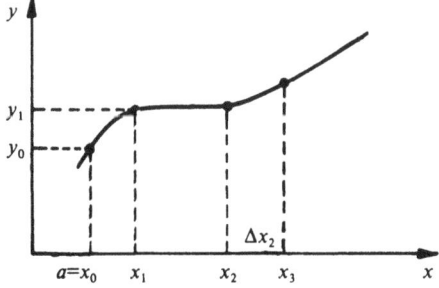

Figure A.5.2. Discretization of an interval and corresponding function values.

The first error will be discussed when we talk about convergence. The fatality of round-off error in a problem of numerical analysis is demonstrated in the following example.

Problem.   Find the value of $\exp(-5.5)$. You have available to you a five-digit decimal computer.

Mathematical analysis
(1) For all x, the Taylor series (A.1.64)

$$\exp(x) = 1 + x + (x^2/2!) + \ldots + (x^n/n!) + \ldots \tag{10}$$

converges.
(2) For $x < 0$ the series has terms of alternating signs.
(3) For all $n > |x|$ the terms in absolute value are monotone decreasing. Using (1)–(3), one can prove that the value of $\exp(-5.5)$ lies between the sum of the first N terms and the sum of the first $N + 1$ terms, for all $N > 7$, and the error is less than the first term neglected, i.e. less than $(-5.5)^N/N!$ (rapidly approaching zero).

Our analysis shows that we can just sum the series until adding more terms does not change the sum. According to theory we then have $\exp(-5.5)$ to five decimal figures of accuracy. Proceeding to do this we end up with the value of 0.0026363 whereas the *true* value is 0.0040868. The reason for this is as follows. Writing down the first few terms in the series we have

$$\begin{array}{r}
1.00|00 \\
-5.50|00 \\
15.12|5 \\
-27.73|0 \\
38.12|9 \\
-41.94|2 \\
38.44|6 \\
-20.76|8 \\
\vdots \\
\hline
0.0026363
\end{array}$$

The sum does not change after about 25 terms. Notice, though, that the final result is approximately 10^{-3}. Thus the figures to the left of the vertical dashed line should sum up to zero, the only contribution to the final result coming from figures to the right of the dashed line. Because of the limited accuracy, the digits 5, 0, 9 etc. in the third, fourth, fifth terms are already rounded-off, and these are the figures which are significant to the final result. This explains the catastrophic result. In numerical analysis, problems of round-off error can be overcome by increasing word accuracy (e.g. using double precision) at places where round-off error may be significant. Nevertheless one must be aware that this might be a problem.

Convergence and stability

We shall say that the approximate solution **converges** to the exact solution of the ordinary differential equation, at x, if (neglecting round-off errors)

$$y_{\text{appr}}(x) \to y_{\text{exact}}(x)$$

as

$$\Delta x_i \to 0. \tag{11}$$

Notice that if $x = a + \Delta x_1 + \ldots + \Delta x_n$, and we let all $\Delta x_i \to 0$, then the number of steps required to reach the same value x approaches infinity. Hopefully, if we can show convergence, then taking small but finite Δx_i would imply that y_{appr} is close to y_{exact}.

As we have seen, generated errors are introduced because of the approximation method, and because of round-off. Crucial to the possibility of convergence is how those errors propagate with continuing computations. We would like these **propagated errors** not to grow. Of course this can only apply to a stable equation, as the difference between two initially close solutions of an unstable equation does grow. We thus make the following definition: if we apply the approximate method to a stable equation and propagating errors do not grow, then the approximation method is **stable**.

Unfortunately, this definition is too general for us to be able to check it, so we utilize the idea of a test equation. To demonstrate this idea let us define a method to be stable if applied to the test equation

$$y' = -\lambda y, \qquad \lambda > 0, \tag{12}$$

the method is stable in the sense of the definition above.

To demonstrate convergence and stability we shall look at **Euler's method**. This is the simplest approximation one may think of, and is not recommended for practical use. It is only considered to exhibit the ideas. We want to solve (1), with condition (2), in the interval $0 \leq x \leq b$. Take equal steps $\Delta x_i = h = t/N$, for some fixed point t inside the interval $(0, b)$. By Taylor's series we have

$$y(x+h) \approx y(x) + hy'(x)$$

and Euler's method just takes these first two terms in the Taylor series as the approximation;

$$y_{n+1} = y_n + h(y')_n = y_n + hf(x_n, y_n), \qquad n = 1, 2, \ldots, N, \qquad (13)$$

with y_0 the given initial value.

One can then prove the following **convergence theorem** for Euler's method. For any $0 \le t \le b$, $y_N \to y(t)$ as $N \to \infty$ (and $h = t/N \to 0$) provided $y_0 \to y(0)$ (and this convergence is uniform in t) provided $f(x, y)$ satisfies the following conditions.

(1) $f(x, y)$ is a continuous function of x and y for all $0 \le x \le b$, and $|y| < \infty$.

(2) $|f(x, y_1) - f(x, y_2)| \le K|y_1 - y_2|$, K independent of x for any y_1, y_2.

The second condition says essentially that f does not grow faster than an exponential in y. If we consider for example the problem

$$y' = f(x, y) = 1 + y^2, \qquad y(0) = 0, \qquad (14)$$

then

$$|f(x, y_1) - f(x, y_2)| = |y_1 + y_2| \, |y_1 - y_2|.$$

Thus K, if it exists, should bound $|y_1 + y_2|$. If the interval $(0, b)$ is such that $b \ge \pi/2$, this bound is impossible, since the solution to (14) is $y = \tan x$.

Let us now check the stability of Euler's method by applying it to our test equation (12). Here $f(x, y) = -\lambda y$, and thus (13) becomes

$$y_{n+1} = y_n - h\lambda y_n = (1 - h\lambda)y_n, \qquad (15)$$

and we obtain

$$y_n = y_0(1 - \lambda h)^n. \qquad (16)$$

Stability is therefore equivalent to the condition

$$|1 - \lambda h| < 1, \qquad (17)$$

and, since we have assumed $\lambda > 0$ and $h > 0$, the condition implies $-1 < 1 - \lambda h$, or

$$h < 2/\lambda. \qquad (18)$$

The region in the λ–h plane where (18) obtains is shown in Figure A.5.3. Notice that the larger λ, the smaller h should be. Thus Euler's method would not be stable for all decaying exponentials. This is a regular feature of approximation methods, in that, only for a range of values of λ and related h, do we have stability. From a practical point of view we see that fast decaying exponentials will have to be followed by very small steps, if one uses Euler's method – a time-consuming procedure.

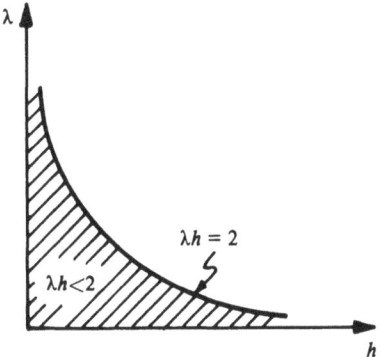

Figure A.5.3. The region in the $\lambda - h$ plane for which (18) holds.

Order of method

The convergence theorem stated above does not provide any information as to how close the solution of the approximation is to the exact solution when we use some small, but finite h. To estimate this error, write the difference method in the form

$$\frac{y_{n+1} - y_n}{h} - G(x_n, y_n) = 0. \tag{19}$$

$G = f(x_n, y_n)$ in Euler's method. The same expression may be written for the exact solution, but the right-hand side will not be zero

$$\{[y(x_{n+1}) - y(x_n)]/h\} - G = F(x_n, y_n; h). \tag{20}$$

Subtracting the two, assuming $y_n = y(x_n)$, will give us an indication not only of how close y_{n+1} is to the exact solution $y(x_{n+1})$, but also how close the difference equation and the differential equations are. More specifically, how close is the exact solution towards solving the difference equation. Practically, this is done by expanding $y(x_{n+1})$ in a Taylor series around $y(x_n)$. The right-hand side of (20), $F(x, y; h)$, then becomes a power series in h. The leading power in h obtained is the **order of the method**. (F is also termed the truncation error.)

For Euler's method

$$y_{n+1} = y_n + hf(x_n, y_n),$$

$$y(x_{n+1}) = y(x_n) + hf(x_n, y_n) + \tfrac{1}{2}h^2 y''(x_n) + \ldots .$$

Thus

$$\left[\frac{y(x_{n+1}) - y(x_n)}{h} \right] - f(x_n, y_n) = \tfrac{1}{2}h y''(x_n) + \ldots . \tag{21}$$

The leading power of h in (21) is one, and Euler's method is a **first-order method**. Clearly, the higher the order of the method, the more accurate it is. Conversely, we may use a larger step h to reach a desired accuracy, thus saving time, *provided* we do not violate the stability condition for the method we are using. An example of a second-order method is the following 'Runge–Kutta-type' method;

$$y_{n+1} = y_n + h\left\{(1-b)f(x_n, y_n) + bf\left[x_n + \frac{h}{2b}; y_n + \frac{h}{2b}f(x_n, y_n)\right]\right\}, \qquad (22)$$

for any $b \neq 0$.

If $f(x, y) = f(x)$ is independent of y, then (22) reduces to the mid-point rule for $b = 1$, and to the trapezoidal rule for $b = \frac{1}{2}$. The widely used (classical) Runge–Kutta method is order 4. It should be noted that the use of higher-order methods presumes the existence of higher-order derivatives, which is not always the case.

Stiff differential equations

In many cases special properties of the equation or equations to be solved demand a special treatment and special methods of attack. Such a property is stiffness, which is quite common in biochemical equations. We shall demonstrate stiffness using the following example. Consider again the system of (3a, b) and (4),

$$u' = 998u + 1998v,$$
$$v' = -999u - 1999v, \qquad (23)$$
$$u(0) = 1, \qquad v(0) = 0.$$

The solution to (23) is

$$u = 2 \exp(-x) - \exp(-1000x),$$
$$v = -\exp(-x) + \exp(-1000x). \qquad (24)$$

We thus see that u and v are composed of two exponentials with exponents orders of magnitude apart. Loosely speaking this phenomenon is termed **stiffness**. As we have seen for Euler's method, stability demands $\lambda h < 2$ and therefore we must have here

$$h < 2/1000, \qquad (25)$$

an impractically small h. The same holds for other methods such as Runge–Kutta-type methods, where the stability condition is always of the form $|\lambda h| = 0(1)$. Condition (25) restricts the step size h because of the existence of exp $(-1000x)$ in the solution. But, for all practical purposes for $x > 10^{-2}$, we have

$$u = 2 \exp(-x), \qquad v = -\exp(-x). \tag{26}$$

Thus, the additional negligible solution restricts the use of Euler's method, and for that matter *all* popular methods. Therefore special methods are needed and have been devised for stiff systems, although they are not always satisfactory. The numerical solution of stiff systems or more generally differential equations with a large Lipshitz constant (K in the convergence theorem above) is the subject of intense research effort at the present time. Many of the papers are being published in *Mathematics of Computation*.

A word of warning. 'Black boxes' that solve initial value problems for ordinary differential equations, such as the CSMP system, are considered reliable for regular 'nice' problems and one can use them confidently. Once specific properties, such as stiffness, are detected or suspected, the routine may not be able to handle it in a satisfactory manner. The best advice one can then give is 'get help' – consult an expert!

Two basic reference books are given below.

References

Acton, F. A. (1970). *Numerical Methods that Work*, New York, Harper & Row.
Gear, S. W. (1971). *Numerical Initial Value Problems in Ordinary Differential Equations*, New York, Prentice-Hall.

Author index

Numbered by sections

Subject index

For EU product safety concerns, contact us at Calle de José Abascal, 56–1°,
28003 Madrid, Spain or eugpsr@cambridge.org.

www.ingramcontent.com/pod-product-compliance
Ingram Content Group UK Ltd.
Pitfield, Milton Keynes, MK11 3LW, UK
UKHW042034080526
470874UK00007B/290